KU-766-330

Zoo Animals

Geoff Hosey
University of Bolton

Vicky Melfi
Paignton Zoo Environmental Park

Sheila Pankhurst
Anglia Ruskin University

Zoo Animals

Behaviour, Management,
and Welfare

OXFORD
UNIVERSITY PRESS

Great Clarendon Street, Oxford OX2 6DP

Oxford University Press is a department of the University of Oxford.
It furthers the University's objective of excellence in research, scholarship,
and education by publishing worldwide in

Oxford New York

Auckland Cape Town Dar es Salaam Hong Kong Karachi
Kuala Lumpur Madrid Melbourne Mexico City Nairobi
New Delhi Shanghai Taipei Toronto

With offices in

Argentina Austria Brazil Chile Czech Republic France Greece
Guatemala Hungary Italy Japan Poland Portugal Singapore
South Korea Switzerland Thailand Turkey Ukraine Vietnam

Oxford is a registered trade mark of Oxford University Press
in the UK and in certain other countries

Published in the United States
by Oxford University Press Inc., New York

© Geoff Hosey, Vicky Melfi, and Sheila Pankhurst 2009

The moral rights of the authors have been asserted
Database right Oxford University Press (maker)

First published 2009

Reprinted (twice) 2009, 2010, 2011, 2012

All rights reserved. No part of this publication may be reproduced,
stored in a retrieval system, or transmitted, in any form or by any means,
without the prior permission in writing of Oxford University Press,
or as expressly permitted by law, or under terms agreed with the appropriate
reprographics rights organization. Enquiries concerning reproduction
outside the scope of the above should be sent to the Rights Department,
Oxford University Press, at the address above

You must not circulate this book in any other binding or cover
and you must impose the same condition on any acquirer

British Library Cataloguing in Publication Data

Data available

Library of Congress Cataloging in Publication Data

Data available

Typeset by Graphicraft Limited, Hong Kong
Printed in Great Britain by
CPI Group (UK) Ltd., Croydon, CR0 4YY

ISBN: 978–0–19–923306–9

7 9 10 8 6

Foreword

There has been a need for a book such as this for some time and it will be welcomed by many: students; zoo professionals; those whose work brings them in contact with zoos and aquariums; and members of the general public, who may simply want to know more about the work and ethos of the modern zoo. It therefore gives me great pleasure to write this foreword and to welcome the publication. The volume contains a huge amount of information that successfully encapsulates the varied range of skills and knowledge that is required to run a modern zoo and aquarium—a range that is often underestimated by the uninitiated. The book also features a comprehensive list of relevant references for those wishing to explore topics in more depth.

Zoos have an amazing ability to change and refocus: from the menageries of old, to the modern dynamic conservation centres of the twenty-first century. But people do not always realize how rapid this change of focus has been—particularly over the past 30 years—or what it involves. Chapter 2 provides a clear résumé of this evolution and the ethos of the modern zoo, and also approaches that 'elephant in the room' question: is it ethical to keep wild animals in captivity? Many anti-zoo organizations, following animal rights philosophy, claim that there is no justification for zoos and that they are morally indefensible. While Chapter 2 considers this hot topic directly, the remaining chapters provide details of the role of the modern zoo, supplying the reader with sufficient information to come to his or her own conclusion on the matter.

There is no escaping the fact that, however laudable the mission of the modern zoo in terms of conservation and education, there is no justification for bad zoos—that is, for keeping animals in inadequate conditions, with poor husbandry and welfare. Therefore, as in any area, controls are necessary, and these are supported and encouraged by national and regional zoo and aquarium associations and their members. Zoo legislation and accompanying standards are complex areas, with which Chapter 3 deals clearly, also covering all of the other regulatory areas that affect the modern zoo. The optimum conditions for housing, nutrition, and veterinary care for different species must be researched, understood, and communicated; important enabling tools are husbandry guidelines, and well-kept and readily available animal records.

Some knowledge of animal behaviour is essential in order to appreciate and enjoy fully the experience of observing animals in zoos and aquariums: why is the animal behaving in that way and how do the individuals of a species communicate with each other? These areas are expertly explained in Chapter 4, which also covers differences in the captive and wild environments, and the appearance of some behaviour in captivity that can be classed as abnormal, in that it either has not been observed in the wild or has been observed there only at a significantly lower frequency. It is important to try to understand which of the incidences of these kinds of behaviour actually indicate poor welfare and how to evaluate empirically activity and behaviour in order to try to establish what comprises good welfare. This is a fascinating area that is producing exciting new ideas and concepts, which are covered in Chapter 7, with Chapter 8 giving information on enrichment—that is, one of the ways in which we can improve the captive environment by trying to simulate similar activity patterns to those seen in the wild. But does enrichment work? This question can only be answered after carefully observing behaviour and subsequent detailed analysis. This chapter touches on the area of perception and reality; we assume that the wild represents optimum conditions, but that it can be a stressful environment within which animals get killed and eaten, or fall sick. In zoos and aquariums, there is excellent veterinary care, and much of the knowledge gained from working with animals in zoos can be transferred to the treatment of animals in the wild, often now a part of field conservation programmes. Through good veterinary care, the welfare of animals in zoos can exceed that of their conspecifics in the wild. All of these issues are fascinating areas for debate and discussion.

Although many animals in the wild—especially those in some national parks—come into regular contact with humans, the lives of zoo animals are lived in closer, and more regular, proximity to the visiting public. But is this a good or a bad thing? Chapter 13 examines the argument that it is good in that it gives the public a unique and exciting experience that may result in them being inspired to learn more about wildlife and its conservation. This chapter examines also the argument that being close to the public for many hours each day may compromise the welfare of some species.

Zoos offer fabulous opportunities for research and training in observation techniques for students. Some aspects of species biology are much easier to study in the captive environment than in the wild, thus adding vital information to our knowledge and understanding. Chapter 14 provides excellent coverage of the range of biological research potential in zoos, much of it for the ultimate benefit of species both in captivity and in the wild.

The evolution of the modern zoo is a fascinating subject, as is the ability of the better zoos to become powerful and influential forces for conservation. Through integrated science and effective communication to their many visitors, zoos are in a unique position to achieve conservation success in many different ways. The future of zoos is exciting and much remains to be done. This volume will aid progress and inspire many to carry on the good work, and, by so doing, help zoos and aquariums to fulfil their potential as conservation organizations of the twenty-first century.

Miranda Stevenson
Director, British and Irish Association of Zoos and Aquariums (BIAZA)
April 2008

About the authors

Geoff Hosey was principal lecturer in Biology at the University of Bolton until his retirement in 2005 and is now an honorary professor at the University. He has a long-standing interest in zoo animals and, particularly, in how their behaviour is affected by the zoo environment. In addition to undertaking his own research, he has also supervised numerous undergraduate and postgraduate research projects on zoo animals. Geoff is an Associate Member of BIAZA and a member of the BIAZA Research Group. He has a keen interest in all animals, but a particular affection for lemurs—especially ring-tailed lemurs *Lemur catta*.

Vicky Melfi is senior research officer with the Whitley Wildlife Conservation Trust (WWCT), working with the animal collections at Paignton Zoo Environmental Park, Living Coasts, and Newquay Zoo; she also undertakes field work in Sulawesi. Vicky started working as a zookeeper when she was 16 years old and, with the benefit of this experience and subsequent academic training, she now leads the behavioural husbandry and animal welfare group at WWCT. Other than engaging in the academic duties of teaching and student supervision, Vicky actively participates in BIAZA and EAZA, notably as gibbon TAG chair, Sulawesi crested black macaque *Macaca nigra* EEP coordinator, and Eastern black-and-white colobus *Colobus guereza* ESB coordinator.

Sheila Pankhurst is a principal lecturer in Animal Behaviour at Anglia Ruskin University in Cambridge and is also a member of the BIAZA Research Group. As a postgraduate student at Cambridge University, she carried out her PhD research at Whipsnade Wild Animal Park in Bedfordshire, investigating the social structure of a large, South American rodent: the mara *Dolichotis patagonum*. She went on to supervise further research into the ecology and parasitology of the mara (in collaboration with the Zoological Society of London), and has also supervised many undergraduate projects at zoos. Sheila has had a special interest in rodents, from all taxonomic groups, since she was a very small child.

Preface and acknowledgements

This is a textbook about how and why zoos manage and maintain populations of animals. It is about how zoos meet the animals' physical needs—such as ensuring that they have good housing, health, and an appropriate diet—and it is also about how their psychological needs are met, how they are kept as free as possible from stress, and how they are given opportunities to perform as many as possible of the behaviours they would perform in the wild. But it is more than that: we have also tried to give some idea of the context within which zoos operate, so there is some consideration here of topics such as zoo research, which is so important to advancing our theory and practice of zoo biology, and zoo history, which helps us to understand how zoos have evolved.

There are several good books about zoos, but none seems to cover all of the different aspects of animal management, behaviour, and welfare that are brought to bear in modern zoo practice. This book arose more out of frustration than inspiration: frustration that we were not able to recommend to our colleagues, or to our students, a single textbook that provided a broad, but accessible and reasonably concise, overview of the zoo, how it operates, and how it fulfils a role within modern society. But we were also very aware of the growing amount of good research in zoo biology and the presence of a large literature that had not really been surveyed in a book of this sort.

All three of us know the sheer joy of being close to animals, and zoos allow that opportunity in a way that few other experiences do. We have tried to capture some of this in the way in which we have approached the topics in this book. Between the three of us, we have clocked up more than 50 years' researching in, working in, and supervising students in zoos. We hope that this, together with the particular mix of zoo-based and university-based experience that we have, makes us a strong team for writing this book.

We are fortunate in having many colleagues who have been willing to read and make constructive (and often critical) comments on various drafts of chapters, and their input has improved the book immeasurably. It goes without saying that any errors or omissions that remain in the book are our responsibility and not theirs. Our particular thanks go to (in alphabetical order): Neil Bemment (Paignton Zoo Environmental Park); Wayne Boardman (Adelaide Zoo); Iain Brodie (Anglia Ruskin University); Julian Chapman (Paignton Zoo Environmental Park); Emma Creighton (University of Chester); Danny de Man (EAZA); John Eddison (University of Plymouth); Harriet Elson; Andrea Fidgett (Chester Zoo); Angela Glatston (Rotterdam Zoo); Sonya Hill (Chester Zoo); Kathy Knight (Whitley Wildlife Conservation Trust); Charlotte Nevison (Anglia Ruskin University); Amy Plowman (Whitley Wildlife Conservation Trust); Kirsten Pullen (Whitley Wildlife Conservation Trust); Pippa Rogerson (University of Cambridge); Stephanie Sanderson (Chester Zoo); Vicky Sandilands (Scottish Agricultural College); Ghislaine Sayers (Paignton Zoo Environmental Park); Andrew Smith (Anglia Ruskin University); Miranda Stevenson (BIAZA); Sarah Thomas (Blackpool

Zoo); Deborah Wells (Queek's University Belfast); and the anonymous Oxford University Press reviewers, particularly Reviewer A, for thoughtful and helpful comments that have improved the text enormously.

We are also grateful to those colleagues who have provided us with help and information when we have been stuck. In particular, we thank Andrea Fidgett and Stephanie Sanderson (Chester Zoo), who spent time with us at Chester helping us to understand the intricacies of zoo animal nutrition and health, and Nicola Marples (Trinity College, Dublin), who provided us with understandable information about statistics as applied to zoo research.

We have tried to identify copyright holders for all of the illustrations we have used and to obtain their permission to use the figures. If we have failed to gain the correct permission for any figure reproduced here, we apologise and will try to rectify this in any future edition. We would like to thank the following publishers for allowing us to use figures originally published in their journals:

- Elsevier for Figures 4.14, 4.16a, 4.16b, 6.21, 13.10, 13.16, and 14.4, all of which originally appeared in *Applied Animal Behaviour Science*;
- John Wiley & Sons, Inc. for Figures 4.17b, 4.18b, 4.19, 4.21, 4.22, 4.23b, 4.29b, 13.2, 13.9, and 13.13, and Tables 4.2, 4.4, and 13.1, all of which originally appeared in *Zoo Biology*, reprinted with permission of Wiley-Liss, Inc., a subsidiary of John Wiley & Sons, Inc.;
- Karger AG for Figure 6.6, which originally appeared in *Folia Primatologica*;
- Universities Federation for Animal Welfare for Figures 4.20, 4.31a, 4.31b, 7.8, 8.7, 8.8a, 8.9a, 8.13, and Table 4.3, all of which originally appeared in *Animal Welfare*;
- World Association of Zoos and Aquaria (WAZA) for Figure 14.6 and Table 14.1, both of which appeared in the *World Zoo and Aquarium Conservation Strategy*;
- Macmillan Publishers Ltd for Figure 4.27, which originally appeared in *Nature*;
- Wiley-Blackwell Publishing Ltd for Figure 6.5, which originally appeared in *Conservation Biology*;
- The Company of Biologists Ltd for Figures 6.1 and 7.9, which originally appeared in *The Journal of Experimental Biology*;
- Cambridge University Press for figures 12.4 and 12.5, reproduced by kind permission from Stevens and Hume (1995).

We would also like to thank the following:

- Frankie Kerridge (University of Bolton), for allowing us to use Figure 4.30b from her unpublished PhD thesis;
- Stuart Semple (Roehampton University), for allowing us to use Figure 14.3 from his paper in the BIAZA Research News;
- Amy Plowman, BIAZA and ICEE, for allowing us to reproduce figures that appeared in the *ICEE Proceedings* and the *Proceedings* of BIAZA Research symposia;

- Graham Franklin (Department of Local Government, Housing and Sport, Australian Government), for permitting us to quote from the Animal Welfare Act 1999;
- Neil Pratt (Secretariat of the Convention on Biological Diversity, UNEP), for giving permission to use Figures 1.3 and 3.3;
- Lee Jackson, for kindly supplying images (Figures 1.2, 2.1, and 2.8) from his website www.victorianlondon.org;
- Miranda Stevenson (BIAZA), Bart Hiddinga (EAZA), and Deborah Martins (ARAZPA), for giving permission to reproduce their respective logos (Figure 3.12);
- Valerie Hare and Karen Worley, for allowing us to use Shape of Enrichment and REEC logos (Figure 8.3).

Many of our colleagues have generously supplied us with photographs to use in this book. Indeed, we have been offered so many excellent pictures that selecting those that have finally appeared in the book has been a difficult task. But we would particularly like to thank: Heidi Hellmuth and Jessie Cohen (both Smithsonian's National Zoo, Washington DC); Tibor Jäger and Amelia Terkel (both Zoological Center, Tel Aviv-Ramat Gan, Israel); Harriet Elson; Nadya Stavtseva (Department of International Co-operation, Moscow Zoo); Johannes Els (Cango Wildlife Ranch, Oudtshoorn, South Africa); Christopher Stevens (Werribee Open Range Zoo, Zoos Victoria, Australia); Nathalie Laurence and David Rolfe (Howletts and Port Lympne Wild Animal Parks); Robyn Ingle-Jones (Pretoria Zoo); Wolfgang Ludwig (Dresden Zoo); Monika Ondrusova (Ostrava Zoo); Leszek Solski and Radoslaw Ratajszczak (both Wrocław Zoo); Olga Shilo (Novosibirsk Zoo); Joy Bond (Belfast City Zoo); Achim Johann (Naturzoo Rheine); Natalie Cullen and Vickie Ledbrook (both Colchester Zoo); Diana Marlena Mohd Idris (Singapore Zoo); Hannah Buchanan-Smith (University of Stirling); Keith Morris (MRC); Georgia Mason (University of Guelph); Vicky Cooper (University of Bolton); Douglas Sherriff (Chester Zoo); Sonya Hill (Chester Zoo); Julian Doberski (Anglia Ruskin University); Julian Chapman (Paignton Zoo Environmental Park); Mark Parkinson (Paignton Zoo Environmental Park); Kirsten Pullen (Whitley Wildlife Conservation Trust); Ray Wiltshire (Paignton Zoo Environmental Park); Mel Gage (Bristol Zoo Gardens); Phil Gee (University of Plymouth); Kathy Knight (Whitley Wildlife Conservation Trust); Barbara Zaleweska (Warsaw Zoo); Andrew Bowkett (Whitley Wildlife Conservation Trust); Olivia Walter (BIAZA); Rachel McNabb and Richard Hezlep (both Zoo Atlanta); Cordula Galeffi, Samuel Furrer, and Edi Day (all Zoo Zurich); Karen Brewer (South Lakes Wild Animal Park); Gillian Davis (Paignton Zoo Environmental Park); Klaus Gille (Archiv Hagenbeck, Hamburg Tierpark); and Jake Veasey (Woburn Safari Park).

We have been able to include some lovely cartoons, which were kindly drawn for us by Phil Knowling (Paignton Zoo Environmental Park). Christine Jackson

(Drusillas Zoo Park) provided the ARKS printout in Table 5.7, and Pierre Moisson (Mulhouse Zoo) provided the SPARKS printout in Figure 5.19. We thank David Price (University of Plymouth) for help with references on water quality in aquariums. We greatly appreciate the support of Miranda Stevenson (Director of BIAZA) and thank her for providing us with the foreword for this book.

We are enormously grateful to Jonathan Crowe at Oxford University Press for his unfailingly good-natured acceptance of our inability to keep to deadlines, but also for his encouragement and wise advice throughout the preparation of this book. Bristol Zoo has been a midway point at which the three of us have met and planned the book, and we thank Brian Carroll and Christophe Schwitzer for their hospitality and for the provision of the zoo library for our meetings.

We owe a huge debt of gratitude to Susie, Julian, and Jonathan, for their patience, support, and encouragement while we were writing the book (and to Sheila's children, Tabitha and Charlie, for their endless enthusiasm for visiting zoos). And finally, we thank the animals themselves—the lemurs, the macaques, and the maras—beautiful animals that were largely responsible for getting us involved in zoo research in the first place, and without which this book might very well not have been written.

Contents

List of boxes

List of acronyms

(see also Glossary)

NOTE: UK or USA in parentheses after an entry denotes the country of origin for legislation, government departments, etc.

AATA	Animal Transportation Association
AAZK	American Association of Zoo Keepers
AAZPA	American Association of Zoological Parks and Aquariums *(Now the AZA—see below)*
AAZV	American Association of Zoo Veterinarians
ABP	Animal By-Products Regulations 2005 *(UK)*
ABS	Animal Behavior Society *(USA)*
ABWAK	Association of British Wild Animal Keepers
ACTH	adrenocorticotropic hormone
ADF	acid detergent fibre
AHA	Animal Health Australia
AI	artificial insemination
AKAA	Animal Keepers Association of Africa
ALPZA	Latin-American Zoo and Aquarium Association
ANCMZA	Advanced National Certificate in the Management of Zoo Animals *(UK)*
ANOVA	analysis of variance
APHIS	Animal and Plant Health Inspection Service *(Part of the US Department of Agriculture)*
APP	African Preservation Programmes
ARAZPA	Australasian Regional Association of Zoological Parks and Aquaria
ARKS	Animal Record-Keeping System
ART	assisted reproductive technology
ASAB	Association for the Study of Animal Behaviour *(UK)*
ASG	Amphibian Specialist Group
ASZK	Australasian Society of Zoo Keeping
AV	approved vet
AWA	Animal Welfare Act of 1966, as amended *(USA)*; Animal Welfare Act 2006 *(UK)*
AZA	Association of Zoos and Aquariums *(Formerly the AAZPA—see above. The organization changed its name first to the American Zoo and Aquarium Association in 1994; there is no 'American' in the current title.)*
BAP	Biodiversity Action Plan *(UK)*
BIAZA	British and Irish Association of Zoos and Aquariums
BMR	basal metabolic rate
BSE	bovine spongiform encephalopathy

BVA	British Veterinary Association
BVS	BSc in Veterinary Surgery
BVZS	British Veterinary Zoological Society
CAZA	Canadian Association of Zoos and Aquariums
CAZG	Chinese Association of Zoological Gardens
CBD	1992 Convention on Biodiversity
CBP	captive breeding programme
CBSG	Conservation Breeding Specialist Group *(Formerly the Captive Breeding Specialist Group—the name was changed in 1994)*
CCTV	closed-circuit television
CHP	combined heat and power
CITES	1973 Convention on International Trade in Endangered Species of Wild Fauna and Flora
CNS	central nervous system
CoPs	Conference of the Parties *(A country that signs up to a UN treaty or convention is known as 'a Party')*
COSHH	Control of Substances Hazardous to Health *(In relation to UK Health and Safety regulations)*
COTES	Control of Trade in Endangered Species (Enforcement) Regulations 1997 *(UK)*
CP	conservation programme
CPS	Crown Prosecution Service *(UK)*
CR	conditioned response
CS	conditioned stimulus
CSF	cafeteria-style feeding
CWD	chronic wasting disease
DCMS	Department for Culture, Media and Sport *(UK)*
DE	digestible energy
Defra	Department for Environment, Food and Rural Affairs *(UK)*
DESD	Decade of Education for Sustainable Development
DfES	Department for Education and Science *(UK)*
DNA	deoxyribonucleic acid
DTI	Department of Trade and Industry *(UK)*
EAZA	European Association of Zoos and Aquaria
EAZWV	European Association of Zoo and Wildlife Veterinarians
EC	European Community *(Confusingly, the European Commission, which is part of the EU—see below—is also sometimes referred to by the abbreviation EC)*
ECAZA	European Community Association of Zoos and Aquaria *(Now the EAZA—see above)*
EEKMA	European Elephant Keeper and Manager Association
EEP	European Endangered species Programme *(The acronym is taken from the German: Europäisches Erhaltungszuchtprogramm)*

EFA	essential fatty acid
EFSA	European Food Safety Agency
EGZAC	European Group on Zoo Animal Contraception
ELISA	enzyme-linked immunosorbent assay
ENG	EAZA *(see above)* Nutrition Group *(Formerly the EZNRG— see below)*
ESB	European studbook
ESU	evolutionarily significant unit
EU	European Union
EZNRG	European Zoo Nutrition Research Group *(Now the ENG— see above)*
FAA	food anticipatory activity
FAO	United Nations Food and Agriculture Organization
FDA	Food and Drug Administration *(USA)*
FMD	foot and mouth disease
FMR	field metabolic rate
FSC	Forest Stewardship Certificate
FSH	follicle-stimulating hormone
GAS	general adaptation syndrome
GE	gross energy
GI (tract)	gastrointestinal (tract)
GnRH	gonadotropin-releasing hormone
GRB	genetic resource bank
GSMP	global species management programme
HPA/G (axis)	hypothalamo-pituitary-adrenal/gonadal (axis)
HSE	Health and Safety Executive *(UK)*
IATA	International Air Transport Association
ICEE	International Conference on Environmental Education
ICEE	International Conference on Environmental Enrichment
ICSI	intracytoplasmic sperm injection
ICZ	International Congress on Zookeeping
ICZN	International Commission on Zoological Nomenclature
ISAE	International Society for Applied Ethology
ISIS	International Species Information System
ISO	International Organization for Standardization
IUCN	International Union for the Conservation of Nature and Natural Resources *(Also known as the World Conservation Union)*
IUDZG	International Union of Directors of Zoological Gardens
IVF	*in vitro* fertilization
IZW	Leibniz Institute for Zoo and Wildlife Research
JMSP	Joint Management of Species Programme
JNCC	Joint Nature Conservation Committee *(UK)*
LARs	IATA *(see above)* Live Animal Regulations

LH	luteinizing hormone
MBA	Methods of Behavioural Assessment
MBD	metabolic bone disease
ME	metabolizable energy
MHC	major histocompatibility complex
MHSZ	*Managing Health and Safety in Zoos*
MSW	Mammal Species of the World
mtDNA	mitochondrial DNA
MTRG	Marine Turtle Research Group
NAG	Nutrition Advisory Group *(of the AZA—see above)*
NAP	National Academy Press
NDF	neutral detergent fibre
NRC	National Research Council *(USA)*
NSAID	non-steroidal anti-inflammatory drug
OIE	Office International des Epizooties *(Also known as the World Organisation for Animal Health)*
PAAZAB	African Association of Zoological Gardens and Aquaria *(Formerly the Pan-African Association of Zoological Gardens and Aquaria, from which title the acronym's initial 'P' remains)*
PETA	People for the Ethical Treatment of Animals
PFA	pre-feeding anticipation
PhD	Doctor of Philosophy degree *(A higher degree than a BSc or MSc)*
PIT	passive integrated transponder
PMP	population management plan
POE	post-occupancy evaluation
PRL	prolactin
PRT	positive reinforcement training
PSM	plant secondary metabolite
RAE	Research Assessment Exercise *(UK)*
RCP	EAZA *(see above)* Regional Collection Plan
RCVS	Royal College of Veterinary Surgeons *(UK)*
REEC	Regional Environmental Enrichment Conferences
RIA	radioimmunoassay
RSG	Reintroduction Specialist Group
RSPB	Royal Society for the Protection of Birds *(UK)*
RSPCA	Royal Society for the Prevention of Cruelty to Animals *(UK)*
SAZARC	South-Asian Zoo Association for Regional Cooperation
SDB	Self-directed behaviour
SHEFC	Scottish Higher Education Funding Council
SIB	self-injurious behaviour
SPARKS	Single Population Analysis and Records Keeping System *(SPARKS is a studbook management system)*
SPI	spread of participation index

SPRG	Scottish Primate Research Group
SSC	IUCN *(see above)* Species Survival Commission
SRP	species recovery programme *(UK)*
SSP	species survival plan
SSSMZP	Secretary of State's Standards of Modern Zoo Practice *(UK)*
SUZI	sub-zonal insemination
SVL	snout–vent length
SVS	State Veterinary Service *(In the UK, the SVS was absorbed into a new government agency, Animal Health, in 2007)*
TAG	taxon advisory group
TB	tuberculosis
TCF	Turtle Conservation Fund
TPR	temperature, pulse, and respiration
TRAFFIC	*Note that TRAFFIC is not an acronym, but rather the proper name of an organization*
TSE	transmissible spongiform encephalopathy (e.g. BSE—*see above*)
TWG	taxon working group *(UK)*
UFAW	Universities Federation for Animal Welfare
UNCED	United Nations Conference on Environment and Development
UNEP	United Nations Environment Programme
UNFCCC	United Nations Framework Convention on Climate Change
UR	unconditioned response
US	unconditioned stimulus
USDA	US Department of Agriculture
UV	ultraviolet
WAZA	World Association of Zoos and Aquariums
WCC	AZA Wildlife Contraception Center
WCS	Wildlife Conservation Society *(Based in New York)*
WIN	Wildlife Information Network
WLCA	Wildlife and Countryside Act 1981 *(UK)*
WNV	West Nile virus
WWF	Formerly, the World Wide Fund for Nature—now known only by the acronym
WZACS	World Zoo and Aquarium Conservation Strategy
WZCS	World Zoo Conservation Strategy
ZIMS	Zoological Information Management System
ZLA	Zoo Licensing Act 1981 *(UK)*
ZOO	Zoo Outreach Organisation
ZSL	Zoological Society of London

Chapter 1 Introduction

It has been estimated that, throughout the world, there are something like 10,000 animal collections that come under the general heading of zoos (WAZA, 2006). We do not know how many visitors they all receive annually, but it is certainly in the order of at least hundreds of millions. Of these zoos, about 1,000 belong to regional or national associations, which endeavour to foster cooperation and to ensure high professional standards among their members. These 1,000 **accredited zoos** alone receive more than 600 million visits every year (WAZA, 2006), which makes going to the zoo one of the most popular of leisure pursuits across the world. In North America, it is often said that zoos attract more visitors than professional baseball, basketball, and football games combined. In the UK and Ireland, more than 18 million people visit accredited zoos every year, which is one in four of the population (BIAZA, 2007).

So what do these statistics tell us? Firstly, they suggest that the motivation to see, at close hand, real, live **exotic animals** is very strong among people. But they also indicate that the worldwide zoo community can have real influence in shaping people's attitudes towards, and knowledge of, animals. This, in turn, can result in more public concern and support for raising welfare standards and promoting conservation. Zoos are in an ideal position to capitalize on this.

Modern zoos are no longer only places that keep a few animals for people to go and look at. They are scientifically run and governmentally regulated institutions, which have a significant role to play in our relationship with the natural world. Many zoos describe their role in terms of four key words: conservation; education; research; and recreation. These roles have been defined more precisely in the last two decades, and this has been accompanied by a great increase in knowledge about the needs of animals and the best ways of maintaining them in captivity. In this book, we review and synthesize much of this knowledge to provide an overview of the functions and operations of modern zoos, how they manage their animals, why they keep them the way that they do, and how they try to provide environments that ensure the best welfare of the animals. We try to show what is currently regarded as best practice in accredited zoos, and the knowledge and scientific research on which that best practice is based.

1.1 Who is this book for?

The short answer to the question of who this book is for is that we hope that there will be something of use and value in here for anyone who has an interest in zoos and the animals they keep.

We did, however, have two particular groups of people in mind when preparing this book. Firstly, we have tried to provide an up-to-date overview of zoos and zoo-related issues for the keepers, vets,[1] curators, education officers, and other zoo professionals who look after exotic collections. These people are highly knowledgeable specialists in their own particular field, but may want to know more about other aspects of how modern zoos work. The second group that we had in mind is the growing

By **accredited zoos**, we mean zoos that are licensed and/or members of one of the regional zoo associations. Much more is said about this in **Chapter 2**.

The term **exotic animals** is used here in a very loose sense. Many zoos keep groups of native species, and some specialize in this. How exotic you think they are probably depends on where you come from.

1 We use the abbreviation 'vets' throughout this book to refer to veterinary surgeons and scientists. The word has a somewhat different connotation in North America.

body of college and university students who study zoo animals as part of their course, and who will almost certainly visit a zoo at some stage in their studies to observe and learn about its animals. Courses that include a consideration of zoo animals have proliferated over the past 10 years, and are often to be found with titles such as 'Animal Management', 'Animal Behaviour', 'Animal Welfare', 'Conservation', and a variety of other terms. For students on these courses, there has previously been no readily accessible text covering the biology of zoo animals at an appropriate level. There is, of course, the classic and excellent—but perhaps rather formidable—*Wild Mammals in Captivity* (Kleiman *et al.*, 1996), which synthesizes a huge amount of literature about mammals in zoos. We do not intend to compete with that tome. Our book is broader, assumes less prior knowledge, and covers birds, **herps**, fish, and invertebrates, as well as mammals.

With such a diverse intended readership, identifying the appropriate level for the book has been no easy task. Students following modules or courses that include consideration of zoo animals may be registered for particular certificates or diplomas, may be foundation or honours degree undergraduates, may be on taught masters courses, or may be beginning their studies for a PhD. Zoo professionals are likely to have one or more of these qualifications already. What we have assumed is that most readers will have a basic knowledge of biology and that, while they may have very good knowledge of one or two of the biological topics covered in this book, they may not have the same level of familiarity with all of the areas covered. This probably means that the book as a whole is located at around Level 2 or Level 3 (roughly, the second and third year of an honours degree), but we say that only as a guide to its style and content; we hope that the text is not too difficult or too patronizing for those who approach it at a different level.

While we have assumed that most of our readers will already have some basic biological knowledge, biology is a very broad discipline and even professional biologists are not usually proficient in all areas of the subject. For this reason, we have included in each chapter a brief review of the relevant background theory in the hope that the rest of the chapter is intelligible to all readers regardless of their previous knowledge.

Because we are familiar with, and work within, the zoo animal management systems that operate in the UK and Ireland, our perspective in this book very much reflects this orientation. We have tried to include a flavour of other perspectives as well, without making the book too cumbersome, and hope that it will also be of use to readers elsewhere in the world—particularly those located in the rest of Europe, and also in North America and Australia.

1.2 Sources of information

There is now a large literature on zoo animals. Despite this, we have become very aware in preparing this book just how many gaps there are in our knowledge. There are large areas of zoo biology that have not been empirically studied at all and plenty of other areas in which there are very few studies. In some of these areas, we have been able to refer to studies on laboratory, farm, companion, or wild animals, but it should always be borne in mind that the zoo environment is very different from these other environments, so we should apply such studies with caution. We have identified these gaps in knowledge wherever possible and hope that this will stimulate fresh research in future.

Although there is a large zoo literature, accessing it is not necessarily very easy. Many of the **empirical studies** about zoo animals are published in a small

The term **herps** has become popular as a general term to describe reptiles and amphibians, and it will reappear else-where in this book. **Herpetologists** are people who study reptiles and amphibians; the science itself is **herpetology**.

number of peer-reviewed journals, notably *Zoo Biology*, *Applied Animal Behaviour Science*, and *Animal Welfare*. These are usually within the full-text subscriptions of university (and some zoo) libraries, but also allow free access to abstracts of the papers on their websites. We have tried to use peer-reviewed sources wherever possible, partly because it ensures the reliability of the source, but also because they are relatively easy to access from most libraries. There is, however, a substantial, non-peer-reviewed zoo literature,[2] some of which contains results from empirical studies, but much of which contains only anecdotal information.

The availability of some of these sources to people outside the zoo world (and perhaps also to some of those within it) can be very poor and, for this reason, we have avoided such sources if alternative sources in the peer-reviewed literature are available. In any case, we have not undertaken exhaustive reviews of the literature in this book. Instead, we have tried to concentrate on notable papers that have advanced theory or practice, together with appropriate examples drawn from those that are available. In doing this, we have attempted to give due prominence to a range of different species, but it remains the case that a large proportion of zoo animal studies are on mammals and that, of the studies on mammals, a large proportion of these focus on primates.

As well as the works to which we refer in each chapter and which are listed in the bibliography at the end of the book (we thought this would be better than having a separate references section for each chapter), we have also tried to identify, in each chapter, sources of further information and appropriate further reading. Of course, websites can come and go, and their addresses can change, so the best we can do is to provide sites that are current at the time of final writing (April 2008), and direct people to the Online Resource Centre that supports this book.

We have also added a selection of questions at the end of each chapter, and hope that these will stimulate readers to think about what they have read and that they will act as prompts for discussion of some of the issues that we raise.

1.3 What do we mean by 'zoo'?

While the question of what we mean by 'zoo' may, at first sight, seem a rather strange question to ask, it is more than just pedantry to ask for some sort of definition of terms. As we shall see in subsequent chapters, it is important for our interpretation of animal behaviour in the zoo to be able to compare it with behaviour in other situations, and, for this, we need to know how we can separate the zoo environment from those other situations. Furthermore, some of the research that informs our understanding of the needs of animals in zoos and the most appropriate ways of managing them has not, in fact, been done in a zoo, but may have been undertaken at some other kind of animal facility. Again, we can only start to identify ways in which these other facilities differ from zoos if we have a clear idea of what we mean by 'zoo'.

So, what is a zoo?

The UK Government, in the legislation that covers zoos, defines the zoo as:

> an establishment where wild animals are kept for exhibition (other than a circus or a pet shop) to which members of the public have access, with or without charge for admission, on more than seven days in any period of twelve consecutive months.
>
> (Zoo Licensing Act 1981)

---

2 The non-peer-reviewed literature about zoo animals is sometimes referred to as the 'grey literature'. Examples range from the newsletters of zoo associations, such as BIAZA and EAZA, to the UK zookeepers' journal, *Ratel*. (See **Box 14.4** for more information.)

(a)

(b)

Figure **1.1** Places such as butterfly houses (a) and aquariums (b) are regarded as zoos within most definitions, and come within the scope of this book if they are within recognized accreditation systems. (Photographs: (a) Geoff Hosey; (b) Sheila Pankhurst)

Circuses and pet shops are excluded, because they are covered by other legislation, but the definition nevertheless permits the inclusion of such establishments as aquariums, butterfly houses, and safari parks, as well as conventional zoos.

Of course, this definition is driven primarily by the requirements of a legal framework, but it does not differ significantly from the view taken by other organizations within this field. For example, the World Association of Zoos and Aquariums (WAZA), in its 1993 *World Zoo Conservation Strategy* (IUDZG/CBSG, 1993), does not give a formal definition of a zoo, but offers two features that characterize zoos:

- they *'possess and manage collections that primarily consist of wild (non-domesticated) animals, of one or more species, that are housed so that they are easier to see and to study than in nature'*;
- they *'display at least a portion of this collection to the public for at least a significant part of the year, if not throughout the year'*.

These definitions are helpful. They allow us, for example, to include small-animal collections in museums, specialist collections, aquariums, bird parks, and the like (**Fig. 1.1**). Significantly, they exclude private and university collections that are not normally open to the public, or are only open for a handful of days each year.

This, then, is the concept of the zoo that we use in this book and, when we use the word 'zoo', we use it to include implicitly the aquariums and other publicly accessible collections of wild animals. But while we use the term 'zoo' to indicate this range of establishments, virtually all of what we have to say in this book is about, and uses information gained from, accredited zoos and aquariums.

1.4 The scope of the book

The book is about how animals are managed in zoos (their housing, husbandry, health, **nutrition**, and breeding) and also the way in which animals experience the zoo environment (their behaviour, welfare, and interactions with people). But to gain a full understanding of the zoo as an environment for animals and people, it is also useful to consider the context in which zoos operate, and their contribution to species and habitat conservation both within and beyond the zoo's boundaries. We therefore also briefly consider legislation, **conservation** (both *ex situ* and *in situ*), record keeping, and research.

Zoos have a major role to play nowadays in awareness raising and education about animals and environmental issues, and we consider this role briefly in so far as it affects the animals and their environment. This book is about zoos today and, with the exception of the brief overview of zoo history in **Chapter 2**, its content is about the health, welfare, and management of animals in modern zoos, and not about any failings in welfare provision, or health care, or husbandry, that may have occurred in the past.

We will now provides a brief summary of the content of each chapter of the book.

1.4.1 The history and philosophy of zoos (Chapter 2)

Collections of wild animals are not a new phenomenon. History records a number of private collections of animals, from the parks of ancient China to aristocratic **menageries**, including the menagerie kept at the Tower of London during medieval times. But the 'modern' zoo, in a form that we would recognize now, dates only from the end of the eighteenth century. Zoos increased greatly in public popularity during the nineteenth century, largely as places that provided entertainment and enjoyment for their visitors (even though the intention of their founders may have been the pursuit of scientific ideals).

Nowadays, we tend not to see the zoos of yesteryear as having any particular educational, scientific, or conservation role at that time, yet the two leading zoos of the nineteenth century in Europe (the Jardin des Plantes, in Paris, and London Zoo—**Fig. 1.2**) were very much scientific institutions. In any case, it is possible that, in a society that was becoming increasingly industrialized and urbanized, zoos passively helped to raise visitors' interest and

awareness of the living world simply by giving them the opportunity to experience animals that would not otherwise be a part of their lives.

The history of zoos shows us the evolving context in which wild animals have been kept, and reflects changing views in both society and the zoo world. This history is considered more fully in **Chapter 2**, which concludes with a brief overview of the philosophy and **ethics** of zoos.

1.4.2 The regulatory framework (Chapter 3)

As with so many organizations, many zoos operate within a regulatory framework that includes both mandatory requirements (that is, those required by law) and procedures seen as reflecting good practice, which are stated in various guidelines and codes of practice (**Fig. 1.3**). In **Chapter 3**, we concentrate on describing and interpreting this framework, largely as it applies to zoos in the UK.

Within the UK, the regulatory framework for zoos is stringent, and generally ensures high standards of welfare and husbandry. Similar regulatory frameworks apply to different extents in other regions of the world and these are also briefly considered in this chapter.

1.4.3 Behaviour (Chapter 4)

Modern zoo housing and husbandry often aim to provide the animals with many of the opportunities for behaviour that they have in the wild. Nevertheless, the zoo environment can be quite different from many wild environments and, if we want our zoo animals to behave in the same way as their wild counterparts (**Fig. 1.4**), then we must look closely at the precise ways in which zoo environments differ from the wild and how these differences affect behaviour.

This chapter examines some of the features of zoos that influence behaviour—an area of study that is growing in research importance. It also looks critically at the notion that the 'wild' provides an appropriate benchmark for evaluating and interpreting all aspects of animal behaviour.

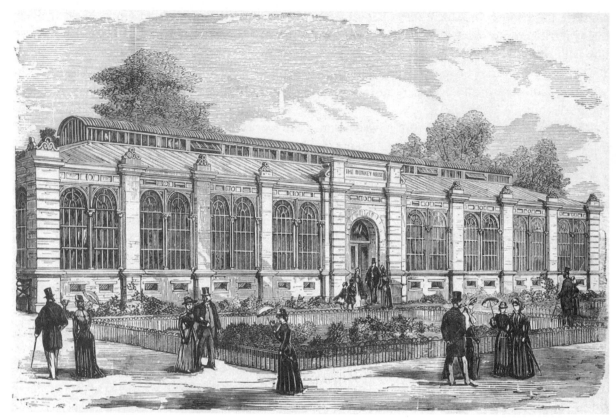

Figure **1.2** The monkey house at London Zoo in late Victorian times (from John Fletcher Porter (1980) *London Pictorially Described*). (Picture: Lee Jackson, `www.victorianlondon.org`)

Figure **1.3** The United Nations Convention on Biological Diversity (CBD) was a major outcome of the 1992 'Earth Summit' in Rio de Janeiro, and forms the framework for much national and international legislation on the role of zoos in conservation. (Picture: Secretariat of the CBD)

Figure **1.4** This young Hamadryas baboon *Papio hamadryas* has just been chased by a more dominant animal, and is showing a submissive facial signal and body posture. Social interactions within groups are an important part of the behaviour of both wild-living and zoo-housed animals. (Photograph: Geoff Hosey)

Figure **1.5** Knowing about your animals requires you to be able to recognize them. The ear tag allows easy recognition of this young bongo *Tragelaphus eurycerus*. (Photograph: Paignton Zoo Environmental Park)

1.4.4 Animal identification and record keeping (Chapter 5)

An essential part of managing collections of zoo animals is knowing your animals (**Fig. 1.5**) and what has happened to them in their lives. For example, a knowledge of previous health issues and veterinary treatments is important in managing an animal's future health, and a knowledge of ancestry and genetic relationships is necessary for planning **captive breeding programmes**.

For most zoos, the days of filing cards have long gone and animal records are maintained on computer databases. This opens up opportunities to look beyond individual zoos and to start to identify trends in whole **taxa**, so these records provide a valuable research database over and above their usefulness in the day-to-day management of the animals.

The use of zoo records for research is considered in more detail in this chapter and also in **Chapter 14**.

1.4.5 Housing and husbandry (Chapter 6)

Housing for zoo animals has changed greatly over the past 150 years and it is rare nowadays—at least in the accredited zoos that are the main focus of this book—to see relatively **barren enclosures** made from concrete, iron bars, and wire. Some ideas of zoo architecture have, in the past, leant more in the direction of what looks good to the visitors rather than what is appropriate for the animals. Today, it is considered necessary to plan housing for zoo animals from multiple points of view, taking into account all of the main 'stakeholders': animals; the public; the zoo staff. This chapter considers these points of view, and examines the different kinds of housing that can be seen in zoos today.

Husbandry is about the day-to-day maintenance of the animals (**Fig. 1.6**), an area that is still very much dominated by experience rather than empirical data, and the extent of knowledge and practice in zoo husbandry is covered in the rest of this chapter.

1.4.6 Animal welfare (Chapter 7)

Of course, zoos have changed over recent years in more than just the design of their enclosures. An increasingly detailed understanding of the physical and psychological needs of animals (see **Fig. 1.7**) has led to their welfare being a priority for zoos, and this welfare priority recognizes that animals have needs over and above simply being comfortably housed and fed. There is now a large body of theory underlying our knowledge of welfare issues and, although this has largely been developed in the context of farm and laboratory animals, considerable advances have been made in zoo animal welfare as well—and this is the subject of this chapter.

The term **taxon** (pl. **taxa**) refers non-specifically to any unit of **taxonomy** or classification, so it can variously refer to a species, a genus, an order, or any other classification category. The system of animal **taxonomy** is briefly described in **Chapter 5**.

Figure **1.6** Housing and husbandry of aquariums can involve getting into the water to clean the tanks. (Photograph: Andrew Bowkett)

Figure **1.7** This giraffe *Giraffa camelopardalis* is showing an oral stereotypy (repetitive licking or chewing movements), which suggests that it has experienced less-than-optimal welfare at some point in its life. (Photograph: Vicky Melfi)

Figure **1.8** Some enrichments are designed to make the animal spend more time in feeding and foraging. (Photograph: Harriet Elson)

1.4.7 Environmental enrichment (Chapter 8)

In some respects, the concept of **enrichment** brings together the behaviour and the welfare of zoo animals. One of the reasons why it is considered important to encourage zoo animals to show similar behaviours to those they display in the wild is the belief that their welfare is compromised if they cannot. Enrichment aims to fill gaps in the opportunities that animals have to display their full range of behaviours (in other words, to change their behaviour—see **Fig. 1.8**), and sometimes to change physiology or other aspects of the animal's biology. What these should be changed to is a debatable issue, which is considered in this chapter, along with descriptions of the main forms of enrichment.

1.4.8 Captive breeding (Chapter 9)

Breeding of zoo animals is often popularly seen as a good thing in its own right, because it implies good welfare and a contribution to conservation. In fact, for many zoo animals, breeding is part of a planned and managed process (**Fig. 1.9**), which is intended to result in self-sustaining zoo populations with high **genetic diversity** and low

Figure **1.9** Captive breeding is an important function of the modern zoo, as illustrated by the successful mating of these spur-thighed tortoises *Testudo graeca*. (Photograph: Warsaw Zoo)

Figure **1.10** Giant waxy tree frog *Phyllomedusa bicolor*. Although this species is not yet of serious conservation concern, it is suffering increasing human pressure because its skin secretions are seen as medicinal. Amphibians as a group are under significant threat of extinction. (Photograph: Douglas Sherriff)

inbreeding (although the extent to which zoos achieve this goal is the subject of much debate and some justified criticism).

Part of this managed breeding process might involve the use of various technologies for assisting (for example, *in vitro* **fertilization**, **artificial insemination**) and monitoring (for example, **ultrasonography**, **hormone assays**) reproduction. These techniques are described in this chapter, which also considers some of the necessary consequences of managed breeding, such as the issue of surplus animals.

1.4.9 Conservation (Chapter 10)

One of the most significant changes in zoos since their formation has been a shift towards an active role in conservation and education, rather than only the exhibition of exotic species. Human involvement in the extinction of species, through hunting and habitat destruction, goes back to prehistoric times and has been implicated in the extinction of large vertebrates in Australia, Madagascar, and the Americas. In historic times, the first extinctions to be noticed were probably the dodo *Raphus cucullatus* and Steller's sea cow *Hydrodamalis gigas* in the eighteenth century. There is now widespread recognition of the threats to species and the habitats in which they live, and most accredited zoos see the conservation of endangered species as one of their highest priorities (**Fig. 1.10**).

This chapter considers both what zoos can do, and are doing, in terms of maintaining self-sustaining populations (the *ex situ* **conservation**) of animals, with possible **reintroduction** to the wild for some; and their increasing involvement with *in situ* **conservation**. It also considers some of the problems that zoos face in deciding how many species can be maintained in zoos and which species these should be.

1.4.10 Health (Chapter 11)

A major component of the management of zoo animals is the need to ensure that they are healthy and in good condition (**Fig. 1.11**), and this is covered in **Chapter 11**.

Figure **1.11** Maintaining health in zoo animals includes a variety of preventative measures, e.g. checking the oral health of animals, as in this case, where a margay *Felis wiedii* is being examined. (Photograph: Colchester Zoo)

Threats to an animal's health can often come from parasites and infections, but may also result from an inappropriate diet or some other aspect of husbandry. Inevitably, much of our knowledge of zoo animal health issues is derived from experience and case studies. This chapter reviews current practice in preventing illness and treating sick animals, and also looks at the role of veterinary staff and keepers within the zoo in relation to maintaining good health of the animals.

1.4.11 Feeding and nutrition (Chapter 12)

Like health, the nutrition of zoo animals is an area of husbandry that relies, to a great extent, on experience and case studies. The natural diets of wild-living animals are often poorly known and, even if they are known, they may contain items that are difficult or impossible for the zoo to provide. Diets must be devised for zoo animals that meet their nutritional and health requirements, but also, if possible, their behavioural requirements, because foraging for and processing food are activities that often occupy a large part of an animal's time in the wild.

This chapter briefly reviews the underlying physiological and ecological theory relevant to zoo animal nutrition, and then considers how this is applied in the zoo setting (**Fig. 1.12**) to provide appropriate and nutritionally correct diets for the animals.

Figure **1.12** Providing a nutritionally balanced diet is only one of the considerations that have to be borne in mind when delivering an appropriate diet to zoo-housed animals. The logistical and practical requirements of delivering food are also important, as with preparing carcasses. (Photograph: Colchester Zoo)

1.4.12 Human–animal interactions (Chapter 13)

The zoo is a setting in which human and non-human animals[3] come into contact with one another (**Fig. 1.13**) in a very intense and intimate way, but the consequences of this had not really been studied empirically until recently. It is important to know what zoo visitors want and get from their zoo experience, and it is also important to know if their presence and behaviour has any effect on the animals. There is also the likelihood that keepers and other zoo staff themselves play an important part in the animals' lives, and, in turn, affect their behaviour and welfare. These issues, together with the zoo's role in educating visitors, are considered in this chapter.

1.4.13 Research (Chapter 14)

Importantly, most of the changes that have led to the modern zoo, and the way in which zoos manage and display their animals, are built on a solid foundation of scientific research. Even as recently as the 1960s, most of what people thought they knew about zoo

---

3 We know that humans are animals too, but to avoid being too wordy, we will refer to non-human animals simply as 'animals' from now on.

Figure 1.13 Interactions between people and zoo animals come in various forms. Here, a penguin *Pygoscelis papua* pecks a zoo visitor's foot. (Photograph: Sheila Pankhurst)

animals and the correct way of keeping them was anecdotal. Early attempts to record other aspects of the zoo than the basics of physical maintenance of animals were attempted by people such as Heini Hediger, director of Zurich Zoo, but, again, his books are largely based on personal observation and experience, rather than on systematic science.

Scientific research in zoos is now a significant undertaking, encompassing areas such as behaviour (**Fig. 1.14**), nutrition, reproductive biology, and **population genetics**, among other things. This chapter examines the current state of zoo research and also considers some of the methodological issues that frequently, but sometimes unnecessarily, worry those about to engage in zoo research.

1.4.14 We hope you enjoyed your visit (Chapter 15)

The final chapter of this book attempts to look ahead into the future of the zoo, and to reflect on the changing role of the modern zoo. With current concerns about climate change and related issues such as **sustainability** now high on the political agenda (see **Fig. 1.15**), we are likely to see zoos trying to become not only conservation organizations, but

Figure 1.14 Good practice in the care and management of zoo animals is based on systematic scientific research. Here, a student collects behavioural data as part of an undergraduate project. (Photograph: Sheila Pankhurst)

Figure **1.15** The next challenge towards which zoos are working is reducing their impact on the environment, which can be achieved by moving towards sustainable operating systems, including recycling. (Photograph: Paignton Zoo Environmental Park)

sustainable organizations in the widest sense, with policies on sustainability affecting all aspects of the day-to-day operation of major zoos. In this chapter, we also provide information about careers in the zoo world, for readers who would like to, or already, work with zoo animals.

1.5 The naming of names

Included within the text of this book are a lot of names of species of animals, of zoos, and of other organizations. Referring fully to all of them each time they are mentioned can result in a very cumbersome text and we have tried to avoid this wherever possible. It is important to identify the **scientific names** of the animals to which we refer so that we all know exactly which species we mean. We may all know what a chimpanzee is—but do we all know to what we are referring if we mention the blue poison dart frog (about which you can read more in **Box 10.3**)? We have therefore used the scientific name of each species when we first mention that species in each chapter, but not at subsequent mentions. (Scientific names of species are also given in the index at the end of this book.)

There is a similar problem in referring to the names of zoos. We have tried to give sufficient information for the reader to know where a zoo is when we mention it, without making a very unwieldy text. We have not used any particular rule about this—just a bit of discretion. Not all readers, for example, might know where Brookfield Zoo is, but we might expect them to know where Chicago is, so we would refer to 'Brookfield Zoo, Chicago' rather than to 'Brookfield Zoo, Chicago, IL, USA'. Where it makes it clearer, however, we have used a country name, rather than only a city name.

We have also tried not to be too cumbersome when introducing technical terms. Throughout the text, we have highlighted what we think are significant words on the first occasion that they appear in each chapter and have endeavoured to explain, or define, them when they first appear, or at appropriate places through the text. At the end of the book is a glossary of some of the terms that are particularly used in zoo world, some of which may not be familiar to all readers. We hope that this, together with the index, will guide the reader through any terms that are new or unfamiliar to them.

1.6 Finally . . . abbreviations

We have tried very hard to make this book accessible and readable, and to avoid any unnecessary jargon. But the zoo world is riddled with acronyms and it is hard to avoid using some of these from time to time when talking about the work that zoos do. For those of you who do not know your 'EEPs' from your 'TAGs' —let alone what 'ZIMS' and 'SPARKS' might stand for—we apologize now. We have tried to explain all specialist terms and acronyms where they first arise in the text and, to make life a little easier for our readers, there is a list of acronyms at the front of the book.

Chapter 2 History and philosophy of zoos

The word **zoo** is now used more or less ubiquitously around the world to refer to a collection of **exotic animals** kept in captivity and on show to the general public. Yet just 200 years ago, the word 'zoo' did not exist. Before about 1800, collections of exotic animals were usually referred to as **menageries**. Many of these were private collections, kept for the amusement of wealthy individuals. It was not until after the opening of the Zoological Gardens in Regent's Park (London Zoo—see **Fig. 2.1**) that the word 'zoo' came into widespread use, in the latter part of the nineteenth century.[1] Nowadays, the word 'zoo' seems to be drifting out of favour again—at least in some parts of the world—with the advent of titles such as 'safari park', 'wildlife park', and **biopark** for the wild animal collections that used to be referred to as zoos.

In this chapter, we will trace the history of zoos around the world and also look at changes in public attitudes towards wild animals in captivity (**Fig. 2.2** provides a timeline for zoos, from the menageries of ancient civilizations to the bioparks and **ecosystem zoos** of today). Zoos nowadays attract very large numbers of visitors,

but can also be the subject of debate and sometimes even criticism. What should be the role of zoos in today's society? Is it acceptable for zoos to keep wild animals for reasons other than conservation? Alongside an overview of zoo history, we will also briefly discuss in this chapter some of the philosophical and ethical issues that arise from keeping wild animals in captivity.

First, however, a caveat: the history of zoos is a very long one, going back at least 4,000 years, and there is consequently a great deal of material available about zoos throughout the ages, only a very small fraction of which can be included here. This chapter does not consider zoos in South America, for example (although there is a rich history of the keeping of exotic animals by the Aztecs and, to a lesser extent, by the Incas), nor does it look in any detail at zoos in India, Russia, or Eastern Europe. The best we can do in a book of this size is to provide suggestions for further reading at the end of each chapter, together with references to key papers and text books.

The main topics considered in this chapter are as follows.

Special topics, such as zoo architecture and the **ethics** of keeping animals in captivity, are included in boxes throughout this chapter, as are brief accounts

of some iconic individual zoo animals, the stories of which mirror a particular period in zoo history. These animals include Jumbo, an African elephant

The word **menagerie** comes from the French word *ménage*, meaning a household or unit of people living together. The word *ménagerie* was used from the 1500s onwards in France to refer to the management of a farm or collection of livestock and gradually came into use to refer to a collection of exotic animals.

1 Kisling (2001) notes that the word 'zoo' first appeared in print to describe Clifton Zoo (now better known as Bristol Zoo), but that the word was popularized as a result of a music hall song about London Zoo, 'Walking in the Zoo on Sunday'.

Figure **2.1** This illustration of the Reptile House in London Zoo in late Victorian times shows the menagerie style of exhibiting animals in closely spaced rows of cages (from John Fletcher Porter (1890) *London Pictorially Described*). (Picture: Lee Jackson, `www.victorianlondon.org`)

Loxodonta africana at London Zoo, and Willie B., a western lowland gorilla *Gorilla gorilla gorilla* housed at what is now Zoo Atlanta in Georgia, USA. There are also short accounts of influential curators and zoo directors, such as Abraham Dee Bartlett, William Conway, and Gerald Durrell.

2.1 What is a zoo?

In **Chapter 1** of this book, we looked at how zoos are defined and also set out the scope of this book, which is largely concerned with **accredited zoos**. So the focus of this book broadly reflects the definition of a zoo provided by Norton *et al.* in *Ethics on the Ark* (1995), which opts for the definition of a zoo or aquarium as:

> A professionally managed zoological institution accredited by the American Zoo

For readers who skipped this bit in **Chapter 1**, **accredited zoos** are zoos that are members of organizations such as the British and Irish Association of Zoos and Aquariums (BIAZA), and which strive to reach high standards in areas such as conservation, education, welfare, and husbandry.

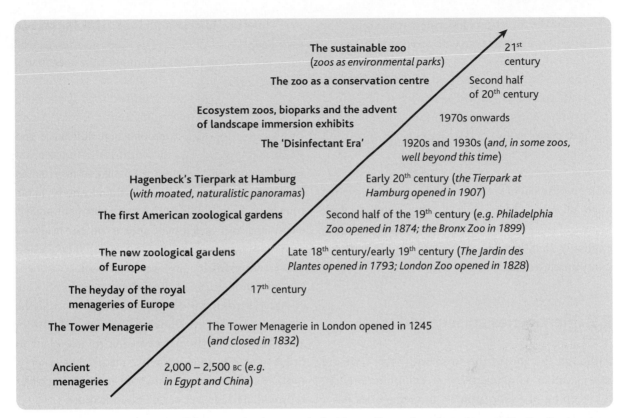

The sustainable zoo   21st
(*zoos as environmental parks*)   century

The zoo as a conservation centre   Second half
of 20th century

Ecosystem zoos, bioparks and the advent
of landscape immersion exhibits   1970s onwards

The 'Disinfectant Era'   1920s and 1930s (*and, in some zoos,
well beyond this time*)

Hagenbeck's Tierpark at Hamburg   Early 20th century (*the Tierpark at
(*with moated, naturalistic panoramas*)   Hamburg opened in 1907*)

The first American zoological gardens   Second half of the 19th century (*e.g. Philadelphia
Zoo opened in 1874; the Bronx Zoo in 1899*)

The new zoological gardens   Late 18th century/early 19th century (*The Jardin des
of Europe   Plantes opened in 1793; London Zoo opened in 1828*)

The heyday of the royal   17th century
menageries of Europe

The Tower Menagerie   The Tower Menagerie in London opened in 1245
(*and closed in 1832*)

Ancient   2,000 – 2,500 BC (*e.g.
menageries   in Egypt and China*)

Figure **2.2** This figure shows some of the major developments in the history of zoos throughout the ages. The timeline is not drawn to scale.

and Aquarium Association and having a collection of living animals used for conservation, scientific studies, public education, and public display.

(We might, perhaps, add a sentence or two about due regard for animal welfare, and delete the word 'American'.)

In contrast, Baratay and Hardouin-Fugier, in their book *Zoo: A History of Zoological Gardens in the West* (2002), offer a harsher definition of a zoo, as '*a place of forced meeting between animal and human*'. But note that this definition could also be used to describe farms, some laboratories, and, indeed, even many **ecotourism** sites.

What a zoo is—and what a good zoo should be—may, of course, be two very different things. Most readers of this book would probably tend towards the Norton *et al.* (1995) definition as a description of what a good zoo should be, but will, at the same time, be aware that there are many zoos in the USA, Australia, Europe, and the rest of the world that fall far short of these standards.

The US Association of Zoos and Aquariums (AZA)[2] provides for its members a detailed and quite prescriptive definition of a zoo or aquarium:

A permanent cultural institution which owns and maintains captive wild animals that represent more than a token collection and,

2 The AZA is the USA's leading accrediting organization for zoos and aquariums, and has its offices just outside Washington, DC, in Maryland. There is no 'A' for American in the acronym.

under the direction of a professional staff, provides its collection with appropriate care and exhibits them in an aesthetic manner to the public on a regularly scheduled basis. They shall further be defined as having as their primary business the exhibition, conservation, and preservation of the earth's fauna in an educational and scientific manner.

It is a definition that allows for the cultural importance of zoos (which tacitly acknowledges their historical role) and also, interestingly, acknowledges explicitly that there is, or should be, an aesthetic component to zoos.

2.2 Menageries ancient and royal

Although the zoo in the form with which we are now familiar is a relatively recent phenomenon, the keeping of exotic animals in menageries has a long history and there is evidence of these kinds of collections from the ancient world dating back at least 4,000 years.

2.2.1 The ancient menageries: Egypt, Mesopotamia, Greece, and China

Egypt

The earliest known records of exotic animals kept in captivity in ancient Egypt go back to around 2,500 BC. Pictures and hieroglyphics at Saqqara cemetery, near Memphis, show that the Egyptians kept many species of antelope in captivity, as well as baboons,[3] hyaenas, cheetahs, cranes, storks, and falcons (Lauer, 1976). Other records from ancient Egypt show that Tutmosis (Thutmose) III kept wild animals in the gardens of the famous temple at Karnak, near Luxor in Egypt (Strouhal,

1992) and the Pharaoh Rameses II, who reigned over Egypt from 1298–1235 BC, kept giraffes, as well as a pet lion that accompanied him into battle (Kisling, 2001).

Mesopotamia

In Mesopotamia (the area between the Tigris and Euphrates rivers, largely within the borders of modern-day Iraq), there are records of lions being kept in captivity by royalty more than 2,000 years BC. Carvings of other wild animals, including monkeys, elephants, and antelopes, appear on the walls of Assyrian royal palaces (Kisling, 2001). The earliest **ecosystem exhibit** may even date back to ancient Mesopotamia. Records show that Sennacherib, king of Assyria from 704–681 BC and believed to be the creator of the famed Hanging Gardens of Babylon, also created an artificial wetlands environment in order to exhibit marsh plants and animals (Dalley, 1993). (Further information about ecosystem exhibits is provided in section **2.3.5** and in **Box 2.6**.)

China

During the Zhou dynasty (c. 1000–2000 BC) in China, walled-in parks were created to house animals. The later Han dynasty (from c. 200 BC) saw the creation of private menageries comprising animals such as birds, bears, and tigers; one of the large royal parks of the Han period contained animals such as alligators, deer, elephants, and rhinoceroses (Schafer, 1968).

Greece

The ancient Greeks kept collections of wild animals, for study and enlightenment as well as for show. Aristotle (384–322 BC), for example, had his own private menagerie and wrote the earliest known zoological encyclopedia, *The History of Animals*, in 350 BC.

---

3 Scientific names of species have only been provided in this chapter where there is reasonable certainty over the identity of the species being referred to. Many historical accounts of early menageries and zoos refer to animals such as 'tigers' or 'the elephant', without giving further details.

(a)

(b)

Figure **2.3** These models show the complex system of lifts and pulleys that was employed to raise animals from underground storage areas up to the arena in the Colosseum, in ancient Rome. (Photographs: Geoff Hosey)

2.2.2 The Roman gladiatorial menageries

The Romans kept many wild animals—not so much for education and contemplation, but rather for slaughter in the gladiatorial arena, often on a grand scale. Any visitor to Rome today can still see clearly at the Colosseum the extensive subterranean holding areas for men and exotic animals. Archaeological evidence points to the use of sophisticated winches and elevators or lifts to bring large animals up from their pens below ground to the level of the arena (**Fig. 2.3**).

Baratay and Hardouin-Fugier (2004) suggest that the Roman practice of killing large wild animals for public entertainment developed from the symbolic slaughter of animals such as elephants captured from their enemies in battle '*to avenge the losses inflicted by military* **pachyderms**'. Whatever the antecedents for this public killing, the scale of the slaughter in arenas such as the Colosseum was immense (Jennison, 2005). The Roman general Pompey, for example, is reported to have financed a single show in which twenty elephants, 500 or 600 lions, and numerous other animals were killed (Kyle, 2001). The logistics of capturing, transporting, and maintaining in captivity 500 or 600 lions are astounding and beg the question of how many more lions died before reaching the Colosseum.

The slaughter of so many animals (and humans)[4] for public entertainment meant that many wild beasts became rare, or even locally extinct, in Roman times. The Roman appetite for contests between man and beast is believed to have contributed to the extinction of the hippopotamus in Nubia, the lion from Mesopotamia, the tiger from modern-day Iran, and the elephant from northern Africa (today, the African elephant is a sub-Saharan animal).

Some did speak out against the killing—notably Cicero, who asked: '*What pleasure can a cultivated man find in seeing a noble beast run through by a spear?*' (See Shackleton-Bailey, 2004, for a translation of Cicero's *Epistulae ad Familiares*, or 'Letters to Friends'.)

2.2.3 From the Dark Ages to the royal menageries of Europe of the seventeenth century

There are very few records of zoos of any kind during the Dark Ages and medieval times (from the fall of the Roman Empire in the fifth century

4 It is unthinkable nowadays that large numbers of human lives could be viewed as being readily expendable in the cause of public entertainment. This serves to highlight the very different attitudes to non-human animals that have prevailed at different times throughout history (and that still prevail, to some extent, in different cultures today).

to around AD 1450). Exceptions are the collections of exotic animals maintained during this time at Baghdad, Cairo, and Istanbul (Constantinople), and also in the parks and gardens of the Yuan and Ming dynasties in China (Kisling, 2001). In Western Europe, Kisling (2001) records that the Emperor Charlemagne (AD 742–814) maintained animal collections at several of his royal estates, with animals such as elephants, lions, bears, and camels among the inhabitants. In Britain, Henry I (the son of William the Conqueror) enlarged an animal collection that had been started at Woodstock in Oxfordshire by his father; this menagerie included lions, leopards, lynx, and camels.

In the thirteenth century, the Holy Roman Emperor Frederick II was the first person in several hundred years to establish a major new zoo in Europe. This was at Palermo in Italy and the collection included elephants, a giraffe, leopards, camels, and monkeys. Frederick II wrote several authoritative books on birds and falconry, and set up permanent zoos in three other Italian cities. He also contributed to the history of zoos elsewhere in Europe as a result of his habit of sending wild animals as gifts to other European heads of state—notably, to his brother-in-law, England's Henry III. Three leopards sent from Italy to Henry III became the first inmates of the infamous royal menagerie in the Tower of London; these animals were later represented on Henry III's coat of arms (Keeling, 1984). Again in Italy, a number of the popes at this time maintained animal collections at the Vatican. Pope Leo X (1475–1521), for example, kept lions, leopards, tropical birds, and other animals, as well as an elephant that had been given to him by the King of Portugal.

During its heyday in the seventeenth century, the animal collection at the Tower of London was typical of the zoos of its time: a private collection of exotic animals kept initially for the entertainment of the king and his court. By the late sixteenth century, nearly all kings and princes in Europe had at least one private menagerie. These collections were rarely open to the public; they were emblems of state, and the animal houses were often elaborate and lavish. The Dutch were a notable exception, building zoos as places of study, which were open to the public for a fee.

This era of zoos in Europe existing largely as private royal collections lasted for around 200 years, extending well into the eighteenth century. One of the most stylish of the royal zoos of the seventeenth century was that of Louis XIII at Versailles, with its formal, symmetrical ground plan, and elaborate sculptures and decoration. Unfortunately, the fate of many of the animals in the Versailles menagerie was slaughter by the proletariat during the early days of the French Revolution. (A few of the Versailles animals did survive and were later transferred to the Jardin des Plantes in Paris, of which more in the next section.)

At the same time—and in stark contrast to the gardens at Versailles, which were accessible only to a small elite—travelling circuses and exhibitions put on by showmen moving from town to town meant that ordinary people could still see exotic animals. From the sixteenth to the eighteenth centuries, there are numerous records of wild animals, including elephants, tigers, and bears, being transported in wagons to the towns and villages of Europe (Baratay and Hardouin-Fugier, 2004).

2.3 The development of the modern zoo

From the start of the eighteenth century onwards in Europe, zoos saw a change not so much in the animals they kept, or how they kept them, but in their audiences. Gradually, through the eighteenth and nineteenth centuries, zoos became public places, rather than private and exclusive menageries usually reserved for royalty or the very rich and powerful. This change in zoo audiences mirrored a wider public interest in natural history, as exploration and the growth of the various European empires saw more and more exotic animals brought back

(a)

(b)

Figure **2.4** These drawings of (a) a chimpanzee *Pan* spp., and (b) an Asiatic rhinoceros *Rhinoceros unicornis* at the Jardin des Plantes around about the middle of the nineteenth century show that the manner of keeping animals in Paris at that time was not unlike that which we saw in London in **Fig. 2.1**. (Pictures from *Cassell's Popular Natural History*, not dated.)

from travels in far-flung corners of the world. By the nineteenth century, the 'modern' zoological garden had been born, with the Jardin des Plantes in Paris and Regent's Park Zoo in London vying for position as world leaders.

2.3.1 The new zoological gardens of Europe of the eighteenth and nineteenth centuries

The nineteenth century is regarded by many as a golden age of exploration and natural history. Specimens brought back from expeditions (not least Charles Darwin's voyages on *The Beagle*) led to a rapid growth in the size and importance of natural history museums and, by association, zoological and botanical parks and gardens.

The Jardin des Plantes, Paris

In Paris, a new public menagerie was created in 1793, on the site of the old Jardin du Roi (renamed during the French Revolution as the Jardin des Plantes). This operated as a division of the Muséum d'Histoire Naturelle (Museum of Natural History), for which two new chairs of zoology[5] were established. There is some evidence that the academics and administrators of the Paris Museum of Natural History had not intended to create a menagerie: the new name of the site was, after all, the Jardin des Plantes. Ownership of a number of wild animals was forced upon the Museum when the police department took action to confiscate exotic animals from the many travelling exhibitors in and around Paris, and delivered these to the Museum (**Fig. 2.4**).

5 One of the new chairs of zoology at the Museum of Natural History in Paris, the wonderfully named 'Chair of Insects and Worms', went to Jean-Baptiste Lamarck, now better known for his controversial theories about evolution.

Figure **2.5** Frédéric Cuvier was probably the first ever curator of a national zoo and held that post for more than 30 years at the beginning of the nineteenth century, during which time he ensured that the Jardin des Plantes became a focus of scientific study. (Picture from *Cassell's Popular Natural History*, not dated.)

The first director of the Jardin des Plantes was Étienne Geoffroy Saint-Hilaire; his colleague and another distinguished professor at the Museum of Natural History was Georges Cuvier, known for his detailed paintings of plants and animals (Kisling, 2001).

It was, however, Frédéric Cuvier (**Fig. 2.5**), Georges' younger brother, to whom the day-to-day responsibility of managing the animals in the new menagerie was given, from 1804 onwards. Cuvier's post, which he held for more than 30 years (to 1838), was one that had never really existed before: curator or scientific superintendent of a national zoo that was open to the public (Burkhardt, 2001). The young Cuvier consequently had to learn quickly how to care for the exotic animals arriving at the Jardin des

Plantes from all corners of the world. As he wrote in his first guidebook to the zoo: '*Nothing has been written, nearly nothing has been seen, everything remains to be done.*'

Cuvier's background was as a chemist and his belief that menageries could become for zoologists what laboratories were for chemists helped to lead to the rapid establishment of the Jardin des Plantes as a place for scientific study, with the live animals in the menagerie as much a focus of research as the dead specimens in the adjacent museum (Burkhardt, 2001). Cuvier was an early pioneer of the study of animal behaviour and demonstrated a keen interest in what we would now refer to as animal welfare and **enrichment**. Of the raccoons in his care, for example, he wrote:

> It would be necessary to see them under other conditions, that is to say sufficiently free and under circumstances sufficiently diverse, so their faculties could be in some measure developed.
>
> (Burkhardt, 2001)

Tiergarten Schönbrunn (Vienna Zoo)

Another notable European zoo founded in the eighteenth century (and one that is still going strong today, under the proud title of 'Europe's oldest zoo') is the Tiergarten Schönbrunn in Vienna (**Fig. 2.6**), founded in 1752[6] by the Emperor Franz Stephan. Collections of exotic animals at Schönbrunn are recorded well before the eighteenth century, but these were private affairs (the menageries of royalty and noblemen), whereas Vienna Zoo was open to the public from the start and, indeed, entrance to the Tiergarten at Schönbrunn was free of charge from its creation until 1918.

Animals for the Vienna Zoo arrived only gradually, bought from the travelling menageries of the

---

6 Some sources give 1751.

Figure **2.6** This cage, in which zoo visitors are currently standing, is one of the old animal enclosures at Vienna Zoo. The cheetahs *Acinonyx jubatus*, which could formerly have been housed inside this cage, are now in a large enclosure outside and the cage has been appropriated for visitor use. (Photograph: Vienna Zoo)

time or captured by Emperor Josef II's expeditions to Africa and the Americas. The first elephant to be exhibited at Schönbrunn arrived in 1770, marking the start of a long tradition of elephant keeping at Vienna that resulted, in 1906, in the first birth of an elephant in captivity in Europe, and, in 2001, in the birth of the first test-tube baby elephant, Abu. In 1924, the Zoo appointed its first scientist, Otto Antonius, as director and, in July 2002, Vienna Zoo marked its 250th birthday by opening an innovative new rainforest exhibit.

Regent's Park Zoo, London

What is now London Zoo, in Regent's Park, was the brainchild of Sir Stamford Raffles (who also founded the port of Singapore). Prompted by the growing reputation of the Jardin des Plantes in Paris as a centre for scholarly activity, Raffles drew up a prospectus for a new academic society and zoo in London. The Zoological Society of London's first council meeting was held in May 1826; the zoo in Regent's Park opened just under 2 years later, in 1828.

The Zoological Gardens were an instant success with the public (see **Fig. 1.2**) and became a fashionable place to be seen (Hancocks, 2001). A song, 'Walking in the Zoo on Sunday' by a music hall artist called the Great Vance became a hit and the word 'zoo' gradually entered popular vocabulary (Bostock, 1993). Sadly, Raffles did not live to see his vision materialize: he died of a stroke, in July 1826, on his 45th birthday (Barrington-Johnson, 2005).

Box **2.1** Jumbo the elephant

David Hancocks opens the first chapter of his book *A Different Nature: The Paradoxical World of Zoos and their Uncertain Future* (2001) with an account of the first elephant to be housed at London Zoo. This was a young male African elephant named Jumbo. Captured from the wild in Ethiopia and subsequently purchased by an animal dealer, Jumbo arrived at London Zoo in 1865. He grew up to be a magnificent specimen, fed on a daily diet that reportedly included bread, biscuits, and cakes, as well as hay and oats (Preston, 1983). Members of the public could, for a small payment, ride on Jumbo, and he was well known and much loved by London Zoo's visitors.

After 17 years in captivity, however, Jumbo started to show periodic outbursts of aggressive behaviour when he became difficult to handle. When, in 1882, the Zoo was offered a large sum of money for Jumbo by Phineas T. Barnum, of Barnum and Bailey's Circus in the USA, the Council of the Zoological Society of London gratefully accepted, relieved to have been given an opportunity to get rid of an animal that the Zoo was no longer confident it could control. The news that Jumbo was to be sold—and worse, sold to a circus in the USA—generated a huge public outcry (Barrington-Johnson, 2005; see **Fig. 2.7**), but the Council stood by its decision and Jumbo was transported to the USA. The massive elephant was a star of Barnum and Bailey's Circus for 3 years, before being killed in a freak collision with a freight train.

Even before his untimely death, Jumbo had become something of an icon. His name came into popular use to denote anything of a grand size (Barrington-Johnson, 2005), from cigars to jars of peanut butter. Passengers travelling on

Figure 2.7 When Jumbo was sold by London Zoo to Barnum and Bailey's Circus in the USA, the resulting public outcry was encapsulated in the British press in cartoons such as this one. (Picture: `www.historypicks.com`)

jumbo jets today probably do not realize that the popular name for these planes has its origins in the name of a zoo elephant. As David Hancocks points out (2001): '*No other animal's name has become so deeply embedded in our language.*' Jumbo's story, as Hancocks reminds us, is of an animal captured from the wild as an infant, sold on to a dealer and then to a zoo, fed an inappropriate diet, displayed as a curiosity in an unnatural setting, used as a plaything for visitors to ride upon, and finally meeting an untimely death. Hancocks ends his account of Jumbo's life by saying: '*Jumbo's display brought out the best and worst in people. Like many other zoo animals, and like many zoos themselves, Jumbo satisfied the curiosity of all, the vanity of many and the greed of a few.*'

There is a postscript to Jumbo's story. The skeleton of this massive animal ended up at the American Museum of Natural History, in New York. More than a century after the elephant started to behave unpredictably and dangerously, the curator of mammals at the museum, Richard Van Gelder, noticed that the molars in Jumbo's jaw were growing inwards rather than erupting properly (Van Gelder, 1991). This abnormal pattern of tooth development was no doubt due, at least in part, to the unsuitable low-roughage diet Jumbo had been fed in captivity. The probable pain from these impacted molars provides a highly plausible explanation of the animal's increasingly aggressive and erratic behaviour.

During the second half of the nineteenth century, London Zoo developed many innovative exhibits (**Fig. 2.8**), including the first reptile house in the world (in 1849); this was followed by the first public aquarium, in 1853, and the first insect house, in 1881 (Hancocks, 2001). The first children's zoo opened in 1938. The Zoo was also remarkably successful at keeping and breeding wild animals in captivity at this time.[7] Its records for the period between 1860 and 1880 show that species bred in the Zoo included the giraffe, the spotted hyaena *Crocuta crocuta*, the chevrotain *Hyemoschus aquaticus*, a species of giant skink, the sunbittern *Eurypyga helias*, and many others (Kisling, 2001).

For a time, London Zoo was a world leader and spawned many imitations around the world. Not until 1907, with the opening of Carl Hagenbeck's zoo in Hamburg, did the pattern of zoo development take another great leap forward.

London Zoo's influence was due to two key factors:

- firstly, it was founded on scientific principles (more on these in a moment);
- secondly, it was created in a large, public open park with informal, naturalistic landscaping.

The prospectus for the Zoological Society of London (ZSL), under whose auspices London Zoo was created, stated that the Society would acquire 'animals . . . from every part of the globe . . . as objects of scientific research, not of vulgar admiration' (Olney, 1980). Another objective of the ZSL was 'the advancement of zoology and animal physiology'. Initially, the zoo in Regent's Park developed along **taxonomic** lines, with collections based on representative members of a single genus.

2.3.2 Hagenbeck and the Hamburg Tierpark

After the Jardin des Plantes and London Zoo, the next great leap forward in zoo design was prompted not by a scientist or an architect, but by a German animal dealer and trainer, Carl Hagenbeck. Although Hagenbeck is best remembered now for the design of naturalistic zoo enclosures, in his day he was also known as a prolific animal collector and trader, counting Barnum and Bailey's Circus among his many customers (Rothfels, 2002).

Hagenbeck's vision was of a zoo without bars, based on images of natural habitats (**Fig. 2.9**). He began with a series of travelling displays, to which he referred as 'panoramas', and then—in collaboration with Urs Eggenschwyler, a Swiss sculptor—he

---

7 The captive breeding of native species in zoos is not only a recent development. London Zoo was successfully breeding British adders *Vipera berus*, grass snakes *Natrix natrix*, and smooth snakes *Coronella austriaca* in the 1850s, soon after the opening of the new reptile house (Kisling, 2001).

Figure **2.8** This is a view of the lion house at London Zoo in 1896. The building was sizeable (234 ft × 50 ft), and was warmed and ventilated. The cages, which were seen as large in their day, were separated from the public by a broad barrier. This was seen as an advanced design for animal exhibitory in Victorian times. (Photograph: Lee Jackson, `www.victorianlondon.org`)

developed a permanent zoo park with concrete and cement rocks and gorges, based on real geological formations (Hancocks, 2001). This was the Tierpark at Stellingen near Hamburg, which opened in 1907 and was, from the outset, a huge success with the German public. Visitors flocked to see the artificial mountain landscape built to house African animals and the 'Polar Panorama' with its Arctic animals (Rothfels, 2002).

Hagenbeck's panoramas marked the beginning of a move away from collections presented on taxonomic lines towards collections representing the fauna of regions of the world, such as the African savannah, or the Arctic. But perhaps the greatest legacy of Hagenbeck and Eggenschwyler is their use of moats and ditches, rather than fences and bars, to separate animals from visitors. In this respect, they took zoos 'beyond the bars' at the start of the twentieth century. Hagenbeck, who was an accomplished animal handler and trainer as well as a dealer, carefully calculated the maximum distance that animals such as lions and tigers could jump before building his moats. He later wrote of the animals in his zoo:

Figure **2.9** This old photograph shows one of the panoramas designed by Carl Hagenbeck for the Hamburg Tierpark. Hagenbeck pioneered the use of moats, rather than cages and bars, to separate animals from zoo visitors. (Photograph: Archiv Hagenbeck, Hamburg)

I wished to exhibit them not as captives, confined within narrow spaces and looked at between bars, but as free to wander from place to place within as large limits as possible.

(Hagenbeck, 1909)

2.3.3 The first American zoological gardens

Prompted by developments in Europe, the USA saw the creation of its first zoological parks towards the end of the nineteenth century. Philadelphia Zoo, with its mission to educate and entertain, opened in 1874 and Cincinnati Zoo in 1875 (see **Box 2.2** for the story of a species extinction that happened at this zoo); other cities soon followed. The New York Zoological Park, better known as the Bronx Zoo,[8] opened in 1899.

By comparison to Europe, the history of North American zoos only stretches back to a little over a century ago. The USA had circuses prior to the nineteenth century, and there were travelling menageries and some private collections of exotic animals, but there were none of the elite menageries associated with wealth, power, and royalty that

---

8 The parent body of the Bronx Zoo is the Wildlife Conservation Society (WCS).

Box **2.2** The last passenger pigeon

Cincinnati Zoo was the home of the last known passenger pigeons *Ectopistes migratorius*. These birds were once found in vast numbers across the eastern USA and Canada, but, by the early part of the twentieth century, the species had become extinct (**Fig. 2.10**). The loss of the passenger pigeon is made all the more dramatic by the historical reports of the size of the flocks of this bird, which probably numbered into the hundreds of millions. Schorger (1973) has estimated that, in the time of the naturalist John James Audubon (1785–1851), one in every four birds in North America was a passenger pigeon. Early settlers reported flocks taking several hours to pass overhead and told of the sky being darkened by the multitude of birds (Kisling, 2001).

Despite some initial success at breeding passenger pigeons, Cincinnati Zoo's small flock dwindled from twenty individuals in 1881 to just three surviving birds by 1907. A female known as Martha, the last of these birds, died in 1914 and, with her death, the species vanished (Schorger, 1973). Martha's body was presented to the Smithsonian Institution in Washington and Cincinnati Zoo has preserved the aviary in which she lived as a memorial to all endangered species (Kisling, 2001).

Figure **2.10** This old print shows a passenger pigeon *Ectopistes migratorius* (with a Carolina turtle dove behind). Within decades of this picture being produced, the passenger pigeon was extinct. (Picture: `www.historypicks.com`)

were typical of European countries. The new North American zoos were civic institutions (**Fig. 2.11**), and unlike the major 'new' zoos in Europe, did not (usually) develop out of links to museums or universities (Hanson, 2002).

A number of the early zoological gardens in the USA were quick to follow Hagenbeck's lead and built naturalistic, moated enclosures; these zoos included those of Denver, St Louis, Chicago, and Cincinnati (Kisling, 2001).

2.3.4 The 'Disinfectant Era'

The 1920s and 1930s saw the start, in some zoos, of a depressing move away from Hagenbeck-inspired naturalistic exhibits towards 'modern' and often minimalist enclosures. In the interests of hygiene,

(a)

Figure **2.11** Some early pictures of American zoos: (a) schoolchildren feeding bears at the Smithsonian's National Zoo in 1899; (b) hyenas (published as 'laughing hyenas') at New York's Central Park Zoo in the nineteenth century. (Photographs: (a) Smithsonian's National Zoo; (b) www.historypicks.com)

cages were designed primarily for ease of cleaning rather than with regard to the needs of the animals housed within them (Hancocks, 2001).[9] The trend for sterile cages with concrete floors and tiled walls persisted well into the 1960s and 1970s. Regrettably, cages of this type can still be seen in many European zoos (see **Fig. 2.12**). This period of zoo design and management is sometimes referred to as the **Disinfectant Era** (Hancocks, 2001) or the **Hygiene Era**. As Hancocks (2001) has pointed out,

the use of tiled walls, concrete floors, plate-glass viewing windows, and steel doors in many of these minimalist enclosures leads to an environment that is not only sterile, but which is likely to be noisy in a way that promotes increased stress, with the sound of steel doors *'reverberating at a painful level in the hard acoustics of the cages'*.

Architects such as Russian-born Berthold Lubetkin (principal partner in the Tecton Group) must take their share of the blame for the sterile,

---

9 Both Heini Hediger and William Conway have referred to architects as the most dangerous animals in the zoo.

(b)

Figure **2.11** (continued)

Box **2.3** Willie B., the story of a lowland gorilla

'*On a rainy day in May 1988, a lowland gorilla named Willie B. stepped outdoors for the first time in twenty-seven years*': Elizabeth Hanson (2002) begins her excellent book about the history of American zoos with an evocative account of the story of an individual animal, a male western lowland gorilla *Gorilla gorilla gorilla* captured from the wild in Africa and acquired in 1961 by the zoo in Atlanta, Georgia. This was Willie B., who was housed, by himself, in an indoors enclosure with a concrete floor. Photographs from that time show an overweight animal, surrounded by bars and tiled walls (**Fig. 2.13a**).

Figure **2.12** This tiled enclosure is a good example of an enclosure that has been built with hygiene as its priority, because its design facilitates easy cleaning. When captive animal management was still in its infancy, it was thought that preventive health care and the elimination of disease risk would ensure healthy and reproductively active animals. (Photograph: Sheila Pankhurst)

Twenty-seven years later, in 1988, a new naturalistic gorilla exhibit was created at the Zoo, now renovated and renamed Zoo Atlanta. For the first time at the Zoo, Willie B. was able to step outside onto grass, amid trees (**Fig. 2.13b**). Willie B. lived on at Zoo Atlanta for another 12 years, during which time other gorillas were introduced into the exhibit. Willie B. even became a successful father, siring five surviving offspring (his second, a female gorilla born in 1996, was named Olympia to commemorate the Atlanta Olympics). He died in 2000, at the age of 41; more than 7,000 people attended a memorial ceremony held by the Zoo, and today, visitors to Zoo Atlanta can see a large sculpture of Willie B. in a special garden near the gorilla enclosure. As Hanson puts it: '*In his lifetime he had journeyed from being an object of voyeurism in a sterile cage to a muscular silverback, foraging for raisins and behaving like a gorilla.*'

The story of Willie B. encapsulates both the depressing era of hygienic, but sterile, zoo cages, with their easy-to-clean tiled walls ▸

Figure **2.13** Willie B., a lowland gorilla *Gorilla gorilla* sp., was one of the best-known and best-loved residents of what is now Zoo Atlanta, in Georgia. The two photographs here show him (a) housed alone in a tiled indoor enclosure and (b) after he was moved into a new, naturalistic outdoors enclosure. Other gorillas were gradually introduced into the new outdoor exhibit and Willie B. went on to father five offspring. (Photographs: courtesy of Zoo Atlanta)

and concrete floors, and also the advent of the new biopark zoo, with naturalistic exhibits designed to encourage animals to behave as they would in their natural habitat. Hanson points out that, during Willie B.'s lifetime, zoos in the USA, Europe, and elsewhere moved from collecting animals from the wild to participating in **captive breeding programmes**. Visitors to zoos no longer expected to see animals behind bars; zoos, in turn, began to take on more of a role in promoting the conservation of species such as gorillas.

minimalist designs adopted in many zoos in the 1930s. For a period of time, form came before function and the animals' needs often came a very poor second to architectural impact.

Lubetkin's penguin pool at London Zoo is perhaps the best-known example of a zoo building or enclosure from this period. Completed in 1934 with the assistance of Danish structural engineer Ove Arup and highly regarded by other architects, the pool is now a Grade 1 listed structure (see **Box 2.4** for further information about zoo architecture

and listed buildings). But it was never a particularly successful home for the penguins and the birds were eventually moved out in 2004 to a more naturalistic pool, where they are reported to be thriving (ZSL, 2005).

It is worth pointing out, however, that public reaction against some of the worst enclosures of the Disinfectant Era may well have speeded up the advent of the modern zoo, with its naturalistic settings and a new emphasis on seeing animals within the context of their ecosystem.

2.3.5 The advent of ecosystem exhibits, bioparks, and wildlife parks

The impetus for change

For a number of US and European zoos, the 1950s, 1960s, and 1970s was a period of stagnation and, in many cases, decline. The beginnings of an environmental movement in the 1960s, and of an **animal rights** movement in the 1970s, saw increasing public concern about poor zoos and even about whether zoos should exist at all (Donahue and Trump, 2006; see section **2.6.2** on anti-zoo campaigns). The growth in television programmes about wildlife emphasized the contrast between animals in sterile, badly designed enclosures, and the same species roaming free in their natural habitat. At the same time, people were being offered greater choice in how they spent their leisure time (Kisling, 2001). To survive, zoos had to compete with a range of other recreation facilities, from theme parks or amusement parks, to better and more-accessible sports facilities.

In the face of increasing criticism and declining visitor numbers, zoos in the latter half of the twentieth century needed to reinvent themselves, or face an uncertain future and the possibility of closure. Some zoos did close; others rose to the challenge of reinvention and emerged from a process of radical, and sometimes painful, change as a new

---

Box 2.4 Zoo architecture and protected buildings

Older city centre zoos, in particular, often have to manage their animals within constraints imposed by old and unsuitable buildings. This job is made more difficult when these buildings have some kind of legal protection, so that they cannot be demolished or even substantially altered without special permission. Within the UK, for example, a 'listed building' refers to a building included in the national List of Buildings of Special Architectural or Historic Interest. Buildings on this list are graded into categories according to their architectural or historical importance (Grades I and II* are the top categories) and are subject to additional planning controls and protection. Listed building consent is required not only to demolish a listed building or any part of it, but also to alter the building in any way that would affect its character, internally or externally.

London Zoo faces a special challenge in this respect (it has a rich architectural heritage, particularly if you happen to like early concrete structures). As well as Lubetkin's famous penguin pool, London Zoo has eleven other listed buildings, in relation to which even the smallest alteration to the fabric of the building requires special permission. These buildings include Hugh Casson's concrete elephant house, built in the early 1960s, and the 1932 gorilla house or 'Round House', another Lubetkin building (Barrington-Johnson, 2005). (Further information about London Zoo's architectural heritage can be found online at www.zsl.org/info/about-us—there is a link to ZSL architecture.)

The constraints imposed by the listed status of these buildings at London Zoo has invariably resulted in the animals that they were originally designed to house being moved elsewhere, to new sites or enclosures in which modern standards of welfare and husbandry can more easily be provided (the elephants have gone to Whipsnade Wild Animal Park near Dunstable; the gorillas are still at London, ▶

▶ but are now housed in the new £5.3m Gorilla Kingdom exhibit, which opened in March 2007). This leaves the Zoo with buildings it almost certainly will not be allowed to demolish, or even substantially to alter, but which are unsuitable for housing many of the species that the Zoo would like to keep.

Other zoos face similar problems: Bristol Zoo's former Giraffe House, as well the main entrance lodge and the south gates, are Grade II-listed buildings; Dudley Zoo, in the West Midlands, has no fewer than twelve Lubetkin and Tecton listed buildings (**Fig. 2.14**), of which five are categorized as Grade II*.

Figure **2.14** This early concrete structure was designed by Berthold Lubetkin's Tecton Group of architects and was used for many years to house polar bears at Dudley Zoo, as shown in this photograph taken in 1960. More recently, it has been accepted that enclosures like this do not meet the needs of the animals for which they were designed. But many such enclosures are now listed buildings and therefore cannot be demolished. (Photograph: Geoff Hosey)

generation of ecosystem zoos, wildlife parks, and bioparks. (**Box 2.5** looks at pioneering zoo directors, such as William Conway at the Bronx Zoo in New York and Gerald Durrell at Jersey Zoo, who were at the forefront of the drive to put wildlife conservation at the top of the zoo agenda.) Many lost, or tried

to lose, the name 'zoo' along the way, with varying degrees of success. For example, the Bronx Zoo became the 'International Wildlife Conservation Park' for a while in the 1990s, but has now abandoned the struggle, and has reverted to the name by which it has always been widely and popularly known.

Box **2.5** Pioneering zoo directors and curators

One of the best-known directors of London Zoo in Victorian times was Abraham Bartlett, who was appointed as Superintendent of the London Zoological Gardens in 1859 (the same year, incidentally, that Darwin published *On the Origin of Species*). He was then paid the princely salary of £200 p.a. (Blunt, 1976). Bartlett held the post of Superintendent for 38 years, until his death in 1897 (Vevers, 1976). During this time, he had to take responsibility for species sent to the Zoo from all over the world, many of which had never before been kept in captivity, and for which he had to develop appropriate diets and housing.

Bartlett proved to be a resourceful and highly successful director of the Zoo (Barrington-Johnson, 2005) and many 'firsts' in the captive breeding of exotic animals were recorded at London Zoo during his tenure. He also kept extensive and detailed notes of the animals in his care, and of their husbandry, which were used as the basis for the posthumous publication of two books, *Life Among Wild Beasts in the Zoo* (1890) and *Wild Animals in Captivity* (1898).

Almost a century later, in 1956, a young curator from the St Louis Zoo, named William Conway, joined the Bronx Zoo as its Associate Curator of Birds (Hancocks, 2001). By 1962, Conway was the Zoo's Director, and by 1966, he was General Director of the New York Zoological Society—which became the Wildlife Conservation Society (WCS) in 1993. He eventually became President of WCS as well, a post that he held until his retirement in 1999 (**Fig. 2.15**).

Together with other 'zoo pioneers' such as Gerald Durrell at Jersey Zoo and George Rabb at Brookfield Zoo in Chicago, Conway was at the forefront of moves to make conservation a central role of zoos. He was also largely responsible for putting the Bronx Zoo onto the world map as a leading example of what zoos could and should be doing, developing innovative and naturalistic exhibits such as The World of Darkness and The World of Birds. But perhaps Conway's greatest contribution to setting the standards for modern zoos was his emphasis on the Zoo's involvement with *in situ* **conservation** projects, ranging from the Freshwater Fish Conservation Program in Madagascar to monitoring populations of

Figure **2.15** William (Bill) Conway, formerly General Director of the New York Zoological Society and President of the Wildlife Conservation Society. (Photograph: AZA)

▶ endangered primates in Africa. By 1999, when Conway retired, WCS was pursuing more than 350 conservation programmes in 52 nations. In the year of his retirement from the presidency of the WCS, the AZA created the William Conway Chair for Conservation Science, in recognition of Conway's enormous contribution to wildlife conservation and, in particular, his lead in showing how zoos could support *in situ* conservation projects.

On the other side of the Atlantic, one of the most influential zoo directors of the twentieth century was Gerald Durrell, who founded Jersey Zoo in the Channel Islands. Known initially as an author (with best-selling books such as *My Family and Other Animals* to his name), the now world-famous zoo at Les Augres Manor in Jersey opened officially in 1959. Like Conway, Durrell was ahead of his time with his emphasis on conservation and the captive breeding of endangered species from around the world. In 1963, Durrell incorporated the Zoo into a new charitable trust, the Jersey Wildlife Preservation Trust. This was followed in the 1970s by the creation of an International Training School for young conservationists. To date, around 1,500 students have benefited from the experience of a placement and training at Jersey Zoo.

With Durrell's death in 1995, the Trust was renamed the Durrell Wildlife Conservation Trust, and Gerald Durrell's work has been continued by his widow, Lee. Today, the Trust is involved in efforts to save a wide variety of endangered species, particularly **island endemics**, ranging from the mountain chicken *Leptodactylus fallax* from Mauritius (this species is, in fact, a frog, not a bird), to the Lesser Antillean iguana *Iguana delicatissima*.

Finally, a brief mention of Desmond Morris, who as a young postgraduate researcher studied under Niko Tinbergen at the University of Oxford, and who was appointed Curator of Mammals at London Zoo in 1959. Morris did not become Director of London Zoo (he left to pursue other interests, particularly in writing and broadcasting), but he did greatly increase public awareness and appreciation both of the zoo world and of animal behaviour, largely as a result of his role as the presenter of television programmes such as the popular series *Zootime*, as well as a number of documentaries on animal behaviour.

Landscape immersion

In the 1970s, the Woodland Park Zoo in Seattle, USA, employed the architects Jones and Jones to help to develop a plan to renovate the zoo. For the first time, both the visitors and the animals were brought together in the same habitat—it was the advent of **landscape immersion** (Hancocks, 2001). Instead of wandering around a city park looking at replicas of desert, or savannah, or rainforest habitats, zoos visitors walked *through* the rainforest, with the

---

An **island endemic** is a species that is native and confined to an island or group of islands. Examples include the aye-aye *Daubentonia madagascariensis* and other lemur species on Madagascar, which originate from Madagascar and are not found in the wild anywhere else in the world.

Landscape immersion is defined on the useful ZooLex website (provided by the World Association of Zoos and Aquariums, WAZA, at www.zoolex.org) as 'a term used to describe zoo exhibitions that provide the illusion of "naturalism" to the point that visitors are perceptually immersed in the environment'.

animals and rainforest plants (or replicas of rainforest plants) all around them.

The move towards more naturalistic exhibits was not always welcomed at first, either by the paying visitors or by the zoo employees. Well into the 1970s, it was believed, for example, that attempts to house gorillas in exhibits with live vegetation were doomed to failure: the animals would simply destroy the plants or would injure themselves falling out of trees (Hancocks, 2001). The pioneering new gorilla enclosure at Woodland Park Zoo, Seattle, which opened in 1978, showed other zoos what could be done: Not only could gorillas safely be housed in large, outdoor enclosures with natural vegetation and even tall trees, but both the gorillas and the visitors

could benefit from the experience (Hancocks, 2001; see also Embury, 1992).

The concept of landscape immersion has now been adopted by other zoos, although not as rapidly and widely as some might have hoped. Hancocks (2001) gives as other examples of **immersion exhibits** the gorilla enclosures at Zoo Atlanta in Georgia, USA, and at Melbourne Zoo in Australia, as well as the African savannah exhibit at Disney's Animal Kingdom in Florida. Other good examples can be seen at Burger's Zoo, Arnhem (**Fig. 2.16**), in the Netherlands; within the UK, examples of landscape immersion exhibits are Bristol Zoo's Seal and Penguin Coasts, and Paignton Zoo's Desert House.

Figure **2.16** This photograph shows visitors at Burger's Zoo, Arnhem, in the Netherlands, exploring an immersion exhibit. This style of exhibit is designed to make visitors feel as though they are immersed in the animals' natural habitat. (Photograph: Andrew Bowkett)

Ecosystem exhibits and bioparks

An ecosystem exhibit is designed to represent an entire ecosystem, rather than a single species or a group of animals displayed on taxonomic or regional lines (see **Box 2.6**). In the USA in particular, the trend towards exhibiting animals within a portrayal of their ecosystem is known as the biopark (or 'BioPark') concept and has been promoted by, among others, Michael Robinson, director of the Smithsonian's National Zoo in Washington DC. Robinson (1996a; 1996b) explains that '*a bioexhibit should portray life in all its interconnectedness*'. In other words, in a biopark or ecosystem zoo, the plants, as well as the animals, are integral to the exhibit and the overall aim is to demonstrate the interdependencies between the living things in a particular ecosystem. Of course, a realistic portrayal of '*life in all its interconnectedness*' would also include predators and predation, but this is not always fully possible in the zoo environment, other than perhaps allowing some natural foraging for invertebrate prey.

Box 2.6 Ecosystem exhibits

An 'ecosystem zoo' or 'ecosystem exhibit' is one that attempts to recreate or to provide information about an entire ecosystem, not only the animals and plants that live within it. An ecosystem comprises both living organisms and their non-living environment, such as the rainfall, and the rocks and soils. A modern example of an ecosystem exhibit is the Masoala Rainforest at Zurich Zoo, in Switzerland. This is an award-winning greenhouse ecosystem within which temperature, humidity, and 'rainfall' are carefully controlled to match, as closely as possible, natural conditions in a Madagascar rainforest (**Fig. 2.17**). The plants are an integral part of the exhibit, not just a backdrop to the animals.

(a)　　　　　　　　　　　(b)

Figure **2.17** The Masoala Rainforest Hall was opened in 2003 at Zurich Zoo and represents a contemporary approach that brings together native flora and fauna in one exhibit. In this exhibit, visitors are transported to a 'real-life' Madagascan experience. (Photographs: Zurich Zoo)

Hancocks (2001) cites as a good example of an ecosystem exhibit the Louisiana Swamp at the Audubon Park Zoo in New Orleans. Here, the plants, animals, and people that live in the swamps of southern Louisiana are combined into an exhibit that extends to more than 2 hectares, with black bears, turtles, raccoons, otters, and alligators dispersed among the swamp cypresses and lagoons. In Europe, the Masoala Rainforest exhibit at Zurich Zoo has been widely praised (see **Fig. 2.17** and **Box 2.6**). This 11,000m^2 greenhouse ecosystem, which opened in 2003, promotes the species diversity of the Masoala National Park in Madagascar and also contributes to

raising funds (currently more than US$100,000 p.a.) for *in situ* conservation initiatives in the National Park (Bauert *et al.*, 2007).

The first ecosystem zoo in the USA (see **Table 2.1** for some other notable 'firsts' in the zoo world) did not open under the name 'zoo'. The Arizona-Sonora Desert Museum, which opened in Tucson, Arizona, in 1952, set out to be a 'living museum' of the plants and animals of the Sonora Desert. Each enclosure represented a particular habitat and signage provided information on the entire ecosystem (Kisling, 2001). These innovations were widely copied by other zoos in the latter half of the twentieth century.

Date	Initiative/development	Zoo/location
1793/1828	First openings of a major public zoo as part of a scientific institution	The Jardin des Plantes, Paris, France, as part of the Museum of Natural History (1793) London Zoo in Regent's Park, London, UK, under the umbrella of the Zoological Society of London (ZSL) (1828)
1847	Word 'zoo' first appears in the *Oxford English Dictionary*	A reference to Clifton 'Zoo', Bristol, UK, and, later, an abbreviation of the Zoological Gardens in London, UK
1849	Opening of first reptile house	London Zoo, UK
1853	Opening of first public aquarium	London Zoo, UK
1906	First birth of an elephant in captivity in Europe	Tiergarten Schönbrunn, Vienna, Austria
1907	First scientific zoo journal launched (*Zoologica*)	Published by the US New York Zoological Society
1938	First oceanarium opens in USA (displaying marine mammals)	Marineland, St Augustine, Florida, USA
1938	First children's zoo	London Zoo, UK
1943	First duck-billed platypus (*Ornithorhynchus anatinus*) bred in captivity	Healesville Sanctuary, near Melbourne, Australia
1952	First ecosystem zoo opens in USA	Arizona-Sonora Desert Museum, Tucson, Arizona
1953	First nocturnal house at a zoo	Twilight World, Bristol Zoo Gardens, UK
1963	First giant panda cubs born in captivity (as a result of artificial insemination)	Chengdu Giant Panda Breeding Research Centre,[*] China
1978	First 'landscape immersion' exhibit opens, at an American zoo	Gorilla World, Woodland Park Zoo, Seattle
2001	Birth of first test-tube elephant (Abu) in captivity	Tiergarten Schönbrunn, Vienna, Austria
2005	First aye-aye born in captivity in UK	Bristol Zoo Gardens, UK
2006	First captive-bred giant panda released back into the wild	Wolong Giant Panda Protection and Research Centre, China

Table **2.1** Some notable 'firsts' in the history of zoos and aquariums

* Now the Wolong Giant Panda Protection and Research Centre.

Zoos in the twenty-first century

So how are zoos likely to evolve from here? What will be the defining characteristics of zoos in the twenty-first century? We have tried to address some of these questions in the final chapter of this book (**Chapter 15**). We are already beginning to see the advent of the sustainable zoo, or at least a greater focus by zoos on **sustainability** in all of their operations. But before considering, in the final sections of this chapter, the philosophy and ethics of zoos, and how modern zoos see their role at the start of the twenty-first century, it is worth pausing to look specifically at the history of aquariums and how this differs from the history of zoos.

2.4 The history of aquariums

Public aquariums are a more recent phenomenon than public zoos (see **Table 2.2** for the dates of opening of some of the world's best-known zoos and aquariums), although brightly coloured ornamental fish have been kept in captivity in ponds and containers (both natural and artificial) for centuries, particularly in Asia (Kisling, 2001).

From the nineteenth century onwards, advances in the manufacture of glass tanks allowed fish to be viewed from the side, rather than only from above when in ornamental ponds (Hoage and Deiss, 1996).

Date of opening	Zoo or aquarium	Location
1245	Tower Menagerie *(closed in 1832)*	London, UK
1800	Exeter 'Change (Exchange) *(closed in 1828)*	London, UK
1752	Tiergarten Schönbrunn *(Vienna Zoo)*	Vienna, Austria
1793	Jardin des Plantes	Paris, France
1828	London Zoo *(part of the Zoological Society of London, or ZSL)*	Regent's Park, London, UK
1830	Dublin Zoo	Dublin, Republic of Ireland
1835	Bristol Zoo	Bristol, UK
1838	Natura Artis Magistra *(Artis Zoo)*	Amsterdam, Netherlands
1843	Antwerp Zoo	Antwerp, Belgium
1844	Zoologischer Garten *(Berlin Zoo)*	Berlin, Germany
1858	Diergaarde Blijdorp *(Rotterdam Zoo)*	Rotterdam, Netherlands
1861	Melbourne Zoo	Melbourne, Australia
1864	Moscow Zoo	Moscow, Russia
1868	Chicago Zoo *(Lincoln Park Zoological Gardens)*	Chicago, IL, USA
1873	National Aquarium	Washington DC, USA
1874	Philadelphia Zoo *(the first zoological park to open in the USA)*	Philadelphia, PA, USA
1889	National Zoo of the United States *(Smithsonian National Zoological Park)*	Washington DC, USA
1899	Bronx Zoo *(Wildlife Conservation Park)*	New York, NY, USA
1907	Tierpark *(Hagenbeck's Tierpark)*	Hamburg, Germany
1913	Edinburgh Zoo *(Royal Zoological Society of Scotland Zoo)* Burgers' Zoo and Safari	Edinburgh, Scotland Arnhem, Netherlands
1916	Taronga Zoo	Sydney, Australia

Table **2.2** Dates of opening of some major zoos in Europe, North America, and Australia*

* After Kisling (2001), who provides a detailed and comprehensive list of chronological information about the zoos and aquariums of the world.

Date of opening	Zoo or aquarium	Location
1923	Paignton Zoo	Paignton, UK
1928	Münchener Tierpark (Munich Zoo)	Munich, Germany
1931	Chester Zoo Zoologicka Zahrada Praha (Prague Zoo)	Chester, UK Prague, Czech Republic
1934	Belfast Zoo (originally known as Bellevue Zoo) Duisburg Zoo	Belfast, UK Duisburg, Germany
1937	Dudley Zoo	Dudley, UK
1956	Dierenpark Planckendael (Planckendael Zoo)	Mechelen, Belgium
1959	Jersey Zoo (Jersey Wildlife Preservation Trust)	Channel Islands
1963	Colchester Zoo Twycross Zoo	Colchester, Essex, UK Twycross, UK
1964	Sea World	San Diego, CA, USA
1970	Vancouver Zoo	Vancouver, BC, Canada
1971	Apenheul Primate Park	Apeldoorn, Netherlands
1972	Marwell Zoological Park	Hampshire, UK
1974	Toronto Zoo (Metro Toronto Zoo) (Riverdale Zoo was open from 1894 to 1974)	Toronto, Canada
1981	The National Aquarium	Baltimore, MD, USA
1984	Monterey Bay Aquarium	California, USA
1986	The Seas with Nemo and Friends (previously The Living Seas)	Orlando, FL, USA
1998	National Marine Aquarium Disney's Animal Kingdom Blue Planet (aquarium)	Plymouth, UK Orlando, FL, USA Ellesmere Port, Cheshire, UK
2002	The Deep (aquarium)	Hull, UK
2011	Biota! (The new ZSL aquarium at Silvertown Quays is scheduled to open in 2011)	London, UK

Table **2.2** (continued)

London Zoo was the first zoo to capitalize on this new technology, with the opening of the world's first public aquarium, known as the 'Vivarium' or 'Fish House', in 1853. The following extract from a newspaper of the day (the *Literary Gazette*) gives some idea of the impact of the opening of the Fish House:

> A living exhibition of the sea-bottom and its odd inhabitants is such an absolute novelty, that we must give our readers . . . some account of the elegant aquatic vivarium just opened to the public.
>
> (Kisling, 2001)

Given that the tanks had to be aerated by hand-operated pumps, the Fish House was home to a remarkable variety of species, with more than sixty species of fish and more than 200 different marine invertebrates on display (Barrington-Johnson, 2005).

The Fish House lasted until the early twentieth century, by which time the popularity of keeping marine and freshwater fish in tanks had waned. London Zoo acquired a new aquarium in 1924 and, in the twenty-first century, ZSL has ambitious plans for another new aquarium venture, Biota!, in London's docklands.

In much the same way that the Jardin des Plantes in Paris had prompted the opening of other 'new' zoological gardens in Europe, the Fish House at London Zoo led to the development of aquariums elsewhere in the UK and in mainland Europe. Among these is the Stazione Zoologica aquarium in Naples, Italy, which opened to the public in 1874. This is one of only a handful of nineteenth-century aquariums still open to the public (it is now part of an internationally important multidisciplinary biological research centre); another aquarium that has partially survived from this time is in Brighton in Sussex, now open as a Sea Life Centre.

The first public aquarium in the USA was opened in 1856 as part of the American Museum in New York, by Phineas T. Barnum (the same Barnum whom, as we saw earlier in this chapter, acquired Jumbo the elephant from London Zoo for exhibit in his circus). The first **oceanarium** in America, displaying marine mammals, opened in 1938 (see **Table 2.2**). This was Marineland, in St Augustine, Florida. Other marine mammal parks followed in towns and cities across Europe and the USA, although a few were short-lived and closed again soon after opening (for example, Marineland of the Pacific, in California, which was open from 1954 until 1987). The 1970s and 1980s, in particular, saw an expansion in the number of centres holding marine mammals for public display, although increasing concern for the welfare of these animals in captivity has seen a decline in recent years in the number of **dolphinariums** and oceanariums in the UK and, to a lesser extent, elsewhere.

In contrast to the decline in oceanariums, the past two decades or so have seen great modernization of aquariums (**Fig. 2.18**), and also the opening of a number of impressive and popular new aquariums, from Monterey Bay Aquarium in California (1984) to Océanopolis in Brest, in France (1990), and The Deep, in Hull (2002) in the UK (see **Tables 2.1 and 2.2**).

2.5 Zoos today

It is almost impossible to generalize about zoos today. Whether or not you believe that zoos are a force for good will depend both on your attitudes towards keeping wild animals in captivity (more on this in the next section) and on what type of zoo you are considering. At one end of the scale, there are zoos such as the Jersey Wildlife Preservation Trust, or Chester Zoo (both in the UK), or the Bronx Zoo in New York, with very high standards of animal welfare and husbandry, and a genuine and strong commitment to wildlife conservation. At the other end of the scale, there are far too many appalling 'roadside zoos' in far too many countries around the world, where animals languish in pitifully small cages, with very low standards of welfare.

As Hancocks (2001) has said:

> The history of zoos is replete with contradictions. People have set up zoos because they wanted to control big strong animals . . . in recent years there are increasing numbers of people who want to work in zoos because they are passionate about wildlife conservation.

But regardless of your personal beliefs, there is no getting away from the fact that zoos are numerous, highly popular, and almost certainly here to stay. So what is—or rather, what should be—the role of the modern zoo?

2.5.1 How does the modern zoo see its role?

In common with many organizations nowadays, zoos attempt to articulate their perceptions of their role and purpose through various positioning statements,

---

An **oceanarium**, or 'marine mammal park', houses marine mammals, such as seals, sea lions, and perhaps sea otters. A **dolphinarium**, strictly speaking, houses only dolphin species, which include the orca or killer whale *Orcinus orca*.

Figure **2.18** In much the same way as we have described advances in terrestrial captive animal exhibitory, similar advances have also been made in the display of aquatic animals. Here, you can see one of large tanks at the Aquarium at Burgers' Zoo at Arnhem, in the Netherlands, which houses a diversity of fish and other marine fauna. (Photograph: Julian Doberski)

aims, objectives, or mission statements. These identify what the zoo thinks its purpose is, what perspective it looks from, and what it particularly values in achieving that purpose. These statements are usually published for all to see on the zoos' websites, and a review of a number of these is sufficient to show that most are concerned with education, conservation, and providing a valuable experience for visitors. (Of course, zoos in many countries now have a statutory obligation to demonstrate involvement in conservation and education, so it would be somewhat surprising if these roles were not reflected in their mission statements.)

A typical example is Chester Zoo in the north of England, which states as its mission:

The role of the Zoo is to support and promote conservation by breeding threatened species, by excellent animal welfare, high quality public service, recreation, education and science.

Similarly, Paignton Zoo Environmental Park, in south-west England, '*is an education and scientific charity dedicated to conserving our global wildlife heritage and inspiring in its many visitors a life long respect for animals and the environment*'.

The Los Angeles City Zoo website has the simple banner heading, 'Nurturing Wildlife and Enriching the Human Experience'.

And Taronga Zoo, in Sydney, has as its mission statement, '*We will demonstrate a meaningful and*

Figure 2.19 This highly complex and large gibbon enclosure provides its inhabitants with the structural features necessary to perform **brachiation**, and many other species-specific behaviours necessary to the maintenance of their general health and welfare. Alongside this enclosure, Sydney's Taronga Zoo highlights the need to conserve this species and provides information to their visiting public about how they can contribute to the zoo's conservation activities. (Photograph: Julian Chapman)

urgent commitment to wildlife, our natural environment and the pursuit of excellence in our conservation, recreation and scientific endeavours' (**Fig. 2.19**).

These are all medium- to large-sized zoos with high attendance numbers, substantial animal holdings and resources that can be channelled into research, and *in situ* conservation (that is, conservation in the animals' habitat countries). But even smaller zoos espouse the same priorities. To consider only one, Linton Zoo in Cambridgeshire states that its emphasis is on conservation and education, while providing an enjoyable family day out (www.lintonzoo.com).

Overall, then, zoos see their primary purposes as being involved in the conservation of endangered species, and helping to foster public appreciation and interest in the natural world.

In many countries, zoos may belong to a national or regional association or federation of zoos. The aims and mission statements of these associations differ in some respects from those of their constituent zoos, in that they emphasize the fostering of between-zoo cooperation and the maintenance of high professional standards. The priorities they stress, however, are the same ones: namely, conservation, education, welfare, and research.

Brachiation is a movement seen in animals such as gibbons, which involves swinging from arm to arm through the trees.

In the UK and the Republic of Ireland the relevant association is The British and Irish Association of Zoos and Aquariums (BIAZA). (There is more information about zoo associations in **Chapter 3**, which looks at the legislative and regulatory framework for zoos.) BIAZA's mission states that it '*represents its members and promotes the values of good zoos and aquariums*'. It leads and supports its members:

- to inspire people to help conserve the natural world;
- to participate in effective cooperative conservation programmes;
- to deliver the highest quality environmental education, training, and research;
- to achieve the highest standards of animal care and welfare in zoos, aquariums, and in the wild.

The equivalent to BIAZA in the USA is the Association of Zoos and Aquariums (AZA). Its mission is very similar to that of BIAZA, as are those of other similar associations throughout the world.

So we might summarize the majority view of modern zoos as being that they see themselves as agencies for undertaking conservation—which they do through their collections and, in an increasing number of cases, in the wild as well—and that they also use their collections for education and raising awareness about the natural world. Of course, the rather cynical response to this would be that zoos in many countries now have a statutory obligation to demonstrate that they are involved in conservation.

But would zoos still exist if the paying public were not able to see exotic animals at first hand? As Michael Robinson, Director of the Smithsonian's National Zoo in Washington DC, points out, zoos are fundamentally:

> places of spectacle and entertainment. Research, education and conservation are functions which, in the last one hundred years or so, have been grafted onto the recreational rootstock of zoos.
>
> (Robinson, 1996a)

And this brings us to philosophical questions, such as whether or not zoos *should* exist and under what circumstances it may be ethically justifiable to keep wild animals in captivity.

2.6 The philosophy and ethics of zoos

Because of the charge that zoos are involved in keeping animals captive for the purposes of human recreation and entertainment, questions may be asked about the ethics of what zoos do and their underlying philosophy (see **Box 2.7** for a brief explanation of what is meant by 'ethics'). For example, on what

---

Box **2.7** What do we mean by 'ethics'?

The words 'morals' and 'ethics' are often used together, and, in some cases, as though they mean the same thing. The *Concise Oxford Dictionary*, for example, defines ethics as '*relating to morals; treating of moral questions; morally correct, honourable*' (Sykes, 1977). But are ethics and morals really the same thing? Before we can consider any ethical or moral concerns in relation to keeping wild animals in captivity in zoos, we need to be clear about exactly what we mean by these terms.

Many people will freely express moral concerns about keeping animals in captivity, whether in zoos, or as pets, or on farms, or in the laboratory. For example, some people may believe that it is wrong to keep dolphins or ▷

killer whales in captivity, but perfectly acceptable to keep tropical fish in a small tank in the living room at home. Most people in the UK would think it was wrong to throw stones at a sleeping lion in a zoo to wake it up, but in other parts of the world such behaviour might be encouraged by other visitors rather than frowned upon.

In other words, moral concerns and beliefs are an individual's view of what is right or wrong, and may result from careful thought about the underlying issues, or may just be based on a 'gut feeling' about whether or not something is acceptable (Straughan, 2003). As Straughan points out, ethics is a narrower concept than morality, and refers to a branch of philosophy that tries to clarify and analyse the arguments that people use when discussing moral questions. This involves probing the justifications that we have for believing that some things are right and other things are wrong.

So a good general description of ethics might be '*a set of standards by which a particular group of people decides to regulate its behaviour —to distinguish between what is legitimate and acceptable in pursuit of their aims, from what is not acceptable*' (Flew, 1979). Of course, different people have different moral outlooks and approach ethical decisions in different ways. For this reason, the recommendation of the UK Zoos Forum is that zoos should set up and utilize ethical committees rather than leave ethical decisions to individual people. The *Zoos Forum Handbook* provides advice on how to set up appropriate ethical panels, in both large and small zoos, together with suggestions on the sorts of members that such panels should have.

grounds can we justify keeping wild animals in captivity (**Fig. 2.20**)? How should zoo directors respond to the charge, from animal rights campaigners and others, that zoos are morally indefensible? And how should zoos approach the ethical paradox that conservation is all about species and populations, but welfare is about individual animals?

2.6.1 Should zoos exist?

The philosopher Dale Jamieson (1985; 1995) has argued that there is a '*presumption for liberty*' and that it is morally wrong to deprive animals of their freedom by keeping them in captivity. As Stephen Bostock (1993) points out, we condone the loss of liberty for humans only under very specific and limited circumstances (such as a prison term as punishment for law-breaking).

Bostock (1993) summarizes the three main arguments in defence of keeping wild animals in captivity as follows:

- keeping wild animals in captivity has advantages for humans (education, conservation, recreation, scientific discovery) and sometimes for the animals themselves too (conservation can be viewed as beneficial for populations of animals, if not always for individual animals kept in captivity);

- wild animals in captivity may not necessarily experience negative welfare and may, in some cases, be better off than they would be in the wild;

- animals are not sufficiently comparable to humans to make meaningful comparisons about the morality of captivity.

The first of these arguments reflects a broadly **utilitarian** approach to the question of whether or not wild animals should be kept in captivity (see **Box 2.8** for an explanation of terms such as **utilitarianism**). The last two arguments are essentially about animal rights.

(a)

(b)

Figure **2.20** When some people argue against the notion of keeping wild animals captive in zoos, it is possible that they may have a preconception that animals are still housed in cages like the one illustrated in (a). This cage housed a clouded leopard *Neofelis nebulosa* at London Zoo in the 1960s and the cage design has clearly not greatly advanced from the sort of cage used to house lynx *Lynx* spp. (b) at the Smithsonian's National Zoo in the early years of the last century. Since this time, and from reading of this chapter so far, it should be clear that zoo design has progressed greatly and that these images are thankfully reminiscent of a past that is no longer present in good zoos. (Photographs: (a) Geoff Hosey; (b) Smithsonian's National Zoo)

Box **2.8** Utilitarianism, holism, and animal rights

Before we can consider the impact of the animal rights movement on the zoo world, we need to understand the language and approaches used by writers and philosophers when discussing this area. The idea behind utilitarianism is that we should act so as to produce the greatest good (or 'utility') for the largest number of individuals (put crudely, 'maximum pleasure for minimum pain'). Utilitarians view the suffering of animals (such as rats and mice undergoing medical experiments in a laboratory) as acceptable, provided that the benefits to humans (for example, a cure for a disease) outweigh the costs to the animals (Appleby and Hughes, 1997). Of course, measuring the costs to the animals in this sort of case is far from straightforward.

In contrast to utilitarianism, animal rights campaigners generally hold that the rights of animals are equivalent to those of humans and that it is never acceptable to sacrifice the interests of one animal to benefit another. Peter Singer, the author of *Animal Liberation* (1990), is a strong proponent of the utilitarian viewpoint; another well-known philosopher, Tom Regan, is a proponent of the animal rights viewpoint. Some philosophers believe that we unthinkingly value benefits to humans more than costs to animals; this human tendency is referred to by Singer as **speciesism** (Appleby and Hughes, 1997).

The animal rights movement is a relatively recent phenomenon. It has developed over the past three or four decades, out of various human rights movements (it is worth noting that the General Assembly of the United Nations set out its Universal Declaration of Human ▷

▶ Rights as recently as 1948). Most animal welfare organizations now hold that animals do have certain rights, or entitlements, but that these are not equivalent to those of humans.

The other ethical viewpoint mentioned in this chapter is **holism**. From this perspective, inanimate objects such as rocks, water, and soil must also be taken into account: the ecosystem or life community as a whole is what matters when debating moral considerations. Regan (1995) cites the example of a fur trapper to illustrate this approach. Holists would have no problem with the trapping of wild animals for their fur, provided that the integrity and sustainability of the ecosystem were preserved.

The animal rights viewpoint: are zoos morally indefensible?

Tom Regan is a leading American philosopher who is the author of two influential books: *The Case for Animal Rights* (1983) and, with Peter Singer, *Animal Rights and Human Obligations* (1976, reprinted 1999). Regan (1995) has considered whether or not each of three main ethical viewpoints—namely, utilitarianism, holism, and the animal rights view (see **Box 2.8**)—can be used to make the case that zoos are morally indefensible. He comes to the, perhaps surprising, conclusion that none of these three ethical theories can be used as the basis of a watertight argument that zoos are morally indefensible. Even the animal rights view, which Regan strongly supports and which, as he says, '*takes a very dim view of zoos*' (Regan, 1995) cannot be used to argue against zoos in the (admittedly limited) circumstances under which it can be demonstrated that it is in the best interests of an animal to be kept in captivity for a period of time. He illustrates this with the example of an animal that would be killed by humans if it were to be allowed to remain in its natural habitat.

This short summary is only a very simplistic representation of Regan's conclusions in relation to zoo animals and does not give a full picture of his views on animal rights. We strongly recommend that readers with an interest in the ethics of keeping wild animals in zoos should read for themselves other books and essays on environmental ethics, several of which are listed in **Further reading** at the end of this chapter.

Other dilemmas for zoos

The animal rights view espoused by Regan and others is, of course, not the only critical view of zoos. Others have argued that zoos give a false and compromised picture of habitats and ecosystems, by concentrating on larger species—and particularly mammals—at the expense of the smaller species that make up a vital part of a functioning ecosystem (Hancocks, 1995). There is certainly much truth in this criticism, but zoos are gradually moving towards collections that reflect a more representative sample of the natural world and are also prioritizing keeping threatened species, not only examples of the (largely mammalian) **charismatic megafauna** (**Fig. 2.21**). This debate is revisited and expanded on in **Chapter 10** on 'Conservation'.

Zoos are by no means the only organizations that face these sorts of dilemmas: ethical concerns can arise from a very wide range of activities in which

Charismatic megafauna refers to the animals—nearly always mammals—that have strong popular appeal, such as tigers, pandas, wolves, and elephants. These animals are often (but not always correctly) assumed to be the species that zoo visitors most want to see.

Figure **2.21** Nothing better characterizes the concept of the 'charismatic megafauna' than a large animal that people find inspiring and exciting. The Asian rhino *Rhinoceros unicornis*, pictured here at Warsaw Zoo, epitomizes a species that would encourage many people to visit the zoo. (Photograph: Warsaw Zoo)

humans come into contact with animals. But this book is about zoos, so what are the other ethical issues that zoos need to address? Zoos face issues about the acquisition, transfer, and disposal of animals: for example, should zoos acquire animals from the wild? Routine management procedures can have an impact on welfare; the training of animals to facilitate routine husbandry tasks may also have welfare implications.

Conservation versus welfare?

The genetic management of zoo animals to reduce **inbreeding** and to maintain **genetic diversity** may require the sterilization of some animals, or their removal from breeding groups. This highlights another moral dilemma: as zoos shift their focus towards achieving conservation goals, they may find that these goals come into conflict with the provision of high standards of welfare for individual animals (the tension between the conservation of species and populations, and the welfare needs of individual animals is discussed further in **Chapter 7**). There are strong moral arguments on both sides, but there is clearly the potential for conflict between these two imperatives. Jamieson (1995) has said that '*conflicts of value are intrinsic to wildlife management*', but goes on to say that, just because these conflicts exist does not mean that we cannot make better decisions, nor that we are excused from reflecting on how they can be resolved.

Ethical guidelines for zoos

Support for zoos in resolving these and other ethical issues comes from bodies such as the Zoos Forum and from zoo associations. BIAZA, for example, has published an Animal Transaction Policy that, among other things, identifies good practice in the acquisition, transfer, and disposal of animals. Similarly, the North American zoo association AZA has published policies on the acquisition and disposal of animals, on animal contact with the public, and on other issues.

2.6.2 Anti-zoo campaigns

The 1960s and 1970s onwards saw, both in the USA and Europe, the emergence of a new generation of animal rights and animal welfare organizations concerned either specifically or peripherally with zoos. The first group of zoo animals to be targeted was the marine mammals. Captive killer whales *Orcinus orca* and other members of the dolphin family attracted much attention in the 1970s, 1980s, and 1990s. More recently, elephants have moved up the agenda. The publication in the UK of a study (Clubb and Mason, 2002) commissioned by the Royal Society for the Prevention of Cruelty to Animals (RSPCA) drew attention to a number of health and welfare issues relating to elephants in captivity in zoos. (There is more on this study, and its consequences, in **Chapter 7**.)

In the USA, activists from organizations such as People for the Ethical Treatment of Animals (PETA) and the Humane Society of the United States have joined forces over the past few years with local pressure groups to try to remove elephants altogether from a number of AZA zoos (Donahue and Trump, 2006). The activists argue that elephants should not be kept in small enclosures or small social groups, but should instead be sent to sanctuaries where they can roam in larger groups over a much larger acreage of land. Of course, the public may not be able to visit these sanctuaries to see the elephants.

Similar campaigns in Europe have been instigated by groups such as the Born Free Foundation, Care for the Wild, and Animal Defenders. Some anti-zoo campaigners argue that all animals should be removed from zoos (see McKenna *et al.*, 1987). In the UK, the Zoo Check programme (part of the Born Free Foundation) states explicitly on its website that it has worked since 1984 '*to phase out zoos*' and operates under the slogan '*Keep Wildlife in the Wild*'. Unfortunately, it is simply not feasible for all of the people who currently visit zoos each year to travel to see the same animals in the wild. As Bostock (1993) has put it: '*millions of people . . . could never, in comparable numbers, visit "the wild" without damaging it irreparably*'. And, to the suggestion that zoo visitors could view wildlife films or virtual exhibits rather than live animals, Donahue and Trump (2006) offer the response that '*their consistently high attendance at zoos indicates that they do not desire to replace zoos with television shows*'.

Summary

- The history of captive collections of exotic animals is a very long one, dating back at least 4,000 years. Some of the earliest records of collections of wild animals in captivity come from Egypt, Mesopotamia, China, and Greece.

- In the nineteenth century, scientific societies set up leading European zoos such as the Jardins des Plantes and London Zoo.

- By contrast, twentieth-century zoos, particularly in the USA, were usually municipal (civic) institutions.

- The so-called 'Disinfectant Era' of the early part of the twentieth century saw the introduction in many zoos of easy-to-clean, but sterile, cages, with tiled walls, smooth concrete floors, and steel doors.

- Zoo exhibits in the second half of the twentieth century have become more naturalistic, with a new generation of bioparks, ecosystem exhibits, and landscape immersion exhibits that do not focus only on a single species, but rather on animals in their natural environments.

- In the latter half of the twentieth century, zoos began to face organized opposition from a new generation of animal rights and animal welfare organizations.

- Many zoos, at least partly in response to criticism from animal rights campaigners, have reinvented themselves as conservation organizations.

- At the start of the twenty-first century, the best zoos are retaining their emphasis on conservation, and are also striving to become sustainable organizations and environmental centres.

Questions and topics for further discussion

1. Under what circumstances is it ethically acceptable to keep wild animals in captivity?

2. Are zoos a force for good in modern society?

3. What brought about the so-called Disinfectant Era or Hygiene Era in Western zoos? What are the advantages and disadvantages of this approach to captive animal management?

4. Why should zoos display ecology rather than taxonomy?

5. Discuss the following statement: '*Landscape immersion exhibits are good for visitors but not for animals.*'

6. Who should own zoos? Are zoos best managed at national or at city level, as public bodies or as private institutions?

7. What should be done about 'roadside zoos'? Should all zoos that do not belong to a formal zoo association such as BIAZA or AZA be closed down?

Further reading

The history of zoos

Accounts of zoo history abound, both in print and on various websites, but be warned that not all of these are accurate. We strongly suggest that anyone with a keen interest in zoo history checks, for themselves, wherever possible, the original sources.

For a detailed and thoroughly researched account of the history of zoos around the world, we recommend both Kisling's *Zoo and Aquarium History* (2001) and Baratay and Hardouin-Fugier's *Zoo: A History of Zoological Gardens in the West* (2004). Kisling's book is more global in its scope

and considers the history of zoos in other parts of the world, such as the Aztec and Inca collections, which, for reasons of space, have not been included in this chapter. It has a comprehensive reference list at the end for anyone wanting to delve still further into the records of the animal collections of ancient Egypt, or Greece, or Rome. For anyone wanting to learn more about the keeping (and killing) of wild animals during the days of the Roman Empire, Carcopino (1991) provides a vivid account of daily life in ancient Rome; George Jennison's classic, *Animals for Show and Pleasure in Ancient Rome,* first published in 1931, was reprinted in 2005 and is now available in paperback.

Another comprehensive account of zoo history is provided by Hoage and Deiss (1996), the editors of *New Worlds, New Animals: From Menagerie to Zoological Park in the Nineteenth Century*. Several of the chapters in this book are based on papers presented at a symposium on zoo history, held at the National Zoo in Washington in 1989. For an account of the history of US zoos, Elizabeth Hanson's *Animal Attractions: Nature on Display in American Zoos* (2002) is well worth reading, and features an interesting chapter on the history of wild animal collectors and zoo expeditions over the past century. Another recommended read is Nigel Rothfels' *Savages and Beasts* (2002). Rothfels traces the birth of the modern zoo to Hagenbeck's influence, and offers much detailed information both about Hagenbeck and about the creation of the Hamburg Tierpark.

In the introduction to *A Different Nature: The Paradoxical World of Zoos and Their Uncertain Future* (2001), David Hancocks provides a good overview of zoo history, as does Stephen Bostock in his book on *Zoos and Animal Rights* (1993). An even shorter, but nevertheless recommended, summary of zoo history is found in the 'Foreword' written by Michael Robinson, the Director of the Smithsonian's National Zoological Park, in Washington DC, for Hoage and Deiss (1996).

Finally, if you are not a herpetologist, do not be put off by the title of James Murphy's *Herpetological History of the Zoo and Aquarium World* (2007). This hefty book covers many aspects of zoo and aquarium history, and, although there is plenty about amphibians and reptiles in the book, there is also much information of general interest.

The philosophy and ethics of zoos

The basics

In an excellent booklet produced for the Biotechnology and Biological Sciences Research Council (BBSRC), Roger Straughan (2003) gives a clear and concise explanation of the difference between moral and ethical concerns, and goes on to provide short summaries of topics such as animal **sentience** and speciesism. Although the booklet is entitled *Ethics, Morality and Animal Biotechnology*, much of the content is relevant in a wider context. Straughan also poses the question 'What is an animal?' and rightly points out that you cannot fully consider animal ethics without first deciding what counts as an animal. The booklet can be downloaded, free of charge, as a PDF file from the BBSRC website, at www.bbsrc.ac.uk/organisation/policies/position/public_interest/animal_biotechnology.pdf.

We also recommend Armstrong and Botzler's *The Animal Ethics Reader* (2003).

Further reading on the philosophy and ethics of zoos

Four books on the philosophy and ethics of zoos stand out. Firstly, although it is a challenging read in places, we recommend *Ethics on the Ark: Zoos, Animal Welfare and Wildlife Conservation*, written and edited by Bryan Norton and his colleagues (1995). This includes chapters from moral philosophers such as Tom Regan and Dale Jamieson (although, frustratingly, this book lacks an index). More challenging still, but recommended nonetheless, is the book edited by Singer and Regan, *Animal Rights and Human Obligations* (1976, reprinted 1999), although some of the examples of poor practice in zoos cited by, for example, James Rachels (1976) in this book are now more than half a century old. Stephen Bostock, a zoologist and philosophy student who went on to work at Glasgow Zoo, has produced a much more accessible account of the history and philosophy of zoos in his book, *Zoos and Animal Rights* (1993). Finally, we recommend David Hancock's thoughtful and well-written text, *A Different Nature: The Paradoxical World of Zoos and Their Uncertain Future* (2001).

Zoo politics

Jesse Donahue and Erik Trump's *The Politics of Zoos: Exotic Animals and Their Protectors* (2006) is essentially a political biography of the American Association of Zoos and Aquariums (now known as the AZA). It is a book that offers detailed and useful insights into how zoos in the USA responded to the growing challenge from animal rights and animal welfare organizations in the latter half of the twentieth century.

Chapter 3 Regulatory framework

Zoos operate within a complex framework of conventions, directives, laws, procedures, policies, regulations, and recommendations. This framework extends from international agreements through to regional, national, and sub-national (for example, state) legislation and guidelines. The extent and influence of the legislative and regulatory framework varies considerably from country to country and from region to region. Much of what will be described in this chapter relates to the legal framework for zoos within Europe and within the UK in particular, but many other countries have similar operating frameworks.

The main topics that we will cover in this chapter are as follows.

Tables and boxes in this chapter provide additional information, such as a summary of key international conservation treaties and organizations (**Table 3.1**) and an overview of devolved legislation within the UK (**Box 3.2**). As in other chapters, acronyms are explained on first use (and are also listed at the start of the book).

Given the dynamic nature of legislation, some of the information provided in this chapter is likely to become out of date quite quickly. Web addresses can also change, as can the names or acronyms for national and international organizations. Other than checking all information at the time of writing, which we have done, there is little else that we can do except to add this caveat: the online resources accompanying this book are reviewed and revised regularly, and we would recommend that readers check these for any updates, particularly in relation to the regulatory framework for zoos.

3.1 The law and legislative processes: a brief introduction

Before considering the laws within which zoos operate, let us first briefly consider the law and legislative processes in general.

Laws are the rules that govern society. These rules set out both the rights and the responsibilities of citizens. Legislation is needed to ensure that people can live together safely and peacefully, and to enforce good practice. Sometimes guidelines and persuasion are not enough to ensure compliance with sensible welfare or safety precautions in zoos.

But a law is only effective if it can be implemented; there are many examples, both current and historic, of 'bad laws' that are difficult to enforce and widely flouted.

3.1.1 Civil and criminal law

In most countries in the English-speaking world, 'the law' is divided into civil law[1] and criminal law. Offences under criminal law can be punished by a

---

1 The term 'civil law' also refers to a type of legal system that is in widespread use today. Also known as European Continental Law, the civil law system is largely derived from Roman law.

jail sentence or a fine and the offender, once proven guilty, will have a criminal conviction. Offences under civil law determine liability for harm, rather than guilt, and are not punishable by a prison sentence, but damages may be payable. Liability in a civil court is decided on the 'balance of probabilities' rather than under the tougher criterion of 'beyond reasonable doubt' that is used for criminal law. (This applies to the law in the UK, USA, and Australia, for example, but not in all countries.)

Enforcement also differs between criminal and civil law. Criminal law is enforced by officers of the state bringing a criminal action. In the UK, criminal law is enforced by the police and, in England and Wales, the Crown Prosecution Service (CPS). (The equivalent of the CPS in Scotland is the Crown Office, and in Northern Ireland, the Public Prosecution Service.) A zoo in the UK could face criminal proceedings, for example, under the provisions of the Animal Welfare Act 2006. Breaches of civil law are enforced by people or organizations with an interest in the case (for example, a person who has suffered harm) bringing litigation in the civil courts. Another example would be if a civil action were to be initiated because an individual or an organization believed that breaches of regulations controlling the operation of a zoo had occurred. Such a civil action may not necessarily be taken directly against a zoo, but could be taken against a third party, such as a local council, for permitting a zoo to continue operation when it has not complied with licensing requirements.

3.1.2 Legislative processes at international and national levels

The laws governing the operation of zoos can be considered at three main levels: global or international; regional (European); and national. Countries such as the USA and Australia also have state or territory legislation at sub-national level, and many countries have local regulatory or licensing arrangements that augment national laws, such as local government by-laws in the UK (Cooper, 2003). Although the regulatory framework for zoos varies considerably between different countries, Cooper (2003) points out that there are some basic elements that are common across much zoo legislation, such as: the requirement for authorization to open and to operate a zoo (for example, via a licence or permit); processes for zoo inspections before a licence or permit is granted; and a set of minimum standards that must be met. Zoos are also often required by law to keep certain types of record.

There is no body that makes international laws that are binding on all of the nation states of the world. Instead, various international organizations (such as the United Nations) propose treaties and conventions, to which individual countries may or may not choose to sign up or 'join'. But once such treaties are incorporated into national law, they are every bit as binding as other national legislation.

Within Europe, the European Union (EU) can pass legislation that is binding on all member states, although this depends on the category of EU legislation that is put forward. There are three main categories of EU legislation:

- EC Regulations are directly binding upon member states;
- EC Directives are not directly binding on member states, and need to be incorporated into national legislation before they can take effect;
- EC Decisions are binding on member states, but are limited in scope (these come from the European Commission or Council of Ministers; see **Box 3.1** for a brief guide to the EU).

In the UK, primary legislation is made via Acts of Parliament. There is also secondary legislation, which is not debated in Parliament and which therefore can be amended more rapidly. EC Directives are often implemented within the UK as secondary legislation (the 2002 Regulations amending the Zoo Licensing Act 1981 (ZLA),[2] which is considered later

---

2 Zoo Licensing Act 1981 (Amendment) (England and Wales) Regulations 2002, SI 2002/3080.

Box **3.1** A brief guide to the European Union

What is now the European Union (EU) started life as the European Economic Community (EEC) in 1957, with the signing of the Treaty of Rome by France, West Germany, Italy, Belgium, the Netherlands, and Luxembourg. After 1967, the EEC became the European Community (or, more correctly, the European Communities), generally known as the EC. The UK joined the EC in 1973. After the Maastricht Treaty of 1993, the EC became part of the wider EU. Confusingly, the European Commission, which is part of the EU (essentially, it is the EU's 'civil service'), is also sometimes referred to by the abbreviation 'EC'. The presidency of the EU rotates between its member states every 6 months (the EC Zoos Directive, adopted in 1999, was a significant achievement of the UK presidency in the preceding year).

It is worth noting that the original Treaty of Rome, founding what has now become the EU, did not include any reference to animal welfare. A revision of the Treaty some 40 years later resulted in the signing of the Treaty of Amsterdam in 1997; this did include a Protocol on animal welfare.

Further information about the EU and about EU law can be found online at `http://europa.eu/`.

in this chapter, are a good example of secondary legislation). Some UK legislation is implemented by setting out specific standards, which can be changed and updated by the Secretary of State. It is also worth noting that not all of the UK Acts mentioned in this chapter have power in all of the UK's constituent countries; **Box 3.2** provides details of devolved legislation in Northern Ireland, Scotland, and Wales.

Box **3.2** Devolved zoo legislation within the UK

Scotland and Northern Ireland, in particular, and, for more recent legislation, Wales as well, often have their own versions of UK and European laws. In 1999, Scotland acquired its own Parliament; the same year saw the creation of the Welsh Assembly, with more limited devolved legislative authority.

Some UK legislation relating to zoos, such as that stemming from the international Convention on International Trade in Endangered Species of Wild Fauna and Flora (CITES) (UN, 1973), is not devolved, but in other areas, it is, and so zoos at different locations within the UK may have to meet different legal requirements for some aspects of their operation. The Animal Welfare Act 2006, for example, applies only to England and Wales. Scotland has its own Act: the Animal Health and Welfare Act (Scotland) 2006 (see `www.scotland.gov.uk` for further information). Where UK legislation is devolved, the full title of the Act makes this clear by naming the constituent countries to which the Act applies. So while the Zoo Licensing Act 1981 applies throughout the UK, the Zoo Licensing Act (Amendment) (England and Wales) Regulations 2002 apply only within England and Wales.

In practice, the devolved legislation is usually broadly similar across the UK.

3.2 The international framework: global conventions, agreements, and regulations

Examples of international conventions that impact on the operation of zoos include the 1973 Convention on International Trade in Endangered Species of Wild Fauna and Flora (CITES) (UN, 1973), the 1992 Convention on Biological Diversity (CBD) (UN, 1992), and the **Berne Convention**, which covers the conservation of European wildlife and habitats. (See Holst and Dickie, 2007, for a review of conservation conventions and agreements that influence zoos.)

A wide range of other international organizations, standards, and guidelines have some influence on how zoos operate, including the Food and Agriculture Organization (FAO), the World Conservation Union (IUCN), and the United Nations Environment Programme (UNEP). (**Table 3.1** explains all of these acronyms and provides brief details about each international organization or convention, together with website addresses for finding further information.)

Acronym	Full name	Role	Further information
CBD	United Nations Convention on Biological Diversity (often referred to as the Convention on Biodiversity)	The CBD was one of the main outcomes of the 'Earth Summit' at Rio de Janeiro in Brazil in 1992.	www.biodiv.org Montreal, Canada
CBSG	Conservation Breeding Specialist Group (part of the IUCN) (Before 1994, the CBSG was the Captive Breeding Specialist Group)	The CBSG, set up in 1979, was created to provide a link between the IUCN and zoos.	www.cbsg.org/cbsg/ Apple Valley, MN, USA
CITES	Convention on International Trade in Endangered Species of Wild Fauna and Flora	CITES is a trade convention, not a conservation convention. The CITES Secretariat is administered internationally by the United Nations Environment Programme.	www.cites.org Geneva, Switzerland
FAO	Food and Agriculture Organization of the United Nations	The FAO, established in 1945, has a mandate to work to '*defeat hunger*'. Its remit includes improvement of agriculture, forestry, and fisheries practices. The work of the FAO covers domestic animal transport and diseases such as avian influenza (H5N1).	www.fao.org Rome, Italy
IUCN	International Union for the Conservation of Nature and Natural Resources (also known as the World Conservation Union)	The IUCN issues guidelines and policy positions and statements (e.g. guidelines on species reintroduction) on wildlife conservation and also on the sustainable use of natural resources.	www.iucn.org Gland, Switzerland
OIE	Office International des Epizooties (World Organisation for Animal Health)	The OIE, created in 1924, monitors and disseminates information on animal disease worldwide via its website and periodicals, and other publications.	www.oie.int Paris, France

Table **3.1** Major international conservation organizations and conventions (in alphabetical order)

NOTE: Web addresses shown in this table were correct at the time of publication—but can change. An online search under the full name of any one of these organizations should take you to its current website.

Acronym	Full name	Role	Further information
SSC	Species Survival Commission (of the IUCN)	The Species Survival Commission is a network of experts who (on a voluntary basis) advise the IUCN on issues relating to conservation. It has a number of specialist groups, such as the coral reef fish group and the equids group.	www.iucn.org/themes/ssc/ Gland, Switzerland
TRAFFIC	TRAFFIC is not an acronym, but the name of a joint programme of WWF and the IUCN	TRAFFIC exists to monitor trade in wild plants and animals, and to reduce the threat of such trade to conservation. It works in close cooperation with the CITES Secretariat.	www.traffic.org Cambridge, UK
UNEP	United Nations Environment Programme	The UNEP, established in 1972, describes itself as 'the voice for the environment within the United Nations system'.	www.unep.org Nairobi, Kenya
WCMC	United Nations Environment Programme World Conservation Monitoring Centre	The UNEP-WCMC is the biodiversity assessment and policy implementation arm of UNEP.	www.unep-wcmc.org Cambridge, UK
WTO	World Trade Organization	The WTO was established in 1995 as the successor to the General Agreement on Tariffs and Trade (GATT) and deals with the global rules of trade between nations.	www.wto.org Geneva, Switzerland
WWF	Since 2000, WWF has been known only by its acronym, which originally derived from 'World Wildlife Fund', and then 'World Wide Fund for Nature'	WWF is an independent wildlife conservation organization and a charity that promotes and supports the conservation of both species and habitats. Founded in the UK in 1961, it is now a major international organization.	www.panda.org (The website for WWF in the UK is www.wwf.org.uk)

Table **3.1** (continued)

3.2.1 The 1973 Convention on International Trade in Endangered Species of Wild Fauna and Flora (CITES)

CITES was set up in 1973 at a meeting of representatives of 80 countries in Washington DC, USA, to discuss the regulation of wildlife trade for conservation purposes. CITES entered into force in 1975. Countries or states ('parties') that agree to be bound by CITES are said to have 'joined' CITES and must implement the Convention via their own national legislation.

CITES is an important international agreement, but it is, first and foremost, a trade agreement, not primarily conservation legislation. There are some highly endangered species, for example, in which

there is little international trade. For these species, CITES is of limited relevance. The major impact on zoos of CITES is the requirement for permits for the movement of animals (or animal parts or tissues) belonging to species listed by CITES as endangered or threatened. As Cooper and Cooper (2007) point out, however, prosecutions under CITES frequently involve offences in other areas as well, such as breaches of animal welfare or animal health legislation in the case of illegal transport of wild animals.

Approximately 5,000 animal and 28,000 plant species are listed in three appendices to CITES (see www.cites.org). These appendices are amended and updated in biennial Conferences of the Parties (CoPs). Appendix I of CITES covers species that

Figure **3.1** The Waldrapp, or northern bald ibis *Geronticus eremita* is listed in Appendix I of CITES. (Photograph: Sheila Pankhurst)

are deemed to be at highest risk—that is, those that are threatened with extinction and which may be affected by trade. For these species, trade is permitted only under exceptional circumstances. **Appendix I species** include the great apes (the bonobo or pygmy chimpanzee *Pan paniscus*, the common chimpanzee *Pan troglodytes*, the gorillas *Gorilla* spp., and the orang-utans *Pongo* spp.), as well as the Asian and African elephants *Elephas maximus* and *Loxodonta africana* (the African elephant populations of Botswana, Namibia, South Africa, and Zimbabwe are included in Appendix II rather than Appendix I). Avian species listed under Appendix I include the Waldrapp or northern bald ibis *Geronticus eremita*

(see **Fig. 3.1**); notable reptilian species listed in Appendix I include the Komodo dragon *Varanus komodoensis*. Appendix II of CITES lists species that are not considered to be under current threat of extinction, but which could become so unless trade is closely controlled.

The operation of CITES within the European Union

For the purposes of CITES, the EU is regarded as a single state and, within Europe, CITES is implemented via Regulation (EC) No. 338/97. This Regulation is legally binding on all EU member states.[3] Within each nation state, there is a CITES Management Authority, which has an administrative role, and which deals with the practicalities of issuing permits for the transport of animals and animal parts. There is also a CITES Scientific Authority in each state; this has an advisory role.

Within the UK, the relevant domestic legislation for CITES is the Control of Trade in Endangered Species (Enforcement) Regulations 1997 (the COTES Regulations), as amended in 2005 to increase the penalties for illegal trading in endangered species. The UK Government's Department for Environment, Food and Rural Affairs (Defra) is the CITES Management Authority in the UK and has responsibility for issuing permits for the movement of animals (and animal body parts and tissues) covered by CITES. The **Joint Nature Conservation Committee (JNCC)** is the CITES Scientific Authority (Fauna) within the UK and has an advisory role.

Because the EU is effectively a single state in relation to CITES, the movement of endangered species within the EU is relatively free (at least in relation to the Convention). There are, however, some EU amendments to the CITES appendices or lists of endangered species. These are published in four EU Annexes to CITES, known as Annexes A, B,

---

3 Within Europe, CITES legislation is implemented under criminal, not civil, law and transgressions can result in imprisonment or a fine.

The **Joint Nature Conservation Committee (JNCC)** is the official or statutory adviser to the UK Government on conservation issues (see www.jncc.gov.uk).

Figure **3.2** The European Union has its own Annexes to the CITES legislation. An example of an Annex A species that is not listed on CITES Appendix I is the black stork *Ciconia nigra*. (Photograph: © Robert Hardholt, `www.iStockphoto.com`)

C, and D. Annex A includes all CITES Appendix I species, as well as some Appendix II and III species for which the EU has adopted stricter measures. (For example, the black stork *Ciconia nigra* is an **Annex A species**—see **Fig. 3.2**—but is listed in CITES Appendix II, rather than Appendix I.)

A permit known as an **Article 10 certificate** is needed by zoos within the EU to authorize the sale or movement of animals listed under Annex A. (The summary of **Websites and other resources** at the end of this chapter gives details of a useful summary from TRAFFIC of EU animal trade legislation and required documentation.) Some European zoos have an **Article 60 certificate** (formerly an 'Article 30 certificate'), which allows them to display their Annex A species and also to move them to other zoos within the EU, provided that the other zoos also have an Article 60 certificate. (There is a useful briefing note on the movement of animals under Article 60 on the Defra website, at `www.defra.gov.uk/wildlife-countryside/gwd/pdf/article60-briefnote.pdf`.)

For animal species listed under EU Annexes A and B, permits require minimum standards for trans-port and housing (and animals moved under this legislation must be microchipped). Annex A animals that are captive bred rather than wild captured (that is, second generation, or **F2**, zoo animals) may, under some circumstances, be able to be moved to Annex B and then be transported more easily. Satisfying the CITES definition of 'captive bred' is not as straightforward as it might seem, however. All of the following criteria must be met:

- specimens must have been born or produced in a controlled environment;
- the parents must have mated (or gametes must have been transferred) in a controlled environment;
- the breeding stock must have been established and maintained in accordance with CITES Resolution Conf. 10.16 (Rev.).

Production of a second generation

Zoos wishing to move captive-bred (F2) Annex A animals under Annex B must be able to prove that they have met all of the above criteria. They can do this, for example, by reference to records maintained by the International Species Information System (ISIS) (see section **5.7.1**). In addition, animals moved under this legislation must be clearly marked, usually with a microchip.

3.2.2 The 1992 Convention on Biological Diversity (CBD)

Another international convention that impacts on zoos is the Convention on Biological Diversity (CBD) (see **Fig. 3.3**). Signed by 150 government leaders at the UN 'Earth Summit'[4] held in Rio de Janeiro in Brazil in 1992, the CBD is dedicated to promoting sustainable development. The CBD is an important convention and is the international umbrella for the 1999 EC Zoos Directive (see next section).

---

4 The Rio 'Earth Summit' in 1992 was a meeting of world leaders at the UN Conference on Environment and Development. Two major international agreements were signed at the Rio meeting: the Convention on Climate Change and the Convention on Biological Diversity.

Figure **3.3** The Convention on Biological Diversity was signed by 150 countries at the 1992 'Earth Summit' in Rio de Janeiro. The aim of the CBD is to promote sustainable development. (Photograph: Secretariat of the CBD)

The text of the CBD refers to *in situ* and *ex situ* **conservation** measures and Article 9 of the Convention specifically requires that:

> Each party shall . . . adopt measures for the ex-situ conservation of components of biological diversity . . . , establish and maintain facilities for ex-situ conservation of plants, animals and micro-organisms, preferably in their country of origin . . .
>
> (EC, 2006)

The EU and its member states, via the EC Zoos Directive, can refer to zoo conservation work to indicate compliance with the CBD.

3.2.3 The International Air Transport Association (IATA) Live Animal Regulations (LARs)

Like CITES and the CBD, membership of the International Air Transport Association (IATA) is voluntary. Airlines, rather than nation states, choose to become members of IATA and, once they have joined, must abide by IATA's Live Animal Regulations (LARs) (IATA, 2007). These regulations are intended to ensure that live animals are transported safely, legally, and with good standards of welfare. Copies of LARs are available both as a book and as a CD-ROM, and provide information and detailed specifications in areas such as crate size and ventilation (see www.iata.org/index.htm). IATA membership now extends to more than 200 international airlines; LARs are enforced within the EU, and are also officially recognized by organizations such as CITES and the World Organisation for Animal Health (OIE),[5] as well as by many individual countries.

Additional guidance on the international transport of live animals is available from the Animal Transportation Association (AATA) (see www.aata-animaltransport.org). This is a voluntary organization that works closely with the IATA. The AATA (2007) manual on the transportation of live animals contains much useful information for zoos.

3.2.4 Other international agreements, guidelines, and regulations

As Holst and Dickie (2007) point out, in their review of national and international regulations influencing zoos, there is '*an overwhelming array of conventions and regulations whose sole purpose*

5 The acronym 'OIE' is derived from the French title of the organization: *L'Office International des Epizooties.*

is to protect wildlife and regulate wildlife trade, constituting a complex legal framework with which zoos must comply'.

Two important areas that we have not yet considered in this chapter are international regulations and standards on animal health and disease, and guidelines on the **reintroduction** of animals back into the wild. The following sections provide a very brief overview of these topics, with details of where to find further information.

Animal health and disease

The World Organisation for Animal Health (OIE) produces standards that are widely used by government veterinary agencies as a key source of information about the prevention and management of outbreaks of animal disease. OIE publications include the *Terrestrial Animal Health Code*, which is published annually, and the *Aquatic Animal Health Code* (see `www.oie.int/eng/en_index.htm`).

Reintroduction of animals into the wild

In 1995, the World Conservation Union (IUCN) approved a set of guidelines on the reintroduction of wild animals and plants (IUCN/SSC, 1998). These guidelines are still widely used today and form the basis for many **taxon**- or species-specific reintroduction plans (reintroduction is considered in more detail in **Chapter 10** on 'Conservation'). The website of the British and Irish Association of Zoos and Aquariums (BIAZA) provides a useful overview to the IUCN guidelines for reintroduction (see `www.biaza.org.uk/public/pages/conservation/reintro.asp`).

Some of the key criteria listed in the guidelines are:

- the species must have been a former inhabitant of the area;
- the causes of extinction must be known and must no longer exist;
- there must be suitable habitat available;
- reintroduction should be monitored carefully.

3.3 Zoo legislation within the European Union

The main legislation governing the operation of zoos within the EU is the 1999 EC Zoos Directive. There is a wide range of other EU legislation that impacts on zoos, including Regulations governing the transport and welfare of animals, and also issues such as carcass disposal (animal by-products legislation).

3.3.1 The EC Zoos Directive

For countries that are members of the EU, the EC Zoos Directive of 1999 (Council Directive 1999/22/EC) was a major step forward in European legislation governing zoos and resulted in the closure in some countries of some very poor zoos. The EC Zoos Directive sets out requirements for the licensing and inspection of zoos, for proper record keeping, and for standards of animal care. The Directive also requires zoos to participate actively in education and conservation. Zoos that are not directly involved in captive breeding or species reintroduction programmes can meet the conservation requirement of the EC Zoos Directive by undertaking or supporting conservation-related research and training (although, as Rees, 2005a, has pointed out, the CBD already requires parties to engage in all of these activities).

It should be noted that not all EU member states (Spain was a notable exception) met the April 2002 deadline for incorporating the EC Zoos Directive into their national legislation.

3.3.2 Other EU legislation governing how zoos operate

Transport

The IATA LARs have already been mentioned (see section **3.2.3**) and these are enforced throughout the EU. In addition, Council Regulation (EC) No. 1/2005 on the protection of animals during transport is directly binding on member states (this is a regulation rather than a directive). In the UK, this EU legislation is enacted via the Welfare of Animals (Transport) (England) Order 2006, and other similar legislation in Wales, Scotland, and Northern Ireland.

The Balai Directive

The so-called **Balai Directive** is an EU Council Directive (92/65/EEC) that governs the transport of non-domestic animals between EU member states, with respect to veterinary screening and animal health. Balai was approved in 2002 as a 'catch-all' Directive for a range of animals not covered by other EU legislation and governs the movement not only of live (wild) animals, but also of semen, ova, and embryos.

The original Balai Directive (92/65/EEC) has since been amended, by Council Directive 2004/68/EC.

Obtaining Balai approval involves a considerable amount of work for zoo vets and managers. Once in place, however, Balai approval is intended to allow easier movement of animals between the zoo and other approved institutions (**Fig. 3.4**). Non-Balai-approved premises within Europe, for example, can no longer import or export any primate species, and approval from the State Veterinary Service (SVS),[6] or its equivalent for each country, must be sought before other animals can be transferred to a non-Balai approved zoo. New animals from non-Balai-approved zoos or other sources must go into isolation for a period of 30 days after importation into an approved zoo.

Although the aim of the Balai Directive is to make it easier to transfer animals between approved premises such as zoos, for legitimate purposes such as conservation programmes, in practice, it is '*difficult veterinary legislation*' (Dollinger, 2007). In a review

(a)

(b)

Figure **3.4** Transport between zoos of animals such as (a) the red panda *Ailurus fulgens* and (b) the gharial *Gavialis gangeticus* is covered in Europe by the Balai Directive; zoos that are 'Balai approved' can exchange animals more easily. (Photographs: (a) Paignton Zoo Environmental Park; (b) Smithsonian's National Zoo)

of the Balai Directive, Peter Dollinger, Executive Director of the World Association of Zoos and Aquariums (WAZA), commented:

> Because of time constraints, Directive 92/65 was prepared hastily and without consulting with the zoo community. It was poorly drafted,

---

Balai is the French word for broom. The **Balai Directive** 'sweeps up' animals not covered by other EU legislation.

6 The State Veterinary Service in the UK merged with other services in 2007 to become a new government agency, known as Animal Health (see www.defra.gov.uk/animalhealth/index.htm).

in particular its English version, and thus unclear, misleading and impractical.

(Dollinger, 2007)

Waste management

Another EU Regulation that impacts on the operation of zoos is the Animal By-Products Regulation (EC) No. 1774/2002. This has been transposed into UK legislation, albeit by separate laws for England and Wales, Scotland, and Northern Ireland. The intention behind the EC Regulation was to protect public and animal health by restricting how animal by-products can be used, and, in particular, to prevent food that is unfit for human consumption from entering the food supply chain. In practice, however, the legislation caused some consternation for zoos that wished to 'feed in' dead animals (for example, using meat from a dead goat as food for lions or tigers). Within the UK at least, under reg. 26.3(a) of the Animal By-Products Regulations 2005, zoos now require an authorization from their local **Animal Health** Office before they can feed animal by-products to other animals on their premises.

At the time of writing this book, however, the EU Animal By-Products Regulations, which were enacted in 2002, were under consultation in member states and may subsequently be substantially revised.

3.4 The regulatory framework for zoos in the UK: legislation and guidelines

The most important legislation governing zoos within the UK is the Zoo Licensing Act 1981 (ZLA), as amended in 2002—but this Act is by no means the only UK legislation regulating how zoos operate. For example, the Animal Welfare Act 2006 impacts

on zoos, as does legislation on animal health, human health, animal transport, and a whole host of other areas. In this section, we will look first at the ZLA and then at other legislation governing zoos in the UK.

3.4.1 The Zoo Licensing Act 1981

Within the UK, the EC Zoos Directive is implemented via the Zoo Licensing Act 1981, as amended by the Zoo Licensing Act 1981 (Amendment) (England and Wales) Regulations 2002 (HMSO, 2002). The ZLA was first passed in 1981, some 18 years prior to the EC Zoos Directive, but has been updated and strengthened more recently to reflect the requirements of the EC Zoos Directive, particularly in relation to conservation and education work by zoos (Kirkwood, 2001a).

Since 1981, all zoos in the UK have been required to hold a licence under the ZLA. (See **Websites and other resources** for details of online access to the full text of the ZLA; **Box 3.3** provides a brief overview of how zoo licensing works in practice in the UK.) The ZLA defines a zoo as a place where non-domesticated animals are kept and which is open to the public for at least 7 days per year. It excludes pet shops (which are licensed under the Pet Animals Act 1951), circuses (which are covered by the Performing Animals (Regulation) Act 1925), and animals kept privately with no public access (exotic species listed as dangerous are licensed under the Dangerous Wild Animals Act 1976).[7]

Despite these exclusions, the scope of the ZLA is very broad and everything from large zoos to small aquariums and wildlife parks comes under its remit (**Fig. 3.5**; see also **Box 3.3**). An animal rescue centre with a few **exotic animals**, for example, which is open to the public on just 7 days a year, will count as a zoo in the eyes of the law.

As a result of the EC Zoos Directive, the ZLA was amended in 2002 to incorporate additional

---

7 The Dangerous Wild Animals Act 1976 applies only to species listed as dangerous (see www.defra.gov.uk/wildlife-countryside/gwd/animallist.pdf).

Exotic species that are kept privately and which are not listed as dangerous fall outside the scope of this Act, but do now come within the remit of the Animal Welfare Act 2006.

Box **3.3** How does zoo licensing work in the UK?

The formal requirements for obtaining a licence for a new zoo are set out in the Zoo Licensing Act 1981 (Amendment) (England and Wales) Regulations 2002 (see **Websites and other resources**). There is also a very useful circular from Defra (Circular 02/2003; see `www.defra.gov.uk/wildlife-countryside/gwd/govt-circular022003.pdf`) explaining the key points of the Zoo Licensing Act 1981 (ZLA), including how to go about applying for a zoo licence.

At least 2 months before a licence can be applied for, notification of intent to open a new zoo must be given to the relevant local authority, in writing. This written notification must include a statement about how the zoo will implement the conservation measures detailed in the ZLA, as well as information about location, the kinds of animals that the zoo intends to keep, and various other details. A summary of this information must also be published in one local and one national newspaper, and displayed at the site of the proposed zoo. The local authority must then consider a zoo inspector's report before reaching a decision on whether or not to grant a licence for the new zoo. Once granted, a first licence runs for 4 years. Subsequent licences (usually a renewal licence) run for 6 years.

For existing zoos, an application to renew a licence under the ZLA must be made at least 6 months before the expiry date of the old licence. Local authorities have the power to direct a zoo to apply for a new licence where they believe that there are good grounds for refusing a straightforward renewal. They also have the power, in exceptional cases, to require a zoo to close to the public for a period of time—or even to close permanently—if licence conditions have been breached.

Zoo inspectors are drawn from a list held by the Secretary of State. This list is in two parts: the first part contains names of veterinarians with experience of treating exotic animals; the second part lists persons judged to be competent (by the Secretary of State) to inspect zoos and to provide advice on keeping wild animals in captivity. (The website of Bristol Zoo contains a very good summary of the ZLA and the process of zoo licensing: `www.bristolzoo.org.uk/learning/facts/keepers`).

The costs of a zoo inspection, which can take up to 2 full days, are usually paid in the first instance by the licensing authority, which then seeks reimbursement from the zoo.

requirements relating to conservation and education. The full title of the legislation amending the Act is the Zoo Licensing Act 1981 (Amendment) (England and Wales) Regulations 2002; this legislation came into force in the UK in January 2003. Because the ZLA was already in existence prior to the EC Zoos Directive, the consequent amendments to the Act took the form of secondary legislation and so were not debated in Parliament. In fact, much of the content of the EC Zoos Directive was already built into the 1981 Act: notably, those parts to do with maintaining standards of animal welfare. The EC Zoos Directive, however, strengthened the requirement for UK zoos to build on those aspects of their work related to conservation. As a result, all zoos in the UK now have to demonstrate active involvement

Figure **3.5** Butterfly houses are often located within larger zoos, but there are also a number of butterfly houses that exist as individual collections and, in the UK, these are defined as zoos under the Zoo Licensing Act 1981. (Photograph: © Vladimir Kondrachov, www.iStock.com)

in conservation and/or research that furthers the aims of zoo conservation. They must also demonstrate active involvement in educational programmes designed to raise public awareness and concern about conservation issues.

At first sight, these legal requirements for active involvement in conservation (**Fig. 3.6**) might appear onerous, particularly for small zoos with limited resources. They are in line, however, with the stated aims and purposes of zoos, and guidelines and support are available to zoos (from organizations such as the Zoos Forum, or BIAZA—see section **3.6.1**) on ways of interpreting and implementing these legal requirements.

Figure **3.6** Reddish buff moth *Acosmetia caliginosa* enclosures at Paignton Zoo Environmental Park. This is a species native to the UK that is being bred in captivity as part of a conservation project run jointly with Natural England (formerly English Nature). (Photograph: Paignton Zoo Environmental Park)

Other aspects of the ZLA, as amended, include the requirement that zoos keep adequate records (see **Chapter 5**), prevent escapes, and maintain a safe environment for animals and humans (in fact, much of the Act is about issues relating to health and safety in zoos). The Act also includes provision for the closure of zoos and the disposal of surplus animals. So the ZLA is more than just a vehicle for licensing zoos; it sets the operating framework within which zoos in the UK function.

As well as being the legislation under which licences to operate UK zoos can be granted, renewed, or withdrawn (see **Box 3.3**), the ZLA also includes a requirement for regular inspection of zoos. 'Regular' means once each year and inspection is the responsibility of the relevant local authority. During the 6-year life of a zoo licence (for an existing zoo), two of the annual inspections should be Secretary of State inspections (Defra, 2004), as part of which local authority inspectors must be assisted and accompanied by one or more special zoo inspectors who are appointed by the Secretary of State (Kirkwood, 2001b).

The purpose of licensing and inspection is to ensure that the zoo operates to required standards. Various guidelines are available to zoos to explain what these standards are, the most important of which are the Secretary of State's Standards of Modern Zoo Practice (SSSMZP) (Defra, 2004).

3.4.2 The Secretary of State's Standards of Modern Zoo Practice (SSSMZP)

Of course, responsible zoos do not need to be required by law to operate in the way specified by the various Acts and Directives already discussed in this chapter. But the fact that legislation such as the ZLA spells out what governments expect of zoos serves as a useful reminder to anyone involved in work with captive exotic animals of exactly what standards and goals they should be achieving. To help in the interpretation of the ZLA and other legislation, the UK Government has produced guidelines on what the expected standards are for a wide variety of aspects of the zoo world. These guidelines are the Secretary of State's Standards of Modern Zoo Practice (Defra, 2004) and are regularly updated (the current version at time of writing is September 2004).

These guidelines are intended to set the standards that UK zoos are expected to achieve in all their work. The SSSMZP therefore also function as an interpretation of the law and provide the criteria against which zoos' compliance with the law will be judged. The standards set out in the SSSMZP are based on the so-called **five principles** of animal care and management, which are themselves derived from the **five freedoms** that were originally applied in relation to the welfare of farm animals (Farm Animal Welfare Council, 1992; see **Chapter 7** on 'Animal welfare').

1. **Provision of food and water** This principle deals largely with health and hygiene aspects of the provision of food and water, but also includes statements about the appropriateness of the quality, quantity, and variety of the food for the particular animal, taking into account features such as its species, sex, age, reproductive condition, and so on.

2. **Provision of a suitable environment** This principle includes statements about physical features of the captive environment and their appropriateness for the species being kept, as well as consideration of minimizing the possibility of animals escaping. Also included here are health and safety considerations (for example, electrical equipment servicing, broken barriers) and aspects of hygiene (for example, rubbish disposal, cage cleaning—see **Fig. 3.7**).

3. **Provision of animal health care** This principle is about the level of veterinary care provided for the animals (**Fig. 3.8**) and also about disease prevention. It also includes reference to other kinds of health hazard, such as minimizing the likelihood of animals damaging each other. Finally, it reaffirms the need to keep

Figure **3.7** Part of routine husbandry at zoos is to ensure that animals have an appropriate, safe, and hygienic environment. (Photograph: Christopher Stevens, Werribee Open Range Zoo)

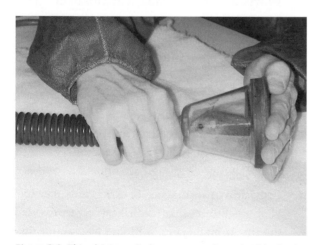

Figure **3.8** This photograph shows a pre-release health check on a common or hazel dormouse *Muscardinus avellanarius*. This species is bred in captivity at a number of UK zoos, as part of captive breeding and reintroduction programmes. (Photograph: Paignton Zoo Environmental Park)

comprehensive records of the animals and their health.

4. **Provision of opportunity to express most normal behaviour**   This principle is to do with the provision of physical and social environments that promote behaviours similar to those seen in the wild. To do this, zoos are expected to be familiar with current knowledge of the behaviour of the animals in the wild. This part of the guidelines also makes statements about the importance of captive breeding and the avoidance of producing between-taxa hybrids.

5. **Provision of protection from fear and distress**   Again, this includes statements about the provision of appropriate physical and social

environments. It also includes reference to zoo visitors and the provision of facilities for animals to avoid human contact if they so choose.

These five principles constitute the main foundation of the standards set for animal care, but the SSSMZP also set standards for other aspects of zoo management, including:

- the transportation and movement of live animals;
- conservation and education measures;
- public safety in the zoo;
- stock records;
- staff and training;
- public facilities.

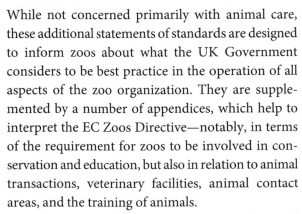 While not concerned primarily with animal care, these additional statements of standards are designed to inform zoos about what the UK Government considers to be best practice in the operation of all aspects of the zoo organization. They are supplemented by a number of appendices, which help to interpret the EC Zoos Directive—notably, in terms of the requirement for zoos to be involved in conservation and education, but also in relation to animal transactions, veterinary facilities, animal contact areas, and the training of animals.

The SSSMZP also require zoos to confront and to resolve ethical issues that may arise during zoo operation, through the establishment of a formal ethical review process.

3.4.3 Guidelines for best practice

In addition to the SSSMZP, zoos in the UK can find advisory guidelines on aspects of their work from sources such as the Zoos Forum and BIAZA.

The Zoos Forum

The Zoos Forum is a panel of specialists in zoo management, education, veterinary science, and animal welfare. It was set up by the UK Government in 1999 to act as an independent advisory body, both to the government itself and to the zoo community.

Thus, it has a dual role of monitoring the licensing system and making recommendations to the government on any necessary amendments or changes, and of encouraging and supporting zoos in implementing research, education, and conservation. The Zoos Forum has regular open meetings and also produces annual reports of its activities.

In its role of supporting zoos, the Zoos Forum has produced chapters for an ongoing *Zoos Forum Handbook* (Defra, 2007b). The purpose of the *Handbook* is to provide further guidance to zoos, over and above that which is available in the SSSMZP, on how to achieve these standards and how to implement the legal requirements, through best practice (**Fig. 3.9**).

At the time of writing, the *Handbook* included chapters on ethical review; conservation, education and research; **sustainability** initiatives; veterinary services; and animal welfare and its assessment. Amongst the appendices to these chapters are real examples from real zoos, showing how **welfare audits**, policies on population management, and so forth, have been set in place. All of these documents are available online, from the Defra website. (The *Zoos Forum Handbook* is currently found at www.defra.gov.uk/wildlife-countryside/gwd/zoosforum/handbook/.)

BIAZA guidelines

The British and Irish Association of Zoos and Aquariums (BIAZA) is another body that provides guidelines and advice on best practice for UK zoos—as do other national and regional zoo associations such as the Association of Zoos and Aquariums (AZA) in North America, the European Association of Zoos and Aquariums (EAZA), and the Australasian Regional Association of Zoological Parks and Aquaria (ARAZPA) (see section **3.6**). BIAZA publishes animal management and **husbandry guidelines** on various species, and also codes of practice for member zoos on subjects as varied as animal transactions and the **euthanasia** of invertebrates. (Details of all of these publications can be found online at www.biaza.org.uk.)

(a)

(b)

Figure **3.9** Some examples of best practice in how zoos look after animals and visitors, what they sell to visitors, and how they encourage environmental awareness: (a) training a spotted-neck otter *Lutra maculicollis* for weighing; (b) 'Zoo-Poo' for sale. (Photographs: (a) Cango Wildlife Ranch, South Africa; (b) Paignton Zoo Environmental Park)

Other guidelines and sources of information for zoos

Defra is not the only UK Government department that produces guidelines for zoos. For example, in its *Learning Outside the Classroom Manifesto* (DfES, 2006), the former Department for Education and Skills (DfES)—now the Department for Children, Schools and Families—addresses education work by zoos. The Association of British Wild Animal Keepers (ABWAK) is another source of information (see www.abwak.co.uk), producing husbandry guidelines for a range of taxa.

3.4.4 Other UK legislation governing how zoos operate

In addition to the Zoo Licensing Act 1981, as amended, other legislation that impacts on zoos includes laws and regulations on areas as diverse as animal welfare, conservation, animal health and diseases, human health and safety (zoo staff and visitors), animal transport, employment, trade, building regulations, charity law and trust law, data protection law, waste disposal (animal by-products),

and disability discrimination (Brooman and Legge, 1997; Kirkwood, 2001a; see **Fig. 3.10**). Even firearms legislation can apply to zoos (vets and keepers using darting equipment, or rifles for **pest** control, require a licence under the Firearms Acts 1968–1997). **Table 3.2** provides a summary of the main Acts other than the ZLA that govern the operation of UK zoos.

Government departments with some responsibility for aspects of zoos in the UK include Defra, the Department for Children, Schools and Families (formerly the DfES), the Department for Trade and Industry (DTI), the Home Office, and the Department of Culture, Media and Sport (DCMS).

The Defra website contains comprehensive details of the legislation governing zoos in the UK.

Animal welfare legislation

The Animal Welfare Act 2006 (AWA) (which is criminal, rather than civil, law) is described on the Defra website as '*the most significant change in animal welfare legislation in almost a century*' (Defra, 2007a), although the Zoo Licensing Act 1981, as amended,

Figure **3.10** Zoos have to comply with a wide range of legislation, including, for example, disability discrimination legislation. Here, wheelchair users are shown enjoying the zoo experience at Rotterdam Zoo. (Photograph: Geoff Hosey)

Topic	Act, Order or Regulation	Further information
Animal health and disease	Animal Health Act 2002	The 2002 Act deals with, among other things, government powers to tackle foot and mouth disease (FMD) and **transmissible spongiform encephalopathies (TSEs)** such as **BSE**.
	Veterinary Medicines Regulations 2005	Most **analgesic drugs** for animals can only be prescribed by a vet who has seen and assessed the animal.
	Misuse of Drugs Regulations 1985	These Regulations cover the use of opioids such as morphine.
Animal welfare	Animal Welfare Act (England and Wales) 2006	The AWA 2006 brings together many existing pieces of animal welfare legislation for farmed and non-farmed animals.
	Veterinary Surgeons Act 1966	
	Animals (Scientific Procedures) Act 1986	This Act covers any invasive research and may apply to some zoos in which veterinary or reproductive research is carried out.

Table **3.2** Other UK legislation affecting zoos

NOTE: This list is not exhaustive. In addition, legislation in Northern Ireland, Scotland and Wales may differ (see Box 3.2 for information about devolved legislation within the UK).

Topic	Act, Order or Regulation	Further information
Building regulations	Building Act 1984	
Conservation	Wildlife and Countryside Act 1981	The WLCA implements European conservation regulations such as the Berne Convention (see online at http://jncc.gov.uk/page-3614 for a useful guide to the WLCA).
Disability discrimination	Disability Discrimination Act 1995	
Employment	Data Protection Act 1998	Zoos must keep records of the details of their employees, clients, and volunteers in accordance with the provisions of the DPA. This also applies to records of student trainees and researchers who may be working temporarily in the zoo.
	Employment Rights Act 1996	
Firearms	Firearms Acts 1968 to 1997	Vets and keepers using darting equipment or other firearms must be licensed to do so.
General animal law	Performing Animals (Regulation) Act 1925	This Act requires that people or organizations who train animals to perform in shows must register with their local authority and be subject to inspection.
General environmental protection law	Environmental Protection Act 1990 Control of Pollution (Amendment) Act 1989	
Health and safety	Control of Substances Hazardous to Health Regulations 2002 (COSHH)	The UK government Health and Safety Executive (HSE) has produced two useful online guides: firstly, *Managing Health and Safety in Zoos* (available at www.hse.gov.uk/pubns/web15.pdf); secondly, a guide to the COSHH Regulations (available at www.hse.gov.uk/pubns/indg136.pdf).
	Health and Safety at Work Act 1974	
Research	Animals (Scientific Procedures) Act 1986	See Box 3.4 for further details about the A(SP)A.
Trade	Control of Trade in Endangered Species (Enforcement) Regulations 1977	COTES is the domestic legislation in Britain implementing CITES (see section **3.2.1**).
Transport	Welfare of Animals (Transport) (England) Order 2006	This Order replaces the older 1997 Order and brings national legislation into line with EU Council Regulation (EC) No. 1/2005 on the protection of animals during transport.
Waste management and disposal (including carcass disposal)	Waste Management Licensing Regulations 1994 (as amended) Hazardous Wastes (England and Wales) Regulations 2005 Animal By-Products Regulations 2005	National legislation was enacted in the UK in 2005 to meet the EU Animal By-Products Regulation (EC 1774/2002).

Table **3.2** (continued)

remains the primary piece of legislation regulating zoos. The AWA came into effect in England and Wales in 2007. The equivalent legislation in Scotland is the Animal Health and Welfare Act 2006; in Northern Ireland, the relevant legislation is the Welfare of Animals Act (Northern Ireland) 1972.

The Act places a legal requirement (a 'duty of care') on the owners and keepers of animals (including zoos) not only to avoid cruelty, but also to ensure that the welfare needs of animals are met. These needs are summarized by Defra (2008) as follows:

Box **3.4** Research work requiring a Home Office licence

In the UK, the Animals (Scientific Procedures) Act 1986 covers **invasive procedure**, which is described as '*any experimental or other scientific procedure applied to a protected animal . . . which may have the effect of causing that animal pain, suffering, distress or lasting harm*'. A protected animal, under the terms of the Act, is any living vertebrate (other than man) and also one invertebrate species: the common octopus *Octopus vulgaris*.

Before invasive research on a protected animal species can be carried out, there is a legal requirement under the Act to obtain a

licence from the Home Office. This requirement might apply to a zoo if, for example, a normally **free-ranging** animal were to be caged temporarily, solely for the purposes of research. Even if a zoo research project is entirely observational, there is still a possibility that procedures such as the capture and/or confinement of animals can cause distress.

The Zoos Forum advises that zoos should seek guidance from the Home Office if there is any uncertainty over whether or not a particular research project requires a licence.

- For a suitable environment (place to live)
- For a suitable diet
- To exhibit normal behaviour patterns
- To be housed with, or apart from, other animals (if applicable)
- To be protected from pain, injury, suffering and disease

The AWA covers all vertebrate animals, not only mammals, and also includes provision to extend its remit in the future to include non-vertebrate animals if, as the Defra website states, '*future scientific evidence shows that other kinds of animals are also capable of experiencing pain and suffering*' (Defra, 2008).

Legislation governing research in zoos

The Animals (Scientific Procedures) Act 1986 may apply to zoos in which research is being carried out, although Kirkwood (2001a) says that little, if any, zoo work is undertaken under this Act. **Box 3.4** provides further information about the Act and the sorts of circumstances under which zoos might be required to obtain a Home Office licence before undertaking research.

3.5 Zoo legislation outside Europe

There is insufficient space in this chapter (or indeed in this book) for a full account of legislation relating to zoos in all countries outside Europe.

The following two sections provide a very brief overview of legislation governing zoos in the USA, and in zoos outside the USA and Europe.

3.5.1 Legislation governing zoos in the USA: an overview

In the USA, the Animal Welfare Act of 1966 (AWA) covers several aspects of zoo operations, from zoo licensing, to animal health, animal purchase, transportation, housing, handling, and husbandry (Vehrs, 1996). Responsibility for monitoring compliance with the AWA rests with the US Department of Agriculture's Animal and Plant Health Inspection Service (APHIS), which administers the granting of licences to zoological parks and aquariums via federal veterinarian services (each US state has a Veterinarian in Charge, whose responsibilities include overseeing licence applications by zoos). The AWA regulations apply only to mammals and not to all species of mammal: farm animals, rats and mice, and most

(a)

(b)

Figure **3.11** There is wide variation between countries in South East Asia in the regulatory framework governing how zoos operate. These photographs are from two leading zoos in the region, and show (a) a capybara *Hydrochaerus hydrochaeris* at Singapore Zoo and (b) the entrance to the Schmutzer Primate Center at Ragunan Zoo, Indonesia. (Photographs: (a) Diana Marlena, Singapore Zoo; (b) Vicky Melfi)

common companion animals are excluded from the provisions of the Act (Vehrs, 1996).

Other US government agencies that are involved with the regulation of zoos are the Food and Drug Administration (FDA) and the US Fish and Wildlife Service (Department of the Interior); the latter enforces international legislation such as CITES. Fowler and Miller (1993), in the third edition of their comprehensive book *Zoo and Wild Animal Medicine*, provide a list of US government departments and agencies, together with a summary of current zoo-related legislation (this useful summary is not included in every edition).

For further information about zoo licensing and legislation in North America, Grech (2004) provides a useful (online) review; the website of the Association of Zoos and Aquariums (AZA) also provides information about federal laws that impact on the operation of zoos and has links to other websites, including that of the Code of Federal Regulations. Gesualdi (2001) looks at zoo licensing and accreditation in Canada and in Mexico, as well as in the USA.

3.5.2 Legislation governing zoos outside the USA and Europe: an overview

The website of the Zoo Outreach Organisation (ZOO) (`www.zooreach.org`) includes much useful information about zoo legislation, standards, and guidelines for zoos in South East Asia (**Fig. 3.11**) and Australia. There is wide variation among countries in this region between their approaches to zoo legislation and, in particular, whether the regulation of zoos is managed predominantly nationally or locally. In India, for example, the zoo licensing system recognizes four different categories of zoo, according to the size of the zoo, and sets different regulatory requirements for each category (Cooper, 2003).

In Australia, zoo regulation is largely a state, rather than a national, responsibility, and legislation on zoo animal health and welfare varies between states and territories.

The same is true of Canada, where zoo legislation varies markedly from one province to another. Ontario, in particular, has come in for criticism in recent years for a lack of adequate legislation covering small 'roadside' zoos (Dalgetty, 2007).

3.6 Zoo associations: from BIAZA and EAZA, to AZA, ARAZPA, and WAZA

Most zoos do not operate in isolation, but belong to zoo associations at a national or regional level (**Fig. 3.12**). These associations are not generally

(a)

(b)

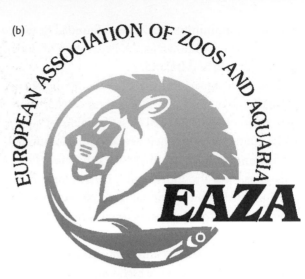

(c)

Figure **3.12** Accredited zoos are zoos that are licensed and which belong to zoo associations at national or regional levels. (Photographs: BIAZA, EAZA, and ARAZPA)

part of the legislative framework within which zoos operate, but they can be thought of as part of the regulatory framework, because part of their purpose is to promote professionalism and good practice within the zoo world, and to help to administer—at an inter-organizational level—the conservation, research, and educational functions of zoos.

The oldest national zoo association is the German Association of Zoo Directors (*Verband Deutscher Zoodirektoren*), founded in 1887; this gave birth to the International Union of Directors of Zoological Gardens (IUDZG) in 1935 at a meeting in Basel (EAZA, 2003). Today, the emphasis is on zoo associations, rather than on zoo director associations, although there are also thriving zookeeper organizations in many countries.

3.6.1 The British and Irish Association of Zoos and Aquariums (BIAZA)

Zoos in Great Britain and Ireland are encouraged to become members of BIAZA, which sees itself as the professional organization representing the zoo and aquarium community. BIAZA was founded in 1966 and was formerly known as the Federation of Zoos of Great Britain and Ireland.

Zoos and aquariums can become full members of BIAZA, but there are also several categories of associate membership of BIAZA available for companies operating within the zoo world, for educational establishments, such as universities that have links with zoos, and for individuals. BIAZA organizes meetings, including an annual research symposium; it also publishes guidelines, promotes research, and manages conservation work. The aims of BIAZA are to promote conservation and public concern for the natural world, and to achieve high standards of animal care in the zoo community. (See section **2.5.1** for more on the published mission and vision of BIAZA.)

BIAZA also produces much useful advice on legislation governing zoos. This is available via the BIAZA website and also in 'Legal Lines', a short section in each edition of *Lifelines*, BIAZA's quarterly magazine.

3.6.2 The European Association of Zoos and Aquaria (EAZA)

Many British zoos are also members of EAZA. Set up in 1988 to represent zoos in the EU, EAZA was originally known as the European Community Association of Zoos and Aquaria (ECAZA). With the fall of the Iron Curtain, the organization was able to expand eastwards and, today, EAZA has members in countries both within and bordering Europe, including Israel, Turkey, and, as of early 2007, Cyprus.

EAZA now has (at the time of writing) more than 300 member zoos spanning 34 countries, with its main offices in Antwerp in Belgium. EAZA's mission comprises four strands: planning and coordinating wildlife conservation through the region; promoting education, particularly about the environment; contributing to discussion at an international level (for example, among the United Nations, EU, etc.);

and advising the European Union and its committees. EAZA places a strong emphasis on captive breeding and involvement in regional **captive breeding programmes**, reflecting its origins in the creation, in 1985, of the first **European Endangered species Programmes (EEPs)**.[8]

To become EAZA members, zoos and aquariums must undergo a rigorous accreditation process and must be members of the International Species Information System (ISIS), managing their collections through the Animal Record-Keeping System (ARKS) (see **Chapter 5**). EAZA members must also comply with EAZA's standards and code of **ethics**.

EAZA has produced a number of policy documents and sets of standards by which zoos are to operate, and was instrumental in formulating the EC Zoos Directive.

3.6.3 The Australasian Regional Association of Zoological Parks and Aquaria (ARAZPA)

ARAZPA, established in 1990, links over 70 zoos and aquariums across Australia, New Zealand, and the South Pacific. The Association's stated mission is '*to harness the collective resources of zoos and aquariums to conserve biodiversity in the natural environment*' (ARAZPA, 2008).

3.6.4 The Association of Zoos and Aquariums (AZA)

With over 5,500 members today, the US-based AZA claims to be the largest zoo and aquarium association in the world. Of this total, however, only around 250 are zoo members of the AZA; others are individual members. And although membership of the AZA is open to individuals and zoos from outside North America, only zoos located within the USA and Canada can become accredited members. Founded in 1924 as the American Association of Zoological Parks and Aquariums (AAZPA), the organization changed its name to the American Zoo and Aquarium

---

8 European Endangered species Programmes are known as EEPs, not as EESPs, because the acronym is taken from the German, *Europäisches Erhaltungszuchtprogramm*.

Association in 1994 and, at the same time, adopted the simple 'AZA' acronym. Today, the Association has dropped the 'American' and refers to itself simply as the Association of Zoos and Aquariums.

The AZA has a variety of membership categories for zoos, other organizations, and individuals, and is worldwide in scope. Its mission is to promote excellence in animal care and welfare, conservation, education and research, and to encourage respect for the natural world. Like BIAZA and EAZA, the AZA publishes policies, guidelines, and codes of practice that are intended to support the work of members and to promote professional high standards.

3.6.5 The World Association of Zoos and Aquariums (WAZA)

Originally the International Union of Directors of Zoological Gardens (IUDZG), this association was renamed the World Association of Zoos and Aquariums (WAZA) in 2000. Membership is open to individual zoos and aquariums (which must, however, already also belong to a regional association), to the various regional and national associations, and to related organizations such as the International Species Information System (ISIS), the Institute for Zoo and Wildlife Research (IZW), and the like.

WAZA states as its mission '*to guide, encourage and support the zoos, aquariums, and like-minded organisations of the world in animal care and welfare, environmental education and global conservation*' (WAZA, 2008).

One of WAZA's best-known achievements has been the production of two important strategy documents relating to wildlife conservation work by zoos. The first of these, the World Zoo Conservation Strategy (WZCS) (IUDZG/CBSG, 1993), was published jointly in 1993 by the IUCN's Captive Breeding Specialist Group (CBSG) and WAZA in its former guise of the IUDZG. The second strategy document is the World Zoo and Aquarium Conservation Strategy (WZACS) (WAZA, 2005). (Both WZCS and WZACS are discussed in more detail in **Chapter 10** on 'Conservation'.)

WAZA has also developed a Code of Ethics (WAZA, 1999), which deals with issues ranging from contraception and euthanasia, to the movement of animals between zoos.

3.6.6 Other zoo associations

National and regional zoo associations

Other areas of the world also have zoo associations that operate in much the same way and with the same sorts of values as those described above. These include, among others:

- the African Association of Zoological Gardens and Aquaria (PAAZAB)—formerly, the Pan-African Association of Zoological Gardens and Aquaria, from which title the 'P' in the acronym derives (**Fig. 3.13**);

Figure **3.13** A keeper at Pretoria Zoo, South Africa, carrying food the traditional way. In Africa, the zoo association is the African Association of Zoological Gardens and Aquaria (PAAZAB). (Photograph: Robyn Ingle-Jones, Pretoria Zoo)

- the Latin-American Zoo and Aquarium Association (ALPZA);

- the Canadian Association of Zoos and Aquariums (CAZA);

- the South-Asian Zoo Association for Regional Cooperation (SAZARC).

 Keeper associations

As well as zoo associations, there are various zookeeper associations worldwide. In the UK, the Association of British Wild Animal Keepers (ABWAK) produces its own magazine, *Ratel*, and other publications such as husbandry guidelines for a number of taxa. ABWAK also offers small grants for work in zoos in two areas:

- any project that is designed to increase knowledge or awareness of factors relevant to the husbandry of captive exotics, particularly Red Data List/CITES-listed taxa;

- any field conservation/education project in the UK or abroad that is principally organized by the applicant.

The ABWAK website (`www.abwak.co.uk`) provides further information and contact details (and is also a good place to start if you are looking for a job as a keeper in a British zoo).

The American Association of Zoo Keepers (AAZK) fulfils a similar role in North America. Among the publications produced by the AAZK is its *Enrichment Notebook* (2004), now in its third edition and available as a searchable CD-ROM (see `www.aazk.org` for further details). The AAZK also produces a CD-ROM on *Zoonotic Diseases*, with a searchable list of **zoonoses** and a useful section on disinfection and hygiene (AAZK, 2005).

In Australia and New Zealand, there is the Australasian Society of Zoo Keeping (ASZK); Africa has the Animal Keepers Association of Africa (AKAA). In Europe, there are various organizations, such as the European Elephant Keeper and Manager Association (EEKMA), as well as national zookeepers' associations.

In 2000, at a meeting in Columbus, Ohio, the idea of an International Congress on Zookeeping (ICZ) was born and, in 2003, the first was held at Avifauna, at Alphen aan den Rijn in the Netherlands. This was followed by a second meeting of the ICZ in Australia in 2006, with Seattle the venue for 2009. The ICZ website (`www.iczoo.org`) lists the main national and regional zookeepers' associations worldwide.

Summary

- Zoos are governed by a wide and complex range of legislation, and non-legislative agreements and guidelines, operating at international, national, and sub-national levels.

- Not all countries have specific 'zoo' legislation. In the USA, for example, the main legislation within which zoos must operate is the Animal Welfare Act of 1966.

- Zoos in the UK are regulated by the Zoo Licensing Act 1981 (ZLA), as amended, which incorporates into UK law the requirements of the 1999 EC Zoos Directive (1999/22/EC).

- The ZLA and the EC Zoos Directive lay requirements on zoos to fulfil certain obligations in order to be allowed (licensed) to operate.

- Zoos in EU countries must now be engaged actively in both education and conservation to meet the requirements of the EC Zoos Directive.

- Help for UK zoos in understanding their legal obligations and achieving best practice is available in government guidelines—the Secretary of State's Standards for Modern Zoo Practice (SSSMZP) —and through bodies such as the Zoos Forum, which publishes an advisory *Handbook*.

- Regional, national, and international zoo associations such as BIAZA, EAZA, ARAZPA, AZA, and WAZA also aim to promote best practice and to coordinate the activities of zoos.

Questions and topics for discussion

1. Is the definition of a 'zoo' within the UK, as set out in the Zoo Licensing Act 1981, too broad to be useful?

2. Should membership of a national or regional association such as BIAZA, EAZA, or AZA remain voluntary, or should it be made compulsory for all zoos as a condition of licensing?

3. Should all UK zoos be required, by law, to participate in conservation and education initiatives, or should some smaller zoos be exempt from this requirement?

4. In addition to the Zoo Licensing Act 1981, as amended, what other legislation governs how zoos operate within the UK?

5. Compare and contrast the legislative framework for zoos in the USA and Europe.

Further reading

For those who want to delve more deeply into the complex regulatory framework within which zoos operate, a good starting point is Chapter 3, 'Importance and application of animal law', in Cooper and Cooper's *Introduction to Veterinary and Comparative Forensic Medicine* (2007). It is also worth reading Chapter 4, 'Animal welfare', in the same book. Margaret Cooper (a lawyer who specializes in animal law) has written a very helpful summary paper on zoo legislation, for the *International Zoo Yearbook* (2003). This contains, as an appendix, a table of examples of zoo legislation for various countries around the world.

For further reading about animal welfare law in Britain, we recommend Mike Radford's *Animal Welfare Law in Britain: Regulation and Responsibility* (2001). We also recommend Brooman and Legge's *Law Relating to Animals* (1997). This features a

section on legislation relating to zoos (as well as a short summary of zoo history). Although published in the UK, the book considers European and international law relating to animals, and also covers animal welfare legislation in Australia.

Within the three volumes of the *Encyclopedia of the World's Zoos* (Bell, 2001), there are several entries about zoo legislation and licensing processes. James Kirkwood (formerly Chief Veterinarian at ZSL and now the Director of the Universities Federation for Animal Welfare) has contributed two useful summaries covering UK zoo legislation and licensing processes (Kirkwood, 2001a; 2001b). Other entries include an overview of legislation and licensing for zoos in Africa (Walker, 2001), and licensing and accreditation for zoos in North America (Gesualdi, 2001).

Websites and other resources

We have found the following websites and other sources of information to be particularly useful, although there are occasional inaccuracies in some of the published website addresses and online information, and we would always recommend that readers double-check against another source. The full text of some of the key documents referred to in this chapter can be downloaded from the following websites.

- **The Convention on Biodiversity (CBD)**
  The text of the Convention is available online at `www.cbd.int/convention/convention.shtml`.

- **The EC Zoos Directive (1999/22/EC)**
  The full text of the EC Zoos Directive is available online at `http://eur-lex.europa.eu/pri/en/oj/dat/1999/l_094/l_09419990409en00240026.pdf`.

- **The Zoo Licensing Act 1981, as amended, and the Zoo Licensing Act 1981 (Amendment) (England and Wales) Regulations 2002**
  The full version of the Regulations amending the Zoo Licensing Act 1981 (© Crown Copyright 2002) is available online at `www.legislation.gov.uk/si/si2002/20023080.htm`. A useful summary of the key points of the Zoo Licensing Act 1981, as amended, is available from Defra (Circular 02/2003; © Crown Copyright 2003) at `www.defra.gov.uk/wildlife-countryside/gwd/govt-circular022003.pdf`.

- **The Secretary of State's Standards of Modern Zoo Practice (SSSMZP)**
  Along with much other useful information for UK zoos, the SSSMZP can be found online at `www.defra.gov.uk/wildlife-countryside/gwd/zooprac/index.htm`.

The legislative framework for zoo licensing in the UK is operated through Defra and a great deal of information is available from that Department either online or by post at: Global Wildlife Division, Defra, Zone 1/16L, Eagle Wing, Temple Quay House, 2 The Square, Temple Quay, BRISTOL, BS1 6EB. Guidance on zoo legislation and the full text of the SSSMZP are available from Defra by post or online (see above). Information about the Zoos Forum, including details of meetings and publications, and copies of the *Zoos Forum Handbook* can be obtained from the same address and website.

The Scottish Government website (`www.scotland.gov.uk`) is a good source of information about devolved legislation, such as animal welfare legislation, in Scotland; the corresponding websites for Northern Ireland and for Wales are `www.northernireland.gov.uk` and `new.wales.gov.uk`.

Information about BIAZA, the work it does, and its meetings can be obtained by post (BIAZA, Regents Park, LONDON, NW1 4RY) or online at `www.biaza.org.uk`. Similarly, information about EAZA and its work can be obtained by post (EAZA Executive Office, PO Box 20164, 1000 HD AMSTERDAM, The Netherlands) or online at `www.eaza.net`. For the AZA, the relevant addresses are AZA, 8403 Colesville Road, Suite 710, Silver Spring, MD 20910–3314, USA, and `www.aza.org`. For ARAZPA, the details are ARAZPA, PO Box 20, Mosman, NSW 2088, Australia, and `www.arazpa.org.au`; for WAZA, the addresses are PO Box 23, CH-3097 Liebefeld-Bern, Switzerland, and `www.waza.org`. If the zoo association in which you are interested is not listed here, the *Encyclopedia of the World's Zoos* (Bell, 2001) contains entries for all of the major zoo organizations throughout the world; there are also links from the WAZA website to other zoo organizations.

TRAFFIC (a joint programme of WWF and the IUCN) has produced a very helpful summary of the documentation needed for trading wild animals (and animal specimens) into and within the EU. This can be found at `www.eu-wildlifetrade.org`. The implementation of the Convention on Biological Diversity (CBD) within the EU is summarized in an EC publication, *The Convention on Biological Diversity: Implementation in the European Union.* This is available online at `http://ec.europa.eu/environment/biodiversity/international/pdf/brochure_en.pdf`.

The website of the Eurogroup for Animal Welfare (`www.eurogroupanimalwelfare.org`) provides access to a useful publication that monitors the implementation of the EC Zoos Directive in member states: the *Report on the Implementation of the EU Zoos Directive* (Eurogroup for Animal Welfare, 2006), available online at `http://eurogroupforanimals.org/policy/pdf/zooreportdec2006.pdf`.

For information about legislation governing zoos in South East Asia, the Zoo Outreach Organisation publishes much useful information online at `www.zooreach.org/ZooLegislation/ZooLegislation.htm`.

Finally, for zoos within the UK, an excellent explanation of the Wildlife and Countryside Act 1981 (WLCA) is available online through the Joint Nature Conservation Committee (JNCC) at `http://jncc.gov.uk/page-3614`.

Chapter 4 Behaviour

The study of behaviour[1] is about understanding what animals do and why they do it. It is about the way in which animals acquire resources (such as food, shelter, or mates) and avoid danger (such as predators or **agonistic behaviours** among competitors), so understanding behaviour is crucial to our understanding of the animal's **life history traits** and its interactions with its environment. How an animal behaves can also help us to interpret its health and welfare, so observing behaviour is an important part of the captive management of **zoo** animals.

For many years, research on animal behaviour was dominated by studies of **learning** and, even then, the importance of the research was primarily seen as what it could tell us about human learning. But in the last 40 or 50 years, there has been a huge growth in studies of animal behaviour from an evolutionary viewpoint and the emphasis now is on understanding how behaviour helps the animal to survive in its natural environment. We now have a substantial body of theory and experimental evidence to guide our understanding of animal behaviour.

In this book, we do not have the space to do any more than briefly summarize this theoretical background in a way that we hope makes the rest of the chapter intelligible to readers, but we would urge anyone interested in finding out more to read one of the many excellent textbooks on animal behaviour that are currently available (see **Further reading**). Our main aim in this chapter is to examine the ways in which living in a zoo can affect the behaviour of the animals, and how we should interpret these changes. This then leads on to more applied uses of our behavioural knowledge in interpreting welfare (**Chapter 7**) and modifying the behaviours of the animals (**Chapters 8**—on **enrichment**—and **13**—on 'Human–animal interactions').

We will cover the following topics in this chapter.

4.1 General principles
4.2 Animal behaviour in the zoo
4.3 Behavioural response to the zoo environment
4.4 Abnormal behaviours
4.5 Comparison with the wild

In addition, boxes will explore further some of the issues raised, with specific examples.

4.1 General principles

At the most basic level of explanation, behaviour can be described in terms of the **responses** animals make to **stimuli** in their environment; stimuli may be aspects of the physical environment (for example, light, temperature, sound) or may come from other animals (for example, signals such as chemicals, postures, calls). At a more complex level of explanation, we must recognize that animals may initiate behaviours in the apparent absence of identifiable stimuli and that behaviours may be guided by

---

1 Animal behaviour involves a number of sub-disciplines, so you may also encounter terms such as 'ethology' (which broadly refers to the study of an animal's natural behaviours), 'comparative psychology' (which is mostly, but not totally, about animal learning), and 'behavioural ecology' (which is about how behaviour helps animals to survive). These are used to represent different traditions and theoretical approaches, but nowadays the distinctions between them are very blurred. **Agonistic behaviours** are those that occur in 'situations involving physical conflict' (Huntingford and Turner, 1987). They consequently include both aggressive and submissive behaviours.

internal (for example, motivational—see section 4.1.3), as much as by external, factors.

Like other aspects of an animal's biology, much behaviour is adaptive—that is, it evolves over generations by the process of natural selection, changing animals so that they are better able to survive and reproduce in their environments. For this to happen, variability in the animals' behaviours must reflect variability in their genes (or, in other words, behavioural differences between individuals must have a genetic basis), because any other differences—such as those arising from an animal's learning—that do not have a genetic basis will not evolve. Thus, behaviour can alter between generations through genetic processes, leading to the evolution of new or modified forms of behaviour that increase the animal's **fitness**.

But behaviour differs from many biological processes in that it is often very flexible, to the extent that it can change dramatically within the lifetime of the individual. Behaviours that change within the lifetime of the individual are usually the result not of genetics, but of experience—a process referred to as learning. Such behavioural change is not usually transmitted to subsequent generations, although it can be under the special conditions of social learning (see section 4.1.2).

Very often the questions we ask about animal behaviour are to do with why the animal is behaving in that particular way at that particular time. The answers to this sort of question—usually referred to as **causal** (or **proximate**) explanations—often require a recognition of which stimuli the animal is responding to, as well as other features, such as genetic differences, hormones, and levels of motivation, which can influence the likelihood and extent of the response. But our explanation of why an animal is behaving in a certain way could also be about what the purpose of the behaviour is and why the animal has evolved to do it this way rather than another way. These are often referred to as **functional** (or **ultimate**) explanations.

A useful way of deciding what sort of question we are asking is to use Tinbergen's 'Four Whys' (see **Box 4.1**). Note that the first 'why' requires a functional answer, whereas the other three are predominantly causal.

Box 4.1 Tinbergen's 'Four Whys'

Our interest in animal behaviour generally revolves around trying to understand why an animal behaves in a particular way that we have observed. But, in this context, the question 'why?' can mean several different things and therefore we can arrive at several different answers. Niko Tinbergen, one of the founders of modern ethology, recognized this and identified four different 'whys' that give us different perspectives on animal behaviour; these still form the framework for behavioural research (Tinbergen, 1963).

Thus, our 'why' can be asking any one of the following.

1. **What is the function of the behaviour?** This is asking about how the behaviour affects the fitness of the animal—that is, how it allows the animal to survive and reproduce in its natural environment.

2. **What is the cause of the behaviour?** This is asking what are the immediate influences that lead to the animal showing that behaviour at that particular time. ▶

Fitness is an indication of how well an individual is adapted to a particular environment and is usually measured as the number of offspring that the individual produces that survive to reproductive age.

▶   These might be internal or external stimuli or states.

3.   **What is the course of development of the behaviour?** This is asking about the stages in the life history of the individual at which the particular behaviour may be shown and what influences the way in which the behaviour might change as the individual matures.

4.   **How did the behaviour evolve?** This is asking about the evolutionary history of the behaviour—that is, how the behaviour has come to be the way it is. Of course, behaviour is rarely captured in the fossil record, so inferences about evolution are usually made comparatively—that is, by looking across closely, and more distantly, related species.

Suppose, for example, we see a small **Carnivore** such as a genet *Genetta genetta* doing a handstand, as in **Fig. 4.1**. What is it doing?

Applying Tinbergen's four whys allows us to understand the behaviour better. In response to 'Why' (1), we find that its function is to apply a scent mark from perineal glands to a substrate; genets use this to recognize individuals and to find out something about their physiological state (Roeder, 1980). In response to question (2), we will identify that a number of stimuli may cause scent marking: for example, anogenital scent marking in the genet may increase in males, but decrease in females, during periods of aggression (Roeder, 1983). In relation to question (3), frequencies of scent marking are low in young genets, but increase as they get older (Roeder, 1984). Finally, in relation to question (4), we find that the handstand position is used for anogenital scent marking in several other viverrid species as well (for example, the dwarf mongoose *Helogale parvula* and the kusimanse *Crossarchus obscurus*), but not in others (for example, the banded mongoose *Mungos mungo* uses anal dragging) (Ewer, 1968), which might allow us to find life history correlates (for example, the need to raise the mark above ground level) to explain why the genet does its marking by handstand rather than any other way.

Figure **4.1** Genet performing an anogenital scent mark using a handstand position. (Photograph: Geoff Hosey)

A **Carnivore** (with an initial capital 'C') is a member of the mammalian order Carnivora, which includes: canids, such as the foxes and wolves; felids, such as lions and tigers; and other animals, such as bears. A carnivore (with an initial lower-case 'c') is an animal that eats the flesh of other animals. So sharks and snakes are carnivores, but not Carnivores, whereas a hyaena is both.

4.1.1 Behavioural repertoires

We have just said that behaviour evolves if it has a genetic basis. But what do we mean by 'genetic basis'? Genes are variable stretches of a substance called deoxyribonucleic acid (DNA) and they provide the instructions for building proteins. Thus, to say that genes 'produce' behaviours is nonsense. But the proteins for which genes code may be involved in the formation and operation of the nervous system, or may be hormones and other chemicals that initiate or maintain physiological states. Thus, genes can influence behaviour through a number of complex pathways and systems. But whatever the **genotype** of the animal, its behavioural **phenotype** is probably always the end result of those pathways and systems interacting with the environment in which the animal lives. The extent of that interaction varies according to things such as the function of the behaviour, the life history of the animal, and the complexity of the environment in which the animal lives.

Animals, then, have a repertoire of behaviours that is based on genetically mediated behaviours, which are the product of evolution; the expression of these behaviours is then modified to different extents by learning—the product of the animal's experience of its environment. It should be possible to compile a catalogue (an **ethogram**) of all of the kinds of behaviours seen in animals of a particular species. Such ethograms have, indeed, been published for several species and even incomplete ethograms (which only catalogue the most frequently seen behaviours, or a particular class of behaviours) are usually the necessary starting point for a behavioural investigation.

An example of part of an ethogram for the mara *Dolichotis patagonum* is shown in **Table 4.1**. Note that each behavioural unit is described in such a way

Behaviour	Description
Grazing	Animal grazing, either from a standing or hunkered-down position. Maras often graze while simultaneously engaged in other activities.
Scent marking	Males and females scent mark in different ways. Males perform an anal drag, moving forward (usually over bare earth or female faeces). Females perform an anal rock, remaining *in situ*.
Refection	Maras practise refection (the ingestion of their own faeces) infrequently and not as a regular daily habit as in the lagomorphs. The reingested faecal pellets are apparently no different in texture or size from normally excreted pellets. Refection involves a head-to-anus movement, often difficult to distinguish from penis licking in males.
Threat	Invariably a male–male behaviour, threat describes an animal facing a conspecific, with its head tilted backwards and mouth open to expose the teeth.
Chase	Chase describes a fast aggressive run towards another animal, often accompanied by threat behaviour.
Bite	Biting of one animal by another, often on the rump. Biting usually occurs only after threat and chase behaviours.

Table **4.1** Extract from an ethogram for the mara*

* From Pankhurst (1998).

NOTE: The descriptions of behaviours in an ethogram such as this should allow any observer to identify the same behaviours when they watch the animals.

Genotype and **phenotype** refer, respectively, to the genetic make-up of the individual with respect to a particular trait and what that trait looks like to an observer. Of course, behavioural traits are generally so influenced by learning that their phenotypes change throughout the life of the animal.

(a)

(b)

(c)

Figure **4.2** (a) Scent marking, (b) refection, and (c) aggressive chase in the mara *Dolichotis paragonum*. Pictures such as these help observers to use the written descriptions in the ethogram. (Photographs: Sheila Pankhurst, from Pankhurst, 1998)

that other observers could also use the ethogram and be confident that they were correctly identifying the behaviours. Often, ethograms are supported by photographs or drawings showing what the behaviours look like. We have included several here for the mara (**Fig. 4.2**) and you can see that they help greatly in describing the behaviours that you are likely to see this animal performing.

The animal's full repertoire of behaviours can be thought of as **species-typical behaviours**, meaning that they characterize the way in which members of that species behave in the wild. The extent to which animals in zoos show species-typical behaviours depends on the opportunities afforded by the physical and social aspects of their environment. For example, prey catching by predatory animals

may not be possible in most zoo environments. Nevertheless, **behavioural diversity** (that is, a wide range of different behaviours) is seen as a good measure of how closely the zoo environment allows animals to display species-typical behaviours and is thus relevant to their welfare.

4.1.2 Learning

Learning can be thought of as the process that takes place when an experience of some sort brings about a relatively permanent change in the way in which an animal responds to a situation (Pearce, 1997). Traditionally, theorists have identified several apparently qualitatively different kinds of learning, although it is possible that these may be more similar than was previously thought. The major kinds are:

- **habituation**, in which the animal reduces its response to a constant or repetitive stimulus;

- **classical conditioning** (also known as **Pavlovian conditioning**),[2] in which the animal learns to associate an existing response with a new stimulus (for example, showing food-related behaviours not to food itself, but to the sound of it being prepared);

- **operant conditioning**[3] (also known as **instrumental conditioning**), in which the animal acquires a new response to an existing stimulus (for example, pressing a lever in order to get food);

- **imprinting**, in which young animals learn about their species identity, sex identity, or relatedness to other individuals.

The significance of learning in animals is that it permits them to detect patterns and relationships between objects and events in their environment. An important aspect of this is associative learning, in which animals learn which events reliably signal consequences that are important for them, and which can be achieved both through classical and operant conditioning. Animals also learn the differences between things (discrimination learning), and there is increasing evidence of their ability to learn routes, topography, categories, numerical quantities, and a variety of other things, often collectively referred to as complex learning, because they require more than only the formation of associations (Atkinson *et al.*, 1996).

It is worthwhile looking a bit more closely at associative learning, because it underpins much of the way animal behaviour is managed in zoos.

Associative learning

Associative learning occurs when there is a change in the animal's behaviour as a consequence of one event being paired with another (Pearce, 1997). Exactly what events are being paired depends, to some extent, on the kind of learning that is taking place, but commonly two stimuli are paired (classical conditioning) or a stimulus is paired with a response (operant conditioning).

In classical conditioning, a neutral stimulus[4]—known as the 'conditioned stimulus' (CS)—is paired with the natural biologically relevant stimulus—or 'unconditioned stimulus' (US)—to produce a response—the 'conditioned response' (CR)—that, in most cases, is very similar to the ordinary response—or unconditioned response (UR)—to the biologically relevant stimulus. The CS–US association tells the animal about events in its environment: the CS is a sign that the US is coming.

In the zoo setting, animals learn these sorts of associations all the time (Young and Cipreste, 2004). The appearance of a familiar keeper, activity in the food preparation room, members of the public starting to appear, and all sorts of other events signal to the animal that something else is likely to happen. Classical conditioning is also a part of training animals, for example, to undergo routine veterinary procedures (see section **13.4**). **Figure 4.3** shows, in diagrammatic form, how this sort of learning takes place.

Operant conditioning refers to the technique of using reward and punishment to modify an animal's behaviour (Pearce, 1997). It therefore differs from classical conditioning in that the animals are learning about their responses (or, more exactly, about

---

2 'Pavlovian' refers to the Russian physiologist Ivan Pavlov (1849–1936), who famously trained dogs to salivate to the sound of a metronome, anticipating that it signalled the delivery of food.

3 Called **operant** because the animal's responses operate on (i.e. affect) the environment, which is why the animal learns them.

4 'Neutral' here is simply meant to imply that the stimulus is not one that the animal would normally encounter and therefore presumably has no particular significance for the animal. Examples might include the noise of a buzzer or a light coming on.

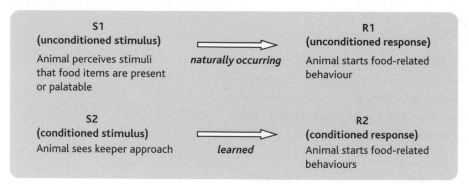

Figure **4.3** Simplified diagram to show the process of classical conditioning. The animal responds naturally to olfactory, visual, and other stimuli from food by showing food-related behaviours, such as salivation, sniffing, etc. In time, the animal may learn that the approach of the keeper is a reliable signal that food will follow, so it sets up an association between S1 and S2. Subsequently, the animal shows responses of food-related behaviours to the keeper's approach (R2), which are very similar to those it naturally shows to the presence of food (R1).

the consequences of their responses) rather than about stimuli. It is achieved by pairing the animal's response with a **reinforcer**, which is simply any event that increases the probability of the response. The reinforcer is referred to as a **positive reinforcer** if it follows the response (for example, the provision of a food reward) and a **negative reinforcer** if it involves the removal of a stimulus (for example, turning off an electric shock). Experimental and anecdotal evidence suggests that animals anticipate the reinforcer, and will learn different responses for different reinforcers, suggesting that the reinforcer acts as a US.

As with classical conditioning, it is likely that operant conditioning of zoo animals occurs informally or semi-formally all the time (for example, keepers 'rewarding' animals for doing the required thing, or perhaps animals learning that certain behaviours draw crowds, and so on). In a formal way, both operant and classical conditioning techniques have long been used in zoos, aquariums, and circuses to train animals to perform shows for the public (see **Fig. 4.4**). Nowadays the main use of these techniques is in **positive reinforcement**

training (PRT), during which animals are trained to undergo veterinary inspection voluntarily, to move willingly from enclosure to enclosure, and to readily undertake other husbandry activities. The way that this is achieved is shown diagrammatically in **Fig. 4.5** and there is more about this technique in section **13.4**.

Before leaving associative learning, we should mention imprinting, which is probably no different in mechanism from the sorts of conditioning we have just considered, but which is usually used to describe the special case of the formation of an attachment between a young animal and its parents (filial imprinting). This is the stage at which the young animal can learn the species to which it belongs and who its relatives are, and this can have consequences for its later choice of a sexual partner (sexual imprinting). In many animals, particularly birds, there is a particularly sensitive period soon after hatching or birth when imprinting is most likely to occur. For this reason, hand-rearing, or rearing in the absence of **conspecifics**, runs the risk of producing animals who fail to imprint appropriately, or, worse still, who imprint on their keeper.[5] Small

---

5 A famous example of imprinting on keepers was Chi-Chi, the iconic female giant panda *Ailuropoda melanoleuca*, who lived in London Zoo in the 1960s and would not mate with An-An, the male who was brought at great cost from Moscow Zoo in the hope of siring a cub.

Figure **4.4** Animals performing shows for the public, like these bottle-nosed dolphins *Tursiops truncatus* at L'Oceanografic, Valencia, are trained through classical and operant conditioning techniques. (Photograph: Vicky Melfi)

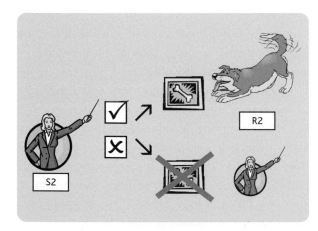

Figure **4.5** (*left*) Training a new behaviour involves building an association between a command—the conditioned stimulus (S2)—and an already occurring behaviour—the unconditioned response (R1). When the keeper gives the command, he or she expects the animal to perform the appropriate behaviour—the conditioned response (R2). If the animal reacts appropriately after the command, it is reinforced with the reward (food). This increases the likelihood that it will perform the behaviour again after the command. Alternatively, if the animal does not perform the appropriate behaviour, it will not receive the reinforcer (so is either ignored or punished), which will reduce the likelihood that it will perform the 'wrong' behaviour after the command in the future. Sometimes, prior to training, a secondary reinforcer (also referred to as a 'bridge') is associated with the primary (naturally occurring) reinforcing stimulus—e.g. a whistle is blown when the animal is eating, so an association is made between the whistle and food. In this case, if the animal reacts appropriately after the command, it is reinforced with the secondary reinforcer (whistle) and then the reward (food). It should be noted that punishments are not necessarily 'harmful' and that an absence of a reinforcer (i.e. time out) can itself be punishing.

Figure **4.6** Sandhill cranes *Grus canadensis* on the nest. Birds reared in captivity require crane-like models so that they imprint appropriately and acquire their species identity. (Photograph: © Walter Spina, `www.iStockphoto.com`)

felids reared without conspecifics, for example, show extreme aggression to potential mates and are unlikely to reproduce (Mellen, 1992).

In birds, especially where artificial rearing is necessary, the use of models may help appropriate imprinting. For example, young sandhill cranes *Grus canadensis* (see **Fig. 4.6**) imprinted on crane-like puppets, or even humans dressed up in crane costumes, learned to associate appropriately with wild cranes when later released into the wild (Horwich, 1989).

Other kinds of learning

As we mentioned above, many animals are perfectly capable of learning much more complex things than only associations between events and, while this ability is not much used or studied in the zoo setting, anyone working with animals should be aware of their potential abilities.

'Discrimination learning' refers to the ability of animals to distinguish between different kinds of events or objects. In some cases, they are able to assign objects to categories. Many of these studies have involved animals such as pigeons, chickens, and monkeys in laboratory settings, but studies in zoos (see **Fig. 4.7**) have included elephants and sea lions—with the sealions needing fewer trials than the elephants to learn the discriminations (Savage *et al.*, 1994). To some extent, we can regard these animals as forming concepts. Pigeons seem to be quite good at this (Herrnstein, 1979), gorillas less so (Vonk and MacDonald, 2002), which warns us

Figure 4.7 Apparatus to test discrimination learning, being used here with Goeldi's monkey *Callimico goeldii*. The animals are rewarded if they successfully go to the black, rather than the white, squares. (Photograph: Paignton Zoo Environmental Park)

stimulus enhancement. Famous examples of these include blue tits *Cyanistes caeruleus* pecking open the foil caps of milk bottles (Fisher and Hinde, 1949) and Japanese macaques *Macaca fuscata* washing potatoes in the sea (Kawai, 1965), both of which appear to be the result of animals watching what others are doing.

Social learning can lead to the establishment of **culture**, one of the most famous examples of which are the different local behavioural traditions (for example, termite fishing, cracking nuts with stones, etc.) of different chimpanzee *Pan troglodytes* populations (Whiten *et al.*, 1999). Again, research on social learning can profitably be undertaken in zoos. The 'artificiality' of zoos presents animals with opportunities for devising novel behaviours that may then spread through the group. An example is tail immersion in ringtailed lemurs *Lemur catta*, which is performed by some individuals in an island-living colony at Chester Zoo, UK (Hosey *et al.*, 1997; see **Fig. 4.8**). Indeed, cultural traditions can be induced experimentally, as in, for example, colobus monkeys *Colobus guereza* at Paignton Zoo Environmental Park, UK, which acquired habits of pushing an apparatus for food in one group and pulling it in the other (Price and Caldwell, 2007).

4.1.3 Motivation

We know both experimentally and anecdotally (and, indeed, personally) that individuals vary from occasion to occasion in their willingness to undertake particular behaviours, including learning tasks. This variability, which also occurs between individuals, is often referred to as 'motivation' and is generally thought of as some kind of internal process that influences the likelihood of whether or not the animal will do the behaviour. In the past,

that the distribution of **cognitive** abilities through the animal kingdom is much more complicated than we might expect.

Other aspects of complex learning include the ability of animals to understand time, number, and the order in which objects are placed, their ability to learn to recognize their image in a mirror, and their ability to form a cognitive map of an area. The evidence for all of these has come predominantly from animals in laboratories, but there is no reason, in principle, why some should not also be studied in the zoo, which would substantially increase the comparative value of learning studies.

Much of the experimental work in animal learning has attempted to describe how an individual learns as a result of its own experience, but recent research has also highlighted the importance to many animals of social learning, through which they can effectively acquire learning from other individuals through processes such as **imitation** and

Imitation refers to an animal copying another's motor pattern. **Stimulus enhancement** occurs when an animal's attention is drawn to a particular stimulus.

Culture, in this context, refers to the non-genetic transmission of behavioural change across generations.

Figure **4.8** A ringtailed lemur *Lemur catta* dipping its tail in the moat that surrounds Lemur Island at Chester Zoo. The lemurs then drink the water from their tails. This behaviour apparently spread in the group through social learning. (Photograph: Geoff Hosey)

particular behaviour can all affect the way in which the behaviour is prioritized.

What happens if the animal is unable to perform the behaviour that it is highly motivated to do? This leads to frustration and, if this is maintained over a long period, it can have adverse effects on the animal's welfare (see **Chapter 7**). Similarly, if the animal is highly motivated to perform two mutually conflicting behaviours—that is, if it cannot do them both, such as, for example, approaching, but also avoiding, a novel object that increases both curiosity and fear—then motivational conflict can result and this too can have welfare implications if it carries on over a long period. In the short term, unusual behaviours such as **vacuum activities** and **displacement behaviours** give us clues about the motivational state of the animal. But frustration and conflict over a long term can lead to the development of **abnormal behaviours** (see section **4.4**), which tell us that the animal has had poor welfare in the past and alert us to the possibility that they are still suffering poor welfare. This is an important area of captive animal management and is considered in more detail in the next chapter.

How an animal responds to stimuli and events is, then, a consequence of both its genetic make-up, and its past experience and learning about these things. Because different individual animals differ from each other in both of these components, it should not surprise us too much that they also may differ from each other in the ways in which they respond. Sometimes, this can lead to consistent individual differences between animals in the ways in which they behave and which cannot be attributed to just their age or their sex. This topic is explored more fully in **Box 4.2**.

motivation has often been seen in terms of some kind of drive or energizer that activates specific behaviours, but while this notion has some intuitive appeal, it fails to account for what is known about variability in responsiveness and has largely been abandoned. We now see motivation more as a decision-making process in which the animal prioritizes the behaviours it must do and switches between them according to these priorities. Factors such as hormones, the attractiveness of a stimulus, and the length of time since the animal last did a

Vacuum activities are those behaviours that are carried out even when the appropriate stimuli are not present, while **dis-placement behaviours** are those that appear to be irrelevant to the stimulus situation in which they are performed.

Box **4.2** Individual differences in behaviour: the influence of personality

Most zookeepers, at some stage, will talk about the animals in their care using the same sorts of terms that people use to describe one another, such as 'aggressive', or 'cautious', or 'shy'. In other words, they are ascribing personality traits to individual animals. But to what extent is it valid to use the term 'personality' (or, better still, 'animality')[6] to describe consistent behavioural responses in animals, and how useful is a knowledge of animal personality in the zoo environment?

We have already seen in this chapter how individual differences in behaviour between animals of the same species can arise as a result of experience, or can be attributed to the animal's age or sex. There is now a growing acceptance of the use of the word 'personality' to describe stable and consistent behavioural styles or temperament traits in a wide range of different **taxa** (see Gosling, 2001, for a comprehensive review). This acceptance has come about largely because of a number of studies providing compelling evidence for a strong genetic basis for personality traits in animals ranging from great tits *Parus major* (for example, in work by Dingemanse *et al.*, 2002, and Van Oers *et al.*, 2004), to vervet monkeys *Chlorocebus pygerythrus* (Fairbanks *et al.*, 2004).

There are a number of areas in which an understanding of personality traits has implications for the captive management of wild animals. Traits such as aggression or a high level of activity, for example, may be more likely to be selected against in zoos, because such animals will be more difficult to manage and may do less well in captivity (McDougall *et al.*, 2006). Carlstead *et al.* (1999a) looked at personality traits[7] in captive black rhinoceros *Diceros bicornis* and found that behavioural scores grouped into the overall trait 'dominant' were negatively correlated with reproductive success for males. Moreover, the higher the female's score in the 'dominant' category relative to her male partner, the greater the breeding success of the pair.

In cheetah *Acinonyx jubatus*, Wielebnowski (1999) used observer and keeper ratings to assess individual behavioural variation in 44 adult captive-born animals. She found that just three temperament components—'tense–fearful', 'excitable–vocal', and 'aggressive'—accounted for 69 per cent of observed variation in behaviour (**Fig. 4.9**).

A suggestion made by two Danish researchers, Hansen and Møller (2001), is that animal welfare could be improved not by modifying the environment to fit the needs of ▷

6 The term 'animality', used to describe personality in non-human animals, was first coined by Geoff Hosey at a zoo research workshop on personality in animals, held at Edinburgh Zoo in July 2004. Animality also featured in the title of a symposium at the University of Stirling in 2005: *Animality! Personality in Animals: An International Symposium on the Importance of Personality and Temperament in Understanding Behaviour.*

7 Carlstead's work was carried out as part of a wider initiative, known as the Methods of Behavioural Assessment (MBA) project. This cross-institutional research initiative by twelve leading North American zoos is discussed in more detail in **Chapter 14** on research in zoos.

Figure **4.9** Personality components such as 'tense–fearful' are derived from keeper ratings and we would expect them to correlate with behavioural measures such as the time taken (latency) to approach a novel object, as seen here when a cheetah was given a traffic cone. (Photograph: Paignton Zoo Environmental Park)

captive animals, but by modifying or selecting the animals kept in captivity, so that they are better adapted to farm, or laboratory, or zoo conditions. In their work on captive mink *Mustela vison* in fur farms, for example, they found that fur-chewing behaviour can be reduced simply by selection, apparently with no negative effects (Malmkvist and Hansen,

2001). But modification of personality traits in captivity, whether intentional or unintentional, may have consequences for animals intended for eventual **reintroduction** back into the wild.

Only a handful of studies have attempted to consider the relationship between personality and survival after reintroduction. One notable example is work by Sam Bremner-Harrison and her colleagues (2004), who looked at personality assessment as a predictor of survival in the wild of captive-bred swift foxes *Vulpes velox*. This study concentrated on an assessment of boldness versus timidity; the results clearly showed that bold foxes are less likely to survive during the immediate post-release period. But while boldness is detrimental to short-term survival in swift foxes, in bighorn ewes *Ovis canadensis* Réale and Festa-Bianchet (2003) found that selection favoured the bold: ewes categorized as non-docile or bold were less likely to be preyed upon by cougars.

So there are no easy conclusions to be drawn about the value of measuring personality in zoo animals. It is becoming increasingly clear, however, that some individual differences in behaviour between animals are consistent, stable, and strongly heritable, and that personality traits in wild animals in captivity cannot be ignored when considering their welfare, captive breeding, and reintroduction.

4.1.4 Functional explanations of behaviour

Functional explanations attempt to understand how behaviours evolve to increase the fitness of the animal—in other words, they increase the animal's survival prospects and reproductive success. The most successful approach to this has been **behavioural ecology**,[8] which examines the evolution of behaviour within the context of the animal's ecology.

---

8 Early in its career, behavioural ecology was often referred to as 'sociobiology'. It came to prominence particularly through Edward Wilson's classic text, *Sociobiology: The New Synthesis* (1975), which is now, 30 years later, still worth a read.

Behavioural ecology is based on a number of key concepts, the most important of which are:

- the gene-centred approach, within which the effects of natural selection on the gene are given prominence, rather than its effects on whole bodies;

- optimality, which recognizes that there are costs, as well as benefits, in performing particular behaviours and that evolving efficient ways of behaving thus involves some sort of trade-off between these;

- the recognition that there are a number of different ways in which behaviours can be done, so that alternative strategies are available to achieve the end of raising fitness.

This is an area of behavioural biology that has grown enormously in the last three decades and, again, we do not have space to do more than give a very brief overview of this subject here. Most importantly, however, we can point out that behavioural ecology has been hugely successful in providing explanations of the evolution of many behaviours that were previously quite puzzling, and that it has led us to view what animals do in a very different way than we did before. The following are some examples.

Altruism and cooperation

Helping somebody else, which potentially lowers one's own fitness while increasing that of the other, is challenging to explain. But although these behaviours (such as alarm calls, division of labour, cooperative hunting, feeding sick and injured members of the group, and many others—see **Fig. 4.10**) look altruistic at the level of the animal that is behaving, at the level of the gene, they can actually be seen to be selfish and therefore we can better understand how they evolve.

At least two explanations are available:

- kin selection—that is, by behaving altruistically to relatives, the animal is furthering the fitness of individuals who have copies of the same genes as itself, through relatedness (originally proposed by Hamilton, 1964);

Figure **4.10** Meerkats *Suricata suricatta* are social mongooses that live in troops made up of family groups, where there is high genetic relatedness. They show a variety of altruistic and cooperative behaviours, including acting as sentinels to watch out for, and warn the rest of the group about, approaching dangers. (Photograph: Sheila Pankhurst)

- reciprocity—that is, behaving altruistically to those individuals who will, at some stage, pay back the good deed (originally proposed by Trivers, 1971).

Because of these explanations, we can start to understand why, for example, helping is often seen in groups of relatives and why an animal may behave more cooperatively to some individuals than it does to others.

Finding food

Finding and processing food is not a simple matter of going looking until something edible is found and then eating it. Animals are confronted with decisions that they must make during their foraging and are also subject to a number of constraints, which limit the sorts of decisions that they can make.

So, what sorts of decisions must foraging animals make? If we apply the 'optimality' concept, we realize that the outcome is not the maximum that an animal can do or achieve, but rather the optimum. In the case of food-finding, this might mean something like the best food return that can be obtained for a given amount of energy expended in finding or processing it. This means that the decisions with which animals are faced are essentially **economic** ones and are to do with optimizing a particular **currency** in the face of particular **constraints**. Examples of currencies are things such as the rate of food intake, the efficiency with which the animal collects food, and the delaying of starvation. Examples of constraints might be the need for minimum daily levels of particular **nutrients**, or the physical size of the animal's stomach.

Now consider an animal that has two kinds of prey: one is small, but can be processed quickly (that is, it has a small 'handling time'); the other is large, but takes a lot of time and energy to process. Which prey should this animal choose? The answer is that it should choose the most 'profitable' prey— that is, the prey with the greatest energy gain per unit handling time. What if the big prey are more profitable, but the animal encounters small prey first? Should it eat the small prey, or should it ignore it and look for bigger prey? Now, the answer is that it should only eat the small prey if what it gains from doing this is greater than that which it would gain from rejecting it and looking for larger prey.

Quantitative models can be constructed to show the circumstances (in terms of search time, handling time, and energy gain) in which this animal should make that decision. These decisions, and others like

them ('Should I stay where the food is, or move out with each piece even though that takes energy, but reduces my risk of being caught by a predator?'), are unlikely to be cognitive decisions that the animal thinks about and works out, but probably involve the application of 'rules of thumb'. But our quantitative models are essentially hypotheses that give us predictions about these rules of thumb, and are therefore a powerful way of understanding how behaviour and ecology have been shaped by evolution.

Mate selection

One of the most direct effects of behaviour on fitness comes when animals are faced with making reproductive decisions. Mating behaviour should not only provide a mate, it should also provide the best possible mate. But what does that mean? Does it mean the best fighter (to defend the young), or the best provider of food (so, someone with a good territory perhaps), or does it mean the most healthy individual (see **Fig. 4.11**)?

Trivers (1972) was one of the first to realize that male animals were likely to have different evolutionary priorities to female animals and that this was likely to lead to conflicts of interest between the sexes. In its most basic form, this conflict starts because males produce plenty of sperm (which are energetically cheap to produce) and therefore can best increase their lifetime fitness by mating with as many females as possible, but deserting them and therefore imposing the costs of raising young on the females. Females, who produce small numbers of ova (which are more costly to produce), can best increase their fitness by caring for a smaller number of young and making sure they survive. Thus, the evolution of mating systems can, to some extent, be seen as a kind of 'arms race' between males and females, each trying to minimize their costs and maximize their benefits at the other's expense.

In terms of an animal's **economic** decisions, **currency** refers to the costs and benefits operating on the animal, while constraints are the behavioural and physiological mechanisms that limit what the animal can do.

Figure **4.11** Why are male peacocks *Pavo cristatus* so gaudy? The most plausible explanation is that bright colourful plumage in a male signals to the female that he is healthy and relatively free of parasites, and therefore a good mating partner. (Photograph: Mark Parkinson)

Out of all of this come what look to us like different sorts of mating systems:

- **monogamy**—a pairing between one male and one female, particularly in those species among which male desertion is not profitable, because females are unable to raise the young alone (for example, many birds);

- **polygyny**—within which one male may mate with many females, particularly in species among which females are able to raise young with little male input (for example, many mammals, in which gestation and lactation facilitate male desertion);

- **polyandry**—in which one female may mate with a number of males (this is rare, but occurs in some species such as the northern jacana *Jacana spinosa*, a species of tropical wading bird);

- **promiscuity**—within which both males and females have multiple sexual partners. Modern molecular techniques have allowed us to see that more species than we expected are promiscuous, even if at surface level they look monogamous or polygamous, because the DNA shows us who an individual's real mother and father are—and often they are not the ones who are bringing the young animal up.

Behavioural ecology proposes fascinating explanations for other aspects of animals' lives, such as the evolution of signals and strategies for aggression and submission. We recommend that, if you are interested in this area, you should read a good textbook such as Krebs and Davies (1993) or Alcock (2005).

4.2 Animal behaviour in the zoo

The body of theory that we briefly reviewed in section 4.1 helps us to interpret what we see animals doing in the zoo and why they are doing it. Behaviour underpins much of the management of zoo animals, from interpreting their welfare (**Chapter 7**) to organizing the structure of breeding groups (**Chapter 9**).

But understanding theory is not enough on its own. We also need basic descriptive information about the behaviours that the animals do. Very often, this information is not available from the wild, and zoo studies represent most of what we know about many species. We can acquire this kind of information by looking at how zoo animals respond to physical and social stimuli in their environment, and also simply by observing and cataloguing what they do.

4.2.1 Stimuli and responses

The physical environment

The environments in which animals live may vary temporally (for example, with day–night cycles, lunar cycles, seasons) and spatially (for example, according to different habitats and microhabitats within the animal's home range). Stimuli associated with these aspects of the environment are used by the animal in choosing appropriate responses. The zoo environment may, or may not, fall within the range of spatial and temporal variability that an animal experiences in the wild, but we would clearly expect that the more overlap that the zoo environment has with the animal's natural environment, the better able the animal will be to show appropriate responses to that environment. It follows, then, that we can provide better environments for captive animals if we know which stimuli are most important for them.

Surprisingly, little research has been done on the stimulus sensitivities of zoo animals, compared with laboratory and farm animals. It has been shown,

for example, that routine procedures and items of laboratory equipment produce **ultrasound** that humans cannot hear, but which many other animals can (Sales *et al.*, 1999)—but virtually nothing is known about sources of ultrasound in the zoo. Clearly, this could have welfare implications, because ultrasound can cause stress in laboratory rodents (Sales *et al.*, 1999).

A good example of the way in which stimulus information can be used to improve the management and welfare of zoo animals is the study by Dickinson and Fa (1997) on the spiny-tailed iguana *Oplurus cuvieri* in Jersey Zoo (**Fig. 4.12**). In these and many other reptiles, ultraviolet (UV) light from sunlight is used in the synthesis of vitamin D, which,

Figure **4.12** Spiny-tailed iguanas *Oplurus cuvieri* in captivity need exposure to ultraviolet light, but prefer to bask under ordinary incandescent white lights because they are warmer. (Photograph: Sheila Pankhurst)

in turn, is necessary for proper bone development and the production of viable eggs. Provision of UV light sources in captivity, however, does not always successfully address this problem. Dickinson and Fa (1997) showed that this was because the iguanas prefer to bask under the ordinary incandescent white lights, which are warmer than the UV lights; this then led to the recommendation that combined UV and hot light sources should be used to ensure that the animals get sufficient exposure to UV light.

The social environment

Social stimuli (for example, stimuli from conspecifics) are also important in guiding behaviour, because all animals in the wild live within a social context, even those traditionally labelled as 'solitary'. When social stimuli from one animal bring about changes in behaviour in another, we usually consider that communication has taken place and, in many cases, the stimuli have become highly ritualized (that is, have been made stereotyped[9] and 'bright' through evolutionary processes) into signals that are obvious even to us. Notice that causal explanations of why an animal gives a signal (for example, a male may give a mating signal because of the combined presence of high levels of hormones internally and a female externally) are at a different level from functional explanations (for example, the male evolves a signal that exploits aspects of the sensory modality, such as colour, that the female favours in an attempt to manipulate her into mating).

Opportunities for social interaction (that is, the exchange of communication) among animals are influenced by the structure and organization of the social group (for example, the number and type of different age/sex classes, and the dispersion of individuals through the habitat). These, in turn, are subject to evolutionary processes. Again, the range of social structures of any particular species maintained in zoos may overlap to different extents with those seen in the wild and social behaviour may therefore be quantitatively—and maybe even qualitatively—different from that in the wild. Because of this, social behaviour is often used as an indicator of whether the zoo environment is affecting the welfare of the animal (see **Chapter 7**) and the manipulation of social behaviour can be used as an **environmental enrichment** to improve welfare (see **Chapter 8**).

4.2.2 Descriptive studies

Some studies of the behaviour of zoo animals are designed to collect basic data on what the animals do, how they distribute their behaviours through time, and how they interact with group members. We can generally call such studies descriptive studies, because they describe the behaviours of the animals in a quantitative way, but the best ones nevertheless follow proper scientific practice in terms of testing hypotheses or answering specific questions. Many of the species held in zoos have never been studied systematically in the wild, so zoos offer important opportunities to find out about the behaviours of the animals. (There is more about this in **Chapter 14** on 'Research'.)

Such studies may take this lack of information as their starting point, so their aim is to find out as much as possible about the animals' behaviours. A good example of this sort of study is the Hutchins *et al.* (1991) description of the behaviours of Matschie's tree kangaroo *Dendrolagus matschiei* (**Fig. 4.13**) at Woodland Park Zoo in Seattle, USA. Tree kangaroos are poorly known in the wild because of their rarity and **arboreality**, making observation difficult. Hutchins *et al.* (1991) started by compiling an ethogram of behaviours, using published work and their own observations. They

---

9 The term 'stereotyped' is used here to describe the process by which ritualized behaviours show an apparently fixed form and are sometimes repetitive. Unfortunately, the same word is used to describe some abnormal behaviours (see section **4.4.3**) that also have fixed form and repetitiveness, but which otherwise have nothing in common with ritualized behaviours.

Animals that are **arboreal** spend most of their lives in trees.

Figure **4.13** Some species, like this Matschie's tree kangaroo *Dendrolagus matschiei*, are difficult to observe in the wild and much of our knowledge of their behaviour comes from zoo studies. (Photograph: Geoff Hosey)

spent 165 observation hours watching a group of four kangaroos and were able to quantify their social interactions (most frequent were approach and nose contact), and to detect quantitative differences in behaviour between males and females.

Other studies may concentrate on one or more specific behaviours that may be of interest as much because they inform behavioural theory as for their impact on animal management. Examples include the investigation of Ralls *et al.* (1987) into how different species of **ungulate** maintain spatial and temporal proximity, Berg's (1983) analysis of the physical structure of elephant vocalizations and the contexts with which they are associated, and Slocombe and Zuberbühler's (2005) investigation of referential features of chimpanzee grunts.

4.3 Behavioural response to the zoo environment

Many behavioural studies in zoos are not designed particularly to tell us basic information about the species, but are more applied, in that they ask questions about how the animal's behaviour is altered by the zoo environment. Such studies are important for a number of reasons:

- they can alert us to welfare issues if the behaviour deviates substantially from what we would expect to see (see section **4.4**);
- they can help us to ensure that behavioural diversity is not being lost through long-term adaptation to captivity;
- they can help us to find ways of providing exhibits that show active, wild-type behaviours, and thus provide a better experience for the public;
- they can help us to evaluate the validity of those studies that do use zoo animals to test behavioural theory.

Many more of these sorts of applied behavioural studies have been undertaken in the last 20 years or so, but it is still the case that most are done on primates (Melfi, 2005), and as we progress through the other mammals, birds, **herps**, fishes, and invertebrates, the studies become fewer and fewer.

We have to start by asking what we mean by the 'zoo environment'. Clearly, there are a great many possible variables here that could influence behaviour. It is therefore convenient to think of them under three subheadings: the physical environment, the social environment, and environmental change.

4.3.1 The physical environment

Enclosures

One potentially important variable in the physical environment is the absolute amount of space in which the animal has to live. Zoo enclosures are mostly smaller than the home ranges that animals occupy in the wild and, at least in Carnivores, home range size is a significant predictor of abnormal behaviours such as **stereotypy** (Clubb and Mason, 2003; see **Box 4.4**), which suggests that restricted space might be a causal factor in stereotypy development. So what effects on behaviour does restricted space have?

One way of answering that question is to compare animals in enclosures of different sizes. This is more difficult than you would expect, because different enclosure sizes usually have different furnishings, and accommodate social groupings of different size and composition. When comparisons have been done, however, they suggest that different kinds of animal are affected differently, as we might suspect. Wolves *Canis lupus*, for example, spend more time resting in large enclosures than in small, but their behavioural diversity appears to be related not to cage size, but to group composition (Frézard and Le Pape, 2003). But in Przewalski's horse *Equus ferus przewalskii*, **activity budgets** differed significantly between groups in different sizes of enclosure, and aggression, grooming, and pacing were higher in the smaller enclosures (Hogan *et al.*, 1988).

Regardless of the absolute amount of space available to them, many animals appear to use the different parts of that space unequally. Some parts of an enclosure may be heavily used, while other parts are hardly used at all. A good example of this is the study by Blasetti *et al.* (1988) of wild boars *Sus scrofa* at Rome Zoo, Italy. They found that nearly two-thirds of the activity of the animals occurred in just two of nine roughly equivalent (in size) sectors. In **Table 4.2**, D is an area with lots of mud and shade, used by the pigs for sleeping and resting, while C is where they were fed. Sectors A and B, which show low usage, were next to a service road, while G, H, and I were next to a visitor road. Thus, the pattern of enclosure usage is partly the consequence of the animals' natural preferences and behaviour, and partly a response to non-natural aspects of the zoo environment.

Similar conclusions can be drawn for cats (pacing mostly at enclosure edges where the animal is at an artificial territory boundary, but where it can see keepers and visitors approaching—Lyons *et al.*, 1997; Mallapur *et al.*, 2002), for great apes (chimpanzees

Area within enclosure	Usage by adults (%)	Usage by piglets (%)
A	2.8	3.3
B	3.1	5.3
C	21.8	29.65
D	44.75	35.75
E	6.1	6.1
F	4.2	5.6
G	7.85	9.3
H	3.3	2.45
I	6.1	2.55
Total	100.0	100.0

Table **4.2** The percentage of utilization of different areas of their enclosure by wild boars at Rome Zoo*

* From Blasetti *et al.* (1988). Reprinted with permission of Wiley-Liss, Inc., a subsidiary of John Wiley & Sons, Inc.

Note how much the pigs used areas C (in which they were fed) and D (an area of mud and shade).

and orang-utans prefer upper areas of the enclosure, whereas gorillas prefer the floor—Ross and Lukas, 2006; Herbert and Bard, 2000), and for Round Island geckos *Phelsuma guentheri* (avoiding vertical glass walls, but preferring hiding places and sunlight—Wheler and Fa, 1995).

The general consensus now (although based mostly on primate research) is that the absolute amount of space is of less importance than the quality of that space, in terms of structural complexity. A multivariate survey of gorillas and orang-utans across 41 different European zoos, for example, found that the activity levels of the animals were related to the number of animals in the group, and the presence of both stationary and movable objects, but were not related to enclosure size or surface area (Wilson, 1982). Laboratory studies show similar effects with rodents and small primates, in relation to which complexity is more important than enclosure size. In many zoos now, the move towards more **naturalistic enclosures** provides an increase in both complexity and absolute size (see **Chapter 6**), and where comparisons have been made, the naturalistic enclosures

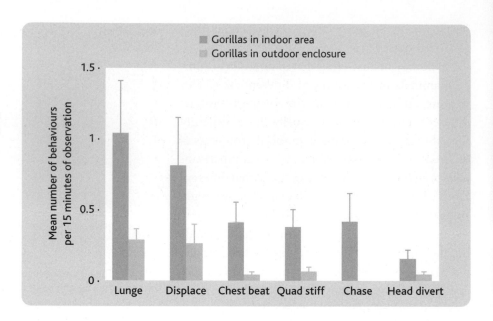

Figure **4.14** Comparison of the frequency of aggressive display behaviours in gorillas when they are in an indoor holding area, compared with when they are in the outdoor enclosure. For each kind of display, the number of behaviours shown by the animals is significantly lower when they are outdoors. (Adapted from Hoff *et al.*, 1997)

are accompanied by more naturalistic behaviours. An example is a study by Hoff *et al.* (1997) of gorillas at Zoo Atlanta, USA, which found that the behaviour of the animals in outdoor naturalistic enclosures differed from their behaviour in the indoor holding area. In particular, aggression was reduced outdoors (**Fig. 4.14**), as were a number of other individual and social behaviours.

Before finishing this section, it is worth saying something about crowding, which occurs when too many animals occupy too small a space. Ethological theories of the 1960s perceived a simple relationship between crowding and stress-induced aggression (see, for example, Desmond Morris, 1969, who saw a parallel between human aggression and that of confined animals in the zoo). Several authors sought to confirm this relationship by studying aggression in zoo primates: an example being a study of rhesus monkeys *Macaca mulatta* in the Monkey Temple at Bristol Zoo, UK (Waterhouse and Waterhouse, 1971; **Fig. 4.15**), which found high rates of aggression.

We now know that there is no simple relationship between crowding and aggression. Early experimental studies showed that social changes had much greater impact on aggression than environ-

mental changes (Southwick, 1967), and more recent studies have demonstrated that chimpanzees *Pan troglodytes* (Nieuwenhuijson and de Waal, 1982; de Waal, 1989), bonobos *Pan paniscus* (Sannen *et al.*, 2004), and rhesus monkeys (Judge and de Waal, 1997) cope with crowding (such as being brought into indoor winter quarters) through behavioural tension-reducing mechanisms, such as an increase in grooming.

Food preparation and presentation

Food is a powerful motivator of behaviour in most animals, so we should not be surprised if the manner and timing of feeding affects behaviour in zoo animals. Timing of feeding can influence behaviour if its predictability allows the animals to anticipate it; this can lead to an increase in aggression—for example, in chimpanzees (Bloomsmith and Lambeth, 1995) and Hamadryas baboons *Papio hamadryas* (Wasserman and Cruikshank, 1983)—or stereotypic pacing—for example, in ocelots *Leopardus pardalis* (Weller and Bennett, 2001). In laboratory stump-tailed macaques *Macaca arctoides*, experimental delaying of routine feeding results in increasing levels of inactive alert, agonistic behaviour, and self-directed and abnormal behaviours during the

Figure **4.15** Rhesus monkeys *Macaca mulatta* in the Monkey Temple at Bristol Zoo, pictured here in 1964 (the structure is no longer used for housing animals). An early study of the effects of crowding on aggression was undertaken here. (Photograph: Geoff Hosey)

anticipated pre-feeding time (see **Fig. 4.16**; Waitt and Buchanan-Smith, 2001). Furthermore, the feeding routine may not be concordant with the natural activity cycles of the animal: mid-morning feeding of brown lemurs *Eulemur fulvus*, for example, disrupts their usual **cathemeral** pattern of activity (Hosey, 1989).

Behaviour is also strongly influenced by the way in which food is presented. For example, primates have traditionally been fed fruits and vegetables that have been chopped into small pieces, usually in the belief that this allows better and more equal distribution of food between individuals. But presenting whole foods to lion-tailed macaques *Macaca silenus* does not result in monopolization by individuals

and individual feeding times increase (Smith *et al.*, 1989). Similarly, multiple feedings of hidden food to leopard cats *Felis bengalensis* result in a decrease in pacing, but an increase in locomotory/exploratory behaviour (**Fig. 4.17**; Shepherdson *et al.*, 1993).

This is, of course, the basis of some kinds of feeding enrichment and this topic is discussed more fully in **Chapter 8**.

The zoo routine

Apart from feeding, few aspects of ordinary zoo routine have been investigated for their effects on behaviour. The following are therefore only a few examples that show the range of behaviours on which the management of animals in zoos can impact.

A **cathemeral** pattern of activity is characterized by sporadic periods of activity through the day and night.

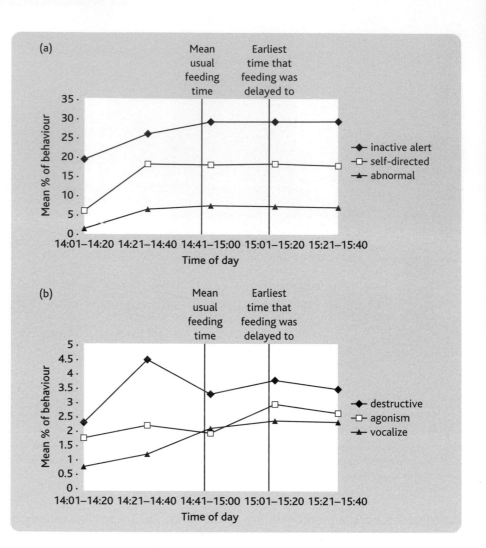

Figure **4.16** Anticipation of feeding in laboratory stump-tailed macaques *Macaca arctoides*. (a) Rates of inactive alert, self-directed, and abnormal behaviours rise in the period leading up to the time at which the animals are usually fed, and remain high if the animals are not fed on time. (b) Vocalizations also increase before the normal feeding time and remain high if feeding is delayed. Destructive and agonistic behaviours, however, rise before the usual feeding time, but then drop at about the time the animals are normally fed. If feeding is then delayed, these two behaviours start to rise again. (Adapted from Waitt and Buchanan-Smith, 2001)

- The application of a contraceptive, which did not affect the physical and behavioural signs of oestrus, to female Hamadryas baboons in an expanding colony at Paignton Zoo did not lead to an expected increase in female agonistic behaviour, even though the frequency of oestrus increased among the females (Plowman *et al.*, 2005).

- Chimpanzees at Chester Zoo confined to their indoor quarters for a month while building work was being undertaken on their outdoor enclosure showed a decrease in aggression, which allowed them to cope with the space restriction (Caws and Aureli, 2003).

- Free-range cotton-top tamarins *Saguinus oedipus* at Jersey Zoo were more vigilant (looking up and looking at building site) when workmen were working at a building site close by (Price *et al.*, 1991).

- Increased noise levels resulted in greater agitation (for example, scratching, vocalization, manipulation of exit door) in giant pandas *Ailuropoda melanoleuca* at San Diego Zoo (Owen *et al.*, 2004).

- Red swamp crayfish *Procambarus clarkii* in aquarium tanks prefer to spend more time near reflective walls, but only if they are dominant (May and Mercier, 2006).

(a)

Figure **4.17** (a) Leopard cats *Felis bengalensis* are often inactive and may show stereotypic pacing. (Photograph: Jessie Cohen/Smithsonian's National Zoo) (b) In this experiment, the ration of food that the cats normally received only once per day was divided into four portions, so that the cats received four feeds irregularly every day. After a month, the four feeds per day were hidden within the enclosure. A significant decrease in pacing, and an increase in exploratory and locomotory activity (as indicated by the asterisks) was seen in the cats in both of these conditions compared with when they received only one feed per day. (Adapted from Shepherdson *et al.*, 1993)

(b)

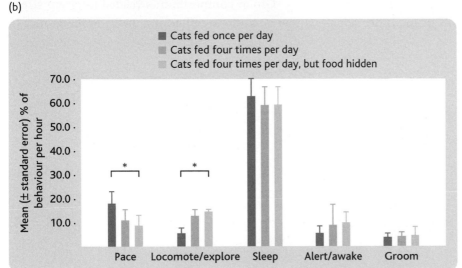

4.3.2 The social environment

The social environment consists of other living things, both human and non-human, whose presence and behaviour impacts on the behaviour of zoo animals. The effects of humans are dealt with in detail in **Chapter 13** and so will not be considered any further here.

Conspecifics

Much of behavioural biology is taken up with trying to understand how conspecifics interact with and influence each other. Here, we are concerned with situations in which the zoo environment may change the ways in which these interactions may naturally occur. An example of this is the way in which the animal is reared: hand-rearing is common in zoos for both mammals and birds, and there are numerous papers (for example, in the *International Zoo Yearbook*) on how this can be done. There are fewer studies, however, on the behavioural effects of different rearing regimes. In its most extreme form, hand-rearing may be accompanied by social deprivation in infancy and this may happen in non-zoo contexts before the animal reaches the zoo. Many zoo chimpanzees, for example, come from circuses, laboratories, and private ownership, and may have a variety of developmental histories. This can affect the social skills of these animals as adults, although some recovery of social skills is possible (Martin, 2005).

Another way in which the zoo environment may influence intraspecific behaviour is through the maintenance of groups that differ in composition from that which usually occurs in the wild. Japanese macaques *Macaca fuscata*, for example, usually live in multi-male, multi-female groups, within which females play an active role in the formation of mating pairs by soliciting some males and rejecting others. But in a group of these animals at Calgary Zoo, within which there was a single male, one female instead prevented all of the other females

from mating, through harassment (Rendall and Taylor, 1991).

One of the best-researched examples of unusual group composition in zoo animals is that of all-male gorilla groups, which are maintained at a number of zoos as a consequence of keeping breeding groups in as naturalistic a form as possible (that is, polygamous harem groups). These 'bachelor' groups can show high group cohesion (see **Fig. 4.18**), particularly in subadult animals (Stoinski *et al.*, 2001). Both silverbacks and blackbacks in these groups may show different intensities of aggression in comparison with males in breeding groups (Pullen, 2005).

Group composition is related to group size—that is, to the number of animals in the group. In

Figure **4.18** Two silverback male Western lowland gorillas *Gorilla gorilla gorilla*, part of a successful bachelor group, engaged in social play. (Photograph: Kirsten Pullen)

Box **4.3** Effects of conspecifics: flamingo breeding displays

Flamingos are very social birds that can form flocks of up to a million birds during the breeding season. In zoos, their group sizes are a great deal smaller than this and this has consequences for their breeding, which is poor in small groups (**Fig. 4.19**; Stevens, 1991; Pickering *et al.*, 1992). So what is it about group size that promotes breeding success in captive birds?

A study of Caribbean flamingos *Phoenicopterus ruber ruber* at the National Zoo in Washington DC (Stevens, 1991) showed that group displays increased by 48 per cent when the flamingo flock size was increased from eighteen to twenty-one, and the synchrony of the displays increased by 100 per cent. Mounts, copulations, and the production of fertile eggs (the first ones in the group's history) also increased. Further addition of birds in subsequent years increased display activity even more (Stevens and Pickett, 1994). Thus, although

(a)

(b)

Figure **4.19** (a) Caribbean flamingo *Phoenicopterus ruber* with young. (Photograph: Ostrava Zoo)
(b) Success at breeding Caribbean flamingos *P. ruber* in zoos depends upon group size. These data, for the period 1983–1988, show that most zoos that keep the birds in flocks of more than twenty-one individuals produce hatchlings, whereas few of those that keep groups of fewer than twenty birds are successful.
(From Stevens, 1991)

▶ there may be other important factors in flamingo breeding, one of the most important is having enough birds in the flock to produce sufficient social stimulation through group displays. Similar effects are seen in other flamingo species, such as Chilean flamingos *P. chilensis* (Farrell *et al.*, 2000).

This example emphasizes the importance of providing a suitable conspecific social environment for zoo animals.

the wild, group size is the consequence of a number of different ecological and behavioural processes, some of which (such as the need to avoid predators and find food) are of less significance in the zoo environment. Consequently, we would expect greater flexibility to be possible in the sizes of captive groups and this does seem to be the case for many species (Price and Stoinski, 2007). Nevertheless, maintaining too many, or too few, animals in a group can have a negative impact on the health and behaviour of the animals. For example, small felids are often solitary in the wild, and housing them with conspecifics can increase stress and reduce reproductive success for the cats; conversely, reproductive success in **callitrichids** is impaired when group sizes are too small, since fewer helpers are available to raise infants (Price and Stoinski, 2007).

Finally, animals may, for various reasons, be kept singly in zoos rather than in social groups. Few studies have investigated the behavioural consequences of this, although pairing primates as a form of social enrichment is well known in the laboratory world. A comparison of the behaviours of pair-housed with singly housed tigers *Panthera tigris* showed that the paired animals displayed a wider variety of more naturalistic behaviours (De Rouck *et al.*, 2005), suggesting that being singly housed affected the behaviour of the animals adversely. Singly housed tigers rolled and played by themselves with objects more than the paired tigers did, perhaps indicating a lack of social opportunities; the greater time that these tigers spent in **flehmen** behaviour (testing for odours from other tigers) supports this interpretation (see **Table 4.3**).

Other species

Animals in zoos may encounter other species within their enclosures, as parts of **mixed-species exhibits**, or they may see, or in some other way detect stimuli from, other species in other enclosures. Mixed-species

Group	Pacing	Rolling	Flehmen	Playing
Pair without neighbours	4.67	0.15	0.19	0.36
Pair with neighbours	21.30	0.16	0.16	0.06
Single with neighbours	23.91	0.31	0.56	0.62

Table **4.3** Average percentages of time tigers spent in those behaviours that differed significantly between housing conditions*
* Data from De Rouck *et al.* (2005).
NOTE: Tigers without neighbours paced less than those with neighbours; single animals rolled on the ground, performed flehmen (an olfactory behaviour), and played more than paired tigers did.

Callitrichid refers to a number of genera of marmosets and tamarins.

exhibits have a long history in zoos and, in general, the mixture of species used is a relatively appropriate one of species that inhabit the same area in the wild and which (hopefully) do not eat each other. A number of descriptive accounts are available in the literature—Pochon, 1998; Young, 1998; Ziegler, 2002—but fewer quantitative studies have been carried out to see how mixed species interact with one another. Popp (1984) investigated aggression between ungulates of different species in mixed groups and found the greatest amount of aggression was between more distantly related species. Events such as births, mating behaviour, and animal introductions were the main triggers of aggression, and male aggression was one of the main features related to the success of the exhibit.

Sometimes, the different species are located in different enclosures, but may still be able to detect each other. Predators and their prey would not normally be housed in the same enclosure, but if they can see each other, we would expect this to be reflected in their behaviour. Stanley and Aspey (1984) studied the behaviour of five ungulate species at Columbus Zoo, housed in an African exhibit where lions *Panthera leo* in a neighbouring enclosure (separated by a dry moat) were visible only if they came to the front of the enclosure. When the lions were visible, the ungulates spent less time with their heads down (for example, feeding, drinking, sniffing the ground) so as to spend more time being vigilant (**Fig. 4.20**).

It is unclear whether situations such as these have welfare consequences for the prey species (in this study, the authors thought not), but laboratory studies with small primates (cotton-top tamarins *Saguinus oedipus*) have shown that even the smell of predator faeces can result in what appears to be anxiety in the animals (Buchanan-Smith *et al.*, 1993; **Fig. 4.21**). On the other hand, tamarins show responses to a model bird being flown over that are similar to those shown for environmental enrichment (Moodie and Chamove, 1990), so this issue is by no means clear.

4.3.3 Environmental change

The management and care regimens at zoos can often require changes to be made that might impact on the animals' behaviours. These changes may affect the physical or social environment of the animals, which, as we have just seen, may already have brought about some behavioural change in themselves. Once again, the literature is dominated by primate studies.

Changes in housing

The laboratory literature shows clearly that cage relocation can be stressful for primates—for example, Mitchell and Gomber (1976); Line *et al.* (1989a); Schaffner and Smith (2005)—but it is unclear how pertinent this is to zoo primates (because the zoo and laboratory environments differ in so many other ways), or indeed to other animal species. Moving leopard cats from barren home enclosures to new, but still barren, enclosures results in elevated urinary **cortisol** and increased stereotypic pacing, both of which can be reduced to some extent by the provision of branches and hiding places (Carlstead *et al.*, 1993).

The adverse effects of translocation in this last study may well be a feature of non-naturalistic enclosures. Indeed, regular routine alternation of two gorilla groups between two naturalistic enclosures at Zoo Atlanta has been used as a form of enrichment for the animals (Lukas *et al.*, 2003). The evidence to support this is that the animals show increased feeding and use of the enclosure, and less **self-directed behaviour**, after being moved to the other enclosure. This sort of manipulation (termed 'activity-based management') has been used elsewhere to move other species (including tapirs *Tapirus indicus*, tigers and babirusa *Babyrousa babyrussa*, as well as primates) routinely between exhibits (White *et al.*, 2003). Enrichment for them appears to come not only from the novelty of the new enclosure, but from stimuli (for example, scents) that have been left in the enclosure by previous occupants.

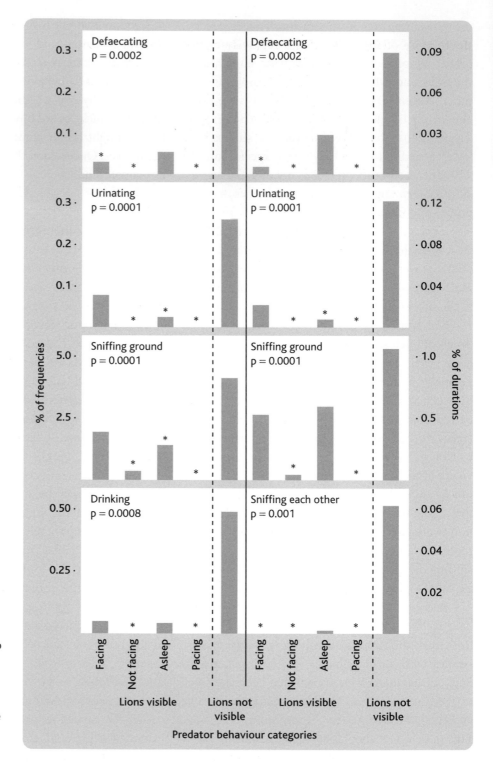

Figure **4.20** When lions were visible to ungulate species in a mixed-species African exhibit, the ungulates showed significantly less (as indicated by asterisks) defaecating, urinating, sniffing the ground, drinking, and sniffing each other compared with when the lions were not visible to them. This was true even if the lions were visible but asleep, and the ungulates stopped all of these behaviours if the lions were pacing. (Adapted from Stanley and Aspey, 1984)

(a)

(b)

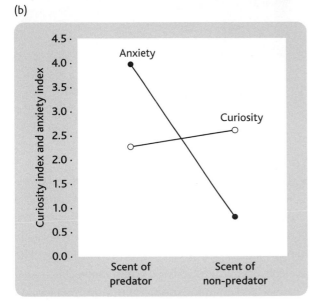

Figure 4.21 Cotton-top tamarins *Saguinus oedipus* (a) are slightly more curious of non-predator faecal scents than those of predators, but predator scents make them behave with higher anxiety (b). The curiosity and anxiety indices are the totals of several individual behaviours in each of those categories. (Adapted from Buchanan-Smith *et al.*, 1993; Photograph: Hannah Buchanan-Smith)

With the increasing trend towards naturalistic enclosures in zoos, a number of opportunities have arisen to monitor behavioural change when animals (again, mostly primates) are moved from traditional cages to naturalistic enclosures and, in some cases, to free range. As expected, the changes seen are usually interpretable as beneficial in terms of producing more wild-type behaviours. For example, the transfer of a group of Hanuman langurs *Presbytis entellus* at London Zoo from a traditional cage to a naturalistic enclosure led to an increase in eating and locomotion, and a decrease in dozing, allogrooming, and aggression (Little and Sommer, 2002; **Fig. 4.22**).

How do we interpret this? At face value, it looks as though enrichment has occurred, but the authors caution us that the animals' activity budgets in both old and new enclosures were within the ranges seen in these animals in the wild (see section **4.5**). Even less clear are the results shown in a study of a gorilla at San Francisco Zoo moved from an old-style concrete grotto to a new naturalistic enclosure (Goerke *et al.*, 1987). In the new enclosure, the gorilla showed less **coprophagy** and **regurgitation/reingestion** (see section **4.4**), but more self-clasping (**Fig. 4.23**) and less play.

Changes in group composition

Changes in group membership can occur because of introductions of new animals to, or removals of animals from, established groups, or through the setting up of new groups. Again, much of the literature on introductions either concerns laboratory primates—for example, Scruton and Herbert (1972); Williams and Abee (1988); Brent *et al.* (1997); Seres *et al.* (2001)—or consists of descriptive accounts of successful procedures—for example, Mayor (1984); Thomas *et al.* (1986); Hamburger (1988). Studies with quantitative data are less numerous.

The following are some examples to show the range of situations in which introductions may occur in zoos.

- Male maned wolves *Chrysocyon brachyurus* at Houston Zoo and Fossil Rim Wildlife Center, both in Texas, were removed from their mates while the latter were giving birth. Later, they were reintroduced in stages, at the end of which they had contact with, and showed affiliative behaviours towards, the pups (**Fig. 4.24**; Bestelmeyer, 1999).

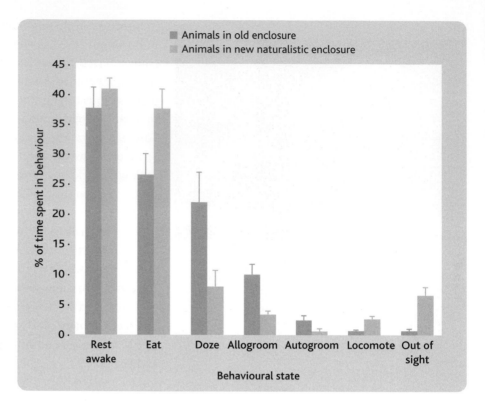

Figure **4.22** Hanuman langurs show changes in their behaviour consistent with enrichment when they are moved from old, traditional-style accommodation to a new naturalistic enclosure, but, in fact, neither is outside the range of behaviours that these monkeys show in the wild. (Adapted from Little and Sommer, 2002)

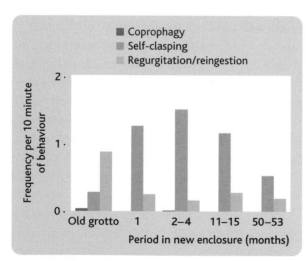

Figure **4.23** When a juvenile gorilla was moved from an old-style concrete grotto to a new naturalistic enclosure, some stress-related behaviours declined (coprophagy and regurgitation/reingestion) while another one (self-clasping) increased. This shows how difficult it can sometimes be to interpret changes in behaviour. (Adapted from Goerke et al., 1987)

- Three female Asian elephants *Elephas maximus* were introduced into an established group at a different zoo; two of them showed increased, the other decreased, stereotypies, which eventually returned almost to pre-introduction levels. The elephants in the established group showed increased social and manipulation/exploration behaviours (Schmid *et al.*, 2001).

- Five hand-reared infant gorillas were introduced into an established group at Bronx Zoo. The infants showed strong tendencies to associate with other infants, but also with the silverback male (McCann and Rothman, 1999).

What happens when animals are removed from groups?

- When two adult Campbell's monkeys *Cercopithecus campbelli* were removed from a laboratory-housed group, there was an increase in social interactions and play in the remaining animals, but they directed more aggression to remaining

(a)

(b)

Figure **4.24** (a) Maned wolf *Chrysocyon brachyurus*.
(Photograph: Ray Wiltshire) (b) Males introduced gradually
to their 7–12-week-old pups show affiliative (friendly)
behaviours towards them. (From Bestelmeyer, 1999)

members of the matriline—that is, the group of
direct female descendents of a particular mother
—from which the females had been removed
(Lemasson *et al.*, 2005).

- Removal of a male giraffe *Giraffa camelopardalis*
at Atlanta Zoo was followed by increased activity,
stereotypies, and contact behaviours, and reduced
utilization of the enclosure by the remaining two
females (Tarou *et al.*, 2000).

Regardless of any welfare considerations, studies
such as these can indicate to us something about
the social dynamics of species in relation to which
detailed field studies are lacking.

4.3.4 What do these examples tell us?

Reading through the previous three sections may
leave us feeling that there is no overall answer to
the question 'How does the zoo environment affect
behaviour?', and that the answer depends on at which
species, at which behaviour, and at which aspect of
the zoo environment we are looking. This is probably a valid conclusion and, given the diversity of
animal life histories and adaptability, we should be
surprised if this were not the case.

There are, however, some general points that
we can make. The first is that the zoo environment
is, indeed, made up of a lot of different variables that
may influence behaviour in different ways. So we
may, for example, be interested in how enclosure
size affects behaviour, but we must recognize that
groups in larger enclosures may be made up of more
animals and may therefore compete more for the
food that is provided, but may also be able to hide
better from each other. So, in this situation, would
we expect a behaviour such as aggression to increase,
decrease, or stay the same? This is not an easy prediction to make, but it is the sort of approach that
we must take if we are to make sense of the zoo
environment and how animals adjust to it.

A second general point is that many animals in
zoos do, indeed, appear to have sufficient behavioural
flexibility that they can adjust to the conditions

imposed on them by the variables of the zoo environment. This may be because the range of environmental variability to which those species are adapted in the wild is as great, or greater, than the variability of the zoo environment; alternatively, it may be that modern zoo environments are not actually all that different from wild environments; perhaps, it is a bit of both.

Either way, this brings us to two related issues that we will consider in the rest of this chapter.

1. How would we know if the zoo environment was outside the range of variability that the animal could accommodate? One possible indicator of an animal's difficulty in coping with its environment is the appearance of abnormal behaviours, which we will consider in the next section, and again in **Chapter 7**.

2. How do we know that the behaviours we are seeing are within the animal's normal range of flexibility? One way is to look at behaviour in the wild and we will consider this in section **4.5**.

4.4 Abnormal behaviours

It has been known for a long time that captive animals sometimes show behaviours that, to us, look unusual and pointless enough for us to label them 'abnormal'. Such behaviours are well known in agricultural animals, among which examples include pacing, weaving, rocking, head-shaking, bar-biting, self-mutilation, feather-pecking, tail-biting, inadequate sexual or parental behaviour, inactivity, hyperactivity, and other apparently odd behaviours (Fraser and Broom, 1990). Examples of these behaviours used to be well known in the zoo world as well, and early descriptions of numerous examples are given by Morris (1964) and Meyer-Holzapfel (1968). The sorts of behaviours that they described included abnormal aggressiveness, stereotyped movements (straight, circular, and figure-of-eight patterns), self-mutilation, and others.

Since those studies, much effort has gone into trying to establish the underlying causes of these behaviours, investigating whether they necessarily constitute a welfare problem, and how best to deal with them. Fortunately, with improvements in housing and husbandry, abnormal behaviours are less of a problem in zoo animals than they were a few decades ago.

4.4.1 What is 'abnormal behaviour'?

What exactly do we mean when we label a behaviour as 'abnormal'? Meyer-Holzapfel (1968) viewed abnormal behaviours as those that were '*uncommon or even absent in **free-ranging** animals*'. A later review used the term also to denote that the behaviour was in some way **pathological** (Erwin and Deni, 1979). Mason (1991) has pointed out that there are two sorts of senses in which the term 'abnormal behaviour' is used:

- that it is rare or unusual;
- that it is apparently lacking in function and may be harmful to the animal, possibly as a consequence of some underlying pathology.

She also points out that, even if a behaviour is seen as abnormal in the first of these senses, it does not necessarily mean that it is also abnormal in the second sense. It is best to use the term, if at all, in a relative rather than an absolute sense, particularly if it is being used to refer to a 'rare' behaviour: rare compared with what? 'Compared with the wild' is a commonly used criterion, on the grounds that these behaviours are not usually seen in wild populations (but see section **4.5** on wild–captive comparisons).

Even if a behaviour is not seen in the wild, it may nevertheless be relatively common in captivity and may not be linked with serious health issues for the animal. Increasingly, some such behaviours are now being referred to as 'undesirable' rather than 'abnormal'. An example of this is regurgitation and reingestion in gorillas (Lukas, 1999)—that is, when the animals bring food voluntarily (not reflexly, as

Figure **4.25** Regurgitation/reingestion in a gorilla. The pool of regurgitated food can just be seen on the floor in front of the animal. (Photograph: Sonya Hill)

Figure **4.26** Feather-plucking has been well documented in an agricultural context and is sometimes also seen in zoo-housed birds such as this blue and yellow macaw *Ara ararauna*. (Photograph: © Petra Jezkova, www.iStockphoto.com)

in vomiting) back up from the stomach to the mouth or a substrate, and then reingest it (**Fig. 4.25**). This behaviour is shown by 65 per cent of captive gorillas and occupies substantial proportions of their activity budgets, and hence, in a relative sense, is 'normal' for captive gorillas. It is possible that this behaviour is performed instead of spending time feeding and it is unclear whether it reflects poor welfare. Using the term 'undesirable', therefore, can avoid some of the negative assumptions that accompany the term 'abnormal'.

The term 'abnormal behaviour' in the zoo context covers a number of apparently different kinds of behaviour, which are generally seen as responses to some aspect of the captive environment. Meyer-Holzapfel (1968) lists the following:

- abnormal escape reactions;
- refusal of food;
- abnormal aggressiveness;
- stereotyped motor reactions;
- self-mutilation (see **Fig. 4.26**);
- abnormal sexual behaviour;
- perversion of appetite;
- apathy;
- abnormal mother–infant relations;
- prolonged infantile behaviour and regression.

Erwin and Deni (1979) distinguish those abnormal behaviours that are 'qualitative'—that is, different in form (for example, stereotypies, self-biting)—from those that are 'quantitative'—that is, when the animal shows elevated or depressed levels of a behaviour that is otherwise normal (for example, hyper-aggression, inactivity).

Many of these behaviours are no longer seen as significant problems for zoos, partly because zoo environments are now much better at meeting the needs of the animals and partly because our knowledge of the extent to which animals show the behaviour has improved. As an example, we can consider **self-injurious behaviour (SIB)**, sometimes referred to as 'self-mutilation' (**Fig. 4.27**). Primates have traditionally been seen as particularly susceptible to this and some laboratory colonies, particularly

Figure **4.27** This tufted capuchin *Cebus apella* has bald patches on shoulder and flank as a result of over-grooming. This type of self-injury is probably best regarded as a stereotypy. (Photograph: Geoff Hosey)

of rhesus monkeys, show high rates of self-injury (Novak, 2003). But a survey of British and Irish zoos by Hosey and Skyner (2007) revealed that, while self-injury was found in a wide **taxonomic** range of primates in zoos, it was nevertheless very rare, with only 24 animals being reported as showing this over a period of about 15 years.

4.4.2 Abnormal behaviour and welfare

The presence of abnormal behaviours is often taken uncritically to indicate that the animal is in some way suffering or that its welfare is, or has been, compromised. This topic is explored further in **Chapter 7**, but, at this point, we can say that abnormal behaviours show such diversity of form, as well as of motivational, physiological, developmental and environmental correlates, that each probably has to be considered separately. In the case of stereotypies, for example, there is clearly a link with welfare, but it is not clear what that link is. Mason and Latham (2004) discuss several possibilities, including that the animal is 'self-enriching', or that repetition of the behaviour helps the animal in improving its own welfare. In the end, it is safest to assume, at least

for stereotypy, that its presence indicates that the animal has been subjected to a suboptimal environment at some stage in its life.

4.4.3 Stereotypies

Stereotypies have probably been of more concern in the zoo context than most other kinds of abnormal behaviour, simply because they are so obvious when present and they cause anxiety in zoo visitors who see them. Most of what we know about their underlying causes, however, is derived from the laboratory and agricultural literature (see **Fig. 4.28**).

Stereotypies can be characterized as behaviour patterns that are repetitive, invariant, and with no obvious function (Mason, 1991). Stereotypies are often seen in environments that appear to be poor for the welfare of the animal, although sometimes they persist in high-quality environments, thus indicating only that the animal has been exposed to a poor environment in its past (Mason, 1991). Often, they are associated with lack of stimulation or with stressful events; stereotypic behaviour may represent the animal's attempts to cope with its environment, but the evidence for this is not as clear as we would wish (Rushen, 1993; Broom, 1998; Mason and Latham, 2004).

Recently, research has indicated a possible link between stereotypies and **perseveration**, which

Figure **4.28** An example of a stereotypy is bar-biting, which is illustrated here in a stabled horse. (Photograph: Georgia Mason)

Perseveration refers to an animal's carrying on of an activity in the absence of the appropriate stimulus.

would suggest that stereotyping animals are unable to inhibit elicited behaviours (Garner and Mason, 2002; Vickery and Mason, 2005). This would imply that the captive environments that produce stereotypies do so through affecting the way brain structures (such as the **striatum**) organize behaviour. This approach may offer ways of explaining why some animals appear to be more prone to developing stereotypies than others.

Recently, Mason (2006) has suggested defining stereotypies as '*repetitive behaviours induced by frustration, repeated attempts to cope and/or **CNS** dysfunction*', a definition that puts more emphasis on possible causal factors than it does on what the behaviour looks like. She sees the first of these (induced by frustration, attempts to cope) as the maladaptive, but reversible, responses of normal animals to abnormal environments, while the second (CNS dysfunction) implies abnormal animals.

In zoo animals, stereotypies may manifest themselves as pacing—for example, in bears (Wechsler, 1991; Montaudouin and Le Pape, 2005)—swaying—for example, in elephants (Wilson *et al.*, 2004)—oral stereotypies—such as licking in **giraffids** (Bashaw *et al.*, 2001; see **Fig. 1.7**)—and other kinds of repetitive movement. Several studies have looked for possible correlates of these behaviours within the zoo environment. A study of primates at St Louis Zoo indicated that rearing history (particularly hand-rearing) was more important in eliciting stereotypies than any of the measured variables of the animals' current housing (Marriner and Drickamer, 1994). A similar conclusion, that early social and environmental deprivation was implicated in the development of abnormal behaviour, was reached by Mallapur and Choudhury (2003) in a survey of 11 species of primate across ten different Indian zoos, and by Martin (2002) in a study of resocialized chimpanzees in

British zoos. Indeed, there is now ample evidence of the link between maternal deprivation and stereotypic behaviours (Latham and Mason, 2008).

But current housing variables can also be important. For example, a survey of nine felid species in eleven different enclosures at Edinburgh Zoo found that, although stereotypic pacing tended to occur at enclosure edges, larger enclosures were not associated with higher levels of pacing; not being fed on alternate days did, however, result in pacing on the non-feed days (Lyons *et al.*, 1997).

Social factors also play a part: paired tigers showed more stereotyped pacing when housed near neighbouring tigers than when not (De Rouck *et al.*, 2005; see **Table 4.3**). It also seems to be the case that some kinds of animals are more prone to developing stereotypies than others, or may be prone to developing particular kinds of stereotypies. Ungulates, for example, are particularly at risk of developing oral stereotypies and this seems to be related to deficiencies in captive feeds (Bergeron *et al.*, 2006). Carnivores, on the other hand, often show locomotory stereotypies ('pacing') that may represent frustrated attempts to escape, because the severity of the behaviour correlates with natural home range size (Clubb and Vickery, 2006; see also **Box 4.4**).

What does all of this mean in terms of the management of zoo animals? Three decades ago, Boorer (1972) concluded that stereotypies were probably not a serious problem, but that they should be dealt with by the zoo because they do not portray the best image of the animals for the visiting public. While the second part of that statement is undoubtedly true, we now know much more about stereotypies and we are now very aware that they might indicate that the animal is in a situation that represents poor welfare. Efforts to deal with stereotypies in zoo animals range from trying to stop them (a common method with

The **striatum** is a part of the basal ganglia, which are masses of grey matter (i.e. neurons or nerve cells) deep within the brain, and appears to have a role in the coordination of locomotion.

CNS is an acronym for the central nervous system, which comprises the brain and the spinal cord.
Giraffids are members of the giraffe family, and therefore include the okapi *Okapia johnstoni*.

Box 4.4 Stereotypy and Carnivores: why are some species worse than others?

We can illustrate our increased understanding of stereotypies by looking at a specific example: the Carnivores. It has long been known that Carnivores seem to be particularly prone to developing stereotypies. Bears, for example, often show pacing (locomotory stereotypy) and may also show high rates of begging (**Fig. 4.29**). In a survey of fifty-eight zoos, Van Keulen-Kromhout (1978) found that high levels of stereotypy and begging were found in the three species for which she had sufficient data: polar bears *Ursus maritimus*, brown bears *U. arctos*, and Himalayan black bears *U. thibetanus*. She also noted that begging was inversely related to

stereotypy, and that polar and Himalayan bears showed more stereotypy than browns, but that browns showed more begging.

Later descriptions of stereotypies in bears concentrated on more detailed description and attempted to correlate the behaviour with underlying motivations. Wechsler (1991) analysed polar bear locomotory stereotypies and suggested that they develop from frustrated appetitive (for example, food-seeking, food-handling) behaviours. Montauduoin and Le Pape (2004) linked stereotypic walking in brown bears to the chance of obtaining food. And Vickery and Mason (2004) considered that

Figure **4.29** Bears often perform unusual behaviours in captivity. (Photographs: Vicky Melfi)

stereotypies were more food-related in sun bears *Helarctos malayanus* than in black bears. Thus, the evidence seems to indicate that stereotypies might develop when animals are prevented from performing behaviours that they are adapted to show in the wild.

This possibility has been explored more fully in carnivorous mammals by Clubb and Mason (2003; 2007), who, after controlling for phylogeny (that is, the likelihood that some correlations will appear in closely related species because of common evolutionary descent), found that the amount of stereotypic pacing that these animals show in captivity can be predicted from their home range size in the wild (**Fig. 4.30**).

Note that polar bears ('PB' in **Fig. 4.30**) seem to fare particularly badly: presumably, because their home ranges in the wild are so vast that even the most generous zoo enclosure cannot replicate them. This result implies that Carnivore stereotypies are not the result of thwarted foraging or hunting behaviours, but are better interpreted as frustrated attempts to escape (Clubb and Vickery, 2006). As far as zoos are concerned, the message might be that better design of Carnivore enclosures is needed to facilitate the animals in performing more of their naturalistic behaviours. Another conclusion might be that there are some species that are unsuitable for keeping in zoos.

This method of comparative analysis, described in more detail by Clubb and Mason (2004), offers promise for giving further insights into the ways in which captive environments affect different species in different ways and, hopefully, will be applied to other taxa as the required data become available.

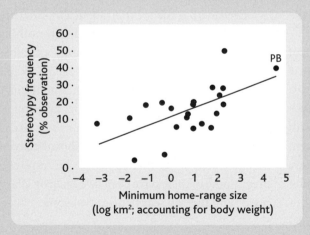

Figure **4.30** Frequency of stereotypy seen in a number of species of Carnivora, plotted against the logarithm of their home range size, with correction for body weight (because bigger animals have a bigger home range). The polar bear is labelled 'PB'. (Adapted from Clubb and Mason, 2003)

agricultural animals), to providing some sort of enrichment. Judging by the agricultural literature, trying to stop an animal from performing stereotypies is unlikely to work and could result in even worse welfare (Mason, 1991). Giraffes do not reduce their stereotypic licking when a bitter-tasting chemical is applied to the substrates that they lick; they just lick elsewhere (Tarou *et al.*, 2003). Increasing dietary fibre, however, can result in a decrease in oral stereotypies in giraffes (Baxter and Plowman, 2001).

Providing some sort of enrichment (see **Chapter 8**) is a common way of attempting to reduce the amount of stereotypies that zoo animals show, as the following examples demonstrate.

- Two Amur tigers *Panthera tigris altaica* at Zurich Zoo were provided with feeding boxes that had to be actively opened by the animals (Saskia and Schmid, 2002). Stereotypic pacing reduced in the female when the boxes were present, both when she was alone and when paired with the male, and also reduced for the male when he was paired with the female.

- In another study, stereotypic behaviours decreased in a female leopard when she was trained to pursue bird sounds, leading to bird parts on a feeder belt (Markowitz *et al.*, 1995).

Much of this effort by zoos is rather 'trial-and-error' —simply trying things out and seeing which will work. For example, in an attempt to reduce the stereotypic behaviour of a pair of vicugna *Vicugna vicugna* at Marwell Zoo, Parker *et al.* (2006) found that, by dividing up the animals' feed, thus increasing their **forage** patch choice, stereotypies could indeed be reduced; providing additional forage, however, which was also expected to reduce stereotypies, actually had the opposite effect.

So how effective is environmental enrichment in reducing stereotypies? Reviews of the literature by Swaisgood and Shepherdson (2005; 2006) have shown that enrichment is, indeed, a suitable technique for this, with enrichment being associated with significant reductions in stereotypy about 53 per cent of the time. They found, however, that in many cases, the form of enrichment was tailored to the particular needs of individual animals. This, and methodological issues to do with sample size, experimental design, and analysis, meant that little in terms of general conclusions about the effectiveness of different treatments could be drawn from the studies.

4.4.4 Birds, reptiles, amphibians, and fish

Up until now, all of our examples of abnormal behaviour have been mammalian. This is not because other animals do not show abnormal behaviour, but rather because there are fewer studies on them in zoos. Studies on birds do, however, show many similarities to the general picture outlined above in relation to mammalian examples. Parrots, in particular, are liable to develop locomotory and oral stereotypies, which may well have the same underlying mechanisms as mammalian stereotypies (Garner *et al.*, 2003). These also can be reduced by appropriate environmental enrichment, such as pair, rather than single, housing (Meehan *et al.*, 2003), and by adding enrichments to promote feeding and locomotion (Meehan *et al.*, 2004).

Again, rearing can affect the development of stereotypies: Hawaiian crows *Corvus hawaiiensis* show higher rates of stereotypy if isolate-reared compared to socially reared (Harvey *et al.*, 2002).

Another kind of abnormal behaviour that has been reported in parrots is self-inflicted feather-picking, which appears to have different environmental and genetic correlates from stereotypy (Garner *et al.*, 2006), but—at least in Crimson-bellied conures *Pyrrhura perlata*—does not appear to be reduced by environmental enrichment (Van Hoek and King, 1997).

Even less has been written about abnormal behaviours in reptiles and fish (and apparently nothing on amphibians). Reptiles are commonly perceived—probably erroneously—as being very adaptable to captive conditions (Warwick, 1990); in addition, they are often regarded as having less behavioural complexity and lower rates of behaviour than birds and mammals. Nevertheless, they do show behavioural changes in captivity that appear to be responses to stress. Some of these have been described by Warwick (1990) and they appear to be of two sorts:

- mobile activities (such as hyperactivity and interactions with transparent boundaries), which perhaps represent attempts to escape;

- hypoactivity/lethargy and anorexia, which perhaps are strategies to 'wait out' the stressful conditions.

Whether these can become stereotypies in the sense described above, and how equivalent they are to abnormal behaviours seen in mammals and birds, is unclear.

Figure **4.31** Surface swimming in a thornback ray *Raja clavata*. This may be a form of stereotypy and is seen in other ray species as well. (Photograph: Geoff Hosey)

Finally, turning to fish, there has been recent discussion about exactly what welfare concerns we should have (Chandroo *et al.*, 2004; Conte, 2004; Ashley, 2007) and behavioural indicators of stress are not clear. Evidence from the **aquaculture** industry suggests that welfare issues for fish can arise from transportation, handling, and inappropriate stocking densities (Ashley, 2007), so we should be aware of these in the zoo and aquarium context as well. Possible stereotypies in fish include vertical swimming patterns in Atlantic halibut *Hippoglossus hippoglossus* at high stocking densities and surface-breaking behaviour in rays in public aquariums (Ashley, 2007; **Fig. 4.31**). There is evidence that captive environments may bring about an increase in aggression in some fish (Kelley and Magurran, 2006), but, in general, how the zoo or aquarium environment affects fish behaviour is unknown.

4.5 Comparison with the wild

At one time, the cages and enclosures used to house zoo animals were predominantly concrete and bars; these bore no resemblance to the natural environ-ment of the animals and the incidence of abnormal behaviour was testament to the difficulty that animals had in coping with these environments. A move towards more naturalistic environments in zoos in the latter part of the twentieth century helped to bring the zoo environment more within the range of environments in which the animals' behaviours were able to operate in (see **Chapter 2**). These developments were, at first, based on perceptions of what made a good captive environment from a human point of view, but gradually became more informed by the growing amount of knowledge of what animals did in the wild (Redshaw and Mallinson, 1991).

This aspect of zoo biology is dealt with in more detail in **Chapters 6**, on 'Housing and husbandry', 7, on 'Animal welfare', and 8, on 'Enrichment'. This section deals with what may be a logical consequence of this philosophy: that behaviour in the wild can be used as some sort of standard with which to compare behaviour in the zoo (see **Box 4.5**). Just what sort of standard we will consider later, but firstly, we will establish some cautions about how to interpret such comparisons.

4.5.1 Interpreting the results

It has been noted, particularly by Veasey *et al.* (1996a), that wild–captive behavioural comparisons pose a number of methodological and technical difficulties that hamper our interpretations of what these studies show.

Veasey *et al.* (1996a) list the following:

- observers may influence the behaviour of wild living animals more than they do those living in the zoo;

- changes in both **biotic** and **abiotic** factors can bring about quantitative and qualitative changes in behaviour in the wild;

Biotic factors are those that are to do with living organisms (such as interactions with other animals and plants), whereas **abiotic** factors are those to do with non-living (physical and chemical) processes in the environment (such as light, temperature, climate).

Box **4.5** Comparing behaviour in the zoo with behaviour in the wild

There are a number of examples of comparisons of behaviour in the zoo with behaviour in the wild in the literature. The two examples given here (hopefully) reflect the differing approaches that researchers have taken to this area of study. (A further example is given in section **8.3.1**.)

Höhn *et al.* (2000) compared the activity budgets and agonistic behaviour of Eastern grey kangaroos *Macropus giganteus* (**Fig. 4.32a**) at Neuwied Zoo in Germany with those shown in the wild in Australia. The same observers collected data in the same way and with the same behavioural measures in the two locations. The group sizes of the animals were similar (fifty-six animals in the zoo group; fifty to sixty in the wild group) and the zoo enclosure was a naturalistic one that resembled the wild.

The observers found that the activity budgets of the animals in the two situations

(a)

(b)

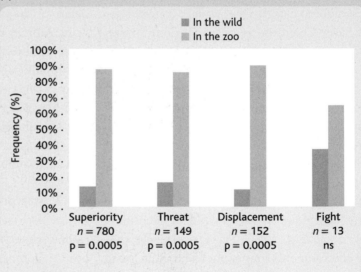

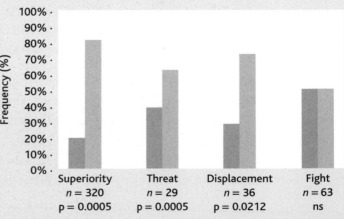

Figure **4.32** (a) Eastern grey kangaroo *Macropus giganteus*. (Photograph: © Sonia Schwantes, www.iStockphoto.com) (b) Frequencies of different aggressive behaviours (expressed as percentages) of large male (top graph) and medium-sized male (bottom graph) Eastern grey kangaroos in the wild and in a zoo. The zoo animals show more low-level aggression (superiority, threat, displacement), but not high-level aggression (fight) in comparison with those in the wild, ns, not significant. (Adapted from Höhn *et al.*, 2000)

were not particularly different, although there were some differences in some behaviours. Agonistic behaviour, however, was consistently more frequent in the zoo kangaroos in three of the four measures of agonistic behaviour (**Fig. 4.32b**). But escalated aggression was not more frequent, which suggested that the animals were using lower intensity aggression more often to avoid damaging conflicts.

Our second example comes from a study by Kerridge (2005) on the activity budgets of black-and-white ruffed lemurs *Varecia variegata* (**Fig. 4.33a**). She collected data in several British zoos and in the wild in Madagascar,

(a)

Figure **4.33** (a) Black-and-white ruffed lemur. (Photograph: Geoff Hosey) (b) Activity budgets for black-and-white ruffed lemurs in the wild and in two different zoos. The zoo-housed lemurs spend less time feeding and moving, and more time in grooming and social behaviour than their wild counterparts. (Adapted from Kerridge, 1996)

(b)

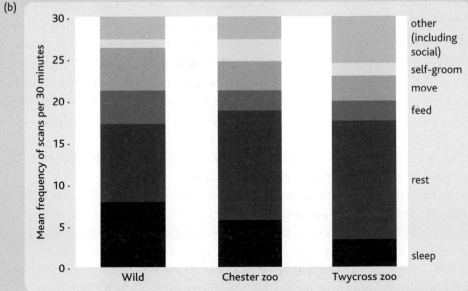

> again using the same behavioural measures. The purpose of this was to identify whether the captive animals had different activity budgets and, if so, whether environmental enrichment could be used to make them more similar to those exhibited in the wild. The results (**Fig. 4.33b**) showed that the captive animals spent more time in grooming and social behaviour, and less time in feeding and moving, than the wild lemurs; they also used fewer different kinds of food manipulation. Environmental enrichment (providing whole rather than chopped food) had the effect of increasing the feeding times of the captive lemurs and manual manipulation of the food, but did not affect time spent in movement.

These two studies illustrate several features of wild–captive comparisons: the observers in the two situations may or may not be the same individuals; particular behaviours may be chosen for comparison, but the use of activity budgets is a common technique; and there is an assumption that other variables in the comparison do not impact significantly on the results. As a consequence, we should be careful how we interpret the results of zoo–wild comparisons.

- many species are rare or extinct in the wild and in limited numbers in captivity, casting doubt on how representative the behaviour is;

- small sample sizes cause difficulties, including the need to pool data, often from different subspecies or even hybrids;

- the zoo population may itself not be representative, because of the zoo survival of animals that may be unfit to survive in the wild;

- there is considerable behavioural variation between individuals and, indeed, between populations in different zoos;

- the same behavioural measures (indeed, the same observers as well) are rarely used in comparisons.

The same authors followed up this critique with a comparison of behaviours of wild and zoo-housed giraffes to try to assess how much these methodological difficulties affected the results (Veasey *et al.*, 1996b). They collected data at four different zoos in the UK, watching giraffes in their paddocks in the day and by night in their sleeping quarters, and also collected data on wild giraffes in Zimbabwe using four different data collection methods. They found (**Fig. 4.34**) that the giraffes in the four different zoos only differed from each other significantly in one behaviour: lying down.

The four different data collection methods used on the wild giraffes gave significantly different results and, when the results of one of the methods were compared with one of the zoo samples, it was found that the zoo giraffes spent less time feeding and also differed from the wild giraffes in several other behaviours. The authors also suggested that the lack of significant differences between zoos was probably a consequence of small sample sizes, rather than a real lack of difference. Clearly, then, comparisons between wild and captive behaviour should attend to these technical difficulties if they are to give an answer in which we can have confidence.

4.5.2 The purpose of wild–captive comparisons

Earlier, we mentioned that behaviour in the wild is used as a kind of standard against which behaviour in the zoo can be compared, without saying what that standard is. In a lot of comparisons, the assumed standard is one of best welfare and the implication is that, where zoo animals differ from their wild counterparts, that difference should be interpreted as lower or impaired welfare in the zoo. This idea

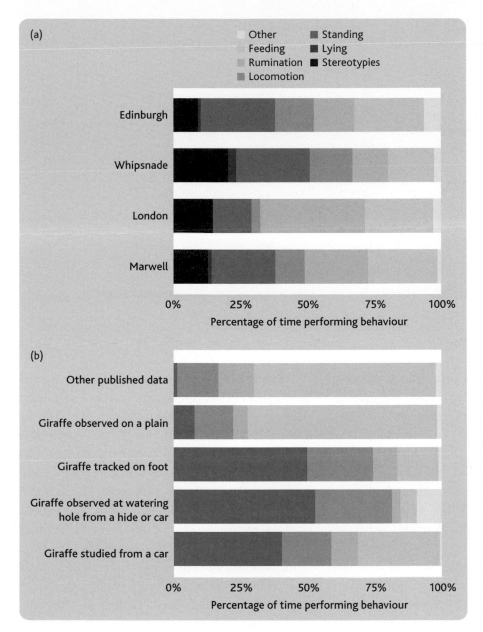

Figure **4.34** (a) Comparison of daytime activity budgets of giraffes in their paddocks at four different zoos. Although there appear to be differences, in fact, only lying is significantly different between the four zoos. (b) Comparison of activity budgets for wild giraffes obtained from five different observation methods. The methods give very different results and are also different from the results for the zoo giraffes. (Both figures adapted from Veasey *et al.*, 1996b)

occurs in the literature on the welfare of farm animals, in which a number of authors have suggested that the wild ancestors of farm animals provide the standard against which to measure the welfare of the captive animals (see, for example, Fraser and Broom, 1990) and has become acceptable in the zoo world as well (see Lindburg, 1988).

Using welfare as the assumed standard, however, has also been criticized by Veasey *et al.* (1996a), because it assumes that the mere expression of a behaviour improves welfare, or, conversely, that not being able to do it decreases welfare, and it is by no means clear that this is actually the case. It is also an assumption that the welfare of wild-living animals is the best that it can be—and this, again, may not be true.

All of this is not to say that the zoo–wild comparison is of no value in our evaluations of

| Environment | Time spent in behaviour (%) | | | |
	Feeding	Moving	Stationary	Social
Zoo-housed (*ex situ*)				
Traditional/indoor	36	6	30	10
Enriched traditional/indoor	52	–	–	20
Naturalistic	66	7	12	6
Confined but in their range country				
Limited provisioning	64	–	–	–
Ad libitum provisioning	28	19	48	5
Unconfined in their range country				
Ad libitum provisioning	22	20	45	13
Ad libitum provisioning	29	25	39	–
No provisioning	63	18	10	9
No provisioning	47	21	22	10

Table **4.4** A comparison of some mandrill behaviours in different conditions across the wild–captive continuum*

* Adapted from Chang *et al.* (1999).

NOTE: The data for unconfined animals are for a different baboon species, as data were unavailable for wild mandrills. (See Chang *et al.*, 1999, for information on the original sources.)

the welfare of zoo animals; only that we should be cautious in making inferences from our comparisons. An alternative use of wild behaviour as a comparative standard is in the area of conservation. If we are concerned that zoo animals should maintain their species-typical behaviours, which is particularly the case if some of them are ultimately destined to be returned to the wild, then how conspecifics behave in the wild is a reasonable guide to what they ought to be capable of doing. Note, however, that the methodological difficulties are still there and should be addressed in any comparison. There is also a potential difficulty of interpretation, in that it is not always clear exactly what we mean by 'wild'.

4.5.3 What is 'the wild'?
While the question of what exactly is 'the wild' may, at first sight, be as absurd a question as 'what is a zoo?', which we came across in **Chapter 1**, it is an important question to try to answer, because human activity nowadays has changed many natural environments that wild-living animals inhabit.

As an example of how these different environments affect behaviour, we can consider a study by Chang *et al.* (1999) of zoo-housed mandrills *Mandrillus sphinx*. These authors studied the behaviour of a small group of mandrills occupying a traditional tiled-and-barred indoor cage at Zoo Atlanta and compared this with the behaviour of the same group after it had been moved to a large, naturalistic outdoor enclosure. The behaviour of the animals changed in the new enclosure: time spent feeding, in locomotion, and out of sight increased, whereas time spent stationary, exploring, playing, and other social interaction decreased (**Table 4.4**). The hope, of course, was that these changes made the mandrills more similar to those in the wild. Unfortunately, no comparable data were available for wild mandrills, but the authors were able to compare their animals with two published studies of mandrills living in captivity, but in their native country in an enclosed area of natural forest. In one of these latter studies, the animals had **ad libitum** provisioning; in the other, they had limited provisioning. The zoo mandrills,

Ad libitum means 'free access'—that is, the animals can eat as much as they want to, whenever they want to.

after release into the naturalistic enclosure, were more like the mandrills with restricted provisioning, but in their native forest. This is probably a good result, but it leads us to ask in what way the forest-enclosed animals differ (if at all) from unrestricted wild animals.

It was pointed out many years ago by Bernstein (1967) that, in order to be considered 'natural habitat', a field site should have space for all of the life history requirements of the animals, little human disturbance, and little human effect on the habitat. Just how many of the field sites that have become well known as places in which to study wild animals actually satisfy these requirements is unclear. More recently, Hosey (2005) has attempted to define the characteristics that distinguish zoo environments from the other sorts of environments in which animals live, including the various sorts of 'wild'. He points out that zoo environments can largely be characterized by the extent to which they confront animals with human presence, require animals to live in what is sometimes a restricted space, and involve management of the animals' life histories. But these three characteristics are not unique to zoo environments: they are found, to a greater or lesser extent, in many 'wild' environments as well. Clearly, again, caution is needed in our interpretations of wild–captive comparisons.

Summary

- Behaviour is of interest because, among other things, it can tell us about how an animal is being affected by the zoo environment and can give us clues about the animal's welfare.

- The behaviour that we observe is moulded by the animal's genetic make-up, but also by its experiences, and this makes behaviour very flexible.

- Zoos have advantages for collecting descriptive information about the behaviours that animals show, which would usually be much more difficult to observe in the wild.

- Aspects of the zoo environment may lead to changes in the animals' behaviours. Both physical (cage size and complexity, food, zoo routine) and social (conspecifics and other species) aspects of the zoo environment may do this, but our knowledge of this—particularly for animals other than primates—is very incomplete.

- If zoo environments are suboptimal for the animals, abnormal behaviours may result. Interpreting these is not easy, but, thankfully, they are much less of a problem for modern zoos, because modern housing and husbandry provide much better environments than were previously available.

- A possible benchmark for interpreting the behaviours that we see in the zoo is how the animal behaves in the wild, but we should be cautious of applying this too simplistically.

Questions and topics for further discussion

1. What can studying animals in zoos tell us about animal behaviour?

2. What evidence is there that the behaviour of zoo animals is changed by their captive environment?

3. Is the term 'abnormal' an appropriate description to use for stereotypies?

4. Is there any value in comparing the behaviour of animals in zoos with that of their wild conspecifics?

5. Do animals adapt their behaviour to the zoo environment?

6. If the behaviours that we see in zoo animals are no different from what we see in the wild, can we assume that captivity has no impact on the animals?

Further reading

There are many good textbooks available on animal behaviour, but several are restricted to functional interpretations of behaviour. Good coverage of all areas of the discipline is given by McFarland's *Animal Behaviour: Psychobiology, Ethology and Evolution* (1999), Alcock's *Animal Behavior: An Evolutionary Approach* (2005), and Manning and Dawkins' *An Introduction to Animal Behaviour* (1998). There are also many good texts on animal learning: a good starting place is Pearce's *Animal Learning and Cognition* (1997). Behavioural ecology is well explained in Krebs and Davies' *An Introduction to Behavioural Ecology* (1993), and texts such as Alcock (2005) are good for bringing this subject up to date. A comprehensive review of the effects of the zoo environment on mammal behaviour is given by Carlstead (1996). An introduction to methods of undertaking behavioural research is to be found in Martin and Bateson's *Measuring Behaviour: An Introductory Guide* (2007).

For specific information on carrying out behavioural research on zoo animals, the best sources are the BIAZA research guidelines, which are available online at `www.biaza.org.uk`. Currently, available guidelines include those on behavioural observation (Wehnelt *et al.*, 2003), visitor studies (Mitchell and Hosey, 2005), surveys and questionnaires (Plowman *et al.*, 2006), statistics for zoo research (Plowman, 2006), and personality profiling (Pankhurst and Knight, 2008).

See also **Chapter 14** on 'Research'.

Websites and other resources

For anybody who wants to know more about animal behaviour, the websites of the Association for the Study of Animal Behaviour (ASAB)—`http:// asab.nottingham.ac.uk`—and the Animal Behavior Society (ABS)—`www.animalbehavior.org`—provide a great deal of information and also links to other behaviour-related websites.

The International Society for Applied Ethology (ISAE) caters for those who are interested in applied aspects of animal behaviour, such as the effects of captivity and human interaction. Its website—`www.applied-ethology.org`—contains relevant information and links.

Chapter 5 Animal identification and record keeping

Being able to identify individual animals within a **zoo** is essential for monitoring their health, promoting good welfare, cooperating with conservation initiatives, and undertaking research. In this chapter, we will provide an overview of the diverse methods used to identify individual captive animals, and will also discuss the relative merits and drawbacks associated with some of them. When animals can be individually identified, the amount of information that can be recorded about them is almost limitless, which means that keeping comprehensive animal records is also an important part of animal management. It probably sounds a little dull and uninspiring, but record keeping is the corner-stone of good captive animal management and informs all of the decisions that we make about animals in zoos. So we will explore in this chapter the need for keeping standardized records that are easily accessible and the systems that are in place to make this possible. These range from record cards, to a computer-based program called the Animal Record-Keeping System (ARKS), and its sister programs, and on to the Internet-based Zoological Information Management System (ZIMS), which promises to revolutionize the ways in which zoos manage their animal records.

The topics that we will cover in this chapter are as follows.

5.1 The importance of knowing your animals
5.2 What species? Nomenclature and taxonomy
5.3 Identifying individual animals
5.4 Temporary artificial methods of individual identification
5.5 Permanent artificial methods of individual identification
5.6 Record keeping: what information could, or should, be recorded?
5.7 Zoo record-keeping systems

In addition, there are several boxes exploring related issues about **scientific names** and the use of molecular techniques, as well as an explanation of the standard units used to measure different kinds of animal.

5.1 The importance of knowing your animals

Successful animal management is dependent on knowing your animals: knowing what they had for breakfast, what has happened to them in their lives, their medical history, how related they are to other animals in their group or the captive population, and so much more. If you know these things, then you can make decisions about their daily management and care, and also plan their long-term management within the captive population, such as their potential breeding prospects. If you do not know, or cannot recognize, your animals at an individual level, then it becomes very difficult to monitor them and you are then likely to miss any changes in them that might reflect their health or welfare status. Because of this, it is a legal requirement for all UK zoos to keep records for individually recognizable animals—or groups, if the individuals cannot be identified—and to keep records accordingly (Defra, 2004; see section **5.6** for more on this) and to produce an annual report that details some of this information (see **Table 5.1**).

As we can imagine, however, while recognition and record keeping is reasonably straightforward

Common name	Scientific name	Group at 1 January 2007	Arrive	Born	Death within 30 days of birth	Death	Depart	Group at 31 December 2007
White-naped crane	*Grus vipio*	2.1.1	0.2.1	0.0.2	0.0.1	1.0.0	0.1.0	1.2.3

Table **5.1** An example of the information that should be provided in annual reports, using the white-naped crane *Grus vipio* as an example*

* It is customary within zoo records to summarize group compositions as the number of 'males.females.animals of unknown sex', e.g. '2.1.1' refers to two males, one female, and one unknown.

for a small number of animals, it gets more difficult as the size of the group or collection gets bigger. For example, Colchester Zoo in the UK is a moderately large zoo that, according to its annual report, held 271 animal species, totalling 3,445 animals, at the end of 2006. This is a lot of animals for which to keep records. Fortunately, there are many processes in place to facilitate the identification of individuals and the keeping of detailed records about them.

There are two main reasons for keeping records for individual animals: to ensure good health and welfare, and to aid conservation. Let us look more closely at each of these.

5.1.1 Animal records for health and welfare

The health and welfare of individuals can be more easily monitored if the animals can be recognized individually. It follows, therefore, that the health and welfare of individually recognizable animals is likely to be greater than that of animals in groups, within which an animal might have to show exaggerated signs of ill health or poor welfare to be noticed among the crowd. By comparing the records of many animals, it is also possible to establish what is 'normal', and thus discern whether any changes that can be seen are usual for an animal of that age, sex, and species (see section **5.6.6**).

Record keeping is also important during the transport of live animals, because transport potentially increases the risk of disease transmission and has been found to be stressful for many species (see section **6.3.2**). There is much legislation and regulation governing the live transfer of wildlife, with regard to health, welfare, and conservation

(Cooper and Rosser, 2002; see also **Chapter 3**). When good records exist for individual animals, it is possible to track the movements of animals, and this enables the authorities (within the UK, the Department for Environment, Food and Rural Affairs, or 'Defra') to map any disease spread and to prohibit any animal transfers that may increase disease transmission. Thus, the risk of transferring disease can be reduced, because movements of animals can be restricted to ensure that healthy and unhealthy animals are kept separate. This has been demonstrated in domesticated animals, where individual identification and the presence of good records greatly enhance disease prevention and control (Disney *et al.*, 2001).

5.1.2 Animal records for conservation

Using records to restrict the transport of live animals is also necessary to safeguard wild populations. As previously mentioned, animals that are listed under the Convention on International Trade in Endangered Species of Wild Fauna and Flora (CITES) (UN, 1973) as endangered or threatened (see section **3.2.1**) cannot be transferred without appropriate permits and licensing, to reduce the likelihood that endangered animals are being traded commercially, which would put pressure on declining wild populations.

Keeping records of animals' rearing and movements is also necessary to facilitate **captive breeding programmes**. Nowadays, animals rarely enter captive breeding programmes from the wild, but, if they do, they need to be classified into the appropriate species or subspecies category (more about this in section **5.2**). For those born in captivity, simply being able to recognize dams and sires of new offspring will greatly increase the accuracy of any genetic and **demographic** analyses (carried out using records) performed to inform recommendations for that captive population.

Comparing information collated between many zoos enables **epidemiological** research to be undertaken and this can identify trends within the captive population, such as the incidence or spread of disease, as well as **fecundity** and/or **mortality** rates. Such studies have included the investigation of **neoplasia** in black-footed ferrets *Mustela nigripes* (Lair *et al.*, 2002) and mortality in cuttlefish *Sepia officinalis* (Sherrill *et al.*, 2000). Results of such studies positively feed back into the management of the animals, and this facilitates improvements in health, welfare, and conservation of the captive population.

5.2 What species? Nomenclature and taxonomy

One of the first things that should be identified about an animal in a zoo is what sort of animal it is. To ensure that an animal is kept in an appropriate social and physical environment, and that it is fed on the most appropriate foods, we need to know to what species it belongs. Indeed, for managing its contributions to the captive breeding and conservation of its species, we often need to be more precise than this and determine to which subspecies it belongs. **Vernacular** names of animals are very variable and tell us little about the animal except that we think we know what it is. For example, the names 'red fox' and 'grey fox' may tell someone that we are referring to something that is small and dog-like, but if we were to say 'flying lemur', we would be talking about an animal that, in fact, is not a lemur (and does not fly either). In Germany, the red fox is *Rotfuchs*, and in France it is *renard*, and given that this animal lives throughout Europe, Asia, and North America, there are doubtless dozens of other words for it as well.

To ensure that we all know precisely what animal we are talking about, we use a system of scientific nomenclature, whereby each species is given a unique **binomial name** in a Latinized form (Latin is no longer widely used as a first language, so it is ideal for this, although some scientific names are derived from the Greek, rather than from Latin, words); these names are then organized within a system of classification or **taxonomy**, which tells us much more about the animal than merely what it is. Taxonomy, for example, can give us information about the distinctiveness and the variability of the species, and can also help us to understand its evolutionary relatedness to other species.

It is worth looking a little more closely at these systems.

5.2.1 Binomial nomenclature

The basis of the scientific naming system for animals (and for plants as well) is that each species has a name made up of two components (which is why it is called 'binomial'), one of which identifies its genus (its generic name) and one that identifies its species (its specific name). The words used for the generic and specific names can come from any language, but are written in a Latinized form. Thus, our domestic cat has the binomial *Felis catus*, in which *Felis* is the Latin word for a cat and *catus* is the Latinized form of the English word for a cat. Note that generic names start with a capital, but specific names start with lower case. The scientific

---

Demography is the term used to describe the characteristics of a population, such as its age or sex structure, birth and death rates, etc.
Epidemiology is the study of the incidence, prevalence, and treatment of disease in the population.

Neoplasia is the growth of new cells resulting in a tumour, which can be benign or malignant.
Vernacular just refers to the word used for something in the local language.

Figure **5.1** Carolus Linneaus (1707–1778) devised the binomial system of animal (and plant) taxonomy. (Photograph: `www.historypicks.com`)

we are to have confidence that the names we use are systematic, comprehensive, unambiguous, and understood throughout the world.

As an example of how this works, let us briefly consider a familiar zoo animal: the lion. Linnaeus recognized that the lion was a kind of cat and gave it the binomial *Felis leo*. But later scientists considered that the lion was sufficiently different from domestic cats, ocelots, pumas, and other assorted small cats to be distinguished by a different generic name. The earliest different generic name given to lions was *Leo*, given by Oken on p. 1070 of his 1816 publication. But it was also clear to most scientists that the leopard and the tiger should be in the same genus as the lion. In this case, the earliest valid generic name is *Panthera*, which Oken used for the leopard. But because he used it on p. 1052 of his publication, it has priority over *Leo*, which is why we now refer to the lion as *Panthera leo*, rather than as *Leo leo*.

When Linnaeus devised his binomial system, species were seen as fixed entities that would not change. We now know that species change over time through evolutionary processes. One consequence of this is that different populations of the same species living in different areas can become different from each other in their **morphology** and behaviour. These populations may be regarded as subspecies, and given a third Latinized name (a trinomial) to distinguish them from the other subspecies. In this case, the population on which the original species description is based is given a trinomial that is the same as the specific name. To return to our lions, Linnaeus' original description was based on the North African lion, so that is now referred to as *Panthera leo leo*, whereas Asiatic and other African lions have different trinomials (for example, *Panthera leo persica* for the Asiatic lion and *P. leo nubica* for the East African lion).

name[1] of an animal is always italicized (or underlined if handwritten).

This system was devised by Carolus Linnaeus[2] (1707–1778; see **Fig. 5.1**), initially for naming plants, but was later extended to animals. The starting point for modern animal nomenclature is the tenth edition of Linnaeus' *Systema Naturae*, which was published in 1758; any Latinized names that date from before this are not considered to be valid.

The process of applying this system is subject to strict rules, overseen by the International Commission on Zoological Nomenclature (ICZN); one important rule is that of priority, whereby the first valid name for a genus or species is the one that must be used. What makes a name valid is subject to informed debate, but decisions on this can lead to animals being renamed (see **Box 5.1**). All of this may seem very pedantic, but it is essential if

---

1 It is more correct to talk about the 'scientific name' of an animal than the 'Latin name', because some of these names are derived not from Latin, but from classical Greek (or even a hybrid of the two). The flightless kiwi, for example, has the generic name *Apteryx*, which comes from the Greek *a*- prefix, meaning 'not' and '*pterux*', meaning 'a wing'.

2 'Linnaeus' was actually Karl von Linné, but we know him better by his own Latinized name. For many years, he was a professor at the University of Uppsala in Sweden, and his house and garden in Uppsala can still be visited.
Morphology refers to the shape, appearance, and structure of living things.

Box **5.1** The naming of animals: who is right?

We have stated in section **5.2.1** that the scientific binomials that we use to identify animal species can sometimes change because of taxonomic rules such as the rule of priority. So how can this happen?

An animal's specific name is most likely to change if it is discovered that what we thought were two different species are only one, or if what were thought to be separate subspecies of the same species are judged to warrant being regarded as distinct species in their own right. Generic names can change if new evidence suggests that a genus is not really distinctive enough to warrant its own name, or, conversely, that the known members of a genus are sufficiently diverse that they really ought to be regarded as belonging to more than one genus. But we have also said that decisions such as these are matters of informed judgement. So, whose decisions are right?

The answer really is that nobody is right, but rather that a consensus of support develops for particular viewpoints. In one sense, giving a species a binomial name is akin to formulating a hypothesis about its evolutionary relationships, so we make our decisions on the basis of how much the evidence supports that hypothesis.

But in that case, which names should we use? Fortunately, consensus views are usually available for people to refer to; these are often called taxonomic authorities and allow us to check that we are using the currently accepted name for a species. We have tried to be consistent and to use accepted taxonomic authorities for the animal names to which we refer in this book. For mammals, the most widely used authority is Wilson and Reeder (2005). This is an enormous book (over 1,300 pages), but, fortunately, it forms the basis of the Mammal Species of the World (MSW) database hosted by the Smithsonian Institution (currently at `http://nmnhgoph.si.edu/msw/`). The equivalent authority for birds is Sibley and Monroe (1990; 1993), which is also available as a database on the Internet (currently at `www.ornitaxa.com/SM/SMOrg/sm.html`). Similar Internet databases exist for amphibians (hosted by the American Museum of Natural History at `http://research.amnh.org/herpetology/amphibia/index.php`) and reptiles (at `www.reptile-database.org/`).

5.2.2 Taxonomy

At this point, we might ask: what do we actually mean by 'genus' and 'species'? Whole books have been devoted to answering this question and there is no unequivocal answer. The most commonly used definition of the species is the **biological species concept** of Ernst Mayr (1942), which states that '*species are groups of interbreeding natural populations that are reproductively isolated from other such groups*'. This reproductive isolation may be the result of living in different areas (allopatry), having different morphologies (for example, reproductive organs that do not fit with each other), or different behaviours (for example, courtship patterns that are not recognized by each other), or may be the consequence of post-copulatory mechanisms (for example, different

The **biological species concept** is that species are populations that are reproductively isolated from, and hence unable to breed with, other populations (Mayr, 1942). This is usually difficult to apply, so morphological or genetic criteria are often used instead.

chromosome shape or number leading to failure of meiosis (see section **9.1.1**) and hence rendering offspring infertile).

This is still a useful definition, not least because it is, in principle, testable—but it is not easy to test and is not of much use to palaeontologists. In practice, judgements based on differences in morphology, reproductive isolation, and how different populations are at the molecular level are used to decide whether similar, but different, populations represent different species or just subspecies of the same species. Similar judgements are used to decide if similar species all belong to the same genus, or whether they are different enough to warrant different genera. Returning to our lion example, we recognize that lion, tiger, leopard, and jaguar are reproductively isolated in the wild and produce infertile hybrids on the rare occasions that they interbreed in captivity. Although they look different, and show ecological and behavioural differences, they show great similarities in their underlying morphology. Molecular evidence suggests that they had a recent common ancestor (Macdonald, 2001), so we place them all in the same genus, *Panthera*. Another large cat, the cheetah, shows more differences from these than they do from each other and does not share the same recent ancestor, so we place it in a different genus, *Acinonyx*. Thus, genera are best regarded as **monophyletic** assemblages of related species.

Both genus and species are grades within a taxonomic system that shows how all species are related to all other species. The main grades in this system are as follows and the system is hierarchical, which means that each grade is ranked higher than the next grade: kingdom; phylum; class; order; family; genus; species.

We can use our lion example again to see how this works (see **Fig. 5.2**).

The way in which we classify animals reflects their evolutionary history, as we have seen in the discussion about what constitutes a genus. Because of this, taxonomy tells us more than only the name of a species; it also tells us who its relatives are (which other species are in the same genus), how variable it is (how many subspecies there are), and how unusual it is (how many others are in the same group).

5.2.3 Relevance to zoos

Although our taxonomic system is intended to reflect what we think is the evolutionary ancestry of different animals, we need to bear in mind, firstly, that it is nevertheless an artificial system. Phyla, classes, and orders have no reality outside our classification system, and they represent our attempts to put order on to our knowledge of the natural world. Secondly, decisions about which animals belong in which taxonomic group are matters of informed judgement, and can therefore change as knowledge and techniques improve.

It is decisions about the species and subspecies grades that are most likely to affect zoos, because these can have consequences for the way in which social groups are set up and the way in which breeding is managed. There is, for example, an issue about whether species or subspecies should be the grade at which conservation imperatives are decided (see **Chapter 10**). For this reason, zoos nowadays need to know not just which species, but which subspecies, they have. And, of course, to do this, they need to have access to the most up-to-date taxonomies (see **Box 5.1**).

5.3 Identifying individual animals

There are lots of ways in which to distinguish individual animals, but these methods vary in their reliability, how much they need trained staff, their costs, and how much impact they have on the animal itself. Some methods have benefited from a fruitful exchange of ideas between field workers and zoo professionals. Many field workers may not even

Monophyletic means having one recent common ancestor.

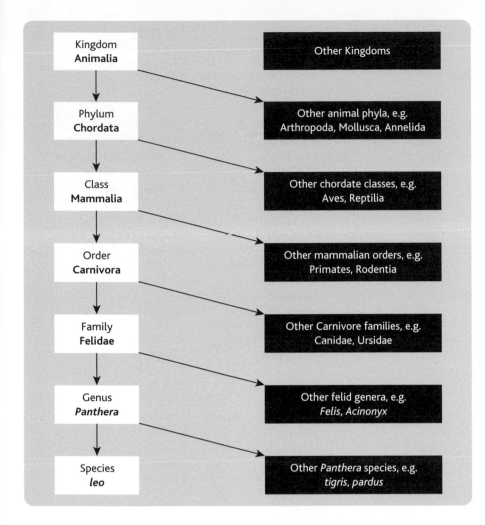

Figure **5.2** The classification of animals is based on a hierarchical system, within which each grade (e.g. class, order, family) encompasses examples of the next lower grade. Here, you can trace the classification of the lion through these grades and also how the lion is related to other species.

be able to see their animals, so they are very reliant on indirect animal identification methods, such as using faecal samples or nests, which may allow the identification of species or, possibly, individuals. In contrast, finding animals in a zoo should not be such a problem and thus it is usually possible to use direct animal identification methods, such as the use of visible characteristics.

There are a number of ways of achieving individual animal identification, but whatever method is used, it should aim to meet a set of basic standards.

1. Pain, suffering, or any change in the animal's behaviour as a direct, or indirect, result of the identification method chosen should be minimal.

2. Identification of the animal should be possible from a distance, to avoid the need to handle it.

3. The method used should ideally be long-lasting, again to avoid recurring handling.

4. The method used should be acceptable to the zoo visitors (see section **6.1.3** for a discussion of the need to integrate the needs of visitors into animal housing and husbandry).

A variety of animal identification methods is summarized in **Table 5.2**. Some of the advantages and disadvantages associated with these different methods are discussed in more detail in the next two sections.

Method	Description/location	Temporary (T) or permanent (P)	Comments	Examples
Natural markings	For example: coat colour or pattern; physical deformity; size	P**	Cheap and easy Requires skilled staff and obvious differences between animals	Okapi: stripes (see **Fig. 5.4**) Tiger: spots
Branding	Flanks; horns	P	Painful	Snakes: freeze branded (see **Fig. 5.15**)
Tattoo	Skin around eyes, rump, or on fingers	P**	Painful	Macaques: ischial callosities
ID cuts	Ears; horns; digits	P**	Potentially painful	Wildebeast: horn notches Rhino: ear holes Lizard: digit removal
Microchips, transponders	Injected under the skin	T	Painful and may require sedation Potential migration under the skin	Everything, from dormice to elephants (see **Fig. 5.3**)
Adornments	Tags; beads threaded through skin	T	Potentially painful Usually attached to young animals	Blue iguana: beads (see **Fig. 5.12**) Mara: ear tags (see **Fig. 5.11**)
	Rings or collars worn by animal	T	Potential risk of entanglement	Ring-tailed lemur: necklaces (see **Fig. 5.6**) Bird: leg rings (see **Fig. 5.9**) Penguin: flipper rings (see **Fig. 5.10**)
Clipping	Hair; feathers	T	May be difficult to see from a distance	Rarely used in zoos
Dye/markers	Ram pads; spray; paint; stuck-on markers	T	May transfer onto other animals Need to ensure non-toxic material used	Beetles: paper numbers (see **Fig. 10.6**) Tortoises: painted dots

Table **5.2** An overview of some methods used to identify zoo animals individually*

* Listed in order of permanence, with more permanent methods first. Considerations are provided that need to be balanced against the need to identify individuals.
** Characteristics may change over time, due to aging or other factors, e.g. pregnancy, new feather growth, injury, etc. This is especially true for colourful or shaped identifiable features.

5.3.1 Using naturally occurring characteristics

When you look at animals, it is often possible to notice natural markings or characteristics that appear to distinguish one animal from another and these can provide a useful tool for individual animal identification. Naturally occurring differences between individuals include: colour or patterns in coat, feathers, or skin; facial and bodily characteristics and features, including deformities (such as a twisted leg or scars, with which the animal may have been born or which may have occurred during its lifetime); size; behaviour.

Visible characteristics are an obvious and practical method of identifying individuals on a day-to-day basis—but this technique is reliant on the observer's skills at consistently recognizing differences between individuals, or identifying the features that characterize a certain individual. Many keepers are able to recognize the animals in their care this way, without the need to use artificial identification methods, but their ability to do this is dependent on their experience and familiarity with the animals, and also on other factors relating to the animals and their housing. As group sizes get bigger, for example, it becomes more difficult to recognize animals individually, because the number of identifying features required also increases and recognizing these can become increasingly complicated. Similarly, enclosure type can also affect individual recognition, because enclosures that enable animals to distance themselves from keepers can make the identifying marks more difficult to see (although binoculars can obviously help in this instance).

Visual characteristics are prominent in a lot of different species, but, when you look at these (for example, in **Fig. 5.3**) it is clear that they do not necessarily make individuals easy to identify. Obviously, the degree to which characteristics vary between individuals will influence how easily they can be recognized. Trying to recognize individual zebra from their stripes, for example, can seem difficult if you look at a herd of zebra within which the individual animals are in close proximity to one another. If you concentrate on only one area of the zebra, however, and compare the pattern you see there with that of the same area in other zebra, the task becomes a bit more manageable (see **Fig. 5.4a**). This can work with other species as well: for example, okapi *Okapia johnstoni* (**Fig. 5.4b**).

Indeed, sometimes, a relative measure of visual appearance is used, within which a process of elimination allows the identification of individuals, as, for example, in 'Fred is the largest in the group'. One problem with doing this is that it requires more than one animal to be in your field of vision at the same time and it also usually needs those animals to be in close proximity to one another, so

Figure **5.3** There is a huge array of colours and patterns present in the animal kingdom, but these do not necessarily help us to distinguish between individuals. Look, for example, at these zebra: how well can you distinguish individuals here? (Photographs: Vicky Melfi)

(a)

(b)

Figure **5.4** If you concentrate on one area of the animal, sometimes it can become easier to distinguish individuals. For example, if you look at the rear patterns of (a) these zebra, or (b) these okapi *Okapia johnstoni*, then it becomes easier to recognize differences between individuals. (Photographs: (a) Paignton Zoo Environmental Park; (b) Vicky Melfi)

that comparisons of size, colour, or some other characteristic can be made. Because of this, there are clear advantages in being able to recognize the identifying feature in each individual independently of what the other animals in the enclosure look like.

On a more practical note, we should remember some limitations on the usefulness of natural characteristics. Firstly, the visibility of some characteristics may be affected by the animal's behaviour: for example, broken digits may only be visible when animals are holding objects or using their hands, or a distinguishing mark or feature may be hidden when an animal sits or lies down. Secondly, some natural markings and characteristics may change over time. For example, as animals get older, their coats may change colour, particularly by losing colour or becoming grey or patchy, and their postures and behaviours may also change. Similarly, the weights, and thus sizes, of animals are also likely to vary over time, because of changes in age, diet, social position, and reproductive status.

It is likely that the individual identification of a large number of animals from their natural characteristics can only really be accomplished by that small group of people who are most familiar with them, such as their keepers. But, in many situations, a wider body of people, such as other keepers, the vet, and researchers, will also need to be able to identify individual animals, and these people will not necessarily have either the time or the expertise to be able to use natural visual characteristics. In these situations, or when direct observation of differences between animals is not sufficiently reliable because of limitations such as those outlined above, then using artificial methods to identify animals individually is probably a more satisfactory method.

5.3.2 Using artificial methods of individual identification

There is a wide variety of artificial methods that can be used to identify animals individually, and these range from leg rings[3] and necklaces, to tattooing and transponders.[4] The choice of method is usually

3 Rings used for animal identification are often referred to as 'bands' in the USA.
4 It should be noted that the term 'transponders' may refer to devices with or without a battery supply, and thus are able to broadcast information or require a device to read

the information directly from the device, respectively. Within the zoo industry, 'transponders' and 'microchips' refer to the latter devices—that is, those that do not have their own battery supply. More explanation is provided in section **5.4.3**.

more informed by keeper experience and tradition than by scientific assessment of the impact or efficacy of the method, mainly because so few studies have been carried out in this area and our knowledge is therefore limited. It is also important that the ethical and welfare implications of using the different methods should be considered. Then, if the method used allows the animal to be identified and the keepers feel that that its behaviour is unaffected, the method is considered successful and its use may subsequently spread elsewhere, through means of positive anecdote.

In determining which method should be adopted, several interrelated factors should be considered: is temporary or permanent identification required? What degree of intervention, invasiveness, or pain is acceptable? What information needs to be portrayed and should it be visual? Which methods have been successful with the target species? What level of security (reliability) is acceptable?

Let us look more closely at each of these in turn.

Temporary versus permanent methods

One of the aims of animal identification is that it should be as long-lasting as possible, so we would expect that those methods that are more permanent would generally be favoured over temporary measures. There are, however, some situations in which temporary methods are sufficient. For example, when newborn animals in large groups need to be identified, marking them with a spot of dye, or shaving or clipping fur or feathers in a known location to create a distinctive coat pattern, can allow them to be monitored. Similarly, animals that have been recently moved into a new group may be marked only until their keepers are able to recognize them, using visual characteristics. Eggs being incubated, a temporary process, are often numbered with a pencil for identification. The lifestyles of some animals can help increase the longevity of temporary measures, if the animals do not interfere with the identifying marker. Examples of this include using paint on iguana skin and tortoise carapaces, or sticking a marker onto a beetle (see **Fig. 10.6**).

Surprisingly few methods of individual identification can be viewed as permanent. Usually, permanent methods of identification are achieved by making physical alterations to the animal's body that will withstand time. These alterations may include identification cuts, **branding**, and tattooing, and are therefore sometimes referred to as 'mutilations' (see section **5.5**). By contrast, temporary methods of identification are achieved either by making temporary physical changes to the animal's body, such as colouring or clipping hair, skin, or feathers, or by adding something to the animal's general appearance using **adornments** such as tags, rings, or necklaces (see section **5.4**). Transponders (sometimes known as 'microchips') are not visible, but can migrate within the animals' body and thus be lost and no longer read, or they can be groomed out by an attentive **conspecific**, so cannot be considered permanent.

The longevity of temporary measures can vary significantly. With luck, some temporary measures of identification can last an animal's lifetime, although it is likely that, over time, they will become indistinct or will be lost. There is a tendency to presume that adornments (temporary) should be chosen over mutilations (permanent), because the latter have negative connotations. But the relationship between the type of identification method used and the animal's welfare is not that straightforward, because the degree of intervention, invasiveness, and potential for pain is variable with both temporary and permanent types of identification.

The degree of intervention, invasiveness, and pain that results

When choosing between temporary or permanent methods of individual animal identification, it is necessary to consider and weigh up the following factors.

1. *The degree of handling necessary*   All methods of individual identification require the animal to be caught up so that the 'identifying feature' can be applied. This can take time and effort, and may be stressful to the animal.

2. *The potential for pain that may be experienced when applying the identifying feature*   This can be highly variable, ranging from the slight discomfort resulting from the application of a leg ring (which is probably negligible compared with the stress of being caught up), to the certain pain that an animal experiences when its ear is pierced for an ear tag, its skin is scalded with extreme cold or heat during branding, or during the amputation of a digit, or the invasiveness of sedating an animal in order to implant a transponder subcutaneously. We will return to this issue when we consider different methods of identification later in this chapter.

3. *The longevity of the identifying feature*   That is, how long it will be before the process needs to be implemented again. It should be remembered that all temporary measures of individual identification are just that: temporary. We have to bear in mind that, at some stage, the process has to start all over again and another 'identifying' feature has to be applied to the animal depending on the longevity of the method chosen. With methods such as colouring or clipping the fur, feathers, or hair, the longevity of the identifying feature may be quite short. Methods that are longer lasting are thus favoured, because the need to catch and restrain the animal can then be minimized.

4. *The impact that the identifier may have on the life of the animal*   This is probably the least easy to answer, so we will now look at this last consideration in a bit more detail.

The question in relation to consideration (4) is whether the marker or identifier will affect the animal behaviourally or physically during its everyday life. Most studies of this kind have been conducted on laboratory-housed animals or those living in the wild, and, sometimes, the effects shown can be quite unexpected. For example, laboratory-housed zebra finches *Taeniopygia guttata* became attracted to the differently coloured leg rings of other finches, with females having a preference for red rings on males and males preferring black rings on females (Burley, 1985). Interestingly, these colour preferences seemed apparent only in laboratory conditions, within which ultraviolet (UV) light is available; we know that UV light plays a fundamental role in bird perception (Hunt *et al.*, 1997). Other studies of wild birds found that some methods of individual identification put demands on the birds that hindered their survival, by increasing mortality or reducing reproduction (reviewed by Gauthier-Clerc and Le Maho, 2001; Jackson and Wilson, 2002). For example, Culik *et al.* (1993) demonstrated that Adelie penguins *Pygoscelis adeliae* with flipper rings used significantly more energy (24 per cent more) when swimming compared with those without.

It certainly appears that some methods of individually identifying animals are associated with problems, while others are not. Leg rings are sometimes reported to be deleterious, but may also be reported as better than alternative methods (such as **wing tagging**), or may even be viewed as having no consequence at all for the animal (Kinkel, 1989; Cresswell *et al.*, 2007).

Similarly, some individual identification methods appear to impact on mammals, while others do not. For example, ear transmitters (transponders powered with batteries, so that they broadcast information) used in moose *Alces alces* calves were thought to be associated with a decreased likelihood of survival, whereas, in fact, the survival of calves with ear tags was no different from that of calves without any form of individual identification

Tagging refers to the addition of a tag to an animal, most commonly seen in ungulates, among which ear tags are used for animal identification. **Wing tagging** operates in the same way, but a tag is fixed to the bird's wing.

(Swenson *et al.*, 1999). Nevertheless, studies on laboratory-housed and wild animals do demonstrate that individual identification methods may have an impact on their lives, so it is something that needs to be considered.

It is likely, however, that captive conditions may mitigate some of these problems, because plentiful food, veterinary treatment, and the removal of many of the other pressures that the animals would have in the wild ensures their survival in captivity. Because of this, zoo populations may be ideal for trying out new individual identification methods or modifications to existing ones, so that they can then be tailored to have as little impact as possible on wild animals. Zoo animals, in turn, gain from these advances in knowledge and technology. An example of this is the study by Simeone *et al.* (2002), who trialled data loggers on zoo-housed Humboldt penguins *Spheniscus humboldti* before using them in the wild.

We started this section by identifying four different aspects on which to judge individual identification methods. Each identification method may rate differently on each of these scales, so the net effect of each method should be considered. For example, feather clipping is not painful, but the quick regrowth of feathers will require frequent catch-ups of the birds—a procedure that can itself be stressful to the animal. Alternatively, applying a tattoo is painful (although this might be mitigated to some extent with **analgesic drugs**), but, once applied, it will then provide a clear individually identifiable feature for the rest of the life of the animal. We should also take account of the species and individual differences between animals, and the housing and husbandry regime in place, when deciding which method is most appropriate. For tame or trained animals, catch-ups may not be stressful, so repeating or replacing 'non-painful' methods may be considered best. For animals that are very wild and become stressed during interactions with keepers, however, the use

of a permanent method of identification that only needs one handling session is ideal.

Portraying information

In most situations, identification methods aim to convey the pertinent information about an individual animal that is necessary for its day-to-day management. Artificial methods of identification can provide that information quickly and effectively to anyone who knows how to interpret it (see 'Codes', below), unlike natural markings, which usually rely on the keepers' skill and knowledge of individual animals. For animal management, the best identifying features are those that can be seen clearly without the need to catch or restrain the animal. Currently, this cannot be achieved with subcutaneous microchips and transponders, even though they are widely used, although technological advances may soon change this (see section **5.4.3**). Indeed, many zoos use more than one method to identify their animals individually, usually using transponders in conjunction with a visible feature, whether natural markings or artificial markers.

At this stage, it is necessary to mention the perceptions of zoo visitors: are their expectations of seeing wild animals in a **naturalistic enclosure** compromised by the sight of bright orange ear tags or other high-visibility identifying features? We do not know the answer to this question, but there are two approaches that can be adopted to deal with this. Firstly, zoos can use identifying features that are less visibly obvious. These are only effective in animals that are comfortable with coming close to their keepers and are obviously useless if the animal is some distance from their keeper. For example, **notching** underneath a **chelonian** carapace can be used to identify individual animals, but requires the keeper to feel along the carapace, so he or she must clearly be very close to the animal.

Secondly, zoos can incorporate the identifying features into the education and interpretation

Chelonian refers to the order of reptiles that includes the turtles, tortoises, and terrapins.

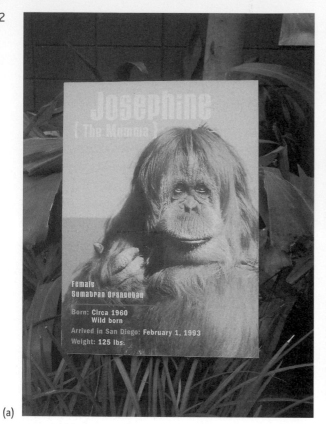

(a)

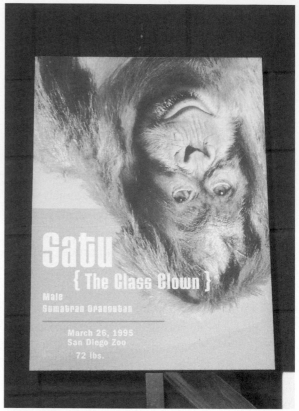

(b)

(c)

Figure **5.5** Illustrating the natural differences of animals to zoo visitors can be achieved by providing them with photographs of the animals. San Diego Zoo provides these pictures of their orang-utans *Pongo* spp. for visitors to be able to identify the animal at which they are looking (a, b, and c are different individuals). (Photographs: Vicky Melfi)

provided to the visitor. An example of this is to give visitors the information that enables them to identify animals for themselves using natural markings and this can be done by providing photographs of individuals (see **Fig. 5.5**), or by highlighting the different patterns of individuals. Another way is to draw the visitors' attention to the artificial identifying features that the zoo uses to recognize its animals. For example, individuals in a large troop of ring-tailed lemurs *Lemur catta* at Edinburgh Zoo were fitted with differently coloured pendants hung from necklaces. A corresponding poster next to the enclosure illustrated to visitors how to use these tags to identify individual lemurs and how they were related to each other (**Fig. 5.6**).

Codes

The information conveyed by the identifying feature is often coded. The simplest way of doing this is to allocate a number or letter code to an individual animal and transfer it directly onto the animal's body, or indirectly via an adornment. Whether putting information directly or indirectly on animals, the space available for this should be kept to a minimum, rather than writing the animal's name or number down the whole side of its body.

This has led to the creation of a number of different coding systems that can portray more information with less space. Often, combinations of colours, numbers, and letters, in conjunction with their position on the animal's body, are used to convey the required information. For example, a zoo may denote an animal's sex through the positioning of an ear or wing tag, with tags on the left denoting males and tags on the right, females. These sorts of code have to be decided upon and adhered to at the different organizational levels in which they are used. At the level of an individual zoo, the position of the marker may be used to denote something like the sex of the animal; at a national level, there may be other

Figure **5.6** To the untrained eye, ringtailed lemurs *Lemur catta* all look very similar. But you can recognize animals individually if they have an identifying feature. In this example, lemurs at Edinburgh Zoo wear necklaces with different-coloured pendants. (Photograph: Julian Chapman)

marking requirements. In the UK, Defra requires all **bovids** to have ear tags with identity codes stamped on them. At a regional, or even global, level, other practices may be in force: the Conservation Breeding Specialist Group of the World Conservation Union (IUCN), for example, recommends using transponders that are ISO-compliant (CBSG, 2004), and that implantation of the transponder must occur subcutaneously at standardized locations (Cooper and Rosser, 2002; see **Table 5.3**).

When physical changes are made to animals as a form of identification, numbering systems are frequently used in such a way that a large number of animals can be distinguished from each other by

The term **bovid** refers to members of the family Bovidae, which includes cattle, sheep, and goats, but also various other species collectively called 'antelopes'.

Taxa	Location
Fish	Base of dorsal fin or within the **coelomic cavity**
Amphibians	Lymphatic cavity; cover wound with tissue glue
Reptiles:	
Snakes and lizards	**Dorsal** side of the tail base
Chelonians	Hind shoulder
Birds	Pectoral muscle or left thigh
Mammals	Behind the ear, at base or between shoulder blades, left of centre

Table **5.3** Globally recognized implantation sites for transponders*

* To increase the usefulness of transponders (so that they can be relocated and read), they should always be implanted on the animal's left side when applicable and at these sites.

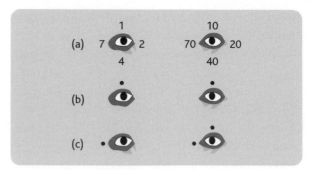

Figure **5.7** A common method of coding information, such as the animal's identification number, can be achieved by using two strings of numbers arranged around an animal's eyes. For example, imagine that an eye has four sides. If the sides on the left eye represent the values 1, 2, 4, and 7 (clockwise, from 1 at the top of the eye) and those of the right eye represent the values 10, 20, 40, and 70, as shown in the diagram (a), then it is possible to identify individually up to 154 individuals. What, then, are the numerical codes of the animals illustrated here?

using permutations of a simple code. For example, tattooing spots around the eyes is one method of identification and, if the numbers 1, 2, 4, and 7 are represented around the left eye and values 10, 20, 40, and 70 are represented around the right eye, then different permutations of spot position will allow a unique number to be given to as many as 154 different individuals (see **Fig. 5.7**).

A similar principle can be used when tattooing an animal's knuckles or digits, removing digits, or when using notching (see section **5.5**). Each toe can represent a number from the right: for example, the first toe represents 1, the next toe, 2, then 3, and finally, 4. The animal's four-figure identification number is read from the front right leg first, then the front left leg, then the back left, and finally, the back right leg. So, for example, an animal identified as number '3024' and marked by digit removal would be marked thus: the third digit removed from the animals' front left leg (3), no toe removed from its front right leg (0), the second toe removed from its back left leg (2), and finally the forth toe removed on its back right leg (4) (see **Fig. 5.8**).

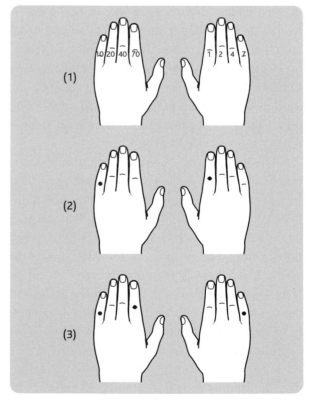

Figure **5.8** The coding system illustrated in **Fig. 5.7** can also be applied by ascribing values to digits or limbs, as shown here.

The **coelomic cavity** is the body cavity in which the internal organs reside. The **dorsal** side is the upper side of the animal.

Method	Fish	Amphibians	Reptiles	Birds	Mammals
Branding	✓		✓		✓
Tattoo	✓		✓		✓
ID cuts			✓		✓ (Ear/horn notching)
Microchips, transponders	✓	✓	✓	✓	
Tags				✓	✓
Beads	✓	✓	✓		
Rings/bands/collars			✓	✓	✓
Coat (fur/feathers) clipping	✗	✗	✗	✓ (Toenail clipping)	✓
Dye	✗	✗	✓	✓	✓

Table **5.4** A summary of the suitability of artificial individual identification methods for different vertebrate groups

Which methods have been successful with the target species?

Despite all of the factors that we have considered so far, the most common reason for choice of method is tradition: that the method has been used before with that species and no deleterious signs have been noted by keepers. **Table 5.4** summarizes some of the diversity in artificial individual identification methods that are used across the different vertebrate **taxa**.

What degree of security (reliability) is acceptable?

There is an obvious expectation that if an individual identification method is applied to an animal, then its identify will be known. In general, permanent methods of individual identification are considered to be more reliable than their temporary counterparts. But we also need to consider the security of the method chosen: can it be removed, changed or replaced? If the answer is 'yes', then the safety of the animal's identify is at risk, so that if it were stolen, it could be untraceable. In addition, an animal's identification can also be in jeopardy if an identifying feature is lost and the animal is held in a large social group within which visual differences between individuals are minimal.

The IUCN recommends that all endangered animals are individually identified, as much to safeguard them, as to 'know' who they are. **Exotic animals**, whether endangered or not, are valuable to wildlife trade. *National Geographic News* noted that animal thefts from zoos worldwide were on the rise (9 February 2007). John Hayward, who runs Britain's National Theft Register, noted that:

> If anybody wants [an exotic animal], chances are they're not going to go into the tropical rain forest to try and find one. It's far easier to break into somebody's private collection or zoo and steal one.

Even if these animals are found and confiscated, if they do not have verifiable identification, their ownership cannot easily be assigned. Sometimes—at least in the UK—informally matching reports of lost animals from a zoo with animals that are available for sale can ensure that animals are returned safely. If the animals are exported, however, then their origins can become far less clear-cut.

Unfortunately, even those methods of identification that seem to be permanent and tamper-proof can still be changed if the need is great enough. Transponder readers, for example, can be purchased by animal dealers operating illegally and the original transponders can be removed or replaced with new ones; brands and tattoos can be disfigured to disguise the original mark, and so on. Because of this, more than one method of identification is sometimes used for some animals.

5.4 Temporary artificial methods of individual identification

Many of the methods of identifying individual animals rely on the temporary addition of something to alter the animal's visual appearance and we can refer to these as adornments. They include such things as tags, beads, rings, collars, or necklaces, and they are most commonly attached to the animals' legs, wings, flippers, ears, or around the neck. The availability and use of adornments in the zoo industry is led by agriculture, field ecology, and companion animal sectors. Usually, the type and position of adornments is determined by each zoo depending on the animal management system it has in place, although, as mentioned above (see **5.3.2**), for some species, there are national or international standards. We can also include transponders in this section, because, although they are not visible, they are nevertheless additions to the animal that do not involve permanent changes to the animal's appearance (these sorts of change are considered in section **5.5**).

Bearing in mind the framework outlined above (section **5.3**), let us now consider the advantages and disadvantages, and hence the suitability, of these different methods.

5.4.1 Rings, collars, and necklaces
Rings, collars, and necklaces are the most benign of adornments, because the animal simply wears them. Of course, the animal needs to be caught and restrained so that the adornment can be secured.

Four main types of ring are available, as follows.

- Closed rings (**Fig. 5.9a**) are slipped over the foot of a young bird with the intention that they will be permanent and never be removed. Because the longevity of these rings is high, they are a preferred option, but care needs to be taken to ensure that the ring is big enough that it does not become restrictive once the bird grows to its adult size. In addition, these rings can be difficult to remove if the bird injures itself and its leg becomes swollen, or if the ring gets caught on something within the bird's enclosure.

- Split rings are flexible (**Fig. 5.9b**), so they can be put onto an animal's leg at any age. These rings are not necessarily very secure, because, depending on the material from which they are made and whether they adequately fit the leg, they can be bitten off or may fall off. Thus, they may have a reduced longevity and this might require the animal being recaught and another ring secured.

- Metal rings (**Fig. 5.9c**) are similar to the split rings, but are secured in place with pinchers (sometimes call 'ringing pliers'). The application of these therefore requires a skilled keeper, because the animal's limb could potentially be damaged when fixing the ring. Metal rings are usually used in long-lived species—for example, on penguins' flippers and cranes' legs—because the rings should be relatively permanent. Unfortunately, over time, the metal can become worn from the animals' movements, and wear and tear to the metal can lead to it becoming sharp, which can then lead to injuries.

- With technological advances, new materials are being used to make rings. For example, a collaborative design team from Bristol Zoo and the University of Bristol has created a new silicone ring, which is designed specifically for ease of application, because they can be stretched onto the penguin's flipper or bird's wing.[5] The robust material should withstand extremes in temperature and movement, and so should be long-lasting, without the associated disadvantage of degradation seen in metal. It should thus also avoid injuries to the bird (**Fig. 5.10**).

---

5 This innovative silicone ring was awarded the 2006 UFAW Wild Animal Welfare Award.

(a)

(b)   (c)

Figure 5.9 Various types of ring can be used to identify animals individually—most notably in birds, but also in bats.
(a) A closed ring, here illustrated on a Dalmatian pelican *Pelecanus crispus*. (b) A flexible ring, shown here on an inca tern
Larosterna inca. (c) A metal ring, here shown on a hammerkop *Scopus umbretta*. (Photographs: (a) Paignton Zoo Environmental
Park; (b) Kirsten Pullen; (c) anonymous)

5.4.2 Beads and ear tags

Beads and ear tags require that the animal's skin is perforated so that the identifying feature can be fixed to the animal's body, so they should be attached by trained keepers. Best practice requires antiseptic be used at the puncture site to prevent infection and a local analgesic to be used wherever practicable. Practical considerations include the

(a)

(b)

Figure **5.10** Bristol Zoo Gardens, in collaboration with the University of Bristol, designed these silicone composite flipper bands, seen here being applied to African penguins *Spheniscus demersus* at Bristol Zoo (a). These bands were awarded the 2006 UFAW Wild Animal Welfare Award (b). (Photographs: Bristol Zoo Gardens)

length of time that the analgesic will take to become effective and whether pain reduction is worth any other 'stressors' on the animal.

Tags

Tags come in a variety of shapes, sizes, and colours. Tags are routinely used in the ears of mammals —mostly **ungulates**—but are also used in bird wings and the appendages of other kinds of animal. Mammals are usually tagged when young—sometimes, when they are as young as 1 or 2 days old —but, in other taxa, tags may be applied at any age. If the animal is very young, care needs to be taken regarding potential aggression from the mother, and the effects of restraint, separation, and time away from the herd or mother need to be considered. It is also important to position the tag correctly, to avoid major veins and cartilaginous tissue, and to prevent excessive damage. In some situations, ear tagging has been known to cause the ear to collapse permanently.

Inserting an ear tag is probably a painful event for the animal, because it involves puncturing the ear so that the tag can be threaded through. Ear tags can also pose some problems in the daily lives of animals, because they can potentially be torn from an animal's ear (**Fig. 5.11**), either by getting

caught on enclosure furnishings or during social events. Appropriate housing and husbandry should minimize these types of incident: for example, by housing socially compatible animals, using well-maintained fencing, and eliminating protrusions into the enclosure. Nevertheless, once an ear tag has been dislodged, it needs to be replaced. An additional problem, which is associated with any identification method that leaves a wound, is the potential for infection. In US cattle farms, parasitic infection of ear-tag sites has been recognized to be a large economic problem that also has deleterious effects

Figure **5.11** Two ear tags have been used to identify this mara *Dolichotis patagonum*, but note that there is some damage to the ear. Sometimes, tags can be torn from the animals' ears during aggressive interactions. (Photographs: Vicky Melfi)

on the cattle's behaviour and welfare (Byford *et al.*, 1992). This has led to the development of various insecticide-impregnated ear tags, which have varied success (see, for example, Anziani *et al.*, 2000). While this is an agricultural example, it demonstrates how important it is, in other situations as well, to keep any wound site clean and to monitor it for infection.

In the UK, all bovids need to have ear tags distributed by Defra and these are part of the animal's 'passport', which ensures that individuals can be monitored and their movements traced. Each animal is given a unique number, combining a prefix representing the zoo and a unique suffix for the individual. Big yellow ear tags are worn in both ears in bovids and each part of the tag (back and front; four pieces in total for the two tags) is stamped with the individual's identification number.

Beads

Threading beads through the skin of reptiles and amphibians is also undertaken as a means of individually identifying animals (for example, Rodda *et al.*, 1988). In much the same way that coloured leg rings are used in birds, combinations of different coloured beads are used to code different individuals (**Fig. 5.12**). We have already mentioned (see section 5.3.2) how birds can develop preferences for sexual partners with particular colours of leg ring

and it is possible that similar preferences for beads may occur in reptiles. It has been suggested that some colours—especially red—should be avoided when using coloured beads with iguanas, because adult iguanas are attracted to this colour and may aggravate the bead site.

5.4.3 Microchips or transponders

The most invasive kind of adornment is the **passive integrated transponder (PIT)**, also referred to as a 'microchip', or simply as a 'transponder' (or, sometimes, a 'PIT tag'). These contain a unique magnetic code and there is a virtually unlimited number of codes available. The transponder is injected just underneath the skin and usually only requires tissue glue, or one or two stitches, to secure the site. Technological advances have enabled even tiny animals such as dormice to have internal transponders (see, for example, Bertolino *et al.*, 2001; **Fig. 5.13**).

Unlike conventional adornments, which aim to provide a highly visible feature to enable individual recognition, the unique numerical code within the transponder can only be identified using a 'reader'. This is a battery-powered handheld device that, when moved close to the site of the transponder, will provide a reading of its unique code. In fact, active transponders are also available: these have a battery unit and are able to broadcast information that can be picked up some distance from the animal carrying the transponder (termed radio-telemetry). Since the 1960s, this latter system has been considered useful to the study of animals, and has subsequently been used to study both wild and captive animals (for example, Essler and Folkjun, 1961; Swain *et al.*, 2003).

Disadvantages with the use of transponders include the need for most animals to be sedated or anaesthetized so they can be inserted (see section **11.6.3** for more on sedation), and the fact that they do not provide a means of identifying an animal that is not 'in the keeper's hands', or in very close proximity to the keeper. At the moment, the readers necessary to identify transponder numbers can only be used when the animal is very close to

Figure **5.12** Individual identification of reptiles is commonly achieved by fixing coloured beads to their dorsal crest. In this example, a green bead has been used to identify individually this Grand Cayman iguana *Cyclura nubila*. (Photograph: Vicky Melfi)

(a)

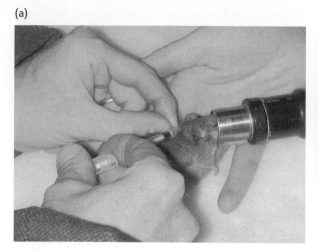

(b)

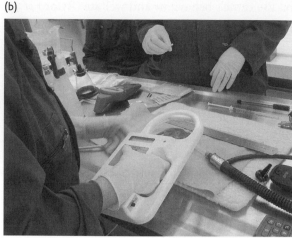

Figure **5.13** (a) Despite their small size, transponders can nevertheless be fitted to small animals, such as this common dormouse *Muscardinus avellanarius*. All transponders should be checked with a reader (b), to ensure they are working properly. (Photographs: Paignton Zoo Environmental Park)

the reader (indeed, sometimes the reader has to be in contact with the animal's skin) and therefore also close to the keepers. But transponder technology is always moving forward, aiming to increase the distance from which the reader can read the transponder in the animal. For example, recent developments have enabled the individual food intake of domestic cattle to be determined by the animal's unique transponder number. As the cow enters a stall, its transponder is read and the appropriate ration of feed is released from the hopper (for example, Sowell *et al.*, 1998; Schwartkopf-Genswein *et al.*, 1999). Unfortunately, however, the creation of readers that can identify transponders in animals while they are still in their paddocks is probably still some time off.

An additional disadvantage at the moment is that several companies produce transponders that can only be read by compatible readers. In practical terms, this can mean that a zoo needs several readers to locate and read the transponders of all of the animals in its collection. Again, however, technology is advancing and, recently, 'universal' readers have been developed that will read more than one transponder—but there is still a need to have more than one reader.

The final disadvantage is that, despite being secured subcutaneously and therefore not likely to

get lost, transponders can, indeed, do so. They can be groomed out, because they are placed very near the skin surface and social groomers often take a lot of interest in recent incision sites. But they can also migrate around the body internally (as mentioned above). Some companies that make transponders are aiming to reduce the likelihood of migration by making them with a coating that binds to the animal's skin after implantation.

Although there are limitations associated with using transponders, therefore, the technology behind them is constantly improving, and they do outperform other temporary measures for longevity and security. Because of this, transponders are usually used alongside other temporary methods of animal identification that provide a visual cue, such as the ear tags or leg rings described above.

5.5 Permanent artificial methods of individual identification

All permanent artificial methods used to identify animals individually change the animal's physical appearance by damaging its body and are therefore referred to as 'mutilations'. In theory, making such permanent changes to an animal's appearance is the only certain way of ensuring that it can be identified indefinitely. At the same time, however, we must recognize that these procedures may cause the animal more pain and stress than many of the other recognition methods that we have mentioned so far, and this should be borne in mind when decisions are made about when they should be used. Fortunately, technological advances in transponder and reader design, as mentioned above, are improving their performance to the point at which their longevity and reliability may be greater than even permanent changes to the animal's physical appearance.

Permanent changes to an animal's appearance can be achieved in two main ways: by the removal of part of the animal, or by the addition of a noticeable mark. The clarity of these identifying measures is determined largely by how well they were executed initially. As with temporary methods, permanent measures are most helpful if they can be viewed from a distance and are obviously distinct from other marks in other individuals. By their very nature, permanent measures are long-lasting, but, unfortunately, their clarity—and thus their efficacy—can alter over time in much the same way as natural visual characteristics (see section **5.3.1**).

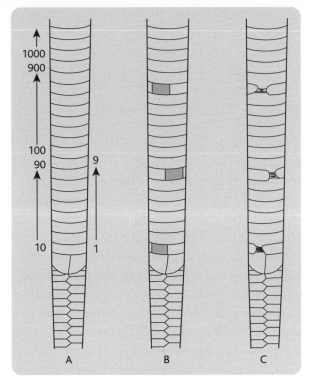

Figure **5.14** One way of identifying individual snakes is to remove a small number of scales, resulting in a unique code. In this example, the numbered code starts from the scale immediately in front of the cloaca (A) so this snake is numbered 819 (B). As the scales heal, the pattern may become less obvious (C).

5.5.1 Removal of part of the animal

There are various ways in which to change the appearance of an animal by removing some bit of it. Most common is a practice called notching (sometimes termed 'puncture'), which involves making carefully placed holes or marks in the animal's horn, shell, scales, or ears. Notches can be made in the undershells of chelonians, along the ventral scales of reptiles, and in the ears of mammals. The number and/or position of these marks allow the animal to be identified, sometimes using coded information (see 'Codes', above).

Another technique, used in snakes, involves the removal of scales so that the underlying muscle is exposed (Ferner, 1979). The removal of **sub-caudal**, rather than **ventral**, scales is thought to reduce the risk of injury, because the possibility of penetration into the snake's abdomen is reduced. Scales are removed from the right and left side, where the first sub-caudal scale is considered to be the one behind the **cloaca**. If the second scale on both the right and left side of the snake were to have been removed, the code would read '2L2R' (**Fig. 5.14**). But this method has many drawbacks, not least

Both **ventral** and **sub-caudal** scales are enlarged scales on the underside of the snake: **ventral** under the snake's abdomen; **sub-caudal** under its tail.

The **cloaca** is the single opening for both the urinogenital and anal orifices in reptiles, birds, and amphibians.

because it is potentially painful and leaves an open wound that might become infected. It is also time-consuming, both to apply, and to read and interpret later, and can lead to ambiguous marks if the scales slowly regenerate or if other injuries occur to disrupt the patterns (Shine *et al.*, 1988).

A final method is to remove parts of digits. Toenails can be clipped as a temporary measure (for example, St Louis *et al.*, 1989), but more permanent marking is achieved by removing whole or part of digits (**Fig. 5.8**)—a practice termed toe clipping, or digit cutting. This method is sometimes used to identify individual amphibians and reptiles, although it is not used often in zoos. We would expect this to be very stressful for the animals, although, counter-intuitively, there is some evidence that the impact of changing enclosures is more stressful than toe-clipping. Of course, this might mean simply that both practices compromise the welfare of the animals.

Unfortunately, even these permanent artificial patterns can deteriorate over time, because other digits can be lost (aside from those purposefully removed), notches can become torn, or additional marks can occur on the animal's horn, shell, or ear. Davis and Ovaska (2001), for example, found that toe regeneration in salamanders *Plethodon vehiculum* meant that toe clipping was lost from 35 weeks onwards.[6] All of this can make the interpretation of these patterns misleading and ambiguous over time.

5.5.2 Branding and tattoos

Branding and tattooing are two popular methods of permanently marking the bodies of animals. In branding, which is widely applied to mark agricultural stock, a shaped metal object known as a 'brand' is rendered either extremely hot or cold and then applied to the animal's skin. Heat branding burns the surface layer of the skin, leaving scar tissue from which hair does not grow and thus a mark is visible. This can be used on a variety of animals: for example,

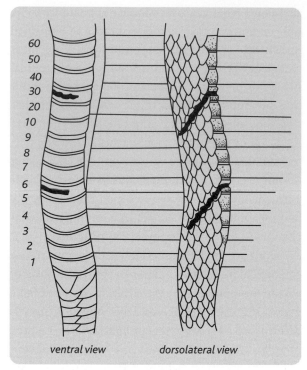

ventral view   *dorsolateral view*

Figure **5.15** Another way of identifying individual snakes is to apply a heat brand to their scales to make a unique mark. As you can see from the illustration, the snake's scales are numbered in a forward direction from the cloaca. In the example we show here, scales 30 and 6 have been branded to identify this individual as #36. (Photograph: *Herpetological Review*)

Winne *et al.* (2006) used medical cautery units to heat brand snakes (see **Fig. 5.15**). For freeze branding, the area where the brand is to be applied needs to be shaved to expose the skin. Freeze branding destroys the pigment in the animal's hair and thus white hair grows back, characterizing the mark.

Both of these processes have been observed to elicit painful **responses** in mammals (for example, Watts and Stookey, 1999). Freeze branding has also been used in a wide variety of taxa, including crustaceans, fish, amphibians, reptiles, and mammals (for example, Lewke and Stroud, 1974; Fletcher *et al.*, 1989; Berge, 1990; Measey *et al.*, 2001). It is possible

---

6 The ability to regenerate lost digits, indeed whole limbs, is widespread in amphibians. Why it is possible in these animals, but not in reptiles, birds, and mammals, is not clear, but it is probably related to their developmental processes (Galis *et al.*, 2003).

that freeze branding leads to slightly less discomfort and pain compared with heat branding, although both processes are associated with a variety of stress and pain-related behaviours, which are displayed both during and after branding. These include the flight response, increased heart rate, vocalizations, and a rise in blood **cortisol** levels (Lay *et al.*, 1992; Schwartzkopf-Genswein *et al.*, 1997).

By contrast, tattooing works by embedding ink under the surface of the skin via an injection. The tattoos can be applied to various parts of the animal's body, depending on the species and on whether or not the mark is expected to be visible. Patterns of tattoo dots are usually used as a visible means of identification, whereby the position and number of the dots is used as a code to identify the animal (see 'Codes', above, and **Fig. 5.7**). Tattoo dots can be placed around the animal's eyes, on its fingers, or wherever there is appropriate skin, such as on the **ischial callosities** of many Old World monkeys. Alternatively, secluded sites, such as inside lips, ears, or the underside of the animal, may be chosen for the location of the tattoo, because these usually have little pigment and the tattoo can therefore easily be distinguished.

The usefulness of artificial marks such as these added to an animal's body relies heavily on their clarity, and this, as with all permanent measures, can decline over time as the animal ages and its appearance changes. This can be exaggerated by the position of the mark. Tattoos, for example, can become hard to read when the skin becomes wrinkled or loose, which is common in aging animals.

5.6 Record keeping: what information could, or should, be recorded?

It has been said that you can never collect too much information—but this is not strictly true. Certainly, it would be ideal to keep as much information as possible about each and every individual animal,

because the information could then be analysed at the level of the individual and/or the population. This analysis could be carried out now, or at some time in the future. The collection and recording of all of this information, however, takes time. Many larger zoos have registrars whose sole responsibility is to manage the zoo's records, but, very often, time is limited, so we need to consider what is the minimum that needs to be recorded about each animal.

The Secretary of State's Standards of Modern Zoo Practice (SSSMZP) state:

Individual records must provide:
a) Individual identity and scientific name.
b) Origin (wild or captive born; identity of parents; previous locations).
c) Dates of entry into, and disposal from the collection, and to whom.
d) Date, or estimated date, of birth or hatching.
e) Sex (when known).
f) Any distinctive markings (i.e. tattoos, rings or transponders).
g) Clinical data, including details of, and dates of treatment given.
h) Behavioural and life history data.
i) Date of death and results of any post-mortem.
j) Any escapes, where to, and any reasons why they occurred; damage or injury caused to, or by, an animal to persons or property; any remedial measures taken to prevent recurrence.
k) Food and diets.

(Defra, 2004)

It also requires an inventory—or annual stocklist—for each animal collection. It is expected that records covering the previous 6 years are kept on site at the zoo and are easily accessible, so that the zoo's local authority can view them.

Ischial callosities is the term used to describe the hardened pink skin on the rump. This skin usually inflates (tumescence) around the time of ovulation.

The main information needed can therefore be summarized as: **life history traits**; daily events; medical information; routine housing and husbandry; and body measurements.

5.6.1 Life history traits

Life history traits are the big events that occur in the life cycle of a zoo animal, such as births, deaths, and transfers between zoos. The amount of detail that can be recorded about these events is largely determined by the record-keeping system implemented (see section **5.7**), although there are some obvious basic bits of information that are required. When animals are born, it is important to know who their parents are, because information about parentage enables the captive population to be genetically managed. This is not always as easy as it might seem: in some situations, for example, it may be difficult to know who the sire (father) is if multiple adult males are present in the group; equally, it may be difficult to know who the dam (mother) is if it is a species within which mothers do not associate with their young after parturition. The dates on which animals are born, moved to other zoos, and die should also be recorded, because this enables the animals to be tracked, which is necessary for disease control and trade prevention. It also permits demographic analyses of the captive population: for example, calculating how long a species lives in captivity, or whether the population is self-sustaining or is not breeding (see **Chapter 9**). As an example, analyses of records kept on cheetah *Acinonyx jubatus* in US zoos between 1987 and 1991 showed that the population was increasing (from 52 to 72 animals) and that infant mortality had declined from 37 per cent to 28 per cent, although this was still highly variable (Marker-Kraus and Grisham, 1993).

5.6.2 Daily events

Unlike life history traits, which are significant events in the animal's life, daily events are observations made of the animal that may, or may not, be of any consequence. For example, what if an animal does not eat one day? Perhaps the animal is ill; alternatively, it might be reproductively active, or might not like the food given to it, or perhaps it found food elsewhere (for example, **carnivores** sometimes catch prey that enters their enclosure), or could be occupied by something else in its enclosure, or there may be something that the keeper has not even considered. Without the collection of more information on a daily basis, it is impossible to know whether the fact that the animal failed to eat is a problem that needs to be resolved or the consequence of a 'reasonable' event.

Information that may be noted on a daily basis varies, but should probably include any observed deviations in the animal's appearance or changes from normal in its behaviour. It is also good practice to note any miscellaneous events that may affect the animal in the future: for example, if any change was made to routine housing and husbandry, such as feeding it later than usual, visitors staying in the zoo late, or the presence of fireworks. By reviewing these daily events, if will be possible to investigate what factors affect the animal's behaviour, health, and reproduction. Patterns may be obvious, as in the above example, or more complicated: for example, a group of animals that will not come in off exhibit several times might be influenced by the behaviour of a neighbouring group of animals.

5.6.3 Medical information

It is obvious that information about the health of animals within the collection needs to be recorded. It is most important that the information is then collated and reviewed, to establish whether animals are healthy and the veterinary care given successful. Medical records are also vital in **preventive medicine** (see section **11.4**), which involves identifying which animals or species seem to have health problems, which sites within the zoo show a high incidence of disease, and whether the disease is spread through the zoo. As with other forms of record keeping, the type and detail of medical information kept is largely determined by the record-keeping system used. (See section **5.7.4** on the Medical Animal Record-Keeping System, or MedARKS.)

5.6.4 Routine housing and husbandry

Keeping routine housing and husbandry information is helpful when investigating patterns in the animals' behaviour and health, because it is likely that the two are linked. Much housing and husbandry, such as the enclosure in which the animals are, the diet given to them, and the basic daily routine (that is, when animals are fed, moved around their enclosure, etc.) is routine and therefore probably varies very little from day to day. This means that records of housing and husbandry can usually remain unchanged for quite long periods of time, usually until some changes occur, such as the introduction of a new diet or modifications being made to the enclosure.

5.6.5 Body measurements

The collection and study of body measurements, such as weights and lengths, is referred to as biometrics, or morphometrics. For most taxa, there are standard measurements that are usually taken, and some of these are described and illustrated in **Box 5.2**.

Box **5.2** Standard body measurements for animals

When taking measurements of animals, the International System of Units (SI units) should be used, so length measurements should be recorded in millimetres or centimetres, and weight recorded in grams or kilograms. A ruler or measuring tape, or string that can later be measured accurately, can be used to determine these lengths. Measurements of small animals, such as many birds, should be taken using callipers, because these are more accurate in this case.

Mammals

There are at least four standard measurements that can be recorded from mammals (see **Fig. 5.16a**), as follows.

- *Total length*   Measure from the tip of the nose to the front end of the tail vertebrae. If possible, the animal should be placed on its back so that the backbone is straightened, but not stretched. Position the head with the nose extending straight forward in the same plane as the backbone.

- *Tail length*   Bend the tail at a right angle to the body and measure the distance from the angle to the far end of the last tail vertebra. Do not include any protruding hairs.

- *Hind foot length*   Measure from the back edge of the heel to the tip of the longest toe, excluding the claw.

- *Ear length*   Measure from the notch at the base of the ear to the furthermost point on the edge of the pinna (external ear flap).

Two additional standard measurements can be recorded from bats (see **Fig. 5.16b**), as follows.

- *Tragus length*   The tragus is a leaf-like structure projecting up from the base of the ear in many bats. Measure from the base to the tip.

- *Forearm length*   Fold the wing and measure from the outside of the wrist to the outside of the elbow. ▶

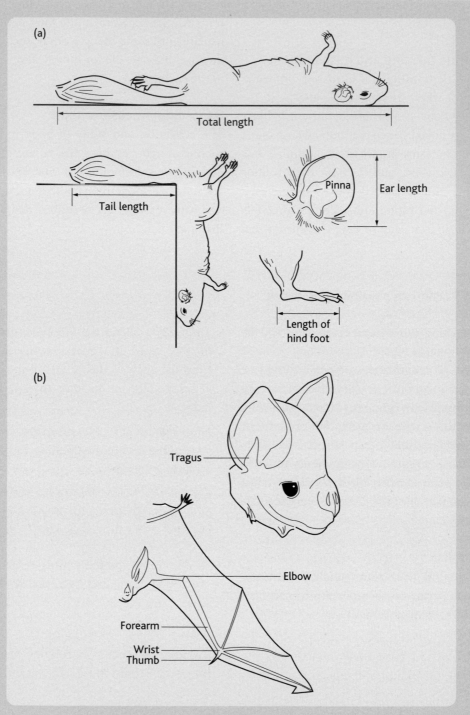

Figure **5.16** (a) An illustration of the standard measurements that can be recorded from a mammal. There are two additional measurements that can be recorded from bats (b).

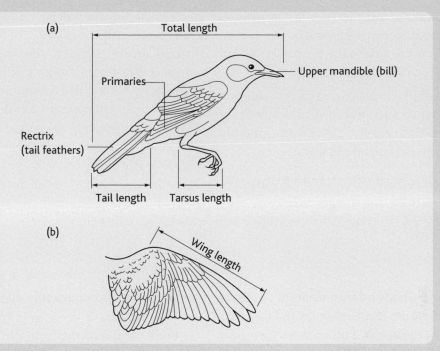

Figure 5.17 An illustration of the standard measurements that can be recorded from a bird.

Birds

There are at least five standard measurements that can be recorded from birds (see **Fig. 5.17**), as follows.

- *Total length*   The distance from the tip of the bill to the tip of the longest tail feather (rectrix).

- *Wing length*   The distance from the bend of the wing to the tip of the longest primary. The curvature is not straightened and the measurement is made from the bend directly to the tip.

- *Tail length*   The distance from the tip of the longest tail feather to the point between the middle feathers at which they emerge from the skin.

- *Tarsus length*   The distance from the point of the joint between the tibia and tarsus, to the point of the joint at the base of the middle toe in front.

- *Bill length*   This is measured from the tip of the upper mandible in a straight line to the base of the feathers on the forehead. In birds with a cere (the smooth featherless skin between the beak and the forehead) or similar structure, the measurement is made from the front edge of the cere to the tip of the bill.

Amphibians and reptiles

Standard measurements in amphibians and reptiles include the following.

- *Snout–vent length (SVL)*   Sometimes referred to as 'total length', this measurement is recorded from the tip of the animal's nose to its cloaca in amphibians, snakes,* lizards, ▶

▶ crocodilians, and tuataras. In salamanders and caecilians, total length is measured from the tip of the nose to the tip of tail (SVL can still be recorded).

- *Measures for chelonians*   Different measurements that can be recorded include: width of carapace (upper shell); shell height; plastron (lower shell) length; and width.

* **NOTE**   Snakes that are artificially stretched can be considerably longer than when they are relaxed. So they should be placed on a soft surface (a towel, a snake bag, etc.) and pressed gently downward, using a piece of transparent material such as Perspex. The outline of the animal is then traced using a soft-tipped pen. Using string, the length is determined from the drawn outline. Alternatively, a novel approach involves photocopying the specimen, preferably in a coiled position. It should be noted that there is a small discrepancy between the size of the actual specimen and the final photocopy, which is around 98 per cent of actual size (Chalmers, 2006).

Knowing the body measurements of an animal is essential to ensure that enclosures and travelling crates are an adequate size. Furthermore, the regular collection of these measurements provides additional information that can be used to monitor the health and welfare of the animals. Thus, for example, regular weighing can help in an assessment of whether the diet is appropriate, and so inform keepers about the animal's health and whether the animal's weight remains constant, increases, or decreases. Actually weighing an animal on scales should be much more accurate than trying to estimate if it has changed weight simply by looking at it. Andrew Fry, of Paignton Zoo, UK, has commented that monitoring the daily weight gain of pregnant female **callitrichids** has been used as a reliable indication of their expected parturition date. In hand-rearing, daily weight gains are seen as a sign that the process is progressing well and the infant is developing (see section **9.5.3**). Very often, wild animals gain and lose weight seasonally, as food becomes more or less available. These fluctuations in weight can affect physiological processes, such as reproductive cycles. So it may be prudent for husbandry regimes to mimic the fluctuation in food quality or quantity throughout the year (see **Chapter 12**).

Body measurements can also be helpful for sexing and aging animals. If data are available from other animals, it is possible to draw growth curves from which estimations of age can be made. For example, the birth dates of young wallaby joeys can be estimated by taking foot measurements and weights (Bach, 1998). Birds, reptiles, and amphibians can sometimes be difficult to sex, because of their lack of external genitalia, and morphometrics can often help with this. In birds, especially, there is a huge wealth of literature documenting the use of morphometric values to distinguish the sexes. In many reptiles, as a general rule of thumb, tail length can be used to help identify the sexes. In snakes, for example, males have broader tails than females due to the presence of the retracted **hemipenes**, whereas in turtles, males have longer tails and the cloacal opening is further down the tail than it is in females (www.chelonia.org; see **Fig. 5.18**).

More recently, molecular DNA (see **Box 5.3**) sexing techniques have been used to establish which morphometric values best predict the sex of an

Hemipenes are the paired intromittent sexual organs of snakes and lizards, usually kept inverted in the body when not being used in copulation.

(a)

(b)

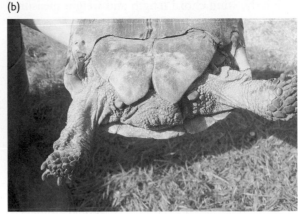

Figure **5.18** Sex determination in turtles cannot be known with complete certainty from body measurements, but a fairly reliable indication can be drawn from a comparison of the tail length, width, and the distance of the cloacal opening down the tail, as in these spur-thighed tortoises *Testudo graeca*. Generally, males (a) have longer tails and the cloacal opening is further down the tail than it is in females (b). (Photographs: Paignton Zoo Environmental Park)

individual (Cerit and Avanus, 2007). For example, the endangered Pohnpei Micronesian kingfisher *Todiramphus cinnamominus* exhibits reversed sexual dimorphism, in which the male is smaller than the female. But the measurements taken from males and females overlap with each other, so no single measurement reliably indicates sex. If a combination of four indices is used—culmen, tarsus and wing length, and weight (see **Box 5.2** for more information on these measures)—**discriminant function analysis** is able to assign correctly the sex of a bird in more than 73 per cent of cases (Kesler *et al.*, 2006).

Box **5.3** DNA: from sampling to interpretation

In recent years, advances in molecular genetics have progressed rapidly (see **Box 10.4**). Using these techniques, it is possible to identify animals from most taxa individually with a variety of samples, the majority of which can be collected non-invasively (see, for example, on non-human primates, Washio *et al.*, 1989; on canids, Ortega *et al.*, 2004; on birds, Burke and Bruford, 1987, and Taberlet and Bouvet, 1991; on reptiles and amphibians, Miller, 2006; on fish, Lucentini *et al.*, 2006). Samples that can be used for this purpose include: fins; scales; blood; saliva; cloacal and buccal swabs; semen; faeces; urine; hair; and feathers.

Other measures that benefit captive breeding and conservation biology, which can be determined using molecular genetics include: species; subspecies; races; parentage; **effective population** size; sex (for example, in the panda—see Durnin *et al.*, 2007); consumed diet (for example, in the pinniped—see Deagle and Tollit, 2007); and much more (Haig, 1998).

Discriminant function analysis is a multivariate statistical technique that is used to determine which of a large number of variables is most successful in classifying objects, or animals, or behaviours (or anything else that can be measured through a number of variables) into groups.

Similarly, wing chord length and weight measured in red-tailed hawks *Buteo jamaicensis calurus* could predict the sex of the bird in over 98 per cent of cases (Donohue and Dufty, 2006).

Sizes and weights of animals are often of interest to zoo visitors. Because of this, some zoos have integrated the process of weighing animals into educational demonstrations. One way of doing this is by connecting a set of animal scales to a large screen that is on display to the public, so that, when an animal steps onto the scales, the public can read its weight. When zoos have trained animals to step onto scales, which are placed in their enclosure, many of the problems associated with catching and handling animals, in order to get weights, can be avoided (Savastano *et al.*, 2003).

5.6.6 Establishing 'normal' values

To monitor and evaluate the health and welfare of zoo animals effectively, it is necessary to have an idea of what measurements are 'normal'. By 'normal', all we mean is the range of values observed in members of a species of the same sex and age, and which do not have any **pathology**. Analysis of pooled data— collected over several years, or from several zoos, or both—can be helpful in calculating the mean, median, mode, and standard deviations for a given parameter. For example, data recorded on captive Arabian oryx *Oryx leucoryx* between June 1986 and April 1992 led to the calculation of various reproductive parameters, including oestrous cycle length (22 days), gestation period (260 +/− 5.5 days), inter-birth interval (295 +/− 42 days), and mean calf weight (6.5 +/− 0.7 kg) (Vié, 1996).

How reliable these 'normal' values are depends on the size of the data set, because the larger the body of information is from which calculations are made, the more accurate the results will be. So, zoo records are not only important in the monitoring of an individual animal in a particular zoo, but also in the monitoring of the captive population in general and of other individuals in other zoos.

5.7 Zoo record-keeping systems

As we have seen, the main aim of keeping records is to use the information contained within them to monitor individuals and populations of animals. It is essential, therefore, that the record-keeping system used should facilitate inspection and analyses of information, so that the health and welfare of individuals can be regularly assessed and maintained to a high standard. Indeed, the SSSMZP (Defra, 2004) expect that zoo records should be kept and filed in such a way that they are easily accessible.

Records can be kept either in a paper format or electronically, and there are advantages and disadvantages associated with both methods. Paper is cheap and low-tech, and there is no restriction about who is able to collect and record information in this way. But paper records easily get lost and damaged. In addition, the collation, organization, and storage of a paper record-keeping system is extremely laborious, compared with how easily a computer software package can undertake a similar task.

Computer programs also have their limitations, however. As a high-tech alternative to paper record-keeping systems, computer-based systems are more expensive to operate and maintain, and they require staff who are specifically trained to use the software. Despite this, they are certainly the easiest method of cataloguing, retrieving, and comparing information, especially for large collections and between collections (for example, in the calculation of 'normals'—see section **5.6.6**).

Given that there are merits associated with both systems, a lot of zoos use a combination of paper-based and electronic record-keeping systems. For example, daily events will probably be recorded by keepers in paper diaries kept close to where they work, whereas less frequently collected information —such as life history, and housing and husbandry data—will be entered into an electronic system.

Paper records can always be transcribed into an electronic form at a later date.

Although there are different computer software programs available for zoo record keeping, most zoos use those that have been formulated by the International Species Information System (ISIS).

5.7.1 The International Species Information System (ISIS)

ISIS is a non-profit organization that was established by Drs Ulysses Seal and Dale Makey in 1973 (Flesness, 2003). They believed that a global database of zoo animal records was necessary to achieve long-term conservation goals. The database started with the records of 51 zoos in Europe and the USA, and now represents almost 650 professional animal collections in over seventy countries on six continents, with records for over 2 million animals of about 10,000 species. By using ISIS software, records can be kept in a format that can be readily uploaded into this ISIS central database. The global database is an invaluable resource, because it provides security—a back-up for individual zoos' records—but it also enables the comparison of information between zoos. The exercise of establishing 'normals' (section **5.6.6**) is much easier when the data are already available in a central database and represents large numbers of animals in many different zoos. These data sets are thus a valuable resource for zoo research, especially epidemiological studies, and investigations of captive breeding and population management (for example, Boakes *et al.*, 2006).

ISIS has produced several different computer programs for recording and analysing information, the most popular of which are ARKS, the Single Population Analysis and Records Keeping System (SPARKS), and MedARKS, at each of which we will now look more closely.

5.7.2 The Animal Record-Keeping System (ARKS)

ARKS is a computer program created to manage all of the types of information that can be recorded about zoo animals and which were described in section **5.6**. A printout from ARKS is illustrated in **Table 5.5** and shows the sorts of information that it can record.

Within the system, there are some basic fields that carry required information and therefore need to be completed for each individually identifiable animal. The information required in these fields includes things such as species, date of birth, birth origin, sire, and dam. There are also additional fields and these are optional, so they can be completed at the discretion of the registrar, or whoever is entering the information. Examples of these additional fields might include descriptions of identifying features, reproductive notes, information from daily report sheets or about the enclosure, or in fact anything that the registrar feels will be helpful in the future and thus worth noting. Keeping some of this information is a legal requirement in the UK (see section **5.6**), but, as we have already discussed, the collation of as much information as possible is important in that it greatly increases our knowledge of the animal and enables us to explore more fully aspects of its life history in captivity.

Information can be exported or printed out from ARKS for only one individual or for a group of animals. If you were to request information on a group, the printout would probably follow a standard format and this can be seen in the example in **Table 5.6**.

ISIS programs were initially biased towards the recording and analysis of data from mammals, but this has now been addressed by the creation of supplementary programs for other taxa. For example, EGGS,[7] designed by Laurie Bingaman Lackey, records information on egg production, incubation periods, **fertility**, hatchability, egg weights, and measurements for breeding and management purposes.

---

7 EGGS is not an acronym, but rather the title of the program.

Specimen Report   Print   Pedigree Report

International Species Information System   10 April 2008

Specimen JERSEY/R991

Names:

Taxonomic:		Common:	Family:	Order:
Paleosuchus palpebrosus		**Dwarf caiman**	Crocodylidae	Crocodylia

Birth Information:

Sex:	Birth Location:	Birth date:	Birth type:	Hybrid:	Rearing:
Male	UNKNOWN	????	Unknown	Not a hybrid	Unknown

Visits:

Date:	Acquisition:	Vendor/LocalID:	Reported by:	Disposition:	Recipient/LocalID:	Date:
7 May 1998	Loan In from	UNKNOWN	JERSEY/R991			

Measurements:

Date:	Measurement:	Value:	Units:	Comments:
7 May 1998	live animal weight	153.00	gram	at JERSEY
28 Aug 1998	live animal weight	280.00	gram	at JERSEY
11 Nov 1998	live animal weight	316.00	gram	at JERSEY
19 Mar 1999	live animal weight	328.00	gram	at JERSEY
7 May 1999	live animal weight	420.00	gram	at JERSEY
30 Jun 2000	live animal weight	817.00	gram	at JERSEY
30 Jun 2000	snout-vent length	300.0	millimeter	at JERSEY
30 Jun 2000	tail length	305.0	millimeter	at JERSEY
15 Aug 2001	snout-vent length	355.0	millimeter	at JERSEY
15 Aug 2001	tail length	335.0	millimeter	at JERSEY
15 Aug 2001	live animal weight	2.20	kilogram	at JERSEY

Special Information

Date:	Note:	Comments:
15 Aug 2001	Sex Modification Log	Old Date: 07 May 1998 Old Sex: Unknown New Date: 15 August 2001 New Sex: Male New Note: Sexed by manual eversion of penis. 28/03/06: sex confirmed by probing. at JERSEY

Sex Information:

Date:	Sex:	Comments:
15 Aug 2001	Male	Sexed by manual eversion of penis. 28/03/06: sex confirmed by probing. at JERSEY

Rearing Information:

Date:	Rearing:	Comments:
7 May 1998	Unknown	at JERSEY

Parents:

Date:	Parent type:	ID:	Location:	Comments:
7 May 1998	Sire	UNK	UNKNOWN	while at JERSEY
7 May 1998	Dam	UNK	UNKNOWN	while at JERSEY

Table **5.5** An example of an ARKS printout*

* A dwarf caiman *Paleosuchus palpebrosus* has been used to illustrate the printout, providing examples of the basic fields of information that are recorded for many species in zoos that use this system.

Report Start Date 07/04/2008	Taxon Report for Nasua nasua	Report End Date 07/04/2008

RTC2	*Nasua nasua*			**Brown-nosed coati**
Date in	Acquisition - Vendor/local Id	Holder	Disposition - Recipient/local Id	Date out
20 Jan 2001	Donation from CHESTER-M00062	ALFRISTON		

Sex-Contraception	Female - Contraception Started	
Hybrid status	Not a hybrid	Birth type:	Captive Born
Enclosure		Birth Location:	North of England Zoological Society
Sire	M00016 at CHESTER	Birthdate-Age:	3 May 2000 - 7Y,11M,4D
Rearing:	Parent	Dam	
Transponder ID:	826098101447110 - Scruff	House Name:	Angel

Additional layout:

RTC2 *Nasua nasua*   **Brown-nosed coati**

Date in	Acquisition - Vendor/local Id	Holder	Disposition - Recipient/local Id	Date out
20 Jan 2001	Donation from CHESTER-M00062	ALFRISTON		

Sex-Contraception Female - Contraception Started
Hybrid status   Not a hybrid   Birth type: Captive Born
Enclosure   Birth Location: North of England Zoological Society
Sire   M00016 at CHESTER   Birthdate-Age: 3 May 2000 - 7Y,11M,4D
Rearing: Parent   Dam
Transponder ID: 826098101447110 - Scruff   House Name: Angel

RTC3 *Nasua nasua*   **Brown-nosed coati**

Date in	Acquisition - Vendor/local Id	Holder	Disposition - Recipient/local Id	Date out
20 Jan 2001	Donation from CHESTER-M00067	ALFRISTON		

Sex-Contraception Female -
Hybrid status   Not a hybrid   Birth type: Captive Born
Enclosure   Birth Location: North of England Zoological Society
Sire   M00016 at CHESTER   Birthdate-Age: 5 May 2000 - 7Y,11M,2D
Rearing: Parent   Dam
                                       House Name: Bebetto

RTC10 *Nasua nasua*   **Brown-nosed coati**

Date in	Acquisition - Vendor/local Id	Holder	Disposition - Recipient/local Id	Date out
19 May 2002	Birth	ALFRISTON		

Sex-Contraception Female -
Hybrid status   Not a hybrid   Birth type: Captive Born
Enclosure   Birth Location: Drusillas Zoo Park
Sire   RTC1 at ALFRISTON   Birthdate-Age: 19 May 2002 - 5Y,10M,19D
Rearing: Parent   Dam   RTC3 at ALFRISTON
Transponder ID: 826098101448231   House Name: Red

RTC11 *Nasua nasua*   **Brown-nosed coati**

Date in	Acquisition - Vendor/local Id	Holder	Disposition - Recipient/local Id	Date out
19 May 2002	Birth	ALFRISTON		

Sex-Contraception Female -
Hybrid status   Not a hybrid   Birth type: Captive Born
Enclosure   Birth Location: Drusillas Zoo Park
Sire   RTC1 at ALFRISTON   Birthdate-Age: 19 May 2002 - 5Y,10M,19D
Rearing: Parent   Dam   RTC3 at ALFRISTON
Transponder ID: 826098101441059   House Name: Little Girl

RTC12 *Nasua nasua*   **Brown-nosed coati**

Date in	Acquisition - Vendor/local Id	Holder	Disposition - Recipient/local Id	Date out
19 May 2002	Birth	ALFRISTON		

Sex-Contraception Female -
Hybrid status   Not a hybrid   Birth type: Captive Born
Enclosure   Birth Location: Drusillas Zoo Park
Sire   RTC1 at ALFRISTON   Birthdate-Age: 19 May 2002 - 5Y,10M,19D
Rearing: Parent   Dam   RTC3 at ALFRISTON
Transponder ID: 826098101442167   House Name: Mandy

Table 5.6 A printout from ARKS for a social group*

* A printout for a social group is usually formatted in a standard form and this has been illustrated using a social group of coatis *Nasua nasua* housed at Drusillas Park, UK.

Report Start Date 07/04/2008	Taxon Report for Nasua nasua	Report End Date 07/04/2008

RTC29   *Nasua nasua*   **Brown-nosed coati**

Date in	Acquisition - Vendor/local Id	Holder	Disposition - Recipient/local Id	Date out
25 Mar 2004	Birth	ALFRISTON		

Sex-Contraception	Female -		Birth type:	Captive Born
Hybrid status	Not a hybrid		Birth Location:	Drusillas Zoo Park
Enclosure			Birthdate-Age:	25 Mar 2004 - 4Y,0M,13D
Sire	RTC1 at ALFRISTON		Dam	RTC3 at ALFRISTON
Rearing:	Parent		House Name:	Elsie
Transponder ID:	826098101487141 - intra scapular			

RTC30   *Nasua nasua*   **Brown-nosed coati**

Date in	Acquisition - Vendor/local Id	Holder	Disposition - Recipient/local Id	Date out
25 Mar 2004	Birth	ALFRISTON		

Sex-Contraception	Male -		Birth type:	Captive Born
Hybrid status	Not a hybrid		Birth Location:	Drusillas Zoo Park
Enclosure			Birthdate-Age:	25 Mar 2004 - 4Y,0M,13D
Sire	RTC1 at ALFRISTON		Dam	RTC3 at ALFRISTON
Rearing:	Parent		House Name:	Mostin
Transponder ID:	981000000257804 - intra scapular		Transponder ID:	981000000438015 - intra scapular

RTC31   *Nasua nasua*   **Brown-nosed coati**

Date in	Acquisition - Vendor/local Id	Holder	Disposition - Recipient/local Id	Date out
25 Mar 2004	Birth	ALFRISTON		

Sex-Contraception	Male -		Birth type:	Captive Born
Hybrid status	Not a hybrid		Birth Location:	Drusillas Zoo Park
Enclosure			Birthdate-Age:	25 Mar 2004 - 4Y,0M,13D
Sire	RTC1 at ALFRISTON		Dam	RTC3 at ALFRISTON
Rearing:	Parent		House Name:	Louis
Transponder ID:	981000000256413 - intra scapular			

Table **5.6** (continued)

5.7.3 The Single Population Analysis and Records-Keeping System (SPARKS)

SPARKS was designed to facilitate the work of **studbook** keepers in managing small captive populations of animals (see **Chapter 9** on 'Captive breeding'). Like ARKS, the program only stores information about the life histories and events of individually identifiable animals—but for one species. Its real purpose, however, is to analyse these data by performing basic demographic and genetic analyses, which can be achieved through SPARKS or, alternatively, through Population Management 2000 (PM2000).[8] Some demographic values calculated include population size, growth rate, sex–age pyramid, and fecundity and mortality parameters. The genetic **fitness** of the population can also be calculated, in terms of the degree of **inbreeding** and **genetic diversity** present both in the population and also in individuals. A printout of some of the analyses that can be calculated with SPARKS is shown in **Fig. 5.19**.

8 PM2000 was designed recently and allows more thorough analyses of the studbook data compared with the SPARKS program.

EAZA Quick Captive Population Assessment for Nomascus gabriellae in EAZA/European region.

Data used in assessment:

Data set:	
Author:	Pierre Moisson
Institution:	Mulhouse
Data Currentness	31st December 2004

EPMAG assessor: Vicky Melfi

Date: 14 December 2007

Age Pyramid EAZA population(graph)

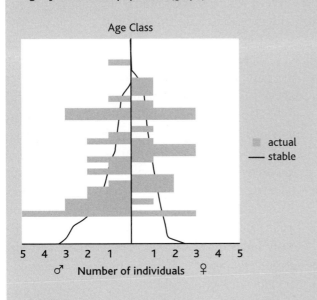

Demographic status of EAZA population

Population size	31.28.4 (63)
Number of institutions	18
No. births/year: 2006	Not complete
(no. of early deaths) 2005	Not complete
2004	1.1.1 (0.1)
2003	3.3.2 (0.1)
2002	4.2 (0.1)
No. births/y needed to sustain population at current number	2.8
No. deaths/year: 2006	Not complete
2005	Not complete
2004	0.1
2003	0.1
2002	1.1
No. individuals in other regions	
Other studbooks	

*At date of evaluation, includes all zoos in EUROPE.

Census EAZA population (graph)

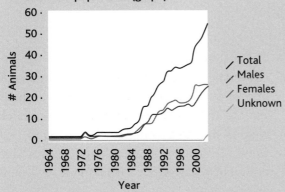

Genetic status of EAZA population

Number of founders (potential founders):	20
Percentage known pedigree:	93.4%
Current Gene Diversity:	95.2%
Founder genome equivalents	10.47 (29.66)
Average Mean Kinship:	0.0478
Average Inbreeding Coefficient:	0
No. founders in intl stubk not represented in EAZA population	

Figure 5.19 Summary of demographic and genetic analyses that can be performed using data entered into the SPARKS program and analysed using the computer software package PM2000. Data provided here come from the Northern buff-cheeked gibbon *Nomascus gabriellae* EEP, managed by Pierre Moisson (Mulhouse). These data outline the sex–age pyramid, population trends, and some calculations of genetic indices (more information to help you interpret these data can be found in **Box 9.2**).

A great feature of these programs is that 'potential' pairings of animals can be suggested and the potential offspring's genetic integrity calculated. Various manuals and guidelines have been published to help with data entry and analyses (Wilcken and Lees, 1998; Wiese *et al.*, 2003; ISIS, 2004; Leus, 2006).

5.7.4 The Medical Animal Record-Keeping System (MedARKS)

MedARKS was designed specifically to record **clinical** and pathological information about individually identified animals. This information, once shared with the global database, facilitates the estimations of 'normals', and thus helps zoo managers and vets to assess the health and welfare of their animals.

5.7.5 Zoo Information Management Systems (ZIMS): a future development

Previously, all ISIS computer software packages have worked independently from each other, which has meant that the same data needed to be entered into each program separately. Thus, for example, the birth of an animal needs to be entered into the ARKS program of the zoo holding it, but also the SPARKS program of the studbook holder, and the MedARKS program of the vet who monitors the animal. What is worse, if the animal is transferred to another zoo, then the new zoo will also need to input all of the animal's data into its ARKS program. There is an obvious duplication of effort here, but, more worryingly, there is an increased potential for errors: each time that data are re-entered into a different program, there is the possibility that a mistake will be made and that the information will no longer match the information current in other software programs.

Thankfully, this worry will soon be a thing of the past. Zoo Information Management Systems (ZIMS) is the newest computer software package to be designed by ISIS and will integrate all of the previous programs into one global Internet database. This means that information will only need to be entered into the database once, in real time, thus reducing the likelihood that errors will be generated. Zoo professionals who need access to information will be able to review zoo records as they did previously, but with additional program features allowing greater analyses of data for population management or epidemiological analyses.

Summary

- The identification of individual animals is essential in a zoo collection, so that they can be monitored, information recorded, and records kept that can be reviewed and analysed.

- Although record keeping may sound a little stale, it is what underpins knowledge of zoo animals, and thus should inform all of our decisions about the animal's daily life and long-term position within the captive population.

- The analyses of zoo records promises a wealth of information that can further inform our knowledge of species, their breeding, mortality, and health, and the impact of our housing and husbandry regimes. In short, being able to recognize individuals is vital and keeping good records about those individuals is the cornerstone of good animal management.

- A number of databases (ARKS; SPARKS; MedARKS) are available for record keeping in zoos and a new Internet-based system (ZIMS) is currently being developed.

Questions and topics for further discussion

1. How might you individually identify an animal, using natural visual characteristics?

2. Name four different artificial methods of individually identifying animals, listing the advantages and disadvantages of each.

3. Should permanent methods of individual animal identification be used? Justify your conclusion.

4. What is 'ISIS' and what does it provide for the zoo community?

5. Name two computer software programs used to keep zoo animal records and explain what records each allows you to keep.

6. Identify two demographic and three genetic details that you can discover about a captive population using SPARKS.

Further reading

Most of the material that we have covered in this chapter is a distillation of accepted practice[9] within the zoo community, together with appropriate references to journal articles where these are available. There is no additional source of information to which we can direct readers, other than Kleiman et al.'s *Wild Mammals in Captivity: Principles and Techniques* (1996), which is particularly concerned with mammals. BIAZA is currently preparing a set of guidelines on record keeping and it is hoped that these will be published in the near future.

Websites and other resources

The are many different manufacturers that make identification tags for animals. A website that is particularly useful is that of Dalton Rototags (www.dalton.co.uk). Dalton produces a range of tags from the 'Minitag' to the 'JumboTag'. These are suitable for many different species and are widely used for tagging livestock in the UK, as well as wild animals from rabbits to seals (see, for example, Testa and Rothery, 1992). Other tag manufacturers in the UK include Allflex (www.allflex.co.uk) and Fearing (www.fearing.co.uk); the online catalogues of all three of these companies include advice on best practice during tagging. In the USA, Biomark (www.biomark.com) is a well-established provider of animal identification tags.

The ISIS website—www.isis.org—contains a wealth of information, including listings of zoos in which different animal species are held and presentations on the new ZIMS software.

---

9 Of the array of animal marking methods reviewed, some of those described are rarely practised, e.g. wing tagging in birds and scale removal in reptiles, but there is no current regulation to prohibit them.

Chapter 6   Housing and husbandry

This chapter is about the accommodation and day-to-day maintenance of animals in **zoos**. Housing and husbandry regimes have to meet the needs of many stakeholders: principally, the animals themselves, their keepers, and the zoo visitors. We will start by outlining the needs of these different stakeholders, and the demands that they place on zoo animal housing and husbandry. Within the construction of zoo animal housing, trends are apparent that reflect the differing popularity of enclosure styles over time. An overview of some of the most prominent housing styles is provided, and their merits and drawbacks are discussed, in terms

of how they meet the needs of different stakeholders. The development of good housing and husbandry regimes requires many variables to be balanced against one another, ranging from financial constraints, the reproductive status of individual animals, or sourcing building materials from sustainable resources, through to things such as ensuring the comfort of visitors while viewing animals in poor weather.

The following key issues relating to zoo housing and husbandry regimes and implementation are highlighted in this chapter.

6.1   The needs of many
6.2   Advances in enclosure design
6.3   Husbandry from birth to death
6.4   Studying how housing and husbandry affects captive animals
6.5   Guidelines and housing and husbandry recommendations

In addition, there are boxes that examine in more detail the related issue of whether it is the physical space itself or the complexity of that space that is of

more importance to animals, and also the technique of **post-occupancy evaluation (POE)**, which can tell us about how that space is being used.

6.1   The needs of many

Enclosure design and good husbandry have developed to incorporate the requirements and needs of many stakeholders, but there are three principal groups whose needs should be considered when designing zoo housing and husbandry regimes:

1. the animals, to ensure that they have good welfare and that they form self-sustaining populations;

2. keepers, so that they can adequately care for the animals;

3. zoo visitors, who want to enjoy their visit to the zoo and who zoos want to educate about conservation and the environment.

6.1.1   Animals' needs

At the very least housing and husbandry should provide animals with the facilities and opportunities to ensure that the **five freedoms** are fulfilled (Webster, 1994; see **Chapter 7**). Different species have evolved to exploit different environmental niches, so there is no 'one size fits all' rule to housing and husbandry that will meet the needs of the whole diversity of species held in zoos. To illustrate this, consider species such as the black rhinoceros *Diceros bicornis* and the Toco toucan *Ramphastos toco*, between which the species differences may be visually obvious, but which may, in fact, both suffer from a similar nutritional problem: iron storage disease (see section **12.8.1**). In contrast, species such as the pig-tailed macaque *Macaca nemestrina* and the stump-tailed macaque *Macaca arctoides* look visually similar, but while providing **enrichment**—in the form of visual barriers—increases aggression in pig-tailed macaques, it reduces it in stump-tailed macaques (Erwin, 1979; Estep and Baker, 1991).

Unfortunately, we know very little about the housing and husbandry needs of many of the species that are held in zoos, which means that very often past experiences (gained through personal knowledge or from zoo colleagues), or knowledge is generalized between similar species in order to create housing and husbandry regimes that it is hoped will meet the animals' basic needs. Because of this, it is important that adequate monitoring of the animals and, in due course, detailed research is undertaken so that a body of knowledge is built up to give us a greater understanding of different species' specific needs. No matter what species we are considering, however, there are various common factors—to which we can refer collectively as 'individual differences'—that will affect their needs.

Individual differences are those characteristics that set one animal (individual) apart from another and affect how that animal functions within its environment. These differences include such characteristics as the animal's age, sex, reproductive status, size and social rank, as well as its temperament (see **Box 4.2** for more about temperament or personality), health (for example, the animal's disease load), past experience, and other factors that contribute to its individuality. These characteristics place biological demands on the animal and thus affect its needs, so they have to be factored into housing and husbandry provision in the zoo. It should also be remembered that many of these characteristics will change during the animals' lifetime. Here, we will look further at how some of these characteristics affect the animals' needs, and at how housing and husbandry regimes should be developed to accommodate them.

Age

In many ways, the needs of old and young animals are similar, because both may have difficulty coping adequately with stressors, but for very different reasons. Old animals undergo **senescence** and young animals have not yet fully developed, so both groups should be considered vulnerable. Animals that are unable to cope with stressors can have compromised welfare due to an increased risk of disease and other associated deleterious consequences (this issue is explored fully in **Chapter 7**).

The period of senescence is also associated with increased **mortality** and reduced **fecundity**. Studies of aging have shown that old animals are also associated with a higher incidence of **hypothermia** and **hyperthermia**, because they are less able to thermoregulate their temperatures physically and behaviourally due to changes in body composition and endocrine function. For example, aged mouse lemurs *Microcebus murinus* were observed to seek out warmer ambient temperatures within their enclosure and thus were shown to thermoregulate behaviourally to compensate for reduced autonomic thermoregulation (Aujard *et al.*, 2006). Old animals may also suffer from **sarcopenia** (Colman *et al.*, 2005).

The sorts of histological change that can be seen in aging human brains can also be found in brains of other mammalian species, suggesting that the process of aging is similar in humans and other animals (Dayan, 1971). Consequently, we might expect

---

Senescence describes the process of aging that occurs after an animal has developed to maturity and is generally characterized by deterioration of physiological function, reduced ability to cope with stress, and greater susceptibility to age-related diseases.

The prefix **hyper-** means that something (in this example – **thermia**, relating to temperature) is excessive or more than expected or normal. By contrast, the prefix **hypo-** means that something (again, in this case, temperature) is deficient or below normal.

Sarcopenia is a decline in skeletal muscle mass and function that is visible as frailty, physical weakness, and increasing disability.

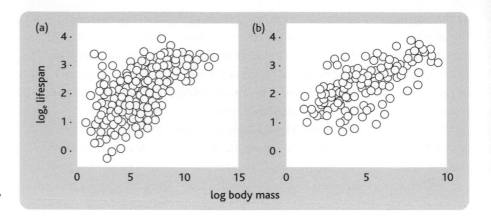

Figure **6.1** Size does matter! Speakman (2005) found that there was a significant positive correlation between an animal's size (body mass) and how long it lived; this is true for (a) mammals (N = 639) and (b) birds (N = 381). (From Speakman, 2005)

cognitive function, including memory and **learning**, to decline in older animals. Studies on aging in mouse lemurs (see, for example, Picq, 2007) and rhesus macaques *Macaca mulatta* (for example, Moore *et al.*, 2006), which are both commonly used species for research on aging, suggest that this is the case.

Young animals are also vulnerable to stressors, just as old animals are, because they too are under-developed both physically and behaviourally, although the effects of these can be mitigated in some species to some extent by parental care.

Size and social rank

Larger birds and mammals live longer and require more energy to survive than those that are small (**Fig. 6.1**). This is probably because larger animals have a proportionately slower metabolism relative to smaller animals—in other words, as an animal gets larger, it requires more energy, but its rate of metabolism declines and thus it is able to use that energy more efficiently (Schmidt-Nielsen, 1997; Speakman, 2005; see section **12.2.1**). It should be noted that there are exceptions to this pattern (Speakman, 2005). Large animals have a low body surface area to volume ratio, and so gain and lose heat slowly, whereas small animals have a higher body surface area to volume ratio, and therefore gain and lose heat more quickly (Schmidt-Nielsen, 1997). Because of these relationships, an animal's size can tell us something about its **longevity**, its **nutritional requirements**, and the extent to which

we need to provide facilities for themoregulation (such as heat lamps; see section **6.2.3**). An animal's size, to a certain extent, also determines the amount of space that it requires, and this is discussed in more detail in section **6.3.3** and also in **Box 6.1**.

Within a species, larger animals are the ones that are more likely to attain positions of high social dominance rank and therefore enjoy priority access to resources. This can have direct implications for many aspects of captive management, such as food presentation (see section **12.7**). It also has implications for breeding success and social stress, because of behavioural and physiological differences between dominant and subordinate animals. These can be the result of mechanisms that are highly variable and complex, of which we need to be aware when managing social groups in captivity.

A good example of this is the African hunting dog *Lycaon pictus*. The majority of cubs born in a pack of these dogs are born to the dominant male and female, both of which show significantly different hormone profiles to other animals in the pack (Creel *et al.*, 1997). Subordinate females have elevated levels of oestrogen compared with dominant animals and this effectively reduces their ability to conceive. On the other hand, subordinate males have lower levels of testosterone than dominant animals, and consequently copulate at lower levels and are less aggressive. This pattern of reduced reproductive success in submissive animals is seen in some other species as well, and it has been suggested that, in primates at least, it results from social stress (Silk, 1989). In the hunting dogs, however, it is the dominant animals that have elevated levels of **cortisol** in comparison with the submissive animals (Creel *et al.*, 1997).

Previous experience

> Somehow a fuzzy, stumbling tiger kitten becomes a monstrously efficient killer. Somehow a big-footed fool of a raven fledgling becomes an aerial acrobat . . . How do baby animals become competent animals? While part of the answer is that cubs and kittens and chicks mature and come into their powers, another part is that they learn what they can do and how to do it.
>
> (McCarthy, 2005)

Previous experience is a term used to describe the sum of what an animal has learnt during its development up to the current day. Early experiences for many species are considered incredibly important and they contribute towards an animal's survival success (see **Box 9.4**).

We have already looked at how animals learn and have pointed out how they may learn to associate with different events in the zoo (section **4.1.2**). Animals may, for example, learn to associate keepers with the provision of food, or the presence of vets with being caught or a painful incident. So, an animal's previous experience in captivity will affect how it perceives and responds to housing and husbandry. How, for example, will the animal respond to novelty: will it be scared or will it be curious? There is some evidence that animals which are exposed to high degrees of environmental complexity (such as **environmental enrichment**) are more able to adapt to novelty (see section **8.6.2**). Keeping detailed records of an animal's previous experiences, including rearing, social groupings, and other aspects of housing and husbandry, will provide a fundamental insight into its behaviour and requirements in captivity.

External environmental factors

Variations in environmental factors such as climate and day length also make demands on an animal's biology and influence its needs in captivity. Many species have evolved seasonal behavioural patterns to reflect changes in resources and climate (Oates, 1989; Menzel, 1991). Tropical and temperate climate zones are characterized by different events, such as heavier rainfall in the tropics and shorter day lengths relative to temperate zones. Because of this, seasonal changes experienced by wild and captive **conspecifics** are frequently different, because captive exotics may live in zoos with dissimilar climates and seasons to those in which they have evolved to live. In addition, some housing and husbandry regimes may change seasonally. For example, in the UK, we experience shorter days in winter, which means that many captive animals are fed earlier and potentially locked into sheltered enclosures for longer periods of time.

Summary of animal needs

It is likely that, within a captive group of social animals, a housing and husbandry regime will need to accommodate both the species-specific and the individual needs of the animals in the group. A good housing and husbandry regime therefore needs to ensure that the five freedoms are met for the group, meeting any species-specific requirements, but must also retain some flexibility, so that the needs of individuals can also be accommodated. This can only be achieved through an understanding of the animals' biology and how it is affected by its captive environment, as well as regular monitoring by knowledgeable staff.

6.1.2 Keepers' needs

The fundamental role of keepers[1] is to ensure that the animals in their care have good welfare. At its most basic, keepers are responsible for monitoring

---

1 Those charged with the daily responsibility of caring for zoo animals are frequently termed 'zookeepers', or 'caregivers', but for the purposes of this chapter, we will refer to them simply as 'keepers'.

Figure **6.2** As many zoos become the focus of television shows, many keepers and other zoo professionals have had to include media skills into the broad array of duties that they may be required to fulfil in a day. (Photograph: Paignton Zoo Environmental Park)

the animals in their care, providing them with food and water, and ensuring that their environment is safe, clean, and enriching. Keepers are also involved in less routine activities that involve close contact with animals: for example, catching and restraining animals, introducing animals to a group, training, administering medical treatments, and many other things. Increasingly, keepers are also involved in educating the public about the animals with which they work, through casual or formal conversations with the zoo visitors, and even through media performances (**Fig. 6.2**).

Keepers are part of a team of zoo professionals who work together to ensure that husbandry procedures are well informed by past experience and also by advances in scientific knowledge. Many of these husbandry procedures are described later in this chapter (section **6.3**) and in other chapters within this book (see **Chapters 8** on 'Environmental enrichment', **9** on 'Captive breeding', and **12** on 'Feeding and nutrition'). This section will consider what the keepers' needs are and what measures need to be considered that might facilitate their work. It should be remembered that the steps taken

to facilitate the keepers' efforts should, in turn, promote the welfare of the animals in their care.

It's all about serviceability

Small details in enclosure design and/or management practice can either facilitate or hinder the efficiency of daily animal husbandry. For example, if drains are placed at the bottom of a slope, enclosure water will run into them, but if they are placed anywhere else, water must be moved towards them, and this requires time and effort that might have been expended elsewhere. There are many such 'little' considerations that can be made to housing and husbandry that greatly facilitate the keepers, and other zoo professionals, in achieving their daily duties.

From the keepers' perspective, some of the most important and influential details that will facilitate husbandry are as follows.

- *Easy-to-clean exhibits* These can be created by providing deep-litter flooring systems, self-cleaning paddocks, the correct type of door (for example, hinged, sliding, or guillotine), correctly placed drains, and so on.

- *Consideration of how staff can safely move animals within the enclosure* It is vital that keepers can view animals easily wherever the animals are in the enclosure, especially when they are moving the animals or catching them up. There should not be any blind spots in which animals can hide from the keepers' view. This can be achieved through features such as good lighting, thoughtful design, viewing windows, and careful consideration in the positioning of enclosure furniture.

- *Providing easy access into enclosures for staff and, where necessary, vehicles* Practical considerations about how keepers and other staff service enclosures should be considered: can all areas of the enclosure be accessed safely? Can large items be moved into and out of the enclosure? Could a keeper escape if necessary? Might enrichment be securely added to the enclosure?

How husbandry can be facilitated through enclosure design is considered in more detail in section **6.2**, but it is important to realize that this can only be accomplished by integrating all staff that will use the facility into the design team (Coe, 1999; section **6.1.5**).

Thinking about how the enclosure is going to be serviced, and ensuring that these activities are facilitated helps to increase the time available to zoo professionals to do other duties. For example, extra keeper time can be devoted to the provision of additional enrichment, husbandry training, or observing their animals.

Staff training

Trained staff are considered beneficial in all types of business, and this is supported and promoted by organizations such as Investors in People (see **Fig. 6.3**). In zoos, the role of staff is varied and they are required to have many different skills, from the catering manager to the conservation officer, and the horticulturist to the head keeper. *Managing Health and Safety in Zoos* (MHSZ) is a document produced by the Health and Safety Executive (HSE). Both the MHSZ (HSE, 2006) and the Secretary of State's Standards of Modern Zoo Practice (SSSMZP) (Defra, 2004) highlight the importance of training staff, and there are different government initiatives to encourage and promote this: for example the Investors in People scheme. Because this book focuses on animal management, we will look only at those zoo employees who work with the animals.

As modern zoos have evolved, so has the role of the keepers, in terms of their expectations of the job,

INVESTOR IN PEOPLE

Figure **6.3** Investors in People is a UK Government initiative that aims to promote staff training in all industries, including zoos.

and the expectations of zoo employers in terms of the keepers' qualifications and experience. All zoo associations offer training opportunities. Within the UK and Ireland, for example, most keepers are expected to complete the BIAZA-accredited Advanced National Certificate in the Management of Zoo Animals (ANCMZA), which is a 2-year course requiring practical 'on-the-job' training, the completion of several pieces of coursework, and attendance of some training workshops. There has also been a proliferation of courses offered that include material relevant to captive housing and husbandry of exotics (see **Further reading**). It is probably unsurprising therefore that, with the increasing competition to become a keeper, many keepers are now required to hold higher qualifications than previously was the case, and to have extensive experience of working with and around animals.

6.1.3 Visitors' needs

To meet the needs of zoo visitors, we really need to know what their motivation is to visit zoos. It is estimated that over 600 million people worldwide visit zoos (IUDZG/CBSG, 1993; see **Fig. 6.4**), but surprisingly few studies have been undertaken to investigate who they are (**demographics**) and why they visit zoos (motivation). Most of those studies that have been undertaken have been completed by sociologists and environmental psychologists, and they suggest that the characteristics of zoo visitors include groups, rather than single visitors, and all ages, although children usually predominate and families are the most frequent type of group (Cheek, 1976; Morgan and Hodgkinson, 1999; Turley, 1999).

We will look in more detail in **Chapter 13** on 'Human–animal interactions' at what people do when they are at the zoo, but, here, it is worthwhile considering briefly some studies on what motivates people to go to the zoo, because this can then help us to understand their needs.

An early study by Kellert (1979) found that the underlying motivation of zoo visitors in the USA was to educate their children (36 per cent), to have

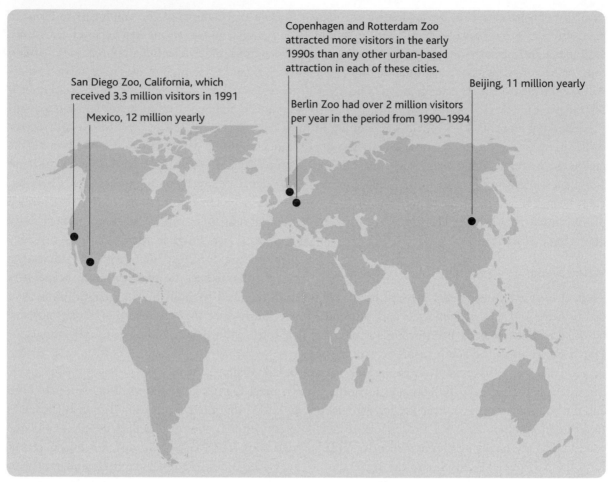

San Diego Zoo, California, which
received 3.3 million visitors in 1991

Mexico, 12 million yearly

Copenhagen and Rotterdam Zoo
attracted more visitors in the early
1990s than any other urban-based
attraction in each of these cities.

Berlin Zoo had over 2 million visitors
per year in the period from 1990–1994

Beijing, 11 million yearly

Figure **6.4** Zoos are the most frequently visited 'museum type' worldwide, with over 100 million visitors recorded to attend zoos in North America and almost 600 million attending other zoos annually worldwide (Sunquist, 1995; Kotler and Kotler, 1998). Some examples of visitor numbers recorded worldwide, from 1990 onwards, are illustrated here (van Linge, 1992; Van der Berg *et al.*, 1995; Shackley, 1996; AZA, 1999).

a fun day out with family and friends (26 per cent), to see animals (25 per cent), and to be around wildlife (11 per cent). Wilson (1984) has suggested that people are inherently **biophilic**: they like nature and want to be around it.

From this perspective, zoos can provide people with much more than just physical access to animals, plants, and (usually) green spaces: they allow people to experience these things with all of their senses, not just to view them, as they would a television documentary. Thus, we can regard zoos as offering an opportunity for **ecotourism** or a 'taster of the wild' (Mason, 2000). But it is important, in terms of tourism, that zoos provide these 'wild' opportunities as a convenient, safe, and comfortable day out (Beardsworth and Bryman, 2001). Morgan and

Biophilia is the concept that people have a love of nature and natural things; it has been promoted by Edward Wilson and Stephen Kellert as the idea that people need contact with nature.

Ecotourism is characterized by seeking out access to nature, wild places, or cultural heritage, but in as responsible and sustainable a way as possible, avoiding many of the negative connotations associated with mass tourism.

Hodgkinson (1999) have cautioned that studies of zoo visitor demographics and motivation should not necessarily be generalized across zoos, because many factors affect people's decisions to visit a zoo, including its location, what animals it holds, and probably other factors as well.

Nevertheless, there does seem to be a consistency in people's motivation to visit zoos. For example, Andereck and Caldwell (1994) found that education and recreation constituted the main reasons why people visited the North Carolina Zoological Park, which is consistent with Kellert's (1979) results. Similarly, the motivation behind visits to UK zoos was investigated by Turley (1999), who found it principally to be recreational, rather than for an educational experience or to support conservation. Indeed, Turley (1999) concluded that most zoo visitors wanted to see and enjoy animals, unlike visitors to museums, who wanted to understand; thus, zoo visitors want '*entertaining, compelling and effectively staged*' events (Beardsworth and Bryman, 2001). There is, then, a need to incorporate educational opportunities into what most zoo visitors consider to be a 'fun day out' (Brodey, 1981). This helps us to understand the statement by the Director Emeritus of the Columbus Zoological Garden, who said that he would rather '*entertain and hope people learn than teach and hope people are entertained*' (Hanna, 1996).

If zoo visitors principally want to be entertained and to have a 'good day out', then there are some basic requirements that they will expect, including the provision of amenities and, above all, good service. If these requirements are not met, the entertainment and educational value of the zoo experience can be undermined:

> The young woman who waited on us [at the snack stand] was so uninterested and the food was so poor that my students talked of nothing else the whole afternoon. The magical opportunity created by the $15 million exhibit investment was quenched by the most junior employee in the whole park!
>
> (Coe, 1996)

But it is the animals that the visitors have come to see. Bitgood *et al.* (1988) considered that a visitor's ideal exhibit was a **naturalistic enclosure** holding a large active mammal, preferably with young. This seems a rather simplistic view, but subsequent research has supported it. Visitors perceive animals in traditional enclosures with bars as having compromised welfare compared with those in greener enclosures (Turley, 1999; Melfi *et al.*, 2004a).[2] Visitors also spend longer observing **free-ranging** tamarins, which are more active than their counterparts in 'traditional' cages (Price *et al.*, 1994), and visitors at the Zurich Zoo spend more time observing larger animals (Ward *et al.*, 1998; see **Fig. 6.5**—but also see section **13.1.2** for an alternative explanation).

Zoos can provide visitors with a magical experience. Vining (2003) describes how interactions with animals can be magical and thus create lasting impressions, which might include medically therapeutic effects, or might encourage an attitude that cares for animals. Fiedeldey (1994) noted that, in the right circumstances, even the tracks of an animal that you do not actually see can be exciting. Close experiences with an animal are particularly rewarding. In many zoos, there have always been opportunities to get close to some species: for example, in the **petting zoo** or in animal-handling sessions. Cheyenne Mountain Zoo has gone a step further, and has built into its zoo animal management and visitor experience philosophy the concept of providing visitors with a 'defining moment' (Chastain, 2005). It provides some visitors with a unique opportunity of close contact, which is not usually possible on a zoo visit. This might entail getting close to a giraffe

---

2 Interestingly, however, Verderber *et al.* (1988) found that elderly visitors to zoos were anxious when animals were not presented behind bars and had 'invisible' barriers.

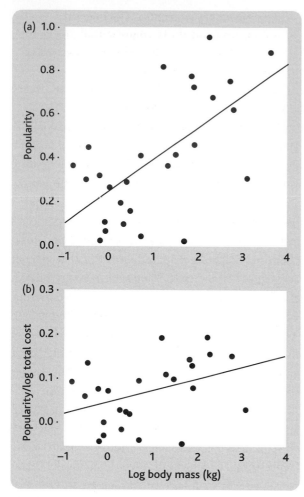

Figure **6.5** It is probably unsurprising that larger animals are perceived by many zoo visitors as the more interesting. (a) Here, you can see that there is a positive correlation between popularity, as rated by zoo visitors, and zoo animal size. (b) It is interesting to note that there is also a positive correlation between the cost of maintaining animals and their overall size (Ward *et al.*, 1998). Issues of animal size in relation to their attractiveness to visitors and their inclusion in zoos is discussed further in section **13.1.2**. (From Ward *et al.*, 1998)

or helping the keeper to scatter food around an enclosure before the animals are let out.

The demand for these 'close encounter' experiences is so high that schemes exists under which visitors can pay to feed animals or even to be a 'keeper for the day', for example, at Taronga Zoo, Sydney, Australia, and Paignton Zoo Environmental Park, UK. In these situations, animals in zoos are no longer only to be viewed by visitors, but they are expected

to perform and/or interact with visitors. It is not in doubt that this provides a more compelling educational message, but it is debated exactly what that education message might be (Beardsworth and Bryman, 2001). Much of this discussion is based on speculation, because there are few **empirical studies** that have been undertaken to explore zoo visitors' attitudes to animals and, more importantly, how 'interactive' sessions may influence these. Robinson (1989) has suggested that the beneficial 'message' gained from seeing animals in zoos can be lost when the animals are taken out of context, away from their naturalistic environment and into a show arena (Robinson, 1989). This can be mitigated, however, by providing a live-animal demonstration within the animals' enclosures. At Zoo Atlanta, for example, husbandry training of the otters was incorporated into the visitor experience (Anderson *et al.*, 2002). Visitors were observed to stay at the exhibit longer, and had more positive attitudes towards the animals and the zoo in general, when they observed husbandry training of the otters.

The topic of visitor education and awareness raising is one to which we will return in **Chapter 13**.

6.1.4 Other interested parties

We have considered how the needs of animals, keepers, and the visiting public have implications for housing and husbandry procedures, but there are other factors that can impact on these as well.

Reducing environmental impact and promoting sustainability

We are all very aware now that many of our natural resources are diminishing and that there is a need to consider the impact on the environment of the way in which we live our lives; the same philosophy applies to how businesses should operate. Many zoos work hard to convey the message that conservation of our precious resources is necessary and the first step that they can take is to 'put their words into actions' by applying those principles to the way in which they operate the zoo.

There are many ways in which zoos can be 'environmentally friendly'. Just like other businesses, they can implement environmentally sound work practices to minimize their impact on the environment. An internationally recognized certificate, ISO 14001,[3] is awarded to those businesses that are able to demonstrate that they implement these methods of best practice. In 2006, two BIAZA-member zoos were awarded ISO 14001 and, in 2006, this rose to a further five BIAZA-member zoos (Stevenson, 2008). A further 55 per cent of BIAZA zoos had environmental policies in 2006 and BIAZA produced documentation to facilitate further the work of their member zoos towards **sustainability** (Stevenson, 2008).

The following are some examples of the many opportunities available to zoos to enable them to operate 'environmentally':

- the sourcing, efficient use, and disposal of resources should be executed in a such a way that it minimizes the zoo's environmental impact, for example, by using sustainable energy and products (such as solar power, reclaimed water, and Forest Stewardship Certificate (FSC)-approved wood);
- ensuring that these resources are used wisely (by turning off power when it is not needed, minimizing packaging on products bought and sold, recycling products, and so on);
- where possible, purchasing products locally or as part of fair trade or **organic** schemes to ensure that the environment is not harmed elsewhere.

This issue of sustainability is discussed further in **Chapter 15**.

Plants: pretty, food, poisonous, or a natural barrier?

Although this is a book about zoo animals, plants must be mentioned, because they play a huge role in animal housing and husbandry, from the exhibition, to the maintenance of zoo animals. Plants are fundamental in the creation of 'naturalistic' enclosures, which are now very common in zoos and popular with the visitors (Jackson, 1996; section **6.2.1**). They can provide natural complexity within the enclosure, providing locomotion and play opportunities, as well as shelter from the elements. If the right plant species are chosen—normally those that are sympatric with the exhibited animals' wild conspecifics—they can provide an ecologically relevant landscape for the animals being exhibited, which serves to facilitate visitor education programmes and the creation of bioexhibits. Unfortunately, this might mean that, in their native habitat, the plants chosen for the exhibit would be a food source and thus vulnerable to being eaten.

When planning which plant species are going to be used in animal exhibits, their function should be identified from the outset. The big question is: does it matter if the animals eat them? To any botanist, the answer is surely 'yes', but in some situations, plants are provided specifically to enable the animals to feed and forage within their enclosure. It could be argued that, unless plants are protected, they are all potential food items. As such, plants need to be chosen that are non-hazardous and non-toxic. The differentiation is necessary because some plants may cause injury, from thorns and other physical defences, while other plants may cause illness, because they contain distasteful or toxic chemicals (see **Box 12.1**). Plant lists are available that provide information on plant toxicity and these should always be consulted before selecting plants for exhibition. For example, BIAZA commissioned a survey of its member zoos to find out which plant species were being fed to which mammal species and whether there were any deleterious side effects. This database can provide some guidance about the variety of plant species that have been provided, and whether or not side effects have occurred during

---

3 A reminder that ISO refers to the International Standards Organization.

the same time period (Plowman and Turner, 2006). There are quite a few examples of plants being provided solely to feed animals on exhibit, with grazing animals (both birds and mammals) being the most obvious. In these situations, it is necessary to have some idea of nutritional value and the extent to which the animals consume the plants within their enclosure (we will return to this issue in **Chapter 12** on 'Feeding and nutrition').

If we do not want the plants to be eaten, because they are endangered, toxic, or serve some other function, then there are various ways in which they can be protected: they can be kept behind barriers, for example, in much the same way as barriers keep animals and visitors separated (Frediani, 2008). Electric fencing is the most frequently used barrier to protect plants from animals, although physical barriers are also used. For example, large trees that provide shelter, in the form of their canopy, can be protected with barriers at their base or halfway up the tree to prevent animals from climbing them. In other circumstances, plants can themselves be used as barriers, especially if they have the defences that we mentioned above (such as thorns or distasteful chemicals). Plants' physical defences, such as thorns, can be enough to discourage both animals and zoo visitors alike from going near them for fear that they might get stung or injured.

Giving researchers somewhere to hide

Because research is now a key goal of zoos (see **Chapter 14**), the needs of researchers are increasingly being integrated into zoo enclosure design. Some zoo exhibits now include areas from which researchers can gain unobstructed views of their research subjects: the animals. This has been achieved through features such as the provision of elevated observation areas, additional 'researcher only' windows and closed-circuit television (CCTV) cameras. At the Lincoln Park Regenstein Centre in Chicago, IL, for example, special windows constructed of three triangular panes protrude into the ape exhibit to give researchers an all-round view. Panda Cam at

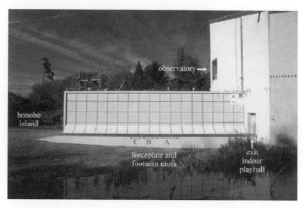

Figure **6.6** Conducting zoo animal research is clearly facilitated at some exhibits, where the needs of researchers have been integrated into the enclosure design. Here, we can see a walkway, which is the main route that bonobos *Pan paniscus* take between their inside and outside enclosure areas, constructed at Planckendael, Belgium. This simple addition has enabled researchers to learn a great deal about locomotion in this species (e.g. D'Aout *et al.*, 2001) (Photograph: *Folia Primatologica*)

San Diego Zoo in California, meanwhile, provides live video footage of the pandas to researchers and visitors, and also on the Internet.

Research into the anatomy and locomotion of animals can also be facilitated through enclosure design. For example, at Dierenpark Planckendael, in Belgium, the study of terrestrial bipedal locomotion in bonobos *Pan paniscus* has been achieved by incorporating force plates into a walkway that the animals use to gain access into and out of a play room (D'Aout *et al.*, 2001). This walkway is surveyed by CCTV cameras and has a backdrop that provides scale (see **Fig. 6.6**).

Building regulations: what can be practically accommodated?

It is necessary to consider all of the stakeholders (including those previously mentioned in this section), but among the most important of these when designing exhibits are the builders. The design of zoo enclosures has to meet with all of the same building regulations that are necessary for 'ordinary' buildings, including environmental and energy-saving considerations. In the UK, building regulations include everything from ensuring disability access to providing sufficient drainage, and are concerned with whether the physical demands on the exhibits can be met: is the building strong enough, for example, to keep in elephants? Will the mesh roof withstand snow? How will 500,000 visitors negotiate their way around the exhibit? Is the glass really shatterproof?

Of utmost importance is that the finished exhibit, and its surrounding area, will be safe for zoo animals, staff, and visitors alike.

Health and safety considerations

The health and safety of keepers—and, indeed, of all zoo professionals—while at work is of the utmost importance. The UK Health and Safety Executive (HSE) has produced a summary document detailing the necessary steps that should be considered and measures that should be employed by zoo operators to ensure safe working conditions for their employees and those who enter the zoo, such as visitors, contractors, and volunteers (HSE, 2006). Much like the SSSMZP (Defra, 2004), this document, *Managing Health and Safety in Zoos* (MHSZ), outlines the measures that zoos should take to adhere to the relevant legislation, particularly the Health and Safety at Work etc. Act 1974 (see **Table 3.2**). The inspection of zoos for compliance with health and safety regulations occurs in addition to the zoo licensing inspection. (See **Chapter 3** for more on zoo licensing.)

There are three main sections to MHSZ: safety issues, health issues, and emergency procedure. Much of the document refers to issues that are common to all businesses (for example, 'slip and trip' hazards, noise at work, and emergency procedures if there is a fire), or are common to work with animals (for example, safety issues related to veterinary procedures, safe animal handling, and public contact with animals), but it also includes some areas that are unique to zoos (safety in large drive-through paddocks and animal escape procedures). For example, this document outlines some recommended precautions that should be taken to reduce the risk of animal escapes:

- nominating a person and deputy to take charge of the situation and make any important decisions;

- raising the alarm and reporting incidents to appropriate personnel as quickly as possible;

- communications with entrances/exits and allocating responsibilities for closure where necessary;

- arrangements for the evacuation or safe confinement of people in the zoo, ensuring that those situated away from buildings receive appropriate assistance as quickly as possible;

- managing crowds safely in an emergency situation and the giving of directions;

- a strategy for recapture appropriate to the various types of animal kept;

- liaison arrangements with senior zoo personnel, vets etc for the recapture plan, which should include the use of radios, equipment, vehicles, firearms;

- identifying essential employees etc;

- briefing staff as to their roles and responsibilities during a recapture operation including the recapture of animals escaped beyond the perimeter of the zoo;

- arrangements to locate the escaped animal;

- arrangements to keep the animal under observation while recapture plans are being formulated, and the movement of key personnel to the area once the escaped animal has been located;

- the provision and location of the necessary capture equipment, eg nets and firearms/darting equipment. Torches will be invaluable for night escapes and should be located in a designated area;

- alerting external emergency services eg police, where necessary;

- stand-down arrangements on the completion of the recapture operation, including notifying all relevant personnel and external organisations involved.

(HSE, 2006)

Like all businesses, zoos should have written health and safety policies. They should also assess the risk to staff, volunteers, and visitors to the zoos that may be incurred by the activities of the zoo in general; this forms the basis of the **risk assessment** (HSE, 2006). All identified risks should be considered and appropriate measures put in place to ensure that these risks are minimized to a safe level. It is also noted that staff should be capable, through having had sufficient experience and/or training, to carry out the work that is expected of them.

Health and safety issues are also dealt with in the SSSMZP (Defra, 2004): Standard 11 is all about visitor health and safety at the zoo.

6.1.5 Incorporating all needs into housing and husbandry

Good zoo housing and husbandry ensures that the needs of all stakeholders are considered, and that, where there is conflict, resolution is sought and agreement achieved on a final decision. A multi-disciplinary team is important for this because:

1. many zoo mission statements imply values and goals that affect a variety of people (see section **2.5.1**);

2. it will ensure that the diverse skills and views of all zoo staff are considered, which will subsequently result in fewer oversights and more people will feel involved, responsible, and thus supportive of the development, across all zoo departments;

3. there are many benefits associated with working in a multidisciplinary team, not least learning what colleagues do and their point of view;

4. innovative ideas usually occur during discussions of professionals with different backgrounds and viewpoints (Coe, 1999).

Nevertheless, some conflicts of interest do occur. The following are some examples of the sorts of situations in which such conflicts might happen:

1. visitors want to be able to view the animals easily, whereas the animals want to be able to hide;

2. visitors want animals to have large, complex enclosures, whereas keepers need to be able to service the enclosures;

3. plants provide a good-looking enclosure, but the animals may want to eat them.

How might we resolve conflicts of interest such as these through good housing and husbandry? In relation to the above, possible resolutions are:

1. visitor viewing can be restricted through various methods, including camouflage netting, which still affords visitors a view of the animals, but allows the animals to feel relatively hidden (Blaney and Wells, 2004);

2. when paddocks or inside enclosure areas are large enough, they can, in a sense, become self-cleaning, because voided faeces represent such a low presence that they rot away into soil or deep litter bedding (see, for example, Chamove et al., 1982);

3. plant species can be selected that are non-toxic, but unpalatable, or which protect themselves with thorns and which animals tend not to eat (Wehnelt et al., 2006).

These and other examples will be expanded upon in section **6.2**.

This leaves only the problem of what happens when a compromise does not present itself, leaving the conflict difficult to resolve. In this case, it seems

---

Risk assessment is an essential part of business and industrial procedures, and involves identifying and grading potential hazards, and the risks they pose, and identifying procedures for minimizing those risks.

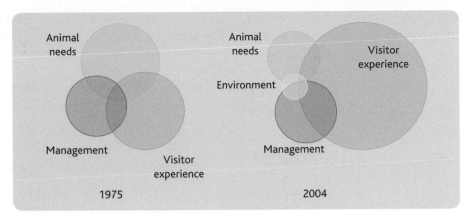

Figure **6.7** There are three main stakeholders whose needs should be borne in mind when designing animal exhibits: the zoo animals themselves; the zoo professionals (notably the keepers); and the visiting public. Melfi *et al.* (2007) suggested that the emphasis placed upon each of these stakeholders has changed over time, which was portrayed by the approximate proportion of content focusing on various elements of design, gathered from the *Proceedings of the First Zoo Design Symposium* (1975) and estimated from abstracts of the *Sixth Zoo Design Symposium* (2004). Notably, emphasis of the animal's needs has reduced, while attention to the visitor experience has risen. It is heartening, however, to see that the inclusion of the fourth stakeholder, the environment, seems to be emerging. (Picture: Whitley Wildlife Conservation Trust)

intuitively right that an emphasis on the animals' needs should be prioritized, because the zoo's primary goal is to conserve species and meet their welfare needs. In recent years, however, there has been a dramatic change in the balance from prioritizing animals' needs to meeting the needs of visitors (**Fig. 6.7**). But does it cause problems if the needs of the zoo visitors shape enclosure design and, ultimately, housing and husbandry in general? Visitors consistently want to see naturalistic enclosures and active animals (playing, climbing, and eating), factors that they also believe represent good animal welfare (Melfi *et al.*, 2004a). The aesthetics of an enclosure, the level of activity displayed, and the appearance of an animal, not whether it has enrichment in its enclosure, are used by visitors as indicators of animal well-being (McPhee *et al.*, 1998).

We should, however, remember that zoo visitors are not trained to understand the complexities of assessing animal welfare. It is also unclear whether they genuinely consider, from an objective standpoint, what housing and husbandry factors promote good welfare, or whether their opinions are shaped to fit what they would most like to see. So shaping zoo enclosures purely to meet the needs of visitors is

potentially perilous for the animals' needs, but also undermines the needs that have been identified of the other parties affected by zoo animal housing and husbandry.

The bottom line in the UK, and for many other zoos worldwide, is that we rely on the admission fees paid by visitors to ensure the zoo's future and thus its activities. Because visitors know what they like, what they want, and what they expect from a zoo, it is up to zoo professionals to ensure that zoo visitors are educated about these issues so that they can make informed judgements about animal welfare, zoo enclosure design, and **exotic animal** behaviour. If this can be accomplished, visitor needs will include many factors that enhance animal welfare, and this will provide overlap between animal, keeper, and visitor needs.

6.2 Advances in enclosure design

Obvious changes have occurred in zoo enclosure design over the past few decades. These can be linked to changes in the way in which we perceive and treat animals within our society (see section **2.3**). They also represent a greater influence of the visitors' needs being incorporated into the design process (see section **6.1.5**). Our knowledge of animals' needs is also greater today than it has ever been, so we are more able to accommodate these within enclosure design—although we should not be complacent about this, because there is always more to learn.

6.2.1 The evolution of enclosure function

This section is not intended to be an exhaustive overview of the methods used to describe and design enclosures, but to provide an overview of those terms that are most commonly used and a balanced perspective from which to assess their merits.

Various terms are used to describe zoo enclosures: hard or soft (of architecture); first, second, or third generation; barren or complex. All of these terms exist on a sliding scale. 'Barren' enclosures, for example, are considered to be those with few, if any, objects or furnishings, but as objects and furnishings within the enclosure increase, so does its 'complexity'. Enclosure complexity, as a concept, is thought to provide the appropriate physical and psychological **stimuli** necessary to provide captive animals with the opportunity to express 'wild' behaviours (Hutchings *et al.*, 1978; Odgen *et al.*, 1993).

Naturalistic enclosures

It is clear that hard, barren, first-generation enclosures have been making way for soft, complex, second- and third-generation zoo enclosures. This evolution in enclosure design has followed an intuitive path towards naturalistic enclosures—that is, aesthetically pleasing enclosures that follow the simple idiom that 'green is good'. Giving the impression of a greener enclosure requires additional planting, but also the removal of bars and other harsh-looking structures within the enclosure. Modifications can be made to 'naturalize' hard enclosures, by adding complexity or natural materials to the enclosure (Coe, 1989).

Polakowski (1987) noted that naturalistic enclosures can be designed and constructed in three different ways, which affect their functional utility

for the animals and thus change the way in which they meet the animals' needs. Naturalistic enclosures could be considered:

1. realistic, reproducing the animals' wild habitat, including land formation and plant life;

2. modified, simulating the animals' wild habitat by using available materials to substitute for the real thing, for example, by using different vegetation and land formations present at the zoo;

3. naturalistic, in which no attempt is made to duplicate the animals' wild habitat and natural materials are used in a stylistic way.

If we take this concept a little further, we arrive at **immersion exhibits**.

Some 30 years ago, Jones *et al.* (1976) coined the term **landscape immersion** to describe a situation in which animals and visitors would share the same setting, but not the same space. The goal of immersion was to continue landscaping beyond the animal exhibit into the visitor area, with the aid of invisible barriers, to provide visitors with the sensation that they were 'in' the animal's habitat (Coe, 1994). When exhibiting zebra, for example, '*the entire setting* [should] *look, smell, and feel as if* [the visitor has] *left the zoo and entered the African savannah*' (Coe, 1985). Immersion exhibits (**Fig. 6.8**) are associated with additional theming, to ensure that educational signage, merchandising, and retail amenities do not detract from the visitor experience of, for example, 'Africa'.

Human–animal contact

The degree to which visitors and animals come into contact with one another also varies between enclosures. All of the enclosures considered so far ensure that animals and visitors are separated by barriers; this is a good idea for many reasons, not least of which is the health and safety of both animals and visitors. For some species, this division has been allowed to lapse and **walk-through exhibits**, which have been in existence since the early 1920s (Olney,

Figure **6.8** One technique employed to increase the visitor experience is to reduce the gap between them and the captive animals that they have to come to see. This can be accomplished through the use of different barrier types (see **Fig. 6.12**), but also by making objects accessible to visitors and animals. Here is a log (a) at the Toronto Zoo and a safari jeep (b and c) in Werribee Open Range Zoo, Victoria, both of which bisect the animal–visitor barrier. (Photographs: Vicky Melfi)

(c)

Figure **6.8** (continued)

1975), have been developed. Recently, there has been a proliferation in the number and the diversity of species that are presented in this format (Kreger and Mench, 1995).

Walk-through exhibits—and their aquatic, swim-through, cousins—still keep animals contained within a fixed boundary, but allow visitors to enter the exhibit. Unlike petting zoos, containing domestic animals, and some **touch stations** in aquariums, walk-through exhibits do not encourage 'actual' contact between visitors and animals. The separation between zoo animals and visitors is completely removed when animals are managed as free-ranging species: these animals are no longer restricted by exhibit perimeters and are more or less allowed to roam wherever they please—although the zoo does not want them to leave the zoo grounds, so appropriate barriers ensure they do not roam too far (see section **6.3**).

There are obviously both advantages and dis-advantages associated with these different kinds of exhibit, and these can best be understood by going back and thinking about the needs of the different stakeholders (which we explored in section **6.1**). There is substantial empirical evidence that demon-strates that complex enclosures meet the needs of animals far more effectively than **barren enclosures** (see **Box 6.1** and also **Chapter 8** on 'Environmental enrichment'). For example, after a group of mandrills *Mandrillus sphinx* was moved from an 'old traditional' enclosure to a 'new natural' exhibit, which was more complex, they were observed to display less stereo-typical behaviour and their activity levels rose (Chang *et al.*, 1999; see section **4.5.3** for more about this study). These changes in behaviour were considered positive, because they brought the captive mandrills' behaviour closer to that of wild baboons.

Box **6.1** Space versus complexity

Zoo visitors sometimes remark that enclosures should be bigger, but how valuable is enclosure size and how big does an enclosure need to be? Enclosure size can compromise animal welfare if it restricts behaviour, or leads to **environmental challenge** or crowding (see **Chapter 7**). It is difficult to decide empirically how big an enclosure needs to be for any one animal and thus how much space should be allotted to a group. There is some evidence that an animal's home range or territory in the wild could predict how restricted it may be in captivity (Clubb and Mason, 2007). This is supported by the generally held assumption that as enclosures become proportionally smaller, they compromise animal welfare more.

This type of analysis is, however, difficult to test empirically and does not necessarily provide a guide to how big zoo enclosures should be. A more pragmatic approach is to ensure that enough space is provided to enable the expression of wild behaviours and, more importantly, those behaviours required to limit stress, such as distancing and comforting behaviours (Hediger, 1955; Berkson et al., 1963). Certainly, an enclosure should be bigger than an animal's **flight distance** to ensure that the enclosure perimeter, and thus visitors or neighbouring animals, do not infringe upon it. Within the laboratory community, there has been a lot of research into optimal enclosure size (for example, Line et al., 1989a; 1989b; 1990; 1991). Some of these studies have highlighted that additional space has not necessarily improved welfare (Woolverton et al., 1989; Crockett et al., 1993a; 1993b; 1994). This suggests that cage size is unimportant, because more or less barren space was of no consequence to captive primates. The contents of the space (that is, its complexity), however,

were important (Reinhardt et al., 1996). This conclusion is supported by many studies that demonstrate that modifications to enclosures (whether zoo, laboratory, or farm) to increase complexity, are associated with many welfare improvements, including reduced **stereotypies** (Whitney and Wickings, 1987; Chamove, 1989; see **Chapter 8** on 'Environmental enrichment').

Space alone, therefore, may only affect captive animal welfare if it falls below a certain threshold; above that threshold, enclosure complexity may contribute more to welfare and behaviour. Indeed, there are a number of studies that show that zoo-housed animals do not fully utilize the space provided to them, but display distinct preferences for certain areas and spent much of their time in those areas (for example, Ogden et al., 1993). In a study by Wilson (1982), enclosure complexity was identified as more important in the determination of gorilla *Gorilla gorilla* and orang-utan *Pongo pygmaeus* behaviour than enclosure space. When Perkins (1992) replicated this study, however, enclosure space and complexity were considered highly interrelated and, therefore, considered to affect behaviour together. This change probably represents a dramatic change in enclosure design, as part of which all enclosures are now much more complex; in older exhibits, enclosure complexity and size were more uniform.

There is a general acceptance that complex enclosures promote good welfare, which has led to the incorporation of complexity into enclosure design, and appears to have resulted in zoo enclosure size and complexity increasing in parallel (Perkins, 1992). To conclude, therefore, zoo housing is considered 'good' if it is complex, dynamic, and large enough to hold appropriate social groups (Mallinson, 1995; Newberry, 1995).

The impact of naturalistic and immersion exhibits on the different stakeholders is a little less clear-cut. In terms of the animals' needs, naturalistic enclosures will only be beneficial if they functionally contribute directly or indirectly to the animals' lives. It probably matters little to the animal whether the additional greenery or structures are just like those of their native habitat or are simulations; what is likely to be more important to them is whether the vegetation provides much-needed shade, or climbing opportunities, or supplementary food. It is always worth considering whether naturalizing the exhibit has increased or decreased enclosure complexity. In terms of the visitors' needs, however, at least one study has shown that the aesthetics of the enclosure —and, principally, how much green there is—greatly affected whether visitors liked the enclosure and also appeared to affect their views about the welfare of the animals (**Fig. 6.9**).

Sometimes, naturalistic and immersion exhibits are considered restrictive by keepers. Trying to ensure that enclosures and fittings are consistent with a naturalistic theme can sometimes go against the interests of the animals, because non-natural fittings and enrichments are abundant and often cheaper than their natural-looking alternatives (see **Chapter 8**). Looking at some of the newer zoo enclosures, it is possible to detect what Beardsworth and Bryman (2001) describe as the 'Disneyfication' of zoos, by which they mean essentially that an emphasis has been placed on the theming of exhibits. There are many advantages associated with this trend: notably, the enhancement of the visitor experience. But whether the expense of some themed exhibits can be justified in terms of achieving the core goals of zoos (conservation, education, and research) is an issue that will surely be debated for some time to come.

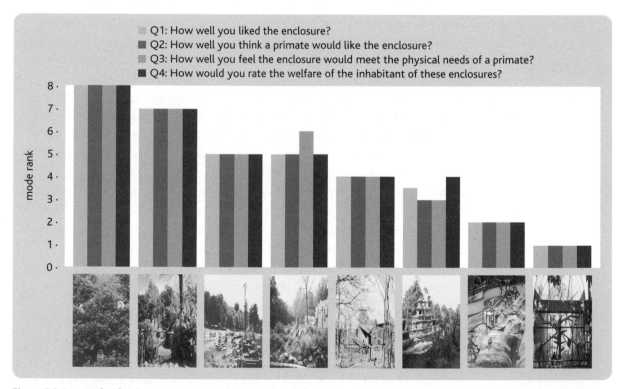

Figure 6.9 In a study of visitor perceptions at Paignton Zoo, UK, visitors were shown several pictures of enclosures and asked to rate them according to a series of questions. As you can see from the graph, visitors preferred enclosures that were more naturalistic, and considered that these enclosures would best meet the needs of the animals and thus promote better welfare in them. (Photographs: Vicky Melfi)

Jon Coe, who has been particularly influential in the design of zoo enclosures and especially the integration of stakeholders into the design process, has also pioneered an integrated approach towards housing and husbandry (Coe, 1987; 2006). In an approach that is termed 'activity-based design', Coe (1997) suggested that the emphasis of the zoo enclosure should be to provide animals with the opportunities to express 'natural' behaviours, and keepers the access to provide enrichment and/or husbandry training, both of which would enhance the visitor experience and increase possibilities for learning. Interestingly, he believed that all of this could be achieved because '*whether* [enclosures] *were simulations of naturally or culturally derived habitats or pure functional facilities, these environments are abundantly provided with appropriate behavioural opportunities for the animals, keepers and zoo visitors*' (Coe, 1997). Getting agreement among all stakeholders is not always easy, however, as illustrated in **Fig. 6.10**.

Figure **6.10** Views of curators and keepers might not always agree. '*We've spent half a million dollars making this zoo habitat exhibit look like a natural jungle to present a message about wild animals. You've stuffed it up with cheap playground equipment!*' exclaimed Mr Hafter B. Green, of the Exhibits department.

'*Gorillas need more than bushes,*' responds Ms Pollyanna Pett, animal keeper. '*These red boomer balls and colourful climbing cables are easily available. Besides, zoos are for the animals first . . . without the animals we wouldn't have a zoo!*'

'*Wrong! Zoos are for people first,*' retorts Mr Green. '*Without visitors, we wouldn't have a zoo.*' (From Coe, 2006) (Picture: Phil Knowling)

6.2.2 Basic housing

There are certain basic requirements that should be met by all animal housing, whether for exotic or domestic species. For example, conditions should be provided to ensure that the five freedoms are met (see **Chapter 7**). Standard 2 and Appendix 8 of the SSSMZP (Defra, 2004) both detail the minimum requirements that zoos should aim to achieve when housing exotic species; standard housing requirements are also thoroughly dealt with in other core texts and guidelines (for example, for domesticated species, Wathes and Charles, 1994; for laboratory-housed animals, Poole, 1999).

In the following sections, we will consider how even basic housing requirements, such as climate control, the provision of water, and ensuring the safety of animals, need to include the exotic animals' species-specific needs. The issue of **stocking density** and how much space should be made available to

animals is covered in section **6.3.3**, and the relative importance of enclosure size versus enclosure complexity is discussed in **Box 6.1**. Some scientific research has been undertaken to study the impact of housing and husbandry practices on zoo animal behaviour and biology (see section **6.4**; see also Kleiman *et al.*, 1996). Unfortunately, however, due to the diversity of species held in zoos, much of our knowledge of the impact of housing and husbandry regimes and its impact on animals comes from anecdotal information and the experience of keepers, although more objective sources of information should always be sought (for example, the SSSMZP; see section **6.5**).

6.2.3 Climate control

The provision of a suitable climate is essential for maintaining good animal welfare and, in many species, can also affect breeding success. Furthermore,

Stocking density is the term used to describe the number of animals in a given area: as the number of animals goes up, so does the stocking density, when the space available remains constant.

appropriate climate control is central to the management of disease. For example, Exner and Unshelm (1997) demonstrated that variations in airborne micro-organism concentrations, recorded over a year in exotic felid enclosures, were affected by climatic conditions.

An animal's preference for resources within its enclosure will vary over the day, or season, so it is good practice to provide animals with variations in climate and other factors within their enclosure (Defra, 2004).

The elements that make up an animal's climate are light, temperature, **relative humidity**, and ventilation, and these affect one another.

Light

Light can be thought of in terms of its strength (intensity) and type (wavelength). Exposure to different wavelengths of light can be important for animals. For example, providing adequate sources of **ultraviolet (UV) light** is necessary for the health and behaviour of several species.

Some mammals (a few marsupials and rodents), birds, reptiles, and insects can actually see **UVA** (400–100nm) wavelengths, which enables them to see markings or patterns on fruits, flowers, and seeds, or within bird plumage and in urine, all of which are invisible to humans (Winter et al., 2003). Thus, UV vision provides an additional channel of information for those species that are adapted to see it and they use it for functions as varied as assessing the ripeness of food or choosing a potential mate (see, for example, Smith et al., 2002; Shi and Yokoyama, 2003).

For example, the UV-reflectance level of the throats of Augrabies flat lizards *Platysaurus broadleyi* has been shown to provide a signal between competing males, because those with the richest UV throats are the best fighters (Whiting et al., 2006). In addition, UVB (320–280nm) is associated with the synthesis of vitamin D, and affects health and behaviour, including calcium metabolism and reproduction (for example, Cole and Townsend, 1977; Regal, 1980; see also **Chapter 12** on 'Feeding and nutrition').

Unfortunately, most windows do not allow the transmission of the full spectrum of light and, as a result, UV levels are usually insufficient if animals do not have access to natural light in outside enclosures (see **Fig. 6.11**). In these cases, animals either need vitamin D supplements in their food or artificial sources of UV, such as lighting (Gehrmann et al., 1991; Bernard et al., 1989). There are many commercially available UV lights, but they should be checked regularly, because the level of UV they provide may be variable. The extent to which animals require UV is unknown and Moyle (1989) cautions that its provision should be restricted. Reptiles from open habitats, for example, may only need 30–45 minutes of exposure daily.

Light is important not only because of its physical properties, but also because its distribution through time (photoperiod) can affect many features of animals' lives. For the majority of species, it is best practice to ensure that the photoperiod incorporates the daily and seasonal variations for which the animal is naturally adapted. For nocturnal species, this can be accomplished by keeping them indoors on a **reversed lighting schedule**, which ensures that when it is light outside, it is dark inside.

Changes in day length can trigger the start of the breeding season, parturition, migration, and many other behavioural changes in animals (see section **9.1.4** for more on triggers for breeding). When animals have adapted to the southern hemisphere and are housed in captivity in the northern

Relative humidity is the result of temperature and the moisture content of the air. Varying either factor or the rate of ventilation will alter the relative humidity of an area.

Ultraviolet (UV) light is light of very short wavelength, shorter than violet light, which is the shortest wavelength of light that the human eye is capable of seeing. The wavelength of UV light ranges from 400 **nanometres (nm)** down to 10nm. Most of the UV radiation that reaches the Earth's surface from the Sun is of wavelength 400–315nm, and is referred to as **UVA**.

Figure 6.11 The provision of light to inside enclosures can be difficult to achieve, because many materials prevent light penetration, especially UV wavelengths of light. In this elephant exhibit at Duisburg Zoo, great efforts were made to increase the light levels and a custom-made polymer fitted to allow UV light penetration. (Photographs: Vicky Melfi)

hemisphere, behavioural changes associated with season will be reversed (and vice versa). Carefully managed lighting systems can negate **circadian** and seasonal discrepancies in daylight, although many animals appear to be comfortable adapting to the lighting conditions of their captive environment. There are many other factors that may vary in a period of 24 hours that affect, or contribute to changes in, observed behaviour and physiology. Many hormone profiles, for example, follow a daily

pattern, such as cortisol, which is naturally higher in the mornings compared with the evenings.

Temperature

Extremes in temperature can have deleterious impacts on animals, ranging from frostbite to heat-stroke, and may also be capable of affecting welfare in more subtle ways. Rees (2004), for example, suggested that low temperatures may influence the performance of stereotypies in Asian elephants, because he found that temperature and the rate of stereotypies were negatively correlated (average temperatures measured were 9°C and 23.2°C on coldest and warmest days, respectively). But for the majority of animals, well-constructed enclosures that offer protection from extreme temperatures can negate this challenge. This protection can be achieved indirectly with the provision of shelters, but also through climate control. Heating large areas, and water, requires specialized heating systems, the environmental impact of which should be carefully considered (see section **6.1.4**), but temperature gradients can also be achieved through localized heating or cooling, or by altering ventilation. Localized heating can be provided with heat lamps or hot rocks, and temperatures can be kept low within inside enclosure areas by refrigerating them.

Some species require hot climatic conditions: for example, the SSSMZP (Defra, 2004) recommend that most reptiles are kept at 20–35°C, and tortoises and crocodilians at 26–32°C. At these sorts of temperatures, extra care needs to be taken with hygiene, because many **pathogens** proliferate under these warm conditions.

Relative humidity and ventilation

Relative humidity—that is, the result of temperature and the moisture content of the air—is a very important part of the environment for many animals. For example, African spurred tortoises *Geochelone*

Circadian is the term used to describe behavioural and physiological patterns that reoccur within a 24-hour period, giving the appearance of a daily cycle.

sulcata develop significantly taller carapace humps (considered a **pathological** condition) when they are maintained in low levels of relative humidity (25.7–57.8 per cent and 30.6–74.8 per cent) compared with higher levels of humidity (45–99 per cent) (Wiesner and Iben, 2003).

Few studies have investigated the impact of humidity on zoo animals, so broad recommendations are suggested by the SSSMZP (Defra, 2004) for providing humidity gradients for animals such as reptiles (50–80 per cent) and amphibians (65–95 per cent). Humidity gradients are important for mammals as well. Anecdotally, a lack of humidity has been observed to result in skin problems, whereas high levels of humidity are associated with increased disease transmission.

To maintain humidity levels, misting the enclosure with water is customary. Again, it is suggested that ventilation be used to create this gradient, so that temperature and the water content of the air can remain fixed.

6.2.4 Safety

Both animals and people need to be kept safe in the zoo. Keeping animals safe is principally achieved through good enclosure design, whether this is to keep animals away from other animals or from people, or vice versa. This involves the careful planning of enclosure perimeters and accessibility to people: both keepers and the public. Because enclosure size varies with species, so will the materials or techniques used for their perimeters or access points; barriers are explored more fully in the next section.

The level of security required is dependent on the animals' abilities (do they present a hazard, or are they thought to be dangerous?) and on the behaviour of the zoo visitors (do they respect the animals space, or will they ignore the signs that advise against certain types of behaviour?). Appendix 12 of the SSSMZP (Defra, 2004) lists all species that are

considered by law to be hazardous and, for some of these, the SSSMZP make specific recommendations. It recommends, for example, that venomous snakes are housed in solid walled or roofed enclosures to prevent their escape, and to stop staff and visitors from reaching them. For many other species, there is a variety of methods that can be used to ensure that the animals can be exhibited safely and that people's safety also remains high.

With all animal enclosures, it is sensible to consider worst-case scenarios, and to build into housing and husbandry appropriate safety precautions to limit these. All enclosures housing dangerous animals should have warning signs, whether on display to visitors or behind the scenes, and all should be individually locked. The staff service area should be free of hiding places, and protocols and staff training for shifting animals between enclosure areas is necessary. Other sensible precautions that can be taken to reduce the likelihood of escape are **double doors** and appropriate enclosure perimeters.

6.2.5 Barriers

Barriers and/or perimeter fencing are not constructed merely to keep animals from getting out of the enclosure, but also to stop other animals and people from getting in. Although this may seem counter-intuitive, consider, for example, waterfowl, which may be at risk of predation at night from foxes and other animals: fencing, in this instance, has to stop animals from getting in. Similarly, the actions of some visitors, such as providing animals with food or taunting them, could potentially harm the animals, and thus the visitors and the animals need to be kept separate.

Stand-off barriers are a simple, but effective, method of increasing the proximity between animals and visitors, the premise of which is to increase the distance between the animal's enclosure perimeter and the zoo visitor. This separation is also necessary if electric fencing is used, so that visitors are not

---

The term **double doors** is commonly used to describe a situation in which two doors, with a vestibule between them, separate animals from the 'outside world'.

harmed. A variety of materials can be used as stand-off barriers, including vegetation and hedges, or other fencing materials, such as metal posts.

Enclosure perimeters can be made from various materials, including horizontal and vertical posts, wire or mesh, electric fencing (or strands), moats (whether dry or filled with water), windows (glass, acrylic, or a combination), or rocks. As with all enclosure design, there are considerations that will determine the material and type of fencing used, including whether:

1. visibility between the animals and their neighbours (animals and/or visitors) is beneficial, and therefore whether solid or partial barriers are used;

2. the animals are able to burrow or climb, which will affect how far above and below ground the perimeter fencing must go, and whether a roof or some kind of deterrent to climbing the fence (for example, electrics) is needed;

3. the barrier is capable of preventing offspring from leaving the exhibit, because these are often a lot smaller than their parents.

Table 6.1 provides an overview of some of the most commonly used barriers, and their associated advantages and disadvantages, from the perspective of the animal, keeper, and zoo visitor.

The use of electric fencing is available in many guises, from single-strand electric tape, to imitation grass and ivy, and electric netting and floor matting (see **Fig. 6.12**). Electric fencing can be used as a perimeter barrier, and also to prevent animals from gaining access to various other resources in their enclosure, such as high trees that they may damage, or which provide them with a vantage point from which to escape. All electric fencing should be regularly checked, should have an alternative power supply (battery) back-up, and should be attached to an alarm system that would be activated if the electric current were to be disrupted or disconnected. It may seem obvious to mention, but electric fencing works because animals get an electric shock when they touch it. Usually, this is because the animals are on the ground and touch a 'live' wire. Some animals, however, may be able to leap onto an electric fence and thus will not be 'earthed', so it is essential, in these situations, that the fencing provides an earth: for example, by providing alternate live and earth strands.

Electric fencing can be very effective at preventing predators (interloper species) from entering enclosures and is recommended by the SSSMZP (Defra, 2004), as long as adequate consideration of zoo visitor safety is adhered to (for example, HSE, 2006). Electric fencing is most effective when animals are trained to recognize it and thus avoid going near it: an electric shock acts as a punishment, which reduces the likelihood that a behaviour will be repeated (see section **4.1.2** for a discussion of such **operant conditioning**). Different species may react differently to electric shocks: for example, **Fig. 6.13** shows how three different rhinoceros species were observed to react to getting an electric shock and thus illustrates, even if anecdotally, how the same barrier method may not be suitable in all situations for all individuals or species.

Many perimeter fencing methods are not used on their own, but in combination with each other. For example, electric fencing is frequently used with mesh-style fences. When the electric fence is positioned at the top, bottom, or all over the mesh, it can prevent animals from escaping over the top of a fence, prevent animals that cannot leap from scaling the fence or from knocking it down (for example, red pandas, binturongs, and equids, respectively), or prevent animals that can leap from making contact with the fence at all (for example, primates).

Electric fencing is also used in conjunction with moats, both dry—that is, a **ha-ha** (see **Fig. 6.14**)—and

A **ha-ha** is a ditch that separates areas of land almost invisibly. Variations on the basic design of the ha-ha will affect the gradient of the slopes used and whether an additional barrier is 'hidden' in the ha-ha.

Table 6.1 The advantages and disadvantages of some of the more commonly used barriers in zoo enclosure design

Type	Advantage			Disadvantage		
	Animal	Keeper	Visitor	Animal	Keeper	Visitor
Solid*	Depending on height, provides safety Prevents disease transmission	Separates animals Prevents visitors from feeding animals	–	Can lead to injury if animals collide with it Can prevent view of surroundings May affect communication between animals	Can prevent view of the animal	Obstructs view of animals
Partial**	Can provide greater useable space	Can aid introductions	Restricted viewing can make a glimpse of an animal more exciting	–	–	Can obstruct view, although new materials are less obvious (e.g. zoo mesh) Considered 'unnatural' Does not prevent human–animal interaction
Bars	As above	Can facilitate keepers' escape from the enclosure, whether the bars are horizontal or vertical	–	–	–	Associated with negative connotations of animal welfare As above
Netting and mesh***	As above	Can accommodate any shape	Can be plastic-coated or painted to reduce visibility	Could become tangled	Metal can rust even under plastic coating UPVC can become brittle in the sun after long exposure Can be eaten through by pest species, e.g. squirrels	–
Electric	Can learn to avoid it	Easily create temporary barriers Cheap	Good visibility	Not visible so can get injured Can get entangled in it A deterrent, not 'fool-proof'	Some body parts do not conduct electricity, e.g. antlers, horns, hair	Needs to be well signed as a hazard and out of public reach
Glass	Prevents disease transmission Laminated glass provides climate control Double glazing provides noise reduction	Using acrylic and glass is very strong	Provides a good view	Close proximity between animals and visitors	Needs regular washing Expensive	Visibility can be reduced, if the glass is scratched, if it is sunny (reflection), or if there is condensation
Moat	Some animals may use the water or objects in it, e.g. eat some of the plants growing it in or wade in	Wet moats can be used as habitats for other species (e.g. plants, fish)	Provides 'naturalistic' view Invisible barrier between species	Animals can fall into the moat and get trapped Water can provide a route for disease transmission A lot of space is required, which cannot be used by the animals in most situations	Dry moats can flood Wet moats can freeze Need method to access animals and/or enclosure safely	Increases the distance between the visitors and animals, which can reduce visibility

* Materials include: wooden fence panels; brick walls; glass or acrylic; hedges, etc.   ** Materials include: metal posts; electric strands; chain link; weld mesh; netting, etc.

*** Materials include: metal; nylon; combinations of the two.

NOTE: A huge variety of barriers are incorporated into zoo animal enclosure design, which fulfil many functions, from preventing animals to escape to restricting visitor access. The materials used will affect both their utility and their appearance.

(a)

(b)

(c)

Figure **6.12** The barrier types employed in zoos are diverse, as are the materials used to construct them. Some of the more common types of barrier include: vertical fencing (a); vertical wire (b); chain-link (c); zoo mesh (d); rock work (e); glass (f); and water, in the form of moats (g). The safety of animals should always be considered when choosing between barriers—i.e. the likelihood of drowning in water moats could be prevented by the provision of gently sloping slides entering the water or ropes (h), which primates can use to prevent them slipping into the water. (Photographs: (a) Achim Johannes, NaturZoo, Rheine; (b–d, f, and g) Vicky Melfi; (e) Leszek Solski, Wroclaw Zoo)

194

(d)

(e)

(f)

(g)

Figure **6.12** (continued)

(h)

Figure **6.12** (continued)

wet. For example, the illusion of a mixed-species paddock can be obtained by using dry moats with electric fencing running at the bottom of them, providing the visitor with a vista of different species sharing the same area. The positioning of the electric fence can determine its function, as illustrated in **Fig. 6.15**.

As with all housing and husbandry features, trade-offs are made between different barrier methods and the needs of all of the stakeholders (the animals, keepers, and visitors). The emphasis on naturalistic enclosures, which was discussed in section **6.2.1**, is particularly evident in the choice of perimeter fencing chosen. It is clear from any casual walk around new zoo enclosures that the most commonly used perimeter fencing types are those that are aesthetically pleasing or invisible to the visitors, such as glass/acrylic, moats, and mesh.

6.2.6 Water

All enclosures, whether aquatic or not, require water (**Fig. 6.16**). Essentially, water is used by keepers to clean enclosures, for animals to drink, and may also function as a swimming pool, or for use as a barrier or aesthetic feature within the enclosure. As might be expected, the care taken on water provision and

(a)   (b)   (c)

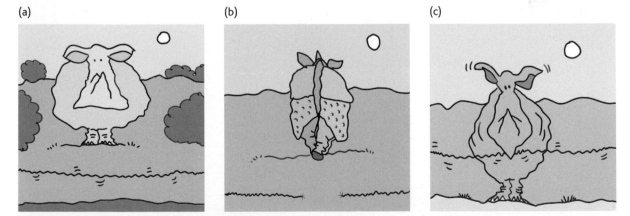

Figure **6.13** The result of coming into contact with electric fencing, having an electric shock, has been observed to lead to various reactions in different species of rhinoceros, according to Julian Chapman of Paignton Zoo Environmental Park. As illustrated in the cartoon above: (a) a white rhinoceros was seen to react as we might expect all animals to and retreat from the fence after receiving an electric shock; (b) an Asian rhinoceros, however, charged straight through the electric fence, regardless of receiving an electric shock; and (c) a black rhinoceros, unfortunately, seemed to be unable to move after being shocked and remained stationary in contact with the electric fence. (Pictures: Phil Knowling)

(a)

(b)

(a)

(b)

(c)

Figure **6.15** Electric fencing can be used to prevent animals from: (a) escaping—at Apenheul Primate Park in the Netherlands, these discreet 'hot wires' are used around the rocky ledges of the perimeter of the barbary macaque *Macaca sylvanus* endosure; (b) using parts of their enclosure—at Disney's Animal Kingdom electric fencing has been designed to look like long grass; and (c) destroying plants within their enclosure. (Photographs: (a) Vicky Melfi; (b and c) Julian Chapman)

Figure **6.14** (*opposite*) Combinations of barrier types are frequently used together; this is especially true when using a ha-ha. In these two examples, the gradient and species housed in the enclosures mean that the animals are able to 'use' the ha-ha area. Escape of the animals is thus prevented by the steep gradient, on the visitors' side, along with (a) a standard vertical fence and (b) a very steep concrete wall. (Photographs: Julian Chapman)

Figure **6.16** Water is an integral part of all enclosure design; this is especially true for aquatic species, for which water makes up their entire environment. (Photograph: Vicky Melfi)

management within aquariums is far greater than when supplying water for many terrestrial species; however, the need for clean and safe provision is paramount to both.

The SSSMZP (Defra, 2004) highlight that some species have specialist needs when it comes to water provision. It recommends, for example, that the total amount of water provided to reptiles and amphibians should allow their full immersion. It also points out that the behaviour of some animals can be dependent on water provision and the degree to which this is true should influence the amount of water provided. A gentle slope should always be provided to ensure that animals can enter and leave water without difficulty; indeed, predominantly aquatic species should be provided with sufficient water to allow them to swim comfortably, but beaching areas should also be provided where necessary.

Water quality is important and needs to be managed differently depending on whether it is an **open system** or **closed system**. With open water systems, it is necessary to monitor the water entering and leaving the enclosure, to ensure that concentrations of chemicals are not toxic or are at non-irritant levels; this is undertaken to secure the safety of

An **open system** describes a set-up within which water comes into the enclosure and leaves it, whereas a **closed** system recirculates the same water around the enclosure continuously.

animals and also people who may be affected by water leaving the enclosure. The precise checks undertaken depend on the species being exposed to the incoming water, but are likely to include salinity, pH levels, and chlorine; a greater array of tests is necessary when dealing with fully aquatic species. The SSSMZP (Defra, 2004) advise that aquarists stringently and regularly monitor water, and have on-site laboratory facilities to undertake this. Water temperature should also be checked for some species, because they may react badly to sudden changes in water temperature.

Water is a very effective mode of disease transmission, so care needs to be taken to ensure that pathogens are kept at low levels and are not allowed to proliferate. In open systems, this is achieved by the removal of 'dirty' water and its replacement with 'clean water'. In closed systems, however, a variety of filters are used to remove any undesirable **organic** matter and noxious waste, and thus ensure that the water quality remains high. Water quality can be especially compromised when housing marine species, because they produce lots of nitrogenous waste, which can react with other chemicals in the water to produce noxious by-products, and these can increase health risks. It is also important, when cleaning ponds and tanks, that toxic disinfectant residues are removed. Finally, the mode of disposal of water from open systems also requires great consideration when designing enclosures. Primarily, this is to ensure that water disposal does not compromise public safety through the transfer of pathogens into public waterways, but also to ensure environmental policy is adhered to (see section **6.2.4**).

Water management is a complex and technologically advanced topic, which has only briefly been touched on here. The SSSMZP (Defra, 2004) have a brief appendix for aquariums, and additional health and safety guidelines have been produced for various managing aquariums (HSE, 2006). The need to oxygenate water, the complexity of the various filters used, and their availability, along with other useful information, can be found in standard textbooks dedicated to this subject, such as Spotte (1992), Moe (1993), Boness (1996), Hemdal (2006), and Adey and Loveland (2007).

6.2.7 Housing designed to aid husbandry

So far, we have considered some basic housing requirements that are tailored to the specialist needs of different exotic species. Another basic requirement for good housing is the inclusion of structural housing features to facilitate husbandry, whether moving, restraining, or simply allowing the separation of animals. Separation areas are necessary if animals require restraint and should be regularly used by the animals so that they are not associated with fearful events. These areas are also helpful to restrict the contact between animal and keeper, because many species will need to be locked away from the area in which the keeper is working.

An appropriately constructed enclosure can also facilitate the provision of enrichment: for example, by providing sturdy uprights from which enrichments can be hung, by anticipating that more furniture will be added in the future, and by making allowances for how extra furniture will be moved into the enclosure (see **Chapter 8**).

Finally, it is important to consider what requirements might be necessary in the future and build them into the design of housing; after all, a zoo may have only a pair of animals today, but, next year, it might have dozens more if the pair reproduce well —or the pair may no longer be compatible and may have to be housed apart from one another.

6.2.8 The furnishings

Technological advances, increased expertise in designing and building zoo animal exhibits, and a greater understanding of animals' needs have shaped changes in zoo enclosure design. It is essential that the 'basics' are included in zoo animal housing, but it is especially important that these enclosures are appropriately furnished. The provision of appropriate furnishings transforms a barren enclosure into a complex one (see **Box 6.1**).

Essentially, furnishings can include all of the modifications made once the structure of the enclosure is completed. There are obviously many changes that can be made to enclosures, including those made to enhance its aesthetics (as viewed by zoo visitors), but, in this section, we will discuss only those furnishings that relate to animal needs.

Surfaces

The physical properties of surfaces within zoo animal enclosures are important in terms of the general maintenance of enclosures, and also because they affect animal behaviour and biology. For example, sea-bird droppings are especially destructive, because they contain high levels of ammonia, which make them corrosive, so the SSSMZP (Defra, 2004) suggest that surfaces need to be durable, non-toxic, non-porous, and to have a waterproof finish. In some circumstances, abrasive surfaces may be provided as an aid to sloughing in reptiles, and to give opportunities to wear down beaks and hooves that could otherwise become overgrown (Defra, 2004; Yates and Plowman, 2004).

Similarly, softer surfaces may be more appropriate for other species, or may be necessary for the expression of certain behaviours. Meller *et al.* (2007), for example, found that rubberized flooring provided to Asian elephants *Elephas maximus* was associated with a reduction in the level of 'discomfort' behaviours expressed and an increase in standing resting behaviour, and this change in behaviour was considered beneficial, because it mimicked the behaviour of wild elephants.

Agricultural housing facilities have used rubberized matting for some time to prevent sleeping sores and bumble foot (ulcerative pododermatitis, illustrated in **Fig. 6.17**), and their use for older, arthritic **ungulates** in zoos is becoming more widespread.

6.2.9 A room with a view

The topography of an enclosure can make a huge difference to how an animal is able to use it.

Figure **6.17** Choice of materials and substrates used in enclosure design affect more than the look of the enclosure. This penguin has developed bumblefoot, an infection associated with abrasive surfaces that cause small cuts in the feet, which then become infected. (Photograph: Mel Gage)

Height is important to many animals so that they can express species-specific behaviours, distancing, or flight **responses**, or simply to gain a better view of their surroundings. Height can be created with perches and platforms, and can also be provided in paddocks through the use of mounds of soil. The SSSMZP (Defra, 2004) suggest that birds of prey should be provided with vantage points within their aviaries so that they can have a view of their surroundings. Providing platforms and perches within enclosures is considered to enhance physical **fitness** (if animals use them); they also enhance welfare. For example, urinary cortisol levels drop in zoo-housed felids when the cats are provided with access to elevated areas (on the clouded leopard *Neofelis nebulosa*, see Shepherdson *et al.*, 2004; on the cheetah *Acinonyx jubatus*, Wielebnowski *et al.*,

2002). A reduction in cortisol levels is generally taken to indicate lowered stress levels and thus improved welfare, although the link between cortisol and stress levels is quite complex (see **Chapter 7** for more discussion of this).

Privacy

Many visitors to zoos will have noticed that, when provided with the choice, many animals will seek out and use areas in which they are out of sight. Why they do this is discussed more fully in **Chapter 13** on 'Human–animal interactions', but it is clear that providing privacy within an enclosure is an important element of enclosure design. Herbert and Bard (2000), for example, found that orang-utans frequently sought out a small area at the top of their enclosure that was out of sight from the public, even though they lived in a large, custom-built enclosure designed specially for them. The degree to which animals seek out privacy varies between species, and it is not surprising that species which are considered shy of people may have problems when they are not provided with enough privacy (for example, Pied tamarins *Sagiunus bicolor bicolor*—see Wormell *et al.*, 1996), although much of the evidence for this is anecdotal.

Degrees of privacy can be achieved through various methods, such as allowing the animals to exit the 'show den' (the area visible to the visitors) or to hide while in the show den, or else by simply obscuring the view of the public and thus reducing the animals' perceptions that they are on 'show'. Dense vegetation, visual barriers, or little crevices within enclosures can be successful in providing areas within an enclosure in which animals can hide.

It is obviously not ideal to have animals hiding, because zoo visitors want to see them, which is why methods developed to reduce the visibility of visitors to animals have flourished. For example, Blaney and Wells (2004) found that the use of a camouflage net, to shield gorillas at Belfast Zoo from the gaze of visitors, significantly reduced aggression and stereotypical behaviours in the gorillas, but did not prevent the public from seeing the animals. Another method, which can be seen in **Fig. 6.18**, shows how Chester Zoo used plants to increase the proximity between zoo visitors and a group of mandrills at their viewing windows. The Zoo noted that there was a reduction in the stress-related behaviours expressed by the mandrills after the provision of these plants.

A particular need for privacy is shown by parent animals when they move their offspring between

(a)

(b)

Figure **6.18** Despite the greatest efforts made to ensure zoo enclosures meet the needs of stakeholders, research can sometimes highlight areas in which this has not been accomplished. At (a) Chester Zoo, it became apparent that visitor pressure was having a deleterious impact on the behaviour of their mandrills. This pressure was successfully alleviated by (b) the use of planters placed between the visitors and the viewing windows. (Photograph: Chester Zoo)

multiple den sites. It is usually thought that moving offspring between multiple dens is the result of disturbance, both from humans and from other species, and could lead to a reduction in the survival of offspring. Habib and Kumar (2007) suggested, however, that Indian wolves *Canis lupus pallipes* moved offspring between den sites as the pups grew and that thus it was not a result of disturbance but a naturally occurring behaviour. Laurenson (1993) also observed that wild cheetah mothers moved their offspring routinely, about every 6.6 days. Thomas and Powell (2006) considered that, among other factors, the provision of multiple den sites to captive African hunting dogs improved their reproductive success. Anecdotally, the provision of multiple den sites to other captive species appears to facilitate breeding: for example, in red pandas *Ailurus fulgens*, binturongs *Arctictis binturong*, and maned wolves *Chrysocyon brachyurus*, according to Julian Chapman of Paignton Zoo Environmental Park. In addition, some parental behaviour may favour privacy from conspecifics. For example, nursing bottlenose dolphins *Tursiops truncatus* at Kolmården Wild Animal Park, Sweden, were observed to suckle their young more frequently when alone (more than 80 per cent of observations) and away from their social group (Mello *et al.*, 2005).

6.3 Husbandry from birth to death

'Husbandry' is the term used to describe the processes undertaken to care for animals and includes many routine, and also some infrequent, events. The most important point to remember when considering zoo animal husbandry is that we, as zoo professionals, are ultimately responsible for the lives of zoo animals, and determine what, how, and when many events occur in their lives, from their conception to their development (which is considered in **Chapter 9**), and, finally, how and when they die. This section will briefly outline some common husbandry practices and illustrate how these, too,

need to meet the needs of the three main stakeholders: the animals, keepers, and visitors.

6.3.1 Daily routine

Keepers meet the animals' basic needs of survival through effective cleaning and feeding, and by monitoring them. Some of these are discussed in more detail in **Chapters 11** on 'Health' and **12** on 'Feeding and nutrition'. Here, we will concentrate on the need for keepers to move animals around within their enclosures routinely, in order to carry out these tasks.

It is dangerous for the keepers to interact, or to go into an enclosure, with many species, so moving the animal around has to be done very carefully. This can be achieved by using a series of different techniques, from moving into the animal's flight distance, to using operant conditioning techniques (training).

When a human walks towards an animal, it will often appear unaffected by the human's presence until he or she gets too close, at which point the animal may stop whatever it was doing. If the human continues walking, the animal will eventually either move away from or threaten him or her, and, at this point, the human will know that he or she has entered the animal's flight distance—so called because invading this space will make the animal take flight. An understanding of flight distance can be used very effectively to move animals—but it needs to be implemented calmly and slowly, because invading an animal's flight distance quickly or aggressively can result in injuries to both animals and keepers alike.

Cues and commands can also be given that the animal learns will be rewarded if it displays the right behaviour. In this way, operant conditioning is achieved and can sometimes take place without either party realizing it. For example, if keepers want to get an animal to move to an inside enclosure, they will frequently provide food or treats inside, so that when the animal goes inside, it is positively rewarded for coming in. This is **positive reinforcement training (PRT)** (see section **13.4.1**; illustrated in **Fig. 6.19**), which will increase the likelihood of

(a)

(b)

(c)

Figure **6.19** All animals are capable of learning and the principles that underpin learning are universal to them all. This means that husbandry training, which is based on learning theory, can be achieved with all species maintained in zoos—i.e. (a) fish, (b) reptiles, (c) birds, and (d) mammals. (Photographs: (a) Phil Gee; (b) www.iStockphoto.com; (c) Jessie Cohen/Smithsonian's National Zoo; (d) Vicky Melfi)

(d)

the animal coming inside in the future when the keeper needs to move it. Alternatively, the keeper could have moved into the animal's flight distance, shouted at it, or physically moved it. In this scenario, the animal would have moved inside to avoid the keeper and so this would represent a punishing episode, and would reduce the likelihood of the animal staying outside when the keeper needed to move it inside. As such, training is, to some extent, inherent in most daily routines.

6.3.2 Infrequent events

There are many planned and unplanned infrequent events that occur during the lifetime of an animal. Again, some of these events are discussed more fully in other chapters: for example, health checks and making provision to identify animals individually (through transponders or tags) are dealt with in **Chapters 11** on 'Health' and **5** on 'Animal identification and record keeping', respectively. In this section, we will consider introducing animals to one another and transporting them, whether within a zoo or between zoos.

Introductions

The movement of individuals into and out of groups is a frequent occurrence in many 'wild' social groups (see, for example, Pusey and Packer, 1987), in which it has evolved, among other things, to reduce the risks of related individuals breeding and having inbred offspring. These are situations that also need to be avoided in captive populations, so recommendations to move animals between groups are routinely made for all animals that are part of **captive breeding programmes** (**Chapter 9**). Trying to simulate natural migrations in captive populations, however, is associated with potential risks due to aggression and injury, even though it enables the genetic management of the population and social stimulation (Visalberghi and Anderson, 1993). Because of this, alongside the genetic merits of potential breeding animals, there are lots of other considerations that also need to be made when considering

the transfer of animals between groups—not least the logistics of doing it (Lees, 1993; Norcup, 2001).

There is no guarantee that, because two animals (or a new animal and a group) are genetically compatible, they will therefore get on with each other when first introduced. In fact, it is always advisable to have a contingency plan in case an introduction has problems and the process should always proceed slowly. Depending on the species, there is a series of steps that can be employed to enable the animals to become familiar with one another before the introduction, such as sharing the bedding and furniture of the animals that are to be introduced, or allowing visual contact and then restricted tactile contact until, eventually, they are mixed together. Kangaroo rats *Dipodomys heermanni* are a solitary species, but compatibility between pairs—and thus breeding success—improves if potential mates are kept in long-term sensory contact with one another (Thompson *et al.*, 1995).

Interestingly—and not particularly conventionally—the scents of two male meerkats *Suricata suricatta* being introduced to one another was reported to be masked by the use of a strong-smelling cold treatment (Vicks vapour rub: *Lancaster News*, 2007). Because no aggression was observed, it was reported that the strong smell facilitated the introduction of these notoriously aggressive and territorial animals. A similar approach, of masking animals' naturally occurring channels of communication, was used in a laboratory facility to improve the success of pairing vervet monkeys *Cercopithecus aethiops sabaeus*. Gerald *et al.* (2006) were able to demonstrate that male vervet monkeys paired together were more likely to be aggressive to one another if their scrotal sacks were similar in colour (pale or dark). As such, they were able to reduce aggression by pairing males of different scrotal sack colour, for example, pale with dark and vice versa.

Although temporary modifications can be made to the way in which animals view each other during introductions, it is essential that, when enclosures are designed, ample facilities are built into the enclosure. At the very least, separate areas should be provided for both new and occupying animals.

Transporting within and between zoos

There are two main considerations involved when moving animals: one is to meet the regulations that govern live animal transport and the other is the physical process of moving the animal.

The regulations that govern live animal transport between zoos are dependent on the species being moved and where it is going; this topic is dealt with in more detail in **Chapter 3**. The practical side of moving animals, however, involves a series of steps that needs to be carefully planned, to ensure keeper and animal safety, to prevent the risk of the animal escaping, and to reduce the likelihood of any deleterious impact on welfare (see **Fig. 6.20**).

There has been an increase in the number of studies on the effects of transport on the welfare of animals, but most of them have focused on domestic breeds (for example, Thiermann and Babcock, 2005). Many of these studies demonstrate that transporting animals, whether by sea, land, or air, can reduce their welfare (Broom, 2005; Norris, 2005), although there is evidence that providing enrichment during transport can mitigate some of the deleterious signs of transport: for example, in pigs (Peeters and Geers, 2006).

Because of the results from these farm animal studies, we would expect that the impact of transport on zoo and other wild animals may also be detrimental, and the limited evidence supports that view. Long-tailed macaques *Macaca fascicularis*, for example, show elevated stress responses during air transport and do not return to pre-transport behavioural profiles in their new homes for some time—certainly, for more than a month—after transport (Honess *et al.*, 2004). Similarly, tigers *Panthera tigris* subjected to a simulated move showed faecal cortisol peaks of 239 per cent above baseline between 3 and 6 days after transport, and took nearly 2 weeks to return to baseline levels (Dembiec *et al.*, 2004). In this latter study, there were indications that prior exposure to some of the aspects of transport could reduce the effects of transport stress. Nevertheless, it is clear that every effort should be taken to ensure that all events that take place before, during, and after transport are carried out with care and with the animal's welfare as the foremost concern.

6.3.3 The management of social groups

It is generally accepted that maintaining successful captive social groups that are similar to those that occur 'naturally' is a goal that zoos should try to attain (Hediger, 1955; Hutchings *et al.*, 1978). But achieving this requires an understanding of the structure and function of animal groups, as well as what factors affect them and how. For example, it is customary in the wild for males or females to leave their natal group upon maturation and find another group to join, because this reduces the risks of **inbreeding**. But carrying out transfers between groups and other social processes in a captive setting may be limited, and could thus lead to problems. Indeed, providing the animals with appropriate opportunities to enable them to express social behaviour and then manipulating this in captivity can be very problematic, because any mistakes that are made could result in a failure to reproduce or lead to fatal fighting (Visalberghi and Anderson, 1993). Animals that do not move from their natal group may become the target of aggression; alternatively, introduced unrelated animals may be the focus or cause of aggression. For example, in a large captive group of collared peccaries *Tayassu tajacu*, high levels of infanticide were recorded (Packard *et al.*, 1990). When aggressive events were observed, it was unrelated females that were seen to attack neonates and related females that defended them. Despite the fact that collared peccaries form large social groups in the wild, it seems that, in captivity, it is necessary to ensure that females are related to reduce the likelihood of aggression.

Managing social groups is frequently considered to be one of the most important, but difficult, tasks to achieve in captivity. This does not only involve

(a)

(b)

(c)

(d)

Figure **6.20** Moving animals is a complex matter, not least because the size of some animals makes the logistics more difficult in every way. (Photographs: (a, b, and d) Paignton Zoo Environmental Park; (c) Leszek Solski, Wrocław Zoo)

large groups of animals; solitary species must have social management as well, because they will need to be brought into contact with other individuals for breeding and will inevitably be housed near other animals. Dysfunctional social groups have negative consequences for the three main stakeholders in housing and husbandry that we identified at the start of this chapter:

- the animals themselves may experience stress, low reproductive output, or impaired immune function, or may develop 'abnormal' behaviour patterns and aggression, some of which can be fatal;

- keepers may find it difficult to carry out daily routine tasks successfully, because some, or all, of the animals may cease to move around the enclosure as normal;

- visitors do not want to see animals fighting, but want to see positive social interactions.

There are a number of factors that limit a zoo's ability to hold natural social groupings. These include space (because, if this is limited, large groups cannot be maintained, and this then affects the extent to which breeding can be promoted and managed within the population), individual (or species) compatibility (because, if animals do not get on and are aggressive to one another, smaller groups may have to be maintained through necessity), and animal availability (because captive populations of some species are sometimes small and so larger groups cannot be created). It is also likely that the social group dynamics will change over time, whether due to animals aging and behaving differently, or because offspring are born. Certainly Baker (2000) noted that older (that is, aged 30–44 years) female chimpanzees *Pan troglodytes* were less aggressive and used their enclosure less than younger (that is, aged 11–22 years) females, while the opposite was true in male chimpanzees.

In the rest of this section, we will look at some of the techniques and processes associated with management of social groups in zoos.

Group size/stocking density

There are some species that live in quite large groups in the wild, but for which, in captivity, attempts to recreate large groups could cause problems if space is limited. The relationship between space and group size is referred to as stocking density, which is the number of animals per unit space. As long as large groups are given appropriate space, they will not suffer the effects of crowding and, in fact, many species thrive in large social groups. Studies have demonstrated that large group sizes are associated with significant elevations in orangutan activity (Wilson, 1982), and increases in reproductive success in Chilean flamingos *Phoenicopterus chilensis*, which require at least forty birds to breed, and Caribbean flamingos twenty *Phoenicopterus ruber ruber*, which require over twenty birds (Stevens, 1991; Pickering *et al.*, 1992).

Clearly, stocking density needs to be carefully monitored, because crowding compromises animal welfare. This can result from increased competition between the animals for resources, which can lead to some animals getting no, or little, food or shelter. Different methods of feeding or resource provision can overcome this problem to a limited extent (see section **12.7** on food presentation).

Crowding can also lead to changes in the physiology and behaviour of the animals, as they attempt to ameliorate increased social pressures, but are unable to distance themselves from one another. For example, a large group of Hamadryas baboons *Papio hamadryas hamadryas* (over the 5-year study period, a maximum of eighty-three and minimum of forty-six individuals) at Paignton Zoo Environmental Park was transferred to a smaller off-show part of their enclosure (measuring 3m³) from the outside area (measuring approximately 35m × 15m × 13m), so that it could be cleaned daily. Stress, measured as the performance of self-directed **displacement behaviours**, rose significantly in this group over the study period as a result of this increased stocking density (Plowman *et al.*, 2005; illustrated in **Fig. 6.21**).

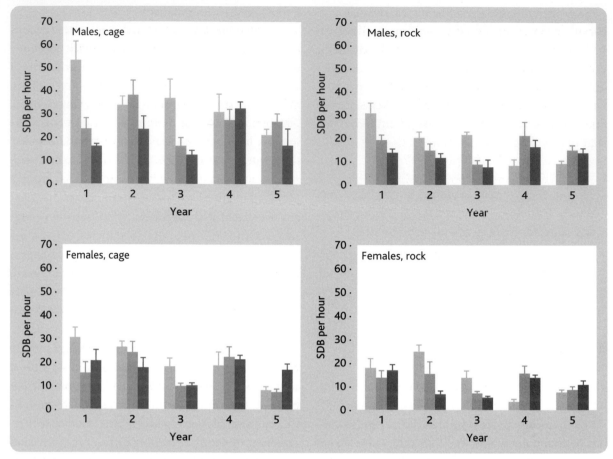

Figure **6.21** To manage a large group of Hamadryas baboons at Paignton Environmental Park, it is necessary to move the animals from their large (rock) enclosure to a smaller (cage) so that the rock area can be cleaned. This move increases the stocking density of the group and, subsequently, an increase is observed in the mean rate of self-directed behaviour (SDB) (Plowman *et al.*, 2005). The relationship between self-directed behaviour and welfare is addressed in section **4.4**. (From Plowman *et al.*, 2005)

Social problems resulting from crowding, such as, for example, when high levels of aggression or stereotypies are displayed (de Waal, 1989), can be mitigated with enrichment. Providing bark as a deep-litter substrate in the small off-show area of the baboon group mentioned above significantly increased levels of play and reduced aggression in the animals.

Finally, the risk of health problems increases in parallel with stocking density (see, for example, Goossens *et al.*, 2005; see **Chapter 11** on 'Health'), not only because the potential disease carriers increase, but also because contamination is facilitated by both the increased proximity between individuals and the greater quantity of faeces being produced. This is true of both terrestrial and aquatic environments.

Single-sex groups

There is a variety of different social systems and group structures in wild animal groups (see **Table 9.1**). Attempting to replicate some of these 'natural' social groups in captivity can, however, contribute to a surplus of animals, such as the offspring from family groups, or maturing males that are excluded from harem systems by more dominant animals (Graham, 1996). Some of the problems caused by surplus animals are discussed further in **Box 9.6**.

One way of preventing surplus animals is forming single-sex groups and thus limiting breeding. An example of this can be seen in the management of the Rodrigues fruit bat *Pteropus rodricensis*, in relation to which the European breeding programme has recommended that many of its member zoos hold either all-male or all-female groups to ensure the population does not continue to grow. Similarly, in the Western lowland gorilla *Gorilla gorilla gorilla*, breeding groups are composed of single adult male harems, so, as males get older, there is no place for them within the group. The European breeding programme for gorillas has recommended that some member zoos hold all-male bachelor groups, as a reservoir for these surplus males.

Single-sex groups have been successfully managed for a variety of species, including blackbuck, ruffed lemur, many reptiles, and flocking birds. Indeed, maintaining a group of male cheetah *Acinonyx jubatus* to present to a potential female mate may even improve the chances of successful breeding (Caro, 1993).

There are, however, some problems that can arise in captive single-sex groups. This may be because the species does not form single-sex groups in the wild, or if it does, they would be transitory or the individuals in the group would tolerate only close relatives (for example, on *Gorilla gorilla*, see Pullen, 2005). Male–male aggression in bachelor groups is of particular concern.

Unfortunately, there do not appear to be any standard 'rules' that can be applied to reduce the likelihood of aggression with all species. For example, levels of male–male aggression have been observed to be high in chimpanzees when there is a large discrepancy in age or social experience (Alford *et al.*, 1995). Differences in coloration can also lead to increased aggression, as described previously in male–male introductions of vervet monkeys (Gerald *et al.*, 2006). Techniques suggested to alleviate the likelihood of this aggression include slow introductions (see above) and matching behavioural profiles to increase the likelihood that animals will be compatible (for example, Kuhar *et al.*, 2006).

Although there is a need for single-sex groups in captivity, the long-term stability of these groups is not always certain, especially when considering non-related animals, the maturation of adolescents, and the movement of animals in and out of the groups (Matthews, 1998; Asvestas and Reininger, 1999; Fàbregas, 2007). In addition to the possible behavioural problems that may compromise bachelor groups, the absence of young animals from single-sexed groups may also make this type of exhibit less attractive to zoo visitors and, consequently, few zoos may be willing to hold all-male groups.

Mixed-species groups

Many animals form mixed-species associations in the wild, where these associations may confer benefits such as a reduced threat of predation, improved foraging opportunities, improved social stimulation, and reproductive enhancement (Noë and Bshary, 1997; Wolters and Zuberbuhler, 2003; Griffin *et al.*, 2005). The benefits of creating **mixed-species exhibits** (thus mixed-species associations) in zoos are considered to include the provision of:

1. dynamic social stimulation (Thomas and Maruska, 1996);

2. a more efficient use of enclosure space, because animals with similar ecological requirements or because different species are able to maximize the use of the available space more efficiently (Dalton and Buchanan-Smith, 2005);

3. a higher-quality educational experience, because biogeographically representative species may be put together, or other combined groups can be assembled of animals that share other characteristics, such as waders that come from similar habitats, but from different continents.

Care needs to be taken, however, when putting different species together. Choosing species with different ecological niches within a similar habitat can reduce the likelihood of competition and thus aggression (Thomas and Maruska, 1996). For example, naturally sympatric species such as **callitrichids** have been housed together in many combinations (for example, Goeldi monkeys *Callimico goeldii* and pygmy marmosets *Callithrix pygmaea*—see Dalton and Buchanan-Smith, 2005). But some naturally occurring associations do not work in captivity, or require special housing and husbandry considerations. For example, when mixing waterfowl and mammals (that is, deer, or other ungulates), an area of land should be provided for the sole use of the waterfowl, so that they are not excluded from using land by more dominant larger species. This can be achieved by excluding the ungulates with fencing (at 30cm above ground—see Defra, 2004).

6.3.4 When should we intervene and what should we do?

Some zoo animal husbandry activities are controversial. Few people would debate the merits of conserving species or ensuring that zoo animals should have good welfare, but the husbandry techniques employed to reach these goals are frequently debated. It seems that the extent to which zoo professionals should intervene in the lives of the animals in their care can be a very contentious issue. The activities that fall under this banner are diverse and include the management of surplus animals through the use of **euthanasia** and contraception (see section **9.7.4** and **Box 9.6**), whether we should hand-rear animals (**Box 9.4**), and the extent to which animals should be modified, intentionally or otherwise, to suit their captive situation.

As we have already seen, the behaviour, morphology, and genetics of captive animals can change over generations in comparison with wild conspecifics. Many of the different housing and husbandry techniques that occur in the lifetime of an animal can affect it and its welfare, either directly or indirectly. For example, physical modifications to a bird's wing, termed **flight restraint** techniques, directly stop it

from flying, but an overly small enclosure can also prevent it from flying.

Whether these activities should take place, when they should occur, and how they should be executed are all topics of debate, and, to some extent, whether they are contentious or not depends on whether or not you agree with them. The most that we can do here is to provide a framework to use when considering different husbandry techniques. It is important to consider the merits and drawbacks of the proposed techniques, what are the alternative techniques, and what the repercussions would be of not intervening.

Let us look in greater detail at the issue of flight restraint.

Flight restraint

Why would we want to keep birds that could not fly in the zoo? One answer is to conserve them, as part of captive breeding programmes, and to prevent them from becoming extinct in the wild. It is argued that it is not always possible to provide a bird species with the adequate enclosure space necessary to enable it to fly: physical space may be limited and so may the funds necessary to build larger 'roofed' enclosures. There are also reports that birds that are able to fly are more susceptible to injuring themselves in some enclosures.

It has been argued that preventing the extinction of bird species through captive breeding programmes is more important than the potential negative ramifications of restricting flight in some birds. The argument is that the impact of restricting flight is dependent on the extent to which the birds would ordinarily fly—or, in other words, the extent to which the bird is motivated to fly. Previous studies have considered that behaviours that occupy a large part of animals' **activity budgets** are necessary and highly motivated (Bubier, 1996). But Mason *et al.* (2001) demonstrated that animals may be

---

Flight restraint refers to a technique that prevents a bird from flying; this can be done temporarily or permanently through housing and husbandry, or through surgery.

highly motivated to perform behaviours that have relatively short durations. This issue is difficult to resolve without adequate empirical data.

The alternative viewpoint is that wilfully amputating a healthy animal (such as tail docking, castration, etc.) and preventing it from performing a natural behaviour (flying, reproduction) is abhorrent and compromises its welfare (see **Chapter 7**).

Discussions of this type often operate on two levels: one that uses information gained as reliably recorded and interpreted facts, and the other based more on conjecture, assumption, or belief. As discussed in **Chapter 7**, the need for objective insight is essential when trying to tease apart the emotive elements of a contentious issue. We need to consider whether the implementation of flight restraint affects bird welfare or conservation. This will, of course, depend on the species and the method of flight restraint used.

There are various methods employed to render a bird incapable of flying. Techniques vary in terms of how invasive or permanent they may be. **Wing management**—especially **pinioning**—is probably the most debated aspect of flight restraint and works on the principal of unbalancing the bird so that it cannot fly, or making it physically unable to use it wings (the more common procedures employed are illustrated in **Fig. 6.22**). Feather clipping is achieved by cutting through the primary feathers from one wing to unbalance the bird. The same is achieved with brails, which consist of a loop of material over the shoulder of one wing and twisted to keep the wing closed.

Surgical methods of wing management are also available and considered by law to be 'mutilations', defined as any '*procedure which involves interference with the sensitive tissues or bone structures of an*

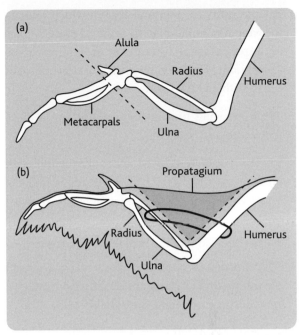

Figure **6.22** Of the different methods that can be used to prevent flight in captive birds, two surgical techniques are illustrated here: (a) pinioning; (b) patagiectomy. Whether zoos should restrict the flight of birds and which methods should be employed is a highly controversial topic, although it should be remembered that, by definition, captivity restricts the movement of all animals, to some extent.

animal, other than for the purpose of its medical treatment' in section 5(3) of the Animal Welfare Act 2006. Within this Act, surgical procedures of wing management are exempt[4] for conservation activities and listed as permitted procedures, if undertaken by a vet or following strict codes of conduct.

There are two main surgical methods:

- pinioning, which essentially involves the amputation of a digit at the wing tip, although there are variations in technique, and which is legal in the UK as long as it is not undertaken on agricultural land;

Wing management refers to methods used to inhibit flying through the physical alteration of the bird's wing, e.g. pinioning, feather clipping.

4 Exemptions to the ban on mutilations—notably, procedures to control reproduction in domestic cattle and wing-tagging of birds, including for conservation activities—are currently being re-evaluated by secondary legislation. (See **Chapter 3** for more information about legislation.)

- patagiectomy (or tendonectomy), which renders the wing functionless by the removal of all, or part, of the propatagium ligament or the tendons of the carpal joint. This method gives the appearance of a bird with an intact wing.

Aside from wing management, flight can also be restricted or prevented by:

- tethering, which is used most frequently with birds of prey and which involves restricting the bird's activity by fixing it via leather straps (jesses) on its legs to a stand or post;

- overfeeding, so that the bird is too heavy to fly;

- enclosure design, because enclosures that are small can inhibit flight in larger birds if they do not have enough space or height to attain lift-off.

The SSSMZP (Defra, 2004) recommend that tethered birds must be flown at least four times a week, unless ill, and should not be tethered all year. It is also recommended that owls and vultures should not be tethered, but trained to fly from pens (Defra, 2004).

Little research has been undertaken that specifically investigates the impact of flight restraint on bird welfare and/or conservation. Hesterman *et al.* (2001) assessed the welfare implications of different flight restraint techniques by considering how they might contravene the five freedoms. **Table 6.2** outlines some of these arguments. But, because of this lack of research, most arguments that are made for or against this husbandry technique are based on the observations of experienced keepers and managers. Indeed, in most situations in which it is necessary to decide whether or not to intervene, there is likely to be little data available to inform any decision, which is why some of these issues become long-standing debates.

Should the ability of birds to fly be prevented or restricted in zoos? If flying is considered a fundamental part of their behavioural repertoire, can they live without it and what effect does that have? These are the types of question that need to be addressed when considering the costs and benefits of implementing this and other husbandry measures. (Cost–benefit analyses are explained more fully in section 7.5.3.)

6.4 Studying how housing and husbandry affect captive animals

Much zoo-based research investigates topics related to housing and husbandry, and whether changes in these affect animal biology, lead to an improvement in conditions for zoo animals, or help the attainment of the zoo's goals (Eisenberg and Kleiman, 1977; Hutchings *et al.*, 1978; Schaaf, 1984; Kleiman, 1992; 1994; Seidensticker and Doherty, 1996). **Chapter 14** discusses what research in zoos entails, but the following section focuses on and outlines some of the different approaches that can be employed to evaluate zoo animal housing and husbandry.

6.4.1 Bottom-up or top-down?

Two different approaches can be adopted to study the effect of housing and husbandry on captive animal biology, and these are best described as 'bottom-up' and 'top-down'. The predominant approach taken is bottom-up, which is based on making speculative changes to housing and husbandry routines until the animal's biology is modified. For example, we might suspect that some element of housing and husbandry is stimulating sterotypies in an animal. In that case, we might make various modifications to housing and husbandry until we can reduce or eliminate stereotypies. This bottom-up approach has been the basis of much environmental enrichment research (Chamove, 1989).

Alternatively, the top-down approach attempts to identify which factors affect captive animal biology and to what extent. This information is then used to make subsequent modifications to housing and husbandry. Using the above example, if we were trying to find the cause of stereotypies so that we could reduce or eliminate its expression, the

Method	Temporary (T)/ permanent (P)	Advantages	Disadvantages
Small enclosure	T	Able to extend full wings and preen	Potential for crowding and lack of space may restrict physical exercise
Feather clipping	T	Easy to implement	Stress from repeated handling and catch-up Irritation Development of self-mutilation Algal growth in waterfowl Excessive bleeding if growing feathers are cut
Overfeeding	T	Able to extend full wings and preen	Health and condition problems associated with obesity
Tethering[1]	T	Given flying opportunities	Lack of space may restrict physical fitness and behavioural expression
Brailing[2]	T	Flight	Long-term muscle atrophy (although apparently reversible) Impaired wing growth
Pinioning[3]	P		
Tendonectomy	P	Wings appear normal	Carrying dysfunctional wing Patagiectomy can take a long time to heal
True for all measures		Large social groups[4] can be more easily maintained in enclosures that are less expensive to construct and which stimulate dynamic expression of social behaviours Might reduce risk of injury from the enclosure and escape of non-native species into habitats surrounding zoo grounds	Increased risk of injury from lack of balance, or from cage mates, because less able to avoid aggression The expression of some behaviours will be restricted, e.g. ability to escape, roost, preen, and potentially mate; the latter may lead to reduced breeding success[5]

Table **6.2** The potential advantages and disadvantages associated with some flight restraint methods, giving examples of worst- and best-case scenarios

[1] Owls and vultures should not be tethered (Defra, 2004).
[2] If brails are used for longer than 2 weeks, they should be alternated between wings to reduce stiffening of the wing (Ellis and Dien, 1996).
[3] Should occur when birds are very young, at about 3 days old (reviewed by Hesterman et al., 2001).
[4] Pickering et al. (1992) found that a minimum of forty Chilean and twenty Caribbean flamingos should be grouped together to ensure breeding success.
[5] Data suggest that flight-restrained flamingos have less successful reproductive lives (Farrell et al., 2000).

NOTE: Temporary measures are reversible; thus, if the conditions under which the bird is housed change, the decision to restrict its flight can be re-evaluated. On the other hand, permanent measures limit the number of times that the animal needs to be caught and handled to maintain flight restriction, but are more likely to be associated with post-operative infection and pain during, and after, the procedure.

top-down approach would be to collect information on a variety of factors, such as genetics, **nutrition**, housing, and husbandry, and then evaluate whether and how these contributed to the expression of the stereotypies observed (Montaudouin and Le Pape, 2005).

The two approaches can complement one another, because the 'top-down' approach can be used to identify factors that affect captive animal biology; the extent to which these factors affect the animals can then be measured by manipulating them until a change in the animals is observed. Both approaches can reveal similar information, but the bottom-up approach is less comprehensive, because the goal is only achieved if and when a change in captive animal biology occurs. As such, the study may be terminated after only one successful modification to housing and husbandry, when, in fact, various housing and husbandry indices are affecting captive animal biology. The 'top-down' approach, on the other hand, provides a greater appreciation of how different housing and husbandry factors affect behaviour—although this approach is more complicated and will require more time to undertake.

6.4.2 Single-site versus multi-zoo studies

Single-site studies are defined as those that are carried out at one location (that is, in one zoo enclosure). Usually, this style of study compares captive animal biology before and after an event (a modification to housing and husbandry) to evaluate its repercussions. A typical method used to evaluate the impact of housing and husbandry changes is **post-occupancy evaluation (POE)** (see **Box 6.2**).

A major benefit of this approach is that it will reveal how housing and husbandry factors affect captive animal behaviour in that enclosure; these studies are also relatively cheap, quick, and easy. In these situations, extraneous variables should not confound the results as they are matched between

---

Box 6.2 Post-occupancy evaluation

Post-occupancy evaluation (POE) was developed to assess how people used and interacted with their environment. Within a zoo setting, POE provides a valuable tool with which to evaluate how different species behave in their enclosures and use the different resources within it. For example, Chang *et al.* (1999) studied the enclosure usage by mandrills *Mandrillus sphinx* before and after major changes to their enclosure. They were able to determine that enclosure refurbishment had led to a significant change in the mandrills' behaviour, which resulted in them expressing behaviour that more closely matched that of their wild conspecifics (see section 4.5.3 for more on this study). Equally, POE techniques can explore an animal's interactions with the environment over time, thus enabling us to evaluate how long an enclosure may stay stimulating after enrichment, refurbishment, or if this enclosure is new. This information is essential when considering the best time at which to make changes to housing and husbandry.

An example is a study of red river hogs *Potamochoerus porcus* at Paignton Zoo Environ-mental Park. It was obvious that these animals had devastated their new enclosure from the moment they were put in it. Initially, it was overgrown with shrubs and trees, but, very quickly, the hogs had chewed, rooted, and foraged away at this vegetation until little was left. POE provided data that demonstrated that, in their first 3 months in the new enclosure, all three animals used the area available to them at significantly lower levels compared with when they first entered the new enclosure (Dayrell and Pullen, 2003). This information enabled a change in husbandry to be introduced: a far-reaching **scatter feed** that increased the hogs' use of the enclosure back to the levels observed when the enclosure was new and interesting (**Fig. 6.23**).

POE can measure how the animal's behaviour or enclosure use changes, or, indeed, how visitors or keepers are able to use the enclosure area. For example, Wilson *et al.* (2003) were able to determine that visitors and keepers using a new giant panda exhibit at San Diego considered the changes to the enclosure favourable.

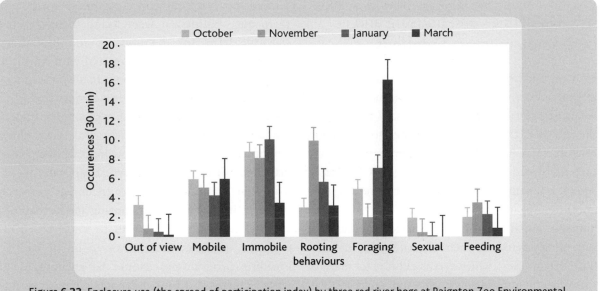

Figure **6.23** Enclosure use (the spread of participation index) by three red river hogs at Paignton Zoo Environmental Park in the first 4 months after they entered the new enclosure. In October, the hogs first entered the enclosure; in March, the hogs were fed using a scatter feed, which increased their use of the enclosure. (From Dayrell and Pullen, 2003)

treatments. Order effects and those contributed by the passing of time can be reduced through design modifications: for example, using a Latin square design or randomization between conditions (Lehner, 1998). Unfortunately, single-site studies usually suffer from having a small sample size, which may be further reduced if subgroups exist within the group (such as sex–age classes; see **Chapter 14** for more on dealing with small sample sizes in research).

A drawback of single-site studies is that they are of limited use when considering how housing and husbandry factors affect animals elsewhere: in other enclosures or in other zoos. The results do not lend themselves to generalizations to other animals or species, although studies of this type cautiously predict ramifications in other circumstances (Carlstead *et al.*, 1991; Baker, 1997). By making changes to food presentation, enclosure structure, and visitor number in one zoo, studies have been undertaken to investigate their impact (McKenzie *et al.*, 1986; Smith *et al.*, 1989; Mitchell *et al.*, 1991a).

Alternatively, multi-zoo studies, which compare one species across several zoos, provide the opportunity to investigate much more about the implications of changing housing and husbandry. For example, in a complex enclosure, the addition of another substrate (for example, a log) may have less effect than if it were incorporated into a more barren enclosure. Other studies have used this method to evaluate how housing and husbandry affect reproductive success, activity levels, and enclosure use (Ogden *et al.*, 1993; Mellen, 1994). Unfortunately, multi-zoo studies are seldom conducted despite their benefits, probably because they require more resources than single-site studies (Mellen, 1994). More details are given about multi-zoo research studies in **Chapter 14**, including the use of surveys and the Methods of Behavioural Assessment (MBA) projects, which have been incredibly useful at identifying the relationships between behaviour, housing and husbandry, mortality, and reproductive success (Kleiman, 1994; Carlstead *et al.*, 1999a; 1999b).

6.5 Guidelines and housing and husbandry recommendations

Housing and **husbandry guidelines** that refer to zoo animals are compiled by a variety of agencies with differing degrees of comprehensiveness. Within the UK, the SSSMZP (Defra, 2004) is referred to by all zoos, because it is required that these recommendations are met in order to obtain a zoo licence (see **Chapter 3**). Also, it is a requirement of most captive breeding programmes, regardless of zoo association, to produce housing and husbandry guidelines. Yet these are surprisingly sparse: the International Species Information System (ISIS), which acts as a library for many of these documents, had only twenty-four on file in 2006.

There are also some concerns about the quality of some of the guidelines that are available. From a rapid survey of guidelines produced worldwide, three out of nineteen were found to have some scientific basis for the housing and husbandry recommendations made, compared with scientific evidence in all veterinary sections in the same guidelines (Melfi *et al.*, 2004a). It seems obvious that good housing and husbandry guidelines should be underpinned by evidence, essentially empirical data, which demonstrates why certain housing and husbandry principles should be employed. It appears, therefore, that there is a need to encourage research investigating the impact of housing and husbandry, so that recommendations can be made based on evidence, and that those involved in the captive breeding programmes should be encouraged to write housing and husbandry guidelines to facilitate zoos in providing the best possible conditions for their animals.

Summary

- Enclosure design is usually a compromise between what is in the best interests of the needs of the three main stakeholders—that is, the animals, keepers, and visitors—although the needs of other parties should also be considered.

- From initial enclosure design through to managing the animals, everything should be done to ensure that housing and husbandry stimulate the animals both physically and psychologically.

- Housing and husbandry should also enable the keepers to maintain and promote good animal welfare.

- Housing and husbandry should provide an experience that is enjoyed by the visitors, who are hopefully able to learn something on their way around the zoo.

- An example of an enclosure in which all of these measures have been incorporated is shown in **Fig. 6.24**.

Questions and topics for further discussion

1. Who are the three main stakeholders in zoo animal enclosure design and which other parties should also be considered?

2. What factors affect animals' needs in terms of their housing and husbandry?

3. What advantages and disadvantages are associated with naturalistic enclosures?

4. What methods might you use to stop an animal escaping?

Figure **6.24** If you have absorbed the contents of this chapter, you may be on your way to being able to design a world-class enclosure, like that shown here for the Hamadryas baboon *Papio hamadryas* at Singapore Zoo, which has been faithfully modelled on their native Ethiopian mountains. (Photograph: Nor Sham/Singapore Zoo)

5. Why is the management of social groups difficult?

6. Discuss, giving reasons to support your argument, whether enclosure size or complexity is more important.

7. Describe how you would evaluate the impact of a change in housing and husbandry on animal biology.

Further reading

Much of the material you have been reading in this chapter has come from the experiences of people who work in the zoo world and a lot of it, where published, has been referred to appropriately in the chapter.

There are few books to which we can direct the reader specifically, although, as always, Kleiman *et al.*'s *Wild Mammals in Captivity: Principles and Techniques* (1996) is a fund of information on the maintenance of mammals in zoos.

A number of journals—notably, *Applied Animal Behavioural Science*, *Zoo Biology*, *Animal Welfare*, *Aquarium Sciences*, and *Conservation*—publish papers of relevance to housing and husbandry. We can also mention the *International Zoo Yearbook*, which always has a number of case studies of housing and husbandry-related experiences from various zoos.

Finally, it is worth reading the proceedings of the Zoo Design Conferences, which are available through the Whitley Wildlife Conservation Trust.

Websites and other resources

In this chapter, we have on numerous occasions referred to the SSSMZP (Defra, 2004). This is available to download and read in its entirety from `www.defra.gov.uk/wildlife-countryside/gwd/zooprac/index.htm`.

We should also direct you to the ZooLex website—`www.zoolex.org`—which is a site endorsed by the World Association of Zoos and Aquariums (WAZA) that provides an extensive zoo animal design bibliography, as well as numerous case studies about zoo exhibits constructed worldwide.

Chapter 7 Animal welfare

All good **zoos** recognize that the animals in their care should have good welfare, regardless of the conservation, education, or research goals or missions that they aim to achieve. Animal welfare is commonly considered to be the subjective state of well-being of an individual, and this is determined by both its physical and psychological condition. This, of course, makes the evaluation of an animal's welfare through the use of objective indices—and thus the determination of which variables affect it—very difficult, because it is not possible for us to appreciate fully animals' subjective states. One way around this difficulty is for us to consider what animals might need during their lifetimes and then judge that, if these needs are satisfied, then, at best, a high level of welfare is attained, but that, at worst, welfare is not compromised.

Understanding and evaluating animals' needs has been pursued via four different approaches, which consider the needs of animals from the perspective of their minds, bodies, and natures, and also by a sort of analogy. This latter approach allows us to suggest that, if we, as humans, are able to experience different emotional states through the physiological stress **response** and if animals have the same physiological apparatus as us, then it may be that they experience the same (or similar) emotions to our own.

In this chapter, we will review some of the circumstances that appear to be associated with compromised welfare, and provide a summary of the objective indices that are used to measure animals' needs and animal welfare. We will also consider how zoos are able to meet the needs of the animals in their care and how they aim to provide the animals with good welfare.

The topics to be covered are as follows.

The accompanying boxes consider some related topics in a bit more detail, including some of the theoretical background that underpins this area (the stress response), methods that are used to assess welfare (non-invasive methods; the use of **cortisol**), and topics that impinge on the subject of welfare (pain; deviations from normal).

7.1 What is 'animal welfare'?

Animal welfare science is the study of an animal's quality of life and more research has been undertaken on this within agriculture than in any of the other situations in which we keep animals. This has led to great advances in our understanding of the welfare of farm animals and has also provided us with general principles that are applicable to animals in other situations, such as the zoo. For this reason, much of the research that we quote in this chapter has come from the studies of agriculturally important animals.

We have to recognize that the welfare of animals does not change simply because we perceive it to be good or bad, or because we condone or deplore the use of animals in certain circumstances within our society. Animal welfare is the subjective state of an animal, which is independent of us and our societal views about welfare. The way in which we view welfare does, however, greatly impact on the way in which we treat animals and therefore the conditions that we provide for them. We should also recognize that welfare is peculiar to an individual. Therefore, welfare may differ between different animals when they are exposed to the same circumstances and, indeed, even the welfare of the same animal may vary over time.

7.1.1 It's all about the individual

Individuals, by definition, are different from one another, which means that they will experience the world differently and have different needs. There is variation in the extent to which even closely related individuals may develop disease, morphological abnormalities, and behavioural **pathologies**, and these may reflect underlying genetic differences. We have already considered how individual differences in factors such as sex, age, and social rank, and thus the needs of individuals, influence how we provide captive housing and husbandry (see section **6.1.1**). In the next section, we will explore the impact on the animals' welfare of, firstly, those individual differences that can best be summarized as differences in personality, and secondly, the shared differences held by individuals of the same species.

Personality

'Personality' is an animal's consistent reaction to different environmental variables, resulting from a synthesis of the animal's genetic make-up and its past experience (see **Box 4.2**). Use of the term 'personality' has sometimes been contentious when applied to animals, and alternative terms such as 'behavioural **phenotype**', 'suites of behavioural traits', 'behavioural syndromes', 'coping style', and 'temperament', have been used instead by some

authors. All of these terms attempt to quantify how individuals differ in their responses to the environment and, with an acceptance that animals do, indeed, act as if they have different personalities, research in this area is now providing great insights into why different individuals react differently to the same environmental factors (Gosling, 2001).

This might be expanded to explain animals' different reactions to **exogenous** and **endogenous stimuli**. Capitanio (1999), for example, was also able to demonstrate that personality dimensions in laboratory-housed rhesus macaques *Macaca mulatta* were biologically valid, in that they correlated with behavioural observations of the individuals, and remained consistent 4.5 years after the initial assessments were made. In this study, animals rated highly on 'sociability' engaged in more affiliative interactions with other animals, whereas those rated highly on 'confidence' displayed more aggressive behaviour. Measures of personality can therefore be considered as objective and repeatable: characteristics that are necessary for measures in the scientific study of animal welfare.

By its very nature, knowledge of an animal's personality can provide us with an insight into how it will perceive and react to factors in its environment, which ought to enable us to assess its welfare under different conditions. In humans, an association between personality differences and resilience or susceptibility to ill health has been demonstrated, and this seems to hold for animals as well. Cavigelli *et al.* (2006), for example, demonstrated that the degree of **neophobia** observed in mice could be linked to the onset and development of pathology, and the eventual death of the animal; this appeared to be due to underlying differences in neuroendocrine function between mice, which was itself linked with different temperaments. **Neophilic** animals, those

---

Exogenous stimuli are those that arise in the environment outside the animal, whereas **endogenous** stimuli arise from within the animal.

The suffixes *–philia* and *–phobia* are derived from ancient

Greek and, added to words, specify an affinity or liking for, or an irrational fear or avoidance of, something, respectively. Thus, **neophilia** is an attraction to novel objects or new situations, whereas **neophobia** is an avoidance of them.

Figure **7.1** The way in which individual animals react to each other and to their environment can be affected by their personality. Studies of personality in non-human animals have concentrated largely on domestic animals, but some work has been done on exotic animals, such as the cheetah *Acinonyx jubatus* (McKay, 2003). (Photograph: Paignton Zoo Environmental Park)

that actively explored their environment, were more likely to die later than their neophobic siblings.

Personality differences in **exotic animals** have been used to predict the survival of reintroduced animals such as swift foxes *Vulpes velox* (Bremner-Harrison *et al.*, 2004) and also to assess the compatibility of breeding pairs and the adaptability of individuals to changes in husbandry and management regimes in species such as cheetah *Acinonyx jubatus* (Wielebnowski, 1999; see **Fig. 7.1**), gorilla *Gorilla gorilla gorilla* (Kuhar *et al.*, 2006), and black rhino *Diceros bicornis* (Carlstead *et al.*, 1999a; 1999b).

Species differences

Within zoos, there are many different species of animals that are adapted to different environmental conditions and thus are visibly different. Although, as we have seen, different individuals perceive

information about their surroundings differently, nevertheless, there are often physical characteristics that are shared by individuals of the same species and which may be different from those of other species. These characteristics affect the animals' perceptions and thus also their responses to **stimuli**.

A good example of this kind of species difference is hearing sensitivity. When Voipio *et al.* (2006) controlled for the difference in hearing sensitivity between rats and humans, rats were found to have much more sensitive hearing than humans. This study demonstrated that the hurried routine cleaning of stainless steel cages in a laboratory generated noise levels that exceeded 90 dB (weighted for the rat's sensitivity) and would not be considered a safe working environment for humans (HSE, 2006). Minor changes to the husbandry routine, using polycarbonate cages in an unhurried fashion, could, however, make a significant reduction of 10–15 dB to the noise generated from this daily husbandry procedure. We know very little about noise levels in zoos and how they affect animal welfare.

A second example of species differences concerns **stereotypies**. We have already discussed these behaviours and looked at some examples from a range of species (section **4.4.3**). There is some evidence that the propensity to perform stereotypies is heritable in species as different as African striped mice *Rhabdomys pumilio* (Schwaibold and Pillay, 2001) and mink *Mustela vison* (Jeppesen *et al.*, 2004).

7.1.2 Changes in our views of welfare

People's perceptions of animal welfare vary according to factors such as the animal's role in society (whether it is considered useful or a nuisance), how familiar we are with it, how attractive we find it, and how rare it is (Appleby, 1999). **Figure 7.2** illustrates how just one species, the rabbit *Oryctolagus cuniculus*, is viewed within our society, ranging from being denigrated as a **pest**, to sharing our home as a pet. The lives of these rabbits will be very different depending on how we see their role within our society, but they nevertheless experience the world in the same way and thus will have very different levels of animal welfare. Many of the species held in zoos are charismatic, rare, and considered attractive, so most people 'like' them and therefore expect them to be maintained in conditions that will result in a very high level of welfare (Appleby, 1999).

People have these different views of animal welfare because of differences in their personal attitudes, emotions, and factual knowledge. But it also appears to be the case that people within different societies often share common views about animal welfare, because these views are shaped by common and shared interests, including culture and religion (Boogaard *et al.*, 2006; Doerfler and Peters, 2006; Heleski and Zanella, 2006; Signal and Taylor, 2006). How we view an animal will impact directly on its welfare, because it will be treated differently as a

(a)

(b)

Figure **7.2** Our perceptions of animals affect the conditions that we impose upon them and the degree to which we are concerned about their welfare. The rabbit can be a family pet (a) or a laboratory animal, or a pest species (b). Which of these animals do you think would have the better welfare? (Photographs: (a) Miroslava Arnaudova, www.iStockphoto.com; (b) Rob Howarth, www.iStockphoto.com)

consequence of our expectations for its welfare. This happens through two routes: firstly, the creation and enforcement of legislation, and secondly, consumer pressure. Animal welfare legislation—which, in the UK, extends back to 1876—has changed continuously over the years in response to societal values (Kohn, 1994; see **Chapter 3** and section **7.5.4**).

'Consumer pressure' is the concept of how societal values can influence an issue (in this case, animal welfare) when consumers follow through on their convictions by purchasing only products that they condone. This can be facilitated and directed through coverage in the mass media. Within the agricultural sector, this concept is termed the **'willingness to pay' principle**; improvements to farming systems that improve animal welfare cost money, so consumers can promote welfare by making informed purchases and by paying more for animals farmed under 'improved' systems. Unfortunately, it appears that people's 'willingness to pay' is not always consistent with societal values, so the cheapest product can still remain popular even if 'societal values' would suggest that it should not be favoured. A survey conducted by Maria (2006) demonstrated that, while many people felt that improvements to farming methods were necessary, this did not influence which farmed products they purchased. In such situations, consumers are not, in fact, exerting pressure to improve conditions, but are supporting the continued use of 'suboptimal systems of farming'.

The impact of consumer pressure in zoos has not been studied systematically in the same way as it has in agriculture. Nevertheless, it does seem to have shaped certain animal housing and husbandry practices, as we have already discussed in terms of a bias by zoos towards meeting the needs of visitors when designing enclosures (see **Fig. 7.3** and **Fig. 6.7**). It is also apparent in the avoidance of certain practices as management tools, such as **euthanasia**, perhaps because, even though they may be the most appro-

priate action to ensure good welfare, they nevertheless gain negative coverage in the media. It is likely that consumer pressure on zoos is effective because they are heavily reliant on the money that visitors bring to the zoo.[1] Furthermore, it is unlikely that many zoo visitors consider their visit to be a necessity, whereas considering buying food or medication would be seen that way. So a decision to stop visiting zoos would probably constitute a relatively small effort on the part of the visitor, but would have a disproportionately large impact on the zoo. Hence zoos seem eager to meet the needs of visitors and to avoid the use of consumer pressure against them.

7.2 Animal welfare science

Science involves the use of standardized repeatable methodologies that give rise to objective data. Many researchers and practitioners in the field of animal welfare aim to follow scientific principles when attempting to understand animal welfare better, because the data collected this way are considered to have a high degree of validity and can therefore be believed. It also avoids simply relying on human perception of the animal's state in order to interpret its welfare. That is not to say that subjective assessments of welfare are of no value, however, and work by Wemesfelder (1999, for example), in particular, has pioneered this approach and has demonstrated the biological validity of some subjective assessments of animal welfare. Nevertheless, the value of many subjective ratings of animal welfare is still unproven and therefore the implementation of results using these types of data are not widely accepted. Objective measures of welfare, on the other hand, ensure consistency, and using them preserves the concept that animal welfare is independent of our perceptions of the animal and its role in our lives, or the lives of others.

---

1 In the UK, for example, many zoos are charities, with no external funding.

(a)

(b)

Figure **7.3** Given that many zoo visitors are attracted to, and spend much of their visit to the zoo watching large and charismatic species, it is not surprising that the lives of these species in zoos and the study of their welfare attracts more attention than that of other species. (Photographs: (a) and (b) Moscow Zoo; (c) Vienna Zoo)

(c)

Figure **7.3** (continued)

When considering what animals may need and thus what we should provide for them to ensure that they have a high level of welfare, four common approaches have been used, described as the study of:

- animal minds, by considering whether animals have emotions or **consciousness**;
- animal bodies, by assessing the ability of animals to thrive and survive in their environment;
- animal natures, by considering the extent of behaviours in wild animals (for example, Fraser and Matthews, 1997; Appleby, 1999);
- ourselves and other animals, in which we draw analogies from ourselves about the needs and abilities of other animals (for example, Sherwin, 2001).

These approaches are not mutually exclusive and overlap considerably, but an awareness of them is useful when we later consider stimuli that may be thought of as compromising welfare, or the indices that we can evaluate to measure welfare (sections **7.3** and **7.4**).

7.2.1 Animal minds

Is the animal 'happy'? This seems to be an extremely appropriate question when considering animal welfare, because the emotions that an animal feels in response to stimuli underpin how it perceives those stimuli and thus whether its welfare is enhanced or compromised. **Box 7.1** explores the concept of pain, which is thought only to compromise welfare when the physiological reaction to withdraw from a painful stimulus is coupled with an aversive emotional sensation (Gregory, 2004).

Box **7.1** Pain

'Pain in animals is an aversive sensory experience caused by actual or potential injury that elicits protective motor and vegetative reactions, results in learned avoidance behaviour, and may modify species specific behaviour, including social behaviour' (Zimmerman, 1986).

As such, pain is considered to be an important part of the **learning** process of animals. It is likely that pain in animals, as in humans, is associated with suffering (reviewed by Rutherford, 2002). Pain is thought to operate on both a physical and an emotional level. As with other kinds of sensory system, there is an underlying physiological basis to pain. Specialized receptors (termed 'nociceptors') detect changes in heat ('thermoreceptors'), pressure ('mechanoreceptors'), and chemicals ('chemoreceptors') from noxious stimuli. The process of transporting the information from these receptors to the central nervous system (CNS), where it is integrated to promote a response, is illustrated in **Fig. 7.4** and is called 'nociception'. Nociception is essentially a reflexive neurological response to aversive stimuli, which does not necessarily involve perception of pain.

Whether animals actually perceive pain is a complex issue, in much the same way as is the study of animals' minds. Much of our knowledge about pain is derived from our human experience and understanding of the phenomenon. Because we are unable to quantify animal emotions, it is difficult for us to appreciate how they feel pain. Flecknell and Molony (2003) pointed out that differences in pain perception between humans and animals should be borne in mind, to ensure that some animal pain is not overlooked. Certainly, animals often show behavioural responses to aversive stimuli that we recognize as similar to our own responses to pain (reviewed by Weary *et al.*, 2006); it also makes evolutionary sense for animals to feel pain, so that they reduce or avoid further damage to already injured parts of the body, which provides time for recovery.

Pain perception can be divided into three different components according to its perceptual and behavioural effects:

1. a sensory component (or nociception), which leads to withdrawal and avoidance of the stimuli;

2. immediate emotional consequences (acute pain);

3. long-term emotional implications (chronic pain).

These three components appear to be located in different areas of the brain (Price, 2000), which are illustrated in **Fig. 7.5**.

Pain is clearly of welfare concern in itself, but also because it may lead to additional problems, associated with incapacitation, for example, hunger, thirst, and social conflict. Although pain can be modified in a variety of ways in humans, *'by opiates, hypnosis, the administration of pharmacologically inert sugar pills, other emotions, and even other forms of stimulation, such as acupuncture'* (Carlson, 2007), prevention is better than cure in animals. Unfortunately, all animals in zoos will experience pain at some time or another, because of illness, injury from the enclosure or a cagemate, transportation, or routine catch-ups. It is essential that the potential for pain is reduced and that, when it occurs, everything is done to make the animal comfortable.

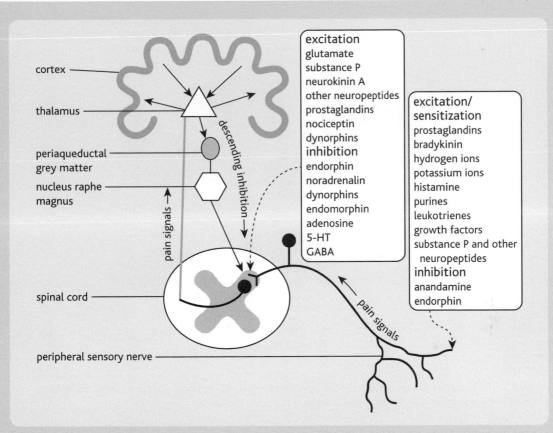

Figure **7.4** Information about noxious stimuli is detected via three different types of nociceptor: chemoreceptors, thermoreceptors, and mechanoreceptors. If any of these receptors is stimulated, information is sent to the central nervous system (CNS) via the sensory nerve fibres. The CNS integrates the information coming from the sensory nerves and initiates a response via motor nerve fibres. Information travels more quickly from mechanoreceptors than from other types of receptor, because their fibres are myelinated (they have a myelin sheath around the nerve fibres).

Of course, this requires that pain can be identified. This can be achieved in three ways (Weary *et al.*, 2006), by measuring: general body functioning; physiological responses; and behaviour.

Some general observed behaviours that have been associated with pain perception in animals include:

- abnormal posture, gait, speed, guarding behaviour;

- vocalizing—groaning, whimpering, squealing, screaming, growling, hissing, barking;

- aggression, withdrawal or recoil during movements, or manipulation;

- licking, biting, chewing, or scratching;

- frequent changes of body posture, restlessness, rolling, writhing, kicking, tail-flicking;

- impaired breathing pattern, shallow breathing, or increased rate of breathing;

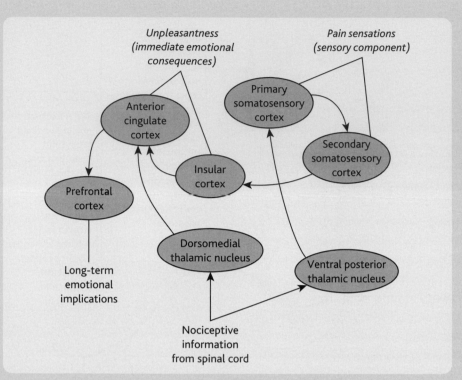

Figure 7.5 The sensation of pain is mediated through different parts of the brain, which determine the sensory (immediate) and emotional (long-term) components of the pain experience.

- muscle tension, tremor, twitching, spasm, straining;

- depression, sluggishness, hiding, lying motionless, seeking cover, sleeplessness (Gregory, 2004).

Many species express different behaviours as a result of pain that can be affected by their age and sex (Flecknell and Waterman-Paterson, 2000). Added to this, many of the species that are kept in zoos have evolved to hide any signs of pain, because showing them would make the animal vulnerable to predation or social pressures. One clear method of assessing whether pain is present is to administer **analgesic drugs**; if indicators of pain stop, then pain was present and has now been alleviated.

For example, Sneddon (2003) demonstrated that fish can feel pain using the following technique. Rainbow trout *Oncorhynchus mykiss* were injected with a noxious substance (0.1ml acetic acid) in their lips, after which they performed pain-related behaviours, rocking from side to side and rubbing their lips into the gravel flooring substrate of their tank. When the trout were administered a morphine injection intramuscularly (morphine sulphate 0.3g/1ml sterile saline), as well as the noxious substance, the expression of pain-related behaviours decreased significantly.

Analgesic drugs, more commonly known as 'painkillers', are compounds used to inhibit pain; taken from the Greek, *an–*, meaning 'without', and *–algia*, meaning 'pain'.

Relatively little is known about animal emotions or the functioning of animals' minds. There are several reasons why this is the case.

1. Animals' minds, like our own, are private and may remain forever subjective. Inferences about animal minds can be made through the collection of suitable indices of behaviour and **cognition**, but, on the whole, this discipline does not lend itself to objective and empirical evaluation. As a consequence, many animal welfare scientists have preferred to collect data following the animal bodies' approach (section **7.2.2**), in relation to which there are clear quantifiable units of measurement (Duncan and Fraser, 1997).

2. Animals' minds and emotions are misunderstood (Midgley, 1983; Welmesfelder, 1999), and so studies in this area are not as popular as they might be. This may be, in part, due to the convergence of different disciplines that all study animals' minds, but nevertheless use their own particular terms. Thus, biologists, sociologists, economists, psychologists, philosophers, and the mass media all have something to say about how animals feel and perceive the world, but use the language of their own discipline. Coupled with this, some of the terms may be used differently in the different disciplines, or may be used in a vague way, or may be considered to be—and thus used as—synonyms. Terms associated with the study of animals' minds, and which often cause definitional problems, include 'affective' (to do with emotion), 'cognitive' (to do with knowledge and the processing of information), 'consciousness' (**self-awareness**), and **sentience** (to do with feeling).

3. Until relatively recently, the study of animal minds was not considered to be scientifically valid. The attribution of emotion to non-human animals was a concept that developed in the eighteenth century, during the Enlightenment, and which has been a contentious issue ever since (Duncan, 2006). The concept went out of favour in the twentieth century, largely due to the influence of **Behaviourism**. This quote from Watson (1928), one of the founders of Behaviourism, neatly sums up its position:

> The behaviourist sweeps aside all medieval conceptions. He drops from his scientific vocabulary all subjective terms such as sensation, perception, image, desire and even thinking and emotion.

Such was the influence of this approach that animals were regarded as automata, or 'black boxes', whose most complex behaviours were considered to be the result of a series of stimuli-response processes, and were neither influenced by 'consideration' nor affected by 'emotion' (Skinner, 1938). Griffin (1992) did a great deal to bring the issue of animal sentience back into the realm of scientific discussion, using extensive reviews of animal behaviour as evidence for his strongly held belief that the highly complex behaviours observed in many species indicated they were, indeed, sentient. More recently, Burghardt (1995) goes so far as to suggest that Tinbergen's (1963) 'four aims of ethology' (see **Box 4.1**) should be expanded to include the exploration of animals' subjective states.

The issue of sentience is still controversial, but even some of those who are unhappy with attributing sentience to animals suggest that we should give them the benefit of the doubt, just in case (Barnard and Hurst, 1996)—a concept that underlies the fourth approach to the study of animal described in this chapter, of welfare by analogy.

Behaviourism was a scientific discipline that viewed mental events as unknowable and, in its most extreme form, denied that they existed. This led behaviourists to study only observable behaviour, so that their view of animals was effectively of automata whose behaviour was shaped by stimuli through a process of **operant conditioning**.

And this brings us back full circle to the eighteenth-century Enlightenment and the sentiments of philosopher Jeremy Bentham who, in 1789, is famously quoted as declaring: 'The question is not, "Can they reason?" nor, "Can they talk?" but, "Can they suffer?"'

7.2.2 Animal bodies

How well does the animal cope in its environment? The 'animal bodies' approach to welfare is based on the concept of **homeostasis**, which is about the processes that maintain the animal's body in a stable state (or a state of balance) so that it can carry out necessary survival functions. For example, when an animal's energy reserves run low (endogenous **stimuli**), hunger is triggered and this motivates the animal to find food; alternatively, hunger might be stimulated by the presence of food in the environment (exogenous stimuli) and so the animal eats. Animal welfare is considered to be compromised and the animal unable to cope either if homeostasis cannot be achieved, or if the physiological and/or behavioural responses that are implemented to achieve homeostasis put undue pressure on the animal, and, as a consequence, it is no longer able to 'cope' within its environment effectively. It is important to note, therefore, that the reaction of the animal to stimuli is not, of itself, evidence of welfare being compromised, if the animal is able to react appropriately and still continue functioning with ease. Welfare is only compromised if the stimuli inhibit the appropriate functioning of the animal.

Unlike the study of animal minds, the animal's reactions to stimuli, and subsequent functioning, can be measured systematically and quantitatively, and, as such, a variety of measures can be collected to study the ease, or otherwise, with which an animal is able to function in its environment. Furthermore, conditions can be experimentally manipulated to investigate how they affect welfare.

This approach is based on the stress concept, which was formalized by Selye (1973) and outlines how an animal's response to exogenous and endogenous stimuli (termed **stressors**) might be interpreted (see **Box 7.2**). The stress concept recognizes that the type of stressor, stress response, and individual differences affect animal welfare (Ewbank, 1985). The physiological process that underlies the animal's reaction to the stressor is mediated by the hypothalamus, pituitary, and adrenal cortex—known collectively as the **hypothalamo-pituitary-adrenal (HPA) axis**—which is illustrated in **Fig. 7.6**, in which you can see that stimulation of the pituitary by a stressor leads to the secretion

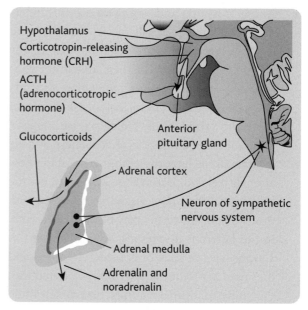

Figure 7.6 The stress response is mediated by the hypothalamic-pituitary-adrenal (HPA) axis. The hypothalamus instructs the anterior pituitary to secrete hormones, which, in turn, exert an effect on the outer part of the adrenal glands (the adrenal cortex); this, in turn, secretes glucocorticoids. Nerve pathways from the hypothalamus affect the inner part of the adrenal cortex (the medulla), which, in response, secretes catecholamine hormones (adrenalin and noradrenalin). Note the similarities to the HPG axis (**Fig. 9.5**).

Stimuli represent any information that the body processes. In terms of **homeostasis**, stimuli can be derived from internal bodily changes, e.g. increases in stomach size after eating, which are detected by mechanoreceptors, or changes external to the body, e.g. increases in temperature, which are detected by thermoreceptors. When discussing stress, stimuli that require homeostatic regulatory changes are often referred to as **stressors**.

Box **7.2** Stress

There is an optimal state that is necessary to ensure that the body can function efficiently (termed a 'stable state'); deviations from this state can reduce the animal's ability to function and can even be fatal. Exogenous (within the body) and endogenous (outside of the animal's body) stimuli put pressure on the animal's body and cause it to deviate from its stable state; an animal's body is considered to be under stress at this time. Stress stimulates various biological regulatory mechanisms to kick in and bring the body back to a stable state; this whole process is termed homeostasis, or the homeostatic control system, whereby the body is maintained at a stable state.

The biological process of stress is therefore one that animals experience a lot of the time and, in most circumstances, it goes unnoticed, because appropriate homeostatic regulation brings the body back to a stable state. This means that stress does not necessarily indicate poor welfare and, hence, some have suggested that the term stress be restricted for use only when pathology occurs—that is, to infer negative welfare—because this is frequently how it is used in practical situations by people working with animals (Fraser *et al.*, 1975). It is important that signs of physiological and behavioural measures indicative of stress are interpreted with extreme caution, which is helped by understanding more about the stress response itself.

Selye (1973) considered that stressors led to a stress response that was stereotyped (always the same), triphasic (made up of three sequential phases: alarm response; resistance; exhaustion), and non-specific (occurs regardless of stressor type). He referred to the whole response as the 'general adaptation syndrome' (GAS). The alarm response is induced by the **sympathetic nervous system** and is responsible for the 'fight or flight' response, during which time the animal is physiologically prepared for muscular exertion as blood supplies are rerouted to supply priority organs, and energy is metabolized and directed towards muscles, away from vital organs. An alarm response cannot be maintained indefinitely, so, if the animal is exposed to a considerable stressor and it is not able to move on to the resistance phase, it will die within a few hours or days. If the stressor continues, the animal will reach the resistance phase, whereby the animal makes adaptations to the way in which it functions, which may be seen as changes in behaviour or physiological processes. Resistance is characterized by overproduction of **glucocorticoids** and a decrease in lymphocyte number, which characterizes impaired immune function and makes the animal susceptible to disease. If the animal is unable to adapt, or if the stressor is extreme, then the animal may enter the exhaustion phase, and it is at this point that the animal's biological functioning is compromised and it can no longer 'cope' in its environment. If the animal's biological functioning is severely compromised, it will die.

▶

The **sympathetic nervous system** is a part of the nervous system, the effects of which include increasing activity, speeding up the heart and circulation, and slowing down digestive processes.

Glucocorticoids are steroid hormones produced by the outer part (cortex) of the adrenal gland and are normally involved in carbohydrate metabolism. One of the best known is **cortisol**.

Type of stress (intensity)*	Phase (according to GAS)[†]	Animals' state[‡]	Biological observations
Innocuous	Alarm (initial)	Eustress	Short-term changes, no compromise to functioning
Aversive	Resistance (continued)	Overstressed	Long-term changes, ability to function impaired
Noxious	Exhaustion (long-term)	Distressed	Unable to adapt to stressor and so damage incurred
Extreme			Death is likely

Table 7.1 The impact of stressors on the biological function of animals

* After Broom and Johnson (1993)
[†] After Selye (1973)
[‡] After Ewbank (1985)

NOTE: The impact of stressors on the biological function of animals is determined by their type, duration, and intensity. The table shows how the different phases of the general adaptation syndrome (GAS) are related to other means of defining stress, i.e. whether the stress is innocuous—termed 'eustress' by Ewbank (1985).

The duration and intensity of stressors, and how they are perceived by the animal, will determine whether an animal moves through all three phases of the stress response or stops at one of them (Dantzer, 1994). **Table 7.1** illustrates how different stressors might affect the animal, and the point at which they might be considered aversive and lead to animal welfare problems. Essentially, if a stressor reduces the animal's ability to function in its environment, animal welfare can be considered to be compromised.

of adrenocorticotropic hormone (ACTH), which, in turn, stimulates the adrenal cortex to secrete **corticosteroids**—namely, cortisol and its metabolites; hence their pivotal role as indicators used to measure stress (this is further explored in **Box 7.3**). These physiological changes facilitate any physiological and/or behavioural responses that should return the body to its normal state.

7.2.3 Animal natures

The third approach to animal welfare considers that deviations from 'normal' or 'wild' reflect compromised welfare, so wild **conspecifics** are held up as examples of animals with good welfare and used as a template from which the welfare of other animals is judged. It follows that, if an animal has evolved as a consequence of performing certain functions,

then the inhibition of these will reduce welfare; this approach is most commonly adopted using behaviour as a welfare index, so a comparison is made between the behaviours of captive and wild animals (see sections **4.5** and **8.3.1**). **Behavioural restriction** is considered to result from situations in which animals are not able to express their full repertoire of behaviours (see section **7.3.2**).

There are some problems in adopting this approach. As Barnard and Hurst (1996) point out, the welfare of many wild animals might itself be considered poor. Indeed, some behavioural patterns, such as aggression and flight responses, occur in the wild when animals are under a high degree of stress (Spinka, 2006). So, in fact, only a selection of the behaviours observed in the wild, which are thought appropriate for captive animals to display, are usually considered in any comparison of wild and captive animals. There are also some methodological difficulties in undertaking wild–captive comparisons (Veasey *et al.*, 1996a; 1996b; see section **4.5.1** for more on this).

In some situations, there may be a deleterious mismatch between the environmental pressures that have shaped an animal in its native habitat and those that it faces in captivity. Historical records show that breeding and survival in zoos has improved in recent years (Kitchener and MacDonald, 2004), which may be due to improvements in husbandry and housing (**Fig. 7.7**; see **Chapter 6**) or may be due to adaptation to captivity. The net welfare impact of adapting to captivity is considered in section **7.3.3**.

7.2.4 Welfare by analogy

Our knowledge of ourselves and other animals is sometimes used to make assumptions about the functioning—and, in this context, welfare—of other animals for which we have little information. This approach believes that we should attribute the 'benefit of the doubt' to animals in the hope that we prevent potential suffering, even when there is no evidence to suggest that they can suffer, and it is a strongly held view. Despite there being very little scientific evidence to support this use of the principle, it is still widely accepted within the scientific community (Dol *et al.*, 1999).

The most rigorous implementation of this principle matches physiological and behavioural processes in humans and other animals, and assumes that there are analogous functions between the two. At its weakest, this is a form of **anthropomorphism**.

7.2.5 Integration

Although these four approaches to animal welfare have been described separately here, they should not be used alone and, indeed, relying upon only one approach has its limitations. Ironically, despite

the study of animal bodies being the approach that dominates the field of animal welfare, most would agree that it really is a knowledge of animal minds that is essential to understanding animal welfare. Rushen (2003) suggests that, without including an animal's subjective experiences alongside measures of physiological and behavioural responses to stimuli, the meaning of animal welfare is lost. Duncan (1993) supports this view, declaring that '*neither health, nor lack of stress nor fitness is necessary and/or sufficient to conclude that an animal has good welfare. Welfare is dependent on what animals feel*'.

Furthermore, an individual's emotional perceptions of the stressor will affect how the stress response impacts upon its welfare (Dantzer, 1994). For example, the physical exertion necessary when either chasing (if a predator) or being chased (if the prey) is extreme and is an example of a high-intensity, short-duration stressor—but we can imagine that the emotional perception of the stressor will be viewed rather differently by the animal being chased than it is by the animal doing the chasing.

The need to include the individual's perception of the stressor is mentioned throughout this section, but, unfortunately, measuring this variable is difficult. And because animal minds are so difficult to interpret, we need to supplement our knowledge of what animals' needs may be on the basis of their natures and through analogy.

7.3 What can compromise zoo animal welfare?

Studies of animal welfare in agricultural and laboratory situations have led us to identify some of the circumstances that are associated with compromised animal welfare, of which, two appear to be particularly significant: **environmental challenge**; and behavioural restriction.

Anthropomorphism refers to the attribution of human qualities to animals when there is no biologically valid evidence to do so.

(a)

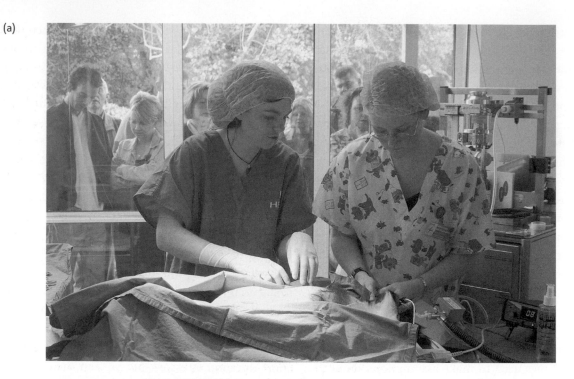

(b)

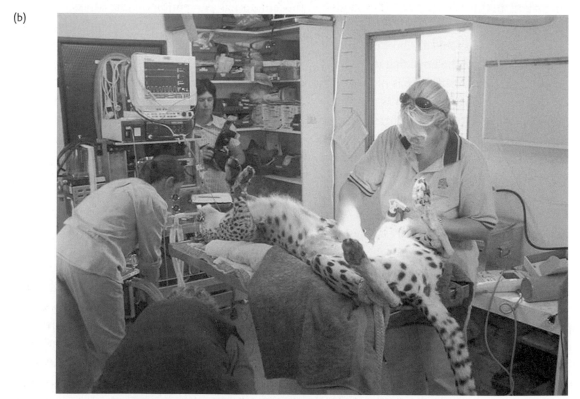

Figure **7.7** Advances in the health care available to exotic animals in zoos, as well as a better understanding of animals' nutritional requirements, are largely responsible for the decline in mortality rates seen in zoos. There is more about this in **Chapter 11** on 'Health'. (Photographs: (a) Rob Cousins, Bristol Zoo; (b) Chris Stevens, Werribee Open Range Zoo, Victoria)

The management of animals in zoos presents additional challenges to welfare that are less of an issue in farm and laboratory animals, such as the steps undertaken for conservation and animal management programmes (for example, transport and introductions), and the factors associated with exhibiting animals (for example, human–animal interactions and housing conditions). All of these factors can nevertheless be described as presenting environment challenge or behavioural restriction, or resulting in adaptation to captivity.

7.3.1 Environmental challenge

The term 'environmental challenge' is used to denote the presence or absence of properties within the environment that may act to overstimulate or understimulate the animal. Morgan and Tromborg (2007) considered that zoo animals were '*confronted by a wide range of potentially provocative environmental challenges*' peculiar to living in captivity. These included artificial lighting, exposure to loud or aversive sound and arousing odours, or reduced retreat space. Whether in captivity or in their native habitat, animals have, of course, evolved to adapt to environmental challenges, including such things as changes in temperature, finding food, orienting themselves within their home range, and living in a complex social system. Given that many zoos maintain animals from all around the world and also transport them worldwide as part of **captive breeding programmes**, zoo animals are likely to experience, at the very least, a climate, day length, and seasonal variation that is foreign to them. (This is reviewed in more detail in **Chapter 6** on 'Housing and husbandry'.)

Many of the issues associated with environmental challenge overlap each other, but, for the sake of simplicity, we will review them in terms of two factors: a lack of control or predictability; and novelty versus disturbance.

A lack of control or predictability

Frustration can develop—and thus welfare becomes compromised—in animals that lose control within

their environment, or when events in their lives are not predictable. This is just as true for positive events in the animal's environment, such as the provision of food, as it is for aversive events, such as being caught up for a health check. Control and predictability are inexplicably linked with each other, because control can be gained in a predictable environment or, from the other perspective, control makes the environment predictable (Bassett and Buchanan-Smith, 2007).

An animal achieves control when its behaviour appears to determine the occurrence of an event or situation (Overmier *et al.*, 1980). In a series of experiments on laboratory rats that had no control or could not predict the occurrence of an electric shock, the rats developed 'stress-related' symptoms, including high corticosteroid plasma levels, stomach wall lesions, and compromised immune systems (Weipkema and Koolhaas, 1993). Rats that could predict the shock, because of a light that came on to precede the shock, or that could control it by turning a wheel that interrupted or stopped the shock, developed no more stress-related symptoms than 'control' animals who did not receive a shock at all (see **Fig. 7.8**).

Bassett and Buchanan-Smith (2007) distinguish two different kinds of predictability:

1. **temporal predictability**, which describes whether an event is provided at a fixed time (and thus is predictable), or at different times (so is unpredictable);

2. **signalled predictability**, in relation to which a signal precedes an event and thus makes it predictable.

In the latter case, the reliability of the signal may vary, from the event always following the signal, to it rarely following the signal. A change in the reliability of the signal changes the predictability of an event.

Animal welfare is greatly improved with signalled predictability, regardless of whether the event being signalled is negative or positive. The study on rats that we have just described (Wiepkema and Koolhaas,

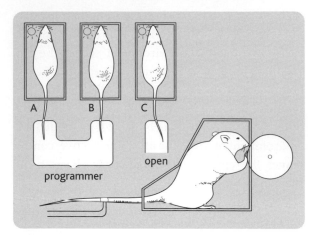

Figure 7.8 Rats housed in identical laboratory cages were positioned so that their tails were left protruding out of the cage. A weak electric shock was applied to the tails of the rats in boxes A and B; rats in box C were control animals and did not receive a shock. The rats in box A saw a light come on just prior to the electric shock, which provided a signal of the shock's imminent arrival, whereas rats in box B saw the light come on at random intervals so that there was no association between the light and the shock. The rats that had a signal (the light) prior to the shock showed a decreased response to the electric shock compared with those rats that were unable to predict its arrival. In a later experiment, rats in box A could prevent or interrupt the electric shock by turning a wheel at the front of the box when they observed the light. (From Wiepkema and Koolhaas, 1993)

1993) supports this interpretation, because knowledge of the onset of an aversive event (an electric shock) reduced stress-related symptoms in the animals. In another study, it was found that **enrichment** had a significantly greater impact on piglets (measured as increased play and decreased aggression) when a signal preceded its presentation (Dudink *et al.*, 2006). On the other hand, predictability within feeding regimes for some zoo-housed species appears to be detrimental to their welfare and has been associated with stereotypies (Carlstead, 1998).

After reviewing the literature, Bassett and Buchanan-Smith (2007) suggested that the provision of food may be an exception to the positive relationship between predictability and improved welfare. They suggest that food should be delivered on a temporally unpredictable schedule, but with a unique reliable signal preceding it. Clearly, more

research is needed to understand fully the implications of predictability and control on the welfare of zoo-housed animals.

Novelty versus disturbance

Wemelsfelder and Birke (1997) suggested that animals in captivity were usually understimulated compared with their wild conspecifics, because their ability to explore and experience novelty was reduced. A lack of challenge is thought to have negative implications for animal welfare, leading to the development of apathy, boredom, and behavioural abnormalities (Stevenson, 1983). In fact, the early development of enrichment aimed to counter this problem and was based on providing novelty to stimulate exploratory behaviours in animals (see section **8.6.2**)—but enrichments that are based on novelty have a very short useful shelf life: once the animals are familiar with the items, they lose their novelty and no longer stimulate exploration.

What if novelty is perceived negatively by the animal—that is, if the animal is neophobic? In this case, we might think of it as a disturbance to the animal and that it might compromise welfare, yet there is some evidence that these novel events may still have positive ramifications. Chamove and Moodie (1990) flew silhouettes of birds of prey over cages that housed cotton-top tamarins *Saguinus oedipus* in order to stimulate '*brief intense* [states of] *arousal*'. This, of course, mimics the sort of environmental challenge posed by real predators that the tamarins might experience in the wild. As a result, the tamarins showed fewer **abnormal behaviours** and more social behaviours, and the authors concluded that the arousal that the model had induced was beneficial for them. Wiepkema and Koolhaas (1993) also thought that short periods of disturbance might stimulate captive animals, because they promoted vigilance.

Despite this, intentionally causing stress or frustration to animals in a captive environment is highly controversial (see, for example, Roush *et al.*, 1992) and, in some situations, this means that the sorts of feeding schedules that the animals would experience in the wild cannot be mimicked in captivity (for example, the provision of live prey—see section **12.3.1**). We are, of course, largely ignorant of how animals perceive negative events in their environment. We might, for example, expect veterinary procedures to be the most fearful events that a zoo animal encounters, and yet Langkilde and Shine (2006) found that moving yellow-bellied water skinks *Eulamprus heatwolei* to an unfamiliar enclosure raised their cortisol levels significantly higher than did **invasive procedures** such as toe-clipping, blood sampling, and microchip insertion (**Fig. 7.9**).

Fearfulness, then, may be caused by the presence of novelty or the absence of familiarity (as in the skink example). Neophobia is generally a sensible, innate precaution to potentially dangerous environmental conditions or objects. For example, wild birds have been observed to avoid eating novel prey items, because they may be poisonous, and rats are notoriously wary of unfamiliar foods.[2] Unfortunately, some husbandry procedures in zoos may inadvertently induce neophobia. For example, species that are highly reliant on odour as their major route of communication spend a great deal of time scent marking and this conveys valuable information about themselves to other animals. It also presumably makes the area in which they live more familiar to them. But the routine use of disinfectants and detergents to remove dirt, and thus reduce the risk of disease transmission, can also remove any olfactory information laid down by the animal inhabitants, thus reducing the familiarity of the enclosure and potentially compromising the well-being of the animals (McCann *et al.*, 2007).

There are other environmental challenges in the zoo environment and these can include factors of which we are aware, but perhaps do not notice, and also some (such as the olfactory signals just mentioned) that we may not be able to perceive. Noise from visitors and construction are both quite obvious disturbances through which zoos animals have to live. Noise generated by visitors standing

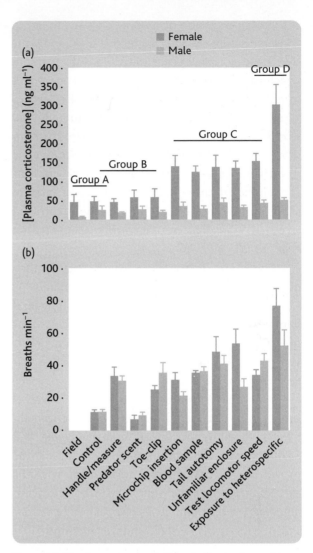

Figure **7.9** The stress response of twenty (ten males and ten females) yellow-bellied water skinks *Eulamprus heatwolei* was measured when they were exposed to different routine husbandry procedures. Plasma cortisol levels (a measure of the stress response), taken 1 hour after an event, are shown in (a); the animals' respiration rates are shown in (b). There were significant differences in plasma cortisol responses for husbandry procedures A, B, C, and D. The greatest elevation in cortisol occurred when the lizards were exposed to a heterospecific (a lizard from another species). (From Langkilde and Shine, 2006)

2 This wariness about novel foods probably explains the evolution of aposematism—that is, the use of warning colours or other conspicuous stimuli to advertise distastefulness or toxicity, as, for example, in ladybirds.

at a gorilla enclosure at Dublin Zoo was found to correlate positively with aggressive displays by the silverback gorilla, as illustrated in **Fig. 7.10**, in which you can see that visitors generated over 70 dB of noise (Keane and Marples, 2003). This level of noise is considered detrimental to humans and working regulations require hearing guards be worn when noise is this loud.[3] Similarly, environmental disturbance caused by construction work and concerts in the grounds of Honolulu Zoo was found to alter significantly the behaviour and faecal corticoid levels of Hawaiian honeycreepers (Shepherdson *et al.*, 2004).

There are other, less obvious, factors than noise that can have an impact on the animals. As an example, the flicker emitted from low-level fluorescent lights affects the mate-choice behaviour of European starlings *Sturnus vulgaris* (Evans *et al.*, 2006); we have little idea how stimuli such as this affect most of the animals housed in zoos.

But not all environmental challenges are negative: some can be considered stimulating to the animals. Meehan and Mench (2007) have suggested that cognitive challenges, for example, which can be provided by **environmental enrichment**, could be extremely beneficial for captive animals. The caveat, however, was that animals should possess the means, skills, or resources to overcome the challenge.

7.3.2 Behavioural restriction

Dawkins (1988) asserted that behavioural restriction —that is, the inability of animals to perform natural behaviours—caused them suffering. As an advocate of this approach, Rollin (1992) suggested that certain behaviours were of inherent welfare benefit to animals: a view that is often summed up as, 'birds gotta fly, and fish gotta swim'.

So what can stop captive animals from expressing their full behavioural repertoire? In general, this can happen because of the direct or indirect impacts

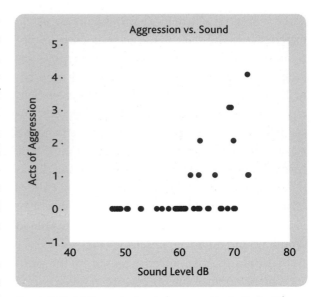

Figure **7.10** Visitor noise levels (measured in decibels, dB) at the gorilla *Gorilla gorilla* exhibit at Dublin Zoo showed a significant positive correlation with the number of aggressive acts expressed by the resident silverback male. (From Keane and Marples, 2003)

of housing and husbandry. Extreme examples are the use of wing-management techniques to prevent flying in birds (see section **6.3.4**), physical restraint, or the application of contraceptives to prevent breeding. Physical restraint is usually a temporary measure and can occur when an animal has been caught for a health check, or for some routine activity within the husbandry regime. For example, a traditional method of housing elephants involves **chaining** (or **picketing**; see **Fig. 7.11**), whereby one front and the diagonal back leg are chained; the elephant is able to move approximately one step forward and one step back. Circus elephants that had previously been chained in this way showed a significant reduction in the level of stereotypies that they displayed when they were provided with pens (paddocks) instead (Friend and Parker, 1999).

Elephant chaining is an extreme example, but there are less obvious aspects of housing and husbandry that can also cause behavioural restriction,

---

3 EC Directive 2003/10/EC on the minimum health and safety requirements regarding the exposure of workers to the risks arising from physical agents (noise).

Figure 7.11 The use of chaining in elephant husbandry routines has decreased markedly in recent times in zoos. This practice is still employed on occasion, however, when the behaviour of the elephant may be unpredictable and when keepers are operating at close quarters: for example, when a female is giving birth, as seen here. (Photograph: anonymous)

1. Animals have evolved over many years to perform these behaviours and are thus highly motivated to express them, so they will experience frustration if the behaviours are inhibited (Stolba and Wood-Gush, 1984; Shepherdson *et al.*, 1993; Lutz and Novak, 1995).

2. The performance of these behaviours is associated with positive affective experiences and long-term benefits such as enhanced adaptation to environmental stressors (Spinka, 2006).

3. A 'void' will be created in their absence and 'abnormal' behaviours will develop in their place (Anderson and Chamove, 1984; Chamove, 1988).

4. Animals will forgo cognitive stimulation if they are unable to express natural behaviours (Reinhardt and Roberts, 1997).

5. Behavioural restriction indicates that environmental challenges of captivity have affected the animal and suggests that the animal has not been able to adapt to those challenges (Hutchings *et al.*, 1978b; Redshaw and Mallinson, 1991).

It should be noted that these explanations are not mutually exclusive and that some of them may only partially explain what is happening to the animals. For example, it seems plausible that abnormal behaviours might fill the gap in the animal's behavioural repertoire (point 3), but it has also been argued that it is just as likely that behaviours of a negligible or beneficial nature could be performed instead (Reinhardt and Roberts, 1997).

7.3.3 Adaptation to captivity

When animals are in captivity, it is inevitable that successive generations will become adapted to the new environmental pressures in the zoo and thus adapt to life in captivity (Frankham *et al.*, 1986). Through this process, the nature of zoo animals can change, indirectly and slowly, and is continuing to

because they do not provide the opportunity for the animal to express some of its behaviours. For example, **flight distances**, which enable animals to get to a 'safe distance' from a stressor, may be severely shortened in enclosures that are too small. When animals in a **petting zoo** were able to retreat fully from visitors, their expression of undesirable behaviours was reduced (Anderson *et al.*, 2002). Restrictions placed on the expression, duration, and diversity of feeding and foraging behaviours are considered to be a principal cause of the development of abnormal behaviours in captive animals (Rushen and Depassillé, 1992).

Why should behavioural restriction impair welfare? Or, alternatively, we could ask why the ability to express natural behaviours promotes animal welfare. It does appear to be the case that benefits are associated with the ability to express natural behaviours and, conversely, that deleterious effects occur when they cannot be performed.

Possible reasons why natural behavioural expression should be facilitated include the following.

An animal's **flight distance** is the nearest distance that an animal will allow a potential danger to come before fleeing.

change. Animals that thrive and breed well in zoos pass on their genes to the next generation. If this process is permitted to carry on, then disproportionately more 'adapted' animals will contribute to the next generation compared with those that do not adapt. As might be expected, the impact of this process grows with each subsequent generation.

If they are adapting to zoo conditions, might we consider these animals to have good welfare, because they are coping with the new captive environmental factors? From an 'animal natures' perspective, any changes to animals that occur as a consequence of captivity should be investigated in terms of their potential to compromise welfare, because they compromise the nature of the animal. This matter is very complex, because adaptations to captivity are usually deemed to improve the welfare of captive animals: animals that do not adapt do not survive, or do so badly (and probably with impoverished welfare); animals that have adapted, however—that is, domesticated animals—usually have successful reproductive lives and show a reduced stress response to captive environmental stressors (Broom and Johnson, 1993). But while their welfare may be improved while the animals are in captivity, if the function of the captive population is to provide a pool of animals for **reintroduction** at a later date, then it is likely that the captive environmental pressures to which they have now become accustomed will not be like those that they will experience in the wild. So, in the short term (while they live in captivity), their welfare may be improved through adaptation, but, in the long-term (if animals are reintroduced back into the wild), their welfare may be compromised.

From a conservation point of view, this process of indirect domestication may present problems. Certainly, captive management programmes attempt to reduce any adaptation to captivity in terms of the genetic integrity of animals (see section **9.2.3**), but they cannot necessarily prevent changes in behaviour (see also section **10.5.6**)—although appropriate husbandry and enrichment might go some way to mitigating it (Shepherdson, 1994; see section **7.5**).

Consider, for example, the process of **socialization**, whereby animals routinely interact with people and become familiar with them (Mellen and Ellis, 1996; see **Fig. 7.12**), leading to changes in the human–animal relationship (**Chapter 13**). Here, we can see that there is a potential conflict between conservation goals and welfare goals. Socialization is usually considered to impair reintroduction success, so attempts have generally been made to limit its impact on captive animals, but especially on animals destined for reintroduction.

The few studies that have investigated the influence of socialization, however, have shown that it can improve the welfare of animals while they are in captivity. In small exotic felids, for example, socialization resulting from increased levels of interaction with keepers was associated with elevated rates of reproduction and a reduction in the cats' stress responses (Mellen, 1991). Similarly, there was a significant positive correlation between breeding success in the cheetah, and the predictability of the keepers' routine and the regularity of the cleaning and feeding regime (McKay, 2003). There can, however, also be negative repercussions associated with changing the human–animal relationship. For example, too much close contact with humans can impede reproduction and put keepers at risk of injury (this is covered in more detail in **Chapter 6**).

This adaptation to captivity has other consequences on the biological functioning of the animals, including more generalized deleterious impacts on their morphology, physiology, and genetics, thus also presumably impacting negatively on their welfare. For example, **inbreeding depression** becomes more likely in successive generations of captive populations, because there is a closed **gene pool**, usually made-up of few **founders**. **Inbreeding** is almost always detrimental to welfare. For example, inbreeding depression was negatively correlated with **longevity** in wolves *Canis lupus* and lynx *Lynx lynx* in Nordic zoos (Laikre, 1999). Similarly, Wielebnowski (1996) found that offspring from closely related cheetah were more likely to be stillborn or suffer congenital

Figure **7.12** All animals in zoos will be exposed to people to varying degrees, whether managed in a 'hands-off' or 'hands-on' style. Typically, contact between keepers and animals occurs during (a) routine procedures, when they are moved between enclosures, fed, or cleaned. More extensive interactions might also occur as a consequence of (b) training, and (c) providing enrichment. Contact with keepers will lead to the process of socialization, as during (d), this penguin parade. (Photographs: (a) and (b) Mel Gage, Bristol Zoo Gardens; (c) Emma Cattell, Howletts and Port Lympne Wild Animal Park; (d) Achim Johannes, NaturZoo, Rheine)

defects, compared with the offspring born to unrelated cheetah. (The management of species to decrease inbreeding is explored more fully in section **9.2.2**.)

There are also morphological changes that have been observed in a variety of species and which have occurred over successive generations in captivity. Some of these have been reviewed by O'Regan and Kitchener (2005), who have attributed the changes to a variety of different factors, including dietary deficiencies, behaviour, and age. Many animals live

Inbreeding depression occurs when related individuals interbreed and the resulting offspring have reduced **genetic** **diversity**, because both parents provide similar genetic material.

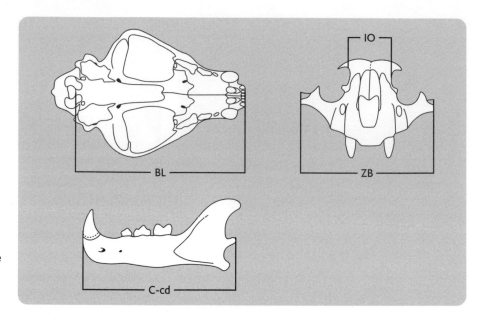

Figure **7.13** A study of the skulls of zoo-housed lions *Panthera leo* and leopards *Panthera pardus* revealed that one cranial dimension, zygomatic breadth (ZB), was significantly larger in captive animals compared with measurements taken from the skulls of wild conspecifics (see Table 7.2). (The measurements of three other dimensions were not significantly different.) (From O'Regan, 2001)

Table **7.2** Significant differences in the zygomatic breadth (ZB) of the skull measurements in zoo-housed and wild carnivores (From O'Regan, 2001.)

NOTE: Mann-Whitney U test results for the difference between modern captive and wild lions and leopards.
* <0.05.

Species	Sex	Wild (*N*)	Zoo (*N*)	Wild median	Zoo median	Alpha
P. pardus	Male	19	5	135.0	137.0	0.6695
	Female	13	4	119.0	137.5	0.0312*
P. leo	Male	11	8	223.0	249.5	0.0289*
	Female	11	9	198.0	210	0.0365*
P. leo	Male	6	5	233.5	261.5	0.0137*

longer in captivity than they would in the wild and some of the morphological changes that they experience could impair their survival in the wild, if they were to be released. **Carnivores**, for example, might have difficulty killing or processing prey. The jaws of American alligators *Alligator mississippiensis* have become shortened in captivity, although their heads are broader and their bodies more robust than wild alligators (Erickson *et al.*, 2003). The skulls of leopards *Panthera pardus*, lions *Panthera leo* and tigers *Panthera tigris*, on the other hand, appear to have become larger (O'Regan, 2001; Duckler, 1998; see **Fig. 7.13**). The cause and impact of these morphometric changes in zoo-housed carnivores is unclear.

It is unlikely that **nutrition** is the cause of this deviation from the wild-type skull, because, in this case, we would expect an increase in the overall size of the skull. Anecdotally, it has been suggested that these changes may have arisen out of 'lack of use', because many captive carnivores do not need to bring down large-bodied prey or mechanically process them as they would in a live-kill situation (although some enrichment does aim to simulate these behaviours).

One pragmatic solution to the potential advantages and disadvantages associated with adaptation to captivity would be to designate individual animals, as far as possible, as being destined either for reintroduction or for a life in captivity. As this

Founders are the individuals from which populations derive. If there are only a small number of founders, then genetic diversity is likely to be low.

is unlikely to be implemented, however, some sort of compromise involving degrees of domestication might be the best recommendation, whereby some natural behaviour is promoted, but the animals still interact with keepers (Newberry, 1995). On this basis, the benefits of adaptations required for life in the wild, as well as those that may better serve life in captivity, can both be enjoyed.

7.4 Indices used to evaluate zoo animal welfare

Many people feel justified in judging the welfare of zoos animals according to what they see when they pass an enclosure. But this glimpse does not usually tell the visitor very much and it certainly does not enable them to interpret whether the animal is suffering from environmental challenge or behavioural restriction, or is adapting to its captive surroundings.

There are many indices that can be used to assess zoo animal welfare, including the diversity of the animal's behavioural repertoire, its enclosure use, the extent of its social interactions, its **life history traits**, and many more. Most zoo visitors are unaware of these indices and, instead, judge welfare by the level of activity that they observe in the animal and whether its enclosure is aesthetically pleasing (Melfi *et al.*, 2004a; see **Fig. 6.9**). Given the complexity of the relationship between activity levels, enclosure style, and animal welfare, however, using only these to assess welfare seems highly unreliable.

Establishing which indices can be used to evaluate zoo animal welfare is not easy. Some of the indices may only indicate that an animal has responded to a stimulus, but this does not necessarily reflect changes in the animal's welfare. Similarly, deviations from normal functioning, whether behavioural or physiological, do not necessarily indicate suffering (see section **7.2**). It is essential, therefore, that any indices that are used to evaluate animal welfare are first validated.

Validation is most commonly accomplished via two main routes.

- Animals can be subjected to an 'almost certain' stressful event or aversive stimulus, and a record taken of which indices change and how. An example of this is taking measurements when animals are being transported, an event that is regarded by many as stressful (see section **6.3.2**).

- Alternatively, new indices can be validated by monitoring them alongside other indices that have previously been validated as accurately monitoring animal welfare. Plowman *et al.* (2005), for example, were interested in studying whether management changes to a troop of Hamadryas baboons *Papio hamadryas* affected their welfare (see **Fig. 6.21**), and to find out they measured **self-directed behaviours** (see **Fig. 7.14**). The expression and frequency of these behaviours had already been validated as a reliable index of welfare, because previous studies (for example, Maestripieri *et al.*, 2000) had shown that they correlated well with other behavioural and pharmacological indices of poor welfare.

The majority of indices currently used to evaluate animal welfare are focused on those that are observed when animal welfare is compromised. It is assumed that, in these cases, the converse is also true and that either an absence of the index or an inverse change may reflect that the animal's welfare is not compromised. More recently, however, parameters associated with good welfare have also been studied, by collecting data during times of positive physiological and social situations that are indicative of good welfare. Seltzer and Ziegler (2007), for example, measured levels of two hormones associated with good welfare

Oxytocin and **arginine vasopressin** are both peptide hormones secreted by the posterior pituitary. Although both have a number of physiological effects in the body, both are also thought to be involved in social interactions and the mediation of positive social behaviours in the brain.

(a)

(b)

Figure **7.14** Self-directed displacement behaviours (sometimes referred to as displacement behaviours) can be a reliable indicator of anxiety in situations such as social tension. Here, you can see Hamadryas baboons *Papio hamadryas* showing some characteristic self-directed displacement behaviours—namely, scratching and yawning (Photographs: (a) Ray Wiltshire; (b) Paignton Zoo Environmental Park)

—**oxytocin** and **arginine vasopressin**—to assess the impact of social isolation on welfare in a troop of common marmosets *Callithrix jacchus*.

Another reason why it is necessary to collect data for many different indices is that zoo animals represent a diverse array of species and they are likely to be affected differently by environmental factors. Interpreting this complexity can be difficult. For example, oral and locomotor stereotypies in giraffe *Giraffa camelopardalis* and okapi *Okapia johnstoni* were found to be affected by different combinations of environmental factors (Bashaw *et al.*, 2001). We strongly recommend that several indices are used whenever trying to assess animal welfare, particularly in the zoo setting.

A final consideration is the time frame over which the welfare of the animal is being assessed. In section **7.3.3**, we said that welfare could be considered in both the short and long term, with regard to the impact of adaptations made by the animal to captive environmental variables. The duration of the stimuli that affect animal welfare can also be long or short term, and thus give rise to chronic or acute responses, as mentioned in section **7.2.2**.

Table 7.3 illustrates some of the physiological and behavioural changes that are associated with acute and chronic stress responses.

7.4.1 Life history variables and events

If we follow the 'animal bodies' approach, we might consider animal welfare to be impoverished if an animal is not able to cope in its environment, which, for our purposes, means that it has not been able to adapt to the captive environment. An animal that has successfully adapted to captivity might be expected to function as well in captivity as it does in its native habitat, thus giving rise to wild–captive comparisons of those parameters that represent success.

Generally, a successful animal is considered to be one that is able to reproduce and live a relatively long time. Thus, measures of **fecundity** and longevity can be used as indicators of an animal's ability to cope in captivity. Fecundity is a measure of reproductive

	Short-term symptoms	Long-term symptoms
Behaviour	Fight or flight Cessation of 'normal' behaviours	Aggression Stereotypies Apathy Self-harming Protective behaviour
Gross biological indices	↑ Heart rate, respiration, and body temperature	↓ Reproductive success, life expectancy, and growth
Endocrine function	Adrenaline Noradrenaline Adrenocorticoids Prolactin	Endogenous opioids Glucocorticoids
Nervous/immune system	Adrenaline Dopamine	Immunosuppression ↑ Incidence of disease
Enzymes/metabolic products in the blood	Renin (from the kidney)	Creatine kinase (from the heart)

Table **7.3** A summary of the symptoms expected according to whether stress is short or long term*

* After Broom and Johnson (1993)

NOTE: The impact of stressors varies according to how long they persist. Short-term stressors are associated with acute stress, whereas prolonged stimulation/stress leads to chronic conditions.

success that can be calculated in various ways, representing greater degrees of rigor: it can be calculated as the number of offspring an animal has produced, or as the number of offspring produced that survive (usually measured up to 30 days, up to 1 year, and then to adulthood if longer), or the number of offspring produced that were themselves able to reproduce successfully. Longevity, measured as the length of time that an animal lives, can also be a useful indicator of how well an animal has been able to survive and thus succeed.

Although animal welfare should be considered from the perspective of the individual, viewing life history traits for a group of animals can provide information about the extent to which the group (due to specific environmental conditions in one location) or whole **taxa** (due to shared species differences and thus ability to cope) are able to succeed. These types of analyses can highlight animals that have specialist lifestyles and which require certain characteristics from their environment, so if they do not gain these characteristics, they are unable to adapt and therefore will not succeed. Equally, these types of analyses can identify ways in which captive environments can, and should, be modified to facilitate success. Clubb and Mason (2007) have used this approach to study the ability of **Carnivores**

to succeed in captive conditions, using measures of longevity and the development of stereotypies to assess their welfare in captivity. They considered that animals with relatively short lives who performed stereotypies had poor welfare, because they were unable to adapt to captivity.

As with all indices used to assess animal welfare, there are some limitations to using life history variables in isolation. The fecundity and longevity of an individual can be affected by many variables, and these may or may not necessarily be related to welfare. It is common for animal management programmes to adopt strategies in which animals are not provided with the opportunity to breed: for example, because of limited space (see section **9.7**). Equally, an animal may die as a result of disease, or socially inflicted wounds, or euthanasia. It is necessary, therefore, that, when using these indices as a measure of animal welfare, details of the animal's life and its captive management are also considered, even though this can make the accurate analysis of these types of variables time-consuming.

7.4.2 Biological processes
The most widely adopted approach to assessing animal welfare is to monitor an animal's biological function, which can provide an indication of why

an animal succeeds, or otherwise, in thriving in its environment. There are many biological processes that underpin biological function, but most studies of animal welfare choose to examine endocrine function, because hormones—particularly cortisol—mediate the animal's reaction to stressful situations (**Box 7.3**). But other biological processes, such as the activity of the immune and nervous systems, also provide information about the animal's ability to cope within its environment. These physiological indices are usually studied in conjunction with observations of behaviour.

Methods used to measure an animal's biological function typically follow standardized protocols that enable the collection of reliable, objective, and repeatable empirical data. These measures are comparable within and between animals, and can be easily scrutinized, which makes them ideal for the

---

Box 7.3 The use of cortisol as a measure of animal welfare

The measurement of cortisol, and its metabolites, has long been recognized as a reliable indicator of adrenal function and thus provides information on an animal's stress response (see **Box 7.2**). The non-invasive assessment of cortisol has been conducted using faecal samples in a wide variety of non-domesticated species, including Alaskan sea otter *Enhydra lutris*, Roosevelt elk *Cervus canadensis roosevelti*, gerenuk *Litocranius walleri*, and Northern spotted owl *Strix occidentalis caurina* (reviewed by Wasser *et al.*, 2000). Other non-invasive techniques include the use of urine (for example, in elephants—see Brown and Wemmer, 1995) and saliva (for example, in gorillas—see Kuhar *et al.*, 2005). Davis *et al.* (2005) provides an example of a zoo study using cortisol as a welfare indicator, in a study of the effect of visitor numbers on spider monkeys *Ateles geoffroyi rufiventris*. They found that increased visitor numbers paralleled an increase in cortisol levels compared with instances in which there were no visitors present in the zoo (see **Fig. 13.10**).

Cortisol is released in increased quantities because of activity of the HPA axis during periods of stress; it is also involved in other physiological mechanisms. Elevations in cortisol associated with a stress response may also occur after a meal has been eaten or when an animal is active, and may be affected by the animal's rank, sex, or age (Lane, 2006). Furthermore, variations in cortisol levels can occur due to diurnal and seasonal rhythms, temperature, humidity, and several other factors relating the animal and its environment (Mormede *et al.*, 2007). Care should also be taken to ensure that non-invasive samples are handled and processed appropriately to ensure accurate estimations of cortisol concentrations. For example, Millspaugh and Washburn (2003) showed that glucocorticoids were not necessarily evenly distributed in faeces, so all voided faeces should be mixed thoroughly and a representative sample taken to reduce the likelihood of biases in the sample.

Despite the many factors that can affect the production and assessment of cortisol concentration, it is still recognized as being an accurate and helpful measure of stress, and thus of animal welfare. When using cortisol as a welfare indicator, consideration of its limitations should be borne in mind and complementary animal welfare indicators measured as well (Millspaugh and Washburn, 2004; Lane, 2006).

scientific study of animal welfare, and this is probably why their use has proliferated in this field. Historically, measuring physiological indices usually required some degree of invasive procedure, through blood sampling, biopsies, or by obtaining information at **post-mortem** after the death of the animal. This has the obvious disadvantage that the method of collection of data to monitor animal welfare may itself compromise welfare, or, at the very least, confound the results. Thankfully, technological advances have now gone a long way towards ameliorating this problem, because many biological processes can now be monitored via non-invasive approaches.

Biological processes are essential for survival, and an animal's responses to stimuli (sometimes termed 'stressors' or 'environmental challenge') do not necessarily represent welfare infringements. They may only reflect routine homeostatic function: glucocoticoids, for example, are important in energy metabolism, as well as in mediating the stress response. Because of this, the interpretation of these indices in relation to animal welfare requires some caution, because changes in different parameters do not necessarily reflect animal suffering per se. There are also many other factors, other than the animal's welfare, that affect changes in biological processes, including time of day, reproductive status, individual differences, etc.

The value of using biological processes as a measure of animal welfare is therefore in combination with data measuring other indices. This is exemplified by the way in which endocrine function is measured and how changes in cortisol concentration are interpreted in conjunction with other welfare indices. For example, behavioural observations and the non-invasive collection monitoring of cortisol have been used to evaluate the welfare of a number of zoo-housed species, including giant pandas *Ailuropoda melanoleuca* (Liu *et al.*, 2006; Powell *et al.*, 2006), clouded leopards *Neofelis nebulosa* (Wielebnowski *et al.*, 2002), and black *Diceros bicornis* and white *Ceratotherium simum* rhinoceros (Carlstead and Brown, 2005).

7.4.3 Behaviour

Observations of animal behaviour are non-invasive and therefore intuitively a method favoured by many animal welfare scientists. Behavioural expression provides a measure of animal welfare that overlaps all four approaches to the study of animal welfare:

- social interactions, problem solving, and the study of animal learning can provide some indication of the animal's mind;

- the development of abnormal behaviours may indicate that the animal's biological processes are unable to function within its current environment;

- an animal's inability to express behaviours seen in its wild counterparts indicate changes in nature;

- in some instances, it can be justified to interpret the function of behaviour by drawing on the knowledge gained from other species and using analogy.

For a long time, the study of stereotypies in zoo-housed animals has been seen as the chief indication of impoverished welfare. The development of stereotypies in zoo-housed wombats *Vombatus ursinus*, for example, was found to be influenced by the amount of time that the animals spent feeding daily (Hogan and Tribe, 2007). But the interpretation of the performance of stereotypical and other abnormal behaviours is not always clear-cut, although the performance of them does reduce **behavioural diversity** and can thus lead to behavioural restriction (Golani *et al.*, 1999). (The topic of stereotypies is explored further in section **4.4.3**.) It should also be noted that whether behaviours are referred to as 'normal' or 'abnormal' is also not straightforward; this is discussed in section **8.3.2**.

As with the use of other measures of animal welfare, behavioural indices should be validated to ensure that their interpretation is justified. Castellote and Fossa (2006) justified their use of underwater vocalizations as a welfare indicator in beluga whales *Delphinapterus leucas*, because they were found to decrease both after the introduction of four harbour seals *Phoca vitulina* into the exhibit and after air transportation to new facilities. Vocalizations have been shown to give a good indication of animal welfare in other species as well.

7.4.4 Health

Poor health can impoverish animal welfare through a variety of routes, not least because it limits an animal's ability to cope with challenges in its environment, can restrict behavioural expression, and can result in pain. There are several main ways in which to evaluate health status in zoo animals and thus infer something about their welfare, and much of this is discussed in **Chapter 11** on 'Health' (see **Fig. 7.15**).

The daily monitoring of animals, which is usually initiated to monitor animal health, can provide a helpful insight into what is considered normal for that animal or species, and deviations from these 'norms' alert zoo professionals to the potential that health and/or welfare may be compromised (see section **5.6.6**). The sorts of information that can be monitored include weight, condition, interactions with conspecifics, feeding behaviour, and fluctuations in these variables. Retrospective information about animal welfare can also be gained from analyses of post-mortem results, which provide invaluable details about the condition and welfare of an animal during its life (see section **11.4.8**) and how they affect its physical functioning. Such studies may not be possible while the animal is still alive, or the conditions may have gone unnoticed because few visible characteristics existed. For example, during post-mortem examination of many zoo-housed carnivores, Kitchener and MacDonald (2004) discovered that

(a)

Figure **7.15** Given that there is evidence from post-mortem results that old Carnivores in zoos can suffer from skeletal problems, it may be prudent to ensure that these animals are provided with plenty of physical exercise, so that that they remain physically fitter for longer (although this will not stave off aging indefinitely). (Photograph: Warsaw Zoo)

(b)

(c)

Figure **7.15** (continued)

most old animals had severe spine deformities that, they suggested, would have been painful during the life of the animals. These data initiated the discussion that old animals in zoos may have poor welfare and that longevity is not necessarily a goal that should be sought, but rather, quality of life.

7.4.5 Asking the animals themselves?

To understand how an animal is feeling and what implications this has for animal welfare, the following types of questions can be asked.

- Is the animal aware?
- Does it know that it is suffering or understand that another animal is suffering?
- Does it remember past events or fear future events?

Answering these kinds of questions and measuring the subjective experiences of animals is difficult, as we explored earlier; **preference tests** have, however, been proposed as a method that can indirectly tell us how an animal feels (Duncan, 2005).

Preference tests involve the simultaneous presentation of different resources to an animal. The resource that the animal chooses is assumed to be that which it prefers and therefore wants. This provides us with a mechanism by which to interpret how an animal perceives its own needs, in terms of what it wants from its environment. In order to evaluate animals' preferences fully, it is very important that we do not underestimate their environmental complexity and/or needs, so we should provide them with sufficient options from which to choose and indicate to us their preferred resource (Fraser and Matthews, 1997). The number and appropriateness of the resources presented to an animal during a preference test determines the value of the results. Simply learning which of two resources an animal prefers may be of little value to us if neither of the resources is what the animal actually 'likes'. The preferred choice may also be affected by previous experience, so an animal's use of a new resource may only temporarily reflect its tendency either to avoid or use novel resources.

Preference testing has been widely used to determine how animal priorities change and are influenced by genetic differences, reproductive status, and numerous different environmental factors. We should, however, be cautious in our interpretation of the choices made by animals in these types of studies, because they may not necessarily promote welfare. Animals may, for example, choose a tastier, rather than a healthier, food. Furthermore, the animals' decisions may only satiate short-term, and not long-term, biological functioning (Duncan, 1978). Therefore, the amount of insight that preference tests give us into the needs of animals is only as good as our experimental design allows.

Animals in zoos have shown a preference to express behaviours that they would perform in the wild. Elson and Marples (2001) found, in several parrot species, that the birds prefer food if it is presented in a manner that reflects their 'natural' feeding habits. Instead of feeding from dishes in their enclosure, or from enrichment devices representing alternative feeding styles, the parrots would use the enrichment devices that required the expression of their natural species-specific feeding behaviours. Thus, golden-shouldered parrots *Psephotus chrysopterygius* selectively used hanging branches with seeds glued to them, which enabled them to employ the sorts of manipulative behaviours that are used by their wild counterparts to gain food from seed pods and/or grass heads. Scarlet-chested parrots *Neophema splendida*, on the other hand, feed mainly on the ground and they showed a preference for foraging trays placed on the ground. This is also an example of **contrafreeloading**—a phenomenon that is explored in more detail in section **8.6.3**.

A variation of the preference test is the **strength of motivation test**, which assumes that an animal's willingness to 'work' for a resource reflects its level of motivation to gain the resource or carry out the behaviour. Elegant experimental designs have been used to calculate how much effort (time or energy) an animal is willing to expend to get access to a resource. This provides a measure of the strength of the animal's preference for a resource, and underpins **behavioural economic theory**. Dawkins (1983), a pioneer in this area, considered there to be a parallel between the choices made by human shoppers (in economic theory) and the choices made by animals when choosing between resources. This method provides a way in which to interpret the animals' preference for different resources, both when they are provided at different times during the day and throughout its life.

There are two main assumptions in this theory: that the resources available to the animal should be ones that it is likely to want, considering its natural history; and that budget constraints will affect choice. The term 'budget' encompasses the pressures on the animal that will determine what choice it makes, such as its reproductive status, or its hunger level. When restrictions are placed on an animal, we would expect that the most important resource will be chosen. This resource would be considered to be a necessity and is referred to as an 'inelastic product'. But in a situation in which there are no constraints on choice, the animal might instead choose a luxury, or 'elastic product'.

There are several further variations to the preference test, developed to measure the strength of an animal's preference, as follows.

1. Will an animal learn a behaviour in order to gain the resource? If not, then the demonstrated preference for that resource is considered weak.

2. An **obstruction test** puts something such as a door between the animal and its resources, and determines how hard the animal is willing to work to get around the obstacle to gain the resource.

3. Animals are offered a choice of resources, but are restricted by the amount of time that they have with them; the assumption here is that, as the time allowed to gain the resources diminishes, the animal will choose the resources that it most prefers or needs.

Figure **7.16** There are many studies of preference tests for mammal and bird species, but few for other taxa. At Paignton Zoo Environmental Park, the preferences of tortoises were measured by providing them with different resources, accessible by climbing up a ramp. The tortoises were willing to walk up a steeper ramp to gain access to their favoured resources. (Photograph: Paignton Zoo Environmental Park)

4. **Aversion tests** study if the animal is willing to choose to undergo something unfavourable, such as an electric shock, to get to a resource.

Some preference testing in zoo animals has been undertaken to determine which foods or enclosure locations are most highly sought after (see, for example, Ogden *et al.*, 1993; Ludes and Anderson, 1996).

More recently, attempts have been made to construct situations in which the motivation of zoo animals can be tested. For example, Nicholls (2003) measured the motivation strength of Hermanns' tortoises *Testudo hermmani* by providing them with a slope, on top of which were different resources. By changing the gradient of the slope, she was able to find out how steep a slope the tortoises would climb in order to get to the resource. She found that the tortoises would climb more quickly and up steeper slopes for sand, compared with twigs (see **Fig. 7.16**).

7.4.6 Cognition

If we follow the 'animal minds' approach to animal welfare, we find that current trends in this field are in favour of some attribution to animals of sentience and of experiencing feelings that matter (Webster,

2006). The extent of their cognitive capacity, which can be measured by the existence of **theory of mind** or self-awareness, is studied in this field to provide a better understanding of the degree to which the animal is able to experience emotion and therefore what influence this might have on welfare.

Gregory (2004) suggested that the complexity of an animal's mind is probably paralleled by its capacity to learn. Thus, learning capacities that go beyond simple stimulus–response connections may be indicative of higher mental processes and potentially suggest a greater capacity for emotion. As we, ourselves, are animals with complex minds and a high capacity to learn, we tend to put primates at the pinnacle of evolutionary intelligence whenever we think of differences in animals' abilities. Interestingly, however, as more data are amassed, the cognitive capacities of many non-primate species are being revealed to rival—or even to surpass—that observed in non-human primates. Sea lions *Zalophus californianus*, for example, have been observed to learn reverse-reward contingency tasks at a more rapid rate than primates (Genty and Roeder, 2006). In this task, the sea lions are required to learn that if they chose a smaller portion of fish when offered a choice, they would, in fact, receive a larger portion of fish to eat as a reward; this is a demonstration of self-control that implies an understanding in the animals about gaining a delayed reward (**Fig. 7.17**).

The complexity of animals' social systems and their interactions with the environment have also been used as an indicator of cognitive ability, because it is suggested that both of these factors would need to increase in parallel. For example, Simmonds (2006) reviewed the literature available on cetaceans and suggested that behavioural measures observed in these animals corroborated the belief that they had high levels of intelligence, which has far-reaching implications for their ability to suffer. Similarly,

Figure **7.17** As the cognitive abilities of more species are tested, the results are challenging the long-held belief that primates are at the 'top of the intelligence pyramid'. For example, sea lions can learn to choose a smaller portion of fish, in order then to receive a reward of a larger portion of fish. This complex task, choosing a smaller portion even though a larger portion is also available, was not achieved by non-human primates during the same trial (Genty and Roeder, 2006).

Douglas-Hamilton *et al.* (2006) used extensive behavioural observations to assert that elephants can show a variety of emotions, and have a general awareness and curiosity about death. This methodology follows closely that adopted by Griffin (1992), a proponent of the 'animal minds' approach to animal welfare, who used a great deal of anecdotal information about animal behaviour as evidence for the existence of animal minds.

7.5 Meeting the needs of zoo animals

Our knowledge of the welfare of zoo animals has advanced greatly in recent years, as the study of their needs, and how they can be compromised in captivity, has been addressed more and more in research. Furthermore, the appropriate monitoring of validated welfare indices permits the identification of factors within the zoo environment that may be compromising welfare and therefore allows remedial action to be taken in an attempt to reduce the impact of these factors.

Theory of mind refers to an individual's knowledge of another individual's mental processes. Self-awareness is having knowledge of oneself as an individual, distinct from other individuals. Both of these are regarded as higher-order cognitive abilities and the extent to which animals possess them is controversial.

Within the field of animal welfare science more generally, advances in knowledge have not necessarily been mirrored by the implementation of appropriate practices to ensure good animal welfare (Millman *et al.*, 2004). Dawkins (1997) suggested that this lag between knowledge and implementation was due to the differing perspectives of the many different parties involved, who included the animal welfare scientists, the public, and the politicians. This was because scientists needed more time to appreciate fully the complex issues underlying the factors that affect animal welfare, while the public were not harnessing their power enough to influence change through consumer pressure and the politicians were not always supportive of legislative change (Dawkins, 1997). On the basis of this, we might suggest that zoo animal welfare could be greatly improved if more dedicated welfare scientists were to study animals in zoos, if there were continued pressure from 'informed' zoo visitors for better standards, and if there were greater harmonization of legislation and regulation promoting zoo animal welfare worldwide.

More pragmatically, and within the scope of those currently working in zoos, a series of steps can be outlined that, if adopted, could greatly enhance animal welfare. These involve identifying what the animal's needs are, minimizing the risks to animal welfare, using a logical and consistent framework to decide how to resolve conflicts between welfare and the other goals that the zoo aims to achieve,

improving our knowledge of animal welfare science and how to best achieve it, and, finally, using this information to inform legislative reform and to support enforcement.

7.5.1 Five freedoms

The **five freedoms** encompass the basic needs of animals and, if they are satisfied, they should ensure an adequate level of welfare (see **Table 7.4** and **Fig. 7.18**). They were originally conceived as a tool to ensure a minimum standard of welfare for farm animals, but they have subsequently become ubiquitous in the assessment of welfare for all animals, whether kept in zoos, in laboratories, or as pets (Webster, 1994).

Although the five freedoms provide a good starting point from which to consider animal welfare, there are concerns that the use of minimum standards can sometimes prohibit advances in knowledge or the promotion of higher standards (Koene and Duncan, 2001). This is because minimum standards can sometimes be viewed as a benchmark towards which to aim, rather than as a standard beyond which to go. In addition, they aim to prevent observed changes in those indices that are associated with poor welfare (for example, the presence of stereotypies or weight loss), rather than aiming to promote signs of good welfare (such as the expression of behavioural diversity, or psychological well-being). Furthermore, Mench (1998) suggested that the

Freedom	Provision	Cross-reference
From thirst, hunger, and malnutrition	Access to fresh water A nutritionally balanced diet	**Chapter 12**
From discomfort	A suitable environment, including shelter and a comfortable resting area	**Chapter 6**
From pain, injury, and disease	Prevention or rapid diagnosis and treatment	**Chapters 11 and 6** **Box 7.1**
To express natural behaviours	Sufficient space, proper facilities, and company of the animal's own kind	**Chapters 4, 8, and 11**
From fear and distress	Ensuring conditions that avoid mental suffering	Section **7.3.1**

Table **7.4** Webster's five freedoms

NOTE: The five freedoms (Webster, 1994) provide a guide to the minimum standards that should be achieved when maintaining animals in captivity. In zoos, the five freedoms can be accomplished through best practice, which has been reviewed in other chapters within this book.

Figure **7.18** The so-called 'five freedoms' of animal welfare can be achieved by implementing best practice in terms of captive animal management. This is illustrated in the examples here: (a) freedom from thirst, hunger, and malnutrition is provided from a mother to her young calf; (b) freedom from discomfort is achieved by providing an appropriate environment for this pipefish (family Syngnathidae); (c) freedom from pain, injury, and disease is ensured by regular veterinary monitoring of this short-beaked echidna *Tachyglossus aculeatus*; (d) freedom from fear and distress is satisfied by allowing this maned wolf *Chrysocyon brachyurus* adequate space in which to distance itself from visitors; (e) freedom to express natural behaviours is stimulated through the provision of cognitive enrichment to these orang-utans *Pongo pygmaeus abelii*. (Photographs: (a), (c), and (d): Paignton Zoo Environmental Park; (b) Warsaw Zoo; (e) Moscow Zoo)

continual study and assessment of animal suffering, rather than animal well-being, would not ultimately lead to major improvements in welfare. Essentially, animal welfare should be not merely the removal of suffering, but the provision of pleasure (Duncan, 2006). Good zoo animal welfare can be achieved through proactive responsibility for enhancing the daily lives of animals in our care and this can be initiated through taking steps to ensure that risks to welfare are not present in the first place. Thus, the aim is to strive to make animals happy and not simply to ensure that they do not suffer.

7.5.2 Minimizing risks to welfare

In section **7.3**, we considered factors that might compromise the welfare of zoo-housed animals and it follows from this that, if conditions can be provided that preclude these factors, then the risk of compromising animal welfare will be reduced. This means that we should provide conditions that give enough environmental challenge to be stimulating and yet not disturbing, and also provide opportunities to ensure that behaviour is not restricted.

So, how can this actually be achieved? Events throughout an animal's life affect its perception of stressors, but those that occur during its rearing, when it is young, have a particularly long-lasting and significant impact on its behavioural expression and development, as well as its perception of stressors. For example, domestic chickens reared in a complex outside environment show behavioural differences as adults, compared with chickens reared inside. They show significantly less fear, as indicated by a shorter **tonic immobility** righting time, and are also quicker to explore and use more of their outside environment (Grigor *et al.*, 1995).

Similarly, deer mice *Peromyscus maniculatus* that are provided with enriched housing while they are young (less than 124 days old) display lower levels of stereotypy than those that are reared in standard housing. This effect persists into adulthood even if the mice are transferred into standard housing (Hadley *et al.*, 2006). Providing appropriate rearing conditions is therefore essential in ensuring that an animal is able to interact effectively in its environment, and so has significant implications for the future life and success of animals (see also section **9.5.1**).

Carlstead (1996) has suggested that captive environments are less complex, more predictable, and afford captive animals less control than the native habitats in which they are adapted to live. Providing zoo animals with choices is a generally accepted method of giving them control in their environment. When giant pandas were given the choice to use different areas of their enclosure (inside and outside areas), their levels of both behavioural agitation and urinary cortisol reduced (Owen *et al.*, 2005). Giving a similar kind of choice to polar bears *Ursus maritimus* led to a reduction in stereotypies and an increase in play behaviour (Ross, 2006).

Choice can also be presented in the form of enrichment, because animals can choose whether or not to interact with it. Sambrook and Buchanan-Smith (1997) suggest that the effectiveness of enrichment can be predicted by the amount of control that it gives to the animals. From this perspective, a puzzle feeder would be more effective than providing scent within the environment, because the animal's actions will affect the function of the puzzle feeder more than they will affect the scent and this gives the animal more control. The animal can explore the feeder, find food in it, destroy it, scent mark on it, defend it, and even ignore it, whereas there is little it can do to change the scent. Although, as Sambrook and Buchanan-Smith (1997) concede, there are many other complex factors that affect environmental enrichment efficacy, nevertheless this hypothesis is supported by the results of studies that demonstrate

Tonic immobility is a natural state of paralysis that occurs in some species (fish, frogs, lizards, birds, rats, and rabbits— see Maser and Gallup, 1974) when they are threatened by an aversive stressor.

that some environmental enrichments are more effective than others (section **8.6**).

Another approach to minimizing the risks to welfare is to carry out a **welfare audit**. As we have seen already in this chapter, good zoos are concerned not only with measuring welfare, but also with assessing whether or not current captive animal management represents best practice—that is, with whether change is necessary to promote better welfare. A welfare audit approach is usually based on a review of the veterinary and husbandry records for individual animals within a collection (for example, incidence of ill health, or reproductive activity). Housing and husbandry parameters may also be recorded and reviewed, to see whether there are common variables that are likely to affect welfare.

Chester Zoo, for example, describes its welfare audit as '*mechanisms for identifying and acting on potential welfare issues*'. The Zoo's welfare audit process informs various activities that may be undertaken by zoo staff on a regular basis, and which relate to the monitoring and assessment of welfare provision. These activities include routine daily checks by keepers on all animals, as well as monthly meetings between keepers, curators, and other zoo staff (such as vets and research staff) to review the previous month's births, deaths, animal movements, husbandry changes, and other ongoing captive management issues.

The Zoological Society of London (ZSL) welfare audit process also makes use of a checklist of issues, covering both the animal's environment and husbandry procedures. There is a section on welfare audits in the *Zoos Forum Handbook* (Defra, 2007b), and details of the systems used by both Chester Zoo and ZSL are provided in Appendix I of Section Four of the *Handbook.*

Other chapters within this book provide ample examples of best practice within zoo housing and husbandry systems that have developed to ensure that welfare is not compromised, especially **Chapters 6** on 'Housing and husbandry', **8** on 'Environmental enrichment', and **12** on 'Feeding and nutrition'.

7.5.3 Making choices

Unfortunately, even in the best zoos, animal welfare is likely to be compromised at some time during the animal's lifetime, whether as direct result of housing and husbandry (for example, during catch-up of an animal for a health check), or indirectly (for example, if receiving injuries from conspecifics). There are also situations in which the zoo's goals and aims for a species might conflict with the welfare of the individual animals, such as conservation efforts that may require animals being moved between groups to ensure appropriate genetic management of the population.

In these situations, decisions need to be made on the basis of the relative 'costs and benefits' of the intended or current conditions in which an animal lives, and their impact on the animal's welfare. It is necessary, in these circumstances, to consider the costs and benefits not only to the individual animal, but also to the animals with which it shares an enclosure, and the animal's captive population and species, as well as the impact for people. The currency used to consider these dilemmas can be highly variable, from empirical indices of welfare, to potential attitude changes in zoo visitors.

Another consideration that should be borne in mind when undertaking cost–benefit analyses is that the welfare of the animal and the impact of the 'topic under discussion' may vary over time. For example, informed discussion may decide that the short-term stress associated with moving an animal out of its natal group is outweighed by the costs incurred if it stays in the group and gets injured as a result of fighting with the resident adults, or the benefits accrued from being able to breed in a new group.

Life for animals in zoos, and in the wild, can be summarized as a complex series of events representing costs and benefits. For example, animals in zoos may be at greater risk of exposure to new diseases, but have access to veterinary care, and may have restricted control over their environment, social groupings, and proximity to people, but do not suffer from predation or starvation. Many more such examples could be given.

7.5.4 Ensuring best practice is implemented

Our increased understanding of animal welfare informs the legislative reform that is necessary to enforce practically the conditions that promote good animal welfare, but animal welfare legislation differs both between and within countries. For example, legislation is different in the USA and Germany, but also between different states in the USA and Germany. This means that, in different countries, animals may or may not be protected by law (see **Chapter 3**) and the extent of that protection is largely determined by the role of animals in society. In the UK, for example, it is unacceptable to feed zoo animals live vertebrate prey, unless there are 'extreme circumstances', whereas the provision of live fish and other vertebrates is allowed and implemented in other countries (Defra, 2004). 'Extreme circumstances' are not frequently encountered, but may include cases such as the potential death of a highly endangered reptile because it will not eat, in which case it might be tempted to eat with live food.

Summary

- Defining, assessing, and thus promoting good zoo animal welfare should be pursued as a scientific discipline.

- Approaches to welfare include the study of animals' minds, bodies, or natures, or by using analogy.

- Our current knowledge of zoo animal welfare is quite limited, because there is a large diversity of species held in zoos and our knowledge of their basic biology is sometimes lacking, making assessments of their welfare difficult.

- It is, however, possible to identify needs in all animals and situations in which welfare may be compromised, and this enables zoos to work proactively to maintain high levels of animal welfare.

- Zoo professionals have a duty of care and are thus responsible for the welfare of the animals in their care, from the time at which they are born to that at which they die.

- The continual monitoring of animals by trained and experienced staff is necessary to assess signs of poor welfare, which should then be swiftly followed by appropriate ameliorative action.

- Housing and husbandry conditions should exceed minimal standards, going beyond what is necessary to ensure an animal's survival, providing conditions under which an animal's behavioural and physical needs are met, and a state of well-being can be attained.

Questions and topics for further discussion

1. Why should animal welfare be approached as a science?

2. What are the four approaches used to study animal welfare?

3. What factors are likely to compromise the welfare of zoo-housed animals?

4. Name some of the main indices used to measure animal welfare.

5. Can stress be considered beneficial?

6. What are the five freedoms?

7. How can zoos aim to maintain animals with good welfare?

Further reading

There are a number of good books that address the topic of animal welfare, but few of these are specific to the zoo industry and/or exotic animal welfare; instead, they are mainly biased towards agriculture, laboratory, or companion animals. The principles and theory of animal welfare are, however, well portrayed and a good general text that discusses factors that might compromise welfare is Appleby and Hughes' *Animal Welfare* (1997), along with Appleby's *What Should We Do About Animal Welfare?* (1999) and Dolins' *Attitudes to Animals: Views in Animal Welfare* (1999), both of which provide easily accessible texts about animal welfare in a wider context. The Universities Federation for Animal Welfare (UFAW) also publishes a series of books on welfare topics, the newest of which is Fraser's *Understanding Animal Welfare* (2008).

More specific texts, which deal in greater depth on topics within the field of animal welfare, include: Mason and Rushen's *Stereotypic Animal Behaviour: Fundamentals and Applications to Welfare* (2006); Moberg and Mench's *The Biology of Animal Stress: Basic Principles and Implications for Animal Welfare* (2000); Flecknell and Waterman-Pearson's *Pain Management in Animals* (2000).

Research undertaken to study the welfare of zoo animals can be found in the following journals: *Animal Welfare*; *Anthrozoos*; *Applied Animal Behaviour Science*; *Journal of Applied Animal Welfare*; and *Zoo Biology*.

Websites and other resources

To gain an insight into the field of animal welfare, it is necessary to use the published literature referenced above. There are, however, a couple of websites that provide useful information about the practical implementation or assessment of animal welfare. The government website that details the regulation that governs practice in the UK can be found at `www.defra.gov.uk/animalh/welfare/default.htm`, and many of the regional zoo associations provide information detailing the expectations that they have of their member zoos, in terms of general animal management.

Finally, a website that you should all visit is `www.vet.ed.ac.uk/animalpain`, which provides brilliant and comprehensive coverage of animal pain.

Chapter 8 Environmental enrichment

Anyone who visits **zoos** regularly, or reads some of the zoo literature, will be aware of the amount of effort that zoos make to stimulate their animals. This effort is generally referred to as **enrichment**. Enrichment is a broad concept, but the term is usually used to describe any change to an animal's environment that is implemented to improve the animal's physical **fitness** and mental well-being. Often, enrichment is provided in the hope that it will stimulate a wide variety of behaviours, and improve the health and welfare of the target animal.

The methods used to enrich are highly variable, but are often categorized and discussed by type: for example, 'food-related'. Unfortunately, the impact of providing an enrichment cannot necessarily be generalized between species, or sometimes even individuals, and therefore a success in one situation might not translate well to another (Maple and Finlay, 1989). This said, there are some changes that can be made to general housing and husbandry that reflect their species-specific needs,

and thus provide opportunities for them to express them: for example, **scatter feeds**, which have been incorporated into the husbandry of many animals (such as lion-tailed macaques—see Mallapur *et al.*, 2007).

On the other hand, changes that appear to provide enrichment can be provided and yet no visible change in behaviour is observed as a result. Spinelli and Markowitz (1985), for example, reported that changes in enclosure size and complexity had no significant effect on the behaviour of some laboratory- and zoo-housed primates. As such, it is necessary, if time is not available to test the efficacy of each enrichment provided, to use some 'rules of thumb' that will increase the likelihood that changes made to the animals' environmental will be enriching (see section **8.6**). There has been a huge growth in this area of husbandry in recent years, with most zoos incorporating enrichment into their housing and husbandry regimes, and education programmes for zoo visitors (see **Fig. 8.1**).

In this chapter, we will cover the following topics.

8.1 What is 'enrichment'?
8.2 The evolution of enrichment as a concept
8.3 The aims and goals of enrichment
8.4 Types of enrichment and their function
8.5 Enrichment evaluation
8.6 What makes enrichment effective?
8.7 The benefits of environmental enrichment

The boxes that accompany this chapter examine some related issues, such as the relationship between enrichment and training, the organization of enrichment, and the scope of animals in relation to which enrichment can be attempted.

A **scatter feed** is said to occur when the provision of all, or part, of the animals' daily ration is spread out around the enclosure. This style of presenting food may be adopted to encourage foraging behaviour and to avoid the monopolization of food by some individuals in the social group.

No Lion' Around Here

To help George and Gracie keep active and alert, keepers installed a spring pull toy, with changeable scents and objects, like deer hides. It's so popular it's been broken twice!

NOTES FROM THE **ZOO**

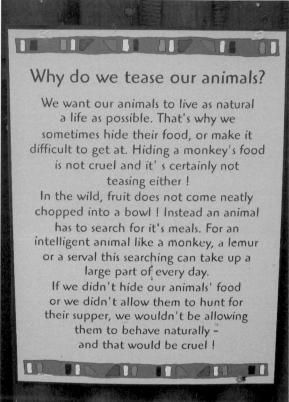

Why do we tease our animals?

We want our animals to live as natural a life as possible. That's why we sometimes hide their food, or make it difficult to get at. Hiding a monkey's food is not cruel and it's certainly not teasing either!
In the wild, fruit does not come neatly chopped into a bowl! Instead an animal has to search for it's meals. For an intelligent animal like a monkey, a lemur or a serval this searching can take up a large part of every day.
If we didn't hide our animals' food or we didn't allow them to hunt for their supper, we wouldn't be allowing them to behave naturally – and that would be cruel!

Figure **8.1** The implementation of enrichment provides an ideal opportunity to educate visitors about the needs of animals, and also to acknowledge sponsors and manufacturers of enrichment devices. Signs at zoos (a and b) explain to visitors why animals need enrichment and (c) acknowledge support from local companies. Often, the provision of enrichment is readily understood by zoo visitors, and provides a spectacle that will be viewed in a positive way, such as the provision of a ball for this lion at Howletts Zoo UK. (Photographs: (a and b) Vicky Melfi; (c) Julian Doberski; (d) KNP/Howletts Zoo)

8.1 What is 'enrichment'?

By definition, there are lots of situations that might be termed 'enriching', and there is a tendency, in some zoos, to call all housing and husbandry changes 'enrichment'. But are they? Would you, for example, consider having an injection enriching? A vaccination may reduce the likelihood that an animal will contract a disease, and therefore improve health and welfare, but do you think an animal may prefer access to a puzzle feeder, which would also confer benefits to its overall well-being? These two methods both achieve improved well-being, but via different routes, which suggests that our simple definition of enrichment requires more clarity and may explain why so many subtly different definitions exist (Young, 2003).

It may be more prudent therefore to consider enrichment to be any change in an animal's life or environment that confers benefits without any negative ramifications. And even this can cause us difficulties, because it raises the issue of whether 'stressful' events should be intentionally employed in zoos (as further discussed in **Chapter 7**).

8.2 The evolution of enrichment as a concept

The suggestion that enhancing the environments, and thus the lives, of captive animals could be beneficial was made by Hediger in the 1950s. But efforts to do this were not really focused until the work of Markowitz, who coined the term **behavioural engineering**, later known as **behavioural enrichment**. Markowitz suggested that behavioural enrichment could successfully stimulate desirable behaviours, increase activity, and provide an easy method of monitoring health in zoo-housed animals (Markowitz *et al.*, 1978; Markowitz, 1982). In his work, animals were required to perform a task for which they would be rewarded: a method very similar to that employed in **positive reinforcement training (PRT)**.

This technique is based on **operant conditioning**, whereby the task requires the animal to express certain behaviours. Operant conditioning is the process whereby a behaviour is more likely to be repeated if it is followed by a reinforcement, or less likely to be repeated if followed by punishment (as explained more fully in section **4.1.2**). If the animal were to show the required behaviours, then it would be rewarded, and this would then increase the likelihood that the animal would express those behaviours again in the future.

Most of the tasks devised by Markowitz involved some kind of mechanical apparatus that could be set to deliver pre-loaded rewards whenever the 'goal' behaviour was performed (for example, touching a lever). Markowitz suggested that behavioural enrichment would lead to the expression of desirable behaviours, because animals were motivated to work for rewards and would thus complete the task. For example, to achieve the aim of increasing activity levels in four Diana monkeys *Cercopithecus diana*, he provided them with a food reward after they had completed a complex series of movements around their enclosure (Markowitz *et al.*, 1978). The monkeys quickly learnt that if they performed this series of behaviours, they would gain food, and so they became more active and were able to obtain more food rewards.

On the face of it, this approach seemed reasonably positive, but it received strong criticism at the time. The main objection was that the behaviour elicited in the animals was not 'natural'. It was considered that the behavioural expression resulting from behavioural engineering was artificial, and that the animals made no association between the behaviour they expressed and the environment. This view was supported by the fact that the behaviour of animals that had been behaviourally enriched was sometimes independent of environmental context, or was performed at abnormally high levels (Hutchings *et al.*, 1978b). But the rate of behavioural expression was determined by the reward schedule—that is, the number and rate of rewards provided to animals after the successful completion of their task—appointed to the behavioural enrichment apparatus, determined by operant conditioning rules. So it would be possible to set the rate of reward to ensure that abnormally high rates of behaviour were not expressed (Forthman-Quick, 1984).

A further criticism (Hutchings *et al.*, 1978) was that the assertion that captive animals were motivated to work for rewards was **anthropomorphic**. We now know, however, that captive animals are frequently seen to work actively for a resource in their enclosure, even when it is also freely available (termed **contrafreeloading**; see section **8.6.3**). Most of the arguments against behavioural enrichment were offered on the basis that an alternative, more naturalistic, enrichment approach was superior to this 'artificial' structuring of zoo animal behaviour.

This alternative, termed **environmental enrichment**, was suggested by Hutchings *et al.* (1978a; 1978b). They advocated that zoo animals should be provided with species-specific opportunities within their environments to enable them to express a diversity of desirable behaviours (see **Fig. 8.2**). For example, to increase activity levels in the same group of Diana monkeys, the environmental enrichment approach would be to provide a large and complex enclosure. This would hopefully stimulate the monkeys to investigate and interact with their environment, and thus result in the desired elevated levels of activity. Creating an environment with vegetation and structural complexity, referred to as a **naturalistic enclosure**, is therefore integral, although not mandatory, to the evolution of environmental enrichment, because it could provide a structural basis for daily enrichment (see section **6.2.1**).

It is now clear that neither of these alternative styles is necessarily particularly natural, nor do they represent the panacea to providing an 'optimal' captive management system. For example, larger, natural-looking enclosures do not necessarily lead to

Figure **8.2** Providing animals with the opportunities to express natural behaviours may be as simple as providing enough water to allow social swimming. (Photograph: Living Coasts)

an elevation in activity levels (Spinelli and Markowitz, 1985) and a feeding device may have limited impact on ameliorating stereotypical behaviour of zoo-housed bears (American black bears *Ursus americanus* —see Carlstead and Seidensticker, 1991; Carlstead *et al.*, 1991). Support for the mechanistic approach of behavioural modification through behavioural enrichment has been overshadowed by the controversy that it has provoked and the overwhelming prejudice for the aesthetics associated with environmental enrichment. It is easy to appreciate how the two techniques can be employed to complement one another: for example, puzzle feeders can be provided in a large, dynamic, and complex enclosure. The two approaches are similar in their underlying mechanisms, but very different in their function (see **Box 8.1**).

More recently, David Shepherdson has worked tirelessly to promote the implementation of enrichment in zoos, through publications, research, and by establishing a biannual International Conference of Environmental Enrichment (ICEE), first hosted by Metro Washington Park Zoo (now Oregon Zoo) in 1993 (Shepherdson, 1998). *Shape of Enrichment* is the official publication of ICEE, edited by Karen

Contrafreeloading describes the phenomenon of an animal choosing to 'work' for a resource, despite it being freely available within the enclosure already.

(a)

(b)

Figure **8.3** (a) Shape of Enrichment is an international organization, the sole role of which is to promote the implementation of environmental enrichment. (b) A working party of this organization is the Regional Environmental Enrichment Committees, which provides 'grass-roots' support at a regional level. (Photographs: (a) Shape of Enrichment Inc.; (b) Regional Environmental Enrichment Committees)

Worley and Valerie Hare, and provides an opportunity for academics and zoo professionals to gain insight into the theory and practical implementation of enrichment. (You may recognize its logos in **Fig. 8.3**.) More recently, Julian Chapman, based at Paignton Zoo Environmental Park, initiated Regional Environmental Enrichment Conferences (REEC) to support a grass-roots approach to enrichment, through the creation of regional committees that host meetings on alternative years to the ICEE to smaller local audiences: for example, the UK and Ireland; and Australasia.

8.3 The aims and goals of enrichment

Most enrichments are implemented with the aim of generating some behavioural change in the animals. That change may be the stimulation or prevention of specific behaviours that are associated with improved fitness and/or animal welfare. The problem is that identifying what those behaviours are depends rather on how you see the goal of your enrichment. In fact, many of these behaviours—and, indeed, the aims behind the provision of enrichment —are interconnected. For example, a frequently quoted aim of enrichment is to enhance natural behaviour expression, by increasing activity and preventing **stereotypies** (for example, in giant pandas *Ailuropoda melanoleuca*—see Liu *et al.*, 2006).

So, what should enrichment be trying to achieve? The behavioural aims of enrichment are usually to be found in one of the following categories.

8.3.1 Wild-type behaviour

The goal of many enrichment studies is to promote wild-type, natural, or normal behaviours, and thus to bring captive and wild animal behaviours more in line with each other. Wild-type behaviours can be considered to be those that are expressed and observed in wild **conspecifics**. Sometimes used synonymously with terms such as 'natural' or 'normal' behaviours, definition of these terms is rarely provided in studies and thus they become meaningless. (Defining 'wild-type behaviour' and a discussion of the value of the concept is dealt with in section **4.5**.)

Although it sounds very worthwhile and theoretically valid to promote wild-type behaviours as a goal for enrichment, in fact, it is very difficult to achieve. Firstly, the premise of the wild–captive comparison is that deviations from the wild type indicate compromises in animal welfare. There are, however, a number of theoretical and methodological reasons to suppose that this premise is not always correct (Veasey *et al.*, 1996a; see section **4.5.2**).

Secondly, in order to promote wild-type behaviours in captive animals, a behavioural comparison of the wild and captive animals is necessary, to establish whether there is a discrepancy and what it might be. An example of such a study is that by Kerridge (2005), who directly compared the behaviour of black-and-white ruffed lemurs *Varecia variegata* through observations of them in their native Madagascan rainforest habitat and in UK zoos (see **Box 4.5** for more on this study). But direct comparisons of this sort, through studying the species in both environments, are time-consuming and expensive, and require, at the very least, comprehensive knowledge of the wild animals' behavioural repertoire. Consequently, few have been undertaken. There is thus a paucity of data for the many species held in zoos, which would enable direct wild–captive comparisons to be performed.

An alternative has been to attempt a more symbolic comparison of wild and captive animal behavioural expression. This approach is characterized by the use of stereotyped assumptions of how captive and wild animal behaviours differ, rather than using actual data. For example, many captive animals are considered to be more inactive than their wild conspecifics and to display 'abnormal' behaviours (also discussed in section **4.4**). Thus attempts are made to increase the captive animals' activity and to reduce any expression of 'abnormal' behaviour.

Although simple and widespread, this method of comparison is, unfortunately, too simplistic and may be based on erroneous ideas of what the animals are like in the wild. It may be true that some captive animals are more inactive than their wild conspecifics, but direct comparisons do not necessarily find significant differences in **activity budgets** between the two populations (in black-and-white ruffed lemurs, see Kerridge, 2005; in Sulawesi black-crested macaques *Macaca nigra*, see Melfi and Feistner, 2002). So, at the very least, if approximation to the wild is what is being sought through enrichment, then the intended behavioural profile should come from real data, not from guesswork.

8.3.2 Desirable versus undesirable behaviours

The promotion of wild-type behaviour is sometimes used as a euphemism for promoting desirable and inhibiting undesirable behaviours. Clearly, we need to consider very carefully which behaviours are assigned to which category, because their stimulation or prevention may have far-reaching ramifications. Many enrichment studies, for example, aim to promote feeding and foraging, and to ameliorate stereotypies and **self-injurious behaviours** (in the fishing cat *Prionailurus viverrinus*, see Shepherdson *et al.*, 1993; in the chimpanzee *Pan troglodytes*, see Baker, 1997). But why should we think that feeding and foraging should be increased? Unless a direct comparison of time spent feeding and foraging has been performed, how do we know how much time the animals should spend performing these behaviours?

Similarly, it is possible that the expression of stereotypies or self-directed **displacement behaviours** provides the animal with a coping mechanism for living within a challenging environment (see section **4.4**). If this is the case, then what does their reduction mean for the welfare of the animal?

8.3.3 Activity versus behavioural diversity

The drive to generate greater activity in captive animals is extremely strong, even though direct comparisons of activity levels of wild and captive animals are rarely performed. For example, wild lions spend many hours resting and yet enrichments have been designed to promote activity in this species when in captivity (Powell, 1995). Part of this drive is almost certainly led by zoo visitors, who wish to see active animals (**Fig. 8.4**; see section **6.1.3**). Certainly, it has been suggested that environmental differences in captivity, compared with the wild, will lead to quantitative changes—that is, changes in the amount of the same behaviour—in animal behaviour (Carlstead, 1996).

Quantitative assessment of behaviours can be made through the generation of activity budgets,

(a)

(b)

(d)

(c)

Figure **8.4** It is easy to see why visitors may want to see active animals: this running (a) sloth bear *Melursus ursinus* at Wrocław Zoo and (b) cheetah *Acinonyx jubatus* at Vienna Zoo both look more interesting than the (c) lion *Panthera leo* and (d) guenon also pictured. (Photographs: (a) Radaslav Rata/Wrocław Zoo; (b) Vienna Zoo; (c) Andrew Bowkett; (d) Vicky Melfi)

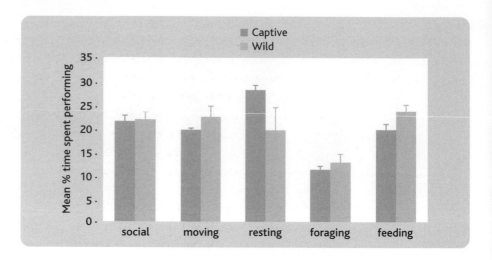

Figure **8.5** Despite the general assumption that wild and captive animals will behave differently, a comparison of activity budgets constructed for captive (*N* = 8) and wild (*N* = 3) Sulawesi crested black macaques found that they were not significantly different. (From Melfi, 2001)

which describe the amount of time that the animal spends performing different behaviours in a day (there is more about activity budgets in section **14.4.1**). But it is not only between wild and captive populations that environmental variation exists, but also within wild and captive populations (Hosey, 2005). For example, variation in activity budgets was higher between three wild Sulawesi macaque groups compared with eight captive groups (Melfi, 2001; **Fig. 8.5**). This might mean that activity budgets are not sensitive enough to detect small differences between different groups of animals, but also suggests that behaviour is sufficiently flexible that the slight differences seen between different captive conditions are not particularly important. Perhaps what is important is whether they fall within the range of variation seen in wild populations.

Qualitative differences in behaviours—that is, different sorts of behaviours—may also occur as a result of environmental differences between wild and captive populations, so there may be behaviours shown in the wild, but not in captivity, and vice versa. There can be practical limitations to comparing specific behaviours such as these between wild and captive situations. As an alternative, **behavioural diversity** (the number of behaviours that an animal

expresses) can provide an indication of how wild and captive behaviours compare qualitatively.

It has long been feared that animals in captivity may not be able to express certain behaviours and may even lose them over time (Frankham *et al.*, 1986). Of particular concern would be the loss of behaviours that are necessary for the animal to survive in the wild (see section **9.2.3**). The loss of these would seriously impede long-term conservation efforts, because the animals would no longer have the skills for surviving in the wild (Box, 1991). For example, Britt (1998) observed that black-and-white ruffed lemurs gained much of their food in the wild by suspending themselves upside-down by their feet, and using their hands and mouths to gather up the food. But he also observed that this method of feeding was not usually possible in captivity, because many animals were fed on the ground from dishes. Thus, an appropriate enrichment for the lemurs would be one that encouraged the diversity of feeding postures and behaviours seen in the wild.

Some behaviours are performed infrequently, but expressing these behaviours may still be very important to the welfare of animals. For example, farmed mink *Mustela vison* have demonstrated that they will push very heavy weights to gain access

Behavioural diversity can refer to an animal's behavioural repertoire, or, more formally, is calculated as the number of behaviours performed in a given period of time.

to water for swimming, even though this behaviour did not occupy much of the activity budget of the animals (Mason *et al.*, 2001). So, the ability to perform all of the different kinds of behaviour that are in the animals' repertoire is clearly of importance to them and enrichment can be designed to promote these behaviours. An additional benefit is that observing animals perform their species-specific behaviours is important in public education and in promoting positive attitudes towards zoos (Forthman *et al.*, 1992). Thus, the promotion of behavioural diversity or species-specific behaviours may be a more suitable goal for enrichment than simply increasing activity levels.

Comparisons of behavioural and environmental enrichment suggest that the latter are more successful at stimulating behavioural diversity. Animals appear to be able to express many different behavioural **responses** to environmental enrichment compared with behavioural enrichment or training (see **Box 8.2**).

Providing enrichment to change behaviour quantitatively or qualitatively has been demonstrated to improve physical fitness, spatial coordination, enclosure use, and reproduction (Chamove, 1989; Shepherdson *et al.*, 1993; in orange-winged Amazon parrots *Amazona amazonica*, see Millam *et al.*, 1995).

8.4 Types of enrichment and their function

Enrichment comes in many shapes and forms. Categorizing enrichment into meaningful groups can help with its practical provisioning and can also give an indication of the enrichment's function by acting as a kind of 'rough guide' to the enrichment's expected outcome. For example, food-based enrichments would probably be expected to promote feeding and foraging behaviours. Much enrichment will affect animals in a variety of ways, however, so it can be difficult to choose a category for an enrichment, because it might seem to fit into more than one class. A puzzle feeder, for example, can be manipulated and could therefore be thought of as a physical, or even a cognitive, enrichment, but it is also food-based.

In these situations, enrichments can be classified according to a hierarchy, whereby classes that have the fewest enrichments attributed to them are filled first. For example, in the case of the puzzle feeder, if there are fewer physical enrichments compared with food enrichments, it could be put in the former class.

We can suggest the following as enrichment categories (see **Fig. 8.6**).

1. *Food-based enrichment* Enrichment that is centred on food: for example, providing food in a new way or providing a new type of food.

2. *Physical enrichment* Any change to the animals' structural environment, whether permanent or temporary (for example, **perching** or climbing frames), or the provision of objects that can be manipulated (for example, flooring substrates or toys).

3. *Sensory enrichment* Anything that stimulates the animals' senses, including what they see, hear, and smell (for example, light reflecting off crystals, a rattle, or blood trails).

4. *Social enrichment* Interactions with other animals or people: for example, De Rouck *et al.* (2005) suggest that housing tigers *Panthera tigris* in pairs is advantageous, because the animals are then able to perform a greater diversity of behaviours compared with tigers housed alone.

Perching is the term used to describe an animal balancing on structures, i.e. ropes, branches, and/or beams. These structures contribute to increasing enclosure complexity and increase the functionality of the enclosure space that is available (see **Box 8.2**). Perching also increases opportunities for physical exercise, which can maintain muscle and joint fitness (e.g. LeVan *et al.*, 2000), and social distancing behaviours (e.g. Appleby and Hughes, 1991).

Figure **8.6** Enrichments can be categorized as: (a) food-based enrichment, i.e. a carcass feed given to a social group of dholes *Cuon alpinus* at Dresden Zoo; (b) structural enrichment, e.g. an underwater platform structure given to sea lions housed in Toronto Zoo; (c) sensory enrichment, e.g. a scent trail is obviously interesting to this fossa *Cryptoprocta ferox* housed at Colchester Zoo; (d) social enrichment, e.g. seen here as species-specific boxing behaviour between two male kangaroos at South Lakes; (e) cognitive, e.g. giving chimpanzees *Pan troglodytes* an artificial termite mound so that they can use tools to fish for its contents. (Photographs: (a) Wolfgang Ludwig, Dresden Zoo; (b) Vicky Melfi; (c) Colchester Zoo; (d) South Lakes Wild Animal Park; (e) Ostrava Zoo)

5. *Cognitive enrichment*   Additions to the environment that require problem solving of differing degrees of complexity to stimulate the animal mentally. Meehan and Mench (2007) suggested that animals' cognitive skills could be challenged by requiring them to use '*navigational, tool-making or cooperative social skills*' within their daily lives.

8.4.1 Food-based enrichment

Many species have evolved to spend a large proportion of their day looking for, **processing**, or eating food. Captive environments may offer few opportunities for **foraging** or feeding, and the food provided frequently requires little processing by the animal, because it has already been processed by keepers.

Several studies have demonstrated that some captive animals spend less time in food-related behaviours and/or express a lower diversity of food-related behaviours than their wild counterparts. To this end, many feeding enrichments aim to prolong the feeding experience, by making the acquisition of food more difficult (for example, by hiding the food, or making it hard to obtain in a puzzle), or by providing food that is low in calories, so that, once found, more food needs to be eaten. The provision of food highly motivates animals, so they are more likely to use the enrichment; as such, it is not surprising that food-based enrichment is the easiest and most abundant type provided.

Foraging can be stimulated simply by hiding food in flooring substrates—a method that has been shown to increase enclosure use, activity, and behavioural diversity, while decreasing incidences of aggression and **abnormal behaviours**. These effects have been shown in various primate species, such as chimpanzees (Baker, 1997), rhesus macaques *Macaca mulatta* (Lutz and Novak, 1995), and white-faced capuchins *Cebus capucinus* (Ludes-Fraulob and Anderson, 1999). Additional surface areas for

foraging can be provided, in the form of foraging boards (in rhesus macaques, see Lutz and Novak, 1995; in the squirrel monkey *Saimiri sciureus*, see Fekete *et al.*, 2000). The use of flooring substrates may change over the course of the day, according to the animals' motivation. Guinea pigs *Cavia porcellus* were found to prefer wood shavings during the day, when they would rest in them, but then prefer cut paper at night time (Kawakami *et al.*, 2003). Food can also be hidden in log piles, which have been shown to increase significantly the amount of time that bush dogs *Speothos venaticus* spend exploring their enclosure (Ings *et al.*, 1997b). It was suggested that the bush dogs learned to be more proficient in searching and finding food with repeated use of the log pile, which was observed as a significant decline in searching behaviour with time (**Fig. 8.7**).

The Equiball™, designed for stabled horses, also increases the processing time required to gain access to food and has been shown to reduce stereotypies

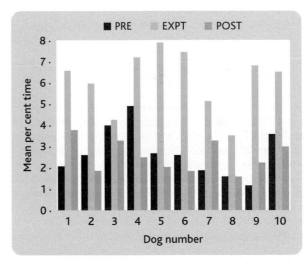

Figure **8.7** Food hidden in a log pile for bush dogs *Speothos venaticus* at Edinburgh Zoo significantly increased the time that they spent searching for food. The mean percentage time that the bush dogs spent foraging for food increased from 2.7 per cent of observations, with 'normal' feeding conditions (before, PRE, and after, POST), to 6.1 per cent when the log pile was used (EXPT). (From Ings *et al.*, 1997)

Processing is a term often applied to the activities that an animal performs to gain access to food—e.g. manipulating food and choosing which parts to eat—and does not necessarily include ingestion. **Foraging** behaviour, however, is often considered to include the location of food, e.g. it could be defined as 'moving slowly, looking towards the ground'.

(a)

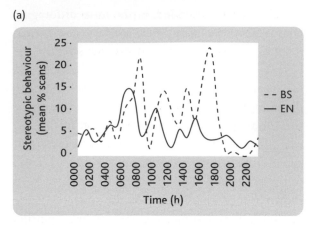

(b)

Figure **8.8** (a) Prior to provisioning with the Equiball™ (BS), six stabled horses were observed to spend 5.27 +/− 8.17 per cent of their time performing stereotypies, which peaked during two feeding periods at about 0800 and 1700. When the Equiball™ was used (EN), stereotypies were reduced in five out of the six horses. This highlights how individuals may differ in their acceptance of changes to housing and husbandry. A variety of feeding balls can be provided to zoo animals, (b) as seen here provided to zebra *Equus burchellii* in Singapore Zoo. ((a) From Henderson and Waran, 2001; (b) Photograph: Low Ai Ping/Singapore Zoo)

in horses *Equus caballus* (Henderson and Waran, 2001; **Fig. 8.8**). The Equiball™ can be set to release high-calorific food concentrates slowly when it is 'nudged' along the ground by the horse. This has two benefits: it increases the time required to feed and also encourages more movement. Similar feed balls are available designed in all shapes, sizes, and materials for a wide variety of animals, and they have also been fixed above ground for arboreal species, which need to use dexterity to gain access to the food.

Food can also be presented in such a way that species-specific behaviours are required to obtain it. The inability of ruffed lemurs to express all of their different feeding postures when they are only fed chopped fruit on the ground was mentioned in the previous section. To promote these postures in zoo lemurs, Britt (1998) provisioned them from the roof of their enclosure and also by using suspended feeding baskets. These enrichments encouraged the lemurs to use the vertical space of their enclosure and to gain food as they would in the wild, using suspensory feeding postures. Indeed, the lemurs were observed to spend as much of their feeding time using suspensory feeding postures when these enrichments were used as did their wild equivalents (wild = 25 per cent; rooftop = 24 per cent; suspended baskets = 30 per cent; standard and scatter feed = 5 per cent).

Providing food unpredictably has also been considered enriching. It has been suggested that **pre-feeding anticipation (PFA)**, also termed **food anticipatory activity (FAA)**, is one of the main contributory factors influencing the development of stereotypies (Howell *et al.*, 1993; reviewed by Mistlberger, 1994; see **Chapter 7**). So it has been reasoned that, if food presentation is less predictable, the development of PFA may also be hindered. Certainly, the amount of time that white-fronted brown lemurs *Eulemur albifrons* and Alaotran gentle lemurs *Hapalemur alaotrensis* at Zurich Zoo spent active and moving increased when they were provided with food boxes, which would open randomly, providing food (Sommerfeld *et al.*, 2006). Similarly, a mealworm feeder that randomly dispensed food to a group of Rodrigues fruit bats *Pteropus rodricensis* led to a significant increase in activity and a decline in aggression, even in bats that were not observed to use the feeder (O'Connor, 2000; **Fig. 8.9**).

Pre-feeding anticipation (PFA), or **food anticipatory activity (FAA)**, is the behavioural expression observed prior to being given food, which can occur due to a variety of signals, including circadian, visual, or odour.

(a)

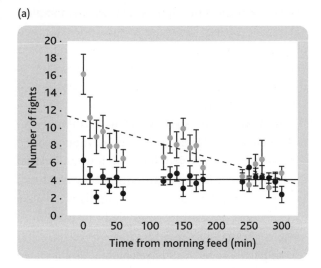

(b)

Figure **8.9** The number (mean +/− SEM) of fights that occurred between Rodrigues fruit bats *Pteropus rodricensis* declined over time after their morning feed when no enrichment was present (grey circles). When mealworms were present in the dispenser (black circles), the number of fights that occurred remained low, regardless of feeding time. ((a) From O'Connor, 2000; (b) Photograph: Paignton Zoo Environmental Park)

Various methods have been used to reduce the predictability of food availability to zoo-housed animals. For example, a hollowed-out coconut was used at the Bronx Zoo to distribute live food (insects) randomly to **dendrobatid** frogs, which led to an observed increase in activity (Hurme *et al.*, 2003).

With so much emphasis placed on food as the motivator for enrichment, care should be taken that the provision of enrichment does not indirectly lead to more food (calories) being provided than is necessary to meet the animal's daily requirement, because this might result in obesity. As a rule, any food-based enrichment should use food from the animal's daily ration. Providing food that is low in calories can be an option if more food is necessary for the enrichment than would normally be given. This mirrors what, in fact, happens for many wild species—that is, that they spend long periods of time feeding, because the food they are eating is of low nutritional quality (see **Chapter 12**). This means that a more 'naturalistic' way of providing food would be to provide fresh foods, which are not as nutritionally rich as many concentrates (manufactured food). For example, provision of browse increased activity levels in African elephants *Loxodonta africana* (Stoinski *et al.*, 2000) and in Livingstone's fruit bats *Pteropus livingstonii* (Masefield, 1999). Similarly, carcass feeding has been associated with many behavioural benefits in zoo-housed **carnivores** (**Box 12.6**).

Some species may have specialized feeding requirements such that they have anatomical adaptations to deal with certain foods. For example, the diet of wild common marmosets, like other marmosets, includes gum exudates from trees, for which they have a specially developed tooth comb designed to aid in the gouging out of gum from trees (see **Fig. 8.10**). Initially used by McGrew (1986), Roberts *et al.* (1999) later showed that, in laboratory-housed common marmosets *Callithrix jacchus*, gum

Dendrobatid frogs are members of the family Dendrobatidae, also known as 'poison dart frogs'. (See **Box 10.3**.)

Figure **8.10** Providing gum feeders, filled with acacia gum, to pygmy marmosets *Cebuella pygmaea*, as seen here, enables them to display species-specific feeding behaviours. (Photograph: Paignton Zoo Environmental Park)

feeders reduced the time spent in stereotypical pacing and sitting. Their use of the device declined over time, however, as did observed changes in their behaviour, towards the end of a 3-hour exposure to the device and after about 3 days, when the device was provided daily.

At Paignton Zoo, straw was provided in wire racks to two elephants (Melfi *et al.*, 2004, unpublished). This not only provided food of low nutritional quality, but in a device that required the elephants to use their highly dextrous manipulative trunks to gain food. Providing straw daily in these wire mesh cages maintained a high level of feeding and foraging in these two captive elephants, as illustrated in **Fig. 8.11**.

(a)

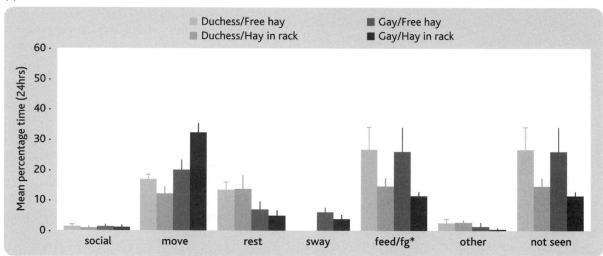

(b)

Figure **8.11** Feeding behaviour of two elephants, Duchess and Gay, housed at Paignton Zoo was increased, compared with baseline, by providing them with their straw ration and other objects in wire cages. The elephants spent about 50 per cent of a 24-hour activity budget feeding and foraging (fg), which is comparable to observations of wild elephants (Shannon, 2005). (Photograph: Julian Chapman)

8.4.2 Physical enrichment

Good enclosure design incorporates structural enrichment anyway, through the inclusion of beams, platforms, ponds, and anchors for fixing perching and objects, as illustrated in **Fig. 8.12**. Many of these features are permanent, but many of them can be modified with some imagination or moved to provide animals with a different perspective. For example, a pond can be filled with water or different substrates, and small bits of food can be provisioned in either the water or the substrates. In addition, perching, and the use of ropes, beams, and branches, can be rearranged or renewed within an enclosure and this can essentially 'rejuvenate' the exhibit. For example, providing climbing structures within a spectacled bear *Tremarctos ornatus* exhibit at Philadelphia Zoo increased the bears' behavioural diversity and use of the enclosure (Renner and Lussier, 2002).

It is well known from laboratory studies that some species show a need for privacy. For example, laboratory-housed rabbits provided with a sheltered area as part of an enriched cage system displayed lower levels of restlessness, grooming, bar-chewing, and timidity, and this was especially true of females compared with males (Hansen and Berthelsen, 2000). Many zoo animals have a similar need for privacy and this can be provided by increasing the amount of cover in the enclosure. Leopard cats *Felis bengalensis* at the National Zoo in Washington were observed to lie in hiding places within their enclosure when their urinary **cortisol** levels were high, indicating that when they were stressed, they would seek out privacy (Carlstead *et al.*, 1993). When hiding places, as well as perching, were provided, the cats showed increased exploratory and decreased stereotypical behaviours relative to those seen in their previously unenriched enclosures. This is further supported by the finding that, when hiding places were provided for clouded leopards *Neofelis nebulosa*, their faecal cortisol levels declined significantly (Shepherdson, 1994; **Fig. 8.13**).

Other species also appear to need a large degree of cover. For example, Chamove (1989) described how groups of common marmosets and cotton-top tamarins *Saguinus oedipus* avoided areas without cover and thus did not use a large open area, although this provided additional space. When dense cover was provided, however, their use of the area became extensive.

Visual barriers have also been used to provide privacy or protective areas within an enclosure. When large groups of domestic fowl ($N = 80$ and 110) had vertical panels installed in their enclosure, the time that they spent resting increased significantly, whereas foraging declined significantly, although feeding itself was not affected (Cornetto and Estevez, 2001). Provision of barriers, however, appears to have mixed results, depending on the species' natural history. Thus, for example, the ability to escape visually from other group members reduces aggressive encounters in stump-tailed macaques *Macaca arctoides* (Estep and Baker, 1991), but increases aggression in pig-tailed macaques *Macaca nemestrina* (Erwin, 1979). Both of these primate groups were maintained as single-male, multi-female troops, but the source of aggression (and thus social tension) was different in the two groups. In the pig-tailed macaque group, the males interrupted female–female aggression, so an increase in aggression was observed when cover was provided, because the male was unable to see all of the females and this prevented him from breaking up aggression between them. In the stump-tailed macaque group, however, the male usually initiated male–female aggression, so the visual barriers reduced social tension and aggression, because the females were able to hide from him.

Qualitative features of the perching can also affect behaviour. Caine and O'Boyle Jr (1992) demonstrated that the orientation of perching in an enclosure for red-bellied tamarins *Saguinus labiatus* affected the type and duration of time spent playing. Vertical substrates stimulated more chasing and grasping behaviours through the expression of a greater variety of play sequences compared with behaviours observed on horizontal substrates.

Objects are frequently used as enrichments. The variety of objects used is vast, ranging from

(a)

(b)

(c)

Figure **8.12** The enclosures at (a and c) the Smithsonian's National Zoo and (b) Disney's Animal Kingdom are both good examples of how structures can be provided that enable animals, here orang-utans *Pongo pygmaeus* and siamangs *Hylobates syndactylus*, respectively, to use species-specific locomotion skills to travel great distances. (Photographs: (a) Mehgan Murphy/ Smithsonian's National Zoo; (b) Julian Chapman; (c) Jessie Cohen/Smithsonian's National Zoo)

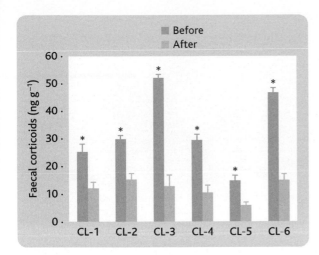

Figure **8.13** Hiding spaces are important to clouded leopards *Neofelis nebulosa*. Prior to the provision of hiding places (before), six leopards all showed some level of behavioural pathology, which was associated with high levels of faecal cortisol. When hiding places were provided within their enclosure (after), there was a significant decline in the faecal cortisol measure in each leopard. (From Shepherdson *et al.*, 2004)

hammocks for laboratory-housed mice, to straw bales for broiler chickens, and novel feeding methods and music for common seals (Grindrod and Cleaver, 2001; Kells *et al.*, 2001; Farlin and Baumans, 2003). Again, their effectiveness may be different in different types of animal. For example, they may stimulate immediate interest, but that interest may then wane over time. This occurred in a group of twenty-eight chimpanzees in a laboratory colony: they spent 41.9 per cent of the day interacting with an uprooted tree on the day on which it was first provided, but thereafter their interaction with it declined to 3.5 per cent (Maki and Bloomsmith, 1989).

On the other hand, equally simple objects, such as ice blocks, significantly increased behavioural diversity in African dwarf crocodiles *Osteolaemus tetraspis*, representing greater interaction with, and use of, their environment (Melfi *et al.*, 2004b). The type and shape of object enrichments are endless, including unwanted items that would otherwise be thought of as 'junk', such as old film canisters and cardboard tubes used as meal-worm feeders or a method for delivering scents, as shown in **Fig. 8.14** (Fry and Dobbs, 2005).

8.4.3 Sensory enrichment

Because we are such 'highly visual' animals ourselves, we can frequently forget that other animals gain information about their environment using many different modalities as well as, or instead of, vision. Visual, olfactory, and auditory channels are all important means of communication for animals, and are routes of information that we are able to manipulate within a zoo environment; we can also modify the tactile information available. With some imagination, it is also possible to affect thermal, humidity, and electromagnetic gradients, and the tastes available to captive animals. We can provide, for example, thermogradients for Round Island geckos *Phelsuma guentheri* (Wheler and Fa, 1995), artificial algae for dolphins *Tursiops truncatus* (Edberg, 2004), and can meet all four senses of **elasmobranchs**, which includes vision, chemoreception (including smell and taste), mechanoreception (hearing and touch), and electroreception (Smith, 2006).

Despite the fact that many zoos use sensory enrichment, described in many *Shape of Enrichment* articles (see section **8.2**), few of these enrichments have been tested empirically. Some reasons and resolutions to providing scent and auditory enrichment are given by Clarke and King (2008), and by Farmer and Melfi (2008).

An example of sensory enrichment is the use of computer-controlled acoustic information in the form of bird sounds (Markowitz *et al.*, 1995). Bird noises played randomly in the leopard cat's enclosure signified that food could be found; towards the end of this study, the leopard cat had learnt that

Elasmobranchs are **cartilaginous fish** of the sub-class *Elasmobranchii*, which includes skates and rays (*Batoidea*), and sharks (*Selachimorpha*).

(a)

(b)

(c)

(d)

Figure 8.14 Junk? Many objects that may otherwise be discarded can easily be transformed into enrichment objects. Here you can see 'spent' film canisters have been made into a variety of mealworm dispensers for (a) Geoldi's monkeys *Callimico goeldii* and (b) meerkats *Suricata suricatta* at Paignton Zoo Environmental Park; (c) large barrels are provided to polar bears *Ursus maritimus* at Central Park Zoo; and (d) a hose is provided to pink river dolphins *Inia geoffrensis* in Duisburg Zoo. (Photographs: (a and b) Julian Chapman; (c) Kathy Knight; (d) Vicky Melfi)

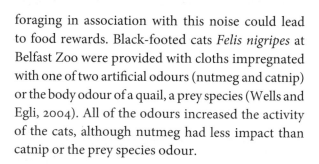

foraging in association with this noise could lead to food rewards. Black-footed cats *Felis nigripes* at Belfast Zoo were provided with cloths impregnated with one of two artificial odours (nutmeg and catnip) or the body odour of a quail, a prey species (Wells and Egli, 2004). All of the odours increased the activity of the cats, although nutmeg had less impact than catnip or the prey species odour.

Television has also been used as a means of visual enrichment. Bloomsmith and Lambeth (2000) provided laboratory-housed chimpanzees with a variety of videotapes showing different scenes (chimpanzees, other animals, and humans). Not surprisingly, single-housed chimpanzees watched the television more than those that were socially housed, although there was a lot of individual variation.

Care needs to be taken when providing sensory enrichment, because you may not always be aware what information you are communicating to your animals. This uncertainty can be reduced by ensuring that you know in what context the information was gained. For example, when playing back vocalizations (**playbacks**), do you know what messages you are sending? O'Brien (2006) used playbacks of conspecific vocalizations with Toco toucans *Ramphastos toco* at Dublin Zoo, but found that they had limited positive impacts on the birds' behaviour. The male bird was observed to feed less during playback conditions compared with baseline days, which was suggested to be a sign of stress.

Equally, when providing olfactory information, do you know what the relationship is between the target animal and the animal responsible for the sample: predator or prey? Buchanan-Smith *et al.* (1993) demonstrated that cotton-top tamarins expressed behaviours associated with anxiety when presented with predator scents. Similar results have also been observed in prey species that are able to see predators. Stanley and Aspey (1984) observed that five African **ungulates** would spend less time expressing several different behaviours (lying down, feeding, drinking, sniffing the ground and each other) when an African lion was visible (there is more about both of these studies in section **4.3.2**).

In terms of biological samples—for example, faeces, urine, and shed skin—used for sensory enrichment, it is necessary to ensure that all samples are from known healthy animals and do not provide a route for the transmission of disease.

8.4.4 Social enrichment

Possible sources of social enrichment are provided by people or other animals (for example, see **Fig. 8.15**) and the potential for humans to act as enrichment is discussed further in **Chapter 13**. Sometimes, the presence of other species (**mixed-species exhibits**,

Figure **8.15** A pair of marbled polecats *Vormela peregusna* at Novosibirsk Zoo. (Photograph: Novosibirsk Zoo)

housing more than one species) can be enriching. In this section, we consider how conspecific social groups may function as enrichment.

Cage mates are the most obvious source of social interaction, whether of the same, or a different, species. Indeed, cage-mates provide dynamic and unpredictable sources of stimulation, which may be one of the most enduring and effective ways of delivering enrichment to captive animals. Most of the evidence for this comes from laboratory studies, probably because it is becoming increasingly unusual for zoo animals to be housed in anything other than social groups. But in the laboratory industry, where animals are routinely housed singularly, there is much evidence demonstrating the importance of social housing (Eaton *et al.*, 1994). For example, Reinhardt and Reinhardt (2000) argue that social enrichment is essential for captive primate welfare and many studies have shown that, even in very small cages, the provision of a cage-mate can be greatly beneficial (Reinhardt, 1994a; 1995; 1998).

In extremely restricted environments, social enrichment has been shown to reduce stereotypies when other environmental changes (such as objects and devices) have failed (Spring *et al.*, 1997). Schapiro

Studies that play sounds—whether vocalizations, music, or other noises—to animals are often referred to as **playback** experiments, which is sometimes shortened and described simply as **playbacks**.

et al. (1997) have suggested that social housing could improve the effectiveness of object enrichment—but social pressures can equally complicate the use of object enrichments, due to the monopolization of resources, resulting in aggression.

Despite the overwhelming benefits of social housing, there are inherent difficulties in managing it, because it can be associated with a myriad of pressures that can result in stress, physical injury, and disease (Visalberghi and Anderson, 1993). Because social housing is the 'norm' within zoos, housing and husbandry regimes have evolved to reduce any deleterious side effects that may occur (see **Chapter 6**).

8.4.5 Cognitive enrichment

Some kinds of enrichment work by providing captive animals with a mental challenge, such as working out a complicated route to food. Indeed, much has been learnt about the cognitive abilities of a wide array of species by providing them with this sort of enrichment (for example, for behavioural traditions in Eastern black-and-white colobus monkeys *Colobus guereza kikuyensis*, see Price and Caldwell, 2007). Enrichment devices are seldom the only source of food available; generally, other sources of food are available in the enclosure concurrently, or else regular meals will ensure that the animal does not 'need' to use the enrichment.

Box **8.1** Is training enrichment?

The use of training, especially husbandry training, is becoming more commonplace in many zoos (see section **13.4**). Training, based on the principles of operant conditioning, is the process of bringing the expression of some of the animal's behaviours under the control of the keepers, who use conditioned cues or commands and reinforcement to elicit them. Training brings several benefits and it has been suggested that one of them is that it is enriching (see, for example, Laule and Desmond, 1998). Indeed, operant conditioning was the basis of environmental engineering, which was the precursor to enrichment as we know and practise it today (Markowitz, 1982; Mellen and MacPhee, 2001). But enrichment is considered by some to have a different function and impact on animals than that observed during husbandry training. This issue is complicated further, because the apparently simple statement that 'training is enriching' seems to be interpreted in a variety of ways.

It is probably easiest, firstly, to consider what we expect enrichment to do and then to consider whether training can achieve this. As we have explored in this chapter, enrichment is an incredibly broad concept. Usually, before a proposed environmental change can be considered enriching, it first needs to be assigned a goal (for example, that it has the effect of reducing stereotypies) and then its behavioural impact should be assessed to see if it achieves this goal. If the environmental change is successful, then it is considered enriching. Put simplistically, then, we may expect that training can function as enrichment—or, in other words, that if enrichment or training is provided, the same positive benefits should be bestowed.

Unfortunately, only one study (McCormick, 2003) has tested this approach. This study aimed to compare the effectiveness of conventional enrichment (logs covered with seeds) and husbandry training at increasing the behavioural diversity of two elephants at Paignton Zoo.

Only the provision of seeded logs increased behavioural diversity and was thus deemed enriching, whereas behavioural training did not and so was not considered to be enriching.

Alternatively, if the underlying mechanisms of training and enrichment are the same, we might be able to suggest that they both have similar impacts on the animal. As yet, although there is a large body of data from enrichment studies, there is little on the impact of training. Hare and Sevenich (1999) proposed that enrichment and training could be thought of as sharing four similar mechanisms:

1. a **stimulus** that triggered the behaviour;

2. a window of opportunity during which the behaviour could be expressed;

3. the behavioural response itself;

4. the connection between the stimulus and behavioural response.

But when they compared these mechanisms in an enriched and a trained tiger, they concluded that the four mechanisms were functionally very different (see **Fig. 8.16**). This suggests that we cannot automatically consider training and enrichment to have the same impact on animals.

There are various other ways in which training and enrichment can be seen to be similar or different. For example, they both provide a change in the animal's day and a cognitive challenge to the animal. While good training should elicit the same behaviour from all animals, however, the same enrichment can elicit behaviours that are very different in form and timing between species, individuals, and even the same individual at different times or on different days.

Currently, there are not sufficient data to support or disprove that training is enriching per se, but we do know that training has an important function in animal husbandry regardless and should be part of animal management programmes (Forthman-Quick, 1984; Reinhardt and Roberts, 1997).

(a)

(b)

Figure **8.16** Here you can see (a) a tiger *Panthera tigris* with a traditional enrichment item at Vienna Zoo, and also (b) another tiger being trained to target at the Bronx Zoo. These are both very different activities: do you think they are both enrichment? (Photographs: (a) Vienna Zoo; (b) Vicky Melfi)

So, if food is available anyway, why do animals take up the challenge of using these cognitively demanding tasks? This phenomenon of contra-freeloading is discussed more fully at section **8.6.3**.

An example of this sort of enrichment is puzzle feeders, which require dexterity (using hands, beaks, or trunks) to manipulate and which have been used with a variety of animals. The observation that some animals use tools to gain access to food has also led to the development of more complicated enrichment tasks. For example, some primate and bird species have been observed using tools to crack nuts. In an attempt to make the use of a tool to crack nuts even more necessary, Visalberghi and Vitale (1990) coated provisioned nuts with a paste made of sawdust and non-toxic glue to make them harder, and thus more difficult for tufted capuchins *Cebus apella* to open. Using food in cognitive enrichment tasks not only increases the likelihood of the device being used, but also requires species-specific behaviours to be expressed to gain the food and prolongs the time spent by the animal attempting to access the food object.

Some zoos have created purpose-built facilities for animal **cognition** research. In 2001, Germany's Leipzig Zoo heralded the opening of the Wolfgang Köhler Primate Research Center, which serves as a state-of-the-art zoo enclosure and also maintains animals that are the subject of high-quality research (Pennisi, 2001). More recently, a similar centre has been built at Edinburgh Zoo, Scotland (see section **14.3.5**).

8.5 Enrichment evaluation

Some previous studies have noted situations in which enrichment has apparently not worked (Hare, 2008). In cases such as these, it is more accurate to consider that the changes made to the life or environment of the subject were not enriching, and therefore not that the enrichment has failed, but that the change provided does not constitute enrichment. It is, therefore, essential that all changes made potentially to enrich animals are undertaken with expected outcomes in mind.

To evaluate the effectiveness of a potential enrichment, you must first consider what you hope it will achieve. That way, you can measure whether the potential enrichment really is an enrichment: did it meet your expected goals? It may be that a potential enrichment is not intended to lead to direct behavioural change, but to provide the animal with choice in its enclosure or to increase enclosure complexity. Both of these are very difficult to measure. Most potential enrichments, however, are provided with the intention of changing behaviour (see section **8.3**) and so data can be collected to evaluate whether this was achieved and thus whether the change to the animals' environment can be considered enriching. In some situations, multiple goals may be set for an enrichment, which makes the process of demonstrating its efficacy more problematic. It also makes comparisons of enrichments more complicated (Newberry, 1995).

Previously, much enrichment has developed on a 'trial and error' basis, whereby stimuli are haphazardly added to an enclosure. But without the resultant behaviour being observed, it is not clear whether these changes to the animals' environment have a positive or negative effect, or, indeed, any effect at all. Past enrichment success probably results from the vast number of potential enrichments provided, rather than from their suitability in meeting any aims set.

Ideally, a systematic method of evaluation is necessary. Simple guidelines to evaluate the effectiveness of enrichment were produced by Plowman (2006). A method developed at Disney's Animal Kingdom, referred to by the acronym 'SPIDER' (see **Fig. 8.17**), can also be helpful in determining the effectiveness of **behavioural husbandry** measures.

> **Behavioural husbandry** is the term used to describe changes in the captive animal's environment that will impact on their behaviour, e.g. enrichment and training.

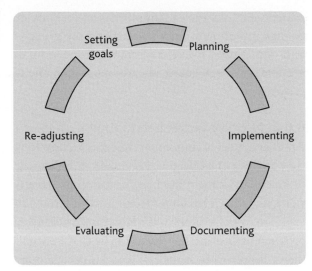

Figure **8.17** 'SPIDER' is an acronym of the stages that should be considered when implementing behavioural husbandry changes, whether they are enrichment, training, or modifications to the animals' housing and husbandry. The level of formality necessary for each stage can be tailored to the needs of the zoo (see, for example, Colahan and Breder, 2003)

We will return to the issue of monitoring enrichment in **Chapter 14** on 'Research'.

8.6 What makes enrichment effective?

Nevertheless, there can be some problems associated with enrichment provision and it is true that they are not all as successful as each other. For example, food enrichments are consistently more effective than other types, including scent, tactile, and object enrichments (in domestic rabbits, see Harris *et al.*, 2001; in domestic cattle, see Wilson *et al.*, 2002). This is why it is important to provide enrichments as part of an enrichment programme using a time-table (see **Box 8.2**).

The likelihood that enrichment will be successful can be increased, and the potential for problems limited, if the following three considerations are borne in mind.

1. The provision of enrichment should not compromise the health and safety of the animals or those working with them. This means that all enrichments should be included in the routine hygiene regime, to minimize the risk of disease transmission.

2. Assessments should be made before enrichments are transferred between different animal enclosures.

3. Finally, all enrichments need to be checked physically to ensure that they will not injure the animals.

8.6.1 Monopolization

Providing enrichment to social groups can be extremely beneficial (Honess and Marin, 2006), but, as with any resource, enrichment in a social situation can also be the cause of competition and the extent of this depends on the natural history of the species. A joystick computer-based system was used at Zoo Atlanta as a cognitive enrichment for pair-housed orang-utans *Pongo pygmaeus*. It was highly sought after by the animals, although one member of the pair monopolized its use; high users spent 48.9 per cent of their time using the system compared with 2.9 per cent in the low users.

The enrichment was considered highly successful, because the level of complexity continually increased and the animals did not **habituate** to it, as shown by the observation that use of it did not significantly decline during its presentation. Unfortunately, however, its presence was associated with increased aggression and anxiety-related behaviours (Tarou *et al.*, 2004; **Fig. 8.18**). The authors consequently suggested that it should only be used with singly housed animals.

On the other hand, when the adult male in a group of Hamadryas baboons *Papio hamadryas hamadryas* at Johannesburg Zoo monopolized a small box filled with food, other group members simply foraged elsewhere in the enclosure (Jones and Pillay, 2004). When a large box filled with food was provided to them, however, the adult

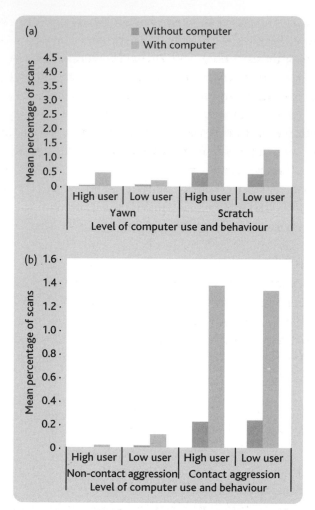

Figure **8.18** A highly complex cognitive enrichment provided to orang-utans required pairs of animals to manipulate a joystick, connected to a computer program. Unfortunately, the enrichment was so highly prized that (a) mean percentage scans of anxiety-related behaviour increased after the introduction of the enrichment, as did (b) aggression. It should be noted, however, that the frequency of these behaviours was still low, even after the introduction of the enrichment. (From Tarou *et al.*, 2004)

male was no longer able to monopolize it and other members of the group tried to feed from it. It was only then that rates of aggression rose.

These two examples demonstrate how differently social structures operate and how monopolization in one context leads to competition, but does not do so in another. It is important to consider how the social structure operates for any group that is given

enrichments. As a rule of thumb, it is best to provide enough enrichment for all animals in the group—and, sometimes, for a couple more, if the animals are able to collect, hoard, or carry more than one of the objects.

8.6.2 Novelty versus habituation

Another matter of contention concerns for how long the enrichment is effective. Previously, the rationale of much enrichment has been based on its novelty value, and thus it has aimed to stimulate interest and exploratory behaviours in the target animal. The big problem with using novelty as the basis for enrichment, however, is that once it is known, it is no longer novel. It will then lose its appeal to the animal and will no longer represent enrichment.

For example, the use of a mealworm feeder by common marmosets was seen to decline steadily in a 3-hour session, despite food being continually dispensed (Vignes *et al.*, 2001). Initial use of the feeder by the animals was associated with declines in pacing and sitting behaviour. The marmosets' loss of interest in the enrichment in this study follows a familiar pattern observed elsewhere, when enrichments are repeatedly provided or left in an enclosure for a prolonged period: their effectiveness lessens. This may have occurred because the marmosets were now satiated, but it may also represent habituation (see section **4.1.2**).

The speed and level of habituation seem to vary. Brent and Stone (1996) observed that singly and pair-housed laboratory chimpanzees, for example, continued to interact, even at low levels, with a variety of enrichments that had been available to them continuously for some months: television (22.75 months); a ball (55.9 months); and a mirror (25.9 months). Sumatran tigers at Paignton Zoo were shown to habituate more slowly to food-related enrichments (Plowman and Knowles, 2003). Habituation was ameliorated if a 3- or 4-week gap was left between repeating enrichment items, and levels of pacing would remain low, indicating that the enrichments were successful (**Fig. 8.19**).

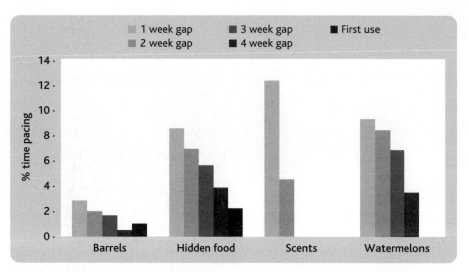

Figure **8.19** The goal of the various enrichments given to two Sumatran tigers at Paignton Zoo was to reduce the percentage of time they spent pacing. As such, when levels of pacing increase, it is considered that the enrichment is no longer effective and that the tigers have become habituated to it. As the gap left between repeating the enrichment (i.e. 1, 2, 3, or 4 weeks) increased, the percentage of time spent pacing reduced, illustrating that the habituation could be reduced. (From Plowman and Knowles, 2003)

Box **8.2** Using enrichment timetables can make provisioning practical

Practical enrichment provision requires a great deal of organization, to ensure that all animals receive enrichment most days and, importantly, receive different types of enrichment. Different types of enrichment timetable have been developed over the years, with two main aims: ease of use by keepers, and ensuring that a diverse array of species-specific enrichments is provided for the animals.

Keeper-friendly timetables have been designed, which are organized on blackboards, with fridge magnets, tick sheets, or specially designed computer software. All of these timetables have to include features that enable all keepers to keep track of which enrichment has been used and which is due, because work schedules mean that not all keepers will work together at all times.

Although it is desirable to study empirically the impact of enrichment or to determine when an animal has become habituated to it (that is, is no longer interested in it), in practice, this is not always possible—usually because of limitations on time or expertise. Equally, it is likely that enrichment programmes will have limited impact if the same few enrichments are repeated frequently, or if the creation of new enrichments cannot be sustained in the long term. So Wojciechowski (2001) suggested that each enrichment could be used for 2 days, by replenishing (with food or fresh scent, etc.) the enrichment or moving it on the second day. These actions were reported to sustain the animals' interest in the enrichment, above that observed if the enrichment was just left in the enclosure.

▶

▶ (a)

(b)

(c)

Figure **8.20** The Sulawesi crested black macaques at Paignton Zoo are provided with enrichment on a 2-month rota, during which time they receive a different enrichment every other day; all enrichments are presented for 2 days. Enrichments are categorized as: (a) manipulative items, e.g. a rugby ball; (b) food-based enrichment, e.g. food in a sack in a wire basket; and (c) scent enrichment, e.g. a herb kong, a scented enrichment device. (Photographs: Paignton Zoo Environmental Park)

For ease of implementation, enrichments can be grouped into different categories and rotated. For example, at Paignton Zoo, enrichments are characterized as food-based, sensory, or manipulative (Dobbs and Fry, 2008). In this timetable, primate species receive diffeent types of enrichment, which is refreshed on the second day, ensuring that an enrich-ment will only be used every 2–3 months (**Fig. 8.20**).

Various other enrichment programmes have been designed and reported in *Shape of Enrich-ment*, the proceedings of the International Con-ference for Environmental Enrichment, and other journals and handbooks (for example, Hooper and Newsome, 2004; Neptune and Walz, 2005).

It has been suggested that habituation can be minimized if the enrichment: is cognitively chal-lenging; provides an outlet for a highly motivated behaviour; provides a contingent link between **appetitive** and **consummatory** behaviours; and is repeated after an interval of time.

Appetitive behaviours are usually thought of as goal-seeking, whereas **consummatory** behaviours are goal-directed, or goal-oriented. Thus, searching for food is appetitive, but handling and eating food are consummatory.

8.6.3 Contrafreeloading: wanting to work for rewards

Animals do, indeed, appear to want to work for rewards, as originally suggested by Markowitz (1982) and as mentioned earlier in this chapter. Despite some authors considering this concept anthropomorphic, many studies have nevertheless shown that animals are willing to 'work'—that is, to perform certain behaviours—in order to get to a reward. More surprisingly, animals have been observed to 'work' for food in the presence of freely available food: animals will choose to use an enrichment device to get food even when there is freely available food present (reviewed by Inglis *et al.*, 1997). For example, thick-billed parrots *Rhynchopsitta pachyrhyncha*, green-winged macaws *Ara chloroptera*, and yellow-backed chattering lorys *Lotius garrulous* preferentially ate from an enrichment device, a wooden log with food hidden in it, even though food was available in a bowl (Coulton *et al.*, 1997). These birds emptied the contents of the enrichment devices within 24 hours and ate from their bowls at significantly reduced rates when these devices were available.

This phenomenon has been termed 'contrafreeloading' and has been observed in a variety of captive animals, which are usually given food-based enrichment while an alternative 'free' food supply is available.

Contrafreeloading may explain why many 'cognitive' or task-orientated enrichments are successful. Enrichment is not usually the only source of food available, so it seems unlikely that animals use food-based enrichment because they are hungry (although the food chosen may be particularly favoured, in which case, the animal may be highly motivated to get access to it). Contrafreeloading also does not necessarily provide more food. Singly caged laboratory rhesus monkeys, for example, could obtain 11.3 primate pellets in 673 seconds when using a puzzle feeder, but this was much less efficient than their 'usual' method of feeding, from which they could obtain 29 primate pellets in 32 seconds (Reinhardt, 1994a).

So why do captive animals choose to 'work' for food? Within **behavioural ecology**, much research has been undertaken to attempt to explain this phenomenon. Possible explanations are that:

- it fulfils a need to gain information about the environment (**information primacy theory**; see Inglis and Ferguson, 1986);
- it allows the animal to express species-specific behaviours (gaining food out of a bowl may not be as natural as cracking a nut; see Elson, 2007);
- there is a behavioural void, because captivity does not provide enough stimulation, so animals perform elaborate behaviours when given the opportunity (Chamove, 1989).

8.6.4 Providing a contingent link

The performance of species-specific goal-seeking (appetitive) and goal-oriented (consummatory) behaviours can establish and reinforce a connection that provides the animal with knowledge that is vital for survival (Misslin and Cigrang, 1986). For example, an animal that is motivated to forage (appetitive) may wander through some long grass and be rewarded by finding insects, which it eats (consummatory). The relationship between the two behaviours here is a contingent one—that is, the two are **causally** connected.

Contingent relationships give an animal predictability about the environment and hence give a certain amount of 'control'. Shepherdson (1994) has suggested that the loss of this contingency or 'control' in a zoo animal's life could have serious consequences to its psychological well-being. Maintaining this contingent link in zoos is sometimes

---

Information primacy theory suggests that animals are highly motivated to learn about their environment and that, as such, they will 'work', expend energy, and behave in such a way that they are able to gain more information about their environment (reviewed by Inglis *et al.*, 1997).

difficult, because animals may be motivated to move away from people and not be able to do so, or may look for food and find none. A lot of research has been undertaken that demonstrates that an animal's welfare is compromised when it lacks control over its environment (reviewed by Bassett and Buchanan-Smith, 2007). For example, providing giant pandas *Ailuropoda melanoleuca* with the control over determining which area of their enclosure they used was associated with fewer signs of behavioural agitation and reduced urinary cortisol (Owen *et al.*, 2005). Providing choice for animals and opportunities for them to gain control over the environment are, therefore, difficult to quantify, but are known to confer welfare benefits (Chamove, 1989; Shepherdson, 1991; see section **7.5.2**).

Enrichment can provide an ideal method of maintaining this contingency in captivity. For example, tigers provided with scents in their enclosure are given opportunities for behavioural expression that can be rewarding in and of itself. The tiger can choose whether to walk away from the scent, or to overmark the scent with its own, to scratch or rub the scented area, or to perform the **flehmen** response.

8.6.5 Motivating the animals

An animal's motivation to undertake any behaviour (such as to interact with an enrichment) will be affected by a great many factors (see section **4.1.3**), including species-specific biological needs and abilities, individual differences (age, sex, past experience, health status, etc.), and temporal circumstances (season, time of day, etc.). A good enrichment will be tailored to an animal, or group of animals, by considering these factors. **Behavioural economic theory**—that is, the study and quantification of an animal's motivation to express behaviour or gain

resources—and other methods can be employed to measure an animal's motivation to perform a type of behaviour or to gain access to a resource (see section **7.4.5**).

A classic example of seasonal variation in motivation is nest building, which is only required prior to parturition or courtship. This has been studied empirically with pre-partum farmed pigs. Gravid pigs show a high preference for straw pre-partum; when this is provided, it stimulates significantly more nest-building and rooting behaviour, and decreases the risk of the mother crushing the piglets (Thodberg *et al.*, 1999).

8.6.6 Providing opportunities for behavioural expression: do not underestimate the animal

It should be clear by now that well-thought-out enrichment will provide opportunities for the animal to perform natural behaviour. In this case, it is essential that we do not underestimate the potential of animals, either because of our lack of knowledge of them, or because our observations of what they do are based on situations in which their housing and husbandry are not ideal. For example, it has been theorized that **ectothermic** reptiles are unlikely to express play. Nevertheless, provision of objects such as balls, sticks, hoses, etc. to a Nile soft-shelled turtle *Trionyx triunguis* at National Zoo in Washington DC led to extensive, vigorous, play-like behaviour and a reduction in self-injurious behaviour (Burghardt *et al.*, 1996).

Intuitively, we may think that older animals may be more difficult to enrich; certainly, they are likely to be less agile than their younger counterparts. Swaisgood *et al.* (2001) found that adult giant pandas were less responsive to enrichments than subadults,

---

Flehmen behaviour may look like an animal yawning or twisting their upper lip, but is an action that draws air into their mouth and past the Jacobson's, or vomeronasal, organ, which is able to detect chemical signals such as pheromones.

Ectothermic animals—sometimes referred to as 'cold-blooded'—are those that regulate their body temperature by acquiring heat from external sources. The alternative is to regulate body temperature through controlling internal metabolic processes and animals that do this are referred to as **endothermic**, or 'warm-blooded'.

although both were more receptive to food enrichments. Novak *et al.* (1993) found that age did not affect enrichment use in group-housed laboratory rhesus macaques: older females would manipulate objects extensively. They observed that social facilitation and avoidance played a role in enrichment use, and suggested that this demonstrated that housing was a larger determinant of enrichment than age. But contrary evidence was provided by Brent and Stone (1996), who found no effect of housing on enrichment use in singly and pair-housed chimpanzees.

8.6.7 People

One of the biggest hurdles to providing effective enrichment is people, or (more diplomatically) the different needs and priorities of people within an organization.

Hare *et al.* (2003) describe how a comprehensive enrichment plan was set up for giant pandas, and how this was successful mainly as a result of considering '*staff time and effort, institutional requirements, resource availability, visitors' experiences and research protocols*' (see **Fig. 8.21**).

(a)

Figure **8.21** Enrichment provision can reflect seasonal changes in availability, and simply sometimes follow holidays to promote visitor interest in the animals and/or associated issues, i.e. conservation. Pictured here are the provision of (a) Christmas parcels to black-and-white ruffed lemurs *Varecia variegata* at Colchester Zoo; (b and c) pumpkins at Halloween to short-beaked echidnas *Tachyglossus aculeatus* and porcupines *Hystrix africaeaustralis* at Paignton Zoo. (Photographs: (a) Colchester Zoo; (b and c) Gillian Davis)

(b)

(c)

8.7 The benefits of environmental enrichment

Full appreciation of the impact of enrichment has not yet received much attention in zoos, which see its key aims and goals to be modifying behaviours or promoting behavioural diversity. But research conducted on laboratory and farm animals, in relation to which a greater array of indices can be measured, has highlighted that there are many other, more far-reaching or 'indirect', benefits of providing enrichment.

Behavioural expression is determined by neural activity, so if enrichment can modify behaviour, it is logical to assume that the brain and associated neural networks are also affected (Carlson, 2007). It is amazing the extent to which enrichment has a direct effect on the morphology, development, and function of the brain, and consequently behaviour (Van Praag *et al.*, 2000). **Neurogenesis** can be promoted through the use of enrichment. Both old (aged 25 months) and young (aged 2 months) laboratory-housed Wistar mice provided with enrichment showed increases in neurogenesis in the hippocampus (Segovia *et al.*, 2006). The hippocampus appears to be involved particularly with the 'encoding' of information about the environment so that it can be stored, elsewhere in the brain, as long-term memories. The cerebral cortex, the area of the brain associated with 'higher' cognitive processing (for example, problem solving or complex movement), appears to be even more receptive to enrichment than other parts of the brain (Diamond, 2001). Neurotrophins—the proteins associated with a variety of neural cell activities—increase within several regions of the rat's brain when the animal is provided with enrichment (Ickes *et al.*, 2000); increased levels of neurotrophins have been associated with enhanced visual cortex plasticity—

that is, the ability to change and be flexible—among other things, which led Prusky *et al.* (2000) to suggest that the enriched mice might have higher visual acuity than their unenriched counterparts, which, indeed, they did.

It has been known for a long time that enrichment enhances **learning** and memory in rodents (Hebb, 1947). More recent studies have demonstrated that enrichment increases brain plasticity and this enables adaptation to environmental changes, which, in turn, influences learning and memory throughout life. For example, Sneddon *et al.* (2000) demonstrated that pigs provided with enrichment (additional space, peat and straw in rack) learned more rapidly than those without. These learning tasks included operant conditioning tasks and navigation around a maze. Learning deficits resulting from lead exposure have also been shown to be reversible with the provision of enrichment (Guilarte *et al.*, 2003).

Enrichment can also affect an animal's interactions with its environment. A number of studies have demonstrated that fearfulness can be reduced in animals that have been provided with enrichment (see, for example, Reed *et al.*, 1993). Orange-winged Amazon parrots reared in enriched enclosures, for example, have been shown to interact with novel objects more quickly and for shorter **latencies** than conspecifics reared under 'normal' conditions (Meehan and Mench, 2002). These results led the researchers to suggest that enriched conditions can successfully modify the fear responses of captive animals.

Furthermore, the physical exercise associated with enrichment has been shown to contribute to recovery from brain damage and the deleterious effects of aging (Jones *et al.*, 1998). Providing enrichment has also increased recovery post-trauma (Jadavji *et al.*, 2006). Thus, animals that need surgery can

> **Neurogenesis** refers to the creation of new neuronal cells.

also benefit from enrichment, as in the study by Coviello-McLaughlin and Starr (1997), which found that mice provided with preferred nesting materials were less likely to 'agitate' wounds.

Enrichment can also promote breeding. Egg laying in captive orange-winged Amazon parrots is more likely if they are provided with enrichment (Millam *et al.*, 1995). Indeed, in a review of the impact of environmental enrichment on the reproduction of captive animals, Carlstead and Shepherdson (1994) concluded that several indirect benefits of enrichment improved reproduction. These included the

modulation of stress, socio-sexual stimulation, and changes in physical and psychological fitness.

At the moment, most of the studies that have identified the non-behavioural (or indirect) benefits of enrichment have been on either laboratory or domesticated species; whether similar benefits occur for zoo animals provided with enrichment is not known. We can, however, hope that, as we continue to study the impact of enrichment, we will discover that its benefits go well beyond what we initially expect, or see, not only in laboratories and on farms, but in zoos as well.

Box **8.3** Enrichment is not only for primates

Much of the peer-reviewed published literature on enrichment focuses on primates, which means that many other mammals—and certainly non-mammals—are sorely underrepresented. Although this is unfortunate and we consequently have a limited understanding of the impact that enrichment has on a variety of species, this does not truly represent the diversity of enrichment that is implemented in zoos (**Fig. 8.22**). In fact, enrichment is provided to many different kinds of animals, from the highly ingenious octopus, to the seemingly unmoving crocodile (Rehling, 2001; Melfi *et al.*, 2004b). In some instances, minor modifications to the enrichments make them suitable for use with other species. For example, many complex puzzle feeders intended for primates are ideal for some bird species, which are certainly comparable in their cognitive abilities (Helme *et al.*, 2008).

Figure **8.22** A variety of enrichments can be provided to all animals in zoos and aquariums, and do not necessarily have to be complex. Here, the provision of a 'plant pot' is used by various fish in a mixed-species tank at Warsaw Zoo. (Photograph: Warsaw Zoo)

Other species, however, deserve and require enrichment that is specially constructed according to their own innate needs and abilities, and a huge diversity of enrichments is reported in *Shape of Enrichment* (www.shape.org) (see secton **8.2**).

Summary

- Enrichment broadly refers to any change to an animal's environment that leads to a positive outcome and thus brings about improved or enhanced welfare.

- Enrichment is thought to stimulate captive animals both mentally and physically.

- Enrichment methods are highly variable and their success cannot necessarily be generalized between species, or sometimes even between individuals.

- Any change made to an animal's environment that aims to enrich should be appropriately monitored.

- Enrichment can sometimes provide a more dynamic environment, in which animals can display more choice and control over their lives, but this can be hard to quantify, because it may not lead to obvious behavioural changes.

- There is good scientific evidence that enrichment is extremely beneficial to animals on many levels and that, if it is incorporated into a wide variety of housing and husbandry regimes, it improves animal welfare.

Questions and topics for further discussion

1. Who was instrumental in the development of the enrichment concept?

2. How do behavioural and environmental enrichment differ from each other?

3. Discuss three enrichment aims and detail their limitations.

4. Identify, giving examples in each category, the five categories into which enrichment can be classed.

5. Why is it important to evaluate the effectiveness of enrichment?

6. Discuss three methods of increasing the likelihood that enrichment will be effective.

7. What non-behavioural benefits may be associated with the use of enrichment?

Further reading

There are various sources available in which you can find out more about the background and theory of environmental enrichment, including the recent and comprehensive book by Young, *Environmental Enrichment for Captive Animals* (2003), and Markowitz's older, but still very relevant, *Behavioral* *Enrichment in the Zoo* (1982). There are also conference proceedings of the International Conferences on Environmental Enrichment (ICEE), which are available from The Shape of Enrichment Inc., and those of meetings held by the Regional Environmental Enrichment Committees. Proceedings from the first ICEE meeting are published in the form of a book edited by Shepherdson, *Second Nature: Environmental Enrichment for Captive Animals* (1998). There is also a huge wealth of information, especially on the implementation of environmental enrichment, the amazing variety of enrichments available, and the numerous species that can benefit from enrichment, in the quarterly publication *Shape of Enrichment*, which is available from the organization of the same name.

Websites and other resources

The Shape of Enrichment Inc.—`wwww.enrichment.org`—is a not-for-profit international organization that promotes, through its quarterly newsletter, training workshops, and international and regional conferences and meetings, the implementation of enrichment. A sub-group of Shape is the Regional Environmental Enrichment Committees—`www.reec.info`—which provide grass-roots support for the international organization.

Chapter 9 Captive breeding

Most captive populations in **zoos** today are largely composed of captive-bred, rather than wild-born, animals. Conway (1986b) noted that more than 90 per cent of mammals and some 75 per cent of birds in **accredited zoos** in North America had been born in captivity. The number of captive-bred animals and the ratio of captive-bred to wild-caught animals recorded in zoos worldwide increased dramatically between 1971 and 1981 (as reviewed for several species in the orders Carnivora, **Perissodactyla** and Primates by Knowles, 1985).

This suggests that our knowledge and our ability to manage **exotic animal** reproduction in zoos and aquariums has greatly advanced in recent times. The incorporation of reproductive technologies, as summarized in section **9.4**, highlights that advances in captive breeding are continuing. It is worth noting that considered and careful management of zoo populations requires that as many, if not more, animals should *not* breed in zoos as those that should—although many zoos refer to the activities related to the captive management of species as 'captive breeding' or **captive breeding programmes**, and this is how we have titled this chapter. This label, however, perhaps implies that all programmes are initiated and managed with the ultimate goal of increasing the population size. This is not always the case, so it is probably more appropriate to refer to all of these activities, including the programmes themselves, as 'captive management'.

Data on the reproductive biology of exotic animals are sparse, because nearly all formal studies of reproductive biology have been of only fourteen domestic species, including man, and domesticated animals such as the cow, mouse, chicken, and cat (Wildt *et al.*, 2003). These studies provide us with a good understanding of the physiology, endocrinology, and genetic underpinnings of reproduction, but much this information is not directly applicable to breeding wild animals in zoos. There are also considerable differences in reproductive biology between seemingly related species. This restricts our ability to generalize across **taxa** and means that many recommendations must be worked out species by species, for example, protocols for **artificial insemination (AI)** (Stone, 2003).

We will start this chapter with a brief summary of the theory underpinning our knowledge of reproductive biology and then identify some of the environmental factors that can affect breeding in zoo-housed exotic animals. We will also summarize the knowledge and techniques available to zoo professionals on which decisions are made about whether animals are allowed to breed with or without intervention, and what these interventions may be. This is followed by an overview of the management programmes that operate under the banner of the main regional zoo associations. Finally, we will discuss the general methods that are implemented with the aim of achieving self-sustainable captive populations.

The order **Perissodactyla** are commonly referred to as 'odd-toed ungulates', of which there are two sub-orders: Hippomorpha and Ceratomorpha. Equidae are the only family of Hippomorpha, and Tapiridae and Rhinocerotidae represent the two extant families of Ceratomorpha.

We will cover the following topics in this chapter.

As with the other chapters in this book, several related issues are considered in the boxes, which include some supporting theory, but also include discussion of some more contentious issues, such as whether or not to hand-rear animals and what to do with surplus animals.

9.1 Reproductive biology

Effective captive management in zoos as part of the conservation of an endangered species requires a multidisciplinary approach, combining knowledge gained from: reproductive physiology; endocrinology; genetics and population biology; ecology and behaviour; veterinary medicine; and **nutrition**. The underlying theory of some of these disciplines is summarized in other chapters, but, here, we review some of the theory that underpins our knowledge of reproduction and forms the basis on which we make decisions about captive management in zoos. As we mentioned earlier, much of this knowledge has been gained from studies of common domestic breeds and we are now learning rapidly that diversity in reproduction biology is as great as the diversity of species we hold in zoos (Wildt *et al.*, 2003). Because the process of reproduction has evolved to transfer our genetic material into the next generation, this is where we will start.

9.1.1 Genetics

Genetics is about the way in which various features of the organism's structure, physiology, and behaviour are inherited. The gene itself is a section of deoxyribonucleic acid (DNA) (see **Fig. 9.1**), which carries the code for making a protein, or part of a protein (that is, a peptide). Because proteins form a large part of the structure of the body (for example, collagen, which holds tissues together; actin and myosin, which form muscle), and are also actively involved in assembling and running it (for example, as enzymes and hormones), then the set of genes that an individual contains—that is, its **genotype**—can be thought of as the assembly and running instructions of the final body—that is, the **phenotype**. At least one complete set of genes is found in each cell of an organism's body; within each cell, this may be only one strand of DNA, but is usually several, with the precise number depending on the species. DNA strands are long and thin, but, when they divide (at the time when the cell divides), they coil up such that they become shorter and fatter, and easily visible under a microscope. At this stage, they are known as **chromosomes**.

Cell division usually occurs to produce more copies of the cell; thus, it is a part of the growth and replacement of cells and tissues. In this sort of division, known as **mitosis**, the DNA also replicates itself, so that each daughter cell contains the same amount of DNA as the original parent cell. In most sexually reproducing organisms (discussed in section **9.1.2**),

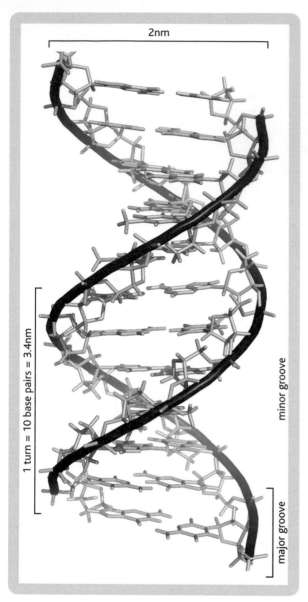

2nm

1 turn = 10 base pairs = 3.4nm

minor groove

major groove

Figure **9.1** Deoxyribonucleic acid (DNA) carries the information that codes for the production of proteins, which make up much of the structure and operating system of the body. Here, you can see that the DNA molecule forms a double helix, with the bases that constitute the genetic code inside the helix.

each cell contains two copies of each DNA strand (and hence, of course, two copies of each gene): one inherited from one parent, the other from the other parent (known as 'diploid cells'). Because of this, cell divisions to produce sex cells (for example, ova and sperm) cannot be mitotic, because this would lead to subsequent generations with four, then eight, then sixteen, and so on, copies of the genetic material. Instead, a different sort of cell division through reduction (**meiosis**, as illustrated in **Fig. 9.2**) occurs, in which the two copies of each DNA strand are separated from each other into different daughter cells, which thus have only one complete set of DNA instead of two (known as 'haploid cells').

During sexual reproduction, an animal inherits one copy of each DNA strand from each parent and hence two copies of each gene are present. The two copies of each gene may not be identical; if they are, the organism is said to be **homozygous** for that gene. But genes commonly occur in several varieties, known as **alleles**, and the organism may thus have two different alleles of the gene (that is, may be **heterozygous**). In some cases, one allele may mask the effect of the other, in which case, the alleles are known respectively as 'dominant' and 'recessive'. In this case, the recessive allele only shows its effects in the organism's phenotype if it is homozygous. An example is fur colour in cats and some other mammals, in relation to which an allele for one of the genes involved in the production of the pigment melanin fails to produce the required enzyme, resulting in animals that, if they are homozygous for that allele, have no skin or fur pigment, and are thus albino.

The ultimate source of new alleles is gene mutation, in which the physical structure of the DNA is changed: for example, through faulty copying during mitosis. New alleles may affect the organism's phenotype in a way that is advantageous or is

Mitosis is a process of cell division in which one cell divides to become two daughter cells, the two cells divide to become four, and so on, and the genetic information in each of the cells remains identical. In **meiosis**, on the other hand, one cell divides to become two, but the genetic information contained within the daughter cells is different in each cell, because of the separation of the two chromosomes of each chromosome pair.

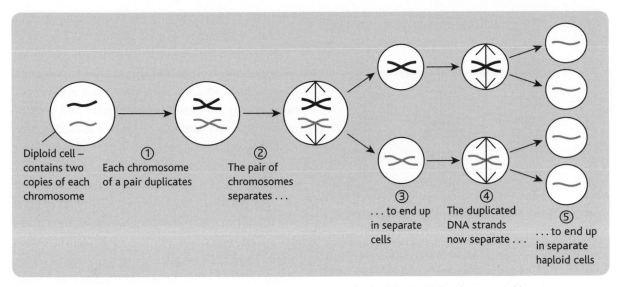

Figure **9.2** During the process of meiosis, the DNA molecules become shorter and fatter so that they are visible as chromosomes. The chromosomes of a pair (one originally from the mother and one from the father) separate from each other and each goes to one daughter cell, so the daughter cells have half of the genetic constitution of the parent cell.

detrimental to the organism's ability to function. Thus, they are subject to **natural selection** and, as a consequence, may increase or may disappear from the population.

Over evolutionary time, then, populations can be thought of as being made up of all of the alleles of their constituent members, referred to as the **population**'s **gene pool**. Different populations of the same species exist that are unable to breed because of, for example, geographical barriers. Even within the same species, different populations will have slightly different gene pools, because they live in slightly different environments in which natural selection will act differentially on the alleles present. Many genes are represented in the gene pool by several alleles (that is, they are 'polymorphic'), and this contributes to the overall **genetic diversity** of the gene pool and hence of the population.

9.1.2 Reproduction or sex?

'Reproduction' and 'sex' are not the same thing. 'Reproduction' is the process of producing offspring. For most vertebrate animals, this involves sexual reproduction, which requires two animals to come together so that their 'gametes' (haploid sex cells) can merge through the process of fertilization to produce a 'zygote' (a single diploid cell). The term 'sex' is commonly used to refer to the act of mating, but, strictly speaking in biological terms, sex is a process that forms new individuals containing genes from more than one source. The gametes of males are called 'spermatozoa' (or sperm) and are produced in the testes (the result of meiosis; the process of sperm formation is known as 'spermatogenesis'); 'ova'[1] (the singular of ova is ovum), or eggs, are the female gametes and are produced in the ovaries (also by meiosis, in a process known as 'oogenesis').

Natural selection is the process, observed in wild animal populations, whereby the alleles of some 'fitter' animals are transferred into the next generation, whereas the alleles of 'less fit' animals are not. (See **Box 9.2**.)

A **population** is a group of animals that are able to breed with each other.

1 Scientific papers about assisted reproduction usually refer to 'oocytes' rather than 'ova': these are female gametes or ova at a particular stage of development.

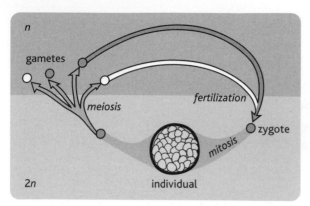

Figure **9.3** Male (sperm) and female (egg/ova) gametes are produced by meiosis, so each only has half of the chromosomes of the parent cell (referred to as haploid, or '*n*'). When they come together and fertilization results, the genetic complement of the new cell (zygote) is restored to two copies of each chromosome (referred to as diploid, or '*2n*').

In mammals, females do not produce ova continuously, but release eggs during oestrous[2] cycles (the process of ovulation), which either occur continuously or seasonally (see section **9.1.4**). Birds differ from mammals in that single eggs or ova are ovulated in succession (this is known as 'serial ovulation'), forming a clutch. In males, once sexual maturity has been reached, the process of spermatogenesis is continual, either throughout the year or throughout the breeding season. Fertilization occurs internally in mammals, birds, and most reptiles, whereas fertilization occurs externally in most fish[3] and amphibians.

As mentioned above, gametes are produced by the process of meiosis, in which the diploid cell, with a full complement of chromosomes, divides to produce haploid gametes, each with only half the complement of chromosomes. When two haploid gametes fuse during fertilization, the resulting zygote is diploid, with a (more or less) equal contribution of genetic material from each parent (see **Fig. 9.3**).

In species where DNA does not determine gender, it is possible to find animals that are **hermaphrodites**. These animals, which include some fish and **gastropod molluscs**, can be born[4] as one gender and develop into the other (a phenomenon known as 'dichogamy', producing **sequential hermaphrodites**), but only one set of sexual organs can function at one time. The process whereby an animal that starts life as a male and changes to female later in life is referred to as 'protandry' (for example, in the clownfish, *Amphiprion* spp.). The reverse, seen for example in wrasse (family Labridae), is called 'protogyny', in which an animal starts as a female, but later turns into a male. It is also possible for animals to be **simultaneous hermaphrodites**; examples include earthworms, some gastropod molluscs, and some fish (for example, hamlets *Hypoplectrus* spp.). Despite having both male and female sexual apparatus, and thus potentially the ability to self-fertilize, in fact, this rarely happens in these species.

Alternatively, some vertebrate (for example, whiptail lizards *Cnemidophorus* spp.) and many invertebrate animals (for example, honey bees *Apis mellifera*) can reproduce asexually—that is, one parent is able to create copies of itself as offspring. Of the various processes that can underlie asexual reproduction, the most common form in animals is **parthenogenesis**. During parthenogenesis, one haploid gamete is transferred to offspring, so in animals among which XY chromosomes determine sex, the offspring tend to be female, and in those among which ZW chromosomes determine sex, the offspring are usually male (see **Box 9.1** for further explanation of this).

2 Oestrous is an adjective, e.g. an 'oestrous female', or the 'oestrous cycle', whereas oestrus without the second 'o' is a noun, e.g. 'oestrus occurs post-partum in this species'. The American spellings are 'estrous' and 'estrus', respectively.
3 Some fish show internal fertilization, notably sharks and dogfish, among which males can be distinguished by their possession of paired modified fins, which are used as intromittent organs (these are often referred to as 'claspers').
4 We know that 'born' is probably not a very good way of referring to an animal that hatches out of an egg, but we are

using it in a very wide sense here to avoid the text becoming too cumbersome.
Hermaphrodites have both male and female sexual organs. **Sequential hermaphrodites** function only as one sex at a time. **Simultaneous hermaphrodites** function as both sexes at the same time.
Gastropod molluscs have a single shell or no shell at all, and include snails and slugs.
Parthenogenesis is a process of asexual reproduction in which the unfertilized ovum develops directly into a new individual.

Figure **9.4** Within the animal kingdom, reproduction can be achieved through sexual or asexual reproduction. Komodo dragons *Varanus komodoensis* have very recently been added to the list of species that can reproduce asexually. (Photograph: Vicky Melfi)

Parthenogenesis can be achieved via different processes. In 'thelytoky', for example, only female offspring are produced and no mating is observed (an example of this is the African Cape bee *Apis mellifera capensis*), but in 'pseudogamy' (sometimes referred to as 'gynogenesis' or 'sperm-dependent parthenogenesis'), mating is required, because the sperm 'activate the eggs', but only the maternal chromosomes are inherited by the offspring (this occurs, for example, in some salamander species). Recently, it was confirmed that the Komodo dragon *Varanus komodoensis*, thought to only reproduce sexually, was able to reproduce via parthenogenesis (Watts *et al.*, 2006)—a discovery that was made from observation of female dragon births at Chester and London Zoos. It has been suggested that producing male offspring parthenogenetically may be a useful adaptation for an island-living female who does not have access to mates; after the production of male offspring, she would then be able to reproduce sexually (see **Fig. 9.4**). Sexual and asexual reproduction have different costs and benefits, but the benefits of sexual reproduction must outweigh the costs, because sexual reproduction is widespread.

In some species, mostly invertebrates, the ability to swap between sexual and asexual reproduction has been observed (a strategy referred to as 'heterogamy'). The Komodo dragon represents one of the few advanced vertebrates in which this has been identified.

9.1.3 Endocrinology

The hormone, or endocrine, system is one of two major control systems in the body (the other is the nervous system). It releases chemical messengers, called 'hormones', which are secreted (in minute quantities) by endocrine glands and then transported in the blood to target tissues elsewhere in the body on which they exert an effect. There are three main types of hormone, classified on the basis of their chemical structure:

- steroid hormones, which include the sex hormones testosterone and the oestrogens (the US spelling is 'estrogen'), and also hormones such as **cortisol** (see **Box 7.3**), are made of rings of carbon atoms;

- protein or peptide hormones, which include oxytocin, a hormone with many effects on the processes of mating and birth;

- amine hormones, which include substances such as adrenaline (also known as 'epinephrine').

The action of all hormones, regardless of their chemical structure, is usually selective rather than general, because the target tissues on which they exert their effects possess receptors that recognize specific hormones.

A key part of the endocrine system controlling reproduction is the **hypothalamo-pituitary-gonadal (HPG) axis**, which is found in all vertebrates (see **Fig. 9.5**).

- The hypothalamus is the main link between the nervous and endocrine systems in the body. It is part of the brain and secretes, in minute quantities, releasing hormones and other neurohormones, including gonadotropin-releasing hormone (GnRH), which is particularly important in reproduction.

- GnRH, in turn, induces hormone release by the pituitary, which is an endocrine gland that sits under

Endocrinology is the study of hormones and how they work.

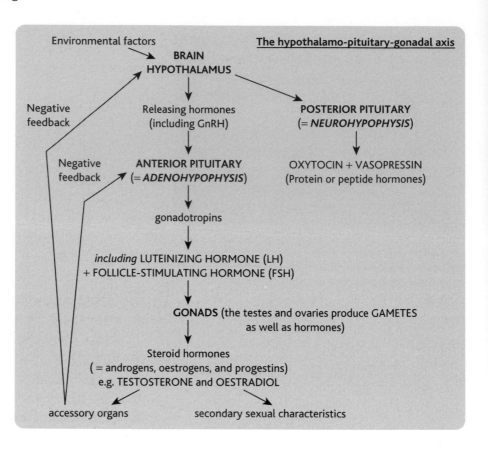

Figure **9.5** There are many endogenous and exogenous factors that influence reproduction. The hypothalamo-pituitary-gonadal (HPG) axis is central to the development and control of reproduction, which is mediated by both the endocrine and nervous system. The hypothalamus influences hormone production by both the anterior and posterior pituitary; these hormones go on to influence other organs in the body, including the testes and ovaries, which, in turn, are stimulated to secrete their own hormones.

the hypothalamus and is connected to the hypothalamus by a structure called the 'pituitary stalk'. The pituitary consists of two sections: the neurohypophysis, or posterior pituitary, and the adenohypophysis, or anterior pituitary, which secretes a wide range of hormones, including the reproductive follicle-stimulating hormone (FSH), which initiates the oestrous cycle, luteinizing hormone (LH), which, among other things, triggers ovulation, and prolactin (PRL), which is involved in lactation.

- These pituitary hormones are transported in the blood and, in turn, trigger a **response** elsewhere in the body, particularly in the gonads (testes and ovaries). As well as producing gametes or sex cells (sperm and eggs), the gonads also produce hormones. The male steroid hormones produced

by the testes are known collectively as androgens, such as testosterone, while the steroid hormones produced by the ovaries are known as oestrogens, and include oestradiol and the progestins (e.g. progesterone). (Incidentally, it should be noted that steroid sex hormones are not exclusive to only one sex, so testosterone, for example, is secreted in females as well as males, albeit usually in much lower concentrations.)

In female mammals, the oestrous cycle is associated with a series of hormonal changes, starting with FSH, which initiates the cycle, and leading to the release of the ova, through the influence of LH, during ovulation. This process is illustrated in **Fig. 9.6**.

If the released ova are fertilized, further hormonal changes take place. Monitoring the hormone

Anoestrous refers to the period of time during which an animal is not cycling and thus is not receptive to mating. This can be because the animal is between cycles, or it may be because it is pregnant, lactating, ill, or old. Many physiological functions diminish in older animals (see section 6.1.1). The duration of an anoestrous period varies according to how many oestrous cycles an animal experiences annually.

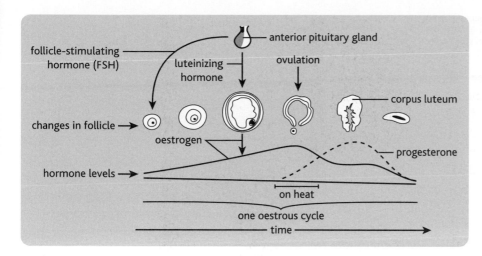

Figure 9.6 The oestrous cycle starts with the formation of a follicle, which is triggered by the secretion of follicle-stimulating hormone (FSH), and, as the follicle matures, it secretes oestrogen. Luteinizing hormone (LH) stimulates ovulation and the subsequent development of the corpus luteum, which produces progesterone. If the ovum is not fertilized, then the corpus luteum degenerates and progesterone levels fall.

profiles of an animal can therefore provide an indication of whether she is cycling, **anoestrous**, or pregnant (see section **9.3.4**), provided that there are enough data available describing the normal reproductive hormone profile—and, for many exotic species, these data can be limited (reviewed by Asa, 1996). In humans and some non-human primates, if the released eggs are unfertilized, the endometrium (the lining of the uterus) is shed (termed 'menstruation'), so, in these species, the oestrous cycles are often referred to as 'menstrual cycles'. In other species, the endometrium is reabsorbed if the eggs are not fertilized during an oestrous cycle and thus no menstrual bleeding is observed.

Females are usually only sexually active for a short period of time during their oestrous cycles. This period of sexual activity is often referred to as an animal being 'in heat' or 'in oestrus'. It is the period of time during which a female is able to mate and conceive, and it is sometimes associated with changes in behaviour or external physical signs (see section **9.3.1**). In contrast, females that have a menstrual cycle can be sexually active regardless of their position with the cycle.

Oestrous cycles start at different ages even within the same species, depending on heredity and diet. Data are mixed about whether cycling starts earlier in captive, compared with wild, animals.

The functions of the different hormones involved in reproduction differ among taxa, even though their chemical structures may not differ much among species. Several progestins, for example, are found in all vertebrates, but carry out different functions. Similarly, the hormone prolactin is involved with lactation in mammals and 'crop milk' production in pigeons and doves, but is active in water retention in fish and has a non-reproductive role (in skin shedding and tail regeneration) in lizards (Nelson, 2000).

Box **9.1** Sex determination and differences between male and female reproductive morphology

The mechanisms of sex determination vary quite widely across different **taxonomic** groups, and whether an individual develops as a male or a female will be determined by genetic or environmental factors, depending on the species. In (nearly all) mammals, the fusion of two X chromosomes gives rise to a female (XX) and the fusion of an X and a Y chromosome gives rise to a male (XY). Ova (eggs) in mammals contribute only an X chromosome, whereas sperm may contribute an X or a Y chromosome. A variation of this system is observed in many ▶

▶ insects: for example, in grasshoppers and cockroaches, males are determined by an absence of one chromosome (XO), although females still retain two chromosomes (XX). In birds, the situation is reversed, with males having two sex chromosomes (known as ZZ) and females having ZW sex chromosomes. Many egg-laying reptiles, including alligators, crocodiles, some turtles, and many lizards, have no obvious sex chromosomes at all. In these animals, the sex of the offspring is not determined at conception, but during a specific period of embryonic development. The factor controlling sex determination is the temperature of the nest in which the eggs are incubating. Some species (hermaphrodites) have both male and female gonads, or can function as both male and female during their life (see section **9.1.2**).

An obvious difference between the sexes, in mammals at least, is in the visibility of the gonads or reproductive organs. While the ovaries in vertebrate animals are always internal, most mammals have testes that are external to the body cavity, contained in a pouch called the scrotum. The mammalian exceptions to this are known as the **testiconid** mammals, where the testes are internal (for example, elephants). In birds, reptiles, amphibians, and fish, the testes are always internal.

Most male mammals possess a penis, but most bird species do not and have an opening known as a 'cloaca', as do the egg-laying mammals, or 'monotremes'; exceptions include the swans, ducks, geese, and ostriches, all of which have a penis. Crocodilians and turtles also have a penis, but snakes and lizards have a pair of hemipenes—although only one hemipenis is used at any one time during copulation (see **Fig. 9.7**). Tuataras (illustrated in **Fig. 9.8**),

Figure **9.7** The sex of all animals is not immediately obvious, because there may be a lack of secondary sexual characteristics and external sexual organs. Although invasive, one of the common methods used to determine the sex of snakes is to probe them, a technique that should only be carried out by a trained and skilled practitioner. (Photograph: Paignton Zoo Environmental Park)

Figure **9.8** Because there are various ways in which animals reproduce, the apparatus that animals use to copulate is also different. Tuataras *Sphenodon punctatus* represent a taxon in which the animals do not have external sex organs, so exchange of gametes is achieved by pressing the cloacas of both animals together. (Photograph: Geoff Hosey)

which are endemic to New Zealand and of which there are currently two extant species, *Sphenodon punctatus* and *Sphenodon guntheri*, do not actually have sex organs, so simply press their cloacas together when mating.

9.1.4 Triggers for breeding

The time when animals are able to breed and have offspring is shaped by several factors. If breeding is synchronized with the maximum availability of environmental resources, **fecundity** (the likelihood of offspring survival) is enhanced. This synchronization can be achieved through a number of mechanisms, both internal and external, including environmental factors (such as day length and food availability), pheromones (chemical signals from other animals), and social interactions.

Pheromones are chemical signals used to communicate within a species. They are detected in a variety of ways, ranging from the receptors on the antennae of male moths, to the olfactory mucosa of mammals. Some mammals, such as felids and equids, possess a specialized organ for this—the vomeronasal organ, or Jacobson's organ (in the roof of the mouth) —and take in air carrying the pheromone molecules through the mouth, using a distinctive **flehmen** behaviour. Pheromones may also be detected from urine or other substances. These chemicals have various functions, from mate detection and selection, to hunting and territory protection, and avoidance behaviours (Wyatt, 2003).

Environmental factors, including temperature, food availability, and predator pressure, help to shape the breeding biology of species. As a result, many animals have restricted periods during which they breed and have young, and these animals are termed 'seasonal breeders'. A few species, however, including some primates, pigs, mice, and rabbits, are considered 'continuous breeders', which means that they are able to breed and have offspring throughout the year. This may be because they experience sequential oestrous cycles, which makes them ready to mate and therefore have offspring throughout the year. Alternatively, it can happen in species among which mate availability is not reliable, so when a potential mate is in close proximity, this is more important in determining the timing of breeding than environmental features such as climate. These species, which include **lagomorphs**, felids, and camels (Lombardi, 1998), have developed into **induced ovulators**, so called because the act of mating or coition stimulates ovulation and therefore makes the female ready to breed whenever an appropriate male is available.

The number of oestrous cycles and the timing of these (whether or not they are seasonal) can vary from having several cycles ('polyoestrous', for example, in felids and bovids) to having two cycles yearly ('dioestrous', for example, canids) or only having one cycle annually ('monoestrous', for example, some **Carnivores**). In fact, polyoestrous animals rarely experience several oestrous cycles, because, if conditions are appropriate, they conceive and become pregnant, and this stops further cycles. But this system does enable females who do not conceive during their first cycle to make further subsequent breeding attempts.

In seasonal breeders, of course, reproductive activity is synchronized with seasonal aspects of the environment and usually this involves day length (photoperiod). The controlling mechanism that synchronizes seasonal breeders to their environment, via photoperiod length, is mediated by the release of the hormone melatonin.[5] Long-day breeders, such as equids, come into oestrus during the longer daylight periods of spring and summer, whereas

---

Lagomorpha is a mammalian order that includes the families Ochotonidae (pikas) and Leporidae (rabbits and hares). The latter includes the critically endangered Sumatran short-eared rabbit *Nesolagus netscheri*.

5 The hormone melatonin is present in all animals and also in algae. In vertebrates, it is secreted from the pineal gland, a body at the base of the brain, into the blood stream. Melatonin is secreted during dark periods, so the amount secreted is dependent on the proportion of dark to light in the day (Arendt and Skene, 2005). It is responsible for linking many physiological and behavioural patterns into a circadian, or seasonal, rhythm (such as daily or seasonal changes in coat growth, changes in colouring and behaviour), and may also be associated with other functions, such as skin colour changes in amphibians (Filadelfi and Castrucci, 1996).

short-day breeders, which include sheep, goats, and deer (remember, we stated at the outset of this chapter that much of our knowledge in this area comes from well-known domesticated species), come into oestrus during the short daylight periods of autumn and winter. It is thought that higher melatonin levels stimulate reproduction in short-day breeders when daylight hours reduce, but repress reproduction in long-day breeders at this time. This effect diminishes when the daylight hours increase, so long-day breeders become reproductively active and short-day breeders cease to be reproductively active. This mechanism cannot, however, be stimulated artificially simply by changing the hours of daylight experienced by the animals, because this process apparently involves a variety of other factors, at least in agricultural animals (Dooley and Pineda, 2003).

9.1.5 Group structure and function, and mating systems

Many variables shape group size and group composition in different species, and these include habitat type, the risk of predation, and the availability of resources. Survival, over generations, is optimized when individuals adapt to the environmental conditions around them and this can result in them developing different social organizations. For example, the formation of family units (of two breeding adults and their growing offspring) can have the advantages of allowing the members of the group to hunt together, defend a territory, and share parenting.[6]

Indeed, group size and composition, in turn, have a large impact on behaviour, affecting such things as mating strategies, feeding and foraging strategies, methods of communication, social behaviour, parenting, and social **learning**. Examples of social organizations include family units made up of parents and offspring and single-male, multi-female harems.

We have already mentioned mating systems in section **4.1.4**, where we identified **monogamy** (male–female pair), **polygyny** (one male mating with a number of females), **polyandry** (one female mating with a number of males),[7] and **promiscuity** (breeding between multiple males and females). Examples of species that have these different mating systems are identified in **Table 9.1**.

These terms describe what these mating systems look like to us. Unfortunately, animals that live in these apparent mating systems do not necessarily behave as we might expect them to, which can make any interpretation of the advantages and disadvantages of the systems problematic, and can also lead to some confusion when housing and managing species in zoos. For example, 90 per cent of bird and 3 per cent of mammal species appear to be socially monogamous in the wild, based on the observation that they form a bond with only one individual of the opposite sex for a prolonged period: a breeding season, or, in some animals, for life. But studies of birds such as the dunnock *Prunella modularis* have shown that, if the DNA of offspring attributed to known pairings is analysed to detect paternity, males, in fact, sire offspring in clutches other than their own (Davies, 1992). Similar results have now been found in other bird and mammal species, and demonstrate that, rather than being monogamous, both males and females may, in fact, engage in 'sneaky matings' or extra-pair copulations.[8] Consequently, these mating systems have been referred to as 'social monogamy', but not necessarily sexual fidelity and 'genetic monogamy'. It has been suggested that as many as 90 per cent of apparently

---

6 Cooperative breeding occurs if individuals other than the parents take part in parenting (Stacey and Koenig, 1990; Soloman and French, 2007).

7 You might also come across the term **polygamy**, which simply means one individual mating with several others of the opposite sex, and thus includes both polygyny and polyandry.

8 Extra-pair copulations are rarely seen and it is only recently that we have become aware of how frequent they are, because their effects can be detected in DNA analyses.

Mating system	Number of: Males	Females	Examples	Implications for captive management
Monogamy	1	1	Gibbons; swans; beavers; the mara *Dolichotis patagonum*	High reliance on finding compatible animals Need to move on offspring
Polygamy				
Polygyny	1	>2	Lions; equids, such as zebras	Potential surplus of males
Polyandry	>2	1	Marsupial mice (*Antechinus* spp.); several species of shorebirds; jacanas; pipefish	Potential surplus of females
Polygynandry	>2	>2	Red fox *Vulpes vulpes*; the acorn woodpecker *Melanerpes formicivorus*; some fish species, including the zebra cichlid *Pseudotropheus zebra*	Require large enclosures to reduce aggression and crowding Determining parentage of offspring can be difficult
Promiscuity	Any male within a group	Any female within a group	Chimpanzees *Pan* spp.; the cuis *Microcavia australis* (a small rodent)	Determining parentage of offspring can be difficult

Table **9.1** A summary of mating systems

NOTE: Various mating systems exist in wild animals, the replication of which can have ramifications for the captive management of the species. Listed here are the main mating structures observed, some examples of which species adopt them, and a summary of what the consequences might be for captive management.

monogamous pairings actually represent 'social monogamy' (reviewed for bird species by Westneat and Stewart, 2003; see **Fig. 9.9**).

Breeding opportunities in other mating systems are also not clear-cut. In some situations, breeding can be suppressed by conspecifics: for example, in singular cooperative breeding species such as common marmosets *Callithrix jacchus* and beavers *Castor fiber*, among which dominant females suppress conception in other females via the secretion of pheromones (Wyatt, 2003). Similarly, in many social species, only the dominant male and female will breed (for example, wolves *Canis lupus*), a mechanism that is reinforced through behavioural assertions of dominance, which are associated with healthier and larger individuals (Peterson *et al.*, 2002). These polygamous societies have high levels of competition between individuals, particularly between males

Figure **9.9** (*left*) Penguins (here, a macaroni penguin *Eudyptes chrysolophus*) are one of many species that have been shown to be socially monogamous. This means that they pair for breeding, but are not necessarily faithful to that mate, and take part in extra-pair copulations. (Photograph: Living Coasts)

(a)   (b)

Figure **9.10** Sexual dimorphism is a term used to describe morphological differences between sexes. Common characteristics of sexual dimorphism are differences in colour and size, and often it is the male of the species that is more colourful and larger, as seen here in the mandrills *Mandrillus sphinx*, among which the (a) male is more highly coloured, than the (b) female. (Photographs: (a) Tibor Jäger, Ramat Gan; (b) Ray Wiltshire)

in polygynous societies and between females in polyandrous societies. As a consequence, the more strongly competing sex may evolve marked morphological differences in size or ornaments (sexual dimorphism) or in colour (sexual dichromatism) (see **Box 9.2** for more information on sex differences). These differences in morphology may make an animal more attractive to a potential mate, or more able to fight competitors, thus gaining territory and mating opportunities (see **Fig. 9.10**).

An alternative sexual strategy that is adopted by individuals of some species is not to appear 'dominant', but to go unnoticed, resulting in different 'forms' of the same sex. For example, orang-utan *Pongo pygmaeus* males gain breeding opportunities with females by securing territories in which females live. They secure these territories through exaggerated sexual dimorphism, becoming very large and developing **secondary sexual characteristics** in the form of cheek flanges. There are, however, only

Secondary sexual characteristics—not to be confused with the primary sexual characteristics of the sex organs themselves—are traits that distinguish between the sexes. These can include size, colour, patterning, or physical attributes that are unique to one sex.

so many territories available over which males can compete, so if a male is neither particularly big nor very successful at competing with other large males, he may end up with no breeding opportunities. In orang-utans, then, an alternative strategy for these males is to remain small and appear 'female-like'; thus, they go unnoticed when entering the territories of other males, and may gain from 'sneaky matings' with the inhabiting females (Scharmann and van Hooff, 1986).

9.1.6 Parenting

The extent to which parental care is provided, who delivers it, and for how long it is provided vary greatly among different species. Parental care is sometimes considered to lead to a bond being formed between parent and offspring, which serves to enhance the survival of the offspring (see, for example, in non-human primates, Maestripieri, 2001; in **ungulates**, Lamb and Hwang, 1982). Parents also increase the likely survival of their young by virtue of their own **fitness**, which results in heritable features increasing or decreasing an offspring's chance of survival (such as 'big people produce big babies, which are more likely to survive').

The extent to which offspring are protected during their early development varies more consistently among taxa. The offspring of **viviparous** animals develop inside one of the parents (usually, but not always, the mother; male seahorses 'give birth' to live young from a pouch on their ventral surface). This strategy allows the parent to give birth to the offspring at a later stage of development when they are better able to survive. Most mammals, with the exception of the egg-laying monotremes (the duck-billed platypus *Ornithorhynchus anatinus* and echidnas *Tachyglossus* and *Zaglossus* spp.) are

viviparous. This contrasts with **oviparous**, or egg-laying, animals, which need to protect their nests if they are to protect their young to the same extent of viviparous parents. Birds, without exception, are oviparous, while reptiles, amphibians, and fish show greater variability (for example, frogs and toads are oviparous, but other amphibians, such as some caecilian species, give birth to live young).

In viviparous species, the long-term health and fitness of offspring can be affected by their parents while they are still developing inside the mother. Prenatal stress in pregnant laboratory rodents and primates, for example, has been identified as causing severe impairments in their offspring's development, learning, and memory (Egliston *et al.*, 2007), and can have long-term effects on temperament, behaviour, and response to stress (Austin *et al.*, 2005).

For those offspring that receive parental care after birth or hatching, the time at which they become independent of their parents varies considerably. For many species, the period of dependence is characterized by the young animal spending a disproportionate amount of time in play compared with adult animals: given the opportunity, juveniles have been observed to spend at least 10 per cent of their time in social play (Bekoff and Byers, 1998). Both mental and physical fitness are achieved through play.

A comparative study of **cognition** has demonstrated that 'cognitive milestones' are achieved by different species at very different times. For example, object permanence, finding objects, and knowing they exist even if they cannot be seen is considered a necessary survival skill for many species.[9] Research has shown that different species may be able to attain different levels of object permanence and at different ages; this is illustrated in **Fig. 9.11**.

---

9 Consider, for example, a cat stalking a mouse, which runs behind a dense plant: even though it cannot see the mouse, the cat knows that it is still there, and can even anticipate where the mouse will reappear.

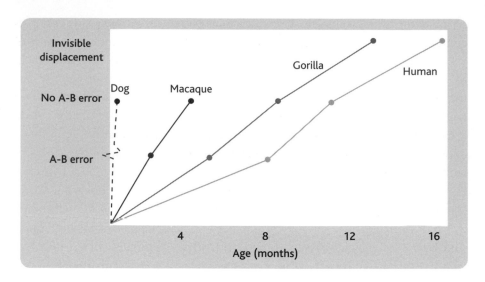

Figure **9.11** Here, we can see the speed and age at which four different species (dog, macaque, gorilla, and human) learned to find objects, and the errors that they encountered in the process. All primates were able to attain object permanence, but at different ages and different speeds, but dogs never reached this level. (From Gomez, 2005)

9.2 Issues and constraints on reproduction in captivity

As we reviewed in the previous section, reproduction and mating strategies are very different between species and there are a great many factors that can influence whether breeding attempts are going to be successful or not. It is not surprising, therefore, that breeding animals in zoos is not necessarily going to be easy and straightforward (although it can be for some species).

As we have already mentioned, fundamental information about the mechanics of reproduction, or the factors that influence it, is lacking for many of the species housed in zoos. How, then, should zoos proceed when they attempt to create and manage such a diverse collection of animals?

In the following section, we have outlined some of the main issues that are considered to be the basic requirements of captive management. These range from the logistical considerations of how to get the animals together, through to the genetic requirements of reducing **inbreeding** between individuals and maximizing genetic diversity, and, finally, to the need for behaviourally competent animals that are able to mate and rear young. Some other factors, such as the impact of stress, health, and nutrition on reproduction, will not be considered here, because they are covered more fully in **Chapters 7** on 'Animal welfare', **11** on 'Health', and **12** on 'Feeding and nutrition'.

9.2.1 Getting animals together for breeding

There are several different approaches that can be followed to bring animals together for breeding in zoos. The obvious choice is to maintain the animals in appropriate social groupings, similar to those observed in their wild conspecifics. Housing and husbandry has been, and can be further, developed to enable these types of groupings to be maintained in zoos (see **Chapter 6**).

Although this can be effective for animals that live in social groups and among which both parents reside in the same group, many species do not live in these types of group. Some species, for example, are solitary, or show a **fission–fusion** strategy, or live

Fission–fusion groupings are seen in various species, including chimpanzees *Pan troglodytes* (Lehmann and Boesch, 2004), among which membership to a group is not fixed, but transient, with animals moving between groups.

in single-sex groups. Furthermore, even in stable social groups, there is natural migration of animals between groups in zoos, which is essential to prevent inbreeding between related animals. This means that, for many species, successful captive management requires much intervention to overcome the complications of getting two animals together. In sections **6.3.2** and **6.3.3**, we addressed some of these complications, including managing the tension expected when introducing animals, and managing the establishment of dominance hierarchies and competition for resources. This whole process can, of course, be avoided by using artificial reproductive technologies such as artificial insemination (AI) (see section **9.4.1**), but many zoos much prefer to bring two animals together physically to breed.

It has been suggested that there are certain housing and husbandry criteria that will ease introductions between males and females, but that an increased likelihood of successful breeding is affected by a more complex set of variables. Those working in zoos will be familiar with situations in which individuals that have been appropriately matched by species, sex, age, and genetic representation do not mate when put together. This is especially a problem for the cheetah *Acinonyx jubatus* and, anecdotally, other species, such as parrots. It is possible that individuals among these species express a high level of mate choice and that, if the potential mate provided does not match their requirements, they will not breed with them.

So what are these animals looking for in a prospective mate? We have already briefly considered sexual selection theory, which suggests that, in many species, individuals of the same sex compete with one another to acquire mates and, in the process, develop phenotypic differences (such as body size, weapons, ornaments, and colours), which may give them an advantage over others or signal something about their fitness. This then makes them appear more attractive to the opposite sex and thus more likely to be selected as a mating partner (see section **9.1.5** for more on sexual dimorphism and sexual dichromatism).

More recently, it has been found that mate choice may also be influenced by the **major histocompatibility complex (MHC)**. The MHC is an array of tightly linked genes (Boehm and Zufall, 2006), found in all vertebrates except the jawless fishes (that is, lampreys and their relatives), and with a similar function in all of them. These genes are in a unique combination of alleles in each individual, so they are able to produce molecules that can recognize invading molecules (such as bacteria), because those molecules have a different MHC composition. Close relatives have MHC genes that are more similar than those of non-relatives, so any animal that can detect MHC products outside of the body (for example, in urine or sweat) can potentially gain information about the other animal's genetic make-up, identity, and relatedness. This detection is achieved through olfaction, often by specialized cells in the nose (Boehm and Zufall, 2006; see **Fig. 9.12**). So, recognizing another individual as highly unrelated

Figure **9.12** Olfaction—that is, using the sense of smell—is a prominent route for communication for many species, including many felids, and can convey a lot of information about the animal that left scent. (Photograph: Paignton Zoo Environmental Park)

The **major histocompatibility complex (MHC)** is an array of linked genes found in most vertebrates. The MHC plays an important role in the immune system, but also in other areas such as mate selection.

(that is, one with a very dissimilar MHC) would reduce the potential for inbreeding and thus indicate the presence of a compatible mate.

9.2.2 Genetic goals

The need to prevent inbreeding and maintain genetic diversity among our captive populations is central to how we currently manage zoo populations. Genetic variability, resulting from sexual recombination of genes, is the raw material for evolutionary adaptation. Even slight variations in genetic code can result in different phenotypes that are more or less able to survive in a given habitat. It is this process that leads to genetic diversity in the population. In small populations, however, it is more likely that closely related animals will breed with each other and, because they have many alleles in common, they may become homozygous for recessive alleles. The recessive alleles, the effects of which would otherwise be masked by dominant alleles (see section **9.1.1**), are then able to affect the animal's phenotype, often in a deleterious way.

Unfortunately, zoo animal populations are especially small, even when zoos combine their efforts via national and international management programmes. Most captive management programmes have an **effective population** (N_E = the number of animals that actually breed and contribute offspring to the population) of fewer than a hundred animals. (See **Box 9.2** for more information on genetic terms used in captive management programmes.) This makes these populations susceptible to problems arising from inbreeding and increases the likelihood that genetic diversity will become reduced.

It is thought that reductions in genetic diversity will impede the long-term conservation goals of captive-bred populations, by potentially reducing the likelihood that these populations will be able to adapt to different wild habitats, and thus survive,

when and if they are reintroduced in the future. The deleterious impacts of **inbreeding depression** are already visible in some captive populations and include:

1. reduced **fertility**, seen as genetic disorders, smaller litter sizes, and reduced sperm viability;
2. reduced reproductive output, through lower **birth rates** and increased infant **mortality**;
3. reduced fitness, due to slower development, smaller adult size, and reduced immune function.

Current thinking among geneticists and reproductive biologists working with zoo management programmes is that '*the great majority of populations forced to inbreed will suffer deleterious and insidious genetic-demographic consequences and increased extinction risk*' (Taylor, 2003). Numerous examples from captive populations of wild animals support this assertion. Examples include:

- decreases in semen volume and sperm motility, both within and across species of inbred gazelles *Gazella dorcas*, *Gazella dama*, and *Gazella cuvieri* (Gomendio *et al.*, 2000);
- reproductive problems and reduced survival of juvenile South China tigers *Panthera tigris amoyensis*, among which inbreeding is increasing in the small captive population, and unfortunately no wild animals exist to bolster the gene pool (Xu *et al.*, 2007).

But inbreeding can also occur naturally in wild populations, especially in island-living species, where breeding opportunities are limited within the one population, and those animals that survive become more and more specialized to the island habitat (Leck, 1980).

The importance placed upon the need to guard against a loss in genetic diversity and prevent

---

Inbreeding depression, usually just referred to as **inbreeding**, occurs when two closely related animals breed and produce offspring, the genetic material of which is homozygous for more recessive alleles than would otherwise be normal. This occurs because the related animals share many genes.

inbreeding has led to most captive management pro- grammes being managed from a theoretical genetic perspective. A widely accepted target for captive management programmes is to maintain 90 per cent (or, for larger populations, 95 per cent) of genetic diversity in a **demographically** stable population over 200 years (Soulé *et al.*, 1986). There are many different factors that affect our ability to achieve these targets, not least differences in the **life history traits** of different species and animal management logistics (that is, those tasks required to move and pair certain animals). Because of this, a rule of thumb is used that requires newly paired animals to be matched according to their **mean kinship** and allowed to breed if their potential offspring would have a low inbreeding coefficient (see **Box 9.2**).

Box **9.2** Genetic terms explained

In recent years, there have been great advances in our understanding and applications in the field of **population genetics** as the study of '*allele frequency distribution and change under the influence of the four evolutionary forces: natural selection, genetic drift, mutation and gene flow*' (Frankham *et al.*, 2002).

Natural selection is the process, observed in wild animal populations, whereby the alleles of some 'fitter' animals are transferred into the next generation, whereas the alleles of 'less fit' animals are not. Fitter animals are considered to possess traits that are adaptive—that is, that make them more able to survive, breed, and rear young more successfully than other animals—hence this process is sometimes referred to as 'survival of the fittest'. Natural selection will be greatly affected by the environ- mental factors at the time, which may result in different allele combinations being 'fit' at one time, but not at another.

By contrast, genetic (allelic) drift is the change in allele frequency observed in a popula- tion of animals from one generation to the next that cannot be explained by environmental factors and, as such, is considered to occur randomly. Of relevance to the captive popula- tion, genetic drift can occur when a few animals from one population become **founders** of a new population, also termed the 'founder effect'. In this circumstance, allele frequencies that were low in the former population come to be the most representative in the new population— that is, the alleles in the new founders may not have been very common or representative in their old population, but, having moved and created a new population with related animals, their alleles are now the more abundant in the new population. Equally, 'genetic bottlenecks' can occur when a large population dramatically reduces in size. If the reduction is due to random events, it is possible that the resulting allele frequency of the smaller population that is left may be very different from the larger popula- tion. This type of genetic drift can have very deleterious implications for the future survival of the species, because its genetic diversity and thus ability to adapt to changes in environmental conditions is greatly limited. A classic example of a species that has been affected by a genetic ▶

An animal's **mean kinship** is a value that describes its genetic representation in the population. Animals with a low mean kinship (close to 0) are related to fewer animals than those with a high mean kinship. Thus, an animal with a mean kinship of 0 is not related to any other animals in the population.

▶ bottleneck is the cheetah *Acinonyx jubatus*, which has been attributed to its vulnerability both in the wild and in captivity (O'Brien *et al.*, 1985).

Mutations, as their name would suggest, are changes in the animal's genetic information, but which can have adaptive, non-adaptive, or neutral consequences for survival, thus contributing to either natural selection or genetic drift. Mutations can occur for a variety of reasons, including errors made when the genetic material is copied within the cell, through exposure to ultraviolet (UV) light or radiation, or infection from viruses.

Finally, gene flow is the term given to the process whereby allele frequency is affected as a consequence of animals migrating into or out of a population.

From this brief overview, it is easy to see that these four forces that affect population genetics are not mutually exclusive and that the consequences of them acting together will determine the genetic make-up of a population.

The study of these four processes has informed our knowledge of many genetically based phenomena observed in the animal kingdom, underpinning the current captive management of zoo animals, which is highly biased towards their genetic management. The founding principles for most captive management programmes are to maximize genetic diversity and prevent inbreeding. Achieving these goals will ensure that the population has the potential to change according to environmental factors (that is, that the population would be able to adapt if it were to be reintroduced into the wild), will represent the wild population from which it was drawn (that is, that founder effects, genetic drift, and gene flow are limited), and will not suffer the deleterious consequences associated with inbreeding (see section **7.3.3**).

So how is this achieved for zoo animal populations? As a rule of thumb when creating potential pairs for breeding, it is necessary to match individuals of a similar mean kinship, but to attempt to ensure that their potential offspring will have a low inbreeding coefficient (r); these calculations can be made using computer software described in section **5.7.3**—for example, SPARKS and PM2000.

The degree to which an individual is related to others in the population can be calculated and the resulting value is termed mean kinship. An animal with a low mean kinship—that is, a value of zero or close to it—represents an individual whose alleles are not present in other individuals in the population (that is, it is not related to other animals in the population); an animal with a high mean kinship is related to many more animals in the population.

An animal's inbreeding coefficient, also known as the 'degree of relatedness', is calculated as the proportion of alleles that are likely to be shared between individuals. For example, an inbreeding coefficient of 0.5 signifies that half of the alleles would be shared and is thus calculated for parent–offspring; an inbreeding coefficient of 0.25 indicates that a quarter of the alleles are shared and is calculated for grandparent–grandchild.

Ensuring the mean kinship values match, but inbreeding is prevented, is more difficult than it sounds, because animals that are closely related are also going to have similar mean kinships, but will therefore have potential offspring with high inbreeding coefficients.

However, once that conundrum is solved, the appropriate animals need to be brought together to breed, because they may not be in the same social group, zoo, or country.

9.2.3 Behavioural competence

Making captive management recommendations on genetic needs alone (that is, on the need to maximize genetic diversity and limit inbreeding) ignores the importance of the 'animals' (as opposed to their genes) in reproduction. **Behavioural competence** is a term that has been used to describe the ability of an animal to express appropriate behaviour in a given situation. For example, we might expect a behaviourally competent animal to express courtship behaviours in response to appropriate **stimuli**, such as being paired or viewing courtship behaviour in a mate (see **Fig. 9.13**). Similarly, we might expect that, after parturition or hatching, animals would perform the necessary parental behaviours, if indeed this is characteristic of these species.

Figure **9.13** Even for animals that are compatible and attempt mating, the physical act of positioning often requires some practice: for example, it is easy to appreciate that a male black rhinoceros *Diceros bicornis* may need some skills to manoeuvre himself physically into the right position. (Photograph: Paignton Zoo Environmental Park)

Frankham *et al.* (1986) suggested that losses in behavioural competence in captive animals were as likely to occur as genetic changes. It is important therefore to recognize the need to preserve behavioural competence as part of the long-term conservation strategy of a species, because it is as important as maintaining their genetic integrity. One way of achieving this goal is through the implementation of **environmental enrichment** (Shepherdson, 1994; see **Chapter 8**).

However, a note of caution needs to be made at this point: just because an animal does not mate or show parenting behaviours, it does not necessarily indicate behavioural incompetence. There may be extremely good reasons why an animal does not express certain behaviours, despite the fact that all of the stimuli appear appropriate. An animal that does not display appropriate breeding behaviour may not like the mate that has been selected for it, and may simply be exerting mate choice (as discussed previously). Anecdotally, it also appears that, sometimes, a lack of parental behaviour in an animal might reflect a problem with the offspring (for example, it may be unwell or deformed; essentially, it is not a fit and healthy animal) and so there is no biological advantage to the parent in looking after it.

9.3 Monitoring the reproductive status of animals in captivity

It is important to monitor the reproductive status of animals in captivity for a variety of reasons, not least to prevent surprise babies that are not wanted (due to captive management recommendations), or for which preparations have not been made. It is also important to monitor reproductive status because this information informs decisions about:

• when to bring animals together for mating. Timing this correctly is essential to increase the likelihood that animals brought together will mate, but also to reduce the time during which potentially

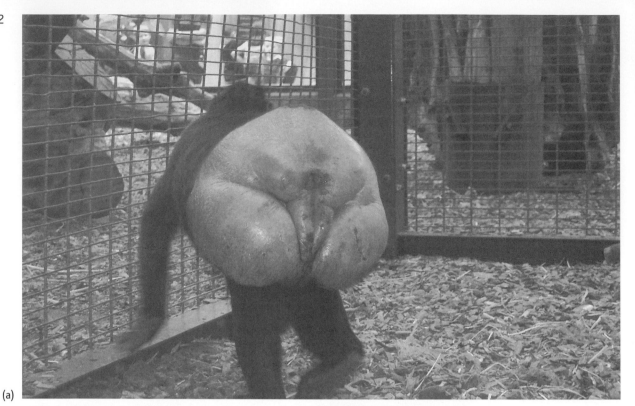

(a)

(b)

Figure **9.14** Secondary sexual characteristics are features that are unique to one of the sexes (and can therefore help us in sex determination), but which can serve to advertise information to the opposite sex—for example, (a) the reproductive status of the female Sulawesi macaque *Macaca nigra* here is advertised by her exaggerated sexual swelling—or to the same sex—for example, (b) the long tusks on this bull Asian elephant *Elephas maximus* serve as an advertisement of his fitness and deter competition from less-fit males. (Photographs: (a) Vicky Melfi; (b) Sheila Pankhurst)

aggressive animals are in proximity to each other (the opposite sexes of many species do not readily coexist together);

- when to make the necessary changes to animals' housing and husbandry, for example, increasing food rations for pregnant and lactating females (see **Chapter 12**) or the provision of nesting materials pre-partum or pre-lay;

- interpreting observations and records made about the animals, for example, their food intake, faeces consistency, behaviour. It is obviously important that changes due to reproductive status can be distinguished from those that represent a change in health and/or welfare;

- implementing breeding recommendations, which may require that a zoo attempt to breed (or not to breed) from particular individual animals.

The first reproductive state that should be recognized is when animals become sexually mature, but there are many other stages of reproduction that can be monitored. Reproductive states can be determined in three ways: by observing external visible signs (such as secondary sexual characteristics); from information gained via non-invasive procedures (such as the analysis of faecal samples); and from **invasive procedures** (for example, the evaluation of hormone profiles from blood). These three methods are used to varying degrees with different species, because our ability to identify reproductive states across taxa varies considerably. For example, for many mammals, external visible signs can be used to determine sexual maturity, oestrus, conception, and also when a female is ready to give birth.

Analyses of data from a large population and **epidemiological** studies can provide good estimations for the likely timing and durations of reproductive states, including the onset of sexual maturity, gestation or hatching length (thus expected birth/hatching dates), inter-birth intervals, and litter/clutch size. (The use of records is explored in more detail in **Chapter 5**.)

9.3.1 External signs

Monitoring external signs is a low-tech, low-cost, and non-invasive method of determining reproductive status, so can be done by observing the animals daily and keeping good records of what is seen. Secondary sexual characteristics are the most obvious visible sign to look for and to monitor, and these can be highly visible in some species, as illustrated in **Fig. 9.14**. The development of secondary sexual characteristics indicates the onset of sexual maturity, while changes in these characteristics over a period of time can indicate when animals are ready to mate.

Equally, in many species, weight gain in wild living males, or those in captivity given adequate access to increased food rations, signifies that they are ready to breed. A well-documented example of this is what has been termed the 'fattening phenomenon', which is observed in male squirrel monkeys *Saimiri* sp. Seasonally, male weights may increase by more than 20 per cent of their pre-breeding season body weight (Boinski, 1987). It has been suggested that this weight gain prepares males for a period of fasting that may occur while they are making breeding attempts. Seasonal changes in hormone levels, testosterone and cortisol, have also been recorded to occur during the breeding periods (Schiml *et al.*, 1996).

The most notable secondary sexual characteristics in females are seen in mammals, which have mammary glands. Some female mammals also have perineal sexual swellings, which become swollen (termed 'tumescent') when the animal ovulates, providing an obvious indicator of their reproductive status. Thus, when sexual swellings are swollen, the females are ready to mate, while the cessation of cyclical swelling indicates that they have probably conceived and become pregnant. Other signs of reproductive status in mammals can include changes in vulva appearance, vaginal bleeding, and swollen mammary glands and/or nipples. In some cases, these visual displays have been shown to coincide with the sequential change in hormones observed

Figure **9.15** Sometimes a change in reproductive status is accompanied by obvious behavioural changes, such as the onset of courtship behaviour. Here, a pair of great white egrets *Ardea alba* are displaying to each other on the nest, as part of courtship behaviour. (Photograph: © Mark Kostich, www.iStockphoto.com)

during oestrus. In Malayan tapirs *Tapirus indicus*, for example, a swollen vulva gives a good indication of lowered progesterone levels, which indicate that the females are ready to breed (Kusuda *et al.*, 2007).

A variety of behavioural changes also occurs during different reproductive states. Obvious behavioural changes related to reproduction include the expression of courtship displays (**Fig. 9.15**), increased aggression and competition for resources, changes in appetite, and mating and mating-related behaviour.

9.3.2 Invasive versus non-invasive monitoring methods

For many species, visible signs of external changes associated with reproductive status are either missing or are not considered reliable enough to inform decisions about their management. This should not be surprising, given that simply sexing many species —for example, many reptiles, amphibians, and fish —can itself be problematic. So, when managing these species, information about reproductive states can be gained by direct assessment of hormone profiles, gonad status, or 'putting animals together and observing what happens'.

There are many advantages to choosing non-invasive methods over their invasive alternatives, primarily because these methods are less disruptive to the animal. In this context, invasive refers to measures that break the skin or penetrate a body cavity, and also those that cause a disturbance to the animal. At worst, invasive methods may require that an animal is caught up and sedated, then a procedure undertaken to monitor its condition (which may be painful), after which it is then reintroduced back into its group. Invasive procedures that may be undertaken to investigate the reproductive status of animals include taking blood samples, sperm collection, and exploratory surgery.

The impact of invasive procedures can be greatly reduced through husbandry training techniques (see section **13.4**), or, in the case of potentially painful procedures, **analgesic drugs** can be given. Husbandry training for invasive procedures entails the animal being trained to approach keepers and/or vet staff. Zoo staff are then able to conduct the necessary procedure while the animal remains in its enclosure and, potentially, in its social group, with minimal disruption. For example, animals can be trained to offer up an arm for a blood draw, or the anogenital region for swabs. Using husbandry training enables comprehensive and reliable monitoring, with less disruption to the animal.

Alternatively, the animal's reproductive status can be monitored using non-invasive methods. These procedures can include the use of external visible signs, as described in section **9.3.1** (reviewed by Asa, 1996), the monitoring of reproductive hormones via analyses of faeces and urine, and, in some contexts, **ultrasonography** (see **9.3.3**). Implementation of the latter can be achieved either invasively, if the apparatus must be entered into the animal's body cavity, or non-invasively, if the apparatus can be moved over the animal's skin.

9.3.3 Ultrasonography

Ultrasonography works by using a probe that emits high-frequency sound waves. As the probe is moved over the relevant area, the sounds waves are reflected from the surfaces of organs and other structures, and the reflected sound waves are received and transformed into an image (**Fig. 9.16**). For some animals —for example, dolphins—this probe can be moved externally over the skin of the animal, but, in some circumstances, greater conductivity may require that hair is clipped or a gel applied to the skin. For other species, a probe may need to be inserted into the animal, via the vagina (transvaginal) or anus (transrectal). Animals can be trained to stand while this procedure is undertaken (for example, elephants and rhinoceroses; see Hildebrandt *et al.*,

2006). Other species require anaesthesia (the big cats) or restraint (birds) before this technique can be used.

The resulting images can be recorded as: two-dimensional (2-D), showing the length and width of the object of interest; three-dimensional (3-D), showing length, width, and depth of the object; and four dimensional (4-D), showing length, width, and depth in multiple images over a period of time, which essentially provides a video of the object and thus enables the observer to watch any movement.

Pictures captured using ultrasonography in zoo-housed animals have been used for four main reasons: to sex animals (for example, penguins, among which the male and female may appear identical); to monitor gestation; to support assisted reproductive

(a)

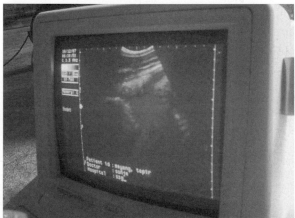

(c)

(b)

Figure **9.16** To ensure the safety of zoo staff at Singapore Zoo, this female Malaysian tapir *Tapirus indicus* is (a) crated before she is examined and (b) an ultrasound scan taken, which provides (c) images of a growing baby, enabling zoo staff to monitor the progress of her pregnancy. (Photographs: Singapore Zoo)

technologies, such as artificial insemination; to investigate disease.

The incorporation of ultrasonography techniques into the captive management of many exotic animals nowadays facilitates breeding and the investigation of breeding problems. It has been described as *'the golden standard for reproductive assessments in megavertebrates'* (Hildebrandt *et al.*, 2006).

9.3.4 Monitoring reproductive hormones

The importance of hormones in reproduction has been described already in **Box 9.1** and section **9.1.3**. Historically, endocrine (hormone) function could only be detected and monitored using blood samples, which can usually only be gained through invasive catch-ups of the animals, although, as we have seen in section **9.3.2**, husbandry training can ameliorate this. Methods of non-invasive monitoring of reproductive hormones, however, are not new (human pregnancy tests using urine samples date back to the 1920s—Cowie, 1948), although their application to the management of wild animals in captivity is relatively recent, dating from the late 1980s onwards (Pickard, 2003).

Today, steroid hormone metabolite concentrations can be measured from urine, faeces, saliva, and sweat, by either radioimmunoassay (RIA) or enzyme immunoassay (ELISA) (reviewed for mammal species by Asa, 1996). The non-invasive monitoring of hormones using assays is described in **Box 7.3**, using cortisol as the example; the principles involved are the same when monitoring sex-steroid hormones.

Research on the non-invasive monitoring of reproductive hormones in exotic wild animals has not only facilitated good animal management, but has also led to a number of useful insights into problems affecting the captive management of some species. In captive elephants, for example, ovarian inactivity affects 15 per cent of Asian elephants and 25 per cent of (female) African elephants (Brown, 2000).

9.4 Providing a helping hand: assisted reproductive technologies

The three main assisted reproductive technologies (ARTs) utilized by zoos are artificial insemination (AI), *in vitro* **fertilization (IVF)**, and **embryo transfer** (for a useful summary of the use of ARTs in the captive management of wild animals, see Loskutoff, 2003). The use of ARTS often also involves surrogacy, which can take several different forms. In interspecies embryo transfer, for example, a fertilized embryo from a wild animal species is transferred into the uterus of a similar domestic species for gestation.

Zoos also make use of other reproductive techniques, such as cross-fostering (see **9.5.3**). Cloning (more correctly referred to as 'nuclear transfer') is now widely used for reproductive management in domestic livestock, but remains rare in the captive management of wild animals[10] (Critser *et al.*, 2003).

9.4.1 Artificial insemination

Before AI can occur, semen must be collected. This is achieved via a variety of means, such as electroejaculation, massage (easier and safer in birds than in large mammals), provision of a 'dummy' female with an artificial reproductive tract, or 'cooperative ejaculation' (falconers, for example, use sexually imprinted birds trained to copulate voluntarily on special devices (Hammerstrom, 1970). In mammals, at least, viable sperm can be obtained up to 24 hours after the death of an animal, providing that the testes are kept cool. This has obvious benefits in the case of the unexpected death of a genetically

---

10 One example of the use of cloning, or nuclear transfer, in a wild animal is reported by Lanza *et al.* (2000), who successfully cloned a calf from a gaur *Bos gaurus*, a species of wild cattle. This was achieved using nuclear transfer and then transfer of the embryo to a domestic cow for gestation. As with many cloned animals, however, the calf did not survive for more than a few days after parturition.

valuable animal from an endangered species. Once collected, sperm (semen) may be used immediately or preserved by freezing (cryopreservation)[11] for future use.

Obtaining viable sperm is only the first stage in successful AI. Sperm must be introduced into the right part of the female reproductive tract, at the right time in the female's reproductive cycle. Again, there is considerable variability between taxa and even between related species, and species-specific AI protocols are needed for successful fertilization to take place.

Wishart (2001) provides a useful review of sperm cryopreservation and AI in birds. A collaborative workshop between the Chinese Association of Zoological Gardens (CAZG) and the IUCN's Conservation Breeding Specialist Group (CBSG) on captive breeding of the giant panda *Ailuropoda melanoleuca* resulted in the development of better methodology for the cooling and cryopreservation of panda sperm (reported in Wildt *et al.*, 2003).

9.4.2 *In vitro* fertilization

In vitro fertilization (IVF) refers to the production of an embryo in culture in the laboratory or clinic, as opposed to *in-vivo* fertilization, which occurs inside the mother's body (in the case of mammals and birds).

IVF can now be accomplished via the injection of a single spermatozoon into an oocyte. This technique is referred to as 'sperm microinjection'. The first mammalian offspring produced in this way were laboratory animals, such as mice and rabbits, using a technique now referred to as 'sub-zonal insemination' (SUZI) (Mann, 1988), in which a spermatozoan is inserted just inside the outer coating (zona pellucida) of an oocyte. A development of this technique, which is now more commonly

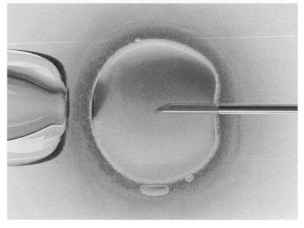

Figure **9.17** This photograph shows intracytoplasmic sperm injection (ICSI), a technique that was developed in 1992 in Belgium. Here, the egg is being held in place by a pipette while a single sperm is injected into the egg using a tiny needle, so that fertilization can occur. ICSI is a particularly useful assisted reproductive technology (ART) when males have a very low sperm count or low sperm motility. (Photograph: © Kiyoshi Takahase, www.iStockphoto.com)

used, involves the injection of a spermatozoon into the cytoplasm of the oocyte. This is known as 'intracytoplasmic sperm injection' (ICSI) (see **Fig. 9.17**). ICSI offspring have now been produced in a number of primate species, including humans (for a review, see Loskutoff, 2003).

9.4.3 Embryo transfer

Whether produced *in vitro* or as a result of **in vivo fertilization**, there have been a number of cases in which embryos from wild mammals have been transferred successfully into a different species for gestation. Pope (2000), for example, documents this form of surrogacy in two endangered felid species —the Indian desert cat *Felis silvestris ornata* and the African wild cat *Felis silvestris lybica*—among which embryos were transferred into domestic cat surrogate mothers and living offspring produced.

The term **in vitro** literally means 'in glass' and refers to the production of 'test tube' embryos in the laboratory.
11 Although cryopreservation of sperm is now widespread in farm animal reproduction and in the treatment of human fertility problems, the success of this technique for wild animals is very variable. Donoghue *et al.* (2003) reported a high degree of variability between even closely related species in sperm tolerance of, for example, rapid cooling.

Other zoo species in which successful surrogacy has been documented include the gaur (via domestic cattle), the bongo (via an eland), and Grant's zebra and Przewalski's horse (via the domestic horse).

Loskutoff (2003), however, points out that there have been many failed attempts at embryo transfer from wild to domestic species, and that most wild animals do not have an appropriate domestic animal species to act as a surrogate host, so the potential of embryo transfer to help zoos to breed endangered species remains limited, at least for the present time.

9.4.4 The advantages and disadvantages to zoos of using ARTs

The advantages of ARTs

There are a number of advantages to zoos of using techniques such as AI. These can be summarized as follows.

- Transferring sperm or embryos between zoos is cheaper, easier, and safer than moving live wild animals, and also has welfare benefits, in avoiding the need for the sedation, capture, and transport of live animals.

- The use of ARTs avoids the risk of fighting and injury during the introduction of a male and female for mating.

- Reproduction can still take place between incompatible animals or where there are physical blocks to natural reproduction.

- The number of offspring from a given female animal can be regulated.

- Genetic material can be exchanged between *in situ* and *ex situ* populations without the need for the transport of live animals.

- The risk of disease transmission from parent(s) to offspring can be reduced, via screening of semen for **pathogens** and/or by embryo screening prior to implantation; some developmental abnormalities can also be avoided by use of embryo screening.

- Cloning or nuclear transfer offers the possibility of producing several identical offspring from a genetically valuable animal.

- Additional offspring can be obtained from a 'valuable' female by using embryo transfer and surrogacy; this offers the potential of rapid population growth (Loskutoff, 2003).

- The sex ratio of offspring can be manipulated, either by flow cytometric sorting of sperm, or by the selection of an embryo of the desired gender prior to implantation. This can help zoos to overcome the problems associated with surplus males in a population, for example (most zoos have a 'bachelor paddock' housing male animals that would fight if they were allowed to remain within the main social group).

The disadvantages of ARTs

The use of ARTs in captive management of zoo animals will undoubtedly continue to grow as the techniques are refined and developed, and as protocols are developed for a greater number of species. But there are two major caveats about the use of ARTs for wild animals.

1. There is enormous variability in the success of these techniques across species and taxonomic groups (ICSI, for example, does not appear to be very effective in bovid species).

2. The use of ARTs has been linked to a range of problems, such as an increase in foetal and chromosomal abnormalities, and a higher rate of spontaneous abortions.

Loskutoff (2003) provides a review of the successes and failures of these techniques, and warns that much more research is needed to develop effective protocols for non-domestic species.

Similar caveats apply to the successful freezing or cryopreservation of sperm, in relation to which even related species show considerable differences in tolerance of freeze-thawing procedures (Donoghue *et al.*, 2003; see **Box 9.3**).

Box **9.3** The frozen zoo

It is unlikely that there will ever be many visitors to a 'frozen zoo', but freezing genetic material from animal and plant species, at very low temperatures, represents a technically viable method for helping to conserve species biodiversity, without having to maintain large captive populations of each organism. The principle behind the so-called 'frozen zoo' is to preserve samples of sperm, oocytes, or embryos —particularly from endangered animals— which, with the appropriate technology, could be used in the future to contribute to maintaining genetic diversity in breeding programmes (Holt *et al.*, 2003). This storage of gametes or embryos is also known as 'genetic resource banking', and efforts by zoos and other organizations to establish genetic resource banks (GRBs) have now been in place for more than 50 years.

Freezing gametes (and other material) at very low temperatures (typically −196°C, or −321°F) is known as cryopreservation. Techniques for freezing spermatozoa were developed more than 50 years ago, with pioneers such as Sir Alan Parkes leading the way in the newly emerging field of cryobiology (Watson and Holt, 2001). Vertebrate eggs or oocytes (called ova when mature) are more difficult to cryopreserve successfully than spermatozoa, or even embryos, and even with recent technological advances in this field, the successful cryopreservation of vertebrate oocytes remains a challenge.

The Frozen Ark Project, set up in 2003, is a worldwide initiative to collect, preserve, and store both DNA and, where possible, gametes and embryos, from threatened species of animal (see `www.frozenark.org`). Currently coordinated from the University of Nottingham in the UK, the Frozen Ark's list of consortium members includes the Natural History Museum in London, the Zoological Society of London (ZSL), Monash University in Melbourne, Australia, and many other universities, zoological societies, and conservation organizations from around the world.

Cryopreserved gametes can be regarded as genetically equivalent to living animals, with the advantage that they will provide valuable genetic material for years to come without the threat of the genetic loss that can occur over generations when living animals are bred.

9.5 Rearing: facilitating the successful survival of offspring

In the earlier sections of this chapter, our focus has been on the physiological and behavioural basis of reproduction, with special emphasis on mating behaviour and conception. In this section, we will consider rearing (see **Fig. 9.18**) and review how the survival of any offspring that hatch or are born can be enhanced through husbandry. This starts with a look at the prudence of ensuring that adequate facilities are made available to animals prior to them giving birth or laying and protecting eggs, and then when the offspring hatch or are born.

There are potentially many reasons why zoo professionals may feel that it is necessary to intervene with the rearing process. Usually, this is when rearing does not appear to be going well, but intervention may also occur to ensure or artificially increase rearing success. Interventions can be broadly divided into those that leave the offspring in the social group, so that the keepers provide supplementary

(a)

(b)

Figure **9.18** Within the animal kingdom, the degree to which parents care for and rear their young is highly variable, although it is common for many mammal and birds species to invest considerably in rearing young, depicted here by (a) babirusa *Babyrousa babyrussa* and (b) white stork *Ciconia ciconia*. (Photographs: (a) South Lakes Wild Animal Park; (b) Andrew Bowkett)

rearing, and those that involve removing the offspring from the social group. Equally, it is necessary that precautionary steps are in place prior to parturition or hatching, just in case there is a hitch.

Of course, for a great many species—especially amphibians, fish, and some reptiles—the process of rearing by the parents can be extremely limited and sometimes non-existent. Eggs are laid, the parents leave, and the survival of offspring is then dependent on their own abilities to survive. In zoos, ensuring the survival of these offspring is sometimes simpler than for those that require rearing, because they can often be provided with the appropriate feeding and climatic conditions, and they then survive to become reproductively competent adults. But the survival and long-term success of animals that require, but do not receive, adequate rearing from their parents can be severely compromised. The outcome of providing supplementary or full-time artificial rearing by keepers is variable (see **Box 9.4**).

9.5.1 Providing adequate facilities

Zoos should provide animals with conditions that best enable them to rear their offspring successfully. Identifying the optimal conditions for any situation is always a difficult task, so it may be more constructive to consider what animals need in three stages: when pre-partum or gravid; post-partum or post-lay; and during rearing.

Animals' needs can be assessed in a variety of ways, as outlined in section **7.2** (see **Fig. 9.19**). Prior to and during rearing, it is likely that the factors affecting the animals' ability to rear their offspring will include the health (including nutritional demands) of the parent and offspring, and housing and husbandry facilities. Zoos should attempt to identify the needs of the animals and what behaviours they might need to express during the three stages of rearing offspring, and ensure that the appropriate conditions are provided. It is also important to make efforts to ensure that the artificial nature of living in a zoo environment does not compromise the animals or restrict their behavioural expression during these stages.

(a)

(b)

Figure **9.19** Given the opportunity, many animals in zoos will breed successfully, as here, for example, in (a) Javan brown langur *Trachypithecus auratus* and (b) red-crowned crane *Grus japonensis*. (Photographs: (a) Joy Bond, Belfast Zoo; (b) Paignton Zoo Environmental Park)

At the moment, much of the general practice implemented for gravid or nesting animals is informed not by empirical data, but by intuition or anecdotal information, based greatly upon 'tradition' and 'word of mouth'. Some of these practices make perfect sense and do not require empirical support: for example, the separation or removal of eggs or young animals that are in danger of being eaten by their parents or by other conspecifics or heterospecifics housed in the same enclosure. More debatable, however—and probably variable among individuals and species —is the need for privacy, during mating, laying, or parturition. Anecdotally, it has been noted that many zoo animals give birth at night, or at least outside of visiting hours, which has been interpreted as a choice for 'quieter' conditions, although many wild animals also give birth at night and this may be related to predation pressures (Rowland *et al.*, 1984).

For some animals that seem especially sensitive to disturbance, some routine husbandry procedures (for example, periods of human–animal interaction, when keepers may enter the animals' enclosure for cleaning, or viewing of the animals by zoo visitors) may be delayed for a short time around parturition or egg-laying. The extent to which animals are susceptible to disruption is a topic that has received little attention from researchers, even though, in a captive environment, it is very likely to influence, among other things, reproductive success (see section 7.3.1; see **Fig. 9.20**). There are data, however, that highlight the need to increase the food availability to gravid and lactating animals (see section 12.4) and that provide some information about what constitutes appropriate nesting materials or structures for some species.

9.5.2 Supplementary parenting

In some circumstances, it is possible to facilitate rearing without the need to take offspring away from their parents or social group. This may simply require that some alterations are made to the routine housing and husbandry of the animals, such as food being chopped up smaller for weaning offspring.

(a)

Figure **9.20** Providing animals with the appropriate conditions is fundamental to: stimulating reproduction, which can be achieved by providing appropriate nesting conditions, as for these Hyacinth macaws *Anodorhynchus hyacinthinus* (a); increasing the likelihood that parents will rear their young successfully, which may require that they are given some privacy and so monitored remotely, as with this Asiatic lion *Panthera leo persica* (b); or that areas within the enclosure provide privacy for the offspring, which was accommodated by long grass for this baby pudu *Pudu pudu* (c). (Photographs: Paignton Zoo Environmental Park)

(c)

(b)

Alternatively, more dramatic interventions can be undertaken to supplement rearing directly. For example, keepers can provide additional food to infants, train animals in parental skills, and undertake health checks and diagnostics on infants without the need to handle them directly (Desmond and Laule, 1994). These and other steps, termed 'supplementary parenting', have been shown to promote the survival of offspring without the intervention of removal of the offspring.

9.5.3 Removing offspring from their parent and/or group

To many people, the removal of offspring from their parents would be seen as a last resort and would only be carried out if the survival of the parent or offspring were in doubt. This might result from inappropriate housing and husbandry, or from ill health or poor parenting skills. For example, Buckanoff *et al.* (2006) noted that the high level of neonatal mortality in pottos *Perodicticus potto* is probably largely due to inadequate maternal behaviour, and thus hand-rearing attempts have been made (see **Fig. 9.21**). As discussed above, it can take several attempts to appreciate fully and to identify the housing and husbandry facilities that are necessary to enable parents to rear their offspring. This might be thought of as a gap in the abilities of zoos, but really it is not surprising, given the large number of species that zoos attempt to breed in captivity.

There are, however, some circumstances under which the removal of offspring from their parents is part of planned captive management. The most notable argument for such an action is to increase the number of offspring produced. In section **9.6.1**, we will discuss the practices of **double-clutching** and decreasing inter-birth interval, as methods used to increase population size. Both of these techniques require that the offspring are removed and reared independently of their parents.

When and for how long an offspring is removed from its parents is highly variable and dependent on the end goal. If offspring are removed with the sole aim of **hand-rearing**, they will obviously remain away from their parents for the duration of their rearing. If removal has been undertaken as a mitigating step, however, to try to ensure survival of either the parent or offspring, then the two could be reunited very quickly indeed: eggs may be removed from some birds and incubated, and then the chicks returned to the parents after hatching. Similarly, problems around the time of birth can be resolved with some immediate and short-lived intervention from keepers, which enables them to get the offspring back to its parents in a very short period of time. Certainly, documentation of successful hand-rearing events has demonstrated that offspring can be successfully reunited with parents (Abello *et al.*, 2007). It should be noted, however, that there are a great many factors that determine the success of these reunions, including the social group size and composition, enclosure design, personalities of the animals, and much more.

When offspring are taken away from their parents, the decision of who takes over the role of rearing is determined by the availability of appropriate surrogates and facilities, and the offspring's age (for example, do the offspring still require milk or are they weaned?). It might also be possible for slightly older animals to forgo the surrogate mother and be grouped socially with peers, and have any additional needs met through some supplemental parenting from keepers. Surrogate mothers can be animals of the same or different species, including people, who are usually keepers. Domestic chickens, for example, have been used for many years to incubate and rear chicks from game birds and waterfowl species (Sutherland *et al.*, 2004).

Whether animals should be taken away from their parents for rearing is considered to be a

---

Hand-rearing refers to a situation in which a person, usually a keeper, takes on the responsibility of rearing an animal.

The more general term of **cross-fostering** implies that rearing might be undertaken by a person or animal.

(a)

(b)

(c)

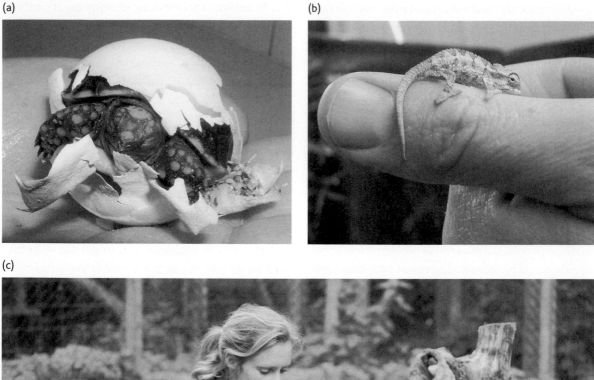

Figure **9.21** Hand-rearing is the most common and successful method of breeding many species of bird and reptile in captivity, but the topic of hand-rearing mammals is often seen as much more controversial, with many zoo professionals split over whether hand-rearing should or should not be undertaken. For some species, the alternative is something that is almost as unacceptable. (Photographs: (a and b) Paignton Zoo Environmental Park; (c) Dave Rolfe, Howletts)

contentious topic both within and outside of the zoo industry. Consider, for example, the discussion and media attention that was attracted by the particularly charismatic Knut, an abandoned polar bear cub, hand-reared at Berlin Zoo. But clarity is required: rarely are the issues of the utility of incubators to facilitate bird and reptile reproduction, or, in fact, methods used to increase the populations of these taxa, discussed. Instead, the controversy is usually about the use of hand-rearing for zoo animals and especially the hand-rearing of mammals.[12]

A summary of some of the costs and benefits associated with hand-rearing is presented in **Box 9.4**.

9.6 Manipulating exotic animal reproductive output

Carefully planned captive management programmes use demographic and genetic analyses to produce long-term projections and goals for the species. Goals made for a captive population will include whether attempts should be made to increase or decrease its size, or to keep the population size stable (see **Box 9.5**). There are various mechanisms that can be used to modify population size and these are reviewed in the next two sections. Any changes made to the size of the population need to be implemented with caution and with consideration of the genetic goals of the captive management programme. These, as explained in section **9.2.2**, usually aim to maximize genetic diversity and avoid inbreeding. Because of this, taking steps to change the population size needs to be coordinated between all of the zoos holding the given species and thus is usually managed by captive management coordinators as part of a captive breeding programme (see section **9.7.2** and **9.7.3**).

9.6.1 Increasing reproductive output

'Reproductive output' refers to the number of offspring born. There are two main ways in which this can be relatively easily increased. In egg-laying species, this is through a process termed **double-clutching**; in mammals, this can be achieved by reducing the **inter-birth interval** or by decreasing the age at which animals have their first offspring.

Double-clutching

Some egg-laying species have been observed to lay more than one clutch in a breeding season. Double-clutching is undertaken in zoos in the hope that this will stimulate the animal to lay another clutch and thus double its reproductive output. For example, Woolcock (2000) comments that successful breeding attempts with keas *Nestor notabilis* at Paradise Park in the UK are largely due to the removal of eggs that have been incubated. In this case, one female laid twenty-two eggs in a 12-week period, which obviously increased her reproductive potential and accelerated population growth.

Reducing the inter-birth interval

Anecdotally, it is assumed that inter-birth intervals have become shorter and that the age of first birth has decreased in zoo-housed mammals, as an unintentional consequence of housing and husbandry. It has generally been thought that reducing these two parameters will also increase the overall fecundity and fertility (reproductive output) of animals. Only recently has the impact of changing these life history strategies been investigated empirically.

Cocks (2007) reviewed various life history traits in the zoo population of orang-utans and discovered that when the age of first birth and inter-birth interval were shortened, compared with the ages

---

12 There has also been some debate about the potential drawbacks of hand-rearing some charismatic bird species, e.g. Californian condor *Gymnogyps californianus* (see section **10.4.5** and **Fig. 10.14**).
Double-clutching is a process whereby some, or all, of the eggs laid in one clutch are removed and incubated

artificially in the hope that the animal will lay another clutch and thus double its output.
Inter-birth interval is calculated as the time between births, which is recorded from the birth of the first offspring until the birth of the second offspring.

Box **9.4** Hand-rearing

Whether zoos should hand-rear animals or not is a highly controversial topic. But it is probably safe to say that much opinion about hand-rearing, and certainly the research undertaken to monitor its impact, are biased towards mammals (for example, Ryan *et al.*, 2002). Published literature is available about hand-rearing other taxa, but these are usually methodological practical papers (for example, on the Anegada island iguana *Cyclura pingluis*, Lemm *et al.*, 2005). This is true of many topics that we have discussed in this book, but the importance of highlighting the species bias in this matter is that hand-rearing is considered standard, and not at all controversial, for many reptile, amphibian, and bird species that would not necessarily gain much parental care in the wild, but which require it in zoos. It is possible that the degree to which hand-rearing will have negative implications, if indeed it has any, will be greater for offspring that are more heavily reliant on their parents and whose **survival skills** and behaviours are formed during this period.

There are three main reasons why a young animal might be removed from its parents and hand-reared, which can best be summed up as providing an advantage to one of three different parties: the young animal—if neglected by its parents and still dependent on their care, it may die or suffer without appropriate attention; the species—the removal of offspring may trigger breeding and increase the population size more quickly compared with natural breeding (double-clutching); the zoo professionals—the process of hand-rearing requires skills that are most effective with practice.

It should also be noted that, in the recent past, animals may also have been hand-reared as a visitor attraction, but this is less common and certainly not usually given nowadays as the primary reason for hand-rearing.

So why should hand-rearing be considered bad? Many animals appear to thrive during hand-rearing and as adults, but there are notable exceptions within the mammal taxa and among some bird species, in which disruption of the mother–infant bond appears to have deleterious ramifications (Cirulli *et al.*, 2003). Hand-rearing has been reported to result in behaviourally incompetent animals, which is observed as animals that are unable to communicate effectively or to behave normally with conspecifics; this has far-reaching ramifications, because it potentially compromises their long-term social interactions and breeding contributions to the population (Martin, 2005). One of the main disadvantages of hand-rearing occurs as a consequence of **imprinting** on the human care-giver, whom the animal then recognizes as a conspecific. This can be partly overcome by, for example, modifications to hand-rearing protocols, which have included the use of puppets to conceal the identity of the keeper (Valutis and Marzuluff, 1999; also see **Fig. 10.14**).

And what if zoos do not hand-rear animals that are rejected from their parents? The answer is simply that they have a duty of care to **euthanize** them, rather than to allow them to die 'naturally', because this will limit any pain or suffering.

Survival skills have been outlined by Box (1991) and include the ability to orientate, find food, avoid predators, and find suitable places to rest.

at which these events are observed in the wild, the **longevity** of the animals was also shortened, so that mortality was earlier. Because there are so few data available on this topic, it is difficult to know whether these results reflect a trend that changing life history variables in captive mammals is deleterious, or whether orang-utans are an exceptional case. But it is certainly an area that requires greater research.

9.6.2 Reducing reproductive output

It may seem counter-intuitive that a section is required in a chapter on captive breeding describing methods of reducing reproductive output in zoo animal populations—but, as outlined in **Box 9.5**, many captive management programmes may have limitations imposed on them, requiring that they do not increase populations, but maintain them at stable levels, or, indeed, reduce them over time.

There are various way in which this can be achieved, the simplest of which is to separate the sexes and maintain groups of singly sexed animals. This approach has been adopted for several species, including Rodrigues fruit bats *Pteropus rodricensis* and red-ruffed lemurs *Varecia rubra*.

Another simple, but very controversial, method is to operate a system termed **breed and cull**. This strategy allows animals within the captive management programme to breed, as normal, but any resulting offspring are culled. Both of these strategies can be implemented simply and easily. There are potentially greater complications in separating sexes than there are in culling offspring, because single-sex groups are not necessarily part of the natural social systems of the species for which this strategy is attempted. Conversely, allowing adults to have offspring enables them to express many of their naturally occurring behaviours, and so could be thought of as an ideal management tool, if viewed from the perspective of the behavioural needs of the adult animals.

The other main method used to reduce reproductive output is the use of contraception. Various types of contraception are available that operate in different ways: for example, by preventing ovulation or the implantation of a fertilized egg. They act on the reproductive biology of either the male or female animal, and thereby ensure that no offspring are born. Contraception can be achieved through physical separation, surgical techniques, and manipulation of endocrine function with synthetic hormones. Some of the most common methods are summarized in **Table 9.2**. Obviously, if it is likely that an animal might be required to breed in the future, then a temporary measure should be chosen. The various methods available are effective through the removal of reproductive organs (or parts thereof), the modification of endocrine control of reproduction, immunization against implantation, and delaying maturation or reducing the survivorship of birds' eggs (the latter is a method used to render eggs unviable).

9.6.3 Sex ratios and their manipulation

As described in **Box 9.1**, there are various factors other than genetic make-up and endocrine function that can influence the determination of sex. Although there are few data that have empirically described how these other factors can affect sex ratio determination, some papers have demonstrated that factors such as nutrition, population density, and the social status of the mother can have a significant influence on sex ratio (for example, Kilner, 1998; Loeske *et al.*, 1999).

ARTs can be harnessed to manipulate the sex ratio of offspring. Firstly, IVF embryos of the desired sex can be directly selected and implanted. Secondly, AI and IVF can be carried out using pre-sorted sperm, so that only male or only female embryos result. Flow cytometric sorting is a technique for separating the X and Y chromosome-bearing

The term **breed and cull** is used to describe a management strategy in which animals are allowed to breed, but the group size is kept a constant size by the selective culling (euthanizing) of individuals with the group.

Method	♂/♀	Reversible (duration)	Latency to effectiveness	Surgical	Behavioural effects	Comments
PHYSICAL SEPARATION						
Single-sex group	Both	Yes	Immediate	No	Abnormal social grouping— aggression may occur	May be able to introduce an adult male to an all-female group for short periods for breeding
SURGICAL						
Vasectomy	♂	No	Immediate	Yes	None (that we know of)	Will not reduce birth rate unless all of the males in the group are treated, but can reduce breeding by over-represented individuals
Castration	♂	No	Immediate	Yes	May reduce aggression between males BUT only if done before puberty	As for vasectomy
Tubal ligation	♀	No	Immediate	Yes	Normal sexual maturation and behaviour	Specialist surgical technique (keyhole surgery) or laparotomy, with concurrent need for post-operative wound management
Ovarectomy/ hysterectomy	♀	No	Immediate	Yes	Acyclic	Laparotomy required, with concurrent need for post-operative wound management
HORMONAL						
GnRH implants (e.g. deslorelin)	♂	Yes, variable (each implant lasts approx 2 years) Implants biodegrade and cannot be removed	Highly variable, can take months	Yes	Delays puberty, by reducing testosterone levels May not eliminate aggression in post-pubertal males (similar to castration)	Can lead to abortion if used in pregnant females No other known contraindications Further evaluation required to determine optimum dose rate and interval

Method		Reversible?	Time		Notes
	♀	Once hormone levels drop below threshold, normal reproductive function resumes (c.f. Depoprovera)	Up to 3 weeks		Delays puberty Can synchronize seasons May induce a season immediately after implant insertion—should then be acyclic for duration of implants
Progestin implants (e.g. melengestrol acetate, implanon, Norplant)	♀	Yes (approx 2 years) Implants can be removed early to initiate oestrus	2 weeks	Yes	Some individuals may still show a sexual swelling, which does not imply contraceptive failure Can be used in pregnancy and during lactation Caution: early removal by grooming may occur
Oral progestins (with or without oestrogens)	♀	Yes, should return to normal within 1–2 weeks	1–2 weeks	Yes	If the progestin is used continuously, no swellings will be seen Administration of a week's worth of placebo tablets will lead to signs of oestrus Can be used in pregnancy and during lactation
Injectable progestins (e.g. Depoprovera)	♀	Yes, in most cases (approx 2–3 months), but it may take up to 2 years for normal cycling to return	2 weeks	Yes	Generally acyclic Useful for short-term contraception Caution: with consecutive treatments, the ability to predict when normal functioning will return becomes difficult

Table **9.2** A summary of available forms of contraception*

* From Sanderson (2005).

NOTE: Contraception can be achieved in a variety of ways, as summarized here, although broadly categorized as separating the sexes, so that they do not have access to one another and therefore cannot reproduce, to performing surgery to impair the proper functioning of the reproductive tract, and finally through the administration of hormones, which causes endocrine dysfunction and reproductive failure.

sperm of mammals. This is possible because the X chromosome is larger than the Y chromosome, and so sperm carrying an X chromosome are slightly heavier than their Y chromosome counterparts. Flow cytometric sorting of sperm is now widely used in farm animal production (for example, to produce cows, rather than bulls, for the dairy industry—see Johnson, 2000), but is not yet widely used in wild animal reproduction.

9.7 Working towards self-sustaining populations in captivity

If the number of births and deaths in a captive population can be balanced, and the overall population can be increased or kept stable, then it can be considered self-sustaining; it does not require bolstering from external sources. Unfortunately, Magin *et al.* (1994) estimated that, although zoos worldwide held 34 per cent of all threatened mammal species, only half of these species (17 per cent) could be con-

sidered to be part of a self-sustaining population. Another analysis (WRI, 1992) suggested that, of 274 species currently listed in captive management programmes, only twenty-six were self-sustaining in captivity. These papers suggest that zoos maintain few recognized threatened species and that those that are housed in zoos may not be there for the future, because they are not being managed as part of self-sustaining populations.

In this final section, we will consider the decision-making processes used to determine which species should be maintained in zoos. The infrastructure that underpins the captive management of zoo-housed species is very similar in the different regional zoo and aquarium associations, and is overseen by **taxon advisory groups (TAGs)** that manage different levels of captive management programmes. The ultimate goal of these captive management programmes is to increase, decrease, or keep stable the captive population for a given species, and this inevitably requires manipulations in order to ensure that the population will be self-sustaining (see **Box 9.5**).

Box 9.5 Captive management or captive breeding?

The goal of good zoos should always be captive management of populations, which can be achieved through captive breeding or from preventing breeding. We have mentioned at several points throughout this book that there are limited spaces available within zoos to hold animals and yet, to manage a successfully self-sustaining population, a large number of animals is required.

It has been suggested that zoos worldwide might need to support the captive management of 2,000 vertebrate species (Soulé *et al.*, 1986), but more recent estimates for specific taxa have suggested that there are only enough

spaces in the North American zoos for self-sustaining populations of 141 bird (Sheppard, 1995) and sixteen snake species (Quinn and Quinn, 1993). Decisions are therefore made to prioritize which species should be maintained and managed with different regional zoo associations (for example, the European Association of Zoos and Aquaria (EAZA)—see section **9.7.1**), and whether these populations should be increased, kept stable, or decreased (see section **9.7.3**).

Common reasons why a species will be managed and bred in captivity include the following.

1. The 'ark principal', whereby the captive animal represents a 'safety-net population' for an endangered species. The long-term goal of these populations is that they could be reintroduced to ameliorate wild extinctions and also to reduce the pressure on wild populations as a source of animals for other captive situations.

2. Allowing animals to breed also provides them with the opportunity to show their full behavioural repertoire, including courtship and parental care.

3. The very powerful attraction that baby animals offer for the public is likely to attract additional numbers of visitors through the gate.

9.7.1 Should the species be maintained in zoos?

Captive management programmes are managed regionally through the appropriate regional zoo association—the European Association of Zoos and Aquaria (EAZA); the American Association of Zoos and Aquariums (AZA); the Australasian Regional Association of Zoological Parks and Aquaria (ARAZPA), etc. (see **Chapter 3**)—although some inter-regional and international management programmes—that is, global species management programmes (GSMPs)—are also created and co-ordinated through the World Association of Zoos and Aquariums (WAZA).

Within regional zoo associations, groups of specialists form taxon advisory groups (TAGs), which review, assess, and recommend which animals should be held in zoos in that region. The membership of a TAG will include a chairperson, possibly a vice chairperson, the management programme coordinators, and invited specialists, such as advisers on genetics, nutrition, and veterinary research. EAZA currently has forty-one TAGs. The number of species included within each of these TAGs is highly variable. For mammals and birds, the TAG may operate at order, family, or even genus level (for example, 'deer' or 'hornbill'), or may be more general (for example,

'marine mammals'). There are three TAGs that cover species that are neither birds nor mammals: namely, 'terrestrial invertebrates', 'fish and aquatic invertebrates', and 'amphibians and reptiles'.

Regional collection planning

A key role of TAGs is to develop and manage **regional collection plans (RCPs)**, which aim to identify whether species should be managed and consequently make recommendations to zoos about the species that they should aim to incorporate in their collections.

As an example, Wilkinson (2000) provides an overview of the activities of the EAZA parrot TAG. He highlights that, of about 130 parrot species classified as threatened or endangered, only a handful of EAZA captive management programmes exist and five international **studbooks** are maintained that aim to coordinate efforts between zoo associations. A lot of information is gathered and discussed by TAGs in order to make these types of decision, including, but not limited to: current captive population; potential capacity for captive population; value to conservation, research, education, or exhibition; inherent uniqueness within the taxon; and the likelihood that populations can be managed successfully and be self-sustaining. These can be influenced by

A **regional collection plan (RCP)** is a plan drawn up by a regional zoo association, such as EAZA or the AZA, to inform member institutions on which species they should be keeping and how they should manage them. The RCP may also identify species that should (preferably) not be kept.

various factors: for example, space availability, previous breeding success, and genetic parameters (for example, genetic diversity and inbreeding coefficients of the population).

Three main decisions can be made on the basis of this information: to manage the species as part of a captive management programme, which could operate at a series of different levels; to phase out the species; and to gather more information before a final decision is reached. Reaching a decision for each species is complicated, because it relies on information about other species within the TAG. For example, all species of gibbon are considered in danger of extinction (Geissmann, 2007), so when space is limited to exhibit gibbons in zoos, factors other than the conservation value of the different species need to be considered when planning which species zoos in that region should hold.

To facilitate this process, some TAGs have developed decision trees to help them to maintain some consistency and logic when choosing whether a species should be included in the RCP. The most common reason why a TAG would recommend that a species should not be managed or held in zoos is that space is limited and that another species is considered to take precedence, because it is more endangered. Alternatively, if the captive population for a species is very low and would require supplementation from wild stocks to ensure its survival in captivity, unless there were extreme circumstances, it is unlikely that it would be recommended for captive management. The removal of animals from the wild to restock captive populations has been undertaken in extreme circumstances, but is not a practice that is undertaken lightly.

Once it has been decided that a species should be maintained in the region's zoos, it is sensible that this is managed by a captive management programme.

9.7.2 Levels of captive management

It is clear that the captive management of species requires coordination between many zoos and cannot be achieved by individual zoos operating alone (**Fig. 9.22**). The concept of **metapopulations** and the genetic management of captive populations relies on cooperation between zoos, including the exchange of animals for breeding purposes where necessary—and this is a large and complex undertaking.

Although there are various different terms used to describe management programmes in different zoo associations worldwide, the underlying goals and the mechanisms used to achieve these are nevertheless very similar. **Table 9.3** outlines the main captive management programmes that exist in three of the regional zoo associations (EAZA, AZA, and ARAZPA).

There is a bewildering variety of names and acronyms used to describe these programmes: European Endangered species Plans (EEPs)[13] and European studbooks (ESBs) in the EAZA area; **species survival plans (SSPs)** and population management plans (PMPs) in the AZA and ARAZPA areas (ARAZPA also has conservation programmes, or CPs); and simply management or preservation programmes elsewhere in the world.

Some of the key acronyms are summarized and explained in **Table 9.4**.

Conservation management within BIAZA

In 2006, a major restructuring within the British and Irish Association of Zoos and Aquariums (BIAZA) was undertaken to allow better coordination of captive

---

The term **metapopulation** refers to a collection or network of sub-populations that are connected in some way. The links between the sub-populations are usually through immigration and emigration from one sub-population to another. Thus, the groups of animals housed in different zoos can be regarded as sub-populations of a larger metapopulation, which also includes the wild population, if there is transfer of animals between sub-populations (see Fig. 9.22).
13 EEPs are colloquially referred to as **European Endangered species Programmes**, because the acronym is taken from the German *Europäisches Erhaltungszuchtprogramm*; they should not be referred to as EESPs.

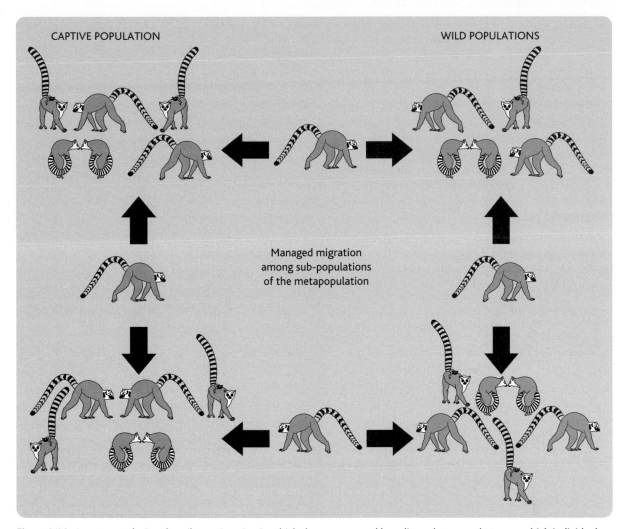

CAPTIVE POPULATION

WILD POPULATIONS

Managed migration
among sub-populations
of the metapopulation

Figure **9.22** A metapopulation describes a situation in which there are several breeding sub-groups, between which individuals will sometimes move. This is exactly how captive management programmes operate, with each of the zoos representing a breeding sub-group and the regional management programme serving as the mechanism through which to move animals between sub-groups/zoos

management and conservation activities within UK and Irish zoos, and similar work was undertaken in mainland Europe. The Joint Management of Species Programme (JMSP) and BIAZA TAGs disappeared, and were replaced with seven broader-based **taxon working groups (TWGs)**, representing: Mammals; Birds; Terrestrial Invertebrates; Reptiles and Amphibians; Aquariums; Plants; and Native

Species. (The organization of these groups is illustrated in **Fig. 9.23**.)

The main aims of these groups are to encourage advances in the knowledge, development, and training of keeping staff, to support and/or actively carry out *in situ* **conservation**, to disseminate information, and to work cooperatively with other groups, including similar groups within EAZA.

Taxon working groups (TWGs) are UK-based groups of zoo professionals who meet to coordinate and plan the captive management and conservation of a particular taxon (e.g. terrestrial invertebrates). TWGs replaced taxon advisory groups in the UK in 2006, to bring the focus of these groups more closely into line with the EAZA TAGs.

	AZA (American Association of Zoos and Aquaria) Established 1924		ARAZPA (Australasian Regional Association of Zoo Parks and Aquaria) Established 1991		EAZA (European Association of Zoos and Aquaria) Established 1988	
Taxon advisory groups	46		16		40	
Captive breeding programmes	SSP: 111	PMP: 324	Conservation Plan: 33	PMP: 56	EEP: 165	ESB: 161
Date initiated	1981	1994			1988	
People involved	Species coordinator and committee	Population manager/studbook keeper			Species coordinator and committee	Studbook keeper
Collect data: births/deaths/moves	✓	✓	✓	✓	✓	✓
Analyse data	✓	✓	✓	✓	✓	✓
Produce studbook	✓	✓	✓	✓	✓	✓
Recommend moves/mates	✓	(✓)	✓	✓	✓	(✓)
Construct long-term management plan	✓	(✓)	✓	✓	✓	(✓)
Housing and husbandry guidelines	✓		✓	✓	✓	
Link with *in situ* conservation	(✓)		✓	(✓)	(✓)	

Table **9.3** A summary of the capture management programmes of AZA, ARAZPA, and EAZA

NOTE: There are many similarities in the way in which captive management programmes are organized between different regional zoo associations, as illustrated here for three of the regional zoo associations, in Australasia (ARAZPA), Europe (EAZA), and North America (AZA).

Acronym	Full name	Area of operation	Other information
CBSG	Conservation breeding specialist group	Worldwide Operates via the IUCN	Formerly 'captive breeding specialist group'
CP	Conservation programme	ARAZPA (Australasian region)	ARAZPA also operates PMPs
IUCN	International Union for the Conservation of Nature and Natural Resources	Worldwide	Better known as the 'World Conservation Union'
RSG	IUCN/SCC reintroduction specialist group	Worldwide Operates via the IUCN's SSC	One of the IUCN/SSC specialist groups
SSC	IUCN Species Survival Commission	Worldwide Operates via the IUCN	Produces the IUCN Red List of Threatened Species
WZCS	*World Zoo Conservation Strategy* (1993)	Accredited zoos worldwide	Produced as a joint initiative between WZO and IUDZG, and CBSG/IUCN/SSC
WZACS	*World Zoo and Aquarium Conservation Strategy* (2005)	Accredited zoos worldwide	Published under the title *Building a Future for Wildlife* Also developed jointly with the IUCN, CBSG, and SSC

Table **9.4** A summary of some of the key acronyms used within the zoo and conservation industries

NOTE: For a comprehensive list of acronyms relating to conservation initiatives and programmes, both within and outside zoos, see the AZA website—www.aza.org.

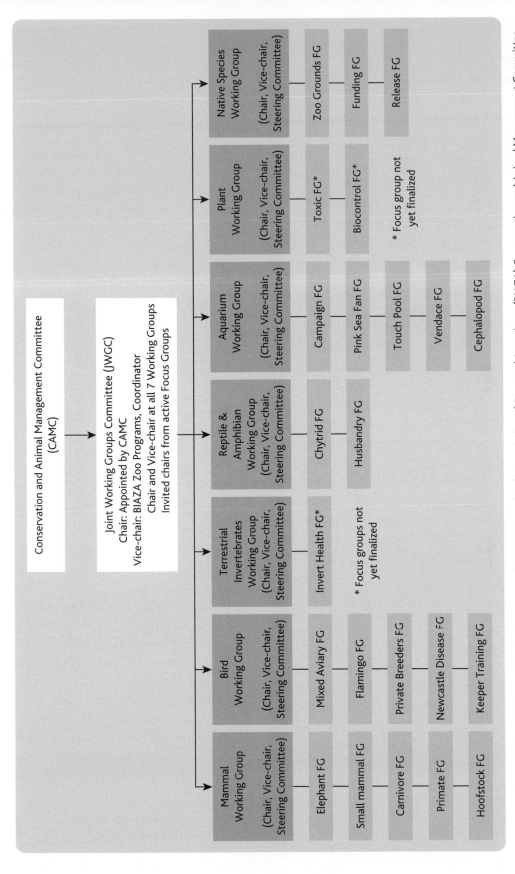

Figure **9.23** Illustrated here is the organizational structure of the British and Irish Association of Zoos and Aquariums (BIAZA) Conservation and Animal Management Committee. The Committee includes seven working groups, each of which specializes in different taxa. This structure is currently under review. (Photographs: BIAZA)

Conservation management within EAZA

The different captive management programmes that are coordinated through EAZA vary from one another in the intensity with which they are managed. An EEP is the most intensive captive management plan. Each EEP has a species coordinator and a studbook keeper (these first two positions are sometimes held by the same person), and they are assisted by a species committee and species advisers, who are specialists in various fields, such as nutrition, research, veterinary, and education. They are thus much like a TAG, but dedicated to one species rather than to a group of species.

Any zoo that is an EAZA member, and holds a species for which there is an EEP, instantly becomes part of, and is required to cooperate with, that management programme. Each year, a studbook should be produced, which provides a summary of the captive population held in EAZA member zoos, and highlights any births, deaths, and transfers that have occurred during that year. On the basis of the information held within the studbook, genetic and demographic analyses are undertaken, on which recommendations are made to all EEP members in relation to whether any of their animals are permitted to breed or need to be transferred to other zoos for breeding.

European studbooks (ESBs) are less intensively managed programmes that essentially monitor the captive European population for a given species—that is, whether it is increasing, decreasing, or remaining stable.[14] Studbook keepers for each ESB can advise on which animals should be exchanged with other zoos, but their advice is not binding on participating zoos in the same way that recommendations from an EEP species coordinator are.

Finally, some species are managed in monitoring programmes, in which there is no established requirement other than that a person is designated to take a 'special interest' in a species; no studbooks need to be generated.

Species management in other regional zoo associations

Table 9.3 summarizes the main species management programmes implemented through AZA, ARAZPA, and EAZA. There are differences between the regional associations in the number of TAGs, because the composition of these TAGs varies. For example, AZA has the most TAGs (forty-six), because it tends to represent small groups of animals, whereas ARAZPA has the fewest (sixteen).

It should be noted that other regional zoo associations also operate and have similar infrastructure to that described here; some of these other associations are still relatively young, however, and so have not attained, to date, the level of infrastructure and administration seen in the zoo associations described. For example, the African Association of Zoos and Aquaria (PAAZAB) initiated African Preservation Programmes (APP) in 1991 and now administers twenty regional studbooks.

9.7.3 Balancing births and deaths with varied lifespans and litter sizes

In order to create a stable self-sustaining population in which the number of animals in the population remains constant, the number of animals that are born each year must be equal to the number of animals that die each year. If, on the other hand, the aim is to create a growing self-sustaining population, then the number of births that occur yearly should surpass the number of animals that die yearly. There are obviously many variables that affect how many animals are born or die in a year, but in a well-organized captive management programme, some of these variables can be controlled by housing and husbandry, and other strategies adopted by zoos (see section **9.7.4**).

---

14 New international studbooks can only be set up with the approval of WAZA and the IUCN/SCC. Global Species Management Programmes (GSMP) are currently being initiated, which aim to integrate more fully all aspects of the *in situ* and *ex situ* conservation of a species.

Which strategy is adopted will, to some extent, be determined through analyses of the data held within studbooks. By looking at the historical population dynamics for a species, it is possible to determine what level of births (fertility) and deaths (mortality) can be expected annually. For example, if you look at **Fig. 9.24**, which illustrates the fertility and mortality of the animals in the Sulawesi crested

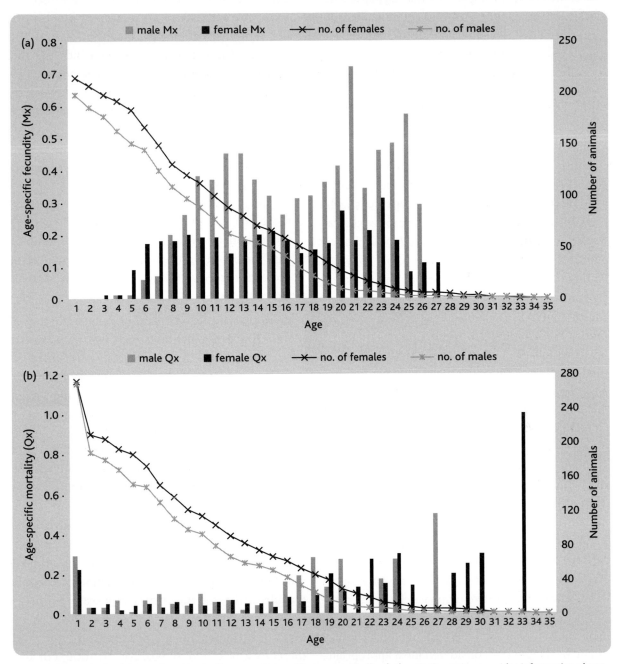

Figure **9.24** Analysis of data collated in the European Sulawesi macaque studbook since its inception provides information about fecundity (Mx) and mortality (Qx) levels that have occurred over time: for example, we can see from these graphs that (a) females start breeding before males and will continue breeding into their later 20s. We can also see from the data compiled for mortality rates (b) that there is a peak of deaths in the first year of life and then as animals get older. (Pictures: Vicky Melfi)

black macaque *Macaca nigra* EEP, you can see that the rate of births far outweighed the number of deaths in the population for the first 30 years or so. These data demonstrate that this species could be maintained as a growing self-sustaining population, when considering the number of animals alone. But this information does not provide an indication of the genetic value of the population.

The ease with which a self-sustaining population can be managed to ensure that it has a high genetic diversity and low inbreeding depends on several life history traits, including the animals' longevity, inter-birth interval, and litter size. Good examples of the type of information that can be gleaned about captive populations are presented by Marker-Kraus (1997) and Blomqvist (1995), who provide information about how the captive cheetah *Acinonyx jubatus* and snow leopard *Panthera uncia* populations have changed over some 30 years. During this period, the snow leopard management programme has increased from ten in member zoos in 1961 to 160 in 1992, 98 per cent of which are captive-born (Blomqvist, 1995).

It is more difficult to manage genetically short-lived species and those that have a large number of offspring, compared with longer-living species or those that have few offspring. Morton (1990) argues that, for some of these former species, such as butterflies, the need to maintain genetic diversity is only part of the conservation conundrum, because these species are highly sensitive to environmental change in the wild and this is reflected in genetic changes.

9.7.4 Common strategies employed in captive management

There are various strategies that can be adopted when managing captive animals. In some situations, the TAG or captive management programme may make recommendations about which strategy should be adopted or considered most appropriate. In most circumstances, however, TAGs or regionally organized management programmes will make

recommendations to a zoo that specific animals should or should not breed, but will leave the decision as to which strategy should be adopted to achieve these recommendations up to the zoo. The strategy adopted will not only be determined by species characteristics (for example, life history traits, reproductive biology, etc.), but also by the legal and ethical framework under which the zoo operates. It is important that a management strategy is identified and adopted, so that its efficacy can be monitored and, if it does not achieve its objectives, an alternative strategy can be sought.

So what strategies are there and how do we determine the most appropriate for a given species? Strategies can be categorized as:

- those that aim to increase the size of the captive population, for example, by providing animals with the opportunities to breed, or by facilitating successful breeding through artificial reproductive technologies;
- those that aim to maintain or reduce the size of the captive population, which can be achieved by maintaining single-sexed groups, by otherwise limiting access between differently sexed animals, by using contraception, or by operating 'breed and cull' strategies.

There are various considerations to bear in mind when choosing the most appropriate captive management strategy and these are summarized in **Table 9.5**.

Some of the issues relating to the captive management of zoo animals—in particular, whether or not they should breed and which strategy is adopted to achieve this—attract more attention from people outside of the zoo industry than most other areas of zoo animal housing and husbandry. This is partly because baby animals are perceived as 'cute' and therefore attract a lot of positive attention, but also because the methods used to prevent babies, or control against surplus animals (see **Box 9.6**), are

	Health	Behaviour	Legislation	Ethics
Allowing breeding	Care needs to be taken that births do not start too early and that inter-birth intervals are not artificially shortened	–	–	Could potentially result in too many animals for which there is no space
Single-sex groups	–	Potential for aggression, lack of social interactions, and reduced opportunities for expression of natural behavioural repertoire	–	Might be considered 'unnatural'
Contraception	Potential detrimental side effects Animals may not be able to breed afterwards (even if contraception considered temporary)	Potential detrimental side effects	Use is regulated by drug licensing, which means that not all formulations are easy to obtain	Pro-life lobby are against this
Breed and cull	–	–	Euthanasia may be restricted by law, e.g. it is illegal in some states of Germany	Pro-life lobby are against this

Table 9.5 An outline of the considerations affecting different captive management strategies

NOTE: No single captive management strategy is ideal, although the limitations inherent in the different strategies vary, from potential health infringements, to controversy over the ethical justification for some methods.

viewed negatively. People's perceptions of using euthanasia as a management tool vary considerably for different species and in different regions (and euthanasia of zoo animals is not even legal in some countries).

There also seems to be a popular assumption that, if zoo animals are not breeding, it is because the zoo is managing them incorrectly (and therefore their welfare is compromised) or the animals are not contributing to the conservation of species. For a long time, zoos advertised their role and their approach to conservation as one of species preservation (Tudge, 1992). The analogy of zoos representing an ark (see section **10.3**), which could conserve animals by keeping them safe in *ex situ* populations, propagating them, and then, when conditions in the animals' native habitats were appropriate, releasing them into the wild, was popularized by many zoo professionals (Durrell, 1976; Mathews *et al.*, 2005). But this approach was recognized by Conway

(1996b), and others, to be rather too simplistic and to have inherent restrictions—that, in fact, conservation efforts would need to be much broader if zoos were proactively and effectively to conserve biodiversity (see, for example, Hutchins and Wiese, 1991; see **Chapter 10**).

Unfortunately, there are many species held in zoos for which the breeding, and life, in captivity is not coordinated and monitored by a regional management programme. In these situations, there is not necessarily a recommendation about whether breeding should take place or not, and so it is even more difficult to know which captive management strategy should be adopted. This can lead to the scenario in which a zoo may consider that they are 'just letting their animals live naturally' and consequently may, or may not, have offspring that, in time, will need to be rehomed. It is these situations—or, indeed, where preventative measures have failed—that create an issue of surplus animals.

Box **9.6** Surplus animals

Animals can be surplus because of a lack of space, and this can be exacerbated by low rates of mortality and unrestricted breeding. Lack of space may occur because priority is given to another animal that is considered to be more important to the captive management of the species. By virtue of considering the types of animal that would benefit from a captive management programme, we can also identify which animals may be considered surplus if space is limited.

Animals that will be prioritized for inclusion in a captive management programme are those that are: not related to other animals in the population (that is, they have a low mean kinship); of the right sex and age to help to create a demographically stable population, which is represented by a sex–age pyramid (see **Fig. 5.19**); reproductively active; fit and healthy; behaviourally competent.

Potentially, surplus animals could be avoided through better management programmes or they can be euthanized (reviewed by Glatston, 1998). It seems an intuitive and simple idea to calculate the number of births required for a year and then to take measures (such as single-sexed groups, contraception; see section **9.6.2**) to ensure that only that number of offspring are born, so as to avoid a surplus. But this is overly simplistic, not only because 'accidents' happen, but also because the offspring born may not be suitable. For example, a surplus of male animals is often a problem with many species that are managed in harem systems, with one male maintained with several females, because many more females are required in the captive population than males. Restricting breeding would limit the number of groups with young animals in them, and consequently deprive them of the positive social implications associated with young and reproductively active animals.

Alternatively, surplus animals can be culled, or euthanized, which is usually practised as part of a breed-and-cull programme. This term is used to describe situations in which animals are allowed to breed, but some animals from the group will be selected and culled to ensure that the group size remains constant. The selection of animals that will be culled can be variable, and is dependent, to some extent, on the group and their life history traits. It may, for example, be appropriate that offspring are culled at the time that they would naturally leave their parents or emigrate from the group. Alternatively, older animals, which may represent post-reproductive or genetically well-represented individuals, or those less able to cope with **environmental challenges**, may be favoured as those chosen to be culled; animals falling in this latter category may have been culled for welfare reasons (see section **7.3.3**). This method enables animals to express their full repertoire of reproductive behaviours, and ensures that fit and healthy animals are selected to contribute to the captive management programme, much akin to natural selection. But public perception and media coverage of euthanasia as a management tool, whether for welfare or conservation goals, is negative and unfortunately puts pressure on zoos to avoid using this measure, despite its many associated benefits for animals (see **Table 9.5**).

Summary

- There is still a great deal to learn about exotic animal reproductive biology, because data collected thus far provide only a basic understanding of the complexities underpinning the various stages required to make a new life. As with all topics related to zoo animal management, the diversity of species managed in zoos compounds our lack of knowledge.

- Despite many of these gaps in our understanding, many zoos have been able to breed animals successfully, using modifications to methodologies developed for animals in other industries: for example, efforts taken to create a self-sustaining population of captive elephants have involved the collection and analysis of data about the elephants' endocrinology and behaviour, the implementation of various assisted reproductive technologies, and, in some circumstances, supplementary rearing (see **Box 9.7**). But all of these efforts have greatly increased our knowledge of elephant reproduction and enhanced the welfare and management of captive elephant populations (Stevenson and Walter, 2002).

Box 9.7 The captive breeding of elephants

Zoo-housed elephants have been the focus of much attention in relation to their welfare and conservation (e.g. Clubb and Mason, 2002; Hutchins, 2006). In terms of the conservation of elephants in zoos, neither of the captive breeding programmes for African *Loxodonta africana* and Asian *Elephas maximus* species are currently self-sustaining. In fact, analyses of data collected from the Asian elephant captive management programmes has shown that it is declining at a rate of 2 per cent per annum (Faust *et al.*, 2006).

To reverse this trend and to work towards self-sustaining captive elephant populations, more offspring are needed. A variety of techniques is currently being explored and implemented to investigate both why zoo populations of elephants are not breeding successfully and also to facilitate breeding. Studies of the elephant's endocrine function have also been conducted to evaluate different reproductive stages in the females—e.g. predictors of ovulation and parturition—and in males—e.g. maturation and musth (Cooper *et al.*, 1990; Brown and Wemmer, 1995). Hormone profiles and **ultrasound** results found that a number of captive elephants were not cycling in 14 per cent of forty-nine respondent zoos that held Asian elephants and 29 per cent of sixty-two respondent zoos that held African elephants (Brown *et al.*, 2004).

To complicate elephant reproduction further, social factors have also been identified that affect the ovarian cycle, including dominance rank (Freeman *et al.*, 2004).

On a positive note, however, initial examinations of female reproductive tracts used to develop AI techniques have proven successful, resulting in some AI babies (Balke *et al.*, 1988; see **Figs 9.25a, b, and c**).

▶

(a)

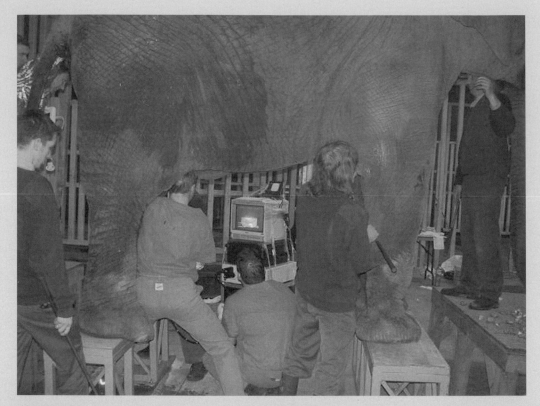

(b)

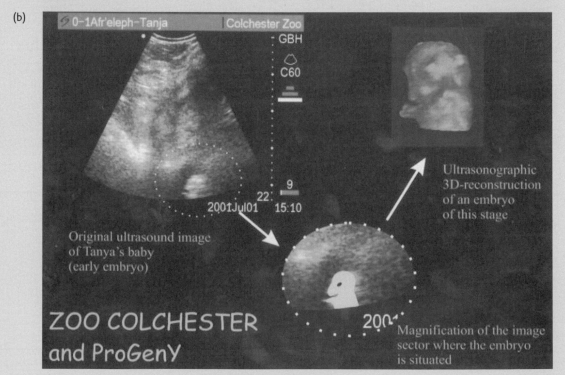

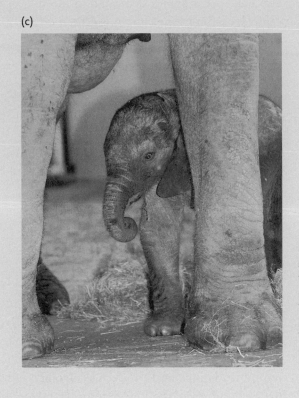

(c)

Figure **9.25** Elephant populations in zoos are still not self-sustaining, but efforts are being taken to reverse this trend, through the use of (a) artificial insemination, (b) proactive monitoring of reproductive state, using techniques such as ultrasonography (which is able to follow the development of the foetus), and, if all goes well, (c) this work is repaid with a successful birth, as pictured here at Colchester Zoo. (Photographs: Colchester Zoo)

- Similar work has also increased our ability to manage the captive populations of a great many other species, although there is yet much to learn and to apply to the successful breeding of other endangered species managed in zoos, which are hoped to represent a safety net and prevent the extinction of the species, even if they are lost in the wild habitat.

Further reading

Unlike some chapters in this book, in relation to which there is very little published information available beyond that cited in the text, the study of reproduction, genetics, and captive breeding of animals has resulted in a wealth of publications: academic, professional, and popular. To ensure, therefore, that this section does not represent too long a list, we have highlighted only a couple of key texts for each of the main topics considered in this chapter, and we have tried, where possible, to refer to texts that have exotic animals and zoos as their central focus.

- *Genetics and captive population management*
Frankham *et al.*'s *Introduction to Conservation Genetics* (2002) provides all of the necessary theory on the basis of which decisions are made about how to manage a captive population, with respect to its genetic integrity. If this is a little too heavyweight, try Frankham *et al.*'s *A Primer of Conservation Genetics* (2004).

- *Reproduction and contraception*
Holt *et al.*'s *Reproductive Science and Integrated Conservation* (2003) provides a very comprehensive

overview of exotic animal reproduction, while Asa and Porton's *Wildlife Contraception: Issues, Methods and Applications* (2005) considers the methods, and their implications, used to prevent offspring in exotic species.

• *Captive breeding in zoos*
Colin Tudge's *Last Animals at the Zoo: How Mass Extinction can be Stopped* (1992) has done a great deal to popularize the concepts behind captive breeding programmes. Quite a few different journals have also had special issues dedicated to the subject of captive breeding: for example, the *International Zoo Yearbook* and *Conservation Biology*.

• *General texts*
Kleiman *et al.*'s *Wild Mammals in Captivity* (1996) provides a very thorough review of many of the topics covered in this chapter, obviously for the mammal taxa. Norton *et al.*'s *Ethics on the Ark* (1995) explores many of the controversial issues raised by the captive management of animals in zoos.

Websites and other resources

It is probably prudent to consider the above texts as the best port of call for further detail about the theory underpinning many of the issues discussed in this chapter, whereas the websites available can provide more information about how animal management programmes operate in practice.

The various zoo association (AZA, EAZA, etc.) websites provide much detail about the captive management programmes that they run, along with varying degrees of additional information. The AZA Wildlife Contraception Center (WCC), a contraceptive advisory group to the AZA, has a very good website—www.stlzoo.org/animals/scienceresearch/contraceptioncenter— featuring recommendations and lots more.

In Europe, a new EU zoo contraception advisory group is under development, referred to as the European Group on Zoo Animal Contraception (EGZAC). The aim of EGZAC is to collect information from European zoos on contraception recommendations and product availability within the EU, and, over the longer term, to merge this information with the AZA's WCC database.

The website of the Frozen Ark Project— www.frozenark.org—has already been mentioned in this chapter; this site has very good short explanations of topics such as 'Why DNA?', and the viability of frozen cells and tissue.

Chapter 10 Conservation

Conservation of endangered species is now a major goal of **zoos** accredited by organizations such as the British and Irish Association of Zoos and Aquariums (BIAZA), the European Association of Zoos and Aquaria (EAZA), the US Association of Zoos and Aquariums (AZA), and the Australasian Regional Association of Zoological Parks and Aquaria (ARAZPA) (see **Chapter 3**). (There are, of course, many non-**accredited zoos** around the world, which may or may not take part in conservation programmes.)

The emphasis on a conservation role for zoos has grown greatly in importance over the past two or three decades, prompted partly by the zoos themselves and partly by external pressures, such as new international treaties and national legislation. To members of the public, the focus on conservation is probably most apparent in the signs and labels at the zoo, which draw attention to breeding successes, particularly of rare species, and also in zoo websites and literature, which prominently feature zoos' conservation priorities and activities. The increasing importance to zoos of an active involvement in conservation has been accompanied by a reduction in the number of different species maintained, more coordinated cooperation between zoos, and a great reduction in the number of animals taken from the wild. Zoos are also expanding their involvement in field conservation projects, supporting and engaging with conservation of animals outside the zoo, in their natural habitats. All of this has taken place alongside a much more science-based and planned approach to conservation than was apparent 20, or even 10, years ago.

In this chapter, we will consider the following.

10.1 What is 'conservation' and why is it necessary?
10.2 The role of zoos in the conservation of biodiversity
10.3 Zoos as arks
10.4 Reintroduction
10.5 Other aspects of zoo conservation work
10.6 How good are zoos at conserving biodiversity?
10.7 Conservation and zoos: looking to the future

Boxes in this chapter provide additional information on special topics and case studies, such as EAZA conservation campaigns and efforts to conserve endangered *Partula* snails.

10.1 What is 'conservation' and why is it necessary?

The term 'conservation' is used in a number of different ways and contexts, and can, for example, refer to the restoration of artistic artefacts in museums, or to saving a historic building from demolition. In the context of species and environments, a typical definition of a conservationist is '*someone who advocates or practices the sensible and careful use of natural resources*' (Hunter, 1995). Implicit in this definition is the idea of human intervention and management to achieve some kind of sustainable usage of resources.

Some conservationists make a distinction between conservation and 'preservation', in which management would be minimal and resources would be kept as pristine as possible. Indeed, 'preservationist' is seen by many conservationists as a rather negative term. Despite these niceties, however, preservation is effectively what conservationists do, although in an active and interventionist way.

Mace *et al.* (2007) offer a simpler definition of conservation as '*the persistence of wild habitats and species*'. This is a useful definition to bear in mind when we talk about measuring the success (or otherwise) of conservation initiatives by zoos and, indeed, is broadly similar to the definition of conservation that zoos themselves have proposed, in the World Zoo and Aquarium Conservation Strategy (WZACS; WAZA, 2005) (see section **10.2.2**). WZACS defines conservation as '*the securing of long-term populations of species in natural ecosystems and habitats wherever possible*'.

The science that underpins conservation is conservation biology, which is heavily based on biology, but draws on other disciplines as well. It may be defined as '*the applied science of maintaining the earth's biological diversity*' (Hunter, 1995).

10.1.1 Biodiversity

Today, the terms 'conservation' and **biodiversity**[1] are often used together, to make explicit the distinction between the conservation of living organisms and non-living structures, such as buildings or books. Another way of defining biodiversity would be as '*the totality of genes, species and ecosystems in a region*' (WRI/MCN/UNEP/FAO/UNESCO, 1992).

So biodiversity is concerned with the variety and variability of different kinds of animals and plants, the genes that make them what they are and give them their adaptability (for example, to changing environments, to evolving diseases, etc.), and the associations and communities in which they live.

For the remainder of this chapter, when the word 'conservation' is used, this will mean the conservation of biodiversity and, particularly, of animal biodiversity, because animals are what zoos (and this book) are all about.

10.1.2 Why is conservation necessary?

Conservation biology, as defined above, is concerned with trying to maintain biodiversity. As we are all now very aware, biodiversity is under threat as a result of human impact on the environment (Wilson, 1988; Reaka-Kudka *et al.*, 1996).[2] Human pressure on the places in which animals live (for building, agriculture, tourism, and sometimes simply wanton destruction) is causing an unprecedented loss of habitats. Habitat destruction and hunting have already resulted in the extinction of many species, and many more will follow in the next couple of decades. It is not only terrestrial habitats that are under threat: pollution and drainage schemes threaten lakes and rivers, and overexploitation and pollution threaten marine species (Jenkins, 2003). The introduction of invasive non-native or exotic species, whether deliberate or accidental, is another major contributor to the extinction of species, particularly of **island endemics**. Climate change (global warming) presents an additional threat to terrestrial and marine habitats alike, with a prediction by Thomas *et al.* (2004) that mid-range warming to 2050 could result in the loss of between 15 per cent and 37 per cent of species.

William Conway (former president of the World Conservation Society and director of the Bronx Zoo) has said that '*we live in an age of extinction*' (Conway, 2007). But just what is the scale of this extinction? Extinction is, of course, a natural part

---

The term **biodiversity** means, in a nutshell, the variety of all living things, from the Greek *bios*, meaning 'life'.
1 'Biodiversity' is shorthand for biological diversity. The term was first used by US scientist Walter Rosen as the title for a National Forum held in Washington DC in 1986. Papers from this forum were subsequently published in an influential book, edited by E. O. Wilson, with the title *BioDiversity* (1988).

2 Reaka-Kudka *et al.*'s *BioDiversity II: Understanding and Protecting our Natural Resources* was published in 1996 as a follow-up to Wilson (1988).
An **endemic** species is a species that is not found anywhere else. **Island endemics**, such as the Mauritius pink pigeon (see section **10.5.4**), are often at particularly high risk of extinction, because they may be restricted to only one or two localities. Genetic diversity is often lower in island species and this can also contribute to a higher risk of extinction.

of the evolutionary process and it is often said that most of the species that have ever existed are now extinct. In the context of the current extinction crisis, we should treat such a statement with some caution. Firstly, many of those species in geological time have not become extinct as such; they have evolved into different species.

Secondly, it is clear that the rate of species extinctions that we are now experiencing is very much greater than that which has happened throughout evolutionary time. An estimate of pre-human extinction rates (that is, a 'natural' rate) gives a figure of about three species per year for described species—twenty-five per year if we accept that most species are unknown to science (Magin *et al.*, 1994). Since 1600 (when records first become available), over 490 species of animals (mostly molluscs, mammals, and birds) are known to have become extinct; if corrections are made for invertebrates (which are generally not in the historical records), then an estimated 18,000 of described species and 140,000 of all species (including those not yet known to science) have become extinct since 1600 (Magin *et al.*, 1994). Even if only birds and mammals are considered, the expected losses to extinction in the period since 1600 would be four species of mammals and nine of birds; the actual known extinctions are sixty mammals and 122 birds.

A more recent listing (Baillie *et al.*, 2004) identifies 784 species (338 vertebrates; 359 invertebrates; 86 plants; one **protist**) that have become extinct in recent history, and a further sixty (twenty-two vertebrates; fourteen invertebrates; twenty-four plants) that have become extinct in the wild, although they still exist in captivity.

The extinct list includes some well-known species, such as the dodo *Raphus cucullatus* (last reported in

Figure **10.1** A drawing of the quagga *Equus quagga quagga*. It is possible that this species survived in the wild until the late 1870s, but the last known specimen died in Amsterdam Zoo in 1883 and is now in the Zoological Museum in Amsterdam. Modern evidence suggests that the quagga belongs to the same species as Burchell's zebra, formerly called *Equus burchelli*, but probably now best referred to as *Equus quagga burchelli*. (Photograph: `www.historypicks.com`)

1662), the great auk *Pinguinus impennis* (last seen alive in 1852), the quagga *Equus quagga quagga* (**Fig. 10.1**; the last known animal died in Amsterdam Zoo in 1883), the passenger pigeon *Ectopistes migratorius* (the last known animal died in Cincinnati Zoo in 1914; see **Box 2.2**) and the thylacine *Thylacinus cynocephalus* (see **Fig. 10.2**). It also includes many that are not widely known, such as forty-eight species of *Partula* snails (see **Box 10.1**) and thirty species of cichlids of the genus *Haplochromis*.[3]

In only 20 years from 1984 to 2004, seven species of amphibian and three birds have become extinct (Baillie *et al.*, 2004), and a further five species of animals (Wyoming toad *Bufo baxteri*, Hawaiian crow *Corvus hawaiiensis*, Alagoas curassow *Crax*

Protists are mostly unicellular organisms, although some are multicellular. Some protists (e.g. **protozoans**) are animal-like, whereas others (algae) are plant-like. These are organisms that do not fit neatly into categories such as animal, plant, or fungus.

3 Most of the now-extinct species of cichlids of the genus *Haplochromis* are restricted to Lake Victoria, where they have been decimated by the introduced predator the Nile perch *Lates niloticus*.

Figure **10.2** The thylacine *Thylacinus cynocephalus*. This large predatory marsupial probably became extinct on the Australian mainland 2,000 years ago (possibly through human activity), but survived on the island of Tasmania until the early years of the twentieth century, when it was hunted to extinction. The last known specimen died in Hobart Zoo in 1936. (Photograph: courtesy of the Australian Museum)

Box **10.1** Small is beautiful: saving the *Partula* snails

When talking of wildlife conservation, many people tend to think of giant pandas, tigers, elephants, and other large, charismatic species. But small species can be just as charismatic, and much effort has gone into saving some of the little animals that might so easily be overlooked.

A good example is the case of the *Partula* snails. These are small, spiral-shelled, land snails, usually smaller than 20mm in length (**Fig. 10.3**). They live on the islands of French Polynesia in the Pacific Ocean, where they have evolved into over a hundred different species through the process of adaptive radiation, just as the more familiar Galapagos fauna has done.

But these little snails have been vulnerable to a number of threats. The chief of these is a predatory snail *Euglandina rosea*, which was introduced into Polynesia in the 1970s in an attempt to control another introduced species, the African giant land snail *Achatina fulica*, which had become an agricultural **pest**. Unfortunately,

Figure **10.3** *Partula rosea*, along with many others snails in this genus, are critically endangered (as listed by the IUCN) in their native habitat of French Polynesia due to the introduction of a predatory snail, which unfortunately preferred to eat them rather than the pest African giant land snail that was its intended target. (Photograph: Doug Sherriff)

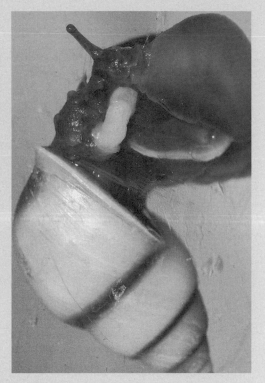

Figure **10.4** As you can see from the mating behaviour of these *Partula faba*, they are currently being managed as part of a very successful captive breeding programme, which is their last hope for survival, because they are extinct in the wild. (Photograph: Doug Sherriff)

Euglandina found the *Partula* snails to be a perfectly acceptable meal and, because partulids reproduce rather slowly (because they give birth to live young), they declined in numbers. By the late 1980s, all partulid species were extinct on the island of Moorea (Murray *et al.*, 1988). By the start of this century, fifty-six out of sixty-one species of *Partula* had become extinct throughout the Society Islands (which include Tahiti, Moorea, and others), primarily because of predation by *E. rosea* (Coote and Loeve, 2003).

Fortunately, there has been a **captive breeding programme** (the Partula Conservation Programme), coordinated since 1994 by the Zoological Society of London (ZSL). This programme involves both captive breeding (for twenty-five species in fifteen different zoos

worldwide; **Fig. 10.4**) and fieldwork within the snails' natural habitat. A successful release of zoo-bred *Partula taeniata* into semi-free range (among Polynesian plants in the Palm House at the Royal Botanic Gardens at Kew) showed that they were capable of displaying the behaviours needed for life in the wild (Pearce-Kelly *et al.*, 1995).

Since then, several have been released into enclosures back on Moorea, where all of the species had previously become extinct (Coote *et al.*, 2004). Unfortunately, *E. rosea* is still there and still partial to partulas, but enclosed areas may provide refuges for reintroduced partulas in the future.

(a)

Figure **10.5** These three species represent animals that are extinct in the wild and the future survival of which depends on their successful captive breeding. Reintroduction of individuals back into their native habitats is being attempted for all three species: (a) the Hawaiian crow *Corvus hawaiiensis*, the last known wild specimens of which died in 2002; (b) Guam rail *Gallirallus owstoni*, which disappeared in the wild during the 1980s as a result of predation by the introduced brown tree snake *Boiga irregularis*; (c) scimitar-horned oryx *Oryx dammah*, which was probably exterminated by overhunting in the 1980s. (Photographs: (a) US Fish and Wildlife Service; (b) Smithsonian's National Zoo; (c) Tibor Jäger, Zoological Center Tel Aviv—Ramat Gan, Israel)

(b)

(c)

Figure **10.5** (continued)

mitu, Guam rail *Gallirallus owstoni*, and scimitar-horned oryx *Oryx dammah*) are now extinct in the wild and exist only in captivity (**Fig. 10.5**).

So, are human-generated extinctions increasing? In general, the answer is 'yes' (Magin *et al.*, 1994), although predictions vary that between 2 per cent and 25 per cent of species will be lost by 2015. Nevertheless, it now looks likely that two-thirds of the world's species could be lost by the end of this century (Raven, 2002). As Balmford *et al.* (1996) have pointed out, the rate of extinction is now an order of magnitude higher than background levels.

Of the land mammals, disproportionately represented in zoos, Ceballos *et al.* (2005) suggest that approximately one quarter of all species are now at risk of extinction. Attention, then, must be on the species that still exist, but which are threatened with extinction. The 2007 IUCN (World Conservation Union) Red List of Threatened Species[4] gives a total of 5,742 vertebrate species and 1,601 insect (**Fig. 10.6**) and mollusc species at risk—that is, categorized by the IUCN as 'vulnerable', 'endangered', or 'critically endangered'.

These, of course, are only the species in relation to which enough is known about their status

4 The IUCN Species Survival Commission (SSC) maintains the Red List database, from which the Red List of Threatened Species is produced. This list is updated at least every 2 years. The IUCN/SCC Red List should not be confused with the many Red Data Books in circulation. The latter are national or regional accounts of endangered species, and usually deal with a specific group of animals or plants (for example, the Red Data Book of European Butterflies).

Figure **10.6** An example is shown here of an endangered insect, the Frégate Island giant beetle *Polposipus herculaneus*, which is endemic to one island in the Seychelles archipelago, where its numbers have been severely reduced through predation by introduced rats. It is now listed as critically endangered by IUCN, and its conservation includes both *in situ* (such as eradication of rats) and *ex situ* (captive breeding) initiatives. (Photograph: Sheila Pankhurst)

10.2 The role of zoos in the conservation of biodiversity

Zoos can play an active role in conservation in four general areas:

- education and campaigning about conservation issues;
- research that benefits the science and practice of conservation;
- maintenance of captive stocks of endangered species (and **reintroduction** of these animals back into the wild where possible);
- support for, and practical involvement with, *in situ* conservation projects.

These conservation roles and activities have been defined in two strategy documents: the *World Zoo Conservation Strategy* (IUDZG/CBSG, 1993) and its successor, the *World Zoo and Aquarium Conservation Strategy* (WAZA, 2005). These two documents are considered in more detail in section **10.2.2**, but let us firstly look more closely at the concepts of *in situ* and *ex situ* **conservation**.

10.2.1 *Ex situ* versus *in situ* conservation

The best place to maintain biodiversity is, of course, in the natural habitats in which animal and plant species live and this is known as *in situ* conservation. It is now apparent to most people, however, that, for various reasons, *in situ* conservation is not a realistic option for some of the most-endangered species and that, if they are to be conserved at all, this can only be done away from their natural habitat, in zoos and wildlife parks. This is *ex situ* conservation.[5] The expectation of modern zoos is

in the wild for this designation to be made; while many of the known species of land vertebrates have been evaluated, very few invertebrates have been (fewer than 4,000 out of over a million named species). Nevertheless, this is an enormous list of species that need active conservation measures if they are to survive. For many of these animals, *in situ* **conservation** is being attempted.

In this chapter, we are concerned with looking at what role zoos can play in this conservation effort.

The term *in situ* refers to the conservation of animals outside zoos, in their natural habitats, whereas *ex situ* conservation takes place in zoos. The origin and use of these terms is discussed in more detail in section **10.2.1**.

5 There is some debate within the zoo community about the practical use of the terms *in situ* and *ex situ*. Animals kept in sanctuaries in their country of origin, for example, may not be considered *in situ* if the sanctuary is fenced and if the animals are not free to migrate out of the immediate area.

that they will be involved in both *ex situ* and *in situ* conservation, as part of an integrated strategy.

The terms '*ex situ*' and '*in situ*', in relation to the conservation of biodiversity, are set out in the text of the Convention on Biodiversity (CBD) (UN, 1992), the international conservation treaty that resulted from the Rio Earth Summit[6] in 1992.

The CBD made some very specific recommendations on *ex situ* conservation and these recommendations have been a major driving force pushing conservation to the forefront of the zoo agenda. The Rio Summit and the CBD have already been discussed in **Chapter 3** (in the context of the regulatory framework for zoos), but it is worth stating again here that Article 9 of this treaty sets out that:

Each Contracting Party shall, as far as possible and as appropriate, and predominantly for the purpose of complementing in-situ measures:

(a) Adopt measure for the ex-situ conservation of components of biological diversity, preferably in the country of origin of such components

(b) Establish and maintain facilities for ex-situ conservation of and research on plants, animals and micro-organisms, preferably in the country of origin of genetic resources

Article 9 of the CBD also lists other measures relating to *ex situ* conservation. These are set out in full in the WZACS (WAZA, 2005), which is discussed in section **10.2.2**.

Within Europe, the EU has looked to zoos to fulfil the CBD obligation to '*establish and maintain facilities for* ex-situ *conservation of and research on . . . animals*'. The CBD provided much of the impetus and direction for the EC Zoos Directive 1999, which, in turn, has impacted on national zoo legislation via measures such as the amendment of the UK's Zoo Licensing Act 1981 by the Zoo Licensing Act 1981 (Amendment) (England and Wales) Regulations 2002 (see **Chapter 3**).

10.2.2 The *World Zoo and Aquarium Conservation Strategy* (WZACS)

The current *World Zoo and Aquarium Conservation Strategy* (WZACS) was published in 2005 under the title *Building a Future for Wildlife* (WAZA, 2005). WZACS builds on the earlier *World Zoo Conservation Strategy* (WZCS) (IUDZG and CBSG/IUCN/SSC, 1993), which was a joint venture of the World Zoo Organization (now WAZA) and the Captive Breeding Specialist Group (CBSG) of the IUCN's Species Survival Commission (SSC).[7]

Although WZCS has now been superseded by WZACS, it is worth looking at how this first zoo conservation strategy document came about and what its goals were. WZCS aimed to identify how zoos could be involved in conservation, and also how they could formulate policies and procedures to achieve conservation goals. Another aim of the document was to raise awareness of, and support for, the conservation efforts of zoos among other agencies and authorities.

WZCS suggested that zoos should support the objectives of the *World Conservation Strategy*[8] by conserving populations of endangered species both *in situ* and *ex situ*, through coordinated programmes, by undertaking conservation-related research and providing facilities for others to do the same, and by

6 The so-called Rio Earth Summit is more properly referred to as the United Nations Conference on Environment and Development (UNCED), which took place in Rio de Janeiro, Brazil, in June 1992. Apart from the Convention on Biodiversity (CBD), another output of the 1992 Rio Earth Summit was the United Nations Framework Convention on Climate Change (UNFCCC). The better-known Kyoto Protocol is an addition to the UNFCCC.

7 For those readers who are reeling from acronym overload and having trouble remembering their 'WZCS' from their 'WZACS', take a moment to re-read the **List of acronyms** at the front of this book.
8 The *World Conservation Strategy* was published by IUCN in 1980 and essentially advocates ecologically sound development in the world's nations.

promoting greater public and political awareness of the importance of conservation. Although these were activities in which many zoos were already involved, by incorporating them into a written document, WZCS effectively formulated a set of standards against which the conservation activities of zoos could be judged (Wheater, 1995).

From WZCS to WZACS

Much of WZCS was informative, outlining the theory and practice of conservation as undertaken by zoos. WZACS, its successor, is not really a new version of the strategy, but rather a further elaboration of the policies and standards that should be reached in achieving the conservation goals of the strategy. Thus, in each of the areas in which zoos have a role to play, WZACS provides a stated vision, supporting information, and a set of recommendations towards which zoos are to work.

An important aspect of WZACS is its advocacy of **integrated conservation**, which involves the bringing together of different activities and different agencies in order to achieve conservation priorities (see section **10.5.4**).

10.2.3 *Catalysts for Conservation*

At the same time as the regulatory framework for zoos has changed to include explicit commitments to *ex situ* conservation, zoos are also being expected to expand their involvement in, and support for, *in situ* work. A symposium on the role of zoos as catalysts for *in situ* conservation was held in London in 2004, co-hosted by the Zoological Society of London (ZSL) and the New York-based Wildlife Conservation Society (WCS).

One of the outcomes of this meeting (which, under the banner of *Catalysts for Conservation*, attracted participants from five continents) was the creation of a list of specific conservation-related

challenges for zoos, including: '*All activities of zoos and aquaria should be conceived and designed to contribute to achieving the overall goal of in-situ conservation.*' In other words, zoos should undertake *ex-situ* work only as part of an integrated strategy, ultimately directed towards the conservation of species in their natural habitats. (The concept of integrated conservation and how to make this work in practice is discussed in more detail in section **10.5.4**.)

Papers from the symposium were published in book form in 2007, under the title *Zoos in the 21st Century: Catalysts for Conservation?* (Zimmerman et al., 2007). Some of its key outputs and recommendations are described in section **10.7**.

10.3 Zoos as arks

Defining the conservation role of zoos in both theory and practice is, as we have just seen, an ongoing process. Just 20 or so years ago, many people would probably have seen the maintenance of captive breeding groups of endangered species as being the major component of this role. From this perspective, zoos are seen as modern arks, keeping species safe from extinction until a time at which it is possible to re-establish them safely in the wild. This activity is still seen as important, although it is only one of a number of important conservation activities in which zoos should be engaged.

10.3.1 Captive breeding in support of conservation

There were some notable early successes in maintaining captive populations in zoos of species that had become extinct in the wild; these successes, with species such as Przewalski's horse *Equus ferus przewalskii*[9] (**Fig. 10.7**) and Père David's deer *Elaphurus davidianus*, were relatively unplanned

---

9 The nomenclature for Przewalski's horse is debated; some authors give *Equus caballus przewalskii*, whereas other sources give *Equus ferus przelwalskii*. What is not under debate is that this species of wild equid has been saved from extinction by captive breeding in zoos.

Figure **10.7** Przewalski's horse *Equus ferus przewalskii*, also known as the Mongolian wild horse, became extinct in the wild during the 1960s. Fortunately, a captive breeding programme had been initiated after the Second World War, which led to the successful propagation of a large number of animals, some of which were reintroduced to Mongolia during the 1990s. (Photograph: © Hien Nguyen, `www.iStockphoto.com`)

by modern standards (see **Box 10.2**), but indicated the potential of zoos as arks, to help save species from extinction.

Much work on quantifying this role—that is, on identifying just how much saving of species zoos could do—and formulating the underlying theoretical basis of captive propagation—that is,

how to manage small fragmented populations— was carried out during the 1970s and 1980s (see, for example, Foose, 1980; Mace, 1986).

The WZCS (IUDZG and CBSG/IUCN/SSC, 1993) pointed out that captive animals represented an important part of the remaining **gene pool** for some species and that, because of this, captive propagation was probably the only option for maintaining stocks of critically endangered species. The role of the zoo gene pool was to supplement wild populations, or to build entirely new ones (for example, through reintroduction), and, to do this, the gene pool had to be managed through cooperative effort.

More details on how zoos manage these gene pools are given in **Chapter 9**. Here, it is worth looking more closely at the scope of the problem in terms of how many species zoos are likely to be able to save and how zoos should select the species on which they focus their attention.

10.3.2 How many species can be saved in zoos?

We saw earlier in this chapter that the IUCN has identified 5,742 species of vertebrates, and 1,601 of insects and molluscs that are threatened in the

---

Box 10.2 Saved from extinction: accident or planning?

There are several species of animal that are extinct in the wild, but still exist in captivity. One of the most famous is Père David's deer, or milu, *Elaphurus davidianus* (**Fig. 10.8**), originally from the marshes of Central China, but probably already extinct in the wild two or three thousand years ago (Whitehead, 1972). When Père Armand David 'discovered' the milu for Western science, it only occurred in captivity inside the Imperial Hunting Park in Peking (now Beijing). This herd was exterminated at the end of the nineteenth century

when the wall of the park was breached by floods and the deer were eaten, the last few being killed during the 1900 Boxer Rebellion. Luckily, the Duke of Bedford kept a small group of milu in the grounds of Woburn Abbey and, realizing the plight of the species, he managed to assemble a group of eighteen, drawn from his own park and other European zoos. These eighteen are the **founders** of all milu alive today.

The species bred well in captivity, with about a thousand animals alive at the start of the 1980s (Cherfas, 1984), distributed among ▶

Figure **10.8** Père David's deer *Elaphurus davidianus* was already extinct in the wild long before it became known to Western science. A small number of captive animals were the ancestors of the modern population and, recently, the species has been reintroduced back to China. (Photograph: Geoff Hosey)

a number of different zoos. By the 1980s, it was possible to start reintroducing milu into the wild in China: firstly, into the Beijing Milu Park (in 1985), and subsequently into the Dafeng Natural Reserve (in 1986), in both of which areas the herd has subsequently grown (Hu and Jiang, 2002).

Surprisingly, analysis of the records of 2,042 individuals born between 1947 and 2000 shows that **inbreeding** is relatively low (Sternicki *et al.*, 2003). This seems almost the perfect example of a species, extinct in the wild, being maintained in zoos and eventually reintroduced back into its former range. In retrospect, however, it is probably only good luck that some specimens were available to form a founder herd, that there was someone with the vision to form the founder herd, and that the subsequent descendents are not too troubled with inbreeding.

Other species for which the last surviving members also lived in zoos—such as the quagga, the passenger pigeon, and the thylacine, were not quite so lucky—and became extinct.

wild (IUCN, 2007). How many of these can be saved from extinction by captive breeding in zoos?

It has been estimated (Seal, 1991) that about a thousand zoos worldwide that are likely to be involved in cooperative breeding programmes may have about 500,000 spaces for animals. Although the spaces available will be limited in terms of the animals for which they are suitable, nevertheless, this seems a reasonable number of spaces—but is it enough?

Unfortunately, the answer is probably 'no'. Maintaining captive populations of rare species involves a lot more than simply putting animals together and waiting for them to breed. As we saw in **Chapter 9**, captive populations are usually small and fragmented, so maintaining **genetic diversity** and avoiding the problems of inbreeding are real problems. Thus, we have the proposal, now encapsulated in WZCS and WZACS, that the goal of captive breeding should be the maintenance of 90 per cent of the genetic variation of the wild population for a period of 100–200 years (Soulé *et al.*, 1986). To do this requires an effective captive population (see **Chapter 9**) of about 200–250 animals for each species.

In this case, there is space for about 2,000 species, provided that the spaces are used to 100 per cent efficiency (Seal, 1991). Of course, different kinds of animal need different amounts and types of space, so evaluations of space (both available and needed) are perhaps better undertaken at class or order level. For example, a survey of North American zoos indicated that sixteen snake species could be accommodated in captive breeding programmes in the forty-four zoos that responded to the survey (Quinn and Quinn, 1993). A similar survey of spaces for birds concluded that fewer than 141 long-term management programmes could be accommodated (Sheppard, 1995).

On a more positive note, for threatened species of amphibians, if each of the 1,200 zoos represented in the current membership of WAZA were to take responsibility for maintaining a captive population of just five species, then, to quote Dickie *et al.* (2007), '*it is possible that zoos could be instrumental in saving an entire group of vertebrates*'. (See **Box 10.3** for more about how zoos are saving amphibians.)

In some ways, this sort of analysis identifies only the minimum that can be achieved. The WZCS points out that more spaces could be made available (for example, by reallocating spaces that are

---

Box **10.3** Saving the amphibians from oblivion

In the case of the amphibians, we have very nearly a whole class of animals in danger of extinction. Recent assessments suggest that 1,856 species, representing 32 per cent of the known species in the world, are threatened with imminent extinction and as many as 168 species may have become extinct during the past two decades (Stuart *et al.*, 2004).

A major problem for these animals is their susceptibility to infection with a **pathogenic** chytrid fungus, and this epidemic is going to worsen because of the effects of global warming (Pounds *et al.*, 2006). But they also suffer other threats, including habitat loss and exploitation, and those species with small geographical ranges are particularly at risk (Sodhi *et al.*, 2008; see **Fig. 10.9**).

So what can zoos do to help save the amphibians? Amphibians are relatively small animals; they also have high **fecundity**, low maintenance costs, and show few behavioural problems in captivity, all of which means that zoos could have a valuable role to play in their conservation (Bloxam and Tonge, 1995). The Amphibian Specialist Group (ASG) of IUCN/SSC has produced a conservation action plan for these animals (Gascon *et al.*, 2007; available online at www.amphibians.org), which includes captive breeding, some of which is to be achieved in zoos. Recently, ASG has joined with WAZA to form an Amphibian Ark (AArk) to coordinate *ex situ* conservation of as many amphibian species as possible and has designated 2008 as 'Year of the Frog' (see **Box 10.5**).

Let us look at one example: the blue poison dart frog *Dendrobates azureus* (**Fig. 10.10**). This species is endemic to a few isolated ▶

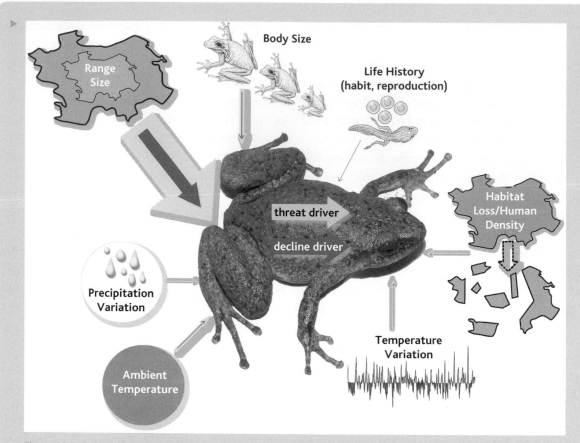

Figure **10.9** Recent studies have highlighted that many amphibians are on the brink of extinction, due to various threats to their survival, including major ecological changes (indicated by light arrows) and an inability to survive in the resulting habitat (indicated by dark arrows), because some of their life history traits are no longer appropriate in their environment. The thickness of the arrows indicates the amount of risk that these variables present: for example, small range size is a greater predictor of risk and decline compared with other environmental and life history variables. (From Sodhi *et al.*, 2008)

Figure **10.10** The blue poison dart frog *Dendrobates azureus* is one of a number of amphibian species, the survival of which currently depends on captive breeding efforts. (Photograph: Doug Sherriff)

areas in Suriname in South America, where it is threatened with extinction. Captive breeding has helped to boost the zoo population of these frogs to 175 first-generation, captive-bred animals in twenty AZA institutions, with fourteen of the original twenty wild-caught founders still reproductive (www.amphibiaweb.org). A **studbook** for this species is held by EAZA, and European zoos are also involved in its captive breeding.

Hopefully, this sort of coordinated action will help save many more amphibian species before it is too late.

currently taken up by animals that are not part of a breeding programme and/or by encouraging many of the smaller zoos that are not part of the international breeding programmes to become involved), that not all species will need even a hundred years of occupation of zoo space, and that maintaining a large, self-sustaining *ex situ* population may only be necessary for the most severely endangered species, whereas for less-endangered species, smaller zoo populations could serve as a reserve in case of problems with the wild populations.

10.3.3 Which species should be saved?

Given that there are not enough spaces to keep self-sustaining populations of all endangered species, then clearly some prioritization is necessary. In the WZACS, several criteria are listed as aids to this prioritization.

- How threatened are the wild populations?
- How unique is the species taxonomically?
- Is the species native to the region?
- Are there established and successful husbandry protocols already in place?
- Is there already a captive population?
- Could the species be regarded as a 'flagship' species?
- Does the species have additional educational or research value?

Setting the priorities is achieved through IUCN and zoo specialist groups (see **Chapter 9**).

An alternative view has been given by Balmford *et al.* (1996), who point out that current breeding programmes are biased in favour of mammals and that, even among mammals, more attention is given to larger-bodied species. They suggest that a more realistic approach would be to concentrate on smaller-bodied animals, which are cheaper to maintain in captivity, use less space, and are not necessarily less popular with the visiting public.

Some support for the current zoo bias in favour of the **charismatic megafauna**, however, is provided by Cardillo *et al.* (2005), who demonstrate that extinction risk rises sharply for animals with a body weight above a threshold of 3kg. This study suggests that the future loss of biodiversity among large mammals could be far more rapid than expected.

Another view (Hutchins *et al.*, 1995) is that zoos should concentrate on **flagship species**—that is, those species that can excite public attention (golden lion tamarins *Leontopithecus rosalia* and gorillas *Gorilla* spp. are often quoted as examples of flagship species), because this can also raise public support and perhaps generate funds for *in situ* conservation, which can benefit many more species.

10.3.4 Species or subspecies?

A further problem for zoos is that of deciding at what **taxonomic** level captive breeding plans should be put in place. As Mace (2004) has pointed out, taxonomy and conservation go hand in hand. The basis of much conservation planning is the **species** list (for example, the IUCN Red Data List), but if the taxonomic unit of choice is the species, then we are still faced with decisions about what constitutes a particular species, because definitions are not easy to apply in many cases. (See **Chapter 5** for more discussion of this.)

Not only is our knowledge of the total number of species on Earth poor (perhaps 1.7 million out of between 7 million and 15 million species—see Mace, 2004), but what we think we do know is liable to change. For example, Roca *et al.* (2001) provide

The so-called **charismatic megafauna** are the large vertebrates that are disproportionately popular with the general public, from giant pandas, to polar bears, tigers, and elephants. The authors' favourite suggestion for a generic term for the very many species not included in the usual list of charismatic megafauna is 'the other 99.99 per cent'.

A common definition of **species** is populations that are reproductively isolated from, and hence unable to breed with, other populations (Mayr, 1942). This is usually difficult to apply, so morphological or genetic criteria are often used instead.

Figure **10.11** The taxonomy of animals has changed over time (see section **5.2.2**) and, coupled with previously misidentified animals, this has led to the creation of hybrids. This is a hybrid of a red ruffed lemur and a black-and-white ruffed lemur, photographed in the 1980s. These two kinds of lemur are now considered to be distinct species (*Varecia rubra* and *V. variegata*, respectively) and are managed in such a way to prevent breeding between the two. (Photograph: Geoff Hosey)

genetic evidence that two species of African elephant should now be recognized,[10] with forest elephants *Loxodonta cyclotis* now classified as distinct from the savannah elephant *Loxodonta africana*. Furthermore, many species consist of a number of geographically separate (allopatric) populations that are distinctive enough to be regarded as subspecies.

So, should different subspecies be maintained as separate captive populations? Until about 30 years ago, different subspecies were routinely allowed to hybridize in zoos. For example, **Fig. 10.11** shows a hybrid of black-and-white and red ruffed lemurs in

a zoo in the 1980s; at that time, the two **taxa** were considered to be subspecies of the same species, but modern thinking is that the two are distinct species: *Varecia variegata* and *Varecia rubra* (Vasey and Tattersall, 2002). Luckily, zoos have maintained these two taxa as separate populations in the last 25 years, because each looks so distinctive, but this does illustrate the difficulties in prioritizing taxonomic levels.

The advent of modern molecular techniques has provided some help in identifying the evolutionary distinctiveness of different populations (Wayne *et al.*,

10 The debate over what is or is not a separate species has even extended to the polar bear *Ursus maritimus*. Analysis of the mitochondrial DNA (mtDNA) of polar bears and brown bears *Ursus arctos* suggests that some populations of brown bears are more closely related to polar bears than to other populations of brown bears (Waits *et al.*, 1998), which raises questions about their species status (Marris, 2007).

1994). For example, African hunting dogs *Lycaon pictus* from southern and from eastern areas of Africa look similar to each other, but, on the basis of mitochondrial DNA (mtDNA) sequences, these two populations are dissimilar enough to warrant sub-species classification. The recommendation is that, in captive breeding and reintroduction programmes, the two populations should be kept separate (Wayne *et al.*, 1994).

In Sumatran rhinoceros *Dicerorhinus sumatrensis*, on the other hand, mtDNA analysis of geographically separate populations from West Malaysia, Sumatra, and Borneo showed that the differences between these populations were slight enough to recommend that the appropriate taxonomic unit for conservation was the species, rather than the local populations (Amato *et al.*, 1995). (See **Box 10.4** for more information about the analysis of mtDNA.)

Note that, in the two examples just given, the molecular evidence is being used not primarily (if at all) to decide on the taxonomy of the species being studied, but rather as an aid in deciding what the significant population unit should be for conservation purposes. This idea is based on the notion of an **evolutionarily significant unit (ESU)**, which was originally formulated by Ryder (1986) to help

---

Box 10.4 The use of mitochondrial DNA in conservation biology

Mitochondrial DNA (mtDNA) is, as its name suggests, DNA that is found in the mitochondria of cells. It is a simple, short loop of DNA that is distinct from the DNA of the cell nucleus in two main ways.

1. It replicates via mitosis, not meiosis, so does not undergo the same 'mixing' of genetic material that occurs in nuclear DNA during recombination. Thus, the inheritance of mtDNA is essentially clonal, with mtDNA passed on virtually unchanged from parent to offspring.

2. mtDNA is nearly always inherited only through the maternal line or matrilineage, via the cytoplasm of the egg or ovum (sperm usually contribute almost nothing but nuclear DNA to the newly fertilized egg or zygote).

Although the mtDNA that is passed on from a mother to her offspring is a copy of her mtDNA, it does not always remain completely unchanged. A feature of mtDNA that makes it particularly useful to conservation biologists is that it undergoes mutation at a markedly higher rate than nuclear DNA. This means that mtDNA, in mammals at least, shows a rapid rate of differentiation among populations in comparison with nuclear DNA (Cronin, 1993).

Conservation biologists can look at mtDNA sequences from different animals and assess the extent to which it has diverged. This provides an estimate of taxonomic relatedness, either between different species or between different populations of the same species.

Most recent textbooks on conservation biology will have information about the use of both mtDNA and nuclear genetic material to support conservation.

An **evolutionarily significant unit (ESU)** can be defined as a population that, for conservation purposes, is considered to be distinct—that is, the minimum unit for conservation management. The usefulness of the ESU concept is that it bypasses difficulties over defining species.

to decide on appropriate conservation units below the species level. He defined ESUs as '*populations that actually represent significant adaptive variation based on concordance between sets of data derived by different techniques*'—in other words, genetic diversity between different populations of the same species was a more appropriate criterion for conservation decisions than formal taxonomy.

Obviously, molecular techniques are again invaluable; indeed, some authors came to define the ESU in terms of molecular criteria. Moritz (1994), for example, defined ESUs as '*populations that are reciprocally **monophyletic** for mtDNA **alleles** and show significant divergence of allele frequencies at nuclear loci*'.

Most authors today view ecological and molecular data as being equally important in defining ESUs (Crandall *et al.*, 2000; Fraser and Bernatchez, 2001).

10.4 Reintroduction

One of the key goals of many captive breeding programmes is the eventual reintroduction of species back into the wild. As we have already seen, however, not all species maintained in zoos are part of a programme that is expected to lead to reintroduction; furthermore, not all individuals that are part of a captive breeding programme which includes reintroduction will themselves be reintroduced (see **Chapter 9**). It is probably also worth pointing out here that reintroduction for most species is an event that is most likely to occur far into the future, simply because the criteria for successful reintroduction (see section **10.4.3**) are not currently satisfied. For all of these reasons, the number of successful reintroductions of zoo-born animals is actually quite low at the moment, but there have been some notable successes, which gives cause for optimism about the prospects of other species in the future. Reintroduction features prominently in both the

WZCS and the WZACS, and usually requires the zoo(s) to be involved collaboratively with other conservation agencies.

10.4.1 What is 'reintroduction'?

Of course, there is a long history of animals (and plants too, come to that) being moved around the planet into new habitats through human agency. So what exactly do we mean by 'reintroduction'?

In 1987, the IUCN defined three different types of translocation that could occur through human agency, and these are the definitions generally adopted by the conservation community:

> *Introduction*: intentional or accidental dispersal by human agency of a living organism outside its historically known home range;

> *Reintroduction*: intentional movement of an organism into a part of its native range from which it disappeared or became extirpated in historic times as a result of human activity or natural catastrophe;

> *Restocking*: movement of numbers of plants or animals of a species with the intention of building up the number of individuals of that species in an original habitat.

> (Stuart, 1991)

The first of these, introductions, are clearly something to be avoided. Many species of island bird, for example, have been rendered endangered or extinct because of non-native fauna introduced to the islands by man (Halliday, 1978). It is reintroduction and restocking with which zoos and other conservation agencies are concerned.

10.4.2 How necessary is reintroduction?

Most individuals and agencies involved in conservation are agreed that populations of endangered species are best maintained *in situ*; this requires

Monophyletic refers to any group of taxa that have evolved from a common ancestor.

habitat conservation and management, which is beneficial to other species as well, and is often relatively less expensive than captive breeding with subsequent reintroduction. Furthermore, translocation of animals from one place to another in the wild has a higher success rate than trying to establish self-sustaining wild populations of captive-bred animals (Stanley Price, 1991).

So why should reintroduction be done at all? To put this into some kind of perspective, only sixty-eight species were recommended for reintroduction in IUCN/SSC action plans by the end of 1992 and, for forty-five of these species, captive breeding was also recommended (Wilson and Stanley Price, 1994). The database of the Reintroduction Specialist Group (RSG) of the Species Survival Commission (SSC) of the IUCN, however, listed 149 past, current, and proposed bird and mammal reintroduction projects, involving 121 species (Wilson and Stanley Price, 1994).

This does not seem very many, so why are there so few? The answer generally, as already mentioned, is that the criteria that need to be satisfied for reintroduction (see section **10.4.3**) are not satisfied. Nevertheless, reintroduction, where feasible, is considered worthwhile because there are public awareness spin-offs and because some species will not survive at all otherwise. With better habitat restoration and the use of gene technologies in the future, prospects for reintroduction may well improve (Stuart, 1991).

10.4.3 The criteria for reintroduction

Establishing new, self-sustaining populations of endangered species in areas where they have become extinct is clearly not likely to be an easy task and a number of general criteria have been used to help to decide whether or not reintroduction is feasible. These criteria are discussed more fully by Kleiman *et al.* (1994) and include the following.

1. There should be a need to augment the size or genetic diversity of the wild population.

2. There should be available and appropriate stock to use in the reintroduction.

3. There should not be any jeopardy to the existing wild population.

4. The original causes of the species' decline should have been removed.

5. There should be sufficient appropriate and protected habitat within the species' former range.

6. The available habitat should have no members of the species (reintroduction) or only a sparse population (restocking).

7. There should be no major negative impact on the local human population.

8. There should be local community support for the reintroduction.

9. There should be support from relevant governmental and non-governmental organizations.

10. The reintroduction should conform to relevant laws and regulations.

11. The reintroduction technique should be known.

12. There should be sufficient knowledge of the species' biology.

13. There should be sufficient resources for the reintroduction.

In 1995, the IUCN approved reintroduction guidelines produced by the SSC Re-introduction Specialist Group (see section **3.2.4**), which are available to view on the IUCN website (see **Websites and other resources**).

10.4.4 How are zoos involved in reintroduction?

Of course, the animals used in reintroduction projects may not have come from zoos; indeed, they may not even be captive-born. So, to what extent are zoo animals involved in reintroduction projects? The answer, at the moment, is that they are involved rather less than we would expect. Of 129 reintroduction projects surveyed by Beck *et al.* (1994), seventy-six

(59 per cent) used zoo-bred animals or their descendents. This involved a minimum of 20,849 individual animals (1,958 mammals; 8,271 birds; 10,620 reptiles and amphibians), which, as the authors point out, is about the equivalent of the collections of only three or four major zoos combined. This may not seem very many, but it is worth remembering that the aim of the captive breeding programme is to maintain 90 per cent of genetic diversity of those species that are part of the programme for at least a hundred years. The assumption behind this is that reintroduction programmes will not be feasible for most species until well into the future. At the moment, there are still sufficient wild stocks in other parts of their range for some of the species in reintroduction programmes and many of the species involved in these programmes are not the most severely endangered. But it is very unlikely that this will be the case in the future.

10.4.5 Some examples of reintroduction projects

Several reintroduction projects have become particularly well known, partly because of the intensiveness of monitoring that has accompanied them and partly because the animals have been regarded as 'flagship' species, the importance of which to conservation goes beyond only being a saved species. The following are three particularly well-known examples, which are often quoted in the literature to demonstrate the effectiveness of reintroduction as a conservation measure.

Golden lion tamarin

Perhaps the best known is the reintroduction programme for the golden lion tamarin *Leontopithecus rosalia* (**Fig. 10.12**) back into the Atlantic Coast rainforest of Brazil (Kleiman *et al.*, 1986; Stoinski *et al.*, 1997). These small primates were severely endangered in the wild and in captivity (where **mortality** rates were exceeding **birth rates** in the 1980s), and the first step in their conservation programme was to identify what was amiss with their captive management and rectify it. As a result, a self-sustaining captive

Figure 10.12 Golden lion tamarins *Leontopithecus rosalia* have been the focus of one of the best-known reintroduction programmes. A successfully managed captive breeding programme, initiated by the Smithsonian's National Zoo, led to the propagation of a captive population and the subsequent reintroduction of animals back into their native Atlantic Coast rainforest of Brazil. (Photograph: Jessie Cohen/Smithsonian's National Zoo)

population was built up at the Smithsonian's National Zoo in Washington DC, and in other zoos, in association with a primate centre in Rio de Janeiro.

Before reintroduction was attempted, the animals went through a pre-release training programme during which the animals learned how to search for hidden food and how to use natural substrates for locomotion. The tamarins were subsequently acclimatized in cages built around natural vegetation at the release site.

The initial set of released animals (fourteen in 1984) fared badly and, 1 year later, eleven of them had

Figure **10.13** The Arabian oryx *Oryx leucoryx* pictured here was considered extinct in the wild by 1972; several attempts have subsequently been made to reintroduce animals into Oman and, more recently, Saudi Arabia. (Photograph: Geoff Hosey)

either died or been removed (Stoinski *et al.*, 1997). In response to this, the training programme was improved—in particular, with a greater emphasis on post-release training—and this led to more successful reintroduction (a reintroduced population of 169 animals, constituting 23 per cent of the wild population, by 1995).

Arabian oryx

In the example of the golden lion tamarins, the captive-born animals were being reintroduced to supplement an existing, although threatened, wild population. A slightly different case is presented by the Arabian oryx *Oryx leucoryx* (**Fig. 10.13**), because this species was extinct in the wild, so the reintroduced animals were part of a programme to establish a wild population where none existed.

This species was exterminated in the wild by 1972, but a captive herd had been established before this. Reintroduction was successfully accomplished

to a reserve in Oman in 1982 and again in 1984, leading to the growth of an independent population that numbered about 280 animals by October 1995, using over 16,000km^2 of the reserve area (Spalton *et al.*, 1999). Unfortunately, poaching, which was the main cause of extinction in the first place, started again in 1996, so that by September 1998, only 138 animals were left, of which only twenty-eight were females. As a result, as many as possible of the remaining animals were caught and brought back into captivity.

Similar reintroduction of the oryx is being carried out in Saudi Arabia and this promises to be more successful (Ostrowski *et al.*, 1998).

Californian condor

Slightly different again is the case of the Californian condor *Gymnogyps californianus* (**Fig. 10.14**), because, here, the last remaining wild specimens were brought into captivity to ensure the survival of a species that was seen as doomed in the wild.

(a)

(b)

Figure **10.14** Attempts made to save the extremely rare Californian condor *Gymnogyps californianus* (a) through captive breeding have resulted in the development of a complex rearing process for the captive-bred chicks, which involves the chick being fed by a puppet (b), representing the head of an adult condor, to help it imprint correctly and learn its appropriate species identity. (Photographs: (a) Scott Frier © US Fish & Wildlife Service; (b) Ron Garrison © US Fish & Wildlife Service)

This species had been rare for many years, but, by 1985, the total wild population numbered only fifteen birds (Toone and Wallace, 1994). Between 1986 and 1987, all of the remaining wild condors were brought into captivity in Los Angeles Zoo and San Diego Wild Animal Park. By 1994, forty-nine condors had been bred at these two zoos; release back into the wild started in 1992. Mortality in the wild continued to be a cause for concern, however, with some birds colliding with overhead power cables and others contracting lead poisoning from eating bullet fragments in carcasses (Meretsky *et al.*, 2000), so the viability of a wild population remains unclear.

As we stated at the start of this section, these three examples are often quoted as reintroduction

success stories, and it is easy to get the impression that the eventual reintroduction of captive-bred animals is what zoo conservation aspires to. The reality, however, is that comparatively few reintroductions have been successfully attempted and the technique is not a common part of conservation practice (Stanley Price and Fa, 2007). More important now is the integrated management of wild and captive populations (see section **10.5.4**).

10.4.6 Is reintroduction good value for money?

The golden lion tamarin project is very well known and has helped to raise awareness of the conservation of endangered species, but it comes at a price.

That price is about US$22,000 for each surviving reintroduced tamarin (Kleiman *et al.*, 1991). This amount is made up of about $1,657 per animal per year in maintenance costs at the zoo (keeper time; food; veterinary care; housing; etc.), plus the reintroduction costs—and that is at 1989 prices.

Clearly, both captive breeding and reintroduction are, at least for this tamarin, costly activities. It has been suggested by Balmford *et al.* (1995) that, for large bodied species, *in situ* conservation is relatively cheaper than captive propagation, can result in equally good population growth, and has the additional benefit that it helps conserve entire ecosystems and not only the target species. But we should bear in mind that these costs quoted for the tamarin project are probably not typical, and that many other reintroduction projects, particularly those involving native species (see section **10.5.5**), are likely to be less costly. Furthermore, it is probably misleading to see captive propagation and *in situ* conservation as alternatives, because, in many cases, both may be needed and, for some species, *in situ* conservation may simply not be an option.

10.5 Other aspects of zoo conservation work

10.5.1 Education and raising awareness

WZACS points out that zoos worldwide reach hundreds of millions of people each year,[11] many of whom come from urban areas in which they have little contact with wildlife.

> Today more and more of us live in cities and lose any real connection with wild animals and plants.
> (Attenborough, 2004 quoted in WAZA, 2005)

Presumably, people visit zoos because they have some interest in animals, so zoos are well placed to raise awareness and knowledge of conservation issues. This can be achieved through enclosure design, signage, keeper talks, interactive education, animal shows, and in other ways. Indeed, it is as important sometimes to keep species of low conservation importance in zoos[12] as it is to keep the high-priority species, because they may be more useful in promoting the conservation message by enhancing people's zoo experience of animals.

Successful education must be targeted appropriately and must take due account of what are sometimes perceived by the public as the more negative aspects of zoos (Whitehead, 1995). The effectiveness of zoo education initiatives remains unclear and it is likely that the most important educational aspect of zoo conservation is simply the caring values shown by zoos in their dealings with animals (Rabb and Saunders, 2005). (More is said about this aspect of zoo work in **Chapter 13** on 'Human–animal interactions'.)

10.5.2 Conservation research

For many years, research (see **Chapter 14**) conducted on zoo animals tended to be concerned primarily with anatomy and taxonomy, but there is a huge potential in zoos to undertake behavioural, genetic, and physiological research that contributes to the *in situ* and *ex situ* conservation of endangered species (Ryder and Feistner, 1995). The importance of research is even more prominent in WZACS, which, among other things, urges zoos to formulate proper research plans within their resources, to promote and carry out research, to contribute to databases, and to disseminate their research findings.

In Europe, the EC Zoos Directive 1999 makes it a requirement that zoos pursue conservation

---

11 The thousand or so accredited zoos worldwide that are affiliated to organizations such as BIAZA, EAZA, AZA, and ARAZPA receive at least 600 million visitors every year. As Achim Steiner (Executive Director of the UN Environment Programme) has said, zoos have immense potential as the *'incubators of the conservationists of tomorrow'* (Steiner, 2007).

12 A good example of a species of low conservation importance that is often a very popular zoo exhibit is the meerkat *Suricata suricatta*. These animals are not considered to be endangered in their native habitat in southern Africa, but meerkats are highly social, and are regarded as an attractive and engaging species that appears to have few welfare problems in captivity.

goals, and undertaking research is one of the ways in which this requirement can be fulfilled (see **Chapter 3**). Undoubtedly, the amount of research undertaken by European zoos has increased markedly in the past decade or so, but how successful this has been in meeting the requirements of the EC Zoos Directive appears to depend on what exactly is counted as research. Rees (2005b), for example, maintains that most of the research studies done in zoos are not directly relevant to conservation (for example, behavioural studies and studies on **environmental enrichment**), and that record keeping, which is admissible evidence of research for zoo licensing requirements in the UK, does not count as research.

The view from the zoos is different (Thomas, 2005; Wehnelt and Wilkinson, 2005), as we might expect given that the research requirement has to be fulfilled by small zoos with few resources, as well as the larger zoos that sometimes have their own dedicated research departments. In any case, it is becoming increasingly clear that behavioural studies are of importance to conservation (Shepherdson, 1994; Wielebnowski, 1998; Festa-Bianchet and Apollonio, 2003; Swaisgood, 2007), not least because of the need to maintain **behavioural diversity** in captive animals that may become candidates for reintroduction.

10.5.3 Sustainability

Chapter 8 of WZACS is about **sustainability** and the measures that zoos can take to reduce their 'environmental footprint'. Sustainability is not the same thing as conservation, but, for zoos to be taken seriously as conservation organizations, they must also be seen to be responsibly managed organizations that do not contribute to the depletion of natural resources (WAZA, 2005). The 'sustainable zoo' is a topic that is considered in more detail in the final chapter of this book (see **Chapter 15**), but it is worth noting here that a number of zoos within the UK (and further afield) have now achieved the environmental management standard ISO 14001.[13]

10.5.4 Integrated conservation

So far, we have considered the role of zoos in biodiversity conservation in terms of education, research, and captive breeding. In fact, these are only convenient labels to apply to a range of conservation-related activities in which zoos are involved. We have, for example, used 'education' to cover such diverse things as raising awareness, raising financial support, and informing the public about the animals. Each of these tasks can itself be approached in a number of ways. Similarly, many zoos may be as involved in research on wild-living animals in their natural habitats as they are with that on captive animals. Bringing all of these things together involves a coordinated approach, whereby different aspects of the zoo's conservation work are integrated with each other and with the external agencies that are also involved in conservation.

The integrated approach to conservation[14] is highlighted in the WZACS, which distinguishes between 'internal' and 'external' integrated conservation. Internal integrated conservation is not only about keeping animals, and providing an enjoyable and informative visitor experience; it is also about making links between the various activities that contribute to these. These links can include embracing sustainability in the zoo's operation, selling crafts from areas with which the zoo has field conservation links, and informing people of the links between the exhibits and the field research and conservation in which the zoo is involved. External

13 ISO 14001, first published in 1996, is the best known of a series of international environmental management standards (the ISO 14000 series) created by the International Organization for Standardization (ISO).
14 The idea of an integrated approach to conservation is not new. More than 10 years ago, in the book *Creative Conservation: Interactive Management of Wild and Captive Animals* (Olney *et al.*, 1994), representatives from zoos and conservation organizations around the world were calling for an integrated approach to managing endangered species.

integrated conservation is about forming links with other agencies to promote field conservation, supporting research on wild populations, and raising money to support *in situ* conservation. Of course, not all zoos will do all of these things, but what the integrated approach is advocating is the most effective use of the zoo's resources in achieving conservation objectives.

Let us look at a few examples.

In the UK, Chester Zoo has conservation programmes in a number of countries, including China, the Philippines, Nigeria, and Mauritius, all of which are run in partnership with local and international conservation agencies, and, in some cases, with universities and local zoos.

In the south east of England, Marwell Zoo is only one part of a much larger body: the Marwell Preservation Trust. This umbrella body also includes Marwell Conservation, which is involved in a number of international conservation programmes and initiatives, with a focus on the animals of Africa. For example, the Marwell Zimbabwe Trust was set up in 1997. This is a conservation and research organization that works with the Zimbabwe Parks and Wildlife Management Authority, and other non-governmental bodies, to help to conserve the country's rhino population.

Marwell Conservation has also been closely involved in the reintroduction of captive-bred scimitar-horned oryx and addax back to Tunisia, and, more recently, to Morocco (Woodfine *et al.*, 2005). The oryx conservation programme combined the captive breeding of this species in zoos, including Marwell, with local *in situ* measures to ensure the survival of the reintroduced animals and the protection of their habitat.

Marwell's work in Zimbabwe also involves another UK Zoo, Paignton Zoo; both are partners in the Dambari Field Station in Zimbabwe (**Fig. 10.15**), and are closely involved in the conservation and management of rhino, cheetahs, and duikers.

Our final example, Jersey Zoo, has an international reputation as a centre of excellence and leader in conservation.[15] The zoo is only one part of the Durrell Wildlife Conservation Trust and the ethos of conservation permeates through all aspects of the zoo's operations. The pink pigeon *Columba meyeri* from Mauritius (**Fig. 10.16**) is one of a number of species facing extinction in the wild that have been bred successfully in captivity at Jersey Zoo.

With only ten birds left in the wild by 1991, the pink pigeon nearly went the way of the dodo. Captive breeding of pink pigeons started in 1977 at Jersey Zoo, alongside measures to protect and manage the remaining few birds in their natural habitat. Captive-bred birds have since been successfully released back into the wild and, today, the pink pigeon population on Mauritius numbers some 350 birds, with the species now listed as 'endangered' rather than 'critically endangered' on the IUCN Red Data List (see **Websites and other resources**).

It is possible that the pink pigeon could have survived in the wild without a captive breeding programme, but it is equally possible that the species could have become extinct. The captive population, now thriving, is both an insurance policy against future threats to this island species, and a reminder to the many thousands of people who visit Jersey Zoo each year of the fragility of island habitats and populations.

Efforts to save the pink pigeon represent just one of a number of conservation programmes run by the Durrell Wildlife Conservation Trust from its base at Jersey Zoo. The website of the Trust lists the Monserrat oriole *Icterus dominicensis oberi*, Mauritius kestrel *Falco punctatus*, Bali starling

15 Like Jersey Zoo, many of the world's leading zoos now see themselves as conservation organizations first and foremost, with the zoo only one part of a wider umbrella organization. The Bronx Zoo in New York is part of the Wildlife Conservation Society; in the UK, Marwell Zoo comes under the umbrella of the Marwell Preservation Trust, and Paignton Zoo Environmental Park, Living Coasts, and Newquay Zoo are all part of the Whitley Wildlife Conservation Trust, which also owns Slapton Ley national nature reserve.

(a)

(b)

Figure **10.15** (a) The Dambari Field Station in Zimbabwe is the location of *in situ* conservation programmes coordinated by the Marwell Preservation Trust and the Whitley Wildlife Conservation Trust, which work with numerous people to preserve local biodiversity including species such as this blue duiker *Cephalophus monticola* (b). (Photographs: Whitley Wildlife Conservation Trust)

Figure **10.16** The Mauritius pink pigeon *Columba meyeri* represents a species that has been reintroduced successfully into the wild as a consequence of a captive breeding programme that was initiated by Jersey Zoo. (Photograph: Paignton Zoo Environmental Park)

Leucopsar rothschildi, Vietnamese pheasant *Lophura hatinhensis*, and many other bird species as the focus of its conservation work, alongside mammals such as the Malagasy giant jumping rat *Hypogeomys antimena*, the aye-aye *Daubentonia madagascariensis*, and the Alaotran gentle lemur, and amphibians such as the blue poison dart frog *Dendrobates azureus* and the Trinidad stream frog *Colostethus trinitatis* (see www.durrellwildlife.org/ for more details). Closer to home, Jersey Zoo manages, within its grounds, native species such as red squirrels *Sciurus vulgaris* and the Jersey bank vole *Clethrionomys glareolus caesarius*.

Many smaller zoos, of course, do not have the resources for this level of activity, although even the smallest zoos can work in partnership with other zoos or with *in situ* conservation projects. Small zoos can also support conservation in many other ways, such as raising public awareness and providing financial support. Again, much of this activity is now coordinated across zoos, a good example being the EAZA campaigns such as Save the Rhinos (see **Box 10.5**).

Box **10.5** EAZA campaigns

Fundraising to support conservation projects is an important way in which zoos can contribute to international conservation efforts. For several years, EAZA has mounted fundraising campaigns targeted at a particular key project, as part of which member zoos attempt to raise money by holding special events, activities, and exhibitions. In 2002–2003 and 2003–2004, the campaign was to support tiger conservation by providing funding towards nine different tiger projects in Russia, Indonesia, Thailand, and India. This was followed in 2004–05 by 'Shellshock', the EAZA Turtle and Tortoise campaign, which aimed to raise

public awareness about the threats faced by these animals, to promote captive breeding of some of the thirty-six most endangered species, and to raise €150,000 to support turtle and tortoise conservation.

The 2005–2006 campaign was 'Save the Rhinos'. Its aims were, again, to raise public awareness of the threats to rhinoceros populations and to raise funds (a goal of €350,000) to support conservation projects within rhino range countries. Save the Rhinos was followed in 2006–2007 with EAZA's Madagascar campaign (**Fig. 10.17**) and, in 2008, EAZA linked with AZA and other zoo organizations in ▶

Figure **10.17** Madagascar was the subject of the EAZA awareness campaign in 2006–2007, in which EAZA member zoos highlighted conservation issues to their zoo visitors and aimed to raise funds to support conservation activities *in situ*, as illustrated here at Apenheul in the Netherlands. (Photograph: Sheila Pankhurst)

a global campaign: 'The Year of the Frog' (see www.yearofthefrog.org/).

EAZA sets specific targets for its conservation campaigns, both in terms of funds raised and other activities, such as raising awareness. The Shellshock campaign, for example, involved 119 EAZA member institutions and also thirteen non-EAZA organizations—notably including the Chinese Association of Zoological Gardens (CAZG). The decision by the CAZG to adopt the Shellshock campaign, and to develop its own long-term turtle and tortoise conservation campaign (under the title 'Careshell'), has the potential for a significant impact in an area of the world in which large numbers of turtles and tortoises are routinely killed for consumption and medicinal purposes.

The final total for funds raised by Shellshock was over €370,000, more than double the initial target of €150,000. In consultation with the Turtle Conservation Fund (TCF) and the Marine Turtle Research Group (MTRG), Shellshock funds have been distributed to a number of *in situ* turtle and tortoise conservation projects worldwide, from Cambodia and Vietnam, to the Philippines.

Further details about Shellshock and other EAZA campaigns can be found online at www.eaza.net.

Figure **10.18** The sand lizard *Lacerta agilis* represents a native British species that has been bred successfully in captivity in several zoos and reintroduced back into the wild. (Photograph: © Karel Broz, www.iStockphoto.com)

Figure **10.19** An endangered British insect, the field cricket *Gryllus* spp., of which more than 14,000 captive-bred individuals have been released back into the wild. (Photograph: © Maks Dezman, www.iStockphoto.com)

10.5.5 The conservation of native species

As well as having to make decisions on the taxonomic level at which conservation projects are carried out, zoos must decide what priority to give to the conservation of native species on their doorstep as opposed to wild animals from further afield. Sand lizards *Lacerta agilis*, **corncrakes** *Crex crex*, and field crickets *Gryllus* spp. may lack the immediate visitor appeal of zebras and tigers, but (in the UK at least) native species often require relatively little space for housing and are better able to tolerate the local climate. Both Chester Zoo and Marwell Zoo, for example, maintain thriving colonies of captive-bred sand lizards (**Fig. 10.18**). Animals from both zoos have been successfully reintroduced back into selected sites in the UK where there is suitable habitat.

Whipsnade Wild Animal Park in Bedfordshire, in collaboration with the Royal Society for the Protection of Birds (RSPB) and English Nature, has been rearing corncrake chicks since 2002 for release back into the wild in the UK. These endangered migratory birds come to the UK from Africa to breed, but have suffered substantial population declines in the UK as a result of changes in agricultural practice in recent years. This collaborative project, which also involved Chester Zoo during the initial stages during which birds imported from Germany were quarantined at Chester before being transferred to Whipsnade, is starting to show positive results; at least one of the birds released at the RSPB's Nene Washes reserve in Cambridgeshire has successfully completed the 3,000-mile migration to Africa and back to its release site.

At London Zoo,[16] a captive breeding programme was established in 1991 for the field cricket *Gryllus campestris* (**Fig. 10.19**), a species listed as 'endangered' on the UK Biodiversity Action Plan (BAP) and restricted to only one site in Sussex. Three new wild colonies of field crickets have now been established, with the release of more than 14,000 captive-bred individuals. The aim of the project is to have ten viable wild populations by 2010.

The costs of projects such as the field cricket breeding programme are modest in comparison with the costs of breeding and releasing into the wild larger animals, such as gazelles or marmosets, and the integration of *ex situ* and *in situ* work is easier within national borders and easier still where animals

---

The **corncrake** is a medium-sized brown bird related to the coot and moorhen. Corncrakes favour grassy meadows, but, where they are present, they are usually heard rather than seen, due to their excellent camouflage.

16 London Zoo and Whipsnade Wild Animal Park both come under the umbrella of the Zoological Society of London (ZSL). Further information about a range of ZSL conservation initiatives can be found online at http://www.zsl.org/field-conservation/.

are being bred for release at sites local to the zoo. To support such initiatives within the UK and Ireland, BIAZA now has a Native Species Working Group, with equal status to the taxonomic working groups such as the Mammals Working Group.

10.5.6 The conservation of behaviour

Much of the theory and practice of small population management that is implemented in zoos is formulated from a genetic point of view, and is concerned with the goals of maintaining genetic diversity and avoiding inbreeding (see **Chapter 9**). We should, however, remember that behaviour is influenced not only by genetic variation, but also by **learning** that takes place in the lifetime of the animal. Because of this, it is important to ensure that behavioural diversity is maintained and that key behaviours that are needed for survival in the wild are not lost from the captive population (Lyles and May, 1987; May and Lyles, 1987; Festa-Bianchet and Apollonio, 2003).

One of the issues here is to avoid **domestication** (Price, 1984), which can happen surprisingly rapidly. For example, farmed silver foxes *Vulpes vulpes* have routinely been selected for tameness, as shown by willingness to be handled and low levels of aggression. After thirty or so generations, these animals show physiological (lower serum cholesterol) and genetic differences, as well as the more obvious behavioural ones, in comparison with unselected foxes (Harri *et al.*, 2003; Lindberg *et al.*, 2005). This sort of selection is not necessarily a totally bad thing to be avoided at all costs in zoo animals. It has, for example, been suggested as appropriate for some common species that are kept for display, but are of no immediate conservation importance, especially if the selection is to maintain a 'classic' species **phenotype** (Frankham *et al.*, 1986).

Maintaining behavioural diversity is particularly important for animals that are being reintroduced to the wild, but potential to survive in the wild should be the criterion of success for all captive breeding programmes (Shepherdson, 1994), because, for many species, reintroduction is a long way in the future and the individuals to be reintroduced may not yet have been born. Appropriate environmental enrichment and training are ways in which animals can be prepared for release into the wild. The purpose of this is to ensure that animals destined for release have the **behavioural competence** to be able to survive in the wild (Wielebnowski, 1998). We have already seen how golden lion tamarins were put through a pre-release training programme so that they could find hidden food and use natural substrates for locomotion.

One particularly important aspect of developing behavioural competence is in training animals to avoid predators, because an inability to express natural anti-predator behaviours is a major cause of failure in animal reintroductions (Griffin *et al.*, 2000). This sort of training is often achieved through **classical conditioning**, in which a predator model is paired with some sort of aversive **stimulus**. This has been used, for example, to train captive-born greater rheas *Rhea americana* at Belo Horizonte Zoo, Brazil, to behave appropriately when confronted by a predator (de Azevedo and Young, 2006a; 2006b).

We are increasingly coming to realize the contribution that animal temperament makes to the success of these initiatives (McDougall *et al.*, 2006; Watters and Meehan, 2007). As an example, Bremner-Harrison *et al.* (2004) found that captive-bred swift foxes *Vulpes velox* (**Fig. 10.20**) previously assessed as bold were more likely to die in the first 6 months after release than those rated as shy, presumably because, as in captivity, they had low fear and were more likely to approach novel objects.

Maintaining behavioural diversity in animals that are not imminently about to be released is largely achieved through the same sort of techniques,

---

Domestication refers to the adaptation by man of a population of animals to a captive environment through a combination of genetic and developmental changes.

Figure **10.20** Personality indicators identified in swift foxes *Vulpes velox*, reintroduced into the wild, were successful predictors of their survival. Animals identified as bold did not survive as well during the first 6 months after release into the wild as did those assessed as shy. (Photograph: Sheila Pankhurst)

although, in this case, the goal is often a more general attempt to achieve a behavioural profile seen in the wild (see **Chapter 4**), rather than the acquisition of particular skills in things such as food finding, predator avoidance, and orientation in the environment. There is perhaps a case to be made that more active management of behavioural development should be undertaken in zoos (Rabin, 2003).

10.6 How good are zoos at conserving biodiversity?

However you define conservation, the ultimate aim is to have self-sustaining populations in natural ecosystems in the wild. While realizing this aim may be far in the future for many species, much work is taking place now to move towards this eventual goal. The integrated conservation ideal (see section **10.5.4**) sees zoos working on a number of fronts in partnership with other agencies to protect habitats and species in the wild.

The answer to the question of how good zoos are at conserving biodiversity is that it depends on the zoo and on what is being measured. This is probably not a very satisfactory answer, but the word 'zoo' encompasses such a wide range of organizations that it is almost impossible to generalize about their achievements, or otherwise, in the conservation of biodiversity. One approach, perhaps, is to look at what the best zoos are achieving in terms of conservation and to consider what can be done to bring other zoos up to the same standards. Let us recall, for a moment, the example of Jersey Zoo's efforts to save the Mauritius pink pigeon, at which we looked in section **10.5.4**. Looking at this example, it is hard to argue that zoos such as Jersey do not make a useful and substantial contribution to the conservation of biodiversity. Of course, not all zoos are like Jersey, but more and more modern zoos are aspiring to the same goals of integrated conservation.

As the example of the pink pigeon shows, conserving species by captive breeding in zoos is not enough. Zoos must also play a role in the conservation of habitats and ecosystems—but how good are they at doing all of this? This is a surprisingly difficult question to answer, mainly because of the sheer diversity of activities that count as conservation, but also because measures of effectiveness are not easy to formulate. Miller *et al.* (2004) identified eight questions that could be asked of zoos to test whether their conservation activities really did contribute towards achieving their mission. The questions included 'is there a functional conservation department?' and 'do exhibits explain and promote conservation efforts?' Approaches such as this are valuable, but yield qualitative data, which can make comparisons between different activities and zoos difficult.

More recently, attempts have been made to find quantitative evaluations of the effectiveness of zoos. One study (Leader-Williams *et al.*, 2007) compared zoos in 1992–93 with what they were doing 10 years on in 2003, using measures such as the number of threatened species held in coordinated breeding programmes (this had increased), how many belonged to each major taxon (there was still an over-representation of mammals), and where they were located (most zoos were in rich countries). They also looked at the activities in which BIAZA zoos were involved and found that those zoos that were run by a charity had more education staff, undertook more institutional research, had more visitors, and were involved in more *in-situ* projects than private, or local authority, zoos. Their conclusion was that many zoos were not fulfilling the conservation aspirations of WZACS.

It is, nevertheless, difficult to compare quantitatively the variety of different activities in which zoos are involved, particularly when the link between an activity (such as an educational display) and a measurable effect on conservation is such a diffuse one. An attempt to do this has been made by Mace *et al.* (2007), using what at first looks to be a very straightforward equation:

$$\text{Overall impact} = \text{project importance} \times \text{project volume} \times \text{project effect}$$

Fortunately, Mace *et al.* (2007) also give an indication of the sorts of measures that can be used to quantify these variables and then apply them to a pilot analysis of forty-one projects undertaken by UK zoos. The calculated impact scores suggested that those projects that were involved in direct action to enhance habitats were more effective than other sorts, with—perhaps surprisingly—educational and training projects being the least effective. They also found that project impact was correlated with the amount of funding it received, but not with for how long the project went on.

This kind of analysis provides a tool that may allow zoos to prioritize their conservation initiatives and then measure how well they are doing. But, for the moment, perhaps the most we can say is that the conservation work that zoos do remains patchy and the effectiveness of that work is still relatively unclear.

10.7 Conservation and zoos: looking to the future

Twenty years ago, the maintenance of breeding groups of endangered species to safeguard their future was often seen as the most important way in which zoos could contribute to animal conservation. Now, although captive breeding is still seen as important, it is only one of a number of integrated activities in which zoos are involved and the emphasis has moved much more towards involvement in *in situ* projects. More and more zoos are now making the transition from being 'zoos that undertake conservation' to becoming 'conservation organizations that run zoos' (Zimmermann and Wilkinson, 2007). This is concordant with the vision of WZCS and WZACS that zoos should become 'conservation centres'.

How do they do this? This question was addressed at the *Catalysts for Conservation* symposium held in London in 2004 (Zimmermann *et al.*, 2007; see also section **10.2.3**) and the main conclusions were that there are three main characteristics of a conservation centre (Hatchwell *et al.*, 2007), as follows.

1. *Bringing about a change in behaviour of zoo visitors and decision-makers*   In the future, zoos should address this by becoming more rigorous in inspiring zoo visitors to care about conservation and measuring their impact in doing this, by defining the highest of standards in animal care and welfare, by becoming involved in conservation locally, and by emphasizing links between humans and biodiversity.

2. *Establishing links between the* ex situ *work of the zoo and* in situ *conservation activities*   There are many ways in which zoos can achieve this, including fundraising campaigns, providing expertise in small-population management, developing zoo exhibits in parallel with *in situ* projects, and making explicit links between exhibits and field sites.

3. *Making direct contributions to* in situ *conservation activities*   Again, there are numerous ways in which zoos can do this, including running their own field projects and raising money for field conservation.

Many zoos, as we have seen, are already following this path and the hope is that these recommendations will form a framework for zoos to plan their conservation role for the future.

Summary

- The roles and priorities of zoos have changed substantially in the last 200 years, but most accredited zoos now cite conservation as their greatest priority.

- In the short term, captive management of breeding populations (*ex situ* conservation) is necessary for many species, with the aim of using them to supplement vulnerable wild populations, or to form new wild populations, if they have already become extinct.

- The ways in which zoos should be involved in conservation are elaborated in two strategy documents, the *World Zoo and Aquarium Conservation Strategy* (WZACS) (WAZA, 2005) and its predecessor, the *World Zoo Conservation Strategy* (WZCS) (IUDZG/CBSG, 1993). Both the WZCS and the WZACS advocate an integrated approach to conservation both within and outside the zoo.

- The WZACS sees *in situ* conservation as the ultimate goal and many zoos are involved in this, either directly through field work or indirectly through activities such as fundraising.

- Some notable successes have been achieved in releasing zoo-born animals back into the wild (for example, the scimitar-horned oryx into northern Africa).

Questions and topics for discussion

1. How good are zoos at conserving biodiversity?

2. How should zoos prioritize which species they keep?

3. Under what circumstances should zoos be permitted to keep and breed species that are not endangered and are not part of a recognized captive breeding programme?

4. 'If zoos did not exist, they would have to be invented.' To what extent, and why, do you agree with this statement?

5. How helpful is the distinction between *in situ* and *ex situ* conservation work by zoos? Can you suggest a more appropriate terminology?

Further reading

Although published more than 15 years ago, a good book about the role of zoos in conservation is Colin Tudge's *Last Animals at the Zoo* (1992), which also gives a lot of information about captive breeding of endangered species. Another book that has stood the test of time, and which is still widely read and cited today, is Olney *et al.*'s *Creative Conservation: Interactive Management of Wild and Captive Animals* (1994). A recommended book for readers interested in the link between animal behaviour and conservation is Festa-Bianchet and Apollonio's *Animal Behaviour and Wildlife Conservation* (2003). For a discussion of what people working in the field of conservation think should be the role of zoos, Zimmermann *et al.*'s *Zoos in the 21st Century: Catalysts for Conservation?* (2007) is not an easy read, but is very thought-provoking.

Websites and other resources

The *World Zoo and Aquarium Conservation Strategy*, published under the title *Building a Future for Wildlife* (WAZA, 2005), is both accessible and readable. This document can be downloaded from the WAZA website—www.waza.org—and is a good starting point for further reading about how accredited zoos are expected to be involved in the conservation of biodiversity.

Conservation activities often move very quickly and are not always reported in the scientific litera-ture, so a good way of finding out more about the projects in which zoos are currently involved is to look at the websites of the individual zoos them-selves, as well as those of the associations such as BIAZA, EAZA, and AZA.

The IUCN website—www.iucn.org—features a considerable amount of information, including a searchable version of the Red Data List.

Chapter 11 Health

Good **zoos** are concerned with both the physical and the psychological health of the animals in their care. Psychological health has already been considered in the chapter on welfare (**Chapter 7**), so this chapter will focus largely on the challenges of monitoring and maintaining physical health in zoo animals. There is, of course, considerable overlap between health and welfare, and signs of disease in captive animals (such as a high parasite load) are often used as measures of welfare. Not all disease will give rise to poor welfare, however: some minor health problems, such as benign fatty tumours, may not give rise to any pain or suffering.

Nutrition also plays an important role in the maintenance of good health. Nutritional diseases, such as metabolic bone disease and iron storage disease (**haemochromatosis**), are mentioned briefly in this chapter; a more detailed account of the relationship between health and diet, along with other aspects of zoo animal nutrition, is provided in the next chapter (**Chapter 12**).

It would not be possible in a book of this size, let alone in a single chapter, to cover all aspects of zoo animal health in anything like sufficient detail to be of practical use to zoo veterinary staff. Instead, the aim of this chapter is to provide, in language that is accessible to non-vets, an overview of the main animal health care issues in zoos. Some of the key textbooks that are most frequently used by zoo vets, keepers, and zoo vet nurses are listed in **Further reading** at the end of this chapter, with a special section devoted to Fowler's and Miller's *Zoo and Wild Animal Medicine*, now in its sixth edition (2007). Most zoo vets have a battered and well-thumbed copy of Fowler and Miller somewhere on their shelves. Websites for organizations such as Defra also provide much useful information on zoo animal diseases, particularly **zoonoses** and **notifiable diseases**.

The main topics that we will be considering in this chapter are as follows.

11.1 What is 'good health'?
11.2 Guidelines and legislation on health of zoo animals
11.3 The role of zoo staff in zoo animal health care
11.4 Preventive medicine
11.5 Diseases of concern in zoo animals
11.6 The diagnosis and treatment of disease in zoo animals

Short accounts of special topics in zoo animal health are included in boxes throughout this chapter. These include the challenges of dentistry for zoo animals and anaesthesia of giraffes, as well as a brief summary of how vaccines work.

11.1 What is 'good health'?

Assessing good health in a wild animal in captivity is not always straightforward. Signs of disease or even injury may not be outwardly obvious and an accurate picture of an animal's health may only be possible after **invasive procedures** (**Fig. 11.1**) such as blood testing for immune function (**Fig. 11.2**). Failure to reproduce can be due to underlying health problems, but can also occur because of a lack of compatibility between individuals (see **Chapter 9**). Weight change can be a sign of disease (for example,

Figure 11.1 Here, a vet is performing an endoscopy on a cheetah *Acinonyx jubatus* at the Cango Wildlife Ranch in South Africa, to try to determine the cause of a health problem. The diagnosis and treatment of health problems in large wild animals presents a number of challenges and, consequently, zoos place a strong emphasis on preventive medicine. (Photograph: Cango Wildlife Ranch)

knowledgeable members of staff who are familiar with the animals in their care. Keepers, and their managers, should have an understanding of species **life history traits** (such as age at sexual maturity) and should maintain detailed, current records of the animals in their care, to identify any recurring health problems or other issues. All UK zoos must (by law) have access to qualified, experienced veterinary support for the prompt treatment of sick or injured animals, together with facilities for the isolation and treatment of sick animals. Good **preventive medicine**,[1] including hygiene, **pest** control, quarantine, and vaccination (see section **11.4**), is a very important part of zoo animal health care; for wild animals in captivity, it is nearly always easier to prevent disease from occurring in the first place than it is to treat it effectively later on.

Husbandry plays an important role in the maintenance of health, via the feeding of an appropriate diet and provision of appropriate housing, including heating and specialist lighting where necessary (see **Chapter 6**). In its broadest sense, husbandry also includes the management of human–animal interactions, such as visitor interactions with animals, to minimize the risk of injury and stress. Ensuring that these measures are in place and that they work effectively is a shared responsibility between keepers, veterinary staff, zoo managers, and support staff.

in non-domestic felids, see Storms *et al.*, 2003; in swift parrots *Lathamus discolor*, see Gartrell *et al.*, 2003), but can also occur for a variety of other reasons, such as a change in reproductive status or due to seasonal patterns of fat deposition.

Maintaining good health in zoo animals requires, as a minimum, routine monitoring of animals by

11.1.1 How can good health be assessed in zoo animals?

Even a small to medium-sized zoo may hold upwards of a thousand individual animals. The zoo keepers, who deal with the animals on their 'section'

Life history traits are the characteristics of a species, such as age at weaning or fledging, age at first reproduction, number and size of offspring, reproductive lifespan, adult survival, and ageing. In a bird species, for example, clutch size and re-nesting rate after clutch failure would both be examples of life history traits. A knowledge of these key maturational and reproductive characteristics is important for managers of captive breeding programmes.

1 Preventive and **preventative medicine** are the same thing; both spellings are correct. 'Preventive' is the more usual spelling in the USA and in many veterinary textbooks, and appears to be gaining ground as the preferred spelling (in the UK, the BBC now uses 'preventive' rather than 'preventative' medicine).
A **pest** can be defined as an animal that is unwanted, or has characteristics that are unwanted.

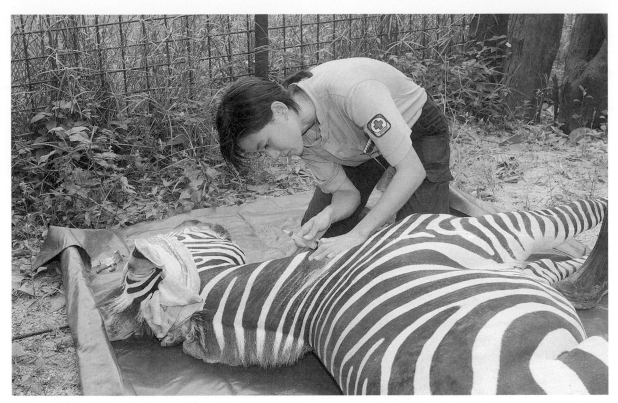

Figure **11.2** Taking a sample of blood for testing from a wild animal in captivity will usually involve sedating or anaesthetizing the animal, as in the case of this zebra *Equus burchellii* at Singapore Zoo. Capture, restraint, and anaesthesia of a large wild animal poses risks both to the animal and to keepers and vets, so most zoos will not take blood samples routinely, but only when an appropriate opportunity presents itself (if, for example, an animal is being anaesthetized for other reasons). (Photograph: William Nai, Singapore Zoo)

on a daily basis, are best placed to report possible health problems. Possible indicators of an underlying health or welfare problem are summarized in **Table 11.1**, and include changes in behaviour (Dawkins, 2003), body condition, gait, weight, or simply the (admittedly subjective) feeling that something is 'not quite right' (**Fig. 11.3**). For further information about indicators of poor health that can be assessed during visual inspection of an animal, section 4 of the *Zoos Forum Handbook* (Defra, 2007b) provides a useful table listing a number of parameters and how to interpret these.

A problem faced by vets and keepers alike is that there are usually no hard and fast definitions of what is 'normal' for any given species; the understanding of what is 'normal' comes with experience, and applies as much to the behaviour of the animal

Figure **11.3** Health problems in wild animals (both in zoos and in their natural habitat) are not always obvious. Spotting a possible health problem requires keepers to have a good knowledge of the animals in their care and sufficient experience to be able to judge when something is not quite right. An animal that has a hunched posture or is isolated from others in a social group may have an underlying health problem, as suggested by the behaviour of this dhole *Cuon alpinus*. (Photograph: Zoo Dresden)

External sign	NOTES
Loss of appetite or refusal to drink	Reduced food or water intake by an individual animal can be difficult to monitor in group-housed species
Changes in behaviour such as coughing, fits, colic, staggering, isolation from group	To notice a change in behaviour, keepers must be familiar with the 'normal' behaviour of an animal
Changes in defecation or urination	Diarrhoea, cloudy urine, etc. can indicate a health problem (but might also result from a sudden change in diet)
Posture and gait	A sick animal may be hunched or unwilling to stand, or may show a change in gait
Appearance of skin, coat, and whole animal	Red, swollen areas of skin indicate inflammation; coats should be glossy not dull; areas of hair loss can indicate health problems such as mange or ringworm
Vomiting	Not all animals are capable of vomiting (e.g. equids cannot vomit)
Signs of pain	See **Chapter 7** for further information about the assessment of pain
Colour and appearance of mucous membranes	Mucous membranes (e.g. inside the mouth) should be pink; pallor can indicate anaemia and a bluish colour can indicate low levels of oxygen in the blood
Temperature, pulse, and respiration (TPR)	Rapid breathing or pulse and/or an elevated temperature (fever) can indicate ill health

Table **11.1** External indicators of possible health problems

as to its body weight and physiological or health parameters. In zoos, it is usually not practical to monitor the temperature, pulse, and respiration (TPR) of animals on a regular basis. Zoos can, however, make use of other methods of monitoring health in their animals, such as **body condition scoring** (see **Box 11.1**) and **gait scoring**. Widely used by livestock farmers (see, for example, Russel, 1984),

Box **11.1** Body condition scoring

Body condition scoring is a method of measuring the general body condition of an animal, as an indication of its nutritional status and energetic state (Millar and Hickling, 1990). Condition scoring has been used in livestock husbandry for some time, but can also be useful for the assessment of **fitness** in zoo animals. Body condition scoring can be determined by visual assessment, or from handling an animal. Russel (1984) describes a commonly used method of condition scoring in livestock, in which the stockperson feels along the spine of an animal and assesses the level of body fat covering the spine. The stockperson (or zookeeper) then records a score for the animal's condition. A score of one, for example, describes an animal with a prominent spine and very little body fat; a score of five indicates a good layer of fat over the vertebrae, under which the joints on the spine can hardly be distinguished.

Body condition scoring is subjective, but can be a useful measure if scores are recorded regularly by the same keeper or stockperson, for the same animal, over a period of time. The Defra website (www.defra.gov.uk) provides detailed (and well-illustrated) information about condition scoring in cattle; these methods can easily be adapted for other bovids and the general principle of condition scoring can be readily applied to a range of other **taxa**.

condition scoring can also be used for a variety of animals in zoos; gait scoring involves looking at the rate, direction, and range of movement of an animal, as well as at the strength or force of its gait. Indicators that something may be awry could include a low-stepping gait, for example, or an animal that is stumbling, or wobbly on its feet.

11.1.2 Why can good health sometimes be compromised in zoo animals?

Animals in a zoo are protected from some of the health risks faced by wild animals, by measures such as vaccination and the provision of an adequate diet. But all animals die eventually and ill health is, at some stage, an inevitable part of zoo animal life. Disease may be spread by contact with **conspecifics**, or with **free-ranging** species (exotic or native), or by pests such as rats and mice, or by keepers or visitors to the zoo. Animals may also be injured in fights with conspecifics, particularly after introduction into a new social group, or during mating. Just as for animals in the wild, complications can occur in female mammals and birds during parturition or egg-laying. Sometimes, trauma (that is, an injury or wound) can occur as a result of poor enclosure design (for example, an animal that falls into a moat) or during capture and transport. Dietary problems can lead to poor health or even poisoning if inappropriate food items, such as toxic plants, are consumed.

There is very little in the peer-reviewed literature on health problems in zoo animals arising from poor enclosure design. We also know very little about the type and extent of harm caused to zoo animals by zoo visitors,[2] either intentionally or unintentionally; there is a need for data from systematic studies in this area of zoo animal health. But there are many anecdotal reports of health problems in zoo animals when litter or human foodstuffs have been ingested, either by accident or as a result of deliberate feeding (for example, soft drinks cans or beer bottles given to chimpanzees; penguins ingesting lollipop sticks or coins).

A particular problem for zoos that are involved in the breeding of rare or endangered species is that **genetic diversity** can be low in captive populations, leading to an increased risk of **inbreeding** and genetic abnormalities (this can also be true of some populations of animals in the wild). (Genetic diseases in zoo animals are discussed in section **11.5.2**.)

11.2 Guidelines and legislation on the health of zoo animals

As has been pointed out already in this book, animals living in the wild are not free from pain, disease, stress, or hunger. Zoos cannot guarantee their animals a life free from disease or pain, nor can legislation ensure this, but zoos can, and should, monitor their animals effectively and ensure that any **clinical** or **pathological** problems are treated promptly and appropriately.

11.2.1 UK legislation on zoo animal health and veterinary care

UK legislation governing the operation of zoos has already been considered in some detail in **Chapter 3**, so this section provides only a brief summary of legislation specific to zoo animal health. The Zoo Licensing Act 1981 does not have a great deal to say about animal health, other than to require zoos (as part of their conservation measures) to keep records '*of the health of the animals*'. The Act also stipulates that zoo inspections must cover the '*health, welfare*

---

2 Heini Hediger, in his book *Wild Animals in Captivity* (1950), records a number of unpleasant instances of zoo visitors behaving badly, including putting pins into an apple before feeding it to an elephant. This shows, sadly, that there is nothing new about a small minority of zoo visitors causing harm to the animals.

The root *patho–* means 'disease'—hence **pathology** is the study of disease and a **pathogen** is an agent that can cause disease. Pathogens can be bacteria, viruses, fungi, or other organisms, such as protozoan parasites. The prions that cause BSE and other spongiform encephalopathies are not micro-organisms; these are pathogenic proteins.

3.1	The condition, health and behaviour of all animals should be checked at least twice daily by the person or persons in direct charge of their care.
3.2	Any animals which give cause for concern must be thoroughly assessed as to whether they are unduly distressed, sick or injured. Where necessary they must receive immediate attention and treatment.
3.3	A daily record must be kept by the person or persons in direct charge of the animals, indicating changes to the prescribed diet, health checks carried out, any unusual behaviour or activity or other problems, and remedial actions taken.

Table **11.2** Minimum requirements for monitoring animals in UK zoos*

* Secretary of State's Standards of Modern Zoo Practice (Defra, 2004)

NOTE: The Secretary of State's Standards of Modern Zoo Practice (SSSMZP) sets out minimum requirements for the routine observation of animals in zoos in the UK, in relation to the health and well-being of the animals.

and safety of the public and animals', and that an experienced vet with **exotic animal** experience must be appointed for special inspections (if these relate to animal health).

For zoos within the UK, section 3 of the Secretary of State's Standards of Modern Zoo Practice (SSSMZP) (Defra, 2004; see section **3.4.2**), entitled 'Provision of animal health care', sets out a number of recommendations relating to health care and veterinary facilities in zoos. These recommendations cover areas such as: routine observation of animals (see **Table 11.2**); enclosure size and design; veterinary care; isolation and containment of sick animals; sanitation and control of disease; and specialist techniques such as **pinioning** the wings of birds (see **Chapter 6**). Other sections of the SSSMZP that have implications for zoo animal health care are section 5, on the provision of protection from fear and distress, and section 6, on the transportation and movement of live animals.

The *Zoos Forum Handbook* (Defra, 2007b) provides interpretation of the SSSMZP recommendations for veterinary services in zoos and considers the options for '*achieving a far better than minimum level of veterinary care in a licensed zoo*'.

In addition to requiring daily checks by keepers of all animals in their care, Appendix V of the SSSZMP sets out recommendations for the frequency of non-emergency visits to zoos by a qualified veterinarian. For large zoos, a routine visit by a vet should occur weekly; for medium zoos, visits should be every 2 weeks; for large bird parks, a monthly visit is recommended. For smaller bird parks and also for large aquariums, the SSSMZP recommend a routine veterinary inspection every 2 months; for small zoos, specialist reptile exhibits, and medium aquariums, visits should be quarterly.

All zoos within the UK, regardless of size, are expected to maintain adequate on-site facilities for the examination and treatment of animals, including a dedicated veterinary treatment room (minimum requirements for veterinary facilities are set out in Appendix V of the SSSMZP; see **Fig. 11.4**). For UK aquariums and zoos that hold aquatic animals, the SSSMZP also set out a requirement for the provision of on-site holding tanks, for the isolation of sick animals.

Zoo animal health: other UK legislation

Zoos must comply with specific animal health legislation, which, for zoos in the UK, can be at national or at EU level. For example, there is specific EC legislation[3] on the monitoring of animals for **transmissible spongiform encephalopathies (TSEs)**, such as **bovine spongiform encephalopathy (BSE)** and **chronic wasting disease (CWD)** (see section **11.5.4**). This requires, among other things, that zoos must test for TSEs in 'fallen stock' (that is, animals that have died) over 2 years of age. In practice, many zoos err on the side of caution, and will send **post-mortem** material from zoo bovids and felids for testing for TSEs even when they are not required to do so by law.

---

3 EC Regulation No. 999/2001 lays down rules for the prevention, control and eradication of certain transmissible spongiform encephalopathies.

(a)

(b)

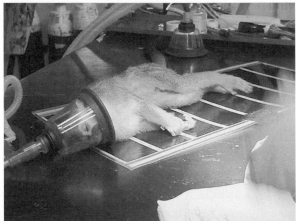

Figure **11.4** (a) Anaesthetic equipment in a zoo veterinary hospital; (b) anaesthetic equipment in use at another zoo, on a squirrel monkey *Saimiri sciureus*. There is a legal requirement for zoos within the UK to maintain on-site health care facilities for their animals, including a veterinary treatment room. (Photographs: (a) Paignton Zoo Environmental Park; (b) Mel Gage, Bristol Zoo Gardens)

Zoos are also legally required (as are farmers and other owners of animals) to report any cases of notifiable diseases in their animals (see section **11.5.3**), under the terms of the Animal Health Act 2002 (this extended the powers of the Animal Health Act 1981). The Animals (Scientific Procedures) Act 1986 may apply in some instances to zoos in which veterinary or other animal health-related research is being carried out (see **Box 3.4**). Another Act that has implications for zoo animal health care provision in the UK is the Animal Welfare Act 2006 (see **Chapters 3** and **7**). Appendix I of section 6 of the *Zoos Forum Handbook* (Defra, 2007b) provides a useful list of the main legislation governing veterinary work relating to zoos in the UK.

11.2.2 Legislation outside the UK on zoo animal health

Zoo animal health care: European legislation

As we saw in **Chapter 3**, all countries that are members of the European Union are bound by the 1999 EC Zoos Directive.[4] This requires zoos to have '*a developed programme of preventive and curative veterinary care and nutrition*'. EC Regulations on the protection of animals during transport also apply to zoo animals; Council Regulation (EC) 1/2005 states that:

> No animal shall be transported unless it is fit for the intended journey, and all animals shall be transported in conditions guaranteed not to cause them injury or unnecessary suffering.

The line here between health and welfare is blurred, but, in practice, the transportation of zoo animals requires considerable veterinary involvement, particularly in relation to carrying out health checks prior to any movement of animals from one zoo to another (**Fig. 11.5**).

4 EC Directive 1999/22/EC relating to the keeping of wild animals in zoos.

Figure **11.5** Moving zoo animals (e.g. as part of a captive breeding programme) is a complex business, and zoos must comply with a wide range of different legislation governing animal transport, animal health, and animal welfare. EC Regulations on the transport of animals, for example, require veterinary screening of the animal prior to transport. Here, an okapi *Okapia johnstoni* is being loaded into a crate prior to being moved. (Photograph: Mel Gage, Bristol Zoo)

The Balai Directive

The 1992 **Balai Directive**[5] has been discussed already in this book (see section **3.3.2**). To recap briefly here, this Directive governs the transport of non-domestic animals between European countries, and also between EU and third-party countries, in relation to veterinary screening and animal health. 'Balai-approved' zoos are required to meet certain requirements, such as the production of a health surveillance plan (which must be reviewed annually) and confirmation that the zoo has enrolled the services of an **approved vet (AV)**. As we saw in **Chapter 3**, obtaining Balai approval entails a considerable amount of work for veterinary staff. Once approval is in place, the Balai Directive is intended to allow easier movement of animals between the zoo and other approved institutions (but see Dollinger, 2007).

Annex A to the Balai Directive lists a number of infectious diseases, but these are largely diseases that do not normally occur in the UK (for example, swine fever, rabies, and anthrax). It is also worth noting that the diseases listed in Annex A of the Balai Directive are diseases of mammals and birds; the Directive does not (at present) cover the diseases of other taxa (with the exception of bees).

Zoo animal health care: legislation outside the European Union

Outside Europe, the picture is mixed. In the USA, zoo animal health falls largely under the banner of the Animal Welfare Act of 1966 (AWA). The AWA covers aspects of veterinary treatment of zoo animals, as well as animal purchase, transportation, housing, handling, and husbandry (Vehrs, 1996). As Spelman (1999) notes, the AWA includes a requirement that licensed animal facilities (including zoos) maintain a pest control programme. As noted in **Chapter 3**, the AWA regulations apply only to selected mammalian species and not to other taxa; farm animals, rats and mice, and most common companion animals are excluded from the provisions of the Act (Vehrs, 1996).

In Australia, the six state and two territory governments have considerable responsibility for legislation relating to animal health and animal

---

5 Council Directive 92/65/EEC of 13 July 1992 laying down animal health requirements governing trade in and imports into the Community of animals, semen, ova and embryos not subject to animal health requirements laid down in specific Community rules referred to in Annex A (I) to Directive 90/425/EEC.

welfare. Animal Health Australia (AHA) is a not-for-profit public company established by the national, state, and territory governments to coordinate animal health-related activities, such as disease surveillance; the AHA website contains much useful information on animal health in general, including the health of zoo animals.

11.2.3 Guidelines and other sources of information about zoo animal health

While section 6 of the *Zoos Forum Handbook* (Defra, 2007b) is entitled 'Veterinary services', much of this information is intended to inform zoo managers about recommended standards of animal health care in zoos, not to tell zoo vets how to do their jobs. Zoo organizations such as the European Association of Zoos and Aquaria (EAZA), the American Association of Zoos and Aquariums (AZA), and the Australasian Regional Association of Zoological Parks and Aquaria (ARAZPA) also set their own standards for veterinary care for member collections.

Husbandry guidelines produced by **European Endangered species Programmes (EEPs)** and (in the USA) **Species Survival Plans (SSPs)** usually include a section on health care. Zoo vets also have access to the conferences and conference proceedings of regional organizations such as the European Association of Zoo and Wildlife Veterinarians (EAZWV). The EAZWV was established in 1996 (`www.eazwv.org`); there is also the American Association of Zoo Veterinarians (AAZV), the website of which (`www.aazv.org`) contains much useful information. These two organizations are the joint publishers of an important journal for zoo vets: the *Journal of Zoo and Wildlife Medicine*. Within the UK, the British Veterinary Zoological Society (BVZS) was founded in 1961 and is part of the British Veterinary Association (BVA).

All of these organizations publish information regularly, in the form of guidelines, newsletters, journals, and proceedings from meetings. There is also a great deal of useful information about animal diseases on the Defra website, at `www.defra.gov.uk/animalh/index.htm`. This includes information about zoonoses and notifiable diseases, which are considered in more detail in section **11.5.3**.

Other guidelines on health care for exotic animals are available from a number of sources, including searchable online resources, such as the veterinary pages of the Intute[6] Health and Life Sciences website (this service was formerly known as Biome and the veterinary website was known as VetGate), at `www.intute.ac.uk/healthandlifesciences/veterinary`. Another very useful online resource is provided by the Wildlife Information Network (WIN):

> a veterinary science-based charity whose mission is to provide information on the health and management of free-ranging and captive animals and emerging infectious diseases to wildlife professionals and decision-makers worldwide.
>
> (WIN, 2008)

Its searchable online encyclopaedia is called Wildpro® and contains many useful links to information about animal health.

11.3 The role of zoo staff in zoo animal health care

Animal health care responsibilities in zoos are shared between zoo veterinary staff, keepers, and other staff, such as zoo animal nutritionists (**Fig. 11.6**). The role of a zoo vet can be very broad and demands knowledge of a wide range of topics, including husbandry and nutrition, legislation, computerized record keeping, statistics, and conservation genetics. The same is true for most zookeepers,

---

6 Intute is a facility operated by a number of UK universities and partnership organizations.

Figure **11.6** Animal health care responsibilities in a zoo are shared between keepers, veterinary staff, and support staff. Here, a team of keepers and veterinary staff at Werribee Open Range Zoo, near Melbourne in Australia, are working together to examine and treat a sedated Przewalski's horse *Equus ferus przewalskii*. (Photograph: Chris Stevens, Werribee Open Range Zoo)

whose responsibilities for zoo animal health care may overlap, to a large extent, with those of the zoo veterinary team, particularly in areas such as preventive medicine and restraint, and the handling of sick or injured animals. Within a zoo, veterinary staff and keepers will be responsible for most, if not all, of the following areas of animal health care provision, depending on the size of the zoo and the range of its activities:

- animal identification, including sexing animals;
- preventive medicine, including **biosecurity**;
- the diagnosis of ill health;
- diet and nutrition;
- the treatment of sick or injured animals, including restraint and anaesthesia;
- post-mortem examinations and carcass disposal;
- reproductive management—that is, support for population management and **captive breeding programmes**;
- the training of other veterinarians, keepers, students, etc.;
- health checks (and paperwork) in relation to the transportation of animals;

Biosecurity simply means taking steps to reduce the incidence and spread of disease.

- the recapture of escaped animals;
- support for conservation initiatives and fieldwork (including *in situ* work);
- human health and safety (for example, in relation to zoonotic diseases, or the use of firearms, or dangerous drugs; general issues relating to the health and safety of zoo staff, and of visitors, are considered in more detail in **Chapter 6**).

These topics are explored in more detail in the following sections of this chapter; see also section 6 of the *Zoos Forum Handbook* (Defra, 2007b), entitled 'Veterinary services', which sets out the main areas of involvement of zoo veterinary staff. Of this list, the diagnosis of health problems in zoo animals presents perhaps the greatest challenge and is the main skill required of a good zoo vet.

There are three main areas in which zoo animal health care can differ from that of farm or companion animals (although there is also considerable overlap in the skills required). Firstly, preventive medicine is of great importance in zoos; treating sick or injured wild animals is particularly difficult and costly, so it is much better to try to prevent them becoming ill in the first place. This is also true, to a certain extent, in livestock practice, but differs somewhat from companion animal practice, in relation to which (other than routine vaccinations, worming, and treatment for fleas), vets may only see an animal once their owner has realized that their pet is ill or injured.

Secondly, a zoo vet will rely heavily on keepers for assistance with capture, restraint, and treatment, and may train keepers to perform minor veterinary procedures such as **venipuncture**. Again, this is not usually the case in companion animal medicine (but may be the case for livestock vets and stockpersons).

Finally, zoos may only know for certain what was wrong with an animal once it has died, and the zoo has been able to carry out pathological tests and a post-mortem examination.

11.3.1 Zoo veterinary staff

A zoo vet may well encounter rare diseases and health problems that a farm animal vet or domestic animal vet has never seen, although many of the challenges of maintaining good health in zoo animals are broadly similar to the challenges of providing health care for other animals kept in captivity (for example, farm animals). Zoo vets are also more likely than livestock or companion animal vets to see an animal from 'cradle to grave', and, even beyond this: a zoo vet may well be involved before and after an animal's life (**Fig. 11.7**).

Within the UK, few zoos have full-time veterinary staff on the premises. Smaller zoos and wildlife parks are likely to use the services of an external veterinary practice within which there is some expertise in the treatment of exotic animals. A larger zoo, however, may have an animal health team comprising one or more full or part-time

Figure **11.7** A young crocodilian hatching from its egg at Cango Wildlife Ranch in South Africa. Zoo vets are involved in an animal's life not only from 'cradle to grave', but often from before an animal's birth to after its death. Pre-birth work may include the monitoring of the reproductive status of the mother; after an animal's death, the veterinary team is responsible for carrying out a post-mortem examination. (Photograph: Cango Wildlife Ranch)

Venipuncture is blood sampling, with a needle and syringe, from a vein.

veterinary surgeons, one or more veterinary nurses, and, sometimes, also an in-house **pathologist** and/or nutritionist. Saint Louis Zoo[7] in Missouri, USA, for example, has a veterinary hospital complex that incorporates six laboratories, a radiology suite, and a surgery suite, as well as a quarantine wing. The veterinary staff team at Saint Louis includes a full-time nutritionist, and an endocrinology team that works within Saint Louis Zoo and also offers support to other, smaller zoos.

Zoo vets may also be actively involved in research, in areas such as new drug treatments, new and safer anaesthetic procedures, or the development of new diagnostic tests. The collection of even basic data on normal values is of use, because such data are incomplete or unavailable for many exotic species kept in captivity, and the collation of data and maintenance of an archive (of, for example, post-mortem records) is part of the role of many veterinary staff in zoos. (Of course, this information is of no use unless it is disseminated more widely, so zoo vets also write papers for publication in appropriate journals.)

11.3.2 Veterinary involvement in captive breeding programmes

Zoo vets may be involved in captive breeding initiatives at both national and international levels, usually via organizations such as the British and Irish Association of Zoos and Aquariums (BIAZA), EAZA, AZA, and the World Assocation of Zoos and Aquariums (WAZA), etc. (see **Chapter 3**). Vets may provide advice to taxon working groups (TWGs) and **taxon advisory groups (TAGs)**, and to **studbook** managers or those involved in breeding programmes such as the European Endangered species Programmes (EEPs) or, in the USA, Species Survival Programmes (SSPs).

A good example of how veterinary support can contribute to the success of a captive breeding programme is provided by Seebeck and Booth (1996), who describe the extent of veterinary involvement in the recovery programme for the endangered Eastern barred bandicoot *Perameles gunnii* in Australia.

11.3.3 Veterinary inspection processes for zoo animal transportation

Vets are involved not only with the health care of animals living in a zoo, but also with the arrangements for the transport of animals between zoos (for example, as part of a breeding programme). In the UK, for example, an animal that is being exported overseas must have an export licence granted by Defra and this licence must be signed by the local Defra Veterinary Inspector on the day of departure of the animal. Even if an animal is from, and is destined for, a Balai-approved zoo, veterinary approval and signature is still required (see section **11.2.2** and **Chapter 3**). Other documents that may be required for transportation and which need veterinary input include any medical notes.

Zoo vets may also be involved in the capture of animals for transport, as, of course, are keepers (**Fig. 11.8**).

11.3.4 Veterinary support for wildlife management outside the zoo

Increasingly, zoo vets (and other zoo staff) are involved in *in situ* wildlife management and conservation projects.[8] A US zoo pathologist, for

---

The work of a **pathologist** involves carrying out diagnostic **necropsies** to determine the cause of disease. Diagnostic tests can be performed on animal tissues, blood, faecal matter, urine, or saliva.

7 Saint Louis is pronounced 'Saint Lewis' not 'Saint Lou-ee'. Widely regarded as one of the leading zoos in the USA, Saint Louis Zoo holds more than 11,000 animals. When the zoo was first created, Missouri state legislation decreed that

'the zoo shall be forever free'—and, to this day, there is no entrance fee for visitors.

8 The Wildlife Conservation Society (WCS), based in New York, provides further examples of the role that vets (not only from zoos) can play in wildlife management and conservation, as well as much useful information for field vets (see the 'Resources and Technical Pages' of the 'Wildlife Health' section of the WCS website, at www.wcs.org).

Figure **11.8** As in **Fig. 11.6**, this photograph shows the team work necessary for successful animal health care in zoos. Here, keepers and veterinary staff at Colchester Zoo are working together to move an anaesthetized tiger (the tail of the animal is just visible on the far right) from its enclosure to the veterinary hospital for treatment. (Photograph: Colchester Zoo)

example, was among the team of investigators who first discovered the link between the rapid decline in numbers of vultures in Asia and the widespread use of the non-steroidal anti-inflammatory drug (NSAID) diclofenac in domestic livestock (Oaks *et al.*, 2004). Keepers and vets from a number of zoos worldwide have since been involved in attempts to reduce the threat posed by diclofenac to critically endangered Asian vulture species, by providing both facilities and captive-bred birds of closely related species for testing of an alternative and less damaging NSAID, meloxicam (Swan *et al.*, 2006).

Vets from the Zoological Society of London (ZSL) are involved in carrying out a variety of health tests on captive-bred animals prior to their release back into the wild, as part of the species recovery programmes (SRPs) operated in the UK by Natural England (formerly English Nature). Common dormice *Muscardinus avellanarius*, for example, are routinely tested for tuberculosis (TB) prior to release (**Fig. 11.9**); captive-bred field crickets *Gryllus campestris* are tested for **protozoan** parasites (Pearce-Kelly *et al.*, 1998); and wartbiter crickets *Decticus verrucivorous* are screened for the presence of a pathogenic fungus (Cunningham *et al.*, 1997).

Protozoan parasites are unicellular organisms that include free-living, as well as parasitic, forms.

Figure **11.9** Common dormice *Muscardinus avellanarius* are bred in captivity at several zoos within the UK, as part of wider conservation programmes for their reintroduction back into selected sites in the wild. Before release into the wild, these animals are screened for diseases such as tuberculosis and for the presence of gut parasites. This photograph shows an anaesthetized dormouse being weighed as part of routine health monitoring. (Photograph: Paignton Zoo Environmental Park)

11.4 Preventive medicine

Preventive medicine and care play a very important role in zoos, because successful diagnosis and treatment of disease in wild animals can be particularly challenging. All zoos, from the smallest to the largest, will be involved at some stage in one or more

areas of preventive medicine, ranging from basic health checks to quarantine procedures, vaccination, foot, beak, and dental care, and post-mortem examinations. Much zoo husbandry is also part of preventive medicine: provision of an appropriate diet, hygiene, pest and parasite control, and enclosure design and management are every bit as important in preventing disease as vaccination or regular veterinary checks of animals.

All of these measures can be grouped under the general heading of biosecurity. This can be as simple as requiring keepers to change into clean uniforms and footwear when they arrive on site, to reduce the risk of transmitting infectious diseases into the zoo. (The Defra website features further information about biosecurity measures; although aimed at livestock farmers, much of this information is also relevant for zoos.)

11.4.1 Health examinations

Requirements for the routine monitoring of the health of zoo animals have been discussed already in this chapter (see **Table 11.2**), but one area of zoo animal health screening that has not yet been considered is dental health. **Box 11.2** provides a brief account of some aspects of dentistry for zoo animals.

Box 11.2 Dentistry for zoo animals

Tooth care poses particular problems in zoo animals, not only because of the physical demands and potential dangers of working with large wild animals, but also because of the very wide variety of tooth sizes and structures that a wild animal dental surgeon may be required to treat, from chimpanzee molars, to elephant or walrus tusks (Glatt *et al.*, 2008). Wiggs and Lobprise (1997), for example, suggest that 3–6-inch drill bits may be needed for canal work on

the canines of large **carnivores** such as tigers; small sledgehammers and industrial power drills may be necessary for **pachyderm** dentistry. Cheek tooth (premolar and molar) removal in large animals may have to be by means of a buccotomy, in which an incision is made in the face to access a diseased tooth from the outside, rather than via intra-oral extraction.

Diagnosis of dental problems in zoo animals is not easy, unless animals have been trained to open their mouths for inspection. Signs of possible tooth problems include weight loss, a refusal to drink cold water, and the animal dropping food from its mouth or selecting only soft items of food. The mouth cavity may also bleed or smell offensive. Because routine oral inspection of wild animals in captivity is not easy, a full dental check-up should be carried out whenever a zoo animal is anaesthetized for any reason (Wiggs and Lobprise, 1997; **Fig. 11.10**).

Braswell (1991) provides a short summary of some of the main problems in exotic animal dentistry, such as tusk and molar problems in elephants, and an oral disease known as 'lumpy jaw', which is relatively common in captive macropods such as kangaroos and wallabies (see also Canfield and Cunningham, 1993). Properly referred to as necrobacillosis, 'lumpy jaw' is difficult to treat once established, although initial signs of gum disease in macropods can usually be treated with antibiotics (Lewis *et al.*, 1989).

(a)

Figure **11.10** The tusks of an elephant are extended teeth (incisors). Dental treatment for elephants in captivity is a major undertaking, as shown in these photographs from Colchester Zoo of an operation on the tusk of Jambo, one of the elephants at the zoo. The successfully treated tusk can be seen more clearly in photograph (b). (Photographs: Colchester Zoo)

▶ (b)

Figure **11.10** (continued)

Further reading on zoo animal dentistry

Glatt *et al.* (2008) provide a useful review, in the *International Zoo Yearbook*, of dental problems and treatment in zoo animals, based on a survey of AZA member zoos. The textbook *A Colour Atlas of Veterinary Dentistry and Oral Surgery* (Kertesz, 1993) is one of only a small number that look in any detail at veterinary dentistry for wild animals (Dr Peter Kertesz is a veterinary—and human—dentist who set up ZOODENT International in 1985, to provide a mobile dental service for zoos and wildlife parks). There is also a chapter on oral disease and dentistry in exotic animals in *Veterinary Dentistry: Principles and Practice* (Wiggs and Lobprise, 1997).

11.4.2 Quarantine

Preventive medicine often starts with the careful selection of new animals and a period of **quarantine** or isolation. Quarantine periods vary considerably, depending both on the species and its country of origin, and on whether or not the animal is being imported into a Balai-approved zoo from another Balai-approved institution (see section **3.3.2**). For non-Balai-approved zoos within the UK, for example, equids are usually quarantined for a minimum of 30 days on arrival, whereas primates are usually kept in quarantine for between 60 and 90 days, largely

The word **quarantine** comes from the Latin meaning '40 days'.

because of the time lapse between tuber-culosis (TB) infection and the development of a positive skin test.

All primates, bats, and Carnivores (excluding pet dogs, cats, and ferrets, which come under different rules) coming into the UK must be quarantined for 6 months for rabies, under the provisions of the Rabies Order 1974 (as amended in 2004). This takes precedence over other legislation such as the Balai Directive (although, at the time of writing, the UK rabies legislation is under review and the quarantine period may well change in the future). Birds and reptiles are exempt from the provisions of the Rabies Order, as are other groups of mammals, such as the lagomorphs (rabbits and hares) and rodents, provided that certain conditions are met. (Further details about current UK rabies quarantine requirements are available from the Defra website.)

The aim of quarantine is not simply to prevent the transmission of disease[9] from newly arrived animals to existing animals (or staff) in the zoo. Quarantine provides a period of time during which zoo staff can assess baseline health parameters, and also appetite and behaviour. Animals that have been moved from one location to another are likely to have experienced stress as a result of translocation; this can trigger the presentation of **clinical signs** for any subclinical disease, making any health problems easier to detect. A period of time in quarantine also allows the animal to recover from transportation, and to adapt to its new surroundings and to a new husbandry regime, before any potentially stressful introduction into a new social group or enclosure.

For further information about quarantine for zoo animals, Hinshaw *et al.* (1996) provide a useful summary, although this source is now somewhat dated in some areas (the forthcoming new edition of Kleiman *et al.*'s *Wild Mammals in Captivity* is eagerly awaited). The review of the Balai Directive undertaken by Dollinger (2007) also provides information about quarantine procedures for zoos, in relation to Balai.

11.4.3 Vaccination

Vaccines are available to zoo vets, but there are almost no commercial vaccines (as yet) that have been developed specifically with zoo animals in mind. Because of this, vaccination is less widely used in zoos than might be expected. Where zoo vets do vaccinate their animals, they must 'make do' with vaccines developed originally for farm or companion animals, and must extrapolate dosage information from data for domestic animals. They can make this procedure more accurate by monitoring the **serological response** of an animal after vaccination. **Box 11.3** provides a brief account of how vaccines work and considers the risks to wild animals of using 'live', rather than inactivated, vaccines.

Some vaccines can be given orally, but most animal vaccines are administered via subcutaneous (below the skin) or intramuscular injection (**Fig. 11.11**), which usually requires the capture and restraint of the animal. Before vaccination takes place, zoo vets must weigh up whether the risk of vaccination (including capture and restraint) is greater than the risk of the animal contracting the disease. This will depend on the severity of the disease and the availability of other means of disease control. For any vaccination programme to be safe and effective, vets need to be able to determine whether or not the vaccination has worked (for example, by monitoring the serological response of the animal) and to know how long a period of protection against disease the vaccination will provide.

---

9 The transmission of disease can be 'vertical', e.g. from mother to offspring, typically during gestation, or 'horizontal', e.g. from one animal to another animal in the same, or a nearby, enclosure.
Clinical signs are objective signs of illness such as a rash; symptoms are what humans may report to their doctor (e.g. feeling tired all the time, or pain in the abdomen), and so are subjective. Animals can present clinical signs of disease, but cannot report their symptoms to a vet.
Serologic, or **serological**, means to do with **serum**, a constituent of the blood. A **serologic test** is simply any diagnostic test made using serum.

Figure **11.11** Vaccination is one of a range of measures that zoos can use in preventive medicine, although it is not used widely in zoos because of the possible risks associated with the use of live, modified vaccines. The use of killed or inactivated versions of domestic cat vaccines, however, is now fairly standard for large cats in zoos, such as lions and tigers. This photograph shows the routine vaccination of a lion cub at Ramat Gan Zoo in Israel. (Photograph: Tibor Jäger, Zoological Center, Tel Aviv)

Some of the more widely used vaccines in zoos are those against rabies (but not within the UK) and canine distemper (but note the risks described in **Box 11.3** associated with the use of live distemper vaccines for some species), polio, tetanus, and measles vaccines for the great apes, and tetanus vaccines for equids and tapirs. Where equine encephalitis is prevalent, zoo equids are also often vaccinated against this disease. Hinshaw *et al.* (1996) provide further information about vaccination for zoo mammals.

Box 11.3 How do vaccines work?

Vaccination makes use of a substance called an antigen, derived from the pathogen (infective agent). This stimulates the immune system of the vaccinated animal to respond, so that it is resistant to natural infection by the pathogen. Vaccination, if effective, protects against disease, but not always against infection (in other words, animals may still be infected with the pathogen after vaccination, even if they do not develop the full-blown disease).

There are two types of vaccine: live and killed. Live vaccines contain an attenuated or changed version of the pathogen that is less harmful than the natural version of the pathogen, but still triggers an immune response in the vaccinated animal. The use of live vaccines, however, can lead to problems such as abortion in mammals and may sometimes trigger the disease that the vaccine is intended to prevent.

This has happened in captive grey foxes *Urocyon cinereoargenteus*, among which animals have died of vaccine-induced canine distemper (Halbrooks *et al.*, 1981). Other non-domestic species in which live vaccine-induced distemper has been reported include the African wild dog *Lycaon pictus* (McCormick, 1983), the red panda *Ailurus fulgens* (Bush *et al.*, 1976), and the black-footed ferret *Mustela nigripes* (Pearson, 1977).

Killed vaccines are less risky (and are also more stable in storage), but can also be less effective; the immunity they induce may be short-lived, necessitating booster vaccination at regular intervals. The disadvantages of traditional vaccines have led to the development in recent years of what are referred to as 'second-generation' vaccines, manufactured using biotechnology such as recombinant DNA or protein engineering.

11.4.4 Hygiene (sanitation) and enclosure management

Achieving a high standard of hygiene in zoo enclosures should be based on good practice (such as regular 'poo picking'), with the use of disinfectants only as a last resort. A wide range of disinfectants and commercial cleaning substances is nevertheless used in zoos, but there is surprisingly little information available either in textbooks or in the peer-reviewed literature about their effectiveness in killing pathogens, or about any possible risks associated with disinfectant use in the zoo environment.

Captive felids appear to be susceptible to poisoning from incorrect use of phenolic disinfectants (Hinshaw *et al.*, 1996) and there is anecdotal evidence of toxicity from improper use of bleach (sodium hypochlorite) to clean enclosures. Most zookeepers rely on manufacturer's literature for information about dilution rates and application methods for disinfectants, and such data that are available tend to be based on farm, rather than zoo, usage.

In the UK, veterinary disinfectants for the control of notifiable diseases are placed on an approved list by Defra once they have passed laboratory tests. This listing process operates under the Diseases of Animals (Disinfectant Approvals) Order 1978. The current approved list is available online from the Defra website, together with a separate list of manufacturers and distributors of approved disinfectants.

Some methods of cleaning enclosures can increase, rather than decrease, the likelihood of the spread of disease (**Fig. 11.12**). Power hoses, for

Figure **11.12** It is tempting to think that thorough power-hosing of an enclosure must be beneficial and will result in high standards of hygiene, but, in fact, the opposite can be true and the use of power hoses can spray faecal matter and other waste over a wide area. Good old-fashioned methods of enclosure cleaning (with a broom, or a shovel and wheelbarrow) are often better in terms of reducing the risk of disease. (Photograph: Sheila Pankhurst)

example, can spray a fine mist of water and faecal material over a wide area. The SSSMZP (Defra, 2004) specifically state that 'Health risks posed by the use of power hoses on animal waste must be minimized'. Overwetting concrete floors can also be a contributory factor in the development of foot problems in some species, particularly elephants.

11.4.5 Pest control

A pest species may compete with the zoo animals for resources such as food or nest boxes, or may transmit disease. Animals regarded as pests in zoos include species such as cockroaches, pigeons, seagulls, herons, and rodents such as rats and mice, all of which can all be vectors of disease. In North America, for example, cockroaches can transmit *Sarcocystis* (a protozoan parasite that can cause disease in birds such as **psittacines**) from opossum faeces (Lamberski, 2003). Rodents can transmit diseases such as Lyme disease, via the tick vector *Ixodes scapularis*, to captive deer (Williams *et al.*, 2002) and leptospirosis to primates, and can contaminate foodstuffs with *Salmonella* bacteria. Mice and rats are also intermediate hosts for parasites such as *Toxoplasma gondii*. At Baltimore Zoo, retrospective analysis of **mortality** in black-tailed prairie dogs *Cynomys ludovicianus* showed that 24 per cent of animals examined post-mortem had hepatic capillariasis caused by the parasitic nematode *Calodium hepaticum* (Landolfi *et al.*, 2003). The authors of this study suggest that wild rats in the zoo may have acted as a reservoir for the disease.

Rodents, such as rats, are not only vectors of disease, but may be direct predators of eggs and nestlings of captive birds, as can snakes, racoons, and opossums in North American zoos (Lamberski, 2003). Free-ranging birds can also spread diseases, such as chlamydiosis and salmonellosis, usually via faecal contamination of perches and food bowls. Effective pest control depends on enclosure integrity and design (preventing rats and mice from gaining

access to exhibits, particularly to feeding areas), active pest control measures, and good hygiene in food preparation areas (see Spelman, 1999, for a useful review of vermin control in zoos). The provision of **scatter feeds**, as part of **enrichment** for captive animals, is likely to increase food accessibility to wild rodents. The use of poisoned bait to control rodents poses the risk of potential ingestion by a non-target species, although careful location of bait and the use of bait traps can help to minimize this risk. Lamberski (2003) suggests a range of pest control measures for aviary birds, including placing food and water pans off the ground, the use of fine mesh along the bottom of aviaries to exclude snakes, rats, and other small mammals, and the use of electric wire to deter free-ranging mammals and birds from entering the aviary.

Zoos also face an ethical dilemma in relation to the control of rodent populations: the use of slow-acting anti-coagulant rodenticides is regarded as an inhumane form of pest control (Mason and Littin, 2003). The use of such rodenticides sits uncomfortably with public statements by zoos of concern about animal welfare. Snap-traps and electrocution devices, if used properly, kill rodents instantaneously and humanely, but some zoos now use biological control measures for pests. At the Burger's Zoo in Arnhem in the Netherlands, for example, biological control methods are employed in the 'Bush' rainforest exhibit (Veltman and van der Zanden, 2000). Bauert *et al.* (2007) examine methods of pest control in their review of the first 3 years of operation of the Masoala rainforest exhibit in Zurich Zoo.

11.4.6 Parasite control

The control of parasites is a very important part of preventive medicine in zoos: many parasites, such as the ascarid worms of carnivores, are hard to eradicate once they are established in a captive population and can cause chronic health problems (Hinshaw *et al.*, 1996). Parasites can be grouped into two categories:

Psittacine birds are parrots, macaws, and lories: the members of the avian order Psittaciformes.

- **exoparasites**, which live on the body surface of the host animal;
- **endoparasites**, which live inside the gastrointestinal tract or the body tissues of the host.

The main exoparasites are flies, lice, mites, fleas, and ticks. Not all exoparasites will be visible to the naked eye; some will require microscopic examination for identification.

Of perhaps greater concern to zoo vets are endoparasites, which can cause serious health problems and even fatalities. The majority of endoparasites are parasitic worms, or **helminths**, and protozoans. Some of the helminth genera with which zoo vets commonly have to deal are *Trichuris*, *Syngamus* (gapeworm), and *Ascaridia*. Protozoan parasites of concern to zoo vets include *Toxoplasma gondii* and *Coccidia*.

Trichuris trichiura, for example, is a commonly occurring helminth species often called the whipworm. It is a major zoonosis and around a quarter of the world's human population is thought to be infected with this parasite. Light infections are often asymptomatic, but uncontrolled infections can result in anaemia and stunted growth.

The presence of helminths in the gut of zoo animals can be diagnosed by examining the animal's faeces for eggs shed by the worm or worms. There are numerous different methods both for detecting and quantifying parasite eggs in faecal matter, but no single method that reliably detects the eggs of all types of parasite. **Box 11.4** provides details of a commonly used protocol, the McMaster flotation technique, for assessing gastrointestinal worm burden by measuring faecal egg counts. This should be used alongside other methods of measuring parasite

---

Box 11.4 The McMaster flotation technique for measuring parasite burden

The eggs of parasitic worms in the gut are shed with the faeces of the host animal, and can be counted and identified under the microscope (**Fig. 11.13**). The result gives an estimation of the number of eggs per gram (epg) of faeces. This, in turn, can be used as an estimation of the degree of infestation of an individual animal, or a group of animals. Egg counts give information about both the prevalence of infection (that is, how many animals are infected in a group or population) and the intensity of infection (that is, how severe or heavy the parasite load is in infected animals).

A commonly used method of measuring epg in faeces is to use a special chambered slide called a 'McMaster slide', to facilitate counting eggs suspended in a known quantity of a salt solution (because the eggs float to the surface of the solution, this method is also sometimes called a 'flotation technique'). The following protocol is a brief outline of how to carry out a faecal egg count using a McMaster slide. For further information, see a good general textbook on practical or diagnostic parasitology, or, if possible, MAFF's *Manual of Veterinary Parasitological Techniques* (1986); Garcia (1999) provides a review of the advantages and disadvantages of flotation methods versus other techniques for measuring parasite egg burden.

The term **helminths** is used to refer to a range of genera of multicellular parasitic worms that inhabit the gastrointestinal tract and, sometimes, other parts of the body. Helminths include trematodes (or flukes), cestodes (or tapeworms), and nematodes (or roundworms).

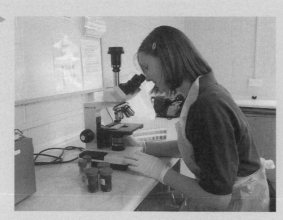

Figure **11.13** Examining a faecal sample under the microscope for the presence of the eggs of gastrointestinal parasites. Laboratory work is an essential part of health care in zoos; samples may be sent out to specialist laboratories, but many larger zoos have their own laboratory facilities. (Photograph: Paignton Zoo Environmental Park)

Disposable latex gloves should be worn for this procedure and all normal hygiene precautions for handling faecal matter observed.

Stage one: the collection of faecal samples

Faecal samples should be collected as soon as possible after emission. Early morning is often a good time to collect faeces. Faecal pellets should be collected into a sterile plastic container, which is then marked with the ID of the animal (if known) and the time, date, and place of collection. Samples can be kept in a fridge for up to 48 hours before analysis. Longer storage is not recommended, because the eggs can hatch; this can lead to a distorted count. Samples should not normally be frozen, because some types of egg may rupture and this will prevent identification.

Stage two: preparation of samples

Place approximately 20g of faecal matter (this is roughly ten large rabbit pellets, or a level tablespoon of other faeces) into a dish.

Break/mash the sample up with a metal fork to mix the faecal pellets. (Parasite eggs are often shed unevenly and one pellet may contain many eggs, while another contains none; mashing several pellets together gives a more accurate overall egg count.)

Place 3g of faecal matter (a level teaspoon) into a plastic tea strainer (aperture of mesh approx. 0.15–0.25mm).

Measure out 45ml of saturated salt solution (370g of NaCl dissolved in 1 litre of hot water) and pour this slowly over the faecal matter, mashing and pushing the sample through the strainer with a small spoon. Collect the filtrate in a dish or bowl.

Stage three: carrying out an egg count using a McMaster slide

Using a Pasteur pipette, fill both chambers of a McMaster slide with the suspension.

Wait for 60 seconds to allow the worm eggs to float to the surface of the chambers.

Using a compound microscope ($\times 10$, $\times 40$, $\times 100$), carry out a count of all visible, intact worm eggs between the marked areas on the two slide chambers.

The number of eggs per gram of faecal matter (epg) is then obtained by multiplying the total number of eggs found in the two squares by 50 (for example, if ten eggs in total are counted, the epg is 500—that is, 500 eggs per gram of faecal matter).

Egg counts can vary from 0 epg to over 5,000 epg in a highly infected animal.

Stage four: egg identification

Using an appropriate guide, identify the eggs observed to genus or family level (where possible).

burden, such as sedimentation or filtration protocols, or a pasture larvae count;[10] the method chosen will depend on what is being looked for.

Not all endoparasites inhabit the gut, however: parasitic worms and flukes can be present in organs such as the heart, lungs, and liver, and detecting their presence in a live animal may require invasive testing, such as taking a tissue sample.

Part of good husbandry in zoos is managing animals in such a way as to minimize the risk of infection with parasites, or to prevent the parasite completing its life cycle (this is a case of 'know thine enemy', or at least knowing something about its life cycle). Helminths can also be treated using **anthelmintic drugs**, which either destroy the worms on contact or paralyse them.

Zoos can administer anthelmintic drugs in a number of ways, such as topically, or in the food stuff given to animals, or in the form of an orally administered drench that can be syringed into the mouth of an animal (**Fig. 11.14**). One of the problems associated with the use of anthelmintic drugs is that parasites can become resistant to treatment after repeated use; a secondary problem is that anthelmintics available to zoo vets have usually been tested and approved for use in livestock rather than wild animals in zoos.

A further problem faced by zoo vets and keepers when administering anthelmintic drugs is that of ensuring that every individual animal in socially (group) housed animals receives an adequate dose of the drug. In a long-term study of *Trichuris trichiura* infection in zoo-housed Abyssinian colobus *Colobus guereza kikuyuensis*, Caine and Melfi (2005) found that individual administration of anthelmintic drugs to the colobus (rather than social drug delivery

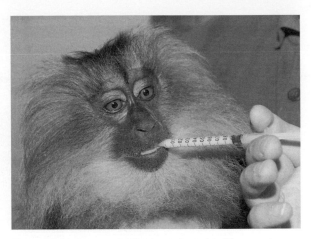

Figure **11.14** Some zoo animals can be trained to accept oral administration of medicine, without the need for anaesthetizing or 'knocking out' the animal. This photograph shows a drug dose being administered to a lion-tailed macaque *Macaca silenus* at Ramat Gan Zoo in Israel. (Photograph: Tibor Jäger, Zoological Center, Tel Aviv)

via communally available food) resulted in a significant decrease in mean egg counts.

11.4.7 Foot care

Foot care is an important part of preventive medicine in zoos, particularly for **ungulates**.

Hoof overgrowth is a common problem in species ranging from zebras and giraffes (**Fig. 11.15**), through to rhinoceroses and elephants, and is often associated with unsuitable enclosure surfaces, genetic factors, and also copper deficiency. The hooves of ungulates are produced from a part of the foot called the corium and grow continuously throughout the animal's life. Normally, hoof growth is worn down as the animal moves around, but sometimes hoof overgrowth can occur. In the even-toed ungulates (order Artiodactyla), hoof overgrowth may lead

10 The infective stage of many helminths is a larval stage that exists outside the host animal on grassland used for grazing by ungulates; these larvae can be detected using pasture larvae counts, in which a specific quantity of grass is cut and rinsed and the rinsing water poured through a series of fine mesh sieves to extract larvae of different sizes.

Anthelmintic drugs (the spelling **antihelmintic** is also correct) are drugs used to treat animals infected with parasitic gut worms, such as roundworms. Commonly used anthelmintic drugs in zoos include ivermectin (in branded drugs such as Oramec®, Ivomec®, and Acarexx®) and fenbendazole (in branded drugs such as Panacur®). Ivermectin is also effective against some ectoparasites, but is not effective against tapeworms or flukes.

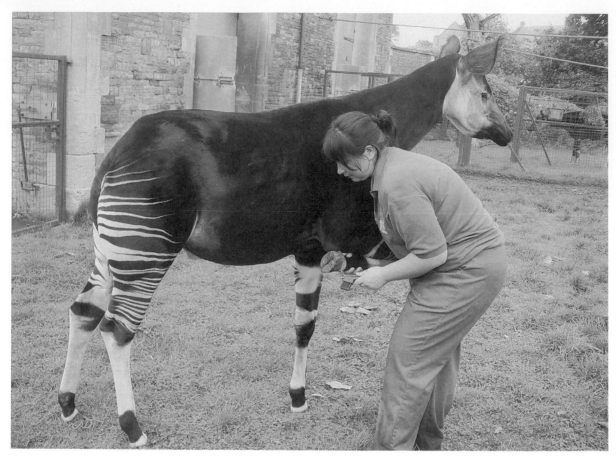

Figure **11.15** Preventive care, such as regular checking and, if necessary, trimming of hooves, is a very important part of the control of foot problems in ungulates in zoos. Here, an okapi *Okapia johnstoni* at Bristol Zoo is having its hooves cleaned. (Photograph: Mel Gage, Bristol Zoo)

to unevenness in weight-bearing on the two claws of each hoof and this can give rise to lameness (as can diseases of the hoof, or dietary deficiencies). Similar problems of hoof overgrowth can occur in the odd-toed ungulates (Order Perissodactyla), such as the equids and rhinoceroses.

In a study of hoof overgrowth in Hartmann's mountain zebra *Equus zebra hartmannae* at Paignton Zoo Environmental Park, UK, Yates and Plowman (2004) found that the zebra were spending less than 8 per cent of their time walking or trotting around the paddock. Of the time spent in locomotion, over 90 per cent was on grass. The hooves of the zebra were not being worn down by sustained contact

with an abrasive surface. A change in diet and enclosure layout, to encourage greater locomotion, resulted both in the animals losing weight and also moving around more, causing their hooves to wear down more normally.

In elephants, Csuti *et al.* (2001) have suggested that foot problems are the most common health care problem in captive animals and are seen in 50 per cent of elephants at some stage in their life. Although African *Loxodonta africana* and Asian *Elephas maximus* elephants differ in their number of toenails,[11] both species have a broadly similar foot structure, with a horny pad covering the sole of the foot (**Fig. 11.16**). Both species of elephant seem

---

11 African elephants have four toenails on their front feet and three on their back feet, whereas Asian elephants (normally) have five nails on the front and four on the back feet (Csuti *et al.*, 2001).

Figure **11.16** Captive elephants are particularly prone to foot problems such as cracks in the sole of the foot or in the toenails. This can allow the foot to become infected, so prevention is much better than cure. This photograph, from the archives of the Smithsonian's National Zoo in Washington DC, shows a keeper filing the toenails of an elephant. (Photograph: Smithsonian's National Zoo)

to suffer from similar foot problems in captivity, although Schmidtt (2003) suggests that these problems are more prevalent in Asian than they are in African elephants. The most common problems are cracks in the sole of the foot or in the toenails or cuticles, foot overgrowth, injuries, and abscesses. Cracks and injuries to the feet can allow the entry of pathogens, particularly if elephants cannot avoid standing in their own faeces. Usually, these problems are treatable, but they can result in disability or even, in severe cases, death (Schmitt, 2003).

Elephants that are housed primarily on hard substrates such as concrete, with little opportunity

for exercise, are most prone to foot problems. In their comprehensive book, *The Elephant's Foot*, Csuti *et al.* (2001) recommend that zoo elephants should be given plenty of opportunity for exercise and should have enclosures with a variety of substrates (sand, mud, grass, etc.) to minimize the development of foot problems.

11.4.8 Post-mortem examinations

Finding out why an animal has died is part of preventive medicine. The cause of death might be an infectious disease, for example, and this obviously has implications for the health of other animals in the collection. Long-term post-mortem records provide useful data on trends in health, both at individual zoos and among the wider zoo community.

Hope and Deem (2006), for example, analysed the veterinary records of 172 jaguars *Panthera onca* at thirty AZA member zoos between 1982 and 2002, including post-mortem records. They were able to determine common causes of **morbidity** and mortality, and found, for example, that musculoskeletal diseases were the most common cause of mortality in captive male jaguars. They were able to use the data from their study to make recommendations for the management of jaguars in captivity.

In the UK, section 3 of the SSSMZP (Defra, 2004) sets out specific recommendations for post-mortem examinations of zoo animals; these recommendations are shown in **Table 11.3**.

3.15	Adequate facilities must be available either at the zoo or within a reasonable distance for the post-mortem examination of all species held at the zoo
3.16	Dead animals must be handled in a way which minimises the risk of transmission of infection
3.17	Animals that die at the zoo should be examined post-mortem in accordance with veterinary advice. Where appropriate, samples for diagnosis or health monitoring should be taken for laboratory examination

Table **11.3** Post-mortem examinations*

* Secretary of State's Standards of Modern Zoo Practice (Defra, 2004)

The Secretary of State's Standards of Modern Zoo Practice (SSSMZP) set out the following recommendations for zoos in the UK, in relation to post-mortem examinations.

11.4.9 Diet

If zoos employ a nutritionist, she or he invariably works as part of the veterinary team; ensuring that animals have a proper diet plays a major role in preventive medicine (see **Chapter 12**). The key aspects of zoo animal diet in relation to health are providing the right **nutrients**, in the right quantities, with the right sort of food presentation.

11.4.10 Exhibit design

Zoo vets, as well as keepers, may be asked to play a role in the development of new enclosures, contributing advice on safety and welfare issues in relation to enclosure design. Poorly designed enclosures can lead to injury or contribute to the spread of **disease** (if, for example, drainage is inadequate). The SSSMZP make a number of health-related recommendations regarding enclosure design, including, for example, a requirement that '*water-filled and dry moats used for the confinement of animals must provide a means of escape back to the enclosure for animals falling into them*' (Defra, 2004).

Decisions on whether or not animals should be housed in **multi-species exhibits** may also require veterinary input, particularly if there may be an increased risk of injury or the spread of disease.

11.5 Diseases of concern in zoo animals

Each zoo will have different 'diseases of concern', depending on its geographical location and the types of animal in its collection. These diseases of concern may vary quite widely from collection to collection, and over time. Foot-and-mouth disease (FMD), for example, was of great concern to zoos in the UK during the outbreak of 2001, but, at the time of writing, blue tongue (a disease that affects **ruminants**) is of particular concern.

Health risk assessment provides the basis upon which zoos decide which diseases are important and is a vital part of health care planning for collections. When drawing up a list of diseases of concern, zoo vets must consider both the probable morbidity and mortality from a given disease, and the likelihood of an outbreak; the latter will be based, in large part, on the species held in the collection and previous patterns of disease.

For all zoos, however, the diseases that zoo vets diagnose and treat can be considered under four broad headings. The first three categories, dealt with here, are:

1. infectious diseases;
2. degenerative diseases;
3. genetic diseases.

The fourth category is nutritional diseases; these are mentioned only briefly here and are considered in more detail in the next chapter (**Chapter 12**).

11.5.1 Infectious diseases

An infectious disease is a disease caused by pathogenic microorganisms such as bacteria, viruses, fungi, or parasites, and which can (usually) be transmitted from one animal to another. The microorganisms that cause disease, along with proteins such as **prions**, are referred to as 'pathogens' or 'infective agents'. The clinical signs of the disease[12] are caused either by the pathogen itself, or by its products (for example, the toxins produced by bacteria such as *Escherichia coli* strain O157:H7), or by the body's response to either of these.

A broad definition of **disease** is '*any impairment of the normal physiological function affecting an organism, producing characteristic symptoms*' (Collins Concise English Dictionary).
A **prion** is a disease-causing agent or pathogen that is not a microorganism, but a protein. Prions do not contain any genetic material (DNA or RNA).

12 A 'clinical disease' has recognizable signs, such as fever, or a rash, or weight loss, whereas a 'subclinical disease' has no detectable clinical signs. Many diseases can be subclinical for a period of time before the signs become obvious; examples are diabetes and arthritis.

Infectious diseases are sometimes also called **communicable diseases**, because they can usually (but not always) be 'communicated', or passed on, between animals or humans. Examples include tuberculosis, rabies, and FMD. Infectious diseases can spread via food and water, or aerosol transmission, **fomite** transmission, or by direct physical contact.

Bacterial diseases

Among the main bacterial diseases that can be diseases of concern in zoos are salmonellosis, shigellosis, anthrax, tuberculosis (TB), yersiniosis, clostridial diseases, and campylobacteriosis. The names of these diseases usually reflect the name of the infective agent: salmonellosis, for example, is the name of the disease caused by infection by any of the species of *Salmonella* bacteria; shigellosis is caused by *Shigella*.

All of the diseases mentioned above are also zoonotic; that is, they can be transmitted to humans. **Table 11.4** provides a list of some of the diseases of concern in zoo animals.

Figure **11.17** Reptiles such as this snake can be asymptomatic carriers of *Salmonella*, which is a zoonosis and can cause serious illness and even fatalities in humans. Sensible hygiene precautions, such as thorough hand-washing before and after handling reptiles, reduce the risk of disease transmission.

Salmonellosis

Salmonella presents two challenges in terms of animal health care. Firstly, *Salmonella* bacteria are unusual, in that they can live and multiply outside a host organism (for example, on, or in, contaminated food). Secondly, animals can be asymptomatic carriers.

There is a very large number of species of *Salmonella* bacteria that cause disease (Ketz-Riley, 2003), but not all *Salmonella* spp. can infect all hosts. Typhoid fever in humans, for example, caused by *S. typhosa*, does not occur naturally as a disease in non-human primates. *S. enteriditis* and *S. enterocolitis*, however, are commonly isolated pathogens in a range of vertebrate zoo animals,

causing diarrhoea, which may be severe. Septicaemia can be a (potentially fatal) complication of infection by *Salmonella* spp.

A number of *Salmonella* spp. are pathogens in reptiles, and salmonellosis is an increasingly reported problem in captive exotic reptiles held outside zoos, as pets. All **chelonid** species, for example, are susceptible to *Salmonella* infection; many reptiles will be asymptomatic carriers of *Salmonella* spp. (Raphael, 2003; **Fig. 11.17**).

Shigellosis

Like *Salmonella*, *Shigella* has a worldwide distribution and the main transmission route is faecal–oral. Along with *Campylobacter*, it is a frequently isolated pathogen in zoo primates; the disease usually presents in primates as an acute gastrointestinal infection.

The most commonly found *Shigella* spp. in non-human primates is *S. flexneri* (Ketz-Riley, 2003).

A **fomite** is an inanimate object that can spread disease from one individual to another. Among humans, coins, towels, or drinking cups can be fomites; in zoos, a fomite could be a contaminated bucket or food bowl, or a keeper's boots.

A **chelonid** species is a member of the reptilian order Chelonia: the tortoises and turtles.

Disease	Causative pathogen	Distribution	Transmission route	Symptoms	Susceptible taxa
Anthrax*	*Bacillus anthracis*, bacterium	Worldwide (rare in the UK and Europe, and more common in warmer climates)	Infected carcasses; ingestion of spores (see de Vos, 2003, for further details)	Ruminants may die suddenly without showing prior symptoms; in other taxa, symptoms include swelling of the face, throat, and neck	Mammals (within zoos, particularly carnivorous species); occasionally birds, e.g. ostriches
Aspergillosis	*Aspergillus* spp., fungi	Worldwide	Inhalation	Usually diagnosed post-mortem (often an opportunistic infection in stressed/immunosuppressed animals)	Frequently occurs in birds
Avian influenza ('bird flu') (*see section 11.5.4*)	Influenza A virus	Worldwide	Direct contact with faeces; aerosol	Highly variable and can be asymptomatic; often clinical signs of respiratory disease	Birds, but can also infect mammals (including humans); waterfowl may be reservoirs for the virus
Campylobacteriosis	*Campylobacter* spp., bacteria	Worldwide (the incidence of campylobacteriosis worldwide is increasing rapidly)	Waterborne (also via milk)		A wide range of mammals, particularly bovids, and birds
Candidiasis	*Candida* spp., fungi	Worldwide	*Candida* spp. are normal residents of the intestine; disease occurs when the usual balance of the gut flora is upset	Usually affects the gastrointestinal tract; infected birds may show weight loss or regurgitation	Birds, mammals (often an opportunistic infection and may follow a period of antibiotic treatment)
Chlamydiosis/ chlamydophilosis	*Chlamydia* and *Chlamydophila* spp. (see Flammer, 2003, for information about changes to the taxonomy of the Chlamydiaceae)	Worldwide	Close contact with an infected animal (may be an asymptomatic carrier); aerosol transmission; contaminated fomites; insect vectors	Weight loss, listlessness; also diarrhoea and respiratory symptoms (the clinical symptoms of chlamydiosis vary widely)	In zoos, chlamydiosis is most common in birds, but it can occur in a wide range of vertebrate and invertebrate taxa (Flammer, 2003)
Clostridial diseases (see also entry for tetanus)	*Clostridium* spp., bacteria, particularly *C. perfringens*	Worldwide	Faeces	Usually enteritis (diarrhoea, dysentery); some forms of the disease can cause rapid death	Most mammals and birds (isolated from some reptiles, but does not appear to cause disease)

Disease	Causative agent	Distribution	Transmission	Clinical signs	Species affected
Colibacillosis	*Escherichia coli*, bacterium (many hundreds of different serotypes exist, of which only a few are pathogenic)	Worldwide (*E. coli* bacteria are a normal part of the intestinal flora of mammals and birds)	Faeces (particularly faecal contamination of foodstuffs)	Enteric disease (diarrhoea, dysentery)	Mammals, particularly ruminants, and birds
Cryptosporidiosis	*Cryptosporidium parvum*, microscopic parasite		Direct contact, particularly with faeces; ingestion of contaminated water or food		Ruminants
Equine encephalomyelitis*	Several strains, virus		Mosquitoes and other biting insects (birds and small mammals are the hosts)		Equids
Glanders and Farcy*	*Burkholderia mallei*, bacterium				Equids (*not in the UK, where glanders has been eradicated*)
Hepatitis (viral)	Various forms of hepatitis, such as hepatitis A		Captive primates can catch hepatitis A from infected humans; other forms of hepatitis can be transmitted via contact with faeces or contaminated food	Lethargy, jaundice, vomiting, but can be asymptomatic; may be fatal (as a result of liver failure)	Primates
Herpesviruses (*see section 11.5.1 on herpes B virus in primates*)	Various herpesviruses	Worldwide	Direct contact including sexual transmission; aerosol	Very variable, from asymptomatic carriers, to severe disease and death	Mammals and birds
Leptospirosis	*Leptospira* spp., bacteria	Worldwide	Direct or indirect contact with infected urine, placental fluids, or milk (Bolin, 2003)	Clinical signs very variable; can include fever, vomiting, diarrhoea, and death	Most mammals
Listeriosis	*Listeria monocytogenes*, bacterium	Worldwide	Ingestion of contaminated food	Abortion; meningitis in young animals	Many mammals; all primates appear to be susceptible to infection
Lyme disease	*Lyme borreliosis*				Deer (via ticks)

Table 11.4 Diseases of concern: common and/or important infectious diseases of zoo animals

* Indicates a notifiable disease in the UK (a notifiable disease is a disease named in s. 88 of the Animal Health Act 1981; not all notifiable diseases are zoonotic). Any outbreak of a notifiable disease must, by law, be reported to the police; in practice, Defra should also be notified.

NOTE: This list is not exhaustive and excludes most of the very large number of diseases caused by endoparasites, such as hookworms, whipworms, lungworms, flukes, etc. See Fowler and Miller (1999; 2003; 2007) for further information about disease in zoo and wild animals.

Disease	Causative pathogen	Distribution	Transmission route	Symptoms	Susceptible taxa
Measles	Rubeola virus (morbillivirus)		Measles can also be transmitted from humans to primates, i.e. is anthropozoonotic	Skin rash, fever; can be fatal	Great apes (see Hastings *et al.*, 1991, for an account of the diagnosis and treatment of measles in wild mountain gorillas)
Mycobacteriosis (see separate entry for *M. tuberculosis*, under Tuberculosis)	*Mycobacterium* spp., bacteria, e.g. *M. avium* in birds	Worldwide	Usually ingestion, also inhalation	Usually a chronic disease	Birds; mammals
Plague	*Yersinia pestis*, bacterium	Parts of Asia, Africa, and the Americas	Transmitted by rodents and their associated fleas	Plague has three main clinical forms: bubonic, septicaemic, and pneumonic (for details of the symptoms of each of these forms, see Williams, 2003)	A wide range of mammalian species
Pox diseases	Poxviruses such as monkey pox, camel pox, etc. (see Greenwood, 2003)	Worldwide	Direct contact; bites and scratches	Skin disease (lesions and tumours); can be fatal	Mammals, birds and reptiles (see Greenwood, 2003, for further information)
Rabies*	*Lyssavirus* spp. (the diseases produced by any *Lyssavirus* infection are indistinguishable from classical rabies—Calle, 2003)	Worldwide, except Australia and Antarctica; 'classic' rabies is absent from the UK, but British *Chiroptera* spp. can carry bat lyssaviruses	Bites from an infected animal; occasionally scratches		A number of vertebrate animals, including felids and canids; also Chiroptera (bats)
Ringworm	*Microsporum* spp. /*Trichophyton* spp., fungal infection	Worldwide	Direct contact with infected animals or fomites	Hair loss; skin crusting; sometimes pruritus	Wide range of vertebrates, particularly canids and other Carnivora; also Chiroptera (bats)
Salmonellosis	*Salmonella* spp., bacteria, e.g. *S. enteritis* (more than 2,000 serotypes have been identified)	Worldwide	Usually ingestion, sometimes inhalation (e.g. of contaminated feather dust); vertical transmission (bird to egg) can occur	Symptoms range from mild enteritis to severe diarrhoea, septicaemia, and sometimes death	Many mammals, birds, and reptiles

Disease	Pathogen	Distribution	Transmission	Clinical signs	Host species
Sarcocyctosis	*Sarcocystis* spp., usually *S. falcatula*, protozoan parasite	N. America	Faeces (opossums are the only known definitive host—Lamberski, 2003)	Variable; infected birds may show pneumonitis	Birds, particularly psittacines
Shigellosis	*Shigella* spp. (Gram-negative rod bacteria)	Worldwide	As with salmonellosis, the transmission route is faecal/oral	Diarrhoea to severe bloody dysentery (some animals may be asymptomatic carriers)	All primates are susceptible, particularly during periods of stress, e.g. introduction into a new social group
Tetanus	*Clostridium tetani*	Worldwide	Spores may enter wounds in the body	Stiffness and spasms; can be fatal	Ruminants, such as cervids
Tuberculosis* (infection with *Mycobacterium* spp. other than *M. tuberculosis* is referred to as 'mycobacteriosis'; see separate entry)	*Mycobacterium tuberculosis*, bacterium	Worldwide	Close contact with infected humans (particularly aerosol transmission via coughing or expectoration)		Primates, elephants, and some other mammals. Other *Mycobacterium* spp. can infect a wide range of vertebrates
West Nile virus (WNV) (see section 11.5.4)	WNV is an arbovirus in the family *Flaviviridae* (some texts refer to WNV as a flavivirus) (Travis, 2007)	Europe, Africa, Asia, Middle East, and N. America from 1999	Blood-feeding biting insects (e.g. mosquitos, sandflies; see Travis, 2007)	Usually diagnosed post-mortem	Birds (some mammals, including humans and equids, are susceptible if bitten by an infected insect)
Yersiniosis	*Yersinia* spp., particularly *Y. pseudotuberculosis* and also *Y. enterocolitica*, bacteria (*Y. pestis* is the pathogen that causes plague)	*Y. pseudotuberculosis* is found worldwide	Contaminated foodstuffs (faecal contamination); insects, such as cockroaches, may also be important vectors for transmission	Listlessness, weakness; death can occur within 48 hours; sudden death with no prior clinical symptoms can also occur	Mammals, many bird species, some reptile and fish species. In zoos, New World primates and some bird species appear to be particularly vulnerable (Allchurch, 2003)

Table **11.4** (continued)

Tuberculosis

Non-human primates are highly susceptible to tuberculosis (TB), which is the collective name given to infection with bacteria in the genus *Mycobacterium* (Mikota, 2007). Tuberculosis is also a zoonosis.

Elephants are also susceptible to both strains of TB, but particularly the human strain, *M. tuberculosis*. A number of cases of TB in elephants in captivity in the USA in the 1990s led to the setting up of a National Tuberculosis Working Group for Zoo and Wildlife Species and the production of *Guidelines for the Control of TB in Elephants* in 1997 (Mikota, 2007; Mikota *et al.*, 2000). These guidelines, which can be downloaded from the website of the US Animal and Plant Health Inspection Service (APHIS), were revised in 2000 and again in 2003, as knowledge of the diagnosis and treatment of TB in elephants improved.

For further reading on TB in elephants, the proceedings of the Elephant Tuberculosis Research Workshop held in Orlando, Florida, in 2005 provide an excellent starting point. The proceedings of this workshop can be accessed via the website of the organization Elephant Care International, which also contains links to the current TB guidelines.

Viral diseases

Viral diseases that can arise in zoo animals include rabies, herpes, and various diseases caused by poxviruses (see **Table 11.4**). (West Nile virus is dealt with in more detail in section **11.5.4** of this chapter.)

Rabies

Rabies is the general term used for a disease of mammals caused by infection with viruses from the *Lyssavirus* genus (Calle, 2003). Once clinical signs of rabies appear, it is nearly always fatal. Worldwide, rabies is a serious zoonosis, causing tens of thousands of human deaths each year. Because of the risk to human life, together with the long incubation period of the disease (weeks to months), many countries have stringent rabies quarantine regulations. In the UK, for example, the quarantine period for rabies is 6 months (see section **11.4.2** for more information about quarantine regulations).

Within North American zoos, free-ranging wild animals have been found to be infected with rabies; *Lyssavirus* infection has been recorded in captive fruit bats in a **walk-through exhibit** at Rotterdam zoo in the Netherlands (Mensink and Schafternaar, 1998).

In Europe, four human deaths from bat *Lyssavirus* infection have occurred since 1977 (Fooks *et al.*, 2003).

Poxvirus infections

Poxvirus infections give rise to generalized skin disease and skin tumours (Greenwood, 2003; **Fig. 11.18**). Poxvirus infections can occur in mammals (for example, seal pox) and in birds (for example, avian pox), and can be fatal in some species, such as orangutans *Pongo* spp. and **callitrichids**. A poxvirus-like infection has also been reported for captive crocodilian species, but the so-called carppox in fish is caused by a herpes virus.

Figure **11.18** There are many different poxviruses that can cause disease in zoo animals (for example, cowpox and sheeppox). This grey seal *Halichoerus grypus* is infected with a poxvirus, which produces unsightly lumps, but usually only a mild infection in these animals. (Photograph: Sheila Pankhurst)

Callitrichids are small New World monkeys. The family Callitrichidae—or, more correctly now, the sub-family Callitrichinae—includes the marmosets and tamarins.

Herpesviruses

Herpes B (sometimes called 'simian herpesvirus') is a zoonotic virus found in monkeys belonging to the genus *Macaca*. Clinical signs in monkeys such as the rhesus macaque *Macaca mulatta* are often quite mild, but the disease is a serious zoonosis that can cause fatal encephalitis in humans (Fowler and Miller, 2003; Richman, 2007). Diagnosis of the disease is difficult, because infected monkeys may be asymptomatic (Richman, 2007) and the virus is difficult to distinguish from the related herpes simplex virus (HSV).

Herpes B virus is classified as a biosafety level 4 pathogen (the highest level) and, therefore, its culture requires a special containment facility. In the UK, for example, only one level 4 containment facility currently exists and, under the Control of Substances Hazardous to Health (COSHH) regulations, animals infected with herpes B must be destroyed.

Richman (2007) reports on a newly recognized disease in elephants—elephant endotheliotropic herpesvirus (EEHV)—which can be fatal. Herpesviruses are also significant pathogens of tortoises, with high mortality rates for infected animals (Johnson *et al.*, 2005).

Fungal diseases

Fungal pathogens that can cause health problems in zoos include *Aspergillus*, *Candida*, *Cryptococcus*, and *Pneumocystis* (see **Table 11.4**). As with bacterial diseases, the name of the disease usually reflects the name of the infective agent, with *Candida* giving rise to candidiasis, *Aspergillus* to aspergillosis, and so on. Fungal diseases, particularly candidiasis, are often associated with general ill health, immunosuppression, or a long period of treatment with antibiotics (Duncan, 2003). Aspergillosis can be a serious problem in captive birds: Flach *et al.* (1990), for example, record aspergillosis as the most common cause of death in gentoo penguins *Pygoscelis papua* at Edinburgh Zoo from 1964 to 1988.

Ringworm

Ringworm (dermatophytosis) is fungal disease of the skin and is also a zoonosis. The infective agents are *Microsporum* and *Trichophyton* spp. in non-human animals, and *Epidermophyton* in humans (Duncan, 2003). Among zoo animals, the felids and canids are particularly susceptible to ringworm infection.

Parasitic infections

The control of parasites in zoo animals has been considered briefly already in this chapter (section **11.4.6**). Health problems caused by ectoparasites in zoo animals include warble fly and scabies (or sarcoptic mange); diseases caused by endoparasites include toxoplasmosis and avian malaria.

Warbles

Warbles, or hypodermosis, is rare in the UK after an extensive eradication campaign, but is still common in other parts of the world. Warble flies lay their eggs on the skin of the host animal. The eggs hatch and the larvae burrow into the animal, remaining under the skin of the host for several months (visible in later stages as lumps, or warbles, on the animal, usually along the back). There are a number of different species of warble fly worldwide; the two species found in the UK both belong to the genus *Hypoderma*.

Bovid species infected with warbles can be treated with the topical application of drugs such as ivermectin.

Scabies (sarcoptic mange)

Scabies is caused by a mite, *Sarcoptes scabiei*, which burrows into the skin of its host to lay its eggs, causing mange. Scabies is most common in ungulates, but can occur in a wide range of mammalian species, including primates. This parasitic infection can also be treated with ivermectin, either topically or orally.

Toxoplasmosis

Most vertebrates, including humans, are susceptible to infection by the **coccidian parasite** *Toxoplasma*

Coccidian parasites are protozoans: small, unicellular animals.

gondii (Wolfe, 2003). Infected animals, however, may be asymptomatic or show only mild signs, and susceptibility varies considerably with both the strain of the parasite and the **taxonomic** group. Australian marsupials and New World monkeys, for example, are highly susceptible to severe toxoplasmosis, whereas Old World monkeys, cattle, and horses are relatively resistant to the disease.

Domestic cats, whether feral or pets, are the definitive host for toxoplasmosis; other infected vertebrates are intermediate hosts. Toxoplasmosis can cause abortion and stillbirth during gestation, and the disease is a significant cause of foetal mortality in sheep and goats; infection in humans during the first trimester can also be lethal (Wolfe, 2003).

Avian malaria

Avian malaria, transmitted by mosquitoes of the *Culex* spp., is a major cause of mortality in outdoor-exhibited penguins in zoos and wildlife parks (Graczyk *et al.*, 1995; **Fig. 11.19**). The infective organisms for this disease are the parasite species *Plasmodium relictum* and, to a lesser extent, *P. elongatum*. In their natural habitat, most penguin species inhabit regions where there are few, if any,

Figure **11.19** Outdoor-housed penguins, such as these Humboldt's penguins *Spheniscus humboldti*, are susceptible to infection with avian malaria, which is transmitted by mosquitoes. This can be a serious, and sometimes fatal, illness in zoo birds, occurring not only in penguins, but in a range of avian species. (Photograph: Sheila Pankhurst)

mosquitoes, so avian malaria is not an important disease of wild penguins.

Avian malaria is not only a disease of penguins; it also occurs in other bird taxa, particularly in non-indigenous falcons (Redig and Ackermann, 2000). It is possible to treat birds prophylactically, with anti-malarial drugs, such as primaquine or mefloquine, administered during the main insect season. A new DNA-based vaccine against avian malaria has had promising results in initial trials, although, at present, an annual booster injection is needed (Cranfield *et al.*, 2000).

11.5.2 Degenerative, genetic, and nutritional diseases

Degenerative diseases

Degenerative diseases, such as arthritis and chronic renal disease, are common in felids, canids (see, for example, Rothschild *et al.*, 2001), bears, primates, and ruminants, all of which are well represented in most zoo collections. These diseases are often age-related and can pose a particular problem in zoological collections, because animals such as bears may live considerably longer in zoos than they would in the wild.

Other health problems associated with ageing animals include dental problems—teeth can become worn down, so that the animal can no longer feed effectively—and diseases such as diabetes and arthritis. These are largely diseases of captivity: wild animals with diabetes or arthritis, or with badly worn-down teeth, would not survive for long.

Genetic diseases and health problems

By definition, zoo populations of some endangered species are likely to be small and the **gene pool** may be restricted. Although the use of studbooks (see **Chapter 9**) limits inbreeding, genetic abnormalities can, and do, occur in zoo animals (Leipold, 1980). In black-and-white ruffed lemurs *Varecia variegata variegata* in zoos, for example, a number of congenital abnormalities have been recorded, including

scoliosis and skull abnormalities (Benirschke *et al.*, 1981). In the grey or timber wolf *Canis lupus*, a hereditary form of blindness appears to be linked to inbreeding (Laikre and Ryman, 1991).

Nutritional diseases

All animals in captivity need to be fed the correct diet, to ensure a strong immune system and to avoid problems such as obesity and tooth decay. Common nutrition-related health problems in zoo animals include metabolic bone disease, iron storage disease (haemochromatosis), and emaciation or, more commonly, obesity. These nutritional problems are considered in more detail in **Chapter 12**.

11.5.3 Zoonoses, anthropozoonotic diseases, and notifiable diseases

Zoo animals can catch some diseases from humans —and vice versa. Diseases that are transmitted from humans to animals are known as **anthropozoonotic diseases** (Ott-Joslin, 1993), whereas diseases transmitted from animals to humans are zoonoses (or zoonotic diseases). The latter form a much longer list than the former, but this is, in part, because we do not yet know as much about the transmission of human diseases to exotic animals. Schwabe (1984) suggests that it is likely, for the **anthropoid apes** at least, that these animals are susceptible to many, if not most, human diseases (**Fig. 11.20**).

A disease that shows very similar pathology in both humans and animals is referred to as an **isozoonosis**. **Table 11.5** lists some of the major diseases (zoonoses) that can be transmitted to humans from exotic animals.

The transmission of zoonoses is via direct contact with an infected animal, or via ingestion or inhalation of pathogens, or by being bitten by an insect such as a mosquito. Most reported cases

Figure **11.20** These lowland gorillas at Apenheul Primate Park in the Netherlands are in good health, but all of the great apes are susceptible to infection with human diseases. Diseases that can be transmitted from humans to animals are known as anthropozoonotic diseases; examples include measles and tuberculosis. (Photograph: Sheila Pankhurst)

of zoonoses arising from zoo animals are among the zoo professionals who come into close physical contact with the animals on a daily basis, rather than among zoo visitors. Good hygiene is the most important control measure for zoonoses (for example, washing hands after handling animals).

Notifiable diseases

A notifiable disease is a disease that, by law (in the UK), must be reported to the police; in practice, Defra should also be notified. Notifiable diseases

---

Scoliosis is lateral (or side-to-side) curving of the spine. The term **anthropoid ape** is rather out of date now (primate taxonomy has moved on). It refers to the 'man-like' apes: that is, the tail-less primates, such as the chimpanzee, gibbons, gorilla, and orang-utan. Current taxonomy puts only the chimpanzee, gorilla, and orang-utan into the family Hominidae, along with humans.

Disease/causative pathogen	Transmission route (to humans)	Animal host(s)	Notes on disease in humans
Anthrax*/Bacillus anthracis, bacterium	Ingestion of spores; direct contact with infected animal (or animal hide) (see de Vos, 2003)	Wide range of species (mainly herbivorous mammals; some bird species). Anthrax infection is very rare in the UK	Anthrax is a rare disease in humans. The most common form of the disease is the cutaneous form, contracted via contact with infected animals or their hides; this form of the disease is readily treatable with antibiotics
Avian influenza (AI)* ('bird 'flu')/influenza A viruses See section 11.5.4	At the time of writing, almost all human cases of AI were the result of direct contact with infected birds or faecal matter from infected birds	Birds, particularly poultry, and domestic cats. Cases of AI have also been reported in non-domestic felids (tigers and leopards) in zoos in Asia	At the time of writing, almost all AI cases in humans have been recorded after there has been close direct contact between humans and infected poultry—but the situation regarding AI is changing rapidly (both the Defra website and the BIAZA website provide updates on the current situation)
Bovine spongiform encephalopathy (BSE)*/ prion (protein)	Consumption of infected meat	Bovids	There is no evidence of any link between cases of Creuztfeldt–Jakob disease (vCJD) (the human form) and zoo animals
Brucellosis*/Brucella spp., bacteria	Consumption of dairy products, such as milk, from infected animals	Bovids, particularly cattle; more rarely, cervids and other mammals	
Campylobacteriosis/ Campylobacter spp., bacteria	Water-borne (also via milk)	A wide range of mammals, particularly bovids, and birds	In humans, infection can cause a severe form of food poisoning. The incidence of campylobacteriosis in humans has increased greatly in recent years
Cryptosporidiosis/ Cryptosporidium parvum, microscopic parasite	Direct contact, particularly with faeces; ingestion of contaminated water or food	Ruminants	Infection in humans generally gives rise to diarrhoea. The disease can be very serious, or even fatal, for immunocompromised patients
Equine encephalomyelitis*/ several strains, virus	Mosquitoes and biting flies	Birds and small mammals are the hosts; mosquitos or biting flies transmit the disease to humans/equids	This is a mosquito-transmitted disease—it is not directly contagious between equids and humans
Escherichia coli/ various strains, bacteria	Faeces	Found in the guts of animals, particularly bovids	Strains such as O157 produce a toxin, called verocytotoxin, which can cause serious illness and even death in humans
Glanders and Farcy*/ Burkholderia mallei, bacterium	Open wounds	Equids	This disease has been eradicated from the UK, but remains a serious zoonosis in other parts of the world, such as the Middle East and China
Hantavirus pulmonary syndrome (HPS)	Airborne—inhalation of aerosolized virus from urine, saliva, or faeces of infected rodents	Rodents	Very rare in the UK and rare in North America, but cases are often fatal
Hepatitis (viral)	Direct contact with body fluids	Primates	Causes fever, anorexia and jaundice
Herpes B/virus (simian herpesvirus) See section 11.5.1	Bites from infected animals	Primates of the genus Macaca (macaques)	Herpes B can cause fatal encephalitis in humans
Hydatid disease/ Ecchinococcus spp., helminth (tapeworm) parasite	Ingestion of parasite eggs	Canids, via sheep (e.g. where dogs or foxes have been fed raw offal)	Various symptoms, including jaundice. Occasionally fatal if cyst ruptures
Leptospirosis/Leptospira spp., bacteria	Skin/wound or mucous membrane contact with infected urine	Most mammals, but particularly associated with rodents, such as rats and mice	Killed vaccines against leptospirosis are available in the USA and some other countries for animal use (e.g. for animals in petting zoos)
Listeriosis/Listeria monocytogenes, bacterium	Ingestion	Many animals, particularly ruminants	Causes fever, nausea and diarrhoea. Can be fatal

Disease/organism	Transmission	Host species	Comments
Lyme disease (Lyme borreliosis)/*Borrelia*, bacterium	Direct contact with an infected animal	Deer (via ticks)	
Monkey pox/virus belonging to the orthopoxvirus group		Primates, such as gibbons, chimpanzees, and orang-utans	Causes symptoms similar to smallpox in humans. Treatment of the disease in humans with cidofovir has been effective
Plague/*Yersinia pestis*, bacterium	Fleas; inhalation; contact with infected animals	Mammals such as rodents (particularly rats); also cats and dogs	Plague is not a major problem in Europe, but still occurs in Central and S. America and in S.E. Asia and southern Africa
Psittacosis/*Chlamydophila psittaci* (formerly *Chlamydia psittaci*), bacterium**	Aerosol; inhalation of dust from faeces or feathers; direct contact	Parrots (psittacines), pigeons, poultry, and waterfowl	*C. psittaci*, found in captive parrots, pigeons, and waterfowl, is responsible for most reported cases of zoonotic infection by *Chlamydophila/Chlamydia* bacteria, although other species can also be zoonotic
Q fever/*Coxiella burnetii*	Aerosol; also via milk	Bovids	
Rabies*/*Lyssavirus* spp., viruses	Saliva of an infected animal, via a bite (or more rarely, a scratch)	A number of vertebrate animals, including felids and canids	Fatal in humans (rabies is responsible for tens of thousands of human deaths worldwide each year, usually after bites from infected dogs)
Rift Valley fever/virus*		Bovids	Rarely fatal
Ringworm/*Microsporum* spp., *Trichophyton* spp., and *Epidermophyton* spp., fungal infection	Direct contact with infected animals or fomites	Wide range of vertebrates, particularly canids and other Carnivora; also Chiroptera (bats)	A common zoonosis, caused by a fungal infection. Rarely serious and usually easy to treat
Salmonellosis/*Salmonella* spp., bacteria	Contaminated food or surfaces; direct contact with infected animals/faeces	Many mammals, birds, and reptiles	A very common source of food poisoning
Shigellosis/*Shigella* spp., bacteria	Faecal/oral transmission (as for salmonellosis)		As with salmonellosis, food poisoning (acute gastrointestinal infection)
Toxocariasis/*Toxocara* spp., parasite (roundworm)	Direct contact with an infected animal	Canids and felids	
Toxoplasmosis/*Toxoplasma gondii*, protozoan parasite	Ingestion (of oocysts)	Mammals, especially cats; also birds	Serious if contracted during pregnancy; can cause abortion/stillbirth
Tuberculosis*/*Mycobacterium tuberculosis*, bacterium	Inhalation (particularly of aerosolized fluids from coughing or expectoration)	Primates, less commonly other mammals	Tuberculosis (TB) infection in humans is of increasing concern, because of the spread of strains resistant to drugs
West Nile virus	Mosquitoes and other biting insects	Birds and equids	Spread by infected mosquitoes
Yersiniosis/*Yersinia* spp., bacteria, particularly *Y. enterocolitica* and *Y. pseudotuberculosis*	Contaminated food including milk	*Y. pseudotuberculosis* is widespread, in mammals, birds, reptiles, and fish (many species appear to be asymptomatic carriers)	

Table 11.5 Major zoonoses of zoo animals

* Indicates a notifiable zoonotic disease in the UK (a notifiable disease is a disease named in s. 88 of the Animal Health Act 1981; not all notifiable diseases are zoonotic). Any outbreak of a notifiable disease must, by law, be reported to the police; in practice, Defra should also be notified.

** There are several *Chlamydia* and *Chlamydophila* spp. of bacteria; these cause disease in animals referred to as chlamydiosis or chlamydophilosis (see Flammer, 2003).

NOTE: This table lists some of the major zoonotic diseases (disease transmissible to humans) in zoo animals. The animal health pages of the Defra website provide much useful information, particularly about notifiable diseases (see www.defra.gov.uk/animalh/diseases/), as does the website of the World Organization for Animal Health (OIE) (www.oie.int).

within the UK are named in section 88 of the Animal Health Act 1981; some, but not all, of these are zoonotic. In the UK, the government agency responsible for managing notifiable diseases is called **Animal Health**; set up in 2007, Animal Health is made up of various bodies including the former State Veterinary Service (SVS). Officially, Animal Health is the agency responsible *'for delivering government policy on animal health and welfare'* (see www.defra.gov.uk/animalhealth/).

Examples of notifiable diseases within the UK include anthrax, blue tongue, and avian influenza ('bird 'flu'). There is a useful table on the Defra website summarizing these diseases (see www.defra.gov.uk/animalh/diseases/notifiable/index.htm).

11.5.4 Five 'headline' diseases

Five very different diseases of domestic or wild animals have attracted considerable press coverage over recent years. These 'headline' diseases are:

- bovine spongiform encephalopathy (BSE);
- foot-and-mouth disease (FMD);
- West Nile virus (WNV);
- avian influenza (AI, or 'bird 'flu');
- chytrid fungus in amphibians.

Bovine spongiform encephalopathy and other transmissible spongiform encephalopathies

BSE is a relatively new disease. It was first diagnosed in 1986, in cattle in the UK (Spraker, 2003), and is believed to have arisen as a result of cattle eating concentrates containing meat and bone meal contaminated with prions,[13] either from sheep with scrapie, or from other cattle infected with BSE. Exotic species can also contract BSE, and the disease has been reported in zoo-housed species such as eland *Taurotragus oryx* (Fleetwood and Furley, 1990),

Figure **11.21** The greater kudu *Tragelaphus strepsiceros* is one of a number of bovid species kept in zoos that are susceptible to transmissible spongiform encephalopathies (TSEs). (Photograph: © John Pitcher, www.iStockphoto.com)

greater kudu *Tragelaphus strepsiceros* (**Fig. 11.21**), and the Arabian oryx *Oryx leucoryx* (Kirkwood *et al.*, 1990).

BSE is not the only transmissible spongiform encephalopathy (TSE) that has been reported in non-human animals. Cats, including exotic species such as the cheetah, can contract the prion disease feline spongiform encephalopathy (FSE) (Kirkwood and Cunningham, 1994; Kirkwood *et al.*, 1995), which appears to be linked to the BSE epidemic. Scrapie in sheep and goats, and chronic wasting disease (CWD) in cervids are also prion diseases (Spraker, 2003; for a review of animal prion diseases other than BSE, see Sigursdon and Miller, 2003).

Prion diseases such as BSE and CWD are characterized by loss of motor control, dementia, paralysis, wasting, and, eventually, death.

Foot-and-mouth disease

FMD is a highly contagious viral disease of ungulates, spread primarily by direct contact (for example, via clothing, vehicle wheels, etc.). Although not usually fatal, the disease is debilitating, and causes fever, lameness, and characteristic blistering (vesicle formation) in the mouths of infected animals. The

---

13 Prion diseases are often called 'spongiform encephalopathies' because of the 'spongy' appearance of the brain post-mortem, with large vacuoles in the cortex and cerebellum.

disease is associated with livestock (sheep, cattle, and pigs), but can affect wild and zoo animals, including deer, camelids, and elephants.

Within the UK, FMD is a notifiable disease. The UK is normally FMD-free, but an outbreak of the disease in 2001 led to the imposition of control measures, such as the slaughter of infected livestock and severe restrictions on the movement of animals from susceptible species. Since the 2001 FMD outbreak, Defra has produced an *Exotic Animal Disease Generic Contingency Plan* (available via the Defra website), covering diseases such as FMD. This document, which is updated annually, sets out plans for the emergency vaccination of zoo animals, where appropriate, as well as providing guidance on biosecurity measures.

West Nile virus

WNV hit the headlines in 1999, when it was isolated for the first time in the USA. The virus, which is zoonotic, was identified by pathologists at the Bronx Zoo in New York, after the death of a number of birds at the zoo (Steele *et al.*, 2000). WNV is transmitted largely by arthropods, such as mosquitoes and other biting insects, but can also be transmitted by direct contact with infected animals or tissue from infected animals (Travis, 2007). Although the disease had been known in Asia, Africa, and southern Europe for many decades, it had not been isolated in animals in the Americas prior to 1999.

Humans[14] and a wide range of vertebrate species—but particularly birds and horses—are susceptible to WNV, which causes potentially fatal encephalitis. Among zoo mammals, cases of WNV have been reported in animals ranging from an Arctic wolf *Canis lupus* to reindeer *Rangifer tarandus* (Travis, 2007).

At the time of writing, two vaccines are currently available against WNV and new vaccines are in development. Protective measures against the disease include spraying with insecticides to kill adult and larval mosquitoes (Travis, 2007).

Avian influenza

AI (or 'bird 'flu') is an infection in birds caused by influenza viruses that are very similar to human influenza viruses. The disease is carried by wild birds—particularly ducks and other wildfowl which may not develop the disease themselves, but can transmit the virus via their faeces to domestic poultry and other captive birds (**Fig. 11.22**). AI is

Figure **11.22** Avian influenza is a disease that can be transmitted to captive birds by wild birds. Preventing wild birds from gaining access to zoo enclosures is an important preventive measure to stop its transmission, but this is very difficult to do in the case of outdoor-housed zoo birds such as these black swans *Cygnus atratus*. (Photograph: Warsaw Zoo)

---

14 Since 1999, more than 700 human deaths from West Nile Virus have been reported in the USA (Travis, 2007).

of concern to zoos not only because of the risk to birds in their collections, but also because the virus can infect a range of mammalian species, from mice, pigs, monkeys, and ferrets, to felids, such as tigers and leopards. At a private zoo in Thailand in 2004, for example, where tigers were fed infected chicken carcasses, forty-five animals died from the disease (Amosin *et al.*, 2006).

Different strains of AI are categorized either as 'low pathogenic' (LPAI) or 'highly pathogenic' (HPAI). The current outbreak (2003 onwards) of bird 'flu is of the H5N1 strain, which is both highly contagious and highly pathogenic; domestic birds, such as chickens and ducks, can die from the virus within 24 hours of clinical signs first appearing.

Information about AI rapidly becomes out of date, so the reader is referred to websites such as that of Defra for guidance on the current situation. Within the zoo community, both the BIAZA and the EAZA websites carry information about the current risk to zoo collections from bird 'flu. EAZA has recommended to its members that they should seek permission to vaccinate birds in their collections; BIAZA members have been provided with advice on appropriate biosecurity and veterinary surveillance measures, as well as advice on how to reduce the risk to zoo staff and to members of the public. The American Association of Zoos and Aquariums (AZA) has produced and circulated to its members detailed guidelines on the monitoring and prevention of bird 'flu. Another website featuring up-to-date information about AI is that of the European Food Safety Authority (EFSA): part of EFSA's role is to '*provide European risk managers with objective scientific advice on the animal health and welfare dimension of avian influenza*' (EFSA, 2008).

Current EU legislation on AI is set out in EC Directive 2005/94 EEC.[15] This Directive covers areas such as monitoring and surveillance for AI in wild birds, as well as control measures to be taken in the event of an outbreak. In the UK, this EU legislation is enforced via the Avian Influenza and Influenza of Avian Origin in Mammals (England) (No 2) Order 2006 and the Avian Influenza (Vaccination) (England) Regulations 2006. (Wales and Scotland have their own versions of this legislation.)

Zoo contingency plans for AI will involve consultation and discussion, decisions on appropriate control measures, and consideration of how to deal with the public perception of an outbreak affecting zoo animals.

Chytridiomycosis in amphibians

Chytridiomycosis, or chytrid fungus, is the name for a fungal disease that has been strongly implicated in the decline of populations of wild amphibians around the world (**Fig. 11.23**). The fungus or pathogenic agent *Batrachochytrium dendrobatidis* belongs to the Class Chytridiomycetes and appears to be a new genus that, until recently, had not been found in vertebrates (Crawshaw, 2003). Infection causes skin problems in anurans (frogs and toads),

Figure **11.23** Wild populations of amphibians such as this White's tree frog *Litoria caerulea* appear to be particularly at risk of infection by a fungus called chytridiomycosis, or chytrid fungus. WAZA has called the current threats to wild populations of amphibians '*one of the greatest species conservation challenges in the history of humanity*'. (Photograph: Paignton Zoo Environmental Park)

---

15 Council Directive 2005/94/EC on Community measures for the control of avian influenza and repealing Directive 92/40/EEC.

ranging from cutaneous erosion to ulceration of the skin; the disease can have a high mortality rate.

But why should a fungus that, at the moment, appears to be a problem mainly in free-living or wild amphibians be of particular concern to zoos and aquariums? In 2006, WAZA produced the following statement:

> Addressing the amphibian extinction crisis represents one of the greatest species conservation challenges in the history of humanity. It is clearly understood that if the global zoo community does not respond immediately and on an unprecedented scale, much of an entire vertebrate class will be lost. This conservation challenge is one that we, the ex situ community, are uniquely capable of addressing. Never before has the conservation community at large charged zoos and aquariums with a task of such large magnitude. This is an opportunity for every zoo and aquarium, regardless of size, to make a vital conservation contribution, and for our community to be broadly acknowledged as a trusted and effective conservation partner. If we were to contain and reverse this situation, the community will have succeeded in our most basic conservation mission as defined in the World Zoo and Aquarium Conservation Strategy . . . Our overall vision is the global persistence of thriving amphibian diversity.

Zippel (2005) provides a useful and comprehensive web-based review of zoo research and conservation initiatives aimed at amphibians, with links to scientific papers, zoo reports, and species information. In the UK, for example, the Zoological Society of London (ZSL) is building a major new centre for the conservation of amphibians. The project will included a new public exhibit at London Zoo, as well as laboratory facilities for further research into amphibian diseases such as chytridiomycosis.

11.6 The diagnosis and treatment of disease in zoo animals

We have seen earlier in this chapter that diagnosis of disease is a key skill for zoo vets. Essentially, the work of a zoo vet involves collecting data and identifying problems, then determining the possible courses of action (a treatment plan, and measures for disease control and prevention).

Clinical signs of illness, such as a fever or rapid breathing, will rarely be specific to a particular disease. A diagnosis is usually based on a suite of characteristics and a zoo vet will often need to call on the services of a pathologist to confirm an initial diagnosis of a particular disease. Once a disease has been diagnosed, treatment will vary, depending on whether or not the condition is **chronic** and also on whether or not immediate veterinary intervention is in the animal's best interests.

11.6.1 To treat or not to treat?

Capture, restraint, and anaesthesia are all stressful procedures for wild animals, and, for any given situation, a zoo vet will have to weigh up the costs and benefits of treatment versus no treatment. It may be better to leave an animal with a superficial injury to heal on its own without treatment, if the only alternative is capture and full anaesthesia. Veterinary treatment may have adverse effects on an animal's reproductive status, or may result in aggression from conspecifics when an individual is removed for treatment and then returned into a social group. A zoo vet will have to consider the practicalities of intervention and treatment in comparison with the long-term prospects for recovery of the animal, and make a welfare assessment based on these factors.

A **chronic** condition is one that persists over a long period of time and which may require long-term management, e.g. diabetes. An **acute condition** is one in which severe or aggressive illness develops, often quite rapidly.

Analgesic drugs for veterinary use	Other methods of pain control
Opioids, e.g. morphine	Distraction (not easy in wild animals)
Non-steroidal anti-inflammatories (NSAIDS), e.g. carprofen, marketed in the UK as Rimadyl®	Application of hot or cold compresses (again, not easy in wild animals)
Alpha-2 agonists, e.g. medetomidine, marketed in the UK as Domitor®	Immobilization of fractures
Local anaesthetics, e.g. lignocaine, marketed in the UK as Lidocaine®	Transcutaneous electrical nerve stimulation (TENS)
Neuroleptoanalgesics, e.g. etorphine, marketed as M99® (These drugs are combinations of tranquilizers and analgesics, with the analgesic component being an opioid.)	Acupuncture

Table 11.6 Analgesia for zoo animals

NOTE: This table lists only some of the more commonly used measures for relieving pain in zoo animals.

Medication that can be administered in food or drinking water may be an option when capture and injection of a drug is not desirable from a welfare perspective, or when it would put veterinary staff or keepers at high risk of injury. **Euthanasia** is also an option (this is discussed further in **Chapter 7**).

11.6.2 The management of pain

The problem of assessing pain in non-human animals has been discussed in **Chapter 7**. It is widely accepted that pain has evolved because it is adaptive —that it has a protective function. Pain helps animals to distinguish between harmful and non-harmful **stimuli**, and provokes animals to avoid harmful stimuli. Pain also inhibits activities that could delay recovery: for example, by stopping an animal from moving an injured limb and so doing more harm.

It is also widely accepted that we have an ethical responsibility to reduce pain and suffering in animals in our care. Vets will act to try to control pain, not only on ethical grounds, but also because pain has serious adverse physical consequences such as changes in behaviour (for example, excitability or self-harm), muscle spasms, vomiting, and even the risk of heart damage.

The relief of pain is termed **analgesia**; drugs or treatments that reduce pain are referred to as analgesics. Some commonly used **analgesic drugs**

and non-drug treatments for the relief of pain in captive animals are summarized in **Table 11.6**.

Analgesic drugs can be administered orally, topically, or via injection. Drugs that can only be administered via injection do not necessarily require the animal to be captured and restrained; intramuscular injections can sometimes be given at a distance, using a dart gun. Before prescribing an analgesic, the vet will consider the age and condition of the animal, the apparent location and severity of the pain, whether or not the animal is in shock or dehydrated, and the possibility of any adverse reaction. Some analgesic drugs are toxic to particular animals, but safe to use in other taxa; other drugs may have adverse side effects in pregnant mammals. The most commonly used analgesics in zoo animals are the non-steroidal anti-inflammatory drugs (NSAIDs), which have useful anti-inflammatory and **anti-pyretic**, as well as analgesic, properties.

Where possible, vets will aim to give pre-emptive analgesia, because this is often more effective at controlling pain (in other words, the vet will try to administer an analgesic before pain is perceived by the animal). This should always be possible for surgical procedures. Veterinary staff will also often give more than one analgesic, from different classes of drug. Multimodal analgesia is often more effective than the administration of only one drug, or type of drug.

An **anti-pyretic** drug is one that reduces fever, i.e. brings down an elevated body temperature.

Figure **11.24** Here, a black rhino *Diceros bicornis* at Ramat Gan Zoo in Israel has been darted with an anaesthetic dart prior to veterinary treatment. The animal is being supported as the drug takes effect, to reduce the risk of injury as it goes down. (Photograph: Tibor Jäger, Zoological Center, Tel Aviv)

Legislation and the control of analgesic drugs

The use of analgesic drugs for animals is strictly controlled within the UK, under the Veterinary Medicines Regulations 2005; most analgesics can only be prescribed by a vet who has seen and assessed the animal.

Opioids, such as morphine, are also subject to the Misuse of Drugs Regulations 1985.

11.6.3 Restraint and anaesthesia

A large part of the health care role of keepers and zoo vets, or vet nurses, will be the **restraint** of animals, either for examination or for veterinary treatment, or both. Physical restraint is usually the responsibility of keepers, rather than the vet, for which there are

two pragmatic arguments: firstly, keepers are likely to 'know' individual animals in their care better than the zoo vet, and can anticipate better how the animal may react; secondly, there is no sense in allowing the attending vet to become injured, so that he or she cannot do his or her job properly.

Restraint and anaesthesia of wild animals can be risky and stressful, both for the animal, and for veterinary staff and keepers. There can be complications as a result of immobilization, particularly in large mammals such as elephants or rhinoceros (**Fig. 11.24**), or giraffes (see **Box 11.5**). One of the first hurdles that a zoo must overcome before anaesthetizing an animal is to try to obtain an accurate weight, so that the correct dose of anaesthetic can be

Restraint means the control of an animal; this can be physical (e.g. use of a crush cage), or chemical, or psychological.

Box **11.5** The anaesthesia of giraffes

The anaesthesia of giraffes in captivity has, historically, been associated with high rates of mortality (Vogelnest and Ralph, 1997; Bush, 1993). Neck muscle spasms can occur as the animal recovers from the anaesthetic. Giraffes can also regurgitate and aspirate their stomach contents while anaesthetized, so fasting the animal for a period of 2 or 3 days prior to the administration of an anaesthetic is recommended by some sources (Bush, 2003). This practice is controversial, however, because of the welfare implications of denying food.

To minimize the risk of complications, the head of the animal should be kept raised above the level of the stomach and the nose pointed downwards (**Fig. 11.25**). This may mean that a catwalk or other raised structure is needed for keepers and veterinary staff to access the head during immobilization. Bush (2003) suggests the use of a long board or ladder to keep the neck straight, and also suggests altering the angle of

Figure **11.25** Anaesthesia of giraffes presents a considerable challenge to zoo vets. Recommendations to reduce the risk of complications include keeping the head of the animal raised above the level of its stomach (Bush, 2003). Here, a team of keepers and veterinary staff at Werribee Open Range Zoo, near Melbourne in Australia, are working together to monitor and treat an anaesthetized giraffe. (Photograph: Chris Stevens, Werribee Open Range Zoo)

the neck every 10–15 minutes to reduce the risk of muscle spasms. Once recumbent, a blindfold and earplugs can be put in place to reduce the impact of external stimuli, such as lights and noise, and so minimize stress.

As a giraffe recovers from anaesthesia, a rope placed around the animal's shoulders can be used (Bush recommends a minimum of three people on each end of the rope) to steady the animal as it attempts to stand.

used. This is easier said than done for animals such as elephants or giraffes, but zoos can make use of industrial weigh bridges or bars (normally used for weighing heavy vehicles), and can position weigh bars under a platform set into the surface of a part of the enclosure, such as a doorway, where the animal will become accustomed to walking over it.

Increasingly, animal training is being used in zoos to reduce the stress of medical procedures such as taking a blood sample or skin swab, or administering eye drops. For example, Neil Bemment of Paignton Zoo Environmental Park, UK, comments that an African elephant in the Park's care has been trained, using **positive reinforcement** (see section **6.3.1**), to present herself to her keeper for the administration of eye drops to treat a recurrent eye problem. This sort of training for the veterinary treatment of elephants is used in other zoos, and for other species, and helps to reduce both the stress of medical procedures for the animal, and also potential health and safety risks to keepers and zoo veterinary staff (see **Box 11.6** for a short summary of the risks to zoo staff from the use of a potentially hazardous drug, Immobilon®).

Ruminants can be prone to complications during anaesthesia, such as the aspiration of regurgitated contents from the **rumen** (the first chamber of the stomach). To reduce this risk, it is better to position an anaesthetized ruminant in **sternal recumbancy** (**Fig. 11.26**) rather than lying flat on its side, and to keep the head raised and the nose down (Flach, 2003).

11.6.4 Medication and self-medication

As Lozano (1998) has pointed out, there is now recognition that some components of diet selection by wild animals are likely to have evolved to help to protect against infection by parasites, or at least to reduce the effects of parasites. As well as the active avoidance of food items and feeding areas

---

Box 11.6 A short note about Immobilon®

Immobilon® (M99™) is a powerful drug containing the opioid etorphine and acepromazine. It is used by zoos to immobilize animals, such as equids and bovids. Immobilon®, however, can be fatal to humans, even if it is only splashed onto the mucous membranes or onto an open wound in the skin and is not washed out immediately. This drug should only be used if the antidote (Narcan®; the generic name is naloxone) is drawn up first and an assistant is on hand to administer this in case of any accident.

Sternal recumbancy means that an animal is lying down with its sternum or breastbone on the ground, usually with the legs folded under the body, but with the body still in an upright position.

Figure **11.26** A group of blackbuck *Antilope cervicapra* at Rotterdam Zoo in the Netherlands. The female on the far right of the photograph shows sternal recumbancy: she is lying down with her breastbone towards the ground, rather than on her side with her flanks on the ground. (Photograph: Julian Doberski)

that may be potential sources of infection by parasites, animals in the wild may seek out and eat plants and minerals (or soil) for prophylactic or therapeutic purposes. This is known as **self-medication** (Glander, 1994).

In an experimental study of leaf-swallowing in captive chimpanzees, Huffman and Hirata (2004) found that the consumption of rough, **hispid** leaves (used by wild chimpanzees to 'hook' and expel parasitic worms from the gut) was, in part, socially influenced. Some enlightened zoos are now offering their animals opportunities for self-medication, by providing a range of plants, such as some herb species, that may have medicinal properties. (For a review of the provision of herbs and medicinal plants within primate exhibits, see Cousins, 2006.)

Summary

- Zoo animal health care is a complex topic: a large zoo may keep upwards of 10,000 individual animals, from more than 500 different species, each of which will be susceptible to different diseases and health problems.

Hispid means covered with stiff, bristly hairs.

- Preventive medicine is very important for zoo animals: treating sick or injured wild animals in captivity is not easy and it is far better to try to avoid the need for treatment in the first place.

- Diseases of concern will vary from collection to collection, depending on the species in the collection and local risk factors.

- Zoo vets may sometimes decide not to treat a sick or injured animal, if the risks of capture and restraint appear to outweigh the possible benefits of intervention and treatment.

- There are far fewer drugs and vaccines that have been developed specifically for exotic animals than have been developed for farm or companion animals; zoo vets face a particular challenge in this respect.

- Zoo vets and keepers also face the challenge of sometimes having to deal with very large animals: dentistry, the treatment of an injury, and anaesthesia pose considerable problems when dealing with an animal the size of a giraffe, or rhinoceros, or elephant.

Questions and discussion topics

1. In what ways does the role of a zoo vet differ from that of a livestock vet or companion animal vet?

2. Why is preventive care of particular importance for zoo animal health?

3. Under what circumstances might zoo staff decide *not* to capture and treat a sick or injured animal?

4. What are the main bacterial diseases of concern in zoos?

5. What are the main zoonoses of concern in zoos?

6. What can be done to prevent the development of foot problems in captive elephants?

Further reading

First and foremost, Fowler and Miller's *Zoo and Wild Animal Medicine: Current Therapy*[16] provides a wealth of practical information about the prevention, diagnosis, and treatment of disease in wild animals. This book, now in its sixth edition (Fowler and Miller, 2007), is regarded by many zoo vets in the UK, the USA, Australasia, and beyond as the first point of reference for information about exotic animal diseases and medicine. It provides a comprehensive guide to the diagnosis and treatment of disease in different taxa, including fish, and also covers topics such as nutrition, the training of animals for medical procedures, and restraint and anaesthesia. Some of the terminology used in the book will be unfamiliar to a reader without a veterinary training, because the book is written by vets for vets—but, while a glossary would be useful to the non-vet reader, it would add to the already considerable length—and weight—of this tome.

If you buy or borrow a copy of Fowler and Miller, you need to check which edition you have got hold of. The content and organization of the book varies

16 Fowler and Miller is an expensive book to buy, and second-hand copies are not that easy to come by, but all the royalties from sales of this book go to animal research, not to the editors and contributing authors.

(deliberately) quite markedly from one edition to the next; the third edition (Fowler and Miller, 1993), for example, contains a useful summary of zoo-related legislation, but this has been omitted from the more recent editions, which have reverted to a taxon-based account of wild animal medicine.

A recommended book on animal health for the general reader is Sainsbury's *Animal Health* (1998). Although this book is about livestock health and does not specifically cover zoo animal health or medicine, it is a useful and readable account of general principles of health care for captive animals.

There are sections on metabolic diseases and on different types of disease, as well as very clear tables listing diseases (including zoonoses) in livestock. Another accessible book for non-vets is Bowden and Masters' *Textbook of Veterinary Medical Nursing* (2003). For the non-vet who wants a clear and easy-to-read account of veterinary diagnostic procedures, or an introduction to infectious diseases, this book is excellent. As with Sainsbury's book, this text is not aimed at zoo staff, but nevertheless there is much useful general information about veterinary medicine.

Websites and other resources

For zoos within the UK (and, indeed, further afield), there is a wealth of useful information about animal health care on the Defra website—ww.defra.gov.uk. This includes guidelines on current legislation, as well as information on topics such as notifiable diseases and updates on the current situation with regard to outbreaks of AI in wild birds.

The BIAZA—www.biaza.org.uk—and EAZA —www.eaza.org—websites are another good source of information about animal health issues relating to zoos. For zoos in North America, the websites of both AZA—see the pages at www.aza.org/AnMgt/AnimalHealth/—and the AAZV —www.aazv.org—provide information about zoo animal health care and related issues such as quarantine regulations, with links to various guidelines and reports. Information about a wide range of animal health issues in Australia, and particularly about infectious disease, can be found on the AHA website —www.animalhealthaustralia.com.au/.

Chapter 12 Feeding and nutrition

For those of us whose experience of feeding animals comprises a trip to the supermarket to buy a pack of dog or cat food, feeding **zoo** animals probably appears relatively straightforward. This is far from the case, however. In the first place, it is by no means clear for a lot of animals exactly what foods we should be giving them, because our knowledge of their natural diets is often quite poor. But even if we know what animals usually eat in the wild, it is unlikely that the zoo will be able to duplicate their natural diet. This is either because of unavailability of the required items, or because it is simply not feasible—or, in some cases such as live vertebrate prey, even legal—to supply food to animals in the form that they eat it in the wild.

These difficulties can be overcome largely by aiming to provide zoo animals with the levels of **nutrients** that they would get in the wild, rather than trying to recreate the wild diet exactly (Dierenfeld, 1997a). This is important, because **nutrition** can have profound effects on other aspects of an animal's life history, such as physical and mental development, reproduction, **mortality**, and emotions, as well as its basic health.

But there is more to this than feeding. Many animals in the wild spend a great deal of time seeking out and processing food, and giving animals the opportunity to do similar things in captivity is an important part of zoo husbandry.

In this chapter, we will consider the following topics.

12.1 Feeding ecology
12.2 Basic nutritional theory
12.3 Guidelines and legislation on feeding zoo animals
12.4 Working out an animal's dietary requirements
12.5 The supply of food
12.6 Food storage and preparation
12.7 Food presentation
12.8 Nutritional problems

Before we can look at how zoos can best meet the nutritional needs of their animals, we need to know something about how different types of animal obtain their food and also about essential nutrients in the diet. The first two sections of this chapter therefore provide a brief overview of feeding ecology and general nutritional theory. If you are already familiar with the terms **carnivore**, **herbivore**, **ruminant**, etc., and have a good basic knowledge of the anatomy and function of the digestive tract, then you may wish to skip sections **12.1** and **12.2**, and go straight to the practical

information about formulating diets for zoo animals, from section **12.3** onwards.

Special topics, such as a brief history of zoo animal diets and nutrition, are included in boxes throughout this chapter. **Box 12.5**, for example, looks at why keepers chop up food for zoo animals, when wild animals manage perfectly well without having their food chopped up for them; other boxes provide information about plant chemical defences against herbivory and on some recommended sources of detailed information on the nutrient requirements for different species.

Nutrition can be defined as the process within organisms of taking in and absorbing nutrients, to provide energy and to meet other metabolic needs.

12.1 Feeding ecology

Wild-living animals have to find food, process it, and eat it. They have to do these things in an environment within which other organisms live: some of these organisms will be potential food items; some will be competitors looking for the same food; some will be predators looking to make a meal of them; and some will be of no food relevance whatsoever. Evolution has equipped these animals with morphological, physiological, and behavioural adaptations for dealing with these eventualities, and although many zoo animals have been removed from this ecological context, it is important that we know something of these adaptations so that they are satisfied as much as they can be by feeding regimes in the zoo.

Figure **12.1** The koala bear *Phascolarctus cinereus* is an example of a folivore—that is, a specialist herbivore that feeds on leaves. The diet of the koala bear is made up largely of eucalyptus leaves. (Photograph: ©pxlar8, `www.iStockphoto.com`)

12.1.1 Feeding categories

A number of different terms are used to describe the sorts of food that animals eat. The most familiar are the very broad terms 'carnivore',[1] to describe an animal that feeds on other animals, 'herbivore', to describe an animal that feeds on plants, and **omnivore**, to describe an animal that eats both plant and animal food. These terms are so broad that they are not always particularly useful, so some narrower terms are also used to describe more specialized diets within these categories. Thus, a carnivore that eats fish is referred to as a 'piscivore', while one that eats terrestrial arthropods is called an 'insectivore'. Herbivorous species might eat fruit (a 'frugivore') or leaves (a 'folivore'; see **Fig. 12.1**).

This sort of naming of categories can go on apparently endlessly. For example, animals such

as galagos[2] (bushbabies) and marmosets, which include a quantity of plant exudates in their diet, are now often referred to as 'gummivores' (Nash, 1986); animals, such as the leatherback turtle *Dermochelys coriacea*, which consume jellyfish, are 'medusivores' (see, for example, Pierce, 2005).

With a few caveats, terms such as 'carnivore' and 'herbivore' apply across all animal **taxa**. Klasing (1998), in his book *Comparative Avian Nutrition*, uses the term 'faunivore' to describe birds that eat food items that are mostly of animal origin, and subdivides this category into carnivores and piscivores. He also adds a few more terms to the lexicon, such as 'molluscivore' and 'planktonivore'. In the scientific literature on avian nutrition, the term 'granivore' occurs quite widely, not necessarily because there is a very large number of species of granivorous, or grain-eating, birds, but because the

---

1 The term 'carnivore', meaning an animal that eats other animals, is easily confused with the taxonomic term Carnivora. The name Carnivora refers to an order of mammals that includes meat-eating animals such as lions and tigers, but also plant-eating giant pandas and omnivorous badgers. Note the lower- or upper-case initial letter 'c' or 'C': this is

the key difference. So a tarantula spider, or a shark, or a snake is a carnivore, but not a Carnivore.
2 Galagos, or bushbabies, are small nocturnal primates from the family Galagidae; marmosets are small New World monkeys that, with the exception of Goeldi's marmoset, belong to the genus *Callithrix*.

nutritional requirements of commercially important granivores such as the chicken *Gallus domesticus* and the Japanese quail *Coturnix japonica* have been extensively studied.

Specialist and generalist feeders

What does become clear from this naming of categories is that some species have much broader diets than others. Thus, we can think in terms of feeding specialists, who take a very narrow range of food items, and generalists, who take a very broad range. These terms should be thought of in a relative, rather than an absolute, sense. For example, lions *Panthera leo* preferentially prey on species within a weight range of 190–550 kg, with a preference for prey of about 350 kg (Hayward and Keeley, 2005). In this respect, they are specialists, but they are more generalist than pandas (both giant and red), who feed on bamboo (Gittleman, 1994; Wei *et al.*, 1999). In the same way, lions are also more specialist than red foxes *Vulpes vulpes*, which eat scavenged food, birds, small mammals, earthworms, other invertebrates, and fruit (Doncaster *et al.*, 1990).

All snakes are carnivores—that is, they all feed on other animals—but, again, there are specialists and generalists. For example, there are two taxa of snakes that have adapted to become specialist feeders on eggs. These snakes (all but one of which are African snakes belonging to the genus *Dasypeltis*) lack both teeth and venom, but have projections from their spinal vertebrae that allow them to puncture and crush an ingested egg. Australia's keel-back snake *Tropidonophis mairii* is a frog-hunting specialist; other snakes, such as the brown tree snake *Boiga irregularis*,[3] are generalist feeders and will eat small mammals, frogs, lizards, birds, and bird's eggs.

Feeding methods

It is also possible to categorize animals according to the way in which they obtain their food. Some animals, for example, filter small food particles that are in suspension in water and are thus known as suspension, or filter, feeders. In the zoo, such feeders are most likely to be seen in the aquarium (sponges, **lamellibranchs**, and other aquatic invertebrates), but some vertebrates, notably flamingos, also feed this way (**Fig. 12.2**).

Carnivorous species are often predators, which hunt, trap, or ambush their prey, but they may also feed on material that somebody else has caught, or that they find dead (scavengers), or may even steal it off someone else (kleptoparasites). Herbivorous species may eat from the ground layer of plants (grazers), or from bushes and trees (browsers). In all of these cases, the different animals have evolved morphological specialisms, usually in and around the mouth (different teeth, different beak forms,[4] etc.), or front limbs (claws, grasping fingers, etc.) for processing the food (for example, biting, tearing, grinding, etc.) before it is swallowed (**Fig. 12.3**).

The terms **nutritional requirements** and nutritional needs are often used interchangeably, but in some texts at least there is a difference in meaning. The term nutritional requirements usually refers to a set of quite specific requirements that have been quantitatively and qualitatively measured, e.g. the publications on nutrient requirements of livestock by the National Research Council (NRC) in the USA. The term nutritional needs is often used in a broader sense and does not necessarily imply that a detailed analysis has taken place.

3 The brown tree snake *Boiga irregularis* is perhaps best known as a pest species on the island of Guam, where it was accidentally introduced. As a generalist predator, it has had a devastating impact on the island's native fauna, particularly bird species.

A **lamellibranch** is a bivalve mollusc belonging to the Class Lamellibranchia. This taxon includes species such as scallops and clams.

4 Klasing (1998) provides an extensive list of the different ways in which birds obtain their food, from dabblers (such as many duck species) and excavators (such as woodpeckers) to probers (such as the curlew *Numenius arquata*), which insert their bill into a substrate, such as mud or sand, to obtain prey.

Figure **12.2** (*right*) This photograph shows a Chilean flamingo *Phoenicopterus chilensis*, a vertebrate filter feeder. There are many invertebrate examples of filter feeders, but relatively few vertebrates feed in this way. (Photograph: Sheila Pankhurst)

(a)

(b)

(c)

(d)

Figure **12.3** Birds show many different specialized beak shapes, such as the probing bill of the Waldrapp or northern bald ibis *Geronticus eremita* (a); the stabbing bill of the fish-eating Inca tern *Larosterna inca* (b); the very large pouched beak of the Eastern white pelican *Pelecanus onocrotalus*, adapted for catching and storing fish (c) and the small beak of the desert-living Gambel's quail *Callipepla gambelii*, adapted for feeding on seeds (d). (Photographs: Sheila Pankhurst)

12.1.2 Gastrointestinal tract anatomy and function

All vertebrate animals have a gastrointestinal (GI) tract or digestive tract, with associated organs and glands that release enzymes and other substances into the gut (Stevens and Hume, 1995). But there is considerable variation between, and even within, taxa in the morphology of the GI tract, depending on whether an animal is a carnivore or a herbivore and how it obtains its food. Not all structural elements of the vertebrate GI tract are found in all species. Stevens and Hume (1995) suggest that comparisons between the digestive tracts of different taxa are best made under broad functional headings such as 'foregut', 'midgut', and 'hindgut'[5] (also reflecting the embryological origin of the different parts of the digestive tract).

Not all invertebrate animals have a gut, but, for those that do, there are functional similarities with the vertebrate gut. A cricket or locust, for example, has a tubular gut with a crop, in which food can be stored temporarily, and a gizzard, in which food is ground up. Insects also have an intestine (in which nutrient absorption occurs) and an anus. The overwhelming majority of zoo animals, however, are vertebrates and so the remainder of this section is about the vertebrate gut.

In its simplest form, the vertebrate GI tract is a long tube with a bulge, the stomach, at one end (some fish do not have a proper stomach). Although simple in structure, the midgut and hindgut of the omnivorous American black bear *Ursus americanus* (**Fig. 12.4**) combine to make a relatively long intestinal tube between the stomach and the anus. In general, carnivores have a short, simple GI tract and herbivores have a much longer GI tract, with the ruminant herbivores possessing the most complex digestive system (**Fig. 12.5**; we shall return to the ruminant digestive system in more detail below).

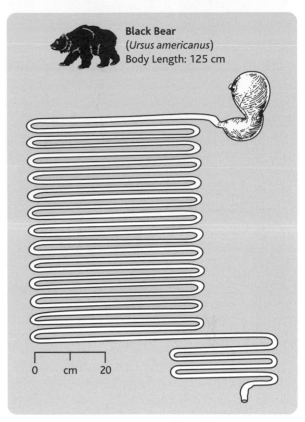

Figure **12.4** This diagram shows the relatively simple gastrointestinal (GI) tract of an omnivore, the American black bear *Ursus americanus*. (Reproduced with permission from Stevens and Hume, 1995)

What are the reasons for these differences between carnivores and herbivores in the structure and length of the GI tract? To answer this question, we need to look at the process of **digestion** and, in particular, the problem that herbivores face in breaking down the tough material that makes up plant cell walls.[6]

Digestion

The macromolecules that animals ingest as food are broken down into smaller molecules in the process of digestion. This breakdown of food during digestion takes place by means of enzymatic hydrolysis

---

5 The terms 'small intestine' and 'large intestine' can be misleading, because these sections of the GI tract may be almost the same diameter in some vertebrate species (Stevens and Hume, 1995). The terms 'midgut' and 'hindgut' distinguish between these sections on the basis of their main function.

6 The cell walls of fruits are not as tough as those of leaves, stems, and other plant material, and are easier to break down with enzymes during digestion. Thus, frugivores face less of a challenge digesting plant material than browsers or grazers.

432

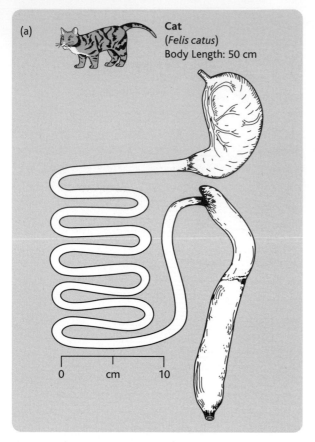

Cat
(*Felis catus*)
Body Length: 50 cm

0   cm   10

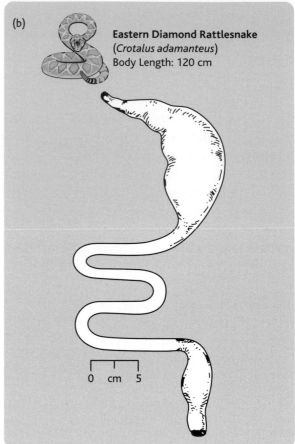

Eastern Diamond Rattlesnake
(*Crotalus adamanteus*)
Body Length: 120 cm

0   cm   5

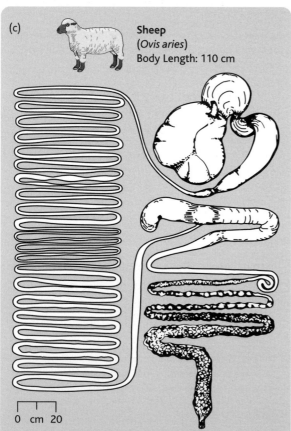

Sheep
(*Ovis aries*)
Body Length: 110 cm

0   cm   20

Figure **12.5** These diagrams show the gastrointestinal (GI) tracts of (a) a domestic cat *Felis catus*, (b) a reptilian carnivore, the eastern diamond rattlesnake *Crotalus adamanteus*, and (c) a sheep *Ovis aries*. The carnivorous cat and snake have a much shorter and simpler GI tract than the sheep, which is a ruminant herbivore. (Reproduced with permission from Stevens and Hume, 1995)

reactions, in which water in the presence of a digestive enzyme splits apart larger molecules. The smaller molecules can then be absorbed and utilized for energy, growth, and tissue repair.

A lot is known about the mechanical and chemical processes involved in digestion, as well as the neural and endocrine mechanisms by which these are regulated. These topics are outside the scope of this book, however, so you are referred to good physiology textbooks such as Schmidt-Nielsen (1997) and Willmer *et al.* (2000). But it is appropriate at this stage to look briefly at the problems involved in **cellulose** digestion.

Figure **12.6** The green iguana *Iguana iguana* is an example of a gut fermenter: an animal that uses microbial fermentation in the gut to break down ingested plant material. (Photograph: ©Stephanie Rousseau, `www.iStockphoto.com`)

The problem of cellulose

Cellulose is a carbohydrate macromolecule used extensively by plants in their structure (for example, as cell walls) and, because of this, it is potentially one of the most widely available sources of energy for animals. But, for some reason, very few animals have evolved an enzyme (a cellulase) for digesting cellulose, so how do they survive on a diet of plant material?

The answer, for many herbivorous animals, is that they rely on **symbiotic** micro-organisms living in their digestive tract. These micro-organisms could not survive independently outside the animal's body and the animal could not derive energy from plant material through its own digestive processes.

The general name for the microorganisms that inhabit the digestive tract is the **gut microflora**.[7] These microorganisms can digest cellulose to produce not simple sugars (which would be the case in starch digestion), but volatile fatty acids, a process

often referred to as fermentation. Examples of gut fermentation to process tough plant materials in the diet can be found throughout the vertebrates. For example, animals as diverse as the llama *Llama glama*, green iguana *Iguana iguana* (see **Fig. 12.6**), and the green sea turtle *Chelonia mydas* use microbial fermentation of plant foods, as do many **gallinaceous** birds.

Probably the most complex, but efficient, evolutionary outcome of this symbiotic relationship is seen in the ruminant mammals (animals that chew the cud). These mammals are all members of the order Artiodactyla[8] (one of the orders of **ungulates**, or hoofed mammals) and most are members of the sub-order Ruminantia; the camels and llamas are an exception, and belong to the family Camelidae, within the sub-order Tylopoda (MacDonald, 2001).

Symbiosis refers to a close structural and physiological relationship between two or more species that could not survive independently of the relationship.

7 All animals have a gut microflora, but there is considerable variation in the extent to which different taxa rely on gut fermentation for their supply of energy and nutrients.

The **gallinaceous** (meaning chicken-like) birds are birds such as pheasants, partridges, grouse, quails, etc., from the

order Galliformes. They are usually ground-feeding birds and many species are hunted as game birds.

8 Mammalian taxonomy is currently undergoing rapid change, with new insights from molecular analysis overturning traditional views on how animals should be grouped. For example, we now know that whales and dolphins (Cetacea) are more closely related to the even-toed ungulates, such as deer and cattle, than to other mammalian taxa (see, for example, Springer and de Jong, 2001).

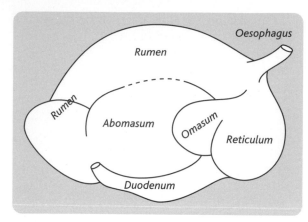

Figure **12.7** This diagram shows the basic structure of the chambered stomach of a typical ruminant. One of the stomach chambers, the rumen, is a major site for microbial fermentation of plant material. (Picture: Geoff Hosey)

The ruminant stomach

In ruminants, the stomach has become more complex by the addition of extra chambers (see **Fig. 12.7**), one of which, the **rumen** is particularly large and houses a substantial microbial population (non-ruminant animals with a single-chambered stomach are sometimes referred to as **monogastric** animals —for example, bears, rodents, pigs, and humans). These microorganisms live in what is essentially a fermentation vat, kept at a temperature of around 37–40°C and buffered from pH fluctuations by the host animal's saliva, which contains large quantities of sodium bicarbonate in solution.

Ruminants often feed very rapidly and swallow their food in a relatively unchewed state. When the rumen is nearly full, the animal rests and starts ruminating, a process of bringing the unchewed food back to the mouth for more chewing. When this chewed food is swallowed, it sinks to the bottom of the rumen, where it is broken down to volatile fatty acids (mostly acetic, propionic, and butyric acids) by the rumen microflora. Importantly, the microflora also synthesize amino acids and vitamin B group compounds (see section **12.2.2**), which are

available for the host animal's use. Because of this, the ruminant gets more from its microflora than only the breakdown products of cellulose.

Foregut and hindgut fermenters

A useful way of considering the nutritional requirements and digestive capabilities of herbivorous animals that rely on gut fermentation is to distinguish between 'foregut fermenters' and 'hindgut fermenters': in other words, to consider where, within the gut, there is space for fermentation. In ruminants, this space is the rumen, which is part of the stomach and so within the foregut (the foregut can be defined, very roughly, as the portion of the gut from the mouth to the stomach exit). Among non-ruminant mammals, sloths, langur monkeys, and kangaroos also have large stomachs with several compartments in which microbial fermentation of cellulose takes place, so these animals too are foregut fermenters.

In other mammals (particularly rodents, lagomorphs, and equids), caecal fermentation takes place: the ingested plant material travels through the alimentary canal, as it does in other mammals, but when it reaches the **caecum**, it is digested by the microorganisms living there. So these animals, along with the non-mammalian herbivores that make use of caecal fermentation (we have already mentioned the green iguana and the gallinaceous birds as examples), can be grouped together as hindgut fermenters.

The location within the herbivore gut of the major site for fermentation of plant materials has important implications for the absorption of nutrients (Oftedal *et al.*, 1996). A substantial proportion of the absorption of nutrients takes place in the midgut. So hindgut fermenters appear to be at something of a disadvantage, in that the breakdown of cellulose occurs after the ingested food material has passed through the main absorptive area of their digestive tract. In a number of taxa, some herbivores

The **caecum** is a large diverticulum (a sac with only one opening) at the point where the small intestine becomes the large intestine. All that is left of it in humans is the sometimes-troublesome appendix.

(for example, many lagomorph and rodent species) that have caecal fermentation also reingest their faeces to extract some of these products a second time around. This strategy, called **coprophagy**,[9] allows better overall absorption of the products of digestion and can play a very important role in the nutrition of species such as rabbits (Hörnicke, 1981).

Given that foregut fermentation allows absorption of nutrients in the midgut without the need for strategies such as coprophagy to improve the efficiency of nutrient uptake, why are so many animals hindgut or caecum fermenters? Rumination is highly efficient, but it carries a cost in terms of the time that it takes for food to pass through the GI tract. For a very large herbivore, such as a horse, or a rhinoceros, or an elephant, the best strategy is a high intake and the faster throughput that can be achieved with a simpler digestive tract. This gives the greatest net intake of nutrients, despite the lower efficiency of hindgut fermentation (MacDonald, 2001).

A consequence of the difference between these two feeding strategies (**Fig. 12.8**) is that rumination is only effective for animals in a certain size range. The largest ruminant is the giraffe, at up to about 1,200kg in bodyweight, but the largest hindgut fermenters can weigh several tonnes (for example, a rhinoceros can weigh more than 2 tonnes and a large male African elephant *Loxodonta africana* more than 6 tonnes).

12.1.3 Food selection: do animals have 'nutritional wisdom'?

Of course, knowing that a species is a herbivore or carnivore, or even what type of herbivore or carnivore, does not always help us very much when formulating diets for these animals in captivity. Both specialists and generalists do not usually simply eat everything that they come across that happens to be within the broad category of 'browse' or 'fruit', or whatever else they are adapted to eat. Most animals are very selective about what they eat and will seek out, or choose, particular preferred food items,[10] while rejecting others that might seem to us to be perfectly acceptable.

The task that animals have in the wild is to satisfy all of their nutritional requirements. At the cellular level, however, the requirements of most animals, whatever their species, are pretty much the same (Moore *et al.*, 2005). The differences that we see in food selection reflect different evolutionary strategies for getting those nutrients.

How do animals choose their diets?

It is possible that animals have an innate **nutritional wisdom** that allows them to select a balanced diet by detecting the nutritional quality of the different foods they encounter: a hypothesis that is termed **euphagia** (Moore *et al.*, 2005). But most evidence of animal food selection contradicts this view: for example, captive capuchin monkeys *Cebus apella* choose a variety of different foods in preference to **monkey chow**, even though the chow has high energy, and is designed to provide a complete and balanced diet for monkeys (Addessi *et al.*, 2005). Even to us, monkey chow does not look particularly enticing, so we could perhaps suggest that palatability is an important component of diet selection. But what does palatability tell the animal about the food?

---

9 Coprophagy, or faeces eating, is by no means confined to lagomorphs and rodents, or indeed to mammals. There have been reports of coprophagy in mammals ranging from koalas (Osawa *et al.*, 1993) to ring-tailed lemurs *Lemur catta* (Fish *et al.*, 2007), and in non-mammalian species such as coral reef fish (Robertson, 1982), hatchling green iguanas (Troyer, 1984a; 1984b), and even one species of cave-dwelling salamander (Fenolio *et al.*, 2006).

10 The processing of food takes time and energy, so animals must take account of this in their decisions about whether or not to eat certain items. But the use of the word 'decisions' should not be taken to imply conscious or deliberate choice: animals have evolved to make decisions about foraging and other behaviours on the basis of what might be thought of as simple 'rules of thumb'—that is, the product of innate behaviours and learning—without necessarily any need for conscious action.

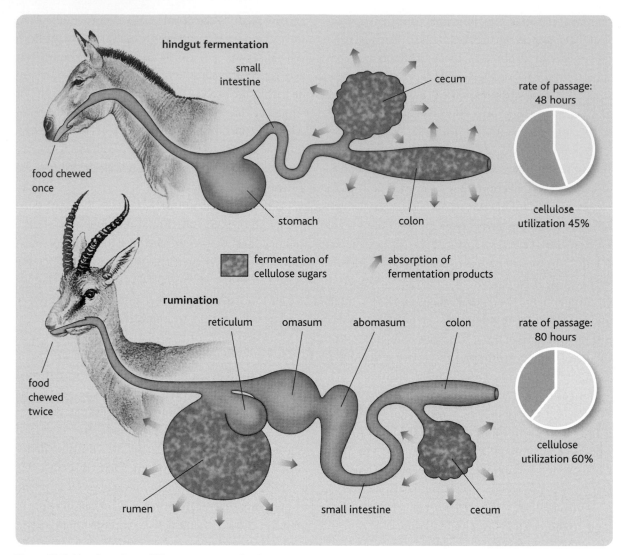

Figure **12.8** Ungulates have different strategies for digesting cellulose. Equids, such as the Przewalski's horse *Equus ferus przewalskii*, have a faster gut throughput, but lower rate of cellulose utilization, whereas ruminants, such as gazelles, have greater digestive efficiency, but a slower throughput of food. (Reproduced with permission from MacDonald, 2001)

One possibility, termed the 'hedyphagia hypothesis' (Moore *et al.*, 2005), is that animals have evolved to find nutritious foods nice tasting and poor quality or toxic foods unpleasant tasting. While this idea has some intuitive appeal, there is little evidence to support the notion that animals have an innate ability to detect high-quality nutrients in their food through sensory cues (Moore *et al.*, 2005). Food selection trials with blue duikers *Cephalophus monticola* and white-bellied duikers *C. leucogaster*,

for example, showed that the animals selected natural foods on the basis of colour over and above features such as size, shape, or chemical composition (Molloy and Hart, 2002).

A major problem with both the euphagia and hedyphagia hypotheses is that, by seeing food selection as an innate ability, we cannot account for the flexibility that animals show when they encounter novel foods, or when they experience adverse effects (such as nausea) after eating something. All of the evidence suggests that **learning** plays a large part in the development of an animal's food preferences and aversions, whether in the wild or in captivity, and helps to explain why many herbivores seek out variety even if nutritional needs can be met by a less varied diet (Provenza, 1996).

Plant defences against herbivory

Plants are not passive participants in the process of herbivory.[11] It is not always in their best interests to be eaten by an animal, so many plants have evolved structures and chemicals to deter animals from eating them. These structures might include things such as spines and thorns. Less obvious are chemical defences, which include a variety of **plant secondary metabolites (PSMs)**,[12] substances that are so-called because they are synthesized from the products of ordinary or primary metabolism. These substances act to make plants distasteful or even toxic, or may make it difficult for the animal to extract the important nutrients.

PSMs are so ubiquitous that herbivores in the wild cannot feasibly avoid them (Moore *et al.*, 2005), and nor can zoos entirely avoid feeding **forage** or browse containing PSMs. So the objectives for zoos are to understand which PSMs may be found in the forage and browse that they feed to their animals, what the effects of these substances may be, and also to keep the amounts of PSMs in the diet of their herbivores within safe limits. Further information about PSMs is provided in **Box 12.1**.

Box 12.1 The plants fight back

Most of the plant chemical defence substances referred to as plant secondary metabolites (PSMs) belong to one of three categories (Harbone, 1991): phenolics (for example, anthocyanins and tannins); nitrogen-containing compounds (such as alkaloids); and terpenoids (or isoprenoids). Tannins, for example, are found in varying amounts in almost all plant species worldwide, with browse containing greater amounts than grasses. Oak bark and acorns—and, to a lesser extent, oak leaves—are particularly rich sources of tannins.

Often, these substances are bitter-tasting, which acts to deter animals from eating them; indeed, some plants advertise their presence, for example, in bitter-tasting secretions that are released when the plant's tissue is damaged. Many herbivores are able to break these compounds down (detoxification), so they are able to tolerate feeding on the plants provided that their intake remains below a certain threshold. The speed at which herbivorous mammals can detoxify these compounds is size-related, with smaller species able to detoxify more rapidly than larger species. But metabolic requirement per unit body mass decreases as body mass increases, so small mammals ingest larger doses of PSMs per unit of body mass than do larger mammals in order to meet their metabolic requirements (Freeland, 1991).

In the UK, in order to address the challenge of which browse could safely be fed to zoo

---

11 Sometimes, it is in a plant's interests to be eaten, as, for example, when seeds need dispersing by animals, or need to be digested by an animal in order to start the germination process.

12 One of the best known plant secondary metabolites is nicotine, found in tobacco plants.
Forage is a general term for any kind of fodder made out of herbaceous plants, so it includes hay, straw, silage etc.

Figure **12.9** Giraffes are browsers. At Colchester Zoo in Essex, members of the public can feed the giraffes with browse supplied by the zoo, during an educational presentation about the diet and nutrition of this species. (Photograph: Sheila Pankhurst)

animals (**Fig. 12.9**), researchers at Paignton Zoo set up a browse database (Plowman and Turner, 2001b), after surveying keepers at a number of zoos to find out which browse species they were using. Perhaps not surprisingly, wildlife parks with extensive grounds fed a greater variety and amount of browse to their animals. The browse species most commonly fed to zoo animals, among the zoos that responded to the survey, were species of willow *Salix* spp. Very few cases of poisoning or other ill effects were reported.

Copies of the browse database, on CD, are available from BIAZA and can be ordered for a small charge via the BIAZA website.

12.2 Basic nutritional theory

Why do animals need food anyway? All living things require energy to operate their cellular and metabolic machinery, and also to fulfil their life history activities, including growth and reproduction. For animals, that energy is obtained by consuming plants (**Fig. 12.10**) or other animals. The correct term for this is 'heterotrophic', which means an organism that obtains carbon for growth from consuming **organic** matter (plants, by contrast, can derive their carbon and energy from non-organic sources, and are referred to as 'autotrophic').

The compounds that supply that energy are mostly carbohydrates and fats, although energy can also be derived from proteins. When eaten, these substances have to be broken down into their constituent units (the process of digestion), which are simple sugars (from carbohydrates such as starch), short-chain fatty acids (from the break-

Figure **12.10** Squirrel monkeys, such as this free-ranging black-capped squirrel monkey *Saimiri boliviensis* at Apenheul Primate Park in the Netherlands, feed on fruits, flowers, and insects, which they obtain while foraging among vegetation. (Photograph: Sheila Pankhurst)

down of fibrous carbohydrates such as cellulose), long-chain fatty acids (from fats), and amino acids (from proteins). These units can then be absorbed into the body, and used for growth and metabolism.

The term **organic** refers to carbon-based compounds, which usually also contain oxygen and hydrogen. These compounds are not necessarily the result of living processes.

Oxidation of these molecules provides most of the animal's energy requirements.

In general, animals have no specific requirements for any particular carbohydrates, fats, or proteins: just about any will do equally well. There are, however, certain specific substances, both organic molecules and chemical elements, that are essential, in the sense that the animal must have a supply of them in its diet, usually because they are a necessary component of enzyme systems and, even if they are organic, the animal may not be able to synthesize them itself. Thus, animals may have specific nutritional requirements over and above the basic need for a supply of energy.

12.2.1 Energy and metabolism

Energy is required by animals to operate their physiological processes, to be active and behave, and to grow and reproduce. Energy enters the animal's body as food and, in its use for physiological and behavioural processes, it is lost again from the body. Some energy is also lost in urine (**Fig. 12.11**), and some more (which has never been assimilated) in faeces.

If food intake exceeds the amount of energy used, the surplus is stored, usually as fat. Conversely, if food intake is less than the amount of energy consumed, stored energy is liberated by metabolizing those fats. We can, then, think of the animal as a system through which energy flows (**Fig. 12.12**). The values can, in principle, be measured and they all come together in an equation that represents the **energy budget** of the animal:

$$C = P + R + U + F$$

(See **Fig. 12.12** for a key.)

In this budget, P represents any energy that the animal has available to put into growth. If the value of this is zero, then the animal can be said to be on **maintenance rations**: the amount of energy entering the animal is equal to the amount leaving the animal. **Figure 12.12** also shows A, or assimilated energy. This is, effectively, that part of the ingested energy that the animal has available for its use.

Figure **12.11** An Asian elephant *Elephas maximus* at Rotterdam Zoo in the Netherlands: some energy is lost from the body during urination. (Photograph: Sheila Pankhurst)

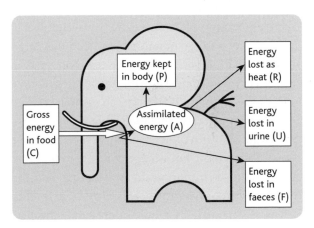

Figure **12.12** This diagram shows energy flow through a typical vertebrate animal. The equation C = P + R + U + F represents the energy budget of the animal. (Picture: Geoff Hosey)

The **energy budget** of an animal is the amount of energy that goes into the animal (from its food), compared with the amount of energy that goes out of the animal (e.g. as heat lost from the animal, and as energy lost in urine and faeces).

Values of A can vary widely according to the kind of food that the animal is eating. Thus, herbivores, for example, usually have a low A:C ratio (about 57 per cent, meaning only just over half of the energy in their food is being assimilated), because cellulose, from which plant cell walls are made, is difficult to digest (see section 12.1.2). Because animal material does not contain cellulose, carnivores, on the other hand, assimilate about 80 per cent of the energy in their food, even without such long or complex digestive tracts as herbivores.

The rate at which animals use energy can be broadly referred to as their **metabolic rate**, which is energy metabolism per unit time. Strictly, this should be measured as the **basal metabolic rate** (which is, effectively, the minimum amount of energy usage required to support life processes), but measuring this in most animals is not feasible, because it has to be done in very particular conditions. Instead, most animal measurements are of the standard metabolic rate, which is the resting or fasting metabolism at a given temperature. If we measure the overall metabolic rates of a range of animals, we find that large animals have higher metabolic rates than smaller animals, but if we standardize metabolic rates per unit mass of the animals, in fact, we can see that smaller animals have a greater metabolic rate per unit mass than larger ones. An understanding of metabolic rate —or at least a basic knowledge of where to find this information—is important in formulating diets for zoo animals: zoos need to know how much to feed, as well as which nutrients should be present in the diet.

12.2.2 Specific nutritional requirements

Nutritional requirements are the particular elements or compounds that an animal must have in its diet in order to satisfy all of its requirements for growth and for building and maintaining its metabolic machinery. The first group of these essential substances is made up of organic compounds such as vitamins or amino acids. The second group is particular elements, which may be required in quite substantial quantities—or, in most cases, in just tiny amounts—such as many minerals.

Essential organic compounds

Proteins are important constituents of animal bodies because they form much of the body's structure, but also because enzymes and some other compounds involved in physiological processes (for example, some hormones) are also proteins. Animals construct their own proteins from the amino acids that they obtain by digestion of plant or animal proteins in their diet.

Of the twenty common amino acids that make up most proteins, animals are able to synthesize about half of them from other amino acids. These are termed 'non-essential amino acids', not because they are not important, but because they do not have to occur in the diet. These non-essential amino acids (based on our knowledge of the laboratory rat) are glycine, alanine, serine, asparagine, aspartic acid, glutamic acid, proline, glutamine, tyrosine, and cysteine. In contrast, ten other amino acids are essential, meaning that animals cannot synthesize them and must therefore ingest them as part of their food. The essential amino acids are lysine, tryptophan, histidine, phenylalanine, leucine, isoleucine, threonine, methionine, valine, and arginine (Schmidt-Nielsen, 1997).

Similarly, there are essential fatty acids (EFAs) that must be present in the diet of mammals (mammals cannot synthesize these substances). Nowadays, fatty acids are often referred to as belonging to one of two main categories: saturated and unsaturated. The unsaturated fatty acids can be further divided into monounsaturated and polyunsaturated fatty acids, and it is in this last category that we find the so-called essential fatty acids, which are alpha-linolenic acid (often represented as 'α-linolenic acid') and linoleic acid[13] (McDonald *et al.*, 2002).

---

13 Texts about human nutrition in particular often refer to the 'omega-6 (n-6)' and 'omega-3 (n-3 families of fatty acids.

Omega-3 fatty acids are derived from alpha-linolenic acid and omega-6 fatty acids are derived from linoleic acid.

In the diet, the so-called oilseeds, such as linseed or rapeseed oil, are particularly good sources of alpha-linolenic acid. Alpha-linolenic acid is also found in fish oils (which is the main reason why we are encouraged to include oily fish, such as salmon or mackerel, in our diet).[14]

Vitamins

Vitamins and related compounds are organic compounds that are only required in small amounts, but which fulfil important metabolic roles within the body, often as coenzymes (McDowell, 1989). Some vitamins (most famously, perhaps, vitamin C, a lack of which causes scurvy)[15] are well known to us because of the dramatic effects that their absence in the diet can cause in human bodies, but, again, we know less about how they are significant to other species. Vitamins were only discovered and isolated in the twentieth century, and the concept of deficiency diseases due to a lack of a specific vitamin in the diet is a relatively new one, although the diseases themselves (such as scurvy and beri-beri) have been known for much longer.

Chemically, the vitamins are a diverse group. Some animals are able to synthesize some vitamins for themselves; others must be present in the diet. For example, amphibians, reptiles, most birds, and most mammals can synthesize vitamin C (ascorbic acid), but primates, some fruit-eating bats, guinea pigs, **teleost** fish, and many invertebrates cannot (McDowell, 1989; **Fig. 12.13**). Similarly, ruminant mammals have no dietary need for some of the B-group vitamins, because these are synthesized by the microbial flora living in the rumen.

Vitamins are often categorized as water-soluble or fat-soluble. An excess of a water-soluble vitamin, such as vitamin C, will simply be excreted. Fat-soluble vitamins (A, D, E, and K), however, will be

Figure **12.13** Teleost (bony) fish, such as this catfish at Warsaw Zoo, are unable to synthesize vitamin C and so must obtain this nutrient from dietary sources. Other animals that are unable to synthesize vitamin C are primates, some fruit-eating bats, guinea pigs, and many invertebrates. (Photograph: Barbara Zaleweska, Warsaw Zoo)

stored in body fat and can sometimes build up to toxic levels in the liver if ingested in excess.

Vitamin D is essential in the body of vertebrates for the absorption and metabolism of calcium (for this reason, it is also called 'calciferol'). This fat-soluble vitamin is unusual in that it does not have to be obtained from the diet, but can be manufactured in the body in the presence of sunlight. Zoo animals that are not exposed to sufficient natural light can become deficient in vitamin D, which, in turn, affects their calcium metabolism (see section **12.8.2**).

Minerals

The greatest part of the animal body (about 96 per cent by weight) is made up of the elements oxygen (O), hydrogen (H), carbon (C), and nitrogen (N), because these are the major constituents of water and of most organic molecules. The remaining 4 per cent is made up primarily of calcium (Ca), phosphorus (P), potassium (K), sulphur (S), sodium (Na),

---

14 Chapter 3 of McDonald *et al.* (2002) provides a detailed account of lipids (including fats) in the diet, including the chemical structure of these substances.

15 Scurvy is not a disease only of humans, but can occur in all animals that cannot synthesize vitamin C. For example, scurvy has been reported in captive capybara (Cueto *et al.*, 2000).

chlorine (Cl), and magnesium (Mg), in that order (Schmidt-Nielsen, 1997).

These six elements are often referred to as 'macroelements',[16] or macronutrients, or macro-minerals (or sometimes the major minerals). In addition, however, there are more than a dozen other elements that are found in the body, often as parts of enzymes, and which appear to be necessary for metabolic function, but only in tiny amounts. These are known as **trace elements**, or sometimes as 'microminerals' (McDowell, 1992). None of these trace elements can be synthesized in the body by living organisms, so any metabolic needs must be met by dietary intake.

The major function of minerals in animals is in skeletal structures, such as bones, teeth, and shells, and eggshells in birds; other functions of minerals are as constituents of enzymes and hormones, and also as electrolytes, helping to maintain osmotic **homeostasis** within the body (McDowell, 1992; **Fig. 12.14**). The main minerals needed for normal life processes are listed in **Table 12.1**, although we should remember that there are still large gaps in our knowledge of what these elements are necessary for within the body, and what knowledge we do have comes largely from studies of laboratory, farm, and companion animals, not of wild animals.

Because health problems can be caused both by too much and too little of these elements (see section **12.8.2**), it is important to know, where possible, how much of these elements are present in diets fed to zoo animals. For example, insects such as crickets and fruit flies commonly fed to insectivorous birds, reptiles, and mammals appear to satisfy requirements for copper (Cu), iron (Fe), magnesium (Mg), phosphorus (P), and zinc (Zn) (based on our knowledge of domestic animal requirements), but some mealworms and waxworms are deficient in manganese (Mn) (Barker *et al.*, 1998; **Fig. 12.15**). A deficiency in either calcium or phosphorus can

Figure **12.14** Calcium and phosphorus are important nutrients for bone and tooth development in vertebrates. These minerals are important for many invertebrate species too, particularly molluscs, such as snails, which need calcium phosphate to build their shells, as in the case of this African land snail *Achatina* sp. (Photograph: ©Kevin Lindeque, www.iStockphoto.com)

result in metabolic bone disorders, such as rickets or osteomalacia, in vertebrates.

Sometimes, knowing how much of a particular mineral is in an animal's diet still does not help us to predict whether the intake for that animal will be sufficient. Minerals can vary greatly in their

Figure **12.15** A leopard gecko *Eublepharis macularius* eating a cricket. Insects such as crickets appear to satisfy vertebrate dietary requirements for some minerals, such as magnesium, phosphorus, and zinc, but some mealworms and waxworms fed to vertebrates are deficient in manganese (Barker *et al.*, 1998). (Photograph: ©Cathy Keifer, www.iStockphoto.com)

---

16 Not all minerals are elements: some are frequently found as salts, which are compounds.

Mineral	Abbreviation	Some metabolic functions
MACROMINERALS		
Calcium	Ca	Major part of the mineral content of bone. Calcium metabolism in the body is very closely linked to phosphorus uptake and metabolism.
Chlorine	Cl	With sodium, makes up common salt (NaCl). Both sodium and chlorine are vital to the maintenance of osmotic homeostasis within the body. Chlorine also plays a key role in regulation of blood pH.
Magnesium	Mg	Diverse functions in body, from tooth and bone integrity to enzyme function.
Phosphorus	Ph	Major part of the mineral content of bone. Usually considered together with calcium.
Potassium	K	Osmotic balance (within cells); transport of oxygen and carbon dioxide in the blood.
Sodium	Na	See chlorine. Vital in maintaining osmotic pressure within the body, particularly in extra-cellular fluids. Also transmission of nerve impulses.
Sulphur	S	Component of amino acids methionine, cystine and cysteine. Also important component of bone, tendon and cartilage structure.
MICROMINERALS		
Chromium	Cr	Important in insulin activity and glucose metabolism.
Cobalt	Co	Major constituent of vitamin B_{12}. Essential in the diet of ruminants for manufacturing vitamin B_{12}.
Copper	Cu	Growth, formation of blood cells. *Toxic in excess. Often considered together with molybdenum.
Fluorine	F	Tooth development. *Very toxic in excess.
Iodine	I	Constituent of thyroid hormones thyroxine and triiodothyronine.
Iron	Fe	Oxygen transport in the blood (iron is a vital component of haemoglobin) and other key biochemical reactions.
Manganese	Mn	Enzyme function; bone growth.
Molybdenum	Mo	Enzyme function. *Toxic in excess. Closely linked to copper metabolism (molybdenum and copper are antagonistic in the body).
Selenium	Se	Antioxidant function (closely linked to vitamin E). *Toxic in excess.
Zinc	Zn	Enzyme function. Zinc deficiency impacts on DNA, RNA and protein synthesis in the body.

Table **12.1** Essential minerals

* All minerals can cause problems within the body both when they are deficient and when they are present in excess, although toxicity problems are better documented for some minerals (e.g. fluorine, molybdenum, and selenium) than others.

NOTE: This list covers the main minerals needed for normal body processes, but is not exhaustive: other trace elements, such as boron (B) and lithium (Li), are known to be required by some animal species. See McDowell (1992) for further information.

bioavailability, which will depend on a variety of factors, such as the chemical form of the mineral or the presence of other minerals that may block absorption—and this leads us to consider interactions between minerals in the diet, because these substances rarely act in isolation within the body and the absorption of one mineral will be influenced by the presence (or absence) of another.

The relationship or ratio between different minerals is often as important, or more important, for growth and health than the absolute amount in the diet (McDowell, 1992). For example, the

Bioavailability means the proportion of a nutrient that is available for physiological functions within the body. The bioavailability of some minerals can be reduced by antagonistic effects between minerals, in which the presence of one blocks the uptake or utilization of another.

relationship between calcium and phosphorus in the body means that animals can be calcium deficient, even if there is enough calcium in their diet, if their diet also contains an excess of phosphorus. Ensuring that the diets of zoo animals provide the correct ratio of calcium (Ca) to phosphorus (P) is a major challenge for zoo nutritionists. The recommended Ca:P ratio in the diet for most mammals lies between 1:1 and 2:1 (the ratio in bone is slightly above 2:1); this dietary ratio should probably be higher in birds, but data are lacking except for poultry (Klasing, 1998). Klasing (1998) also points out that smaller birds need proportionately more calcium for egg-laying than larger birds. This is for two reasons: firstly, small birds lay proportionately larger eggs; secondly, smaller eggs have proportionately more shell. Achieving the right Ca:P ratio in the diets of zoo animals is not always easy without calcium supplementation. Widely available grass hays, for example, such as Timothy hays, have a very low calcium content in comparison with legume hays; cereal grains are also very low in calcium (McDowell, 2003).

Other minerals that have documented antagonistic effects include copper (Cu) and molybdenum (Mo), with high levels of Mo in the diet limiting the retention of Cu in the body. A Cu deficiency, for example, can give rise to a **wasting disease** in grazing ruminants, which is sometimes called 'salt sick' (McDowell, 1992).

12.3 Guidelines and legislation on feeding zoo animals

In **Chapter 3**, we looked at the overall regulatory framework for zoos and particularly for zoos in the UK. In this section, we provide a brief overview of the main legislation covering the provision of food and water for zoo animals, as well as non-legislative guidelines and other sources of information on zoo

animal nutrition. It is worth noting, however, that, in many countries, the legislation relating to the provision of food for zoo animals generally only covers the minimum requirements: that food and water should be made available, and that zoos cannot allow animals to starve. Good zoos will be concerned about providing optimum nutrition, which goes beyond simply the provision of enough food and water.

12.3.1 UK legislation on zoo animal feeding and nutrition

In the UK, as we saw in **Chapter 3**, the primary legislation regulating the activity of zoos is the Zoo Licensing Act (ZLA) 1981, as amended by the Zoo Licensing Act 1981 (Amendment) (England and Wales) Regulations 2002. The ZLA itself does not say very much about zoo animal nutrition, but, in England, zoo inspectors granting licences to zoos under the Act are required to take account of the Secretary of State's Standards of Modern Zoo Practice (SSSMZP) (Defra, 2004), which do make a number of detailed recommendations on feeding zoo animals.

The first of **five principles** of zoo animal care and welfare set out in section 2 of the SSSMZP relates to the provision of food and water, and describes in some detail the standards that are expected of zoos in the UK in relation to feeding their animals.[17] For example, section 2.1.1 of the SSSMZP states that:

> Food provided must be presented in an appropriate manner and must be of the nutritive value, quantity, quality and variety for the species, and its condition, size and physiological, reproductive and health status.
>
> (Defra, 2004)

Other recommendations in section 2 of the SSSMZP deal with the health and hygiene aspects of the provision of food and water, including the requirement

---

17 Legislation in the UK does not prohibit visitors feeding zoo animals, but the SSSMZP say that 'Uncontrolled feeding of animals by visitors should not be permitted.'

that 'Preparation of food, and, where appropriate, drink should be undertaken in a separate area suitably designed and constructed, and used for no other purpose' (Defra, 2004).

The SSSMZP also state that 'Veterinary or other specialist advice in all aspects of nutrition must be obtained and followed.' While vets do provide advice on nutrition at most zoos, only one zoo in the UK (at the time of writing) has its own full-time, in-house nutritionist: Chester Zoo, in the north of England. (Zoos in North America fare a little better, with in-house nutritionists at a number of leading zoos.)

Other UK legislation that impacts on how zoos feed their animals includes regulations resulting from the European Animal By-Products Regulation (EC) No. 1774/2002. In England, for example, the relevant national legislation is the Animal By-Products (ABP) Regulations 2005. **Chapter 3** provides further information and online sources for much of this legislation (largely from the Defra website).

Feeding of live prey

In the UK, it is illegal to feed live vertebrates and **cephalods** to other animals, but acceptable to feed other live invertebrate species, such as locusts, crickets, and mealworms (Defra, 2004). No other country has legislation restricting the feeding of live prey in zoos (see **Box 12.2**).

Box 12.2 Feeding live prey

Outside the UK, there is no country that has legislation explicitly banning the feeding of live vertebrate prey to zoo animals. In practice, however, the feeding of live vertebrate prey in zoos is not widespread. The provision of live fish is probably one of the few exceptions to this generally held rule: live fish are routinely fed (for example, to species such as fishing cats) in some US, Australian, Asian, and mainland European zoos. At Singapore Zoo, visitors can watch from an underwater viewing gallery as polar bears dive into their pool to hunt and catch live fish.

Studies have been undertaken that demonstrate that there are benefits to predatory species in being able to hunt live fish as prey (Shepherdson et al., 1993). In some rare examples, larger vertebrates are fed as live prey to zoo animals (live goats, chickens, and even cows are still routinely fed to large carnivores in at least one Chinese zoo; this practice has not yet been banned at all Chinese zoos).

Animals in the wild capture and eat live prey, so why not in zoos? A survey of visitors to UK zoos suggested that there were few objections from the general public to the feeding of live invertebrate prey, even on-exhibit rather than off-exhibit, but more concern about feeding live vertebrate prey (Ings et al., 1997a). There also are various ethical arguments informing the decision, within the UK, to prohibit the feeding of live vertebrate prey to zoo animals; these are largely concerned with the welfare of both the potential prey and the potential predator. The first concern for the potential prey is that, in a zoo enclosure, it cannot escape the predator, as it might if the situation were played out in the wild. Although the prey may be able to hide or to elude the predator initially, all evasive behaviours within the confines of a zoo enclosure are likely only to prolong the prey's potential state of stress before it is eventually caught and killed. There is good evidence that prolonged pursuits do severely compromise the welfare of prey species (Bateson and Bradshaw, 1997). It is also easy to appreciate that well-fed ▶

► captive-bred predators in the zoo may not be the efficient killers that their wild counterparts are and that any live prey species put into their exhibit may therefore be at a greater risk of prolonged pursuit, and failed attempts at capture and killing, than would be the case in the wild.

There are also potential risks to the welfare of the predator when feeding live prey. It is possible that, in its attempts to survive, the prey may injure the predator by, for example, kicking or biting (see, for example, Frye, 1992). Given the great lengths to which zoos go to preserve the health and well-being of (valuable) animals in their collection, it seems rather perverse to threaten this routinely in order to deliver live food.

There are exceptions to this in relation to which there are clear benefits from feeding live prey and in which these benefits outweigh any potential costs. The obvious example is that of the potential predator being an animal due to be released into the wild as part of a **reintroduction** programme and the provision of live prey being a part of training before release. But even for animals that are part of a reintroduction programme, there are alternatives to the provision of live vertebrates as prey. One of these is carcass feeding, which is explored in more detail in **Box 12.6**.

12.3.2 Legislation outside the UK on zoo animal feeding and nutrition

In the USA, the Animal Welfare Act (AWA) of 1966 requires, under the licensing arrangements for zoos, minimum standards of provision of food and water. But the AWA applies only to certain species of mammals and not to all animal taxa (Vehrs, 1996; Gesualdi, 2001; **Chapter 3**).

In Australia, legislation varies between the different states and territories, but current animal welfare (animal protection) legislation generally includes the requirement that animals are treated humanely and are not neglected. Some territories have more specific legislation covering the provision of food and water. For example, the first part of section 8 of the Northern Territory of Australia Animal Welfare Act 1999 states that:

(1) A person in charge of an animal must provide the animal with food, drink and shelter—
    (a) that is appropriate and sufficient; and
    (b) that it is reasonably practicable for the person to provide.

12.3.3 Guidelines and other sources of information about zoo animal nutrition

In addition to meeting statutory requirements, zoos have access to a range of non-binding guidelines, and other sources of information on nutrition and food presentation. Feeding protocols are available online to members of the Association of Zoos and Aquariums (AZA), via the nutrition sections of **Species Survival Plans (SSPs)**. The AZA also set up, or rather formalized, its own Nutrition Advisory Group (NAG) in 1994 (Dierenfeld, 2005).

The NAG has proposed a format for standardizing diet and nutrition information within **husbandry manuals** (see Fidgett, 2005, for more details). The NAG has also produced a series of technical papers on the nutrition of captive wild animals, which are available online at www.nagonline.net. Fact Sheet 014, for example, is entitled *Fruit Bats: Nutrition and Dietary Husbandry* (Dempsey, 2004); Fact Sheet 006 evaluates hay and pellet ratios for feeding zoo ungulates (Lintzenich and Ward, 1997).

Another useful service provided by the NAG to AZA members is a current list of advisers on the nutrition of different species (again available via the

NAG website), from tree kangaroos to the Komodo dragon *Varanus komodoensis*. These advisers are drawn from the relevant TAGs and SSPs; some are zoo nutritionists, but most are senior keepers and curators with specialist knowledge of particular species or taxa.

A further useful source of information is the series of publications on nutrient requirements by the National Research Council (NRC) of the US National Academy of Sciences (**Box 12.3**).

Within Europe, the equivalent zoo organization to the AZA in North America is the European Association of Zoo and Aquaria (EAZA). Zoos can obtain information on nutrition for their animals from what is now called the EAZA Nutrition Group (ENG); this started life as the European Zoo Nutrition Research Group (EZNRG) at a meeting in Rotterdam in 1999. EAZA does not produce or distribute diet sheets for particular species, but, like AZA, does provide information on nutrition within husbandry manuals. So an EAZA member zoo wanting to find out how to feed a species new to their collection would be likely, as a first step, to consult the most recent **taxon advisory group (TAG)**

Box **12.3** National Research Council publications on nutrient requirements

The importance of including major nutrient groups, such as fats, proteins, and carbohydrates, in the diet of captive animals has been recognized, if not fully understood, since the nineteenth century. But it is only relatively recently that detailed information about the range of essential nutrients has become known and more widely available (Oftedal and Allen, 1996), although there are still considerable gaps in our knowledge. One of the main sources of information about the nutrient requirements of different species is a series of publications by the National Research Council (NRC) of the US National Academy of Sciences.

NRC publications such as the book *Nutrient Requirements of Dairy Cattle* (first published in 1945 and now in its seventh, 2001, edition) provide detailed information on diets, with tables setting out the estimated nutrient requirements for animals at different stages of their life cycle. The NRC guidelines are published in the USA by the National Academy Press (NAP), which provides much useful material free of charge online (see the 'Free resources' section under each publication title online at www.nap.edu). Other titles in the NRC guidelines include *Nutrient Requirements of Swine* (1998), *Nutrient Requirements of Mink and Foxes* (1982), and *Nutrient Requirements of Cats and Dogs* (2006).

Despite the focus on livestock, fur-bearing, laboratory, and companion animals, much of the information in these publications is also relevant to the nutrition of wild animals in captivity. But there is an important caveat: there is quite a bit of the information in the NRC guidelines that is not relevant to the care of zoo animals, because livestock nutrition is generally based on achieving maximum productivity in minimum time, with little consideration for long-term health. Most animals reared on farms will be slaughtered at a young age, unless they are being kept for breeding. The goals of zoo animal nutrition, however, are to achieve good health and to allow breeding, for a maximum lifespan. Therefore, diets based on information for the domestic relatives of zoo animals can lead to problems such as obesity in zoo animals (see section **12.8.2**).

manual. There is also much useful information about zoo animal nutrition online at `www.eaza.net`; this website was substantially revised and updated in early 2008. There are links on the EAZA website to the ENG newsletters and also to conference proceedings.

Zoos are, rightly, sometimes wary of sharing or disseminating diet sheets for their animals. Local conditions may dictate that different diets are needed in different areas (for example, some areas may have a mineral deficiency in their grazing, such as a lack of selenium). It is also risky to assume that a generic diet will be suitable for all members of a **taxonomic** group. Among monkeys, for example, vitamin D requirements are very different for Old World and New World monkeys. Vitamin D comes in two forms: D_2 (ergocalciferol) and D_3 (cholecalciferol). Old World primates can utilize either form, but New World primates can only utilize vitamin D_3 (Hunt et al., 1967). So zoos need to know which type of vitamin D has been added to a commercial primate diet.

12.4 Working out an animal's dietary requirements

We have already said (section **12.2.2**) that the nutritional requirements of an individual animal will depend on its physiology, behaviour, metabolism, and anatomy. These requirements are not static within an animal's life cycle, but will change with age, level of activity, health, and reproductive status. Birds that moult, for example, will need additional nutrients at that time and female birds will need extra nutrients prior to egg-laying. Within the same species, different sexes may have different nutritional requirements. Female mammals will require additional food during gestation and, in particular, during lactation, when energy requirements in many species may rise sharply. (For a review of energy

requirements during reproduction in mammals, see Gittleman and Thompson, 1988.)

The diets[18] fed to zoo animals must meet both the general energy or metabolic requirements of the animal and also its specific nutrient requirements. In other words, zoos have to work out both how much food to provide in total and what the nutrient composition of that food is, or should be, so that nutritional requirements are met. In practice, it is often easier to start with the basic nutritional requirements of an animal (for example, 10 per cent protein) and then work out food quantities that meet these requirements. This, of course, is not always as straightforward as it sounds: formulating diets for wild animals in captivity is a complicated business and there are many factors that need to be taken into account.

Crissey (2005) has provided a useful nutrition matrix that can be adapted to different species and different circumstances. This takes into account not only factors such as nutritional requirements and food quantities, but also the health status of the animal and any management constraints. But even with tools such as Crissey's matrix, a recurring problem is that zoos often do not have adequate information about what the basic nutrient requirements of wild animals are, particularly with regard to trace elements. We will return to this problem in section **12.4.4**.

It is also worth noting here that zoo animal nutrition has only relatively recently become the focus of specific scientific investigation. It was not until 1994, for example, that the American Association of Zoos and Aquariums (now the AZA) first recognized nutrition as a 'scientific advisory speciality' (Dierenfeld, 1997a; **Box 12.4** provides a brief overview of the history of zoo animal nutrition). This is perhaps surprising, given the long history of zoos. But as Oftedal and Allen (1996) point out, in the past, many zoo animals managed to survive for considerable periods of time on what we would now regard as inappropriate diets and, because the

---

18 Diet is not the same thing as nutrition. The word 'diet' can mean slightly different things in different contexts, such as the usual food given to an animal, or a restricted selection of foods given for medical reasons.

Box **12.4** The history of zoo animal nutrition

The importance of minerals such as calcium and phosphorus in animal diets was not widely recognized, or was not considered important, until the early twentieth century (Ammerman and Goodrich, 1983). Vitamins were not 'discovered' until the twentieth century and, even today, there are still considerable gaps in our knowledge of essential nutrients for many animal species.

The incidence of diseases such as rickets in captive animals prompted some early experimental analysis of the role of nutrition in maintaining zoo animal health. In the 1930s at Philadelphia Zoo, for example, cebid monkeys were allowed to range freely and to self-select dietary items. The resulting dietary recommendations published by Dr Ellen Corson-White, of the University of Pennsylvania, were based on a combination of what the animals chose to eat and known nutritional requirements (Corson-White, 1932; Dierenfeld, 1997a).

In the 1960s, Dr Hans Wackernagel of Basel Zoo in Switzerland developed a range of formulated pelleted diets for various taxa of exotic species and also provided details of the nutrient composition of these manufactured diets. But it was not until the 1970s and 1980s that zoos in the USA, and then in Europe, first started to employ in-house professional nutritionists.

animals were not dying, or at least were not dying in huge numbers, there was little pressure on zoos to investigate whether or not diets could be improved.

Even in good zoos today, the diets provided for the animals can be driven more by tradition than science, and the *Zoos Forum Handbook* (Defra, 2007b) warns that diets can suffer from **dietary drift** and can be changed incrementally over time by keepers (not always for the better). The *Handbook* counsels that diets should be subject to regular review, and also recommends that zoos should make use of the services of a nutritionist to support keepers and veterinary staff in delivering appropriate diets for the animals in their care.

12.4.1 Working out how much food to feed

To calculate the general energy requirement of an individual animal, one of the first things that a zoo needs to know is the animal's weight (we shall see why shortly). Working out the weight of a zoo animal is not as easy as it sounds, particularly for an elephant or a giraffe, although some animals can be trained to step onto scales (**Fig. 12.16**). Many zoos build weigh bars into low platforms within an enclosure, often across a doorway, so that the animal becomes accustomed to stepping onto the platform

Figure **12.16** At Singapore Zoo, a giant tortoise *Geochelone gigantea* is being trained to step onto scales for weighing. Knowing the weight of an animal is a vital first step in calculating its general energy requirement and hence how much food to provide. (Photograph: Diana Marlena, Singapore Zoo)

(the website of San Diego Zoo describes its use of the California Highway Patrol's portable truck scales to weigh its elephants).

In addition to knowing an animal's weight, the zoo needs to know how much energy the animal needs to sustain its basic metabolism. A term often used in relation to an animal's energy requirement is **metabolizable energy (ME)**. Confusingly, ME refers to the energy content of the food (along with gross energy and digestible energy), not to the energy requirement of the animal.

Terms used in relation to energy requirements are:

- gross energy (GE), which is the total energy in the food;

- digestible energy (DE), which is GE minus the energy lost in faeces;

- metabolizable energy (ME), meanwhile, is DE minus energy in urine and gaseous products of digestion.

Therefore, ME is the energy available to fulfil the animal's metabolic requirements. This is the energy that the animal needs to sustain its basic metabolism or basal metabolic rate (BMR), plus energy for activity, plus energy needed for growth, thermoregulation, and for any other processes (such as egg-laying in birds or reptiles; see, for example, Klasing, 1998).

The basal metabolic rate has been measured for many mammal and bird species (see, for example, Robbins, 1993) and, for any given taxonomic group, this value is closely correlated to body weight. Birds have a higher metabolic rate than mammals of equivalent size, whereas reptiles have a lower metabolic rate than mammals of the same size.

Once the zoo knows the weight of an animal, there are published equations for mass-specific BMR (see Robbins, 1993), which can be used to calculate the minimum energy requirement per day for that animal (although it is worth noting that there is still considerable debate about the detail and usefulness of some of these equations). Of course, the BMR is only the starting point and the animal will need

additional energy for activities such as running or flying. These sorts of activities can 'cost' between five and ten times BMR.

The software package **Zootrition**™ (see section 12.4.3) has an energy requirement module that will calculate BMR on the basis of several different equations. The user can then pick the one that he or she feels is most appropriate for the species. Zootrition™ will also calculate the field metabolic rate (FMR), or the energy that an animal needs to survive and be active, based on various levels of activity and life stages. In practice, however, it is fairly unusual for a zoo to be confronted with an animal new to its collection that does not arrive with an accompanying diet sheet, or at least some basic dietary information.

How much food is actually given also depends on an assessment of activity and on factors such as whether the animal is solitary, or housed within a social group, in which case other animals may eat some of its food. For taxa such as **herps** and fish, zoos often feed reactively—that is, they feed slightly more than the calculated amount and then adjust feeds on a daily basis depending on how much is left uneaten. Homeothermic animal energy requirements also depend on external temperature: these animals need more food when it is cold just to maintain body temperature. In practice, energy requirement estimation is just that—estimation. It is vital to monitor body weight (ideally directly, but, if this is not possible, then through alternatives such as **body condition scoring**; see **Box 11.1**) and to adjust the diet quantity if animals become underweight or overweight.

As we have seen already in this chapter, however, not all of the content of food provided for an animal will be available to meet its energy requirements. Any foodstuff will comprise digestible matter, indigestible matter, and also material that can be broken down by fermentation in the gut. The relative proportions of these three components will determine the overall availability of energy from the diet.

The digestible matter comprises the carbohydrates, fats, and proteins that we discussed in section 12.2. Fermentable materials are not broken down by the animal itself, but by the symbiotic gut microflora, particularly in ruminants. These materials include substances such as cellulose, from plant walls. Finally, some compounds (for example, lignin) are not broken down at all in the gut and cannot be used as a source of energy; this matter simply passes through the gut and is excreted in the faeces.

But how can zookeepers find out how much digestible material there is in a particular food item? And how can nutrient intake be measured accurately if an animal does not eat all of the food that is provided? Oftedal and Allen (1996) summarize this problem nicely when they talk about distinguishing between '*the diet as presented*' and '*the diet as consumed*'.

12.4.2 How can the nutrient content of the diet be measured?

Questions about the amount of digestible material in an animal's diet can be answered by carrying out food intake studies, within which the total amount of food provided is carefully weighed, and then uneaten food and faeces are also weighed and analysed (see the BIAZA *Research Guidelines on Nutrition*, forthcoming at the time of writing this book, for information on how to go about doing this sort of analysis).

Establishing the nutrient content of a food item, such as a grass hay or another type of forage, is done by laboratory analysis. A typical analysis produces data on the crude protein, ash, fat, and fibre content in the food, expressed as proportions of the total **dry matter** (that is, the weight of the food without water). **Table 12.2** provides a brief explanation of some of the terminology used in a basic laboratory analysis of the nutrient content of a foodstuff.

Thus, different types of study (food intake studies and laboratory analysis of the nutrients in particular food items) can help zoos to decide how

much food to provide and whether the diet is providing all of the necessary nutrients. But both of these sorts of studies can be laborious and time-intensive to carry out. A food intake study for a bird fed on a proprietary seed mix, for example, would involve separating out each type of nut or seed in the mix before weighing each of these, and then the even more laborious job of picking up the uneaten seeds from the enclosure afterwards, and sorting and weighing the leftovers. Nutrient assays in the laboratory are also relatively expensive and a zoo that wanted to sample all of its forages on a regular basis, or wanted a detailed analysis of vitamin and mineral content of its foodstuffs, would very quickly run up a large bill for laboratory assays.

Fortunately, zoos now have access to computerized databases that can provide much of this sort of information. The best known of these databases, and the one most widely used by zoos in Europe and the USA, is Zootrition™.

12.4.3 Zootrition™: dietary analysis software for use in zoos

Zootrition™ is a nutritional software package and database developed by Dr Ellen Dierenfeld, who is a leading zoo animal nutritionist now based at St Louis Zoo, Missouri. Dr Dierenfeld was previously at the Wildlife Conservation Society (WCS) and Bronx Zoo. Support for the development and use of Zootrition™ is provided by the World Association of Zoos and Aquariums (WAZA).

The Zootrition™ software package was developed specifically for use by zoos[19] to help to ensure that the nutritional composition of the diet that they give to their animals is correct. Zootrition™ does this by providing a database of nutrient composition values for more than 2,500 feeds, including whole prey items such as mealworms, crickets, and mice. This database allows the nutritional content

---

Nutrient values are usually reported on a **dry matter** basis to remove any variability between feeds due to differing moisture content.

[19] At its 2001 meeting in Prague, the EAZA Research Committee formally endorsed the adoption of Zootrition™ as the standard dietary analysis software for EAZA member zoos.

Component or fraction of foodstuff	Nutritional terminology explained
Crude protein	Essentially, the protein content of a food is determined by analysing the nitrogen (N) content, because all proteins contain approximately 16 per cent nitrogen. But there will be some N in foods that is not from proteins, so the value obtained from an analysis of the N content of a foodstuff is referred to as the **crude protein** fraction, rather than the **true protein** fraction.
Ash	This is the inorganic content of a foodstuff (e.g. calcium, phosphorus, copper, iron, and other minerals). It is determined by heating (burning) a weighed sample of the foodstuff at a very high temperature, to remove all of the carbon. The ash is the residue left over after this process.
Fat	The fat content of a foodstuff can be given as **crude fat**, which is all of the material that will dissolve in a solvent such as ether, or **total lipids**, which involves a different extraction method using acid hydrolysis and gives the lipid content, excluding substances such as resins or waxy materials.
Neutral detergent fibre (NDF)	This is the entire non-soluble fibre content of the foodstuff (and includes cellulose, hemicellulose, and lignin). The term 'detergent' refers to the chemicals used to separate out the non-soluble fibre from other components of the foodstuff. **Total dietary fibre** is the entire fibre content (soluble and insoluble) of the foodstuff and is a more useful term for animals with minimum gut fermentation.
Acid detergent fibre (ADF)	This is the NDF minus the hemicellulose. It is partially digestible to varying degrees, depending on species. Here, acid, as well as detergent, is used to separate out the indigestible fibrous material.

Table **12.2** Nutritional assay terminology

NOTE: For further information on nutritional analysis and on animal nutrition in general, we recommend McDonald *et al.*'s *Animal Nutrition* (2002).

of specific food items to be compared and also the overall nutritional make-up of an existing or proposed diet to be calculated.

 Further information about Zootrition™ can be found on the St Louis Zoo website, at www.stlzoo.org/animals/animalfoodnutritioncenter/zootrition .

12.4.4 Working out what nutrients an animal needs

Even with databases such as Zootrition™ available to zoos, there is still a paucity of information about the precise dietary requirements of wild animals. As we have seen earlier in this chapter, nutrient requirements have been determined in considerable detail for domesticated animals such as the cow, sheep, horse, pig, cat, dog, and chicken, and also for a very small number of wild animal species: for example, the white-tailed deer *Odocoileus virginianus* (see, for example, Thompson *et al.*, 1973; **Fig. 12.17**). But how similar, for example, is a tapir to a horse in its nutritional requirements?

The problem of a lack of data on baseline nutrient requirements for zoo animals is more acute for birds than for mammals. Nearly all data on avian nutrient requirements come from studies of granivorous domestic species (particularly the chicken). Domestic poultry species do not provide an adequate nutritional model for non-granivorous birds such as carnivorous storks or birds of prey,[20]

20 For birds of prey or other avian carnivores, the domestic cat may provide a more appropriate nutritional model than the granivorous chicken.

Figure 12.17 The white-tailed deer *Odocoileus virginianus* is one of very few species of wild animal for which detailed nutrient requirements have been determined. In general, zoos must use nutritional models developed for domesticated animals, such as the cow, sheep, goat, and horse. (Photograph: © Paul Tessier, www.iStockphoto.com)

or for birds that are fruit eaters, or nectar feeders, or true herbivores (Klasing, 1998).

The problem of a lack of data is even more acute for non-mammalian and non-avian vertebrates: there are some data on nutrient requirements available for fish, from the **aquaculture** industry, but virtually no data on baseline nutrient requirements are available for most reptiles and amphibians.

Studies of the diet of wild animals do exist, albeit for a very limited range of species, and these can provide valuable pointers to zoos for the development of appropriate diets for their animals. The findings of some of these studies should be treated with caution, however, because they may be based on data from only one season; this is not particularly helpful when there is strong seasonal variation in food supply.

Another problem is that the nutrient content of wild versus domestic varieties of fruits and vegetables is not standard. For example, an animal that is known to eat figs in the wild may be given figs in captivity. But a study by Silver *et al.* (2000) of the foods consumed by black howler monkeys *Alouatta*

pigra in Belize showed that the calcium content of wild figs may be seven times greater than that of figs from a domesticated cultivar. Similar findings were reported in a study of bat feeding preferences in Panama (Wendeln *et al.*, 2000).

12.5 The supply of food

Having looked at the theoretical background to feeding zoo animals, what about the practicalities? Where do zoos obtain food for their animals, and what are the main considerations that zoo managers and keepers must take into account when sourcing food items?

12.5.1 Where does the food come from?

There are three main categories of foodstuffs consumed by zoo animals, generally cited as:

- produce[21]—food such as fruit, vegetables, meat, fish, dairy produce, and tinned or frozen foodstuffs;
- commercial formulations—foodstuffs such as pelleted food or concentrates, as well as formulated diets (for example, seed mixes for birds);
- browse—plant material other than grass or hays (for example, willow branches and leaves).

Produce

Produce is generally supplied by companies that also supply the human food industry, so this is usually food that is considered 'fit for human consumption' and, as such, it is subject to the same strict health and safety regulations as the food that we buy in supermarkets.

Ideally, the dietary requirements of the animals in a collection will be calculated and appropriate produce purchased accordingly from wholesalers. Some zoos, however, also incorporate, to a greater or lesser extent, surplus produce items into their

---

21 Produce and other terms used to describe foodstuffs vary among countries. Zootrition™ uses US terms that may not always be familiar to zookeepers in the UK and the rest of Europe.

animals' diets. Surplus produce is those items that are given away freely by wholesalers or supermarkets at the end of the working day or week. This can offer considerable cost savings to zoos and allow limited zoo finances to go further, but caution should be exercised when using surplus food. Although such produce is fit for human consumption, the type of surplus food items that are available may not be ideal for the animals and may not contribute very much to meeting the animal's optimal nutrient requirements. For example, many supermarkets have surplus bakery goods that are readily available, but bread and buns are generally not ideal foodstuffs for most zoo animals.

Commercial formulations

Foodstuffs are also purchased by zoos from dedicated commercial animal food suppliers: for example, those supplying livestock farmers and the pet industry. There are also some specialist zoo foodstuff supply companies, such as Mazuri Zoo Foods® and Nutrazu®. Foodstuffs from these sources are either specific food items, or specially processed or formulated foodstuffs.

Specialist food items purchased by zoos include live food (for example, invertebrate prey items such as crickets or mealworms) and other foodstuffs, such as nectar. Formulated foodstuffs are mixes of ingredients (for example, seed mix) or a combination of ingredients that are processed into pellet form; the latter are usually referred to as concentrates, but can variously be called nuts, biscuits, dry diets, or chow (for example, monkey chow). Because all of the ingredients that go into these formulations are known, it makes it easier to assess the nutritional composition of the diet. In other words, if an animal is fed solely on a commercial formulation, then the zoo will have easy access to information about the nutritional intake (protein, vitamins, minerals, etc.) of that animal.

Another benefit to zoos of using food concentrates is that non-nutritional ingredients can be added to formulations. For example, pelleted food may contain additional roughage, to improve the digestibility of the diet, or an **anthelmintic drug**. The addition of such drugs to formulated diets allows zoos to treat their animals for parasitic worms without having to capture the animals to dose them.

Browse

Zoo grounds are a great source for animal feed. Zoo enclosure design should, ideally, ensure that grazers and browsers have sufficient vegetation within their enclosure to meet their needs, but browse can be supplemented by plant material from elsewhere within the zoo grounds, or with material that is brought in from outside the zoo (for example, from a local park that needs to prune its trees). Finding sufficient browse is a greater challenge for smaller city centre zoos than it is for larger wildlife parks. The provision of browse is also a challenge for some specialist feeders that require vegetation that cannot be sourced locally, such as bamboo or eucalyptus leaves for giant pandas *Ailuropoda melanoleuca* and koala bears *Phascolarctos cinereus*, respectively (see **Fig. 12.18**).

Even when zoos do have access to ready sources of browse, information on the appropriate amount of browse to feed is hard to come by. Keeper's notes may record, for example, that two or three branches of browse are supplied to each animal each day. But this does not describe how much edible material per branch is available to the animal. Large branches cannot easily be weighed, and neither weight nor volume gives much information about the amount of edible material (leaves, smaller twigs, and bark) available to the animal. In a detailed study of the feeding behaviour of moose *Alces alces*, Clauss *et al.* (2003) found that branch diameter at the point of cutting was closely correlated with the amount of foliage and edible

---

An **anthelmintic** (or **antihelmintic**) **drug** is a drug that is used to treat parasitic worms, such as roundworms or tapeworms. The word **helminth** is often used to refer to parasitic worms found in the gut.

Figure **12.18** Giant pandas *Ailuropoda melanoleuca* are specialist feeders, adapted to a diet consisting mainly of bamboo. Most zoos cannot obtain sufficient bamboo all year round to feed giant pandas, so must substitute nutritionally equivalent plant material. (Photograph: © Klass Lingbeek-van Kranen, www.iStockphoto.com)

twigs on the branch. This correlation held true for nine different tree species, including commonly fed browse species such as willow, hazel, and oak.

12.5.2 Other considerations

As well as finding appropriate supply sources for foodstuffs, zoos must take into account other considerations when sourcing food for their animals, such as the nutritional composition of the food, **biosecurity**, seasonal variation in supply, and ethical issues relating to feeding zoo animals.

Variation in the nutritional composition of food

It is highly likely that the nutritional composition of produce from a supplier local to a zoo will be different to that of the same produce found in the animal's natural habitat. The nutritional composition of plant foods will also vary according to species or variety (**Fig. 12.19**), the climate and/or season when it was grown, the soil in which it was grown, and the size and age of the fruits, leaves, or shoots.

Plant material also varies in nutritional composition from one part of the same plant to another: younger grass shoots are more nutritious than older stems, for example.

Luckily, the study of **agronomy**, and the economic importance of farm animals such as cattle and sheep, means that we know a lot about the physiology of grazing livestock, and also about the impact of different types of cereals and forage on the development, growth, and reproduction of these animals. Data from studies of domesticated grazing animals inform us that, although grass and grass-related crops may all look broadly similar, there is considerable variation in nutritional content between raw grass and forages, such as **hay** and **straw**. Other forms of stored forage include silage, which is fermented fodder, and haylage, which is a semi-fermented forage with a moisture content that is higher than that of hay, but lower than that of silage.

Supplements

It is one thing to work out what an animal's nutrient requirements are and quite another to be sure that all of these nutrients are provided in the correct quantity in the animal's diet. Sometimes, even if the nutritional value of foodstuffs has been carefully calculated, zoos will still need to add supplements such as vitamins and minerals to the produce fed to their animals. Calcium powder, for example, is often sprinkled onto fresh produce fed to reptiles, to reduce the risk of the development of metabolic bone disorders (see section **12.8.1** on nutritional diseases).

Often, commercial formulations are available that already contain added quantities of vitamins and minerals. This helps to avoid the problems inherent in administering supplements, for example, in tablet form. Supplements can also be added to invertebrate prey items such as crickets; a number of suppliers

Biosecurity is a term used to describe the measures taken to prevent the transmission of disease.
Within agriculture, the study of cereals and crops is called **agronomy**.

Hay is produced by drying grass; **straw** is the dried stems of cereal crops, such as wheat or barley.

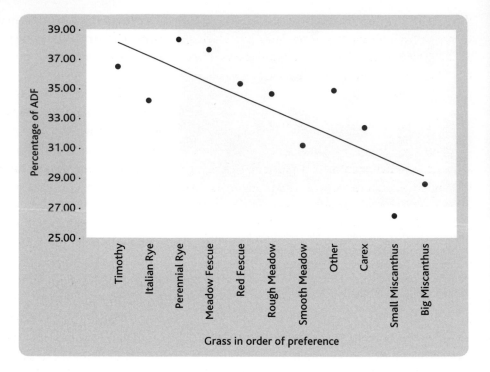

Figure **12.19** This scatterplot shows a significant correlation between the acid detergent fibre (ADF) content of different grass and other plant species, and the grazing preferences of captive zebras in zoos in UK and Irish zoos. The zebras in this study preferred grass species with a low ADF. (Reproduced with permission from Armstrong and Marples, 2005)

of foodstuffs for **exotic animals** offer 'gut-loaded' crickets or locusts containing additional calcium.

Biosecurity and animal health

Part of the process involved in sourcing food is ensuring that the food is safe for zoo animals to eat. For many of the foodstuffs purchased by zoos, this is redundant, because the produce is considered 'fit for human consumption' and, as such, is governed by rigorous standards of hygiene. But for zoo foodstuffs that do not come under the banner of 'fit for human consumption', zoos must ensure that their biosecurity is not compromised when bringing produce into the zoo.

The need for zoos to maintain a high level of biosecurity at all stages of food sourcing and preparation cannot be overstated. For many foodstuffs, this means ensuring that zoos make their purchases from reliable wholesalers, whose methods of storage and transport meet high standards so that foods do not deteriorate or become contaminated before they reach the zoo. Even if zoos are careful to use reliable suppliers, there are still additional steps that can be taken at the zoo to reduce the adverse consequences of any lapses in biosecurity. For example, freezing food items can reduce the likelihood of parasite transfer from meat and associated by-products to zoo animals, because many endo- and exoparasites cannot survive the freezing process (see, for example, Deardorff and Throm, 1988). Keepers and other staff involved in food preparation also need to check foodstuffs for non-biological contaminants, such as stray pieces of wire or baler twine in hay.

Transport costs and environmental impact

The term 'food miles' is often used nowadays as an indicator of how much energy is required to transport a foodstuff from its source to its place of consumption. For example, for a UK zoo, the food miles for apples or pears from a local orchard would be very low in comparison with the food miles for pineapples flown in from Africa, but the nutritional composition of these fruits may not be substantially different.

If zoos wish to be seen as sustainable and environmentally responsible organizations, they need to take into consideration the costs of transporting their foodstuffs. Importing bamboo or eucalyptus leaves, or even pineapples or bananas, can clock up a high total of food miles and have an adverse effect on the zoo's environmental impact and carbon footprint (see **Chapter 15**).

Seasonal variation in food supply

Obtaining sufficient browse for zoo animals outside the summer months is a particular challenge in temperate climates, where deciduous trees shed their leaves in winter. Some zoos are now experimenting with freezing browse, although, given the large volume of freezer space required, this is an energy-intensive option (and may also affect palatability). Another alternative is the production of browse silage for use in the winter months. This offers an inexpensive alternative to freezing browse and has been used at Zurich Zoo for some years now (Hatt and Clauss, 2001).

At Woburn Safari Park in Bedfordshire, UK, both freezing and silage production are used to preserve browse for the winter months. Fresh leaves taken from trees in the Park are compacted and bailed in airtight bags prior to freezing; leaves are also cut and compressed for silage, in airtight barrels (J. S. Veasey, personal communication). In addition, more than 10,000 saplings have been planted over recent years at Woburn, to help to ensure a good browse supply for the Park's herbivores. This includes the planting of fast-growing tree species able to tolerate frequent browsing, to allow the Park's giraffes to browse directly in a strip-grazing (rotation) system.

Browse is not always easy to come by, even in the summer months in the UK, particularly for city zoos. On top of these challenges, zookeepers may not be certain which types of browse can safely be fed to a particular species. BIAZA has produced a browse database, available as a CD-ROM, which provides information about which browse species are not toxic to animals (see **Box 12.1**); Defra has produced a comprehensive guide to poisonous plants and fungi in the UK (Cooper and Johnson, 1998).

Ethical considerations in sourcing animals as foodstuffs

When sourcing animals as food items for other animals, zoos have a duty of care to ensure that the production (life) and death of the food animals is such that it does not compromise their welfare (Spencer and Spencer, 2006; see **Box 12.2** on feeding live prey).

Before the implementation of the Animal By-Products Regulations 2005, it was common practice in the UK for some animal species to be raised in zoos that would then be used as food items for other animals in the zoo. For example, the practice of hatching chickens' eggs and rearing chicks was considered beneficial, because it provided the opportunity to teach zoo visitors about egg hatching and the life cycle of the chicken, as well as providing a food source for other animals in the zoo. In cases such as this, the zoo was directly responsible for the rearing of the chickens (or other animals) as food items and was able to ensure that they had a high level of welfare.

It is still common practice in some zoos outside the UK to breed and rear larger ungulates (for example, deer) partly for exhibit, but also to use 'surplus' animals as food for carnivores within the collection. Again, the zoo is directly responsible for the welfare and humane culling of these animals.

Although the legislative framework within the UK prohibits the feeding of live vertebrate prey, zoos can, and do, feed live invertebrate prey items. The most common invertebrate species fed in zoos are crickets, locusts, mealworms, and daphnia. Many of these species will be bought live by zoos (from specialist breeders) and fed soon after, but some zoos will choose to rear these animals themselves. Zoos still have a duty of care for these animals, to ensure that their welfare is not unduly compromised during their lifetime.

There is still relatively little published research into the welfare of invertebrates and thus our knowledge of how these animals perceive physiological or psychological **stimuli** (if, indeed, they do) is still quite poor. We do know, however, that invertebrates will avoid certain toxic and noxious stimuli, such as extremes in temperature and light (Sherwin, 2001). It follows therefore, that, while storing these animals prior to being fed, they are provided with the appropriate conditions necessary to ensure that the **five freedoms** are satisfied (see **Chapter 7**). As such, they should be given adequate food, water, and shelter, in an area that is kept at a suitable temperature and humidity.

12.6 Food storage and preparation

Foodstuffs purchased by a zoo are rarely used straight away, but will need to be stored for varying periods of time. This may be for only a few hours for some fresh produce, or up to several months for dried or frozen foods. While food is in storage, the zoo has to ensure that it remains fresh and palatable, safe to eat, and does not lose a significant proportion of its nutritional value.

12.6.1 Avoiding contamination of food

Measures to prevent the contamination of food during storage and preparation, and to reduce the risk that foodstuffs may be vectors for disease, should be an integrated part of good husbandry (**Fig. 12.20**). This includes everything from the keepers' personal hygiene (washing hands, not wearing 'work' clothes outside the zoo, etc.) to cleaning food preparation areas and equipment thoroughly and regularly.

All of this may sound rather obvious, but these sorts of measures are essential components of good biosecurity, and are just as important for the health of keepers and other zoo employees as they are for the health of zoo animals.

Zoos also need to take measures to exclude **pests** such as rats, mice, and cockroaches from food storage and preparation areas. Pest control in zoos has already been discussed in some detail in the

Figure **12.20** Keepers at Paignton Zoo Environmental Park preparing a variety of fresh foodstuffs for the zoo animals. (Photograph: Paignton Zoo Environmental Park)

chapter on health (**Chapter 11**), but it is worth pointing out here that plants, as well as animals, can be vectors for disease. Plants can also be toxic (see **Box 12.1**): common ragwort *Senecio jacobaea*, for example, contains alkaloid compounds that are poisonous to a wide range of vertebrate species.

Pest control in relation to food preparation and storage can be achieved through a variety of steps, all of which reduce the impact of pest species on other captive animals. It should be remembered that many pest species are opportunists and thus will take advantage of resources that are easy to access. So zoos need to make access to food, and to other resources that pests may find attractive, as difficult as possible. This can be as easy as ensuring that

Pest species are simply those species (plant or animal) that occur where they are unwanted.

tightly fitting lids are replaced on food containers, doors to food storage and food preparation areas are kept shut, and food is not left in 'non-secured' areas (for example, if food is prepared in advance of feeding, it should not be left in open buckets from which pest species may be able to access it).

12.6.2 Maintaining palatability and nutritional value

The palatability and nutritional value of food can be greatly affected by the conditions in which it is stored. Fortunately, food science and technology is a well-established discipline that has taught us a great deal about good food hygiene and the processes underpinning decay in foodstuffs. (There is a huge literature on food storage and food decay; although this is largely about food intended for human consumption, much if not most of this information is also relevant to the preparation and storage of food for animals.)

The process of decay in foodstuffs is complex, and varies considerably depending on the type of food and the conditions in which it is stored. Heldman (2003) provides much useful information about food processing and storage, and on the deterioration of foodstuffs over time. In general, however, harmful microorganisms in foodstuffs will proliferate much more rapidly in warm, damp conditions, so optimal food storage conditions will usually be found somewhere that is both cool and dry. Fresh foods are most susceptible to deterioration, and the process of denaturing or decay is exacerbated by exposure to extremes of light, heat, and moisture, and by exposure to microorganisms such as bacteria and moulds. Storing fresh foodstuffs in a refrigerator at temperatures of around 1–4°C (34–40°F) is generally regarded as optimal. Below this temperature range, fresh fruits and vegetables will freeze, and their texture and palatability will change; above this temperature range, microorganisms, such as bacteria, will proliferate much more rapidly.

Zoos also make extensive use of freezers for food storage, particularly of meat and fish. Here, foodstuffs are generally stored at temperatures below −18°C (0°F); for fish, lower temperatures (−28°C or −19°F) are recommended by some sources to reduce deterioration and loss of nutrients (for example, Heldman, 2003). Care should be taken that frozen foods are thawed thoroughly before being provisioned, and that foodstuffs are not frozen, thawed, and then refrozen. Frozen fish should be thawed overnight in a refrigerator rather than by immersion in cold or running water, to avoid the loss (leaching) of water-soluble nutrients (Whitaker, 1999).

There is a particular risk to animal health associated with the use of frozen fish, due to the action of enzymes (thiaminases) produced during thawing. However carefully fish is thawed, there will be some degradation and loss of thiamine (vitamin B_1). This means that thiamine needs replacing in the diet of fish-eating animals and, along with vitamin E (which is also lost during storage), thiamine is routinely given as a supplement to piscivores fed on a diet containing frozen fish, to reduce the risk of the development of neurological disorders (Geraci, 1986).

Storage and preparation can also alter the nutritional content of the foodstuffs fed, which means that the animals may not be consuming the nutrients that we hope they are. Compared with other nutrients, vitamins are not particularly stable and can lose their activity over time, particularly if exposed to high temperatures. Vitamin C, for example, can be lost more or less completely from foodstuffs stored for 6 months or more (Klasing, 1998). On the plus side, however, most vitamins are not expensive to manufacture, so supplementation is not particularly costly.

To reduce the risk of deterioration in food quality, zoos need to minimize any lag time between the preparation of food and its consumption. Food that is left in buckets or containers after preparation will be susceptible to desiccation, loss of nutrients such as vitamins, and potential contamination by bacterial activity or pests. There is also a risk of the opportunistic consumption of food by species such as rats, pigeons, seagulls, or herons (**Fig. 12.21**),

Figure **12.21** Grey-crowned cranes *Balearica regulorum* at Warsaw Zoo, with feral pigeons attempting to gain access to the food provided for the cranes. This sort of access by avian pest species is difficult to prevent completely, even within aviaries, but pest species such as pigeons, or rats and mice, can transmit diseases to zoo animals. (Photograph: Barbara Zaleweska, Warsaw Zoo)

before the food reaches the zoo animals for which it is intended.

12.7 Food presentation

After appropriate food has been sourced and stored, it needs to be provided to the animals in the zoo. It may seem an obvious point to make, but if foods are provided to zoo animals with the intention that they should eat them, then the food needs to be accessible (**Box 12.5**).

12.7.1 Access and availability
There are numerous differences between individual animals of the same species (for example, age, social rank, health status, and intelligence) and these factors will affect the ability of different animals to gain food. Not all animals may be able to gain food from all areas of the enclosure, or from complicated feeding devices. The method used to provide food therefore needs to make allowances for these individual differences within a group of animals, to ensure that each animal in a socially housed group can gain access to a nutritionally complete diet. Understanding how individual animals use space and are able to move around their enclosure can enable keepers to provision individuals, or groups of animals, differently. For example, many **mixed-species exhibits** will hold animals that have different dietary requirements and yet share an enclosure. By considering favoured areas or areas to which some species can gain access and others cannot, different diets can be provided within the same environment.

Box **12.5** Why do keepers chop up food for zoo animals?

Many keepers will lament that they spend much their day chopping up food for animal feeds. But why is it necessary to chop food for animals in zoos? After all, wild animals do not have this assistance in accessing and processing their food.

For some species, it may be necessary to chop food so that it can be mixed up, to try to prevent the animals choosing between food-stuffs offered and so ensure that they consume a nutritionally balanced diet (see section **12.7.2**). Similarly, if food supplied to socially housed animals is not chopped up, then keepers may be concerned that dominant animals will get priority access to favoured foods and that the subordinates will not get a nutritionally balanced diet. In fact, a study by Plowman *et al.* (2008) demonstrated that when groups of Sulawesi crested black macaques *Macaca nigra* and red-ruffed lemurs *Varecia rubra* were fed two different diets, one composed of chopped foods and the other of the same foodstuffs presented whole, the overall food intake of the individual animals did not differ.

For many species in zoos, it seems that food is chopped or peeled out of habit. We have suggested already in this book that much housing and husbandry practice in zoos is not influenced so much by research, but more by tradition. Sometimes, this can be invaluable, but sometimes it can be nonsense. Providing whole foods to zoo animals has been demon-strated to be enriching, because the animals are stimulated cognitively and physically in order to handle and access the food items appropriately (see, for example, Smith *et al.*, 1989). In addition, when food is chopped or peeled, it is more susceptible to desiccation and to loss of nutrients, especially vitamins (Lamikanra *et al.*, 2005). Chopping food also increases the surface area available for exposure to bacteria and other contaminants.

So should zoos chop or peel food? Only when there is a good reason to do so; other-wise, the animals will almost certainly benefit more from receiving whole foods.

One of the most widely used methods of feed-ing wild animals in captivity is usually referred to as **cafeteria-style feeding (CSF)** (Marqués *et al.*, 2001). This is where a variety of foodstuffs is offered (for example, a mixture of fresh fruits and vegetables) and the animals are allowed to choose what they consume. (The opposite of cafeteria-style feeding is **complete feed-style feeding**, which is essentially the provision of a processed diet that meets the nutrient requirements of the animal, but does not offer any choice). But, as we have already seen in section **12.1.3**, animals do not necessarily have nutritional wisdom and may not consume a balanced diet if food is presented in a free-choice CSF approach.

Some plant foodstuffs are also freely available within an enclosure, because they grow there: for example, grass, other vegetation, fruit trees, etc. In these instances, the issue may not be that foodstuffs need to be available to animals to eat, but rather the opposite: that the animals' access to the plants may sometimes need to be restricted. This can be for aesthetic reasons (to ensure that attractive trees and shrubs survive), or to limit consumption of potentially toxic plant matter (see **Box 12.1**).

How often should animals be fed?

Many zoo animals are fed once or twice a day (**Fig. 12.22**). But because animals in the wild

Figure **12.22** A giant anteater *Myrmecophaga tridactyla* with a bowl of food, at Ramat Gan Zoo in Tel Aviv, Israel. In the wild, giant anteaters spend a large part of the day foraging for prey species, such as ants and termites. In zoos, giant anteaters are often fed a gruel-like diet containing a mix of foodstuffs, such as canned cat food and fruit, as well as insects, when available. (Photograph: Tibor Jäger, Zoological Center, Tel Aviv)

often forage for a large part of the day, there is a widespread assumption that it is somehow 'better' to feed more often and that providing a greater number of smaller feeds throughout the day may be a useful form of **enrichment**, at least for some species (see **Chapter 8**).

12.7.2 Intake

It is important to know what foodstuffs animals like and dislike, because this will determine which foods they choose to eat first and which foods are likely to be monopolized by dominant animals. Also, if too much food (quantity) is provided, then animals may choose to eat only favoured food items and leave behind those foods that they do not like. This can be very frustrating for keepers and zoo nutritionists who have put much effort into providing a nutritionally balanced diet.

One solution to 'food favouritism' is to provide animals with a diet that has less daily variety, so that all animals will have access to, and indeed will have to eat, certain food items (or go hungry). At Johannesburg Zoo, for example, the animals may receive a lot of apples and no oranges at a feeding session; the next day, oranges will be provided, but no apples (ICEE, 2007).

Alternatively, different foods can be mixed together, so that the animals cannot easily avoid consuming all foodstuffs providing. This food-mixing method was adopted for a colony of Rodrigues fruit bats *Pteropus rodricensis* at Chester Zoo after it became apparent that a particular feeding presentation style used for the bats was having a deleterious impact on the health of the animals (Sanderson *et al.*, 2004). The bats had been fed a diet comprising mashed-up food, combined with primate pellets for additional protein. Larger chunks of fruit were also provided on spikes, to provide feeding enrichment for the bats. Because of the perceived success of the feeding enrichment, and its attractive appearance in comparison with a 'mush' of fruit and pellets, keepers started to provide the primate pellets separately to the fruit when feeding the bats. The dry pellets on their own were not consumed by the bats, however, so keepers reduced the quantity of pellets provided.

In the following breeding season, mortality levels in the Rodrigues bat colony rose, with a disproportionate increase in mortality among female bats that had lactated the previous year. An investigation by the zoo veterinary staff and nutritionist found that the food presentation style, intended as enrichment, in fact allowed the bats to select a nutritionally deficient diet and this had affected lactating females in particular (as well as demonstrating that fruit bats do not appear to posses nutritional wisdom—see section **12.1.3**). Once Chester Zoo returned to its previous feeding regime for the fruit bats, providing fruit and primate pellets mixed together, mortality levels fell and the bat population recovered.

Presentation style can also affect the nutritional composition of the diet fed. We have seen already that food that is left exposed to the elements (for example, in buckets without lids) can start to lose some of its nutrient content. Food that is scattered directly onto the ground can also undergo a change in nutrient value as a result of direct contact with the soil in an enclosure and animals may also ingest soil with their food, particularly if the food (and soil) is moist. For example, iron intake can be increased if food comes into contact with iron-rich soil (McCormick *et al.*, 2006); this change in iron intake may be valuable if iron is deficient in the diet, or undesirable in species at risk of **haemochromatosis** (see section **12.8.1**). Similarly, iron intake in zoo (and farm) animals can rise when water used to fill troughs, or for nipple feeders, is supplied via pipes that are rusty.

Carcass feeding

Only supplying meat to carnivorous zoo animals, such as storks, hawks, lions, and tigers, is not enough; carnivores need to be fed whole prey, including blood, guts, organs, fat, and bones (**Box 12.6**). This is important both to ensure that their nutritional requirements are met and to maintain good dental health. Carnivorous and piscivorous birds and mammals, for example, may become calcium deficient if they are fed only soft tissues and not bones (Howard and Allen, 2007).

12.7.3 Other functions of feeding

In many instances, the impact of feeding zoo animals goes far beyond merely the provision of nutrients. Depending on how food is presented, feeding can facilitate husbandry, improve health, provide mental and physical exercise for the animals, and also have a considerable impact on the education of zoo visitors. For example, providing food at the end of the day may enable keepers to move animals between areas of their enclosure—notably, from outside areas to inside areas, when keepers want animals to be in secure indoor accommodation overnight.

There are also many ways in which food can be presented to zoo animals that can be considered enriching. In **Chapter 8**, we saw that one of the primary goals of much enrichment used for zoo animals is to boost the amount of time the animals spend foraging, so that their **activity budget** more closely resembles that of their wild **conspecifics**. For wild birds in particular, obtaining food is often their main activity and non-migratory bird species, in the winter months, can spend more than 80 per cent of daylight hours foraging (Klasing, 1998). Replicating this level of activity in captive birds poses a particular challenge for zoos, but the use of feeding enrichment can go some of the way towards providing a solution (see, for example, Vargas-Ashby and Pankhurst, 2007).

Haemochromatosis is the correct term for iron storage disease.

Box **12.6** Carcass feeding: what to feed carnivores instead of live prey?

As mentioned in section **12.3.1**, it is illegal in the UK to provide live vertebrate prey to carnivorous (and piscivorous) zoo animals, such as storks, hawks, lions, tigers, and snakes; neither is the feeding of live vertebrate prey routinely practised in zoos elsewhere around the world (see **Box 12.2**).

So what are the alternatives? The obvious alternative to providing live prey is to provide meat (that is, all or part of the body of a dead animal), but there are various forms that this 'meat' can take. At one end of the spectrum, meat fed to zoo animals can be in the form of a dead, but otherwise intact or whole, animal; this is generally referred to as 'carcass, or whole prey, feeding' (**Fig. 12.23**). At the other end of the spectrum, zoos can feed 'meat products', which may contain non-meat material such as cereals (much like commercial dog or cat food).

The advantages and disadvantages of feeding different types of meat to zoo animals have been the subject of considerable debate, centring on issues such as nutritional value, the impact on the health of the animals being fed, and the acceptability to zoo visitors of carcass feeding. In terms of the nutritional content of a carnivore's diet, we know that carnivorous animals in the wild generally eat much more than only the muscle on their prey; they may eat, and gain nutritional benefits from, the blood, guts, organs, bones, and hair/feathers of their prey. This could lead to the recommendation that zoo carnivores should be fed carcasses, because all of these additional body parts would then be presented to the animals. But all of the nutritional requirements of most carnivores can also be met by feeding meat products, because these products are specifically formulated to

(a)

(b)

Figure **12.23** These photographs show zoo animals feeding on carcasses or whole prey. The meerkat *Suricata suricatta* shown in (a) is feeding on a whole dead chick, supplied within a cardboard tube to provide feeding enrichment. Photograph (b) shows a tree boa *Corallus* sp. at Colchester Zoo in the UK, feeding on a whole dead rat. (Photographs: (a) Paignton Zoo Environmental Park; (b) Colchester Zoo)

ensure that they contain all of the necessary nutrients, including vitamins and minerals (which, in the case of vitamins in particular, can be lost during storage of carcasses). In fact, the ability to monitor an animal's nutritional intake when it is fed a formulated diet is easier than trying to quantify the nutritional composition of naturally occurring foods, which may vary in many ways and thus differ in their nutritional composition.

Carcasses may be undesirable foodstuffs in other ways: the fat content may be too high, or the carcass might be a vector for disease—for example, parasite transmission. Other health risks that have been suggested in relation to carcass feeding include gut impaction or obstruction, or gut perforation by bones, but these potential health hazards do not appear to be supported by empirical evidence (Houts, 1999). A recent survey of zoos worldwide shows, however, that feeding carcasses does not increase the risk of health problems in zoo carnivores when compared with feeding commercial diets (Knight, 2006). Moreover, there are other data that associate the provision of carcasses, or meat on the bone (**Fig. 12.24**), with health benefits to zoo carnivores, such as improved muscle and body condition, and oral health (Fitch and Fagan, 1982; Bond and Lindburg, 1990; Houts, 1999).

Figure **12.24** Even if whole carcasses cannot be provided, feeding meat on the bone rather than only meat, or meat-based products, is beneficial to the health of zoo animals. Here, a lion cub and adult male lion *Panthera leo* feed on meat on the bone at Ramat Gan Zoo, Tel Aviv, Israel. (Photograph: Tibor Jäger, Zoological Center, Tel Aviv)

Similarly, concerns that feeding carcasses may lead to increased aggression (if only one or two food items are provided for a social group of carnivores) are not supported by any firm scientific evidence. There is anecdotal evidence that carcass provision may strengthen social bonds in wolves (Houts, 1999; Ziegler, 1995) and bush dogs (MacDonald, 1996). A study by Bond and Lindburg (1990) showed that carcass provision improved the appetites of captive cheetah, which fed for longer and also displayed species-characteristic possessive behaviour over their food, compared with their behaviour when fed commercial diets. Other studies have shown that carcass feeding, or the provision of meat on the bone, promotes natural foraging and food-processing behaviours, and can reduce the expression of **stereotypies** (see, for example, Carlstead, 1998; McPhee, 2002; Bashaw *et al.*, 2003).

These data suggest that zoos should provide carcasses—or at least meat on the bone—for their carnivores. A survey of zoos worldwide, however, suggests that there are considerable regional variations in the type of meat provided for zoo carnivores (Knight *et al.*, 2005; **Fig. 12.25**). It would seem that, in US zoos, commercial 'meat diets' predominate,

whereas carcass feeding is favoured elsewhere in the world. So if all of the available evidence demonstrates that carcass provision does not negatively impact on animal health and is beneficial in terms of behavioural expression, why do zoos not universally implement this form of feeding?

It is possible that the ability to monitor more carefully the nutritional content of a carnivore's diet may bias a zoo towards choosing a commercial diet over carcass feeding. It has also been suggested that the logistics and regulations involved in sourcing, storing, and preparing meat may vary across regions, and that this may discourage zoos from providing carcasses. But it is the perceived views of zoo visitors that seem to be a key factor in determining whether or not zoos feed carcasses to their carnivores.

A survey of zoo visitors in the UK, Belgium, Australia, and New Zealand indicated, however, that they would not object to seeing animals being fed carcasses (Melfi and Knight, 2008). We await, with interest, data on the views of visitors to zoos in the USA; it is possible that the low rate of carcass feeding in US zoos is due to public objections (or the perception of public objections) to this practice.

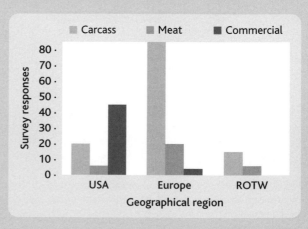

Figure **12.25** This graph shows regional differences in the feeding of carcasses (whole dead prey), meat, and commercial meat diets. (From Knight *et al.*, 2005)

Finally, we need to make some mention of the ubiquitous use of feeding time at the zoo as a visitor attraction. Most zoos make a point of advertising to their visitors when they are going to feed certain species within their collection. This provides an almost guaranteed opportunity for many visitors to see animals that usually hide away, or to see animals doing something that they usually are not, or to learn more about the animals from educational presentations given during feeding time, or a combination of all of these.

Figure **12.26** Reptiles, such as this green iguana *Iguana iguana*, require a source of UVB light in order to be able to metabolize calcium properly. The tail of this animal is permanently distorted as a result of metabolic bone disease, due to an inadequate diet earlier in its life. (Photograph: Sheila Pankhurst)

12.8 Nutritional problems

Zoo animals can experience a variety of nutritional problems, from the accidental ingestion of litter or deliberate feeding of unsuitable food items by zoo visitors, to obesity or **malnutrition**. Vitamin and mineral deficiencies can lead to health problems, as can these substances in excess in the diet. Species categorized as browsers rather than grazers (for example, moose, giraffe, tapirs, and black rhinoceros *Diceros bicornis*) are often regarded as difficult to keep and to breed in captivity, and seem to be more susceptible to diet-related health disorders (Clauss and Dierenfeld, 2007).

Nutritional diseases have been mentioned briefly in **Chapter 11**, but two diet-related diseases that give cause for concern in many zoos and at which we will now look in more depth are metabolic bone diseases and iron storage disease.

12.8.1 Nutritional diseases

Metabolic bone diseases

As Ullrey (2003) points out, the term 'metabolic bone disease (MBD)' is often used as though it is only one disease, but, in fact, there are several differ-

ent bone diseases[22] or disorders that can arise when something goes wrong with an animal's metabolism (**Fig. 12.26**). Bones can lose density or bone mineral (osteopenia or, in severe cases, **osteoporosis**) or can become softened (rickets or osteomalacia).

Rickets, with its characteristic bowing of the long bones such as the femur or humerus, is perhaps the best-known bone disorder and can occur in a wide range of taxa. In fact, rickets and osteomalacia are essentially the same disorder. The term 'rickets' is generally used for the disease in young animals, when the bones are still forming. Both disorders are associated primarily with vitamin D deficiency, but can also occur as a result of insufficient calcium or phosphorus in the diet.

All of the MBDs mentioned here can arise as a result of poor nutrition and dietary deficiencies, although, in several cases, there is also a genetic component to the disease. Bone disorders are of serious concern to zoo vets and nutritionists, because the damage to the bone is often irreversible and can sometimes be fatal. Fidgett and Dierenfeld (2007), for example, report on the potential for the rapid

---

22 It is not only bones that can be affected by MBD: tortoises, for example, can show MBD in carapace development (Hatt, 2007).

Osteoporosis means, literally, porous bone. It is a condition in which there is a marked loss of bone density.

onset of severe MBD in stork species when there is a calcium deficiency in the diet of these birds.

There is a complex relationship between vitamin D$_3$, ultraviolet B light (UVB), calcium, and phosphorus in the body. A number of exotic species lack the ability to synthesize vitamin D$_3$ in the body and dietary deficiencies in this vitamin can lead to bone disorders such as rickets. Other animals, including reptiles such as the green iguana, require a source of UVB light in order to be able to metabolize calcium properly (McWilliams, 2005a). In Komodo dragons *Varanus komodoensis*, for example, captive dragons kept in indoors facilities in zoos showed significantly lower levels of vitamin D$_3$ than those of either wild dragons or captive Komodo dragons kept in exhibits with outdoor access (Gillespie *et al.*, 2001). This study demonstrated that the implementation of corrective measures (such as the use of UV lamps, direct exposure to sunshine, or the installation of UV-permeable skylights) could bring vitamin D$_3$ levels in indoor-housed animals up to levels that were comparable with those found in wild animals.

There are two useful papers on calcium homeostasis in Vol. 39 of the *International Zoo Yearbook*, with feeding and lighting recommendations for lizards and freshwater turtles, respectively (McWilliams, 2005a; 2005b). The Nutrition Advisory Group (NAG) of EAZA also provides a fact sheet (Fact Sheet 002) entitled *Vitamin D and Ultraviolet Radiation: Meeting Lighting Needs for Captive Animals* (Bernard, 1997).

Iron storage disease

Iron storage disease, or haemochromatosis, is the term used to describe the **pathology** associated with deposits of iron in the body tissue, most often in the liver (other organs such as the kidneys and heart may also be affected). Iron storage disease has a known genetic basis in humans and a genetic susceptibility is also indicated in a number of exotic species (for

Figure **12.27** Provision of an adequate supply of browse is particularly important for the health of black rhinoceros *Diceros bicornis* in captivity (Clauss and Hatt, 2006). (Photograph: Kirsten Pullen)

example, the mynah bird *Gracula religiosa*—see Mete *et al.*, 2003). Among mammals, many primate species (such as lemurs—see Spelman *et al.*, 1989) are susceptible to haemochromatosis. Black rhinoceros *Diceros bicornis* and Sumatran rhinoceros *Didermocerus sumatrensis* are susceptible to iron storage disease (Dierenfeld *et al.*, 2005; **Fig. 12.27**), although **haemosiderosis** can occur in black rhinoceros without any **clinical signs** of disease. There is also increasing evidence that iron storage disease is a contributing factor in **morbidity** and failure to breed in zoos in a number of exotic bird species (Taylor, 1984; Cork, 2000; Sheppard and Dierenfeld, 2002).

Iron storage (particularly in the liver) appears to be a problem for those rhinoceros species that are browsers rather than grazers (Clauss and Hatt, 2006). Black rhinoceros feed mainly on browse in the wild, but in captivity are often managed as grazers, with a diet supplemented with high-energy concentrates. Overfeeding of black rhinos appears to be a contributing factor to iron storage disease and can also lead to obesity. Provision of browse and adequate roughage in the diet is also very important for the health of these animals in captivity (Clauss and Hatt, 2006).

Haemosiderosis refers to the accumulation of iron in the body. This can be a precursor to haemochromatosis, but may not have any pathological effects.

Figure **12.28** The most common group of birds to suffer from haemochromatosis, or iron storage disease, are the toucans and hornbills (such as this rhinoceros hornbill *Buceros rhinoceros*), and their relatives. (Photograph: Paignton Zoo Environmental Park)

Haemochromatosis is well documented in captive hornbills (**Fig. 12.28**), and also in toucans, toucanets, and their relatives. The disease is also often found in some mynah birds, tanagers, and birds of paradise in captivity, but not in these species in the wild (Klasing, 1998). Rather puzzlingly, many of the susceptible bird species are frugivores, but fruit is generally a poor source of iron. Consumption of ascorbic acid (vitamin C), however, increases the body's ability to take up iron, so citrus fruits and other foods high in vitamin C are not always recommended for bird species that may be prone to iron storage disease, or are fed separately (Sheppard and Dierenfeld, 2002).

12.8.2 Nutritional disorders

The two main nutrition-related disorders are obesity and malnutrition. Malnutrition is not due to a shortage of food per se, but rather to a deficiency or deficiencies in nutrients in the diet.

Various deficiencies, particularly of vitamins, can give rise to disease.

Malnutrition

Malnutrition arises because of the lack of a specific nutrient (or nutrients) in the diet: the total quantity of food consumed by the animal may be adequate, but an essential component (such as a vitamin or mineral) is lacking. Nearly all macrominerals and vitamins can cause health problems when deficient in the diet (and also when present in excess—see below). We have provided here only a few selected examples.

For general information about vitamin and mineral deficiencies, and their impact on growth and health, we recommend McDowell for mammals (1989; 1992; 2003) and Klasing (1998) for birds. Chapter 5 of McDonald *et al.* (2002) also provides a useful account of vitamins and nutrition, and includes a table summarizing the disorders caused by vitamin deficiencies. Information on vitamin and mineral deficiencies in non-mammalian and non-avian taxa is harder to come by, and is rather more scattered throughout the literature, although Fowler and Miller (see in particular Fowler and Miller, 1999) is a good starting point.

Vitamin deficiencies

Examples of vitamin deficiencies that are of particular concern in zoos include deficiencies in vitamins A, C, and E. Vitamin A is important because of its role in immune system function; a deficiency of this vitamin can make animals much more susceptible to infection. Several studies have reported that death from infection can occur well before signs of a vitamin A deficiency become apparent (for example, in poultry, see Sklan *et al.*, 1994).

Vitamin C deficiency can present a problem in non-human primates, particularly if commercial feeds are stored for long periods of time, because levels of this vitamin decrease markedly in foodstuffs after more than a few months in storage (Lamikanra *et al.*, 2005).

Vitamin E is believed to play an important role in the function of cell membranes. As we have seen already in this chapter, vitamin E is readily lost from fish during storage and, for this reason, is often given as a routine supplement (along with thiamine or vitamin B_1) to fish-eating mammals and birds. Vitamin E supplements (by injection or as an additive to food) are also sometimes given to exotic animals in captivity prior to capture and handling, in an attempt to reduce **capture myopathy**, although the evidence to support the value of this practice is mixed (see, for example, Graffam *et al.*, 1995).

Mineral deficiencies

Mineral deficiencies are rather like the diseases of concern that we discussed in the previous chapter (**Chapter 11**) and will vary from collection to collection. For example, in some areas, the pasture on which grazing animals feed can be deficient in selenium,[23] due to the mineral composition of the soil. This effect can be highly localized, but can have a major impact upon hoof growth and other aspects of health.

The situation is further complicated by inter-relationships between minerals and other nutrients. Selenium, for example, works in conjunction with vitamin E as an antioxidant within the body and, to some extent, these two substances can compensate for each other (McDowell, 2003), so that animals with an adequate intake of vitamin E are better able to cope with very low levels of selenium in their diet, and vice versa.

The signs of mineral deficiencies in animals are not always obvious, and can be similar to the signs of other diseases and disorders. So-called 'wasting diseases' or 'wasting syndromes' (in which an animal shows progressive loss of weight over a period of time) can be related to mineral deficiencies, but can also be due to infectious agents. Bone or antler eating can indicate a deficiency in calcium in the diet. Animals suffering from a mineral deficiency may also exhibit **pica**, which, in livestock, is a term generally used to refer to animals eating items not usually found in their diet (see, for example, Golub *et al.*, 1990).

A deficiency in zinc in captive birds fed a grain-based diet is not uncommon, and can have adverse effects on growth and egg production (Klasing, 1998). Zinc deficiency in ducks, for example, can lead to the development of dermatitis on the webs of the feet and/or the fraying of new feathers after moulting (McDowell, 1992).

Toxicity

As well as health problems caused by nutritional deficiencies, an excess of a particular nutrient or substance can also lead to diseases and disorders that can, in some cases, be fatal. As already outlined in section **12.1.3** and in **Box 12.1**, many plant species are poisonous to the animals that eat, or may attempt to eat, them. Zinc toxicity has been reported in captive birds that have gnawed at the galvanized wire of their cages (on fatalities in **psittacines**, for example, see Howard, 1992).

Obesity

Body weight is often used by keepers and vets as a measure of health in zoo animals, with comparisons to data from wild animals of the same species (see, for example, Terranova and Coffman, 1997). In ruffed lemurs, for example, Schwitzer and Kaumanns (2001) looked at a sample of forty-three animals from thirteen European zoos, and found that more than 46 per cent of individuals were obese.

In general, obesity is a much greater nutritional problem in zoos than malnutrition. This is for a variety of reasons.

Myopathy refers to the build-up of lactic acid in muscles; this can cause stiffness, paralysis, or even death. **Capture myopathy** has been reported in a wide range of taxa and can cause death almost immediately, or up to a few weeks later.

23 The mineral selenium can cause health problems in animals both when it is deficient in the diet and when it is overabundant. Selenium toxicity can cause animals to lose hair or even hooves (McDowell, 2003).
Pica in humans is often used to refer to the consumption of non-food items, from coal to soap, but it can also mean cravings for particular food constituents.

- The legislative framework in the UK does not allow zoos to withhold food (this is set out in the SSSZMP; see section **12.3.1**), so there is a relatively high **plane of nutrition** for most zoo animals.

- In captivity, animals may not have as many opportunities for activity as they would in the wild and their motivation to exercise may be reduced if they are overfed.

- Obesity can also arise in dominant animals in a group if these individuals are highly successful at obtaining food intended for more than one animal. Zoos may deliberately provide surplus food for group-housed animals, to ensure that subordinate animals receive enough food, but this can result in dominant animals consuming much more food than they need and becoming obese. (The converse is also true: subordinate animals may not get their fair share of food and can lose condition.)

- Another factor that can contribute to obesity in zoo animals is the feeding of energy-dense commercial feeds: feeds developed for livestock will be formulated for 'high production', or rapid weight gain, and may not be appropriate for zoo animals.

- Zoo diets also may not show the seasonal variation with which animals are adapted to cope in the wild: many animals are physiologically adapted to lay down fat reserves before the winter, but these animals can then gain excess weight if their food supply is not reduced in the winter months.

- Finally, keepers may overfeed animals in the same way that some (misguided) pet owners overfeed their pets.

Obesity is of concern to zoos because it can lead to other health problems, such as reduced **fertility**, or limb problems. In Indian rhinoceros *Rhinoceros unicornis*, for example, obesity appears to be linked to other health problems such as foot disorders (Clauss *et al.*, 2005). Obesity is also linked to the development of diabetes in captive primates (for example, in orang-utans *Pongo pygmaeus*—see Dierenfeld, 1997b).

Reproductive disorders

Nutritional deficiencies can have a profound effect on reproduction. As we saw in the previous section, there is evidence that iron storage disease is a contributing factor in the failure to breed in zoos in a number of exotic bird species (Taylor, 1984; Cork, 2000; Sheppard and Dierenfeld, 2002). We also saw in **Chapter 9** that factors such as maternal nutrition can have a significant influence on sex ratio (see, for example, Kilner, 1998). Overfeeding can also lead to the development of large foetuses in vertebrates, which, in turn, can lead to health problems at parturition or egg-laying.

A number of mineral and vitamin deficiencies have been documented to have an adverse effect on reproduction. In livestock, for example, a deficiency in vitamin A can cause congenital abnormalities in foetuses, and a copper deficiency leads to reduced egg production and 'hatchability' in poultry (McDonald *et al.*, 2002). Less is known about the effects of nutritional deficiencies on reproduction in exotic animals, but there is still a body of data demonstrating that poor nutrition can have an adverse effect on reproduction (for reviews, see Allen and Ullrey, 2004; Howard and Allen, 2007).

Howard and Allen (2007), for example, have reviewed nutritional factors affecting semen quality in non-domestic felids, and point out a recurring link between poor diets and poor reproduction. This is highlighted in a study of South American felids in captivity by Swanson *et al.* (2003), who found that, in a sample of 185 adult male cats from eight species (including ocelots *Leopardus pardalis* and the jaguar *Panthera onca*), kept in forty-four zoos in Latin America, only around one-third of the cats received nutritionally adequate diets and only 20 per cent of the males were classified as 'proven breeders'.

Among birds, Fidgett and Dierenfeld (2007), in a recent review of minerals and nutrition in stork species, concluded that nutritional factors are likely to be a key component of successful captive breeding of these carnivorous birds.

Dental disorders

As well as the risk of nutritional diseases or disorders developing, provision of an inappropriate diet can have serious consequences for the oral health of zoo animals. The consistency of food has an effect on the rate at which dental plaques form and a diet composed largely of soft foods will lead to greater plaque formation than will a diet of hard foods (Braswell, 1991). This can be a particular problem for carnivores fed only on meat or commercial meat diets, rather than on carcasses or meat on the bone.

Diets may also need to be adapted for elderly animals in zoos that have lost teeth, or have other dental problems.

Summary

- The aim of 'captive feeding' in zoos is to provide optimal nutrition.

- Legislation governing zoo animal feeding and nutrition requires that zoos provide sufficient food and water, but the provision of an optimal diet falls largely outside the legislative framework of most countries (although within the UK, the SSSMZP provide quite detailed recommendations on zoo animal feeding and nutrition).

- Zoos cannot hope to match exactly the wild diets of all of the animals in their care, but they can try to match the nutritional content of wild diets.

- Much of what we know about the detailed nutritional requirements of animals is based on models of domestic animals, such as cattle, sheep, poultry, and cats and dogs, but these nutritional models may not always be appropriate for zoo animals.

- Zoos today have access to an increasing range of guidelines and tools to help them to develop diets providing optimum nutrition for their animals. Foremost among these tools is the nutritional software package and database, Zootrition™.

- Nutrition can have profound effects on other aspects of an animal's life history, such as its health, physical and mental development, reproduction, and mortality.

- Obesity is a greater problem in zoos than malnutrition (by law, zoos must provide sufficient food for their animals); nutritional diseases of particular concern to zoos include metabolic bone disorders and iron storage disease, or haemochromatosis.

Questions and discussion topics

1. Do animals have 'nutritional wisdom'?

2. What is Zootrition™, and how can it be used by zoos to improve the health and welfare of their animals?

3. What are the main nutritional problems and disorders that are found in animals in zoos, and what steps can be taken by zoos to prevent these occurring?

4. Which broad group of animals poses a greater challenge for zoos in terms of 'captive feeding', carnivores or herbivores?

Further reading

For anyone wanting to learn more about the basic principles of animal nutrition, we recommend McDonald *et al.*'s *Animal Nutrition* (2002), now in its sixth edition. Although largely about mammalian livestock species, this book provides much useful general information on nutrients and digestion, and has chapters with titles such as 'Hay, artificially dried forages, straws and chaff'. Another good general textbook is Pond *et al.*'s *Basic Animal Nutrition and Feeding* (2005), now in its fifth edition. The diagrams of gastrointestinal tract structure in this chapter are all taken from Stevens and Hume (1995): their book *Comparative Physiology of the Vertebrate Digestive System* is a very useful source of information on digestive physiology across all vertebrate taxa.

In 1999, Edinburgh Zoo hosted a symposium on *Nutrition of Wild and Captive Wild Animals*. The plenary speaker at this meeting was Dr Ellen Dierenfeld, then at the Wildlife Conservation Society/ Bronx Zoo, who gave a lecture on the historical perspective to the developing science of zoo animal nutrition. Her lecture is published as a paper in the *Proceedings of the Nutrition Society* (Dierenfeld, 1997a) and it is well worth getting hold of a copy, not least for the comprehensive reference list.

In a later paper, Ellen Dierenfeld (2005) also reviews the global dissemination of information about zoo animal nutrition, and catalogues the main national and international developments in nutrition as a discipline within zoo biology over the past three or four decades.

Proceedings from the first European Zoo Nutrition Conference, held in the Netherlands in 1999, have been published in book form under the title *Zoo Animal Nutrition* (Nijboer and Hatt, 2000). Volumes II and III of *Zoo Animal Nutrition* followed in 2003 and 2006, respectively, with Dr Andrea Fidgett (nutritionist at Chester Zoo) as the leading editor (Fidgett *et al.*, 2003; 2006). These useful books contain research papers and reviews on a wide range of topics relating to zoo animal nutrition.

Five chapters in the excellent and detailed book *Wild Mammals in Captivity*, edited by Devra Kleiman and her colleagues, are devoted to the nutrition of mammals in zoos (Kleiman *et al.*, 1996). Recommended texts that consider taxa other than the mammals include Klasing's *Comparative Avian Nutrition* (1998), Robbins' *Wildlife Feeding and Nutrition* (1993), Lovell's *Nutrition and Feeding of Fish* (1998), and (also on fish) Halver and Hardy's *Fish Nutrition* (2002).

Several editions of Fowler and Miller provide detailed accounts of nutrition-related diseases and disorders, across a wide range of taxa (including fish): the fourth and sixth editions (Fowler and Miller, 1999; 2007), for example, contain a great deal of information about the impact on health of various vitamin and mineral deficiencies in the diet of wild animals in captivity.

Finally, three volumes of the *International Zoo Yearbook* have zoo animal nutrition as their main theme: Vol. 6 (published in 1966); Vol. 16 (published in 1976); and, most recently, Vol. 39 (published in 2005).

Websites and other resources

Guidelines and other sources of information about zoo animal nutrition have already been discussed in some detail in section **12.3.3**; we particularly recommend the nutrition pages on the EAZA website— www.eaza.net —as a first port of call.

In addition to these guidelines and to databases such as Zootrition™, another widely used source of information is the series of publications from the National Research Council (NRC) of the US National Academy of Sciences. The NRC publications are consulted not only within zoos, but also by livestock farmers, laboratory animal managers, and the pet food industry. There is more information about NRC nutrient requirement publications in **Box 12.3**.

Also from the USA, the US Department of Agriculture (USDA) has an extensive searchable database of nutrient values for a very wide range of human foodstuffs, which is available online at `www.ars.usda.gov`.

Finally, the website of the Comparative Nutrition Society—`www.cnsweb.org`—has details of published proceedings of the meetings of the Society. (The interests of the Society are wide-ranging, and span nutrition, physiology, and biochemistry across all animal taxa.)

Chapter 13 Human–animal interactions

Zoos are not only full of animals, they are also full of people. Many of these people work in the zoo and are a relatively permanent feature of the zoo environment; others, such as building contractors, maintenance engineers, etc., may be transient, but may nevertheless create a great deal of disturbance; in addition, there are the zoo visitors, several thousand of whom may pass through on a typical day. Humans are thus a significant part of the zoo animal's environment, and are likely to have some impact on the behaviour and welfare of the animal. While this influence may sometimes be positive or neutral, it can also be negative, and detrimental to the welfare of the animals, so we need to measure the extent and type of that influence. Reciprocally, the animals are likely to have some impact on the humans, so we can imagine a sort of 'zone of contact' within which humans and animals come into sufficient proximity to be able to influence each others' behaviours.

In all of these cases, we can refer, in a general way, to 'human–animal interactions', using the word 'interaction' in the sense that it was used by Estep and Hetts (1992) to mean some kind of behaviour performed by one individual that influences the behaviour of another, thus making it a kind of communication. This kind of interaction can be used by the zoo to increase the public's knowledge of, and concern for, the animals and their conservation, while at the same time giving them an enjoyable experience. The ideal is to create a zoo environment in which the animals' and humans' experience of each other is as positive as possible.

In this chapter, we will consider some of the ways in which animals and humans influence each other in this zone of contact. We will cover the following topics.

13.1 Zoo visitors: what do we know about them?
13.2 Education and awareness raising
13.3 Humans in the zoo: how do they affect the animals?
13.4 Training

We are therefore looking in this chapter not only at human–animal interactions in the zoo, but also, a little more widely, at the consequences of this for people's experiences of the zoo.

Related topics to do with the interface between the public, the keepers, and the animals are considered in boxes throughout the chapter.

13.1 Zoo visitors: what do we know about them?

It is self-evident that zoo visitors are fundamentally important to the operation and success of zoos, in terms of their stated objectives. Visitors are the targets of zoo educational initiatives and are, ultimately, the primary funding source for the conservation work that zoos do. So we would expect zoos to find ways of maximizing their attractiveness to visitors, and to provide an experience that is consistent with welfare, educational, and conservation priorities, but is also enjoyable for the visitors.

13.1.1 Who visits zoos?

It has been estimated that, at the start of the 1990s, something in the region of 619 million visits to zoos were made worldwide annually (IUDZG/CBSG, 1993; see **Fig. 13.1**): an astonishing figure, given the

Figure **13.1** Zoos worldwide are popular venues for recreation and education, and also serve as conservation institutions. (Photograph: Tibor Jäger, Zoological Center Tel Aviv—Ramat Gan, Israel)

number and variety of other attractions with which zoos compete.

In the face of that competition, zoos appear to be holding their own extremely well (Turley, 1999); indeed, in the UK and the USA, zoo attendances have been increasing over the past few decades (Davey, 2007).[1] According to *The Manifesto for Zoos*, over 14 million people currently visit a zoo, aquarium, or wildlife park in the UK each year, compared with 15.2 million who regularly attend a professional football match. The number of visitors to Chester Zoo alone exceeds that for many major attractions such as Stonehenge and St Paul's Cathedral (1,060,433 compared with 677,378 and 837,894, respectively, in

2001—information from *Statistics on Tourism and Research*, www.staruk.org.uk). And such figures are not only seen in UK zoos: the zoos in Copenhagen and Rotterdam attracted more visitors than any other urban attractions in those cities during the early 1990s (Mason, 2000).

We may well ask, therefore, who these visitors are, what they do when they are at the zoo, and what they think of their experience. There is, however, not as much information available as we would like in order to answer these questions. What information there is may be difficult to locate, because it appears in unpublished reports with limited circulation or because it is submerged in a wider study that

1 Although Davey's analysis shows that elsewhere there has been a slight decline in zoo attendance since the 1960s.

treats the zoo as a kind of museum[2] (see, for example, Bitgood, 2002), and thus appears in the museum, rather than the zoo, literature.

No clear overall patterns have emerged from these studies, but, in general, it appears that there are no particular distinguishing characteristics that define the zoo visitor. This is probably encouraging, because it implies that all sections of society are likely to visit zoos and therefore be exposed to animals and conservation messages. Similarly, it appears that people visit zoos for a variety of reasons, not all of which are to do with animals (there is more on this in section **6.1.3**). For some, for example, zoo visiting may be a social event (Morgan and Hodgkinson, 1999) and more time is spent in conversation than in looking at animals; for others, the experience of encountering live animals can be an emotional event (Myers *et al.*, 2004).[3]

13.1.2 What do zoo visitors do at the zoo?

The answer to the question of what zoo visitors do when they are at the zoo is, unfortunately, that we do not really know—although there have been studies that have observed the way in which people behave in zoos and some of the things that they say (Davey, 2006a). Probably the most consistent observation that has come out of these studies is that people spend remarkably small amounts of time actually looking at the exhibits.

In one study, Marcellini and Jenssen (1988) observed nearly 600 people visiting the reptile house at the National Zoological Park in Washington DC. They tracked these people through the house, recording how long they spent looking at each exhibit. The mean time spent in the reptile house per visitor was only 14.7 minutes, of which 8.1 minutes were spent looking at the exhibits rather than walking around. This same study found clear differences in the time spent looking at different kinds of animal in the

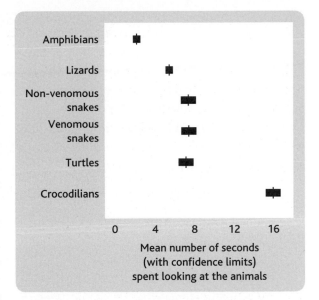

Figure **13.2** Visitors to the Reptile House at the Smithsonian's National Zoo in Washington DC spent most of their time looking at large-bodied species such as crocodilians and least time looking at small animals such as amphibians, but did not actually spend very long looking at any of them. (From Marcellini and Jenssen, 1988)

reptile house, with crocodilians being watched for longest and amphibians for the least amount of time (see **Fig. 13.2**). In fact, it was the relative sizes of the animals that primarily determined for how long people would view them.

This relationship—that larger animals are, in some way, more attractive or popular than small animals—has been claimed for other zoos and other kinds of animal as well (Ward *et al.*, 1998), although it is quite possible that it depends very much on how popularity is measured. In the study by Ward *et al.* (1998), popularity was measured as the proportion of visitors at an exhibit who spend more than 10 seconds looking at it. Balmford *et al.* (1996; Balmford, 2000), however, used a measure of the proportion of people passing an exhibit who actually looked at it rather than simply walked past, and

---

2 Indeed, it has been said that the zoo *is* a kind of museum, differing from other museums only in that its exhibits are alive (Mason, 2000).

3 The emotional dimension of contact with wild animals is expressed in Wilson's (1984) 'Biophilia' hypothesis, that humans have an '*innate tendency to affiliate with life*'.

found no relationship between popularity and body size of animals.

More recently, a study at Recife Zoo in Brazil found no relationship between popularity of a number of mammalian exhibits and a variety of exhibit features, including whether the species was exotic or native (da Silva and da Silva, 2007), so the most we can probably say is that we have little idea of exactly what makes a popular exhibit.

Another factor that might influence how long visitors spend at an enclosure is the visibility of the animals. At Zoo Atlanta, Georgia, visitors complained about difficulties in seeing tigers in a **naturalistic enclosure**, so signs were put up to point out to visitors where the cats were likely to be. Nonetheless, visitors generally did not use the signs, or even read them, and still came away thinking that they could not see the tigers (Bashaw and Maple, 2001).

Box 13.1   Visitor behaviours and preferences in the zoo: the view from Asia

The majority of studies on human behaviour in zoos, and how it impacts on the behaviour and welfare of the animals, have been undertaken in zoos in Europe, North America, and Australia. Yet it is very likely that there will be strong cultural differences among zoo visitors in different parts of the world, and that these differences may affect how visitors perceive and behave towards the animals. Very few studies have looked at this aspect of zoo biology, but there are a few recent studies that have investigated the extent to which zoo visitors in Asian zoos are similar to those in Europe and North America.

Firstly, Mallapur *et al.* (2005) investigated the effects of zoo visitors on lion-tailed macaques *Macaca silenus* in eight different zoos across India. The results were similar to those seen with macaques in Western zoos, in that abnormal and aggressive behaviours increased when visitors were present, and the animals appeared to be stressed by the visitors. The amount of increase in these behaviours, moreover, was quite large in comparison with many of the European/North American studies (20 per cent more **abnormal behaviour** as a short-term effect and 30 per cent as a long-term effect). As the authors point out, there are few well-established conservation and welfare awareness

programmes in Indian zoos, and, consequently, it is common for animals to experience high levels of disturbance (such as shouting, teasing, feeding, and even physical harm) from zoo visitors. This is certainly an aspect of visitor behaviour that deserves further study.

Our second example is from China. Davey (2006b) investigated visitor interest, as measured by viewing times, at a mandrill *Mandrillus sphinx* enclosure at Beijing Zoo. He found that visitors spent more time at the enclosure after it had been transformed from a **barren enclosure** to a more naturalistic one, even if there was no animal visible. This is an encouraging result at a time when Beijing (and probably other Chinese zoos) is faced with the long task of replacing its old concrete enclosures with naturalistic ones.

Finally, Puan and Zakaria (2007) undertook a questionnaire survey of visitors to three Malaysian zoos (National Zoo, Malacca, and Taiping) to find out how they perceived the role of zoos. Most of the respondents (80 per cent) were local, living in the same state as the zoo, and most visited with their families to see the animals. Encouragingly, the majority of respondents were well aware, and supportive, of the conservation role of zoos.

Visitor attendance times can be increased if the exhibits are more active or if they involve the public more. For example, attendance time at an otter exhibit at Zoo Atlanta was greater for animal training sessions than when the public only passively observed (Anderson *et al.*, 2003). People particularly like to see active animals, as has been demonstrated with small felids (Margulis *et al.*, 2003) and free-range tamarins (Price *et al.*, 1994).

13.1.3 How do zoo visitors perceive the animals?

Valuable information can be obtained by watching what zoo visitors do and measuring how they spend their time while they are in the zoo. To understand their perceptions, attitudes, and quality of experience, however, it is usually necessary to use a questionnaire-based, rather than an observational, approach. Several studies have adopted this sort of approach both to find out about people's perceptions of zoo animals in general, and also to evaluate whether various welfare, education, and conservation initiatives have helped to make people's attitudes more positive.

One of the earliest attempts to measure public perceptions of zoo animals was a study by Rhoads and Goldsworthy (1979), in which subjects were asked to rate pictures of animals in three different settings in terms of adjectives such as 'freedom', 'happiness', 'loneliness', 'dignity', 'friendliness', and 'naturalness'. Across several animal species, there were more negative ratings of pictures of caged zoo animals compared with 'semi-natural' (but still zoo) and 'natural'. Because the subjects were first-year psychology undergraduates[4] and they were rating pictures rather than real animals, it is not easy to generalize these results to the visiting public as a whole. Other studies, such as that by Finlay *et al.* (1988), have used similar methodology and a similar cohort of subjects, and have obtained similar results.

In contrast, Reade and Waran (1996) interviewed zoo visitors within the zoo and compared them with a control group of people interviewed in the street. They found that, although there were negative perceptions of zoo animals (for example, that the animals were bored or sad) among the general public, attitudes among those actually in the zoo were more positive. With the trend towards naturalistic enclosures, we might hope that people's perceptions of zoos and their animals would become more positive, and this does, indeed, appear to be the case. Although visitors still spend very little time at the enclosures, they do rate naturalistic or enriched enclosures more favourably (Wolf and Tymitz, 1981; Tofield *et al.*, 2003). This appears to be because people assume that animals in naturalistic enclosures have the best welfare, even though they may have contradictory notions about what best welfare actually is (Melfi *et al.*, 2004a). In this case, the preference among the people interviewed was to see tigers being active and they thought that this would indicate good welfare, even though they accepted that this was not what tigers spent their time doing in the wild.

13.1.4 Do zoo initiatives change attitudes?

The more naturalistic enclosures that zoos have been providing in recent years ought to promote more positive views among zoo visitors about the welfare and naturalness of the animals (Coe, 1985), and studies such as those described in the previous section suggest that this is, indeed, the case. But other changes that are made to the way in which the animals are maintained may also influence people's perceptions and therefore need to be monitored. The hope, of course, is that changes that are instigated to improve animal welfare will also be seen more positively by the zoo-going public, but this cannot be taken for granted.

McPhee *et al.* (1998) tested the effects of **environmental enrichment** on public attitudes by

---

4 Psychology undergraduates are routinely used in psychological research, with the result that we probably know more about them than most other groups of people.

Type of feeding	On-exhibit	Off-exhibit	χ^2 (df 1)
Live insects to lizards	96% (192)	100% (200)	4.04, p<0.05
Live fish to penguins	72% (144)	84.5% (169)	9.18, p<0.01
Live rabbit to cheetah	32% (64)	62.5% (125)	37.32, p<0.001

Table **13.1** Percentage of people interviewed at Edinburgh Zoo who agreed with the idea of feeding live prey to zoo animals on-exhibit and off-exhibit*

* Data from Ings *et al.* (1997a).
NOTE Numbers in brackets indicate actual numbers of people.

interviewing people in front of each of four exhibits: a barren outdoor polar bear *Ursus maritimus* grotto; a vegetated outdoor tiger grotto; a traditional barred outdoor lynx *Lynx* spp. cage; and an indoor **immersion exhibit** for a fishing cat *Prionailurus viverrinus*. Each enclosure also had either a natural object (for example, wood), a non-natural object (for example, a blue plastic barrel), or no object present. They were testing a common assumption that the public see non-natural objects provided for **enrichment** purposes as artificial, and that this leads to more negative perceptions of the zoo and the animals. In fact, they found that the type of enrichment provided hardly influenced attitudes at all, regardless of enclosure type, and that visitors generally understood the function and importance of the enrichment objects.

On the other hand, a study at Beijing Zoo by Davey *et al.* (2005) found that visitor behaviour changed when viewing an enclosure in which enrichment had been added. In particular, viewing and stopping times increased in comparison with those at the enclosure prior to enrichment. The enrichment in this case included addition of more naturalistic items to a previously barren mandrill enclosure.

Similarly, Blaney and Wells (2004) found that the addition of camouflage netting to a gorilla *Gorilla* spp. enclosure (put there to try to reduce the stressful effect of zoo visitors) improved the public's perception of the animals and their enclosure.

Not all kinds of enrichment are as readily acceptable to the public: in principle, if it is important for zoos to give their animals the opportunity to show **species-typical behaviours**, then zoo **carnivores** should be afforded the opportunity to hunt live prey—but how would the public view this? Ings *et al.* (1997a) attempted to find out by interviewing visitors to Edinburgh Zoo. Their results (see **Table 13.1**) imply that the public—or at least those who go to zoos—may not be as opposed to this idea as we might expect, particularly if the feeding of live prey is done off-exhibit. At present, of course, it is not legal to do this with vertebrate prey in the UK (see **Chapter 12**).

13.2 Education and awareness raising

Educating the public about animals, and raising their awareness and support of conservation, are a fundamental part of the modern zoo's mission, so we can justifiably ask how successfully zoos achieve this. In this context, 'education' can mean a sort of passive acquisition of knowledge that comes from seeing the animal and reading the notices (**Fig. 13.3**), and this is often referred to as **informal learning**. Increasingly, however, zoos are using more structured forms of instruction, ranging from keeper talks and demonstrations, through to school and college visits (**Fig. 13.4**) and short courses, all of which we can refer to as **formal learning**.

Education departments have now become widespread in zoos (Woollard, 1998) and zoo educators undertake a variety of activities, including giving talks, volunteer training, and outreach work (Woollard, 1999). Surveying all of the educational

Figure **13.3** A good example of how zoos can provide information to their visitors and create passive learning experiences is through the provision of informative signs, which can be produced to interest visitors of all ages. (Photograph: Sheila Pankhurst)

activities and initiatives that zoos have introduced in the last couple of decades is outside of the scope of this book, but we can briefly consider what impact these activities have, particularly on people's knowledge of, and attitudes towards, conservation.

13.2.1 Informal learning at the zoo

Kreger and Mench (1995) have pointed out that people are actually highly motivated to view animals and that the opportunities for interaction that are afforded by zoos (for example, shows, demonstrations, children's zoos, etc.) can have a profound impact on education and conservation awareness in people (**Fig. 13.5**). In this case, it is probable that not all parts of the zoo experience contribute equally to public knowledge and perception, but that, overall, the experience is important.

Figure **13.4** Active learning in zoos can be achieved by providing informed guides to zoo visitors. In this case, information about animals at Apenheul Primate Park in the Netherlands was presented to a visiting group of university students by a trained guide. (Photograph: Sheila Pankhurst)

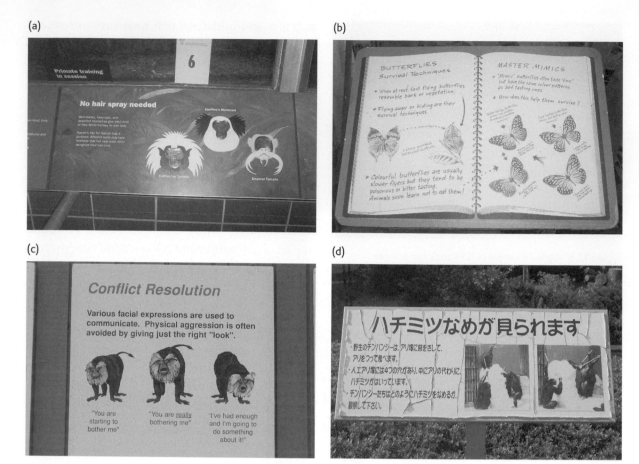

Figure 13.5 Signage in zoos, represented here by the Bronx Zoo in New York (a), Toronto Zoo (b & c), and Nagoya Primate Zoo, Japan (d), is highly variable, both in terms of the information that is being conveyed and also in the manner in which it is depicted. These are some examples that illustrate signs designed to inform visitors about the basic natural history of the animals and the way in which they are managed in the zoo. (Photographs: Vicky Melfi)

Do people learn anything passively from the exhibits? Given the short viewing times described in section **13.1.2**, we would perhaps be surprised if they did. This is borne out by Altman's (1998) study of visitor behaviour at three bear exhibits at Philadelphia Zoo. The conversations of the subjects were predominantly human-centred, rather than about the bears, and only tended to be more about the bears when one of the exhibits (a polar bear) increased its activity, but not when the others (a sloth bear *Melursus ursinus* and spectacled bear *Tremarctos ornatus*) increased their activity.

There is, of course, always the problem of confidence in what our measures are actually measuring in studies like this.

13.2.2 Formal learning at the zoo

Most zoos offer structured educational opportunities, both to ordinary visiting members of the public (such as keeper talks, close encounters, etc.), and also to school and college students on organized visits (**Fig. 13.6**). In the UK and Ireland, 760,000 school pupils use the educational services of British and Irish Association of Zoos and Aquariums (BIAZA) zoos each year, with 400,000 of these receiving some kind of formal tuition (figures from BIAZA).

School and college visits often include 'hands-on' sessions with live animals and with animal parts (for example, fur and bones), and teachers rate these sessions as the most valuable part of the visit

Figure **13.6** School visits provide children with the opportunity to learn about the role of zoos and how they function, but also about the animals that they maintain, and a wider range of issues relating to nature and the environment. (Photograph: Sheila Pankhurst)

(Woollard, 2001). It is widely thought that childhood experiences of zoos are important in fostering positive attitudes to the environment in adults (Holzer and Scott, 1997), which would emphasize the value of school visits—but comparison of the content of conversations at London Zoo between, on the one hand, children on a school visit and, on the other, family groups showed no difference between the two, leading the authors to suggest that schools are failing to make full use of the educational potential of zoos (Tunnicliffe *et al.*, 1997).

What about the less-structured events, such as keeper talks (**Fig. 13.7**)? Given that visitors prefer to see active animals, then the time they spend at an exhibit should be greater if there are animal presentations accompanied by keeper talks (see section **13.1.2**). Does this also increase educational opportunities?

Broad (1996) used a questionnaire-based study at Jersey Zoo to evaluate the relative effectiveness of three different educational media (guidebook, signs, and keeper talks) in enhancing people's understanding of the zoo's conservation work and their knowledge of endangered species. She found that keeper talks were the most effective of the three, as we would probably expect. Given that a large proportion of the respondents had a high level of education, belonged to one of the higher socio-economic groups, and were visiting the zoo because they were interested in wildlife, it is not clear how applicable these results are to other zoos and other countries.

A similar study at Point Defiance Zoo in Tacoma compared visitors watching clouded leopards in a traditional exhibit or in an interpretive presentation, which consisted of encountering a keeper walking the leopard around the zoo and answering visitors'

(a)

(b)

(c)

Figure **13.7** Keeper talks can include opportunities for the public to meet and question the keepers (a), while handling sessions with some of the animals (b), which can bring them 'up close and personal' with animals, and demonstrations of what the animals are able to do (c) can all leaving lasting impressions on the zoo visitor. These methods contribute to educating visitors and providing them with a lasting memory of their visit (see section **6.1.3**). (Photographs: (a) Vicky Melfi; (b) South Lakes Wild Animal Park; (c) Geoff Hosey)

questions (Povey and Rios, 2002). Visitors meeting the interpretive presentation spent longer watching the animal and sought more knowledge than those only seeing the traditional exhibit.

13.2.3 Do zoos raise conservation knowledge and awareness?

Zoos have an important role to play in educating the public about animals and nature in general, but their role in conservation education is particularly important. The term 'conservation education' refers to '*the principles of environmental education and education for sustainability*' (WAZA, 2005).

Zoos are uniquely well placed to undertake effective conservation education (Whitehead, 1995; Sterling *et al.*, 2007) by capitalizing on the live animal collections, and the *in situ* and *ex situ* work in which the zoos are involved. This involves raising awareness of conservation issues, but also demonstrating the links between the animals in the collection and the conservation of their wild counterparts, so that the zoos effectively become interactive and entertaining conservation centres (Tribe and Booth, 2003). But how effectively do zoos achieve this?

To some extent, we might expect that the mere act of visiting a zoo, and encountering the animals and information about them, would help make visitors more conservation-oriented and this seems to be, at least partly, the case. At the National Aquarium in Baltimore, USA, for example, the conservation knowledge, attitudes, and behaviour of visitors were measured on arrival at the aquarium, again at departure, and then again 6–8 weeks later (Adelman *et al.*, 2000). The results of the study indicated that a raised knowledge and understanding of conservation persisted in people after their visit to the aquarium, but did not result in them initiating any conservation-related actions. At the Bronx Zoo, a new 'Congo Gorilla Rainforest' exhibit, which visitors had to pay extra to visit, included not only the simulated African rainforest with over 300 animals, but also an exhibition building with interpretative and interactive panels and displays. Extensive monitoring and questioning of visitors revealed an increase in their knowledge and concern about conservation after touring the exhibit (Hayward and Rothenberg, 2004).

Other studies have reported slightly different results. A survey of attitudes and knowledge amongst zoo visitors entering and leaving several UK zoos found no differences between these two groups in any of the measures used except one: on leaving, visitors had more idea about what differences they could make to conservation (Balmford *et al.*, 2007). This result is consistent with the suggestion that education through facts about conservation is not enough in itself: people want to know what they can do personally (Gwynne, 2007; Sterling *et al.*, 2007).

A number of other techniques have been used to try to enhance the environmental knowledge and awareness of the zoo-visiting public. At Zoo Atlanta, Georgia, for example, an exhibit about the bushmeat crisis was installed in the Willie B. Conservation Centre,[5] consisting of photographs and accompanying text of live animals, dead animals, logged areas, and hunting pictures. Although the pictures of dead animals were disturbing, 97 per cent of visitors who were asked about them thought it was appropriate for the zoo to show them to adult visitors. Eighty three per cent of the visitors had not heard of the bushmeat trade, so the exhibition did raise awareness, but visitors' conservation-related knowledge was not increased by the exhibition (Stoinski *et al.*, 2002).

Improving knowledge and awareness is one thing; bringing about a change in behaviour is, of course, something entirely different. It is also very difficult to measure, because people might not recognize small changes in their behaviour as a result of zoo education to be anything to do with conservation at all. Studies at Monterey Bay Aquarium in California showed that conservation education was most successful if courses of action were suggested to visitors that they were actually able to take, as mentioned above, and which were still consistent with their lifestyle. For example, visitors could pick up at the aquarium a 'Seafood Watch' pocket guide, which listed sustainable seafood to buy in restaurants and shops, and also an 'Ocean Allies Card', which listed conservation organizations that they could join. Visitors turned out to be more interested in changing their eating habits than in joining a conservation organization (Yalowitz, 2004).

The educational impact of zoos has been investigated by an American Association of Zoos and Aquariums (AZA) 3-year project, the results of which (Falk *et al.*, 2007) are available to download from the AZA website. During this project, the authors surveyed the literature, held public forums with a selection of AZA institutions, and interviewed zoo visitors. The study showed that visits to zoos and aquariums do, indeed, raise awareness and a sense of connection to nature, as well as promoting knowledge about, and caring for, conservation. It was also the case, however, that visitors came with more ecological knowledge than expected, so the zoos were supporting and reinforcing the existing values and attitudes of the visitors. But, ultimately,

---

5 Willie B. was a famous gorilla who lived at what is now Zoo Atlanta, Georgia, for nearly 40 years (see **Box 2.3**).

we probably need to recognize that people come to the zoo primarily for a good family day out and to see some animals rather than to be educated, even if, when asked, they think that zoos ought to be educating people about conservation (Reading and Miller, 2007). The challenge is to deliver the conservation message within that enjoyable experience.

Finally, we can note that most educational initiatives that zoos put on are aimed at children or at general zoo visitors. These, however, might not be the most appropriate target. Conway (2007), for example, has urged that more should be done to target policy-makers, with a view to influencing their attitudes and actions. As Conway points out: *'Today's children will not be decision-makers until billions more children have been born and much more wildlife has been lost.'*

13.3 Humans in the zoo: how do they affect the animals?

It is time now to turn our attention away from the zoo visitors themselves and to consider what the presence of people means to the animals in the zoo. As we have already pointed out, humans are a significant part of the environment in which zoo animals live; those humans may be daily visitors, or they may be people who work in the zoo and therefore, to differing degrees, become familiar to the animals. It is reasonable for us to ask if, and how, the animals are affected by these people.

13.3.1 Zoo visitors
One of the first people to try to identify how animals in zoos might perceive the humans with whom they came into contact was Heini Hediger,[6] whose evidence was anecdotal, but which nevertheless helped lay the foundations of modern zoo biology. Hediger (1965) considered that humans could be significant to animals in five different ways:

- as an enemy (that is, to be avoided);
- as prey;
- as a symbiont (that is, as a partner working towards a common goal);
- as of no consequence (that is, part of the background to be ignored);
- as a **conspecific** (for example, a rival or a sexual partner, both of which would cause problems).

Clearly, the particular way in which an animal views people will depend upon its species, its prior experience of human interaction, and the people involved (and probably other variables as well). Of all of these factors, the one that probably impacts most on the welfare of the animals is when they perceive humans as enemies, but systematic studies are necessary to find out if they do view us this way.

Do human visitors affect zoo animals at all?

The first question to ask is whether human zoo visitors really do affect the behaviour of the animals (**Fig. 13.8**). It is, after all, possible that the animals

(a)

Figure **13.8** Zoo visitors behave towards captive animals in all sorts of different ways, ranging from (a) climbing the railings and trying to attract the animal's attention, and (b) teasing the animals, in this case with a pet dog, to (c) passively standing or sitting and watching the animals. (Photographs: (a) Sheila Pankhurst; (b and c) Geoff Hosey)

6 Heini Hediger (1908–1992) was, for a long time, Director of Zurich Zoo and collected together years of observations on the lives of animals in zoos. He can be regarded as one of the founders of modern zoo biology.

(b)

(c)

Figure **13.8** (continued)

perceive visitors as being (in Hediger's words) 'of no consequence'. Many studies have shown, however, that there is an association between visitor numbers and activity, and changes in behaviour of animals. One of the earliest studies, for example, showed that zoo primates were more active, changed their dispersion in the cage, and displayed more visitor-directed behaviour at large, active groups of humans than they did at small, non-active ones (Hosey and Druck, 1987). Mitchell *et al.* (1992b) pointed out that these results could, however, be explained in terms of more active animals attracting bigger, more interactive crowds of visitors.

Hosey (2000) has consequently referred to the 'visitor effect hypothesis' and 'visitor attraction hypothesis' as two alternative explanations of the relationship between zoo animals and their audiences: much of the evidence for zoo primates is consistent with the 'visitor effect hypothesis' (see below), but there is also evidence for the 'visitor attraction hypothesis'. Margulis *et al.* (2003) showed that small cats attracted more visitors when they were active, but the cats were generally unaffected by human audiences. As this paper points out, the two phenomena probably have some mutual effect and are not independent of each other.

Negative effects of zoo audiences on animals

If zoo visitors do affect the behaviour of the animals, then, clearly, we need to know the nature of that effect. Human visitors are a significant component of the zoo environment, and zoos are only really able to fulfil their missions and achieve their aims by attracting more visitors. If those visitors change the behaviour of the animals, then we need to know how.

Most of the studies on zoo visitor effects have been done on primates and the evidence overwhelmingly points to the conclusion that the effect is generally a negative one—that is, that the presence of zoo visitors results in the animals displaying behaviours that are usually associated with a stress **response**. Because of this, there is a risk that prolonged exposure to human visitors could adversely affect the welfare of the animals.

Much of this literature has been reviewed by Hosey (2000). An early study by Chamove *et al.* (1988) at Edinburgh Zoo showed that **agonistic behaviour** increased and affiliative behaviour decreased in three different primate species (Diana monkey *Cercopithecus diana*, cotton-top tamarin *Saguinus oedipus*, and ringtailed lemur *Lemur catta*) when human audiences were present (see **Fig. 13.9**).

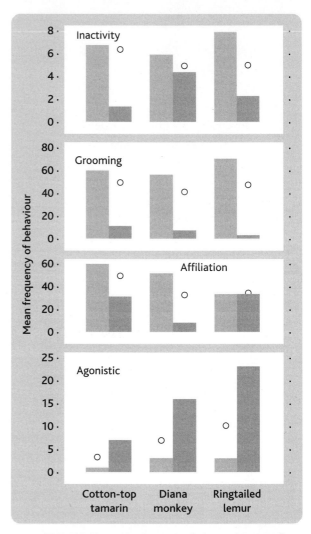

Figure **13.9** This shows the frequency (taken at 10-second intervals) of four different behaviours of three primate species at Edinburgh Zoo. The pale bars indicate when there were no human visitors present, the dark bars show when people were present, and the circles show the animals' behaviours when the visitors crouched down so that only their heads were visible to the animals. When the visitors crouched, the change in the animals' behaviour (compared with when no visitors were present) was less pronounced. (From Chamove *et al.*, 1988)

Similar results were reported at Rotterdam Zoo by Glatston *et al.* (1984) in relation to cotton-top tamarins housed on-show, compared with those housed off-show, and by Mitchell *et al.* (1991a) in relation to golden-bellied mangabeys *Cercocebus galeritus chrysogaster* at Sacramento Zoo, USA. In this study, animals were moved between enclosures that attracted high or low visitor attendance because of their location in the zoo and the animals changed their behaviours accordingly.

More recent studies have shown broadly similar effects. For example, gorillas *Gorilla gorilla* at Belfast Zoo show more intraspecific aggression and abnormal behaviours on summer weekends when visitor numbers are high than during the winter, when visitor numbers are low (Wells, 2005); orang-utans *Pongo* spp. at Chester Zoo are affected by noisy visitors in particular and try to cover their heads with paper sacks (Birke, 2002), and lion-tailed macaques *Macaca silenus* in Indian zoos show up to 20 per cent more abnormal behaviour when confronted with zoo visitors (Mallapur *et al.*, 2005).

All of these studies infer, from behavioural data, that the animals are being stressed by the presence of human visitors. These behavioural indicators of stress include increased abnormal behaviours, particularly **stereotypies** (see **Chapters 4** and 7), increased intra-specific (that is, between cage-mates) and inter-specific (that is, towards humans) aggression, increased activity (sometimes measured as decreased inactivity),[7] and, sometimes, decreased affiliative behaviours, such as grooming. A comparison across studies does, however, indicate some variability and inconsistency in the behavioural responses of different primate groups to zoo visitors. Even within the same species, different individual animals may respond differently to humans (Hosey, 2008; Kuhar, 2008). This is probably because the way in which different animals change their behaviours in response to stressors is itself affected by a number of other variables, such as cage space and complexity, species, and visitor behaviour. Wood (1998), for example, found a quite complex inter-relationship between the responses of chimpanzees *Pan troglodytes* at Los Angeles Zoo to human audiences, the sizes of audience groups, and whether or not the environmental enrichment supplied to the animals was new or not. It is also likely that the previous history of interactions that animals have had with people affects how they subsequently respond (Hosey, 2008).

Of course, inferring stress from behaviour is something that should be done with caution (see **Chapter 7**). An alternative approach is to use physiological measures, which perhaps provide a more direct measure of the animal's welfare. Davis *et al.* (2005) used urinary **cortisol** measurement to look for correlations between visitor numbers at Chester Zoo and the response of Colombian spider monkeys *Ateles geoffroyii rufiventris* to stress. They took advantage of the temporary closure of the zoo during the 2001 foot-and-mouth disease outbreak in the UK to be able to get a 'no audience' condition. Their results (see **Fig. 13.10**) show that urinary cortisol significantly increases as visitor numbers to the zoo increase, indicating that, in this case, the human visitors are indeed stressful to the animals.

Can stressful effects of human audiences be reduced?

If human visitors are stressful to zoo animals, then we can reasonably ask if there are ways of reducing their stressful effect. In the study by Chamove *et al.* (1988) referred to above, visitors were asked to crouch, rather than stand, at the viewing windows of the three primate enclosures, on the grounds that the animals might perceive them as less of a threat. This did, indeed, have the effect of reducing the impact of the visitors on the animals' behaviour, although the behaviours of the primates were still different to their behaviours when no visitors at all were present. The suggestion from this was that enclosures could

---

7 Activity and inactivity are not necessarily the reciprocal of each other, because each may or may not include certain other behaviours.

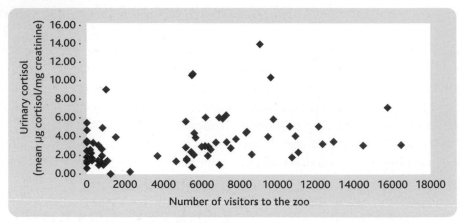

Figure **13.10** Here, we can see the mean urinary cortisol levels in a group of spider monkeys *Ateles* spp. at Chester Zoo, plotted against the number of visitors in the zoo. For part of the study, the zoo was closed to the public (because of a foot-and-mouth disease outbreak in the UK), which provided the opportunity for a 'no visitors' condition. The relationship is significant, with the monkeys excreting higher concentrations of cortisol when human visitor numbers were high. (Adapted from Davis *et al.*, 2005)

be designed with lowered public walkways, to achieve a similar effect.

Existing enclosures can be modified by the addition of screening (**Fig. 13.11**), or additional places (refuges) for the animals to hide from view. A camouflage net barrier was added to the gorilla enclosure at Belfast Zoo by Blaney and Wells (2004) to reduce the visual impact of the human visitors. It resulted in less conspecific-directed aggression and stereotyped behaviour by the gorillas than when

Figure **13.11** It is not uncommon for zoos to use measures in an attempt to reduce the visitor pressure on some of their captive animals. In this case, a camouflage barrier was erected to try to make the presence of human visitors less obtrusive to the animals. (Photograph: Sheila Pankhurst)

the barrier was absent; furthermore, the barrier was rated positively by members of the public, who considered it made the gorillas look more exciting and less aggressive. A similar study by Keane and Marples (2003) at Dublin Zoo, again looking at gorillas, found that a barrier was more effective in reducing the stress behaviours of the animals than the use of carpets in the public walkway (to reduce noise) and signs asking visitors to be quiet.

The use of signs to modify visitor behaviour has not been evaluated to any great extent, but has been tried in an aquarium setting (**Fig. 13.12**), by Kratochvil and Schwammer (1997). They attempted to reduce the amount of banging on the glass fronts of aquarium tanks by the public at Tiergarten Schönbrunn in Vienna. Such knocking sets up high sound pressure levels, and leads to startle responses and avoidance in fish. To try to reduce it, they tried out three different signs saying:

1. 'Knocking kills fish';
2. 'Only loonies would knock';
3. 'Please don't knock on the glass'.

The second of these signs had the greatest success (**Fig. 13.13**), reducing knocking to less than 10 per cent of baseline; the third had the least effect.

(a)

(b)

(c)

Figure **13.12** Signage in zoos is not always only to educate visitors, but also sometimes to try to change their behaviour, for example, by discouraging them from teasing or feeding animals. Here are some examples from (a) Burgers' Zoo, Arnhem, in the Netherlands (b) Entebbe Zoo, Uganda, and (c) Schmutzer Primate Centre, Ragunan Zoo, Indonesia. (Photographs: (a) Sheila Pankhurst; (b and c) Vicky Melfi)

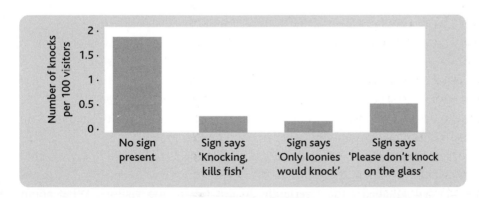

Figure **13.13** Signs at the aquarium in Schönbrunn Zoo in Vienna were effective in reducing the amount of knocking on the glass by zoo visitors. Simply asking people not to knock, however, was not as effective as implying that they would be 'loonies' if they did. (Adapted from Kratochvil and Schwammer, 1997)

A more active way of reducing the stressful effects of human visitors might be to improve the animals' perceptions of people by fostering positive relationships with them. So how can this be achieved?

One way might be through **positive reinforcement training** (see section **13.4.1** for more on this). In a study at Paignton Zoo, interactions between Abyssinian colobus monkeys *Colobus guereza* and zoo visitors declined significantly after the monkeys were trained to undergo oral examinations (Melfi and Thomas, 2005).

Can zoo visitors be enriching?

The suggestion that zoo visitors might be enriching to zoo animals was probably first made by Desmond Morris (1964), who considered that zoo animals were bored for much of the time and that human visitors might be a welcome source of variability in what was otherwise a monotonous environment. Unfortunately, evidence in favour of this suggestion is sparse. A study of green monkeys *Cercopithecus aethiops sabaeus* at Mexico City Zoo by Fa (1989) failed to find any increase in agonistic behaviour on days when the public had access compared with days when the zoo was closed. Instead, the monkeys spent much time and effort attempting to get food from the visitors, which, from their point of view, might have meant that they were enriched rather than stressed. Similarly, chimpanzees *Pan troglodytes* at Chester Zoo are willing to engage in quite long interaction sequences with human visitors, apparently in the hope of being thrown food (Cook and Hosey, 1995).

Moving away from primates, a long-billed **corella** *Cacatua tenuirostris* at Adelaide Zoo made greater efforts to interact with zoo visitors on quiet days when there were fewer people present (Nimon and Dalziel, 1992), which suggests that the bird found people stimulating.

Some of the enrichment attempts initiated by Hal Markowitz in the 1970s, which involved the use of operant techniques to shape the animals' behaviours, could also involve zoo visitors. An example is the apparatus set up for gibbons at Portland Zoo, in which visitors could insert a coin into a machine, thus turning on a light that signalled to the gibbons that a lever press would bring about a food reward at the other side of the cage (Markowitz and Woodworth, 1978). This raised both activity levels of the gibbons and revenue for the zoo, and might be

regarded as a more positive response by the animals to human visitors.

But to what extent can we regard the presence of human visitors as enriching in these studies? Fa (1989) has suggested that feeding by the public can result in overfeeding, with long-term adverse consequences for health and reproduction. He also points out that the food-directed behaviours of the monkeys in his study were done at the expense of other behaviours, such as social behaviour. Similarly, the sorts of techniques used by Markowitz, generally referred to as **behavioural engineering**, have been criticized on the grounds that they promote 'artificial', rather than 'naturalistic', behaviours (Hutchings *et al.*, 1978). Whether or not these changes in behaviour are the result of enrichment probably depends on how you measure enrichment (see **Chapter 8**). Other circumstances in which audiences might be enriching for zoo animals might be through increased opportunities for inter-specific interaction (see section below), but there has been little work on this.

Zoo visitors as of no consequence

We started this section on zoo visitors by noting Hediger's (1965) suggestions of how people may be perceived by animals. One of his categories was that animals might consider people to be of no consequence: just part of the background to be ignored. Is there any evidence to support this view?

Until recently, it was quite a widely held view that zoo animals **habituate** to members of the public—that is, that they get used to them and no longer respond to their presence—but there is actually very little data for us to be able to say whether or not this is true. As we have seen above, primates certainly do not ignore zoo visitors. What about other kinds of animal?

A study at Fota Wildlife Park, Ireland, by O'Donovan *et al.* (1993) failed to find any significant changes in behaviour of a group of cheetah *Acinonyx*

> **Corellas** are Australian birds belonging to a subgroup of the cockatoo.

Figure **13.14** The assumption that all animals are negatively affected by zoo visitors seems unfounded. Felids, such as these lions at Paignton Zoo, do not seem to respond aversively towards zoo visitors as many primates do. (Photograph: Paignton Zoo Environmental Park)

zoo visitors are themselves primates, and share many communicative signals with monkeys and apes. But there is increasing evidence that, even after taxon differences have been taken into consideration, there are still individual differences in the way in which zoo animals respond to humans and we will return to this in section **13.3.3**.

Inter-specific communication

In the cases at which we have looked so far, an effect of human visitors on zoo animals, and vice versa, has been inferred on the basis that the behaviour of one party has been changed by the presence and behaviour of the other, but we have not really examined the form that inter-species communication can take. Few studies have looked at the signals themselves, even though this can potentially give us a fascinating insight into the extent to which animals can discriminate and classify people.

A good example is the study by Mitchell *et al.* (1992a) on the golden-bellied mangabeys at Sacramento Zoo. They found that human visitors did numerous things that resulted in the monkeys threatening them, such as climbing the railings, leaning over, yelling, throwing ice from soft drinks, jumping up and down, making faces, and so on. These behaviours, collectively termed 'harassment', were performed more by boys and men than by girls and women; furthermore, the boys and men directed their harassment more at the male than the female mangabeys. In return, the male monkeys threatened the audience more than the females did and directed most of their threats to the male members of the audience. (Hediger's categories of how animals may perceive humans seem to work in the reverse direction as well.)

Another primate study is that of Cook and Hosey (1995), which looked at the exchange of signals between zoo chimpanzees and human visitors at Chester Zoo. Here again, male visitors tended to direct their behaviours at male chimpanzees, but harassment was less evident and much of what the chimpanzees did was best explained as

jubatus in response to the presence of zoo visitors. More recently, Margulis *et al.* (2003) scored the behaviours of six species of felid in seven different enclosures at Brookfield Zoo, Chicago, during both spring and summer periods. They found no significant effects of visitor presence on any behaviours of the cats, although, as we noted previously, they also found that, when the cats were active, they attracted more interest from the visitors. Other studies on felids seem to agree that these animals do not respond to zoo visitors in the way that many primates do (**Fig. 13.14**), although there may be visitor-related changes in stereotypy (Mallapur and Chelan, 2002; Sellinger and Ha, 2005).

Why should felids fail to respond to zoo visitors while primates show such a clear and obvious response? Margulis *et al.* (2003) suggest simply that some animals are more reactive than others and that this may be **taxon**-specific. In addition, of course,

soliciting food. Several chains of behaviour (interaction sequences) were seen and, in general, it was the human visitors whose behaviour was influenced by the preceding chimpanzee behaviour, rather than the other way around.

We might expect such interchanges between human and non-human primates more than between humans and other animal taxa, because of the similarity in morphology and signal form. Sheep and goats in a **petting zoo**, for example, may show aggression or avoidance of humans (Anderson *et al.*, 2002), which is consistent with the possibility that they perceive humans as potential predators or danger rather than as rival conspecifics, which seems to be the case with the mangabeys.

13.3.2 Human disturbance

Of course, not all of the unfamiliar humans at the zoo are visitors who spend time looking at the animals. Others who come to carry out building, maintenance, or delivery work are probably more noticed by the animals as a result of the disturbance that they cause than because of their presence at the cage.

Again, it is remarkable how little the effects of this disturbance on the animals has been studied. One study investigated changes in both behaviour and cortisol excretion in two giant pandas *Ailuropoda melanoleuca* at the Smithsonian National Zoological Park, Washington DC, while an adjacent cage was being demolished (Powell *et al.*, 2006). The behaviours shown by both animals were characterized by the researchers as 'restless' while demolition was going on and cortisol excretion increased, although this may not have been related to the building work, because the fluctuations in secretion were different in the two animals.

A similar study at Honolulu Zoo, Hawaii, measured changes in behaviour and faecal corticoids in two species of Hawaiian honeycreepers (the 'Apapane *Himatione sanguinea* and the 'Amakihi *Hemignathus virens*) during ordinary days when routine maintenance work might or might not happen (Shepherdson *et al.*, 2004). During the study, various disturbances occurred, such as evening concerts, people entering the cage to do maintenance work, people working on the roof, and machine noise. Faecal corticoid levels increased significantly when concerts or machine noise occurred compared with those on days on which nothing much happened and these disturbances were also associated with behavioural changes, such as a reduction in foraging and perch hopping or flying.

Finally, Hutchings and Mitchell (2003) observed two groups of black-and-white ruffed lemurs *Varecia variegata* at Marwell Zoo, UK: one on show to the public; the other off-show, but subjected to noise disturbance from vehicle movements, a refreshment kiosk, and a miniature train station, none of which the animals could see. The latter group showed higher levels of sniffing, scent marking, locomotion, and vigilance; the on-show group only approached similar frequencies of these behaviours in the afternoons, when visitor densities were greater.

13.3.3 Keepers

Keepers are probably perceived by animals in quite a different way from the visiting public; because of their longer and more intimate contact with particular animals, keepers or caretakers can develop more permanent relationships with the animals (**Fig. 13.15**).

All that is meant by 'relationship' in this context is that the two individuals have a history of interactions between them that lead to a greater predictability about the outcome of future interactions; in other words, they get to know what each other is likely to do. Such relationships can vary in quality, but the most positive kinds of relationship often lead to the formation of a 'bond' or 'attachment' between animal and human (Estep and Hetts, 1992). The main characteristics of such a human–animal bond have been suggested (Russow, 2002) as the involvement of a relationship between a human and an individual animal that it is reciprocal and persistent, and that it promotes an increase in well-being

Figure **13.15** Keepers have longer and more intimate contact with their animals than is possible for members of the public, so more permanent relationships can be established between keepers and individual animals. (Photograph: Julian Chapman)

for both parties. Relationships can, of course, be of poor quality as well; the quality of interactions and, ultimately, of relationships between animals and their caretakers is an essential component of stockpersonship (often just **stockmanship**)—that is, the skill of managing and caring for animals.

Can animals discriminate between individual people?

For anecdotal evidence of whether animals can discriminate between individual people, you need only speak to a pet owner, farmer, zookeeper, or scientist, or indeed anyone who has worked or lived with animals for any length of time. Experimental evidence is harder to find.

In laboratory studies, using techniques such as **preference testing**, **operant conditioning**, and habituation, Davis (2002) has demonstrated discrimination between individual humans in rats, chickens, llamas, rabbits, sheep, cows, seals, emu, rhea, penguins, and honeybees. The literature on farm animals contains numerous studies that show discrimination of different people by the animals, often based on previous experience of being handled (for example, de Passillé *et al.*, 1996; Boivin *et al.*, 1998). There is even some evidence from other species, including some reptiles (Bowers and Burghardt, 1992) and the octopus (Mather, 1992).

In any case, the results of studies on the animal–human relationship imply that animal discrimination of individual people is widespread.

Consequences of human–animal relationships

Once again, surprisingly few studies have been undertaken on human–animal relationships in the zoo. There is, however, a lot of evidence from agricultural research that stockpersons differ in the way they interact with farm animals, leading to stockperson–human relationships (**Fig. 13.16**) of variable quality. The quality of these relationships has consequences for both animal production and animal welfare.

Much of the relevant literature is reviewed by Hemsworth (2003) and by Boivin *et al.* (2003). This research, which has particularly been done with cattle, pigs, and chickens, shows that some stockpeople have good relationships with their animals, the result of positive interactions with them, such as patting, stroking, and talking to the animals, and generally doing things rather slowly. On the other hand, negative interactions such as slaps, hits, shouting at the animal, and fast movement lead to increased fear of humans in the animals. The resulting stress adversely affects the animal's welfare and can lead to a decline in productivity (for example, milk yield).

Among the most powerful predictors of the quality of stockperson–animal relationships are the

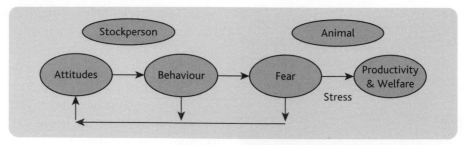

Figure **13.16** Hemsworth's model of human–animal interactions in agricultural animals is illustrated here. The stockpersons' attitudes towards the animals affect the way in which they behave towards the animals and this, in turn, can increase or decrease the fear that the animals have of people. This fear itself feeds back to affect the stockperson's attitudes and affects the productivity of the animals. (From Hemsworth, 2003)

person's attitudes and personality traits; the fearful response of animals to their interactions feed back to and reinforce the stockperson's attitudes.

More recently, the importance of understanding human–animal interactions has been increasingly recognized among animal experimenters as well. Here, of course, part of that importance comes from a worry that a human–animal relationship can be an unaccounted-for variable (a **confounding variable** —see section **14.4.2**) in experimental results, but evidence shows that relationships between experimenters and their animals can have beneficial, as well as aversive, effects (see, for example, on laboratory rats, Dewsbury, 1992).

In fact, the evidence from laboratory animals is very similar to that from farm animals. In a study of laboratory stump-tailed macaques *Macaca arctoides*, for example, Waitt *et al.* (2002) found that the quality of the caretaker–animal relationship had an influence on the behaviour of the animals: those macaques who were considered friendly by their caretakers received more positive interactions from their caretakers during routine laboratory procedures (**Fig. 13.17**), and consequently were less disturbed by those procedures, and more likely to approach and accept food from their caretaker. Similarly, when caretakers spent more time with chimpanzees in a primate centre, the animals showed more grooming of each other and fewer abnormal behaviours, and interactions with the observer moved from 'mostly aggressive' to 'mostly friendly' (Baker, 2004).

Figure **13.17** Laboratory stump-tailed macaques *Macaca arctoides* that receive positive interactions from their keepers are less disturbed by routine laboratory procedures and are more friendly with their caretakers. (Photograph: Keith Morris)

Keeper–animal relationships in the zoo

We would expect that what is true of farm and laboratory animals is also true of zoo animals and, from the very limited amount of research that has been done, this does, indeed, appear to be the case. Golden-bellied mangabeys at Sacramento Zoo, for example, were found to threaten members of the public significantly more often than they threatened keepers (Mitchell *et al.*, 1991b). Furthermore, adult male mangabeys directed the same number of threats to non-human primates in neighbouring cages as they did to keepers (adult females and juveniles directed more), allowing the conclusion to be drawn that the monkeys treated keepers as familiar conspecifics.

Do different keepering styles result in different animal–human relationships, with resulting consequences for well-being, in the zoo as they do on the farm? Hardly any research has been done on this, but a study by Mellen (1991) suggests that they do. She was investigating the inconsistent breeding success of small cats in zoos and measured fifteen variables in eight different zoos to see which ones correlated with breeding success. One of the findings was that reproduction was more successful in cats the keepers of which spent time talking to them and interacting with them (but see **Box 13.2**).

Again, as with farm and laboratory animals, the management and handling of animals can be influenced by the stockperson skills of the keeper. For example, positive relationships of black rhinoceros *Diceros bicornis* with their keepers at Paignton Zoo resulted in low **latencies** of the animals doing what their keepers ask them to do—that is, the animals responded more quickly to their keepers (Ward and Melfi, 2004).

Box **13.2** What can happen when the human–animal relationship goes wrong

We know hardly anything about the relationships between keepers and their animals, or the consequences for the well-being of either party. Keepers are, however, well aware that even the smallest and most innocuous-looking of animals can cause them some damage if they choose to, so possibilities for establishing relationships are constrained to some extent by safe working practices.

This is particularly important with large animals, such as elephants. In relation to this animal, training is important, not only for welfare reasons, but also for safety reasons— but, because of the sheer size of the animals, some element of punishment may have to be included for the training to be successful (Veasey, 2006). Can this damage the relationship with the keepers?

Once again, we do not know the answer to this question. We do know, however, that attacks by elephants on keepers do occur, sometimes with fatal consequences. Gore *et al.* (2006) have analysed the details of 122 elephant-related injuries to keepers and visitors between 1988 and 2003, along with other reports from the literature. Fairly obvious features, such as the animal being in pain or a bull elephant being in musth, only accounted for a very small number of cases. Similarly, the inexperience of the keeper or the procedures that the keeper was doing at the time could only account for a small number of cases. Attacks were mostly initiated by female elephants, but that may be because there were more female than male elephants in the zoos. Nevertheless there were indications that females older than 20 years could be a particularly dangerous group, because they would not have shown previous aggression and keepers would therefore not anticipate an attack.

Gore *et al.*'s (2006) study looked at a number of animal-related, management-related, and keeper-related factors that could have influenced the likelihood of attack, but no single factor was clearly implicated. This sort of study is rare, but important, if we are to properly understand the keeper–animal relationship and its consequences for the well-being not only of the animals, but of the keepers as well.

Figure **13.18** If we assume that zoo animals set up particular relationships with their keepers (familiar people), but have a generalized relationship with unfamiliar people (such as zoo visitors), then we could predict that the quality of those relationships (how positive or negative they are) has consequences for how the animal responds to the presence of people. Clearly, the best scenario for the animal is positive relationships with both familiar and unfamiliar people. (Figure based on Hosey, 2008)

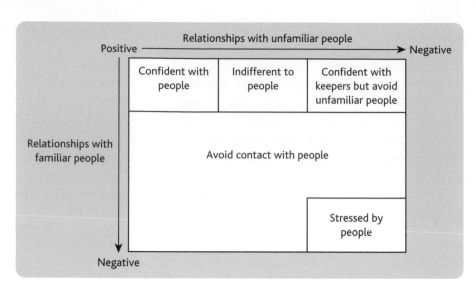

The combination of familiar and unfamiliar people

When we look at the agricultural literature, we can see that the quality of the relationship set up between the animals and their stockperson is of great importance, and can affect the behaviour of both, and ultimately, at least for the animals, their welfare and life history. We would expect this to be true of zoo animals as well and, although there is very little evidence, what there is suggests that this is, indeed, the case.

But zoo animals, as we have seen, also encounter plenty of unfamiliar people who come as zoo visitors and their responses to these are quite variable. Could it be that the quality of relationships that zoo animals have with their keepers influences the way in which the animals respond to visitors? Hosey (2008) has suggested that this is the case. Effectively, zoo animals develop a relationship with the generalized group 'zoo visitors' and this interacts with the relationship they have with their keepers to influence the way in which the animal responds to both. Agricultural animals that have good-quality (positive) relationships with their stockpersons have low fear and some confidence with people, whereas those with poor-quality (negative) relationships show high fear of humans and avoidance of contact (Waiblinger *et al.*, 2006). In between these is a neutral relationship, in which the animal has low fear of humans, but still avoids contact. If we apply this to the zoo setting and incorporate the two different sorts of humans (familiar and unfamiliar) that these animals encounter, then we get the sorts of characteristics that we see in **Fig. 13.18** (Hosey, 2008).

If this way of looking at human–animal interactions is supported by future studies (it has not yet been tested), it may help us to improve the relationships that zoo animals have with visitors as well as keepers. As an example of how this might work, we can point to a study that we have already mentioned: that undertaken by Melfi and Thomas (2005) of colobus monkeys at Paignton Zoo. These animals were trained (see section **13.4**) to undergo oral examination. After training, their interactions with the public declined. Positive reinforcement training is widely thought to increase the quality of the animal's relationship with its keeper (Bassett *et al.*, 2003; McKinley *et al.*, 2003), so its use in zoos might yield the additional benefit of making the animals less responsive to visitors.

13.4 Training

Of course, all captive animals have to be handled occasionally, for purposes such as medical treatment, routine veterinary procedures, transportation to other housing, and so on. The studies on farm and laboratory animals suggest that such handling is generally stressful for the animals.

Stress related to handling can be reduced in various ways, although the simplest method is for handlers to be calm and quiet during handling. If the animals associate positive consequences (such as gentle patting or stroking) with being handled, this can, over time, reduce the degree of fear and aversion they feel towards handling (LeNeindre *et al.*, 1996). Effectively, the animal learns through experience that handling is positive or neutral. By this means, positive or neutral relationships can be set up between the animals and their stockperson (Hemsworth, 2003; Waiblinger *et al.*, 2006), as discussed in section **13.3.3**. For zoo animals, direct handling is not usually a feasible course of action, because, for many species held in zoos, this would be potentially dangerous for both the keepers and the animals. There are also some concerns that too much handling or exposure to humans could have a detrimental effect on the behaviour and/or the breeding of zoo-housed species (see **Chapter 6**).

This effect of handling is probably one of the simplest forms of training. Training is considered to occur when we (humans) determine what animals learn (Mellen and Ellis, 1996). As explained in **Chapter 4**, **learning** occurs when an event increases or decreases the likelihood of a behaviour being repeated in the future. In zoos, we are constantly manipulating the animals' environments and thus influencing which behaviours they will perform in the future. This happens in an informal way, through daily husbandry, and is sometimes referred to as 'passive training', because the animals' behavioural expression is not predetermined by trainers, but occurs as a consequence of the animals learning something about human activity. Zoo animals are continually learning about their environment: for example, an animal may learn to associate the noise of keys with the approach of a keeper, or the sight of a keeper with a wheelbarrow of food with feeding time (Young and Cipreste, 2004).

A more formal approach to changing animal behaviour, or 'active training', can be achieved by eliciting the desired behaviour through appropriate and planned changes to the animals' environment or human–animal interactions. For example, if we were to want an animal to come inside at night, we could start by ensuring that all keepers showed the animal some food in the evening and then left the food inside. The animal would quickly learn to come inside to get access to the food.

13.4.1 Positive reinforcement training

There are many ways in which humans can manipulate what an animal learns, not least by influencing the environment and thus the **stimuli** to which the animal is exposed on a daily basis. The basic underlying principles of learning theory are explained in **Chapter 4** and, when these are understood, they can be used to modify zoo animal behaviour (see, for example, Ramirez, 1999; Pryor, 2002; Laule *et al.*, 2003). The most frequently used methods to train animals in zoos are based on operant conditioning techniques; simply reinforcing a behaviour will increase the likelihood that it will occur again in the future, and this is known as positive reinforcement training (PRT). If, however, a behaviour is ignored or punished, it will be less likely to occur. Steve Martin, an influential and experienced trainer in this field, talks about '*setting animals up to succeed*', by which he means that, if you act appropriately, you should only need to reinforce your animal with a reward, rather than punish it.

The degree of complexity inherent in training varies. One of the simplest forms of training is to capture a behaviour that you want to see again in the future; when you see the behaviour that you want the animal to show, you reinforce it. If the behaviour

Figure **13.19** Shaping an open-mouth behaviour in an Abyssinian colobus *Colobus guereza* so that a dental check could be performed without catching, handling, or separating the animal from the group. The first step was to reward the animal for stationing—that is, staying at a designated stop; then, the animal was rewarded for touching a target—the wooden spoon; the animal was then rewarded for keeping in the same position and moving its other hand to the target; eventually, a grape was shown to the animal, which opened its mouth, and the behaviour was captured. Each behaviour requires a cue that the animal understands and which is always provided consistently. Verbal commands can be added while training or afterwards. (Photographs: Vicky Melfi)

you want to see does not currently exist in your animal's behavioural repertoire, you can reinforce successive approximations to the behaviour: a process called 'shaping' (**Fig. 13.19**). Thus, any slight change in the animal's behaviour that is a bit more like that which the trainer wants is reinforced. Alternatively, if you want to deter a behaviour from being expressed or if the behaviour expressed does not take the animal in the required behavioural direction, you can withhold the reinforcement (punish), or else reinforce a mutually exclusive and alternative behaviour.

Reinforcers and punishments can take many forms, depending essentially on the animal's preferences. A good reinforcer is something that the animal likes a lot and for which it will 'work'. Unsurprisingly, food is often used as a reinforcer, but, for some animals, access to something they like (for example, a toy) or simply being given a scratch can work just as well. Similarly, a 'punishment' only needs to be something that the animal will avoid. Despite the term, punishments are not necessarily painful, but can simply be a 'time out': for many animals, removing attention from them is something that they will try to avoid—although some individuals may, in fact, find this reinforcing.

To be able to train, you need to understand the subtleties of the individual that you are training, because individual, as well as species, differences will

affect training. At Bronx Zoo, for example, it was found that tamarins in general (not lion tamarins) were more rapid in approaching trainers and learning behaviours than some other South American species, such as marmosets, grey titis *Callicebus* spp., and pale-headed sakis *Pithecia pithecia* (Savastano *et al.*, 2003). Similar results were found when training **callitrichids** at Paignton Zoo (Jago *et al.*, 2006).

Many trainers use a secondary or conditioned reinforcer, such as a whistle or a clicker (a sound or cue that is not used in the animal's daily routine), which tells the animal that it has done the correct response and that the primary reinforcer (reward) is on its way. Again, the theory behind this is explained in **Chapter 4** (see particularly **Fig. 4.5**). The secondary reinforcer is only effective, however, if it is used consistently, honestly, and immediately after the behaviour that it is intended to reinforce.

13.4.2 Should we train zoo animals?

Whether zoo animals should be trained or not is highly debated, with each side of the argument suggesting that there are potential benefits and problems associated with its practice; some of these are listed in **Table 13.2**. There are, however, few **empirical studies** that have evaluated the impact of training on zoo-housed animals, and much of this debate is based on anecdotal evidence (for example, Desmond and Laule, 1994).

Benefits	Disadvantages
1) Facilitates husbandry	1) Increases domestication of zoo animals
2) Improves animal health and welfare	2) Can be too invasive, e.g. ultrasonography
3) Is enriching	3) May alter the animals' behaviour outside of the training session:
4) Enhances human–animal interactions	a) reduces the conservation value of the animal;
5) Improves reintroduction success of released animals	b) increases animal–human interactions;
	c) affects animal–animal interactions.

Table **13.2** Some arguments given for and against training zoo-housed animals

Training has been used in zoos to modify the behaviour of animals for many reasons, including husbandry, conservation and **reintroduction**, research, presentations, and education (see **Box 13.3**). The application of training, in any field, should only be attempted by competent staff, and when a need and the required goal behaviour have been identified. There are also some other notes of caution, which are expanded upon in **Box 13.4**.

There are no boundaries to which species can be trained: there are reports of success in training most zoo-housed species, from primates to pinnipeds, crocodiles to cetaceans, and beyond. For example, giant pandas have been trained to move within their enclosure (Bloomsmith *et al.*, 2003) and Aldabra tortoises *Geochelone gigantea* have been trained to stand while blood samples are taken from them (Weiss and Wilson, 2003).

Box **13.3** What can be achieved through training?

Historically, the goal of training animals has been for performances to entertain people. Nowadays, changes in visitor attitudes have meant that many 'traditional' animal performances, in which animals are trained to act like 'little people' or to do tricks, are not so well received and may portray a stereotype of the animal that is inappropriate in today's culture. Anderson *et al.* (2003) found that simply performing husbandry training while an interpreter was present to explain to visitors what was happening (that is, a performance) improved their perceptions of the zoo, provided them with a greater insight into the animals, and meant that they actually stayed longer at the zoo. Other presentations that highlight animals' innate abilities, or portray a conservation or environmental message, are now commonplace in many zoos. Lukas *et al.* (1998) also point out that training sessions in

zoos provide a great and effective opportunity for teaching students about the principles of operant conditioning theory.

A broad array of training is implemented for husbandry purposes. Intuitively, we would expect that taking blood samples from an animal that offers up its arm voluntarily would be less stressful for the animal than having to catch and restrain it, and then take the blood forcibly. This does, indeed, appear to be the case. Reinhardt (2003) found that the mean cortisol concentrations for six laboratory-housed rhesus monkeys *Macaca mulatta* were significantly higher when they were not trained, compared with when they were trained (see **Table 13.3**).

Similar results have been described in bongo *Tragelaphus euryceros*, among which trained animals had significantly reduced cortisol levels when restrained in a box for 20 minutes, ▸

Blood sampling procedure	Mean cortisol concentrations		Difference (significance)
	First sample	Second sample	
Traditional	$20.1 \pm 4.5 \, \mu g \, dl^{-1}$	$33.8 \pm 5.3 \, \mu g \, dl^{-1}$	$p < 0.001$
Trained	$19.6 \pm 3.0 \, \mu g \, dl^{-1}$	$22.3 \pm 5.0 \, \mu g \, dl^{-1}$	$p < 0.1$

Table **13.3** Mean cortisol concentrations measured in six laboratory-housed rhesus macaques, after blood sampling using the traditional restraint method and after the macaques had been trained to 'cooperate' with blood sampling

Source: Data from Reinhardt (2003)

compared with untrained wild **ruminants** and even domestic cattle, which are unused to handling (Grandin, 2000). Training has been used to medicate individuals within a social group: for example, providing a wormer to Abyssinian colobus *Colobus guereza* (Melfi and Poyser, 2007); to facilitate **artificial insemination** in Asian rhinoceros *Rhinoceros unicornis* (Schaffer *et al.*, 2005); to do the same in gorilla *Gorilla gorilla gorilla* (Brown and Loskutoff, 1998); and to teach parenting skills in the giant panda *Ailuropoda melanoleuca* (Zhang *et al.*, 2000).

Operant conditioning techniques have also been used to aid conservation goals, especially those of reintroduction.

Box **13.4** Cautionary notes on implementing training in zoos

Although training has become widely implemented in many zoos, there are few studies that have been undertaken specifically to test its impact on the behaviour and biology of the animal, other than the behaviour that is being trained. We have records of how many sessions it may take to train callitrichids in different tasks (Savastano *et al.*, 2003), or which behaviours can be trained, but we have little understanding of how the training process may affect these individuals or the dynamic of the social group. This does not necessarily mean that training should be avoided or that it has deleterious effects, but it does mean we do not fully appreciate all of its consequences and should therefore work hard to address this.

It is important, then, that research is undertaken to explore and test the advantages and disadvantages associated with training zoo-housed animals. Melfi and Thomas (2005) were able to show that husbandry training did not significantly alter the **activity budget** or social behaviour of Abyssinian colobus monkeys. But it did affect the frequency of monkey-initiated interactions with people over the period during which they were trained. **Figure 13.20** illustrates that, as the training process progressed, they interacted less and less with people (see also section **13.3.3**). So training can clearly have effects on behaviour other than those that are planned.

The majority of training implemented in zoos is for husbandry purposes and requires

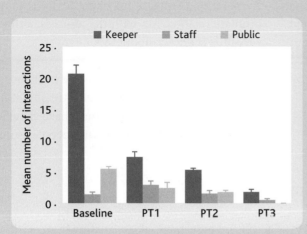

Figure **13.20** A training programme at Paignton Zoo Environmental Park, which aimed to achieve oral examinations of Abyssinian colobus monkeys *Colobus guereza* without the need for catch-ups, had very little effect on the behaviour of the animals—but it did greatly affect how they interacted with people, whether zoo visitors or staff. The frequency of colobus-initiated interactions towards people, whether staff or visitors, declined significantly after training started. Baseline represented the period of time before training started, then data collection continued for the first month (PT1), the second month (PT2), and the third month after training (PT3). (From Melfi and Thomas, 2005)

the animals to learn that certain procedures are not to be feared. This can be particularly beneficial when the alternatives can be dangerous: for example, training animals to station (that is, to remain immobile in a particular spot) can negate the need for physical and/or chemical restraint (Wardzynski *et al.*, 2005). Modification of 'undesirable' behaviours is also attempted in some situations. Bloomsmith *et al.* (2007) have suggested that operant conditioning techniques could be used to limit stereotypies and **self-injurious behaviours** in non-human primates.

In other situations, non-human primates have been trained to react to social situations in an uncharacteristic way. For example, high and low socially ranked macaques were trained to initiate fewer and more social interactions (Shapiro *et al.*, 2003), respectively, and a dominant male chimpanzee was trained to remain still while the rest of his group ate freely around him (Bloomsmith *et al.*, 1994). In these situations, the need for data to clarify what impact training has on the animal is very important. It is possible that training can modify an animal's behaviour, but, if the environmental stimuli or the animal's internal motivation do not change, then the animal may be frustrated between the behaviour that it is 'motivated' to perform and the behaviour that it has been trained to perform (see section **4.1.3** for more on motivation). If this is the case, training may function as a mental barrier to behaviour performance, in much the same way that physical barriers have previously been used to stop animals pacing.

Summary

- Zoos worldwide are visited by huge numbers of people, who appear to come from all sectors of the population; they mostly visit the zoo for enjoyment and to see animals, and are generally positive about zoos.

- Zoos attempt to educate their visitors and change their attitudes through both informal (such as signs) and formal (such as keeper talks) means, but these seem to have limited success in changing attitudes and behaviour.

- There is evidence that zoo visitors affect the behaviour of some of the animals and, in some cases, this appears to be stressful for the animals; on the other hand, some animals are not affected at all and some even seem to be enriched by human presence.

- Keepers, in contrast, are likely to have a more intimate relationship with the animals, although very little is known about this.

- Training of animals allows routine husbandry to take place in what may be a less stressful atmosphere for the animals and may have beneficial effects on their relationships with people as well.

Questions and topics for further discussion

1. What educational messages should zoos be trying to convey to their visitors?

2. Should zoos use some of their animals for demonstrations and displays?

3. 'When zoo animals have been trained or used in public demonstrations, they are no longer representative of their wild counterparts.' Is this true—and does it matter?

4. If keepers have close relationships with zoo animals, does this help or interfere with their ability to undertake routine husbandry procedures with those animals?

5. Should zoos ensure that paying visitors have good views of the animals or should the animals be provided with opportunities to hide from people?

Further reading

Heini Hediger's books (for example, 1955; 1970), now long out of print, are still worth reading as an anecdotal account of the complex relationships between humans and zoo animals. A good account of the role of animal learning and training in the zoo environment is given by Mellen and Ellis (1996).

Websites and other sources of useful information

If you are interested in studying human–animal interactions further, there is a society devoted to that cause: the International Society for Anthrozoology —www.vetmed.ucdavis.edu/CCAB/isaz.htm— organizes conferences and publishes the journal *Anthrozöos*.

Those who are sufficiently motivated to undertake their own research in the area of human–animal interactions in the zoo are advised to read the research guidelines by Mitchell and Hosey (2005), which are available at the BIAZA website—www.biaza.org.uk.

For those wishing to take their knowledge and expertise of training further, the Animal Behaviour Management Alliance (ABMA)—http://theabma.org/home.asp—provides information and help to members.

Zoo education is now a large and innovative field, and a good place to see what is being done is the web page of the International Zoo Educators Association—www.izea.net/index.htm—on which news, information, and online access to the Association's journal are available.

Chapter 14 Research

Research is about finding out new things and is also about solving problems. **Zoo** research can lead to important new information that stretches the boundaries of science, but, equally, it can be about finding the best way in which to deal with difficulties that have arisen in the day-to-day running of the zoo. Research in zoos is not only about animals; researchers may be interested in collecting data on the economics of zoos as commercial enterprises, or on the ability of zoo grounds to support local biodiversity. Indeed, zoo research covers a spectrum from biology, through physics and chemistry, to psychology and sociology, and beyond. But because this book is about zoo animals, rather than zoos per se, this chapter concentrates on research carried out by and within zoos on the biology and behaviour of captive animals.

Zoo animal research poses some particular challenges in comparison with other areas of biological research. These include problems of **confounding variables**, small sample sizes, and a possible lack of independence of data points, which are of wider interest in their own right, because of the methodological problems that they present. Other aspects of zoo research can experience similar problems to research on laboratory or farm animals, or animals in the wild.

In this chapter, we will consider the following points.

14.1 Why is zoo research important?
14.2 What is 'research'?
14.3 What is 'zoo research'?
14.4 Some methodological difficulties in zoo research
14.5 Problems in the analysis of data
14.6 Multi-zoo studies
14.7 Disseminating the results of zoo research
14.8 What zoo research still needs to be done?

There are also boxes dealing with some of the conceptual and methodological issues that arise from zoo research.

The chapter deals primarily with behavioural research, which is probably the area in which most student research is carried out, and in which there are particular issues to do with sampling and data analysis. The chapter should provide a framework for understanding and interpreting the results of zoo research, and also a basis for developing research skills in the zoo environment.

14.1 Why is zoo research important?

Research in any scientific discipline is often characterized as being either 'basic' (sometimes also referred to as 'pure' or 'fundamental'), which is to do with advancing theory and discovering knowledge for its own sake, or 'applied', in which case the research is designed to solve a practical problem or to help a process to perform better. Both of these are of enormous importance in the zoo context.

The zoo environment is a unique one in which to study animals. Unlike the wild, the animals are

Confounding variables are those that you have not measured, but which nevertheless may affect your results, such as a period of very severe weather or a large increase in visitor numbers.

Figure **14.1** Zoos give researchers access to many species of animals that would be difficult to achieve elsewhere. Here, a student gathers vocalization data at an otter enclosure. (Photograph: Whitley Wildlife Conservation Trust)

relatively accessible to the researcher (**Fig. 14.1**), so data can be collected from zoo animals that would be logistically very difficult to get from their wild counterparts. Frans de Waal (1982) has commented on this benefit of zoo research in relation to his studies of the chimpanzees at Arnhem Zoo in the Netherlands. Furthermore, zoo research is a considerably less expensive way of getting data on exotic species than fieldwork in the animals' natural habitat and yet the results of research on animals in zoos can give useful insights into the biology of their wild-living counterparts.

A good example of this is the recently published discovery that Komodo dragons *Varanus komodoensis* can reproduce asexually, via parthenogenesis, as well as sexually (see **Chapter 9**). A team of zoo vets and university researchers collaborated on this study, which was published in *Nature* (Watts *et al.*, 2006). Parthenogenesis is rare in vertebrates and had not previously been reported for this species; this discovery also has implications for the captive management of this endangered species, because

asexual reproduction will reduce **genetic diversity**. Hutchins (2001) cites other examples of zoo-based research that has shed light on the basic biology of species, such as the reproductive physiology of killer whales *Orcinus orca*, in relation to which data would be very difficult to collect in the wild.

Again unlike the wild, some manipulations may be possible in the zoo to take research beyond purely observational and into experimental approaches. The zoo environment is similarly different from that of the laboratory. In most cases, the animals studied in the zoo live in much more naturalistic groups and enclosures than any that might be studied in a laboratory. Because of this, the data collected from zoo animals can have greater biological relevance than that obtained from the laboratory and can tell us more about what these animals are like outside the captive environment. Add to this the fact that zoos are full of interesting animals, many of which have not been extensively studied and many of which are endangered, and you can argue that the zoo is a suitable place in which to conduct basic biological research.

In fact, zoos have a long history of research into aspects of animal biology, even if this was not the primary purpose for which they were established (Hutchins, 2001). For example, the Zoological Society of London's founding charter refers to the need for a '*collection of animals . . . to be applied to some useful purpose, or as objects of scientific research*'. On the other side of the Atlantic, the New York Zoological Society founded the Department of Tropical Research in 1916 and the Institute of Animal Behaviour Research in 1966.

What about applied research in zoos? Problems that may affect the welfare, **nutrition**, health, or other features of the animals' lives in the zoo benefit from the application of scientific research methodology. By this, we mean a systematic approach that makes **hypotheses** about the possible causes of the

---

A **hypothesis** is a sort of 'working theory' to explain what is being observed. Data collected may then support this hypothesis or, alternatively, show that it is false.

problem and then collects data to identify which of the possible hypotheses is most likely to be correct. This empirical approach is far more powerful than unsystematic trial-and-error approaches, because it provides information from studies that are repeatable elsewhere, rather than only anecdotes. This, then, provides data that can be used by the team of animal managers in the zoo to decide on appropriate courses of action to solve or alleviate any problem occurring. For example, is the **enrichment** provided effective? How much of the food provided is digested? As such, applied research is an essential component of zoo management.

14.2 What is 'research'?

The question of what 'research' actually is is not as easy a question to answer as we might at first think. One definition, used by the UK Government in its evaluation of research[1] undertaken by universities and other higher education institutions, is that research is *'original investigation undertaken in order to gain knowledge and understanding'*. This is largely concordant with the description given in the World Zoo and Aquarium Conservation Strategy (WZACS) (WAZA, 2005), that research in zoos and aquariums should be aimed at accruing new knowledge to help the zoo to improve its operations or to further science (basic and applied animal research).

None of this seems particularly contentious, but it is important because zoo critics interpret research in a rather stricter way than do members of the zoo community. The *Zoos Forum Handbook* (Defra, 2007b), for example, which aims to help zoos interpret just such things as this (see **Chapter 3**), tells zoos that research *'need[s] constitute no more than collecting and collating information for statistical*

purposes'. But critics such as Rees (2005b) maintain that this routine record keeping is not research.

This might be true in a technical sense, but the records that are kept can also be seen as data and the statistical analysis of these data[2] is research (see **Chapter 5**). For many small zoos and aquariums, this may be a realistic way in which to contribute to research and the data collected can, and does, get analysed by others to contribute new knowledge (**Fig. 14.2**). For example, Whitford and Young (2004) analysed the records of ten representative UK zoos over a 10-year period to see what trends there were in the captive breeding of endangered birds, and were able to show that there was an increase during this period in the percentage of threatened species kept and an increase in the number of species in each World Conservation Union (IUCN) Red List[3] category that were being maintained in zoos. Pizzi (2004) was able to analyse records of over a thousand **post-mortem** examinations of penguins at Edinburgh Zoo, from a database spanning over 90 years, to identify major causes of **mortality** and any possible correlates. He found that *Aspergillus* infections,

Figure **14.2** Penguins are very popular in zoos and have been maintained in collections for a long time. This means that extensive databases exist that can be used to carry out research to inform husbandry. (Photograph: South Lakes Wild Animal Park)

---

1 This refers to the Research Assessment Exercise (RAE), periodically undertaken by the Higher Education Funding Councils for England, Scotland, Wales, and Northern Ireland.

2 A reminder here that the word 'data' is a plural, so 'data are' is correct, rather than 'data is'.

3 There is more about the IUCN Red List in **Chapter 10**.

causing fungal air sacculitis (a disease of the air sacs), were the most common cause of death (38 per cent of all penguin deaths) and that male birds had twice the likelihood of dying from this as females.

14.3 What is 'zoo research'?

It is probably best to remain relatively flexible about the definition and scope of zoo research, and to regard it as any activity that increases our knowledge and understanding of animals and their environment, whether it be zoo or wild. So, what are the characteristics of zoo research?

14.3.1 What is zoo research for?

As we mentioned above, WZACS (WAZA, 2005) sees the function of zoo research as being to help the zoo or the researcher to achieve his or her goals. What they mean by this is that research should be conducted to help in furthering the operations of the zoo, to increase knowledge and improve practice in things such as captive breeding, conservation, welfare, and so on, and also that zoos should facilitate research (by providing access to animals, biological samples, and such things, as long as this does not interfere excessively with husbandry, reduce welfare, or infringe legal requirements) for university and other researchers to do comparative biological work. Another way of looking at this is that we would expect a lot of the research undertaken in zoos to be applied—that is, designed to provide results that can be used to improve the captive management of animals—but that basic research in zoos, designed to advance knowledge for its own sake, should also be possible and encouraged.

14.3.2 How much research is done in zoos?

The undertaking of research is an important part of zoos' conservation strategies (see **Chapter 10**). Indeed, it is now a legal requirement in the UK that zoos are involved in research[4] and, particularly, research that helps to achieve the conservation objectives of zoos. So how much research is actually done in zoos, and is it increasing in response to strategies and Directives such as these?

This is not an easy question to answer, for a number of reasons. A great deal of research in zoos is conducted by students (for example, as undergraduate and postgraduate research projects or as part of placements) and volunteers. Chester Zoo, for example, may have as many as 200 researchers, many of whom are undergraduates, working there annually and Paignton Zoo often has forty or fifty students every year. This provides an excellent training opportunity for the students, and results in the collection of data and other outcomes that can be valuable to the zoo—but these studies are rarely published in the peer-reviewed journals (see section **14.7**). So simply looking at the published outcomes of zoo research is likely to underestimate the amount of research-related activity that actually goes on.

An alternative way of finding out how much research zoos do is to ask them. Here, again, we are likely to underestimate the true amount of research: we have only to think of the analyses of routinely kept records referred to in the previous section; many zoos may consider that this is not an example of their research, simply because the analysis is done by someone from a different organization. Then again, researchers from universities who use zoo facilities are not necessarily engaged in collaborative research and so may not be counted by a zoo as part of their research effort.

Some surveys of zoo research have, however, been undertaken. One of the earliest was that of Finlay and Maple (1986), who analysed replies by 120 American zoos and aquariums to a questionnaire. Seventy per cent of these reported that they had recently been involved in research, and 46 per cent said that their research programmes were expanding.

---

4 This requirement comes from the EC Zoos Directive 1999/22/EC (see **Chapter 3**).

Only 27 per cent of them had a research committee or research department, however, and an even smaller 21 per cent of them had written research guidelines. Most of the research was being undertaken by larger zoos or aquariums, mostly on behavioural and reproductive biology, and mostly on non-primate mammals. Those zoos and aquariums that were not involved in research tended to give 'lack of funds' and 'lack of trained staff' as the principal reasons for this fact. Nevertheless, for the time, this was a respectable level of research activity.

A repeat of this survey 12 years later (Stoinski *et al.*, 1998) found that research activity had increased (88 per cent of 123 responding zoos and aquariums), but many of the findings of the earlier survey were still being reported. Thus, it was still the larger zoos and aquariums that were involved in research, and lack of funds and staff were still being cited as the major reasons for no involvement in research. But 64 per cent of the zoos surveyed now had written guidelines and 90 per cent had either a research committee or research department (or both). These authors concluded that research was 'alive and well' in zoos and aquariums, but that there was still room for improvement, because the majority had no staff employed solely for research and fewer zoos than in the 1986 survey reported that their research was expanding.

A brief survey by Nogge (1997), which received replies from fifty-seven European zoos, found that 73 per cent of these were involved in research, again mostly in ethology (34.1 per cent of projects), reproduction (15.5 per cent), conservation (12.4 per cent), and veterinary medicine (12.4 per cent). Surveys as detailed as that of Stoinski *et al.* (1998), however, do not appear to have been published for the UK, nor for Europe as a whole, nor indeed the rest of the world, so it is difficult to identify any trends. What we can say is that, in the past 10 years, research outcomes that have appeared through the work of the Research Committee of the British and Irish Association of Zoos and Aquariums (BIAZA; see **Box 14.5**) have increased enormously, suggesting again that there is a sizeable (growing?) research effort in zoos within the Association's area.

14.3.3 What sort of research is done in zoos?

As mentioned earlier, zoos offer opportunities for research in many different scientific disciplines, as well as the social sciences. Almost all areas of biological investigation can be studied in the zoo setting, and **Table 14.1** shows the range of topics that can be covered and the inter-relationships between them.

In general, these can be divided into three categories:

- basic and applied studies on topics such as behaviour, physiology, genetics, nutrition, welfare, wildlife medicine, and so on;
- field-based conservation research;
- research about the zoo itself, and how its roles and procedures can be improved.

But although all of these things can be and are being investigated, we can nevertheless ask where most of zoo research is actually being directed.

The two American surveys mentioned earlier asked the zoos and aquariums what were their major topics of research. In the 1998 survey (Stoinski *et al.*, 1998), the top areas of study were behaviour (85 per cent of respondents), reproduction (75 per cent), visitor studies (67 per cent), and husbandry (66 per cent). An alternative way of getting this sort of information is to look at the profile of zoo-based publications in major scientific journals. Kleiman (1992) analysed the papers published by the journal *Zoo Biology* between 1982 and 1990, and found that only 28.4 per cent of them were behavioural, with the rest divided between management (26.2 per cent), reproduction (19.6 per cent), biomedical (18.2 per cent), genetics (5.1 per cent), and nutrition (2.6 per cent). She also found that, during this period, applied research papers increased and basic research papers decreased in number in the journal.

Topic	Anatomy and morphology	Biogeography	Ecology	Education	Ethology	Genetics	Nutrition	Physiology	Population biology	Social science	Systematics and taxonomy	Veterinary medicine
Ageing	X		X		X	X	X	X	X			X
Animal welfare	X		X		X			X	X			X
Behaviour			X		X	X	X	X				X
Biomaterial banking		X				X				X	X	X
Biotechnology	X					X	X	X				X
Contraception	X				X			X				X
Dietary studies	X		X		X	X	X	X	X		X	X
Disease	X	X	X		X	X	X	X				X
Domestication	X		X		X	X	X	X	X			
Environmental enrichment	X		X		X			X	X			
Husbandry	X		X		X	X	X	X	X			X
Identification	X				X	X		X			X	
Life history	X	X	X		X	X	x	X	X			X
Population management		X	X		X	X		X	X		X	X
Reproduction	X		X		X	X	x	X	X			X
Taxonomy	X	X			X	X		X			X	
Visitor studies				X	X					X		

Table **14.1** The range of academic disciplines that can contribute to research in zoos*

* From the WZACS (WAZA, 2005).

NOTE: The topics in the left-hand column are those that much of zoo research covers; the crosses indicate the primary disciplines that are involved in researching those topics.

In a similar analysis, Hosey (1997) found that the proportion of behavioural papers in *Zoo Biology* had risen to 43 per cent in the period 1991–1994, and only about 40 per cent of these were non-applied.

Behavioural studies also dominate in a survey by Semple (2002) of 904 research projects in the Zoo Federation (now BIAZA) database. In this survey, behaviour and enrichment studies made up the majority, whereas topics such as nutrition (4.5 per cent of projects) and genetics (0.9 per cent) were significantly under-represented (see **Fig. 14.3**).

These results do not appear to be particularly concordant with the American surveys. There are probably many reasons for this. Firstly, *Zoo Biology* is only one of several journals that routinely publish results from zoo studies (see **Box 14.4**). Secondly, opinions sometimes differ as to into which categories particular kinds of research should go: for example, is an enrichment study to be regarded as research into behaviour, welfare, or husbandry? Thirdly, a large proportion of zoo research never gets published in the peer-reviewed journals: Semple's (2002) survey, for example, includes a large number of student projects; in their survey, Stoinski *et al.* (1998) found that zoos preferred to publish their results in their own in-house publications or in conference proceedings. (We will return to this issue in section **14.7**.)

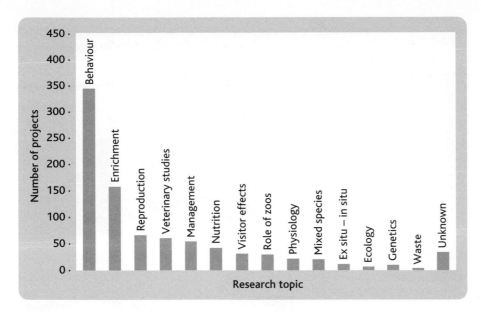

Figure **14.3** Topic areas covered by 904 research projects undertaken in BIAZA member zoos were collated in a database in 2002. There appears to be a bias towards studies of animal behaviour and enrichment. (From Semple, 2002)

A further question that we can ask is about which **taxa** are most frequently studied. In their survey, Stoinski *et al.* (1998) found a heavy emphasis on mammals. Thus, 63 per cent of their respondents were involved in research on **carnivores**, 50 per cent on **ungulates**, 41 per cent on great apes, and 47 per cent on other primates. Birds (53 per cent of respondents), reptiles (50 per cent), amphibians (21 per cent), and fish (26 per cent) received less attention. What this means in terms of amount of research is not clear (after all, the zoos involved in carnivore research might have only one project each, but those involved in birds might have a lot more), but there does seem to be an emphasis on research on the great apes, considering how few species there are in comparison with other taxa.

This is also the conclusion reached by Melfi (2005), who surveyed projects on primates in the BIAZA and American Association of Zoos and Aquariums (AZA) research databases (see **Box 14.5**) and found that hominids[5] accounted for a disproportionately large number of research projects in both the USA, and the UK and Ireland (**Fig. 14.4**).

One possible message for would-be zoo researchers is that there is a need to diversify: there are plenty of taxa that are simply not being studied enough.

14.3.4 Who does zoo research?

The American surveys discussed in section **14.3.2** found that research tended to be concentrated in the larger zoos and aquariums, which had higher visitor attendance figures. It is likely that the same is true in European zoos. Research can be an expensive pursuit and requires investigators who are trained in the skills of research methodology. Potentially, this puts research out of the reach of many smaller establishments.

For these, and probably other reasons, many zoos undertake their research in collaboration with each other and with other agencies, such as universities and conservation agencies. It is, of course, also the case that some academics in higher education institutions are interested in zoo-based research and initiate the collaboration themselves. Unfortunately, there are still rather few of these, probably because many university-based behavioural biologists

---

5 Hominids may sound as if they are people, but recent molecular analyses have led to the great apes being reclassified in the same family as ourselves, Hominidae.

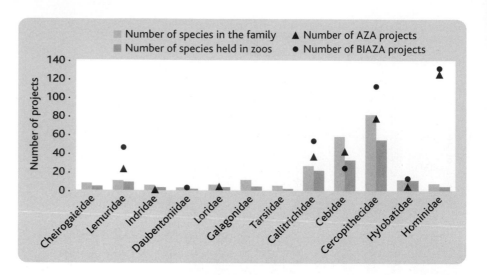

Figure **14.4** In this breakdown of 706 primate research projects on the AZA and BIAZA databases, it is clear that the number of great ape (Hominidae) projects is disproportionately high compared with the number of species, while some other families, such as South American primates (Cebidae) and gibbons (Hylobatidae), are under-represented. (Adapted from Melfi, 2005)

(physiology and anatomy have a longer history of zoo research) do not believe that it is possible to do 'good science' in the zoo setting.

Of 344 studies published in the journal *Animal Behaviour* in 1993 and 1994, for example, 163 involved captive animals, but only three of these were undertaken in zoos (Hosey, 1997). Possible reasons why non-zoo behavioural biologists avoid research in zoos are perceptions that zoo animals constitute abnormal populations, that **functional** theories of behaviour cannot be tested in zoos, and that there are just too many methodological difficulties (Hosey, 1997). Hopefully, a reading of **Chapters 6–9** will convince the reader that the first of these objections is not necessarily true, but we will look more closely at this in **Box 14.1**.

In relation to the second objection, we can point to studies that have tested functional theories in zoos. An example is the study by Cooper and Hosey (2003), which used colour-manipulated photographs to test theories of female mate preference in a number of lemur taxa in UK zoos (**Fig. 14.5**). Such a study would have been extremely expensive and difficult to achieve in the wild.

In fact, even the third of these three objections (methodological difficulties) does not pose

Figure **14.5** Here, you can see a female brown lemur *Eulemur fulvus* looking at a photograph of a male brown lemur in an experimental viewing box, as part of a research project to investigate whether females prefer brightly coloured males. (Photograph: Vicky Cooper)

Functional theories are those that try to explain why particular behaviours have evolved (see **Chapter 4**).

Box **14.1** Are zoo animal populations 'abnormal'?

Suppose we are interested in a particular species of **exotic animal**, or are interested in a particular theoretical problem in something such as behavioural biology. If we were to do our research at a zoo, to what extent would our results be rendered worthless, or at least suspect, because of the unnaturalness of the animals' environment?

This is a difficult question to answer. There is a long tradition of anatomical and physiological research being undertaken in zoos, presumably on the assumption that anatomy and physiology are not affected by captivity. This may not be the case, however: tigers, for example, have different skull morphology in zoo specimens compared with those from the wild, probably because of a different diet and more grooming (Duckler, 1998). But it is in behaviour that we would expect the greatest change (see **Chapter 4**, in which deviations from 'normal' behaviour are addressed more fully). So, at worst, our research data on the behaviour of zoo animals may be suspect because the animals live in an unnatural enclosure, are in groups with an unusual age or sex structure, may show **abnormal behaviours**, and may interact with people. At best, zoo research represents the study of a diverse array of animals in complex environ-ments in semi-natural social settings, providing a unique insight into their biology and behaviour that would be otherwise restricted in their natural habitat, or adversely affected by the standardized and relatively barren conditions of laboratory settings.

Perhaps the best answer at the moment, then, is that zoo enclosures have become much more naturalistic in the last couple of decades and that, in some studies, no differences have been found between animals living in these enclosures and those living in the wild (see **Chapters 4, 7**, and **8**). We are also now amassing much more data about the precise ways in which individual components of the zoo environment affect behaviour (see **Chapter 4**), so there is more opportunity now to build these into the experimental design as possible variables.

Finally, we can point out that even apparently 'wild' populations of many species today are actually also affected in various ways by human activity (Hosey, 2005), so the zoo is by no means unique in potentially changing animals' behaviours. The message is, therefore: go ahead and do the zoo studies, but be as informed from the literature as possible about zoo environments and use this to interpret results appropriately.

a real problem to zoo research, as we will see in section **14.4**.

Indeed, there is much to be gained by university-based psychologists and biologists undertaking research in zoos (Maroldo, 1978; Moran and Sorensen, 1984). They have access to a range of exotic species that could only otherwise be studied at enormous cost and the results of their research can be of great importance in furthering the conservation of endangered species. Indeed, universities and zoos can complement each other in what they can bring to research (**Fig. 14.6**). It is probably also worth pointing out here that a great deal of non-behavioural research on zoo animals, notably veterinary and animal health-related, is published every year in the scientific journals.

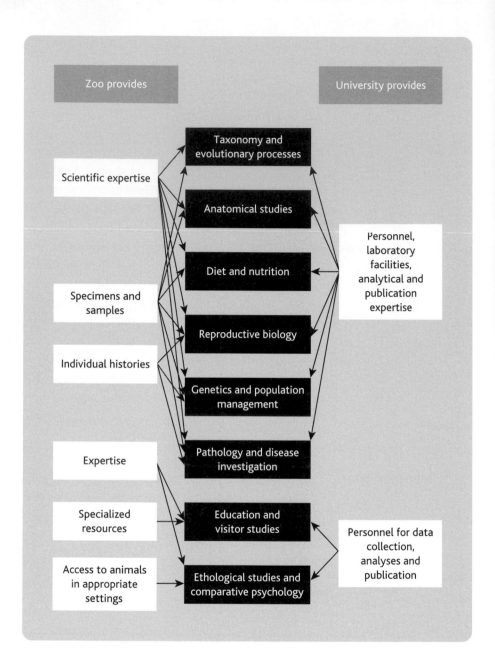

Figure **14.6** Zoo research can be facilitated by support from both zoos and universities, each of which provides different types of expertise, knowledge, and resources. (From WAZA, 2005)

14.3.5 Research departments in zoos

More and more zoos are hiring dedicated scientists onto their staff, to facilitate and conduct research. This research may extend beyond the zoo grounds, to reserves and field conservation, and include a myriad of topics, already explored in section **14.3.3**. Each zoo will prioritize projects accordingly; this will be affected by the animals in the collection, the interests and expertise of staff, links to other organizations or field sites, what can be practically and financially accommodated, global priorities, and many other factors.

Much of the research coordinated by zoo research departments is undertaken by university students and provides training opportunities for new scientists. Training in research methods is also provided to other zoo staff, who are then also able to collect data and conduct research. Increasingly, links between zoo and university-based researchers are expanding, with university departments being sited on zoo grounds, or zoos gaining university funding, or simply providing joint supervision or teaching opportunities.

For example, the Royal Zoological Society of Antwerp conducts research through its Centre for Research and Conservation, a zoo-based research team that has university partners and financial assistance from the Flemish government, and won the European Association of Zoos and Aquaria (EAZA) 'Best Research' award in 2006. In the UK, collaboration between four Scottish universities (Abertay, Edinburgh, St Andrews, and Stirling) and Edinburgh Zoo has resulted in the establishment at the zoo of a purpose-built centre (the Living Links to Human Evolution Centre), for which the Scottish Higher Education Funding Council (SHEFC) is providing £1.6m. Primatologists from these universities, who form the Scottish Primate Research Group (SPRG), will be able to use the centre to study primate behaviour.

These kinds of activities and collaborations further support research in zoos.

14.4 Some methodological difficulties in zoo research

In principle, undertaking research in zoos is no different to any other kind of biological investigation. Any research project should include clear, testable hypotheses or research questions, defined variables and measures, and explicit procedures for collecting and analysing data. Nevertheless, zoo environments do pose some challenges to the researcher, and, in this and the next section, we will try to identify the major ones and how they can be resolved.

For those who are about to undertake zoo research, we strongly recommend following the advice given in the various BIAZA research guidelines, which currently cover:

- project planning and behavioural observations (Wehnelt et al., 2003);
- statistics for typical zoo data sets (Plowman, 2006);
- monitoring stress (Smith, 2004);
- effects of human visitors (Mitchell and Hosey, 2005);
- questionnaire design (Plowman et al., 2006);
- behavioural profiling (Pankhurst and Knight, 2008);
- how to get zoo research published (Pankhurst et al., 2008).

All of these are available to download, free of charge, from the BIAZA website.

In addition, anyone intending to attempt behavioural research would be advised to consult a guide such as Martin and Bateson (2007).

14.4.1 Questions and hypotheses

Any research study should clearly identify what hypothesis it is designed to test and this hypothesis should relate to a body of existing knowledge that effectively provides a theoretical context for the study.[6] Without this, the results of a study can be difficult to interpret.

Zoo research is hypothesis-driven, like any other scientific research, but there are several types of project in which identifying the research hypothesis can be difficult. An example is in enrichment studies (**Fig. 14.7**), in which, potentially, there may be a number of different hypotheses. Enrichment may be provided to try to achieve a broad behavioural profile in the animal that is more like that seen in the wild; it may also be provided to bring about a more specific change, for example, in an attempt to reduce **stereotypies**, or because the animal

---

6 Note that statistical hypotheses, the null hypothesis (H_0), and alternative hypothesis (H_1) do not fulfil this role, because they only predict an effect, but are not related to the underlying scientific rationale of the study.

(a)   (b)   (c)

Figure **14.7** Enrichment is a popular topic for undergraduate research projects; indeed, it is appropriate that enrichment provided is monitored and its effectiveness measured (see section **8.6**). It is necessary that these, as with all research projects, are hypothesis-driven. For example, we might suggest that providing food for this tiger up the pole will increase its activity. (Photograph: South Lakes Wild Animal Park)

is believed to be too inactive, or for all sorts of other reasons (see **Chapter 8**). Whatever the reason for the enrichment, there should be good theoretical reasons for believing that they will achieve their aim and this should be incorporated into the research hypothesis of the study. Otherwise, it will be difficult to be sure that the enrichment is enriching, even if it achieves its planned effect.

Another area in which hypotheses can be difficult to formulate is in the analysis of questionnaires and animal records. Often, the data in these have been collected with no particular question in mind, especially if they are part of routine record keeping. But statistical analysis is only of use if it tests a hypothesis that is derived from theory.

Questions in behavioural research

Research answering any of the four kinds of behavioural 'whys' (see **Box 4.1**) can be undertaken in zoos to find out about the function, causes, development, and evolution of behaviour in particular species. Often, however, the questions are more applied

than is the case with wild-living animals and may be concerned with how the zoo environment affects behaviour. It is important to understand that the 'zoo environment' is actually made up of a number of different variables, such as enclosure size, group size, group composition, feeding schedule, and so on, and that each of these may influence behaviour in different ways. Because the scope for manipulating these variables is limited in most zoos, the researcher needs to take particular care in planning a behavioural study in a way that attempts to account for these. Further advice is available in the BIAZA research guidelines (see **Further reading**).

Examples of the sorts of questions that underlie much zoo behavioural research are:

- how does the animal spend its time?

- how does the animal use its enclosure?

In both of these cases, the purpose of the question is not usually to have that knowledge for its own sake, but because knowledge of these things can help to answer other questions about how the animal is coping with the zoo environment.

To answer the first question, data are often collected to produce an **activity budget** (see **Figs 4.33** and **4.34**). To collect data for this, the behaviours that the animal performs are collapsed into a small number of mutually exclusive categories. Often, these categories are things such as 'resting', 'moving', 'feeding', 'social behaviour', 'sleeping', and 'out of sight', but different categories from these can be used if, for example, the animal has particular behaviours of interest that are hidden within one of these categories. The data collected in this instance are frequencies or durations of behaviour in each category over standardized observation periods, with care being taken to sample all relevant times of the day appropriately. The result shows what proportion of an animal's time is spent in each activity; differences in activity budget between individuals, or between zoos, or between zoos and the wild, can then be examined and, hopefully, interpreted.

To answer the second question—how the animal uses its enclosure—we need to know how often an animal is in each sector of the enclosure. In principle, we would expect all sectors to be used equally relative to their size, but, perhaps surprisingly, many zoo animals do not use all of the space available to them. If this is the case, it might indicate that the enclosure is not as well designed for that species as it could be, or that there are things within the cage environs to which the animal is attracted or which it is trying to avoid. But simply knowing how much time the animal spends in each sector is not enough to tell us about its enclosure use. It is likely that the different sectors that we identify are themselves of different area and that even a randomly moving animal will spend more time in a large part of the cage than in a small one (**Fig. 14.8**).

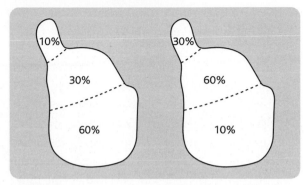

Figure **14.8** The way in which we divide up an enclosure for behavioural sampling may leave us with zones of different areas. We would not be too surprised if an animal spends 10 per cent of its time in the small zone at the top, but if it spends 30 per cent of its time there, we would suggest that perhaps something is attracting it to that zone. Likewise, if it only spends 10 per cent of its time in the large zone at the bottom, we might suspect that there is something there that the animal finds aversive. The SPI helps us to make these sorts of judgements about an animal's enclosure use.

A suitable measure of cage use is the **spread of participation index (SPI)**, which compares the observed frequency of use of different cage sectors with the frequency that we would expect, given their relative sizes. The calculated SPI is a number between 0 (indicating all sectors used equally—that is, maximum use of the enclosure) and 1 (indicating only one sector used, hence very great underusage of available space). A method for calculating the SPI for zoo animals is given by Plowman (2003) (see **Box 14.2**).

14.4.2 Data collection

Again, methods of data collection should, in principle, be no different in zoo-based studies from those of any other kind of biological research. Thought should be given to the form, amount, and timing of sampling, particularly in studies of behaviour (see **Box 14.3**), and the **dependent** and **independent variables** should be defined.

An **activity budget** shows the proportion of time that animals spend in each of the activities that they perform. The **independent variables** are those that might be manipulated or compared, such as cage size, presence or absence of enrichment, etc., which may change the **dependent variable**—that is, a behaviour, a physiological measure, or something similar.

Box **14.2** The spread of participation index SPI

Several investigators have used the spread of participation index (SPI) to quantify how animals use the space available to them. Traylor-Holzer and Fritz (1985), for example, applied the measure to a group of chimpanzees at a primate centre. They found that adult SPI scores ranged from 0.32 to 0.75 (the index ranges from 0, meaning that all zones are used equally, to 1, meaning that only one zone is used), whereas juvenile scores ranged from 0.28 to 0.49, showing greater use of cage space by the juveniles. Shepherdson *et al.* (1993) used the SPI to show increased use of space in a fishing cat *Felis viverrina* on days after live fish had been put in a pool in the enclosure (SPI of 0.4, compared with 0.84 before the arrival of the fish).

As mentioned in the text, the reality of zoo enclosures is that, physically and biologically, they are not made up of equally sized zones, and Plowman (2003) has modified the index to allow for this. Her formula is

$$\text{SPI} = \frac{\sum |f_0 - f_e|}{2(N - f_{e\,min})}$$

in which f_0 is the observed number of observations of the animal in a zone (that is, the frequency of its occupation of the zone) and f_e is the expected number of observations for that zone, calculated to account for differently sized zones, but assuming that the animal uses the whole enclosure evenly. N is the total number of observations in all zones, and $f_{e\,min}$ is the expected number of observations for the smallest zone. The resulting SPI is a number somewhere between 0 (greatest use) and 1 (least use).

A particular problem with zoo research is that there are quite a few variables present over which the researcher has little control and these act as confounding variables. This can happen particularly if the data are being gained from animals in more than one cage, or more than one zoo, but can also cause difficulties in interpreting the results of studies on only one group of animals. How do you know that the effects you are finding are due to the independent variables you are measuring rather than a confounding variable?

As an example, consider how you would investigate whether cage size affects behaviour. This would involve getting data from at least two groups of animals, but it is not only in cage size that they will differ; they might also have different group sizes and compositions (different numbers of males, females, juveniles, infants, adults), the enclosures may be qualitatively different (moated, glass-fronted, indoor, outdoor, etc.), and the husbandry regimens may be different. So what is the solution?

For small research projects, it is often possible simply to acknowledge that these problems exist, although good experimental design can often minimize them (for example, match the groups as precisely as possible). In larger-scale studies, the multi-zoo approach can allow more independent variables to be measured (see section **14.6**).

14.4.3 Practicalities

No matter how well planned a piece of zoo-based research is in terms of its experimental design, it must still take into account several practical issues that make the zoo rather different from other places in which biological research is carried out. Most of these differences stem from the fact

Box **14.3** How do we sample behaviour?

It is tempting, when studying behaviour, to try to write down everything that every animal does, but it is usually not possible to do this, let alone to derive any meaningful conclusions from what we have written down. Inevitably, we must take samples, and this is more of a challenge for behavioural samples than for other kinds of biological samples (such as faeces, blood, physical measurements), because behaviour changes from moment to moment, but also shows daily and seasonal patterns. We can only give a brief introduction here and anyone wishing to carry out behavioural research is advised to consult more detailed guides: particularly, Altmann (1974), Plowman (2006), and Martin and Bateson (2007).

The most detailed data come from 'continuous sampling', in which actual durations and frequencies of individual behaviours are recorded over a fixed period of time. This is usually done to look for all occurrences of a particular behaviour, or to measure the behaviour of each individual animal in turn, or to look for the components of a sequence of behaviours whenever they happen, as, for example, in an interaction. But continuous sampling needs a lot of concentration and can be very time-consuming, so many researchers use time sampling instead.

'Time sampling' involves recording the behaviours seen at or between particular points in time (that is, at sample points, or in the sample interval between successive sample points), rather than whenever they occur. This can mean recording the behaviour that occupies more than 50 per cent of the sample interval ('predominant activity sampling'), or behaviours that occupy the whole of the sample interval ('whole interval sampling'). Alternatively, a behaviour can be recorded as present if it occurs at all (regardless of how often or how long) in the sample interval ('one–zero sampling'), or if it occurs at the sample point ('instantaneous sampling' if relating to one animal; 'scan sampling' if relating to several). Time sampling can generate a lot of data rather less intensively than continuous sampling and is particularly suitable for recording long-duration behaviours (that is, 'states', such as sleeping, resting, feeding) rather than those of short duration (that is, 'events', such as biting, pecking, scent marking).

Each of these methods has advantages and disadvantages, depending on what sorts of data are needed, and how closely they are required to approximate to real frequencies and durations (Tyler, 1979).

that the experimental subjects (usually the animals) are manipulated and influenced in other ways by people (which makes it a little different from the field), but not necessarily by the people who are doing the research (which makes it rather different from the laboratory). Many zoos are understandably unwilling to allow manipulations only for research purposes, but will usually accommodate them if the research is worthwhile, legal,[7] and not too disruptive.

7 The main legislation covering animal research in the UK is the Animals (Scientific Procedures) Act 1986. This is of greater relevance for laboratories and universities than for zoos, but some zoo-based research may require a Home Office licence under the terms of this Act. There is more about this in section **3.4.4** and **Box 3.4**.

By 'manipulations', we mean the sorts of procedure that are often carried out in laboratory research, such as the addition of apparatus to the animals' cages, or changes to cage size or furniture. Such manipulations do take place in zoos, often as part of routine husbandry or enrichment trials, and it is worth the researcher's while building a good relationship with the zoo so that they know when these things are going to happen and can take advantage of the possibilities that they offer.

The perceptions of research held by those zoo personnel who are not directly involved in research are another important factor that must be taken into account during the planning stage of any research project. Sometimes, it is necessary for an animal to be moved (even to a different zoo), or its cage is changed, or various other sorts of husbandry changes are introduced during the course of a study, although most zoos will do all they can to avoid disrupting research of which they are aware and in the planning of which they have been involved. It is important that the researcher who wants to work in a zoo establishes a good relationship with the keepers, as well as with the zoo's research department (if it has one), because this can help to minimize these sorts of problems. It is also important for the researcher to realize that whatever they do must be undertaken within the zoo's health and safety procedures (see **Chapter 3**) and ethical guidelines.[8]

One final practical issue is worth mentioning. Much zoo research is undertaken opportunistically, or as a number of small projects by different people at different times, and perhaps using different sampling regimes. If keepers are collecting data, then, often, the data collection has to fit in with the keepers' schedules of other work, which, again, can be quite challenging for sampling regimes. So how much does this matter when we consider the validity of the data being obtained? Fortunately, the answer is that, in many cases—particularly for the more common behaviours—the answers given by different sampling methods are usually not significantly different from each other (Margulis and Westhus, 2008), which means that fitting in with the constraints posed by zoo and university schedules does not mean sacrificing worthwhile data.

14.4.4 Information that is already available

As we have already seen, there are a lot of data that are already available through the record-keeping processes of zoos (**Chapter 4**). Many zoos use the same computerized databases to store their records, and access to these records is possible for approved projects. Other information is stored in **studbooks** and in the International Species Information System (ISIS) database. Investigating these sources of data first is recommended for many research projects, such as those to do with animal health, **demographics**, and genetics.

14.5 Problems in the analysis of data

Most textbooks on data analysis and statistics assume that the research scientist is able to design **controlled** experiments and have large numbers of subjects. Many of the best-known statistical tests are designed to work with data collected in this way. Thus, parametric tests, which include analysis of variance (ANOVA) for finding differences between conditions and Pearson's correlation (r, for detecting association between variables), are widely used in

---

8 It is important that research is undertaken within the framework of ethical guidelines and most **accredited zoos** have ethical committees to ensure that all of the work of the zoo is conducted ethically. Ethical guidelines for behavioural research are published regularly by the Association for the Study of Animal Behaviour (ASAB) and the Animal Behaviour Society (ABS), and are also available from their websites. There is more information about ethical guidelines for zoos in **Chapter 3**.

Control in scientific investigations refers to a group of subjects that do not receive the experimental treatment and therefore show what would have happened had the experimental manipulation not been done.

experimental biology, but they require that the data points that they are testing are normally distributed, measured on an interval or ratio scale, independent of each other, and show **homogeneity of variance**. So what does this imply for zoo research?

Many research studies in zoos gather behavioural data, which may or may not conform to these requirements. But, provided that the data points are independent of each other, non-normally distributed data can still be analysed with non-parametric equivalents of these tests, such as Kruskal–Wallis (a non-parametric equivalent of the ANOVA) and Spearman's test (a non-parametric correlation analysis). What we have to be aware of is the possibility that the data collected in the zoo are not independent, because this renders these tests inappropriate. We also have to cope, in many zoo studies, with only a small number of animals (sometimes only one), which again makes these tests unsuitable.

But before we look at how to deal with these problems, let us remind ourselves that the important driver of research is the question that we are asking, or the hypothesis that we are testing. There is nothing intrinsically wrong with having a small number of subjects, or, for that matter, only one: this may well be all that is required to answer the question. So if, for example, we want to find out why a particular animal is performing an abnormal behaviour and how we can reduce it, then issues of sample size are pretty meaningless. Indeed, as Kuhar (2006) has pointed out, inferential statistics have been developed to help the researcher extrapolate from a sample to the population as a whole, but if we are only concerned with our animal and are not trying to say something about other animals in other zoos, then statistical analysis may not be necessary.

It is also worth bearing in mind that non-independence and small sample sizes also occur frequently in field studies of animals, and also in psychology. Provided that they are dealt with appro-priately, they need not be a problem. The important thing is that our experimental design and statistical analysis allow for these sorts of issues and give us the answers we need. Ultimately, this might even mean that we break one or two minor statistical rules, which may be better than sacrificing our research, provided that we are honest in our reports of the research and try to discuss whether this affects our conclusions (Kuhar, 2006). One way of approaching this dilemma is to realize that biological significance may sometimes be more important to us than statistical significance. For example, an increase in a stereotypy from 0 per cent to 15 per cent of the time budget of an individual animal is very likely to be biologically significant, but might not be testable statistically because of the zero.

Let us look now at these two problems of non-independence and small sample size.

14.5.1 Lack of independence of data

We think that data are independent if we believe that the value of one data point does not affect the values of any others—but there are three common situations in zoo studies in which this is unlikely to be true: in time sampling, in using animals in the same enclosure, and in constructing activity budgets.

1. *Time sampling* (see **Box 14.3**)   This is a convenient way of gathering behavioural data and, if the sample points are close together (that is, at short sample intervals), they allow us to make an estimate of the duration of a bout of behaviour. Short sample intervals also make it more likely that we will not miss behaviours that have a very short duration. Unfortunately, they also risk a problem of **autocorrelation**: imagine you are scoring the behaviour of a red panda *Ailurus fulgens*, which is at the top of a tree asleep (**Fig. 14.9**); if your next sample point is ten seconds later, it is highly likely that the panda

---

Homogeneity of variance refers to when the variance of the samples are not significantly different from each other.

Autocorrelation means that the value of one sample point is likely to influence the value of the next one.

Figure **14.9** Suppose that you are time sampling this red panda's behaviour: if you sample again in 10 seconds' time, what is the panda most likely to be doing? (Photograph: Geoff Hosey)

will still be asleep at the top of the tree, but even if it is not, there are certain behaviours to which it could not switch in that short time (for example, eating from a food dish on the ground). The best solution is to have sample intervals that are long enough to give you confidence that the animal could, in principle, switch in that time to any of the other behaviours it does. This misses most short duration behaviours, however, and does not usually allow you to calculate bout durations. But, if these are important, there are better ways of obtaining them than through time sampling.

2. *Using animals in the same enclosure*   Animals in the same enclosure may show the same behaviours at the same time, because they are responding to the same variables (**Fig. 14.10**)— for example, they may all get fed at the same time and they may all receive visitors at the same time —but they also influence each other's behaviours by interacting with one another. Either way, it means that data collected from different animals are not independent. One simple solution is to ensure that samples from the different individuals are taken at different times. If this cannot be done, it may be more realistic to treat the

enclosure as the sampling unit, rather than the individual animals, although this leaves you with a small sample size (see section **14.5.2**), unless you replicate your study in a lot of zoos. Again, which of these solutions you adopt depends very much upon your research question. For some research projects, it is important to concentrate on individuals and, very often, their social non-independence is exactly what you are interested in: for example, when looking at social interactions in a group of animals.

3. *Constructing activity budgets* (see **Chapter 4**) Activity budgets are very useful as descriptions of how the animal partitions its time between the different activities that it does. A problem arises, however, if statistical analysis is carried out to compare different activity budgets, because the activities are not independent of each other: activity budgets, by definition, account for 100 per cent of the animal's time; if the amount of time spent in one activity increases, something else has to decrease to provide that extra time. One solution to this problem is to avoid collecting activity data on all of the things that the animal does and, instead, to concentrate on changes in only one or two key behaviours. Alternatively, statistical tests such as the G-test can be used to compare activity budgets (rather than standard parametric and non-parametric ANOVAs), because they do not require independent data categories.

For further information on all aspects of independence problems, including the use of G-tests, see Plowman (2006).

14.5.2 Small sample sizes

Small sample sizes are a frequent occurrence in zoo research. They can occur simply because few animals are available (they may be rare, or big, or may live in small groups), or because the effects of a manipulation (such as adding an enrichment) are being studied and the manipulation is only being

Figure **14.10** If you treat each of these animals as an independent subject for behavioural research, how can you be sure that they do not influence each other's behaviour? Is the gerenuk *Litocranius walleri* on the right about to start feeding because the animal next to it is already feeding? (Photograph: Geoff Hosey)

implemented in one enclosure (maybe even for only one animal).

As pointed out above, this need not in itself be a problem. It can cause some problems of **ecological validity** if you are trying to extrapolate to the whole population or taxon, but much zoo research is about the particular, rather than the general. It can also cause some statistical problems when you analyse the data. With small sample sizes (that is, a small N), it is usually not possible to tell whether or not the samples satisfy the assumptions of the test (such as

normality of distribution, homogeneity of variance), but, in any case, the low N may give the statistical test low power. This means that the test is unlikely to detect meaningful differences unless the pattern of the data is very strong indeed.

A reasonable rule of thumb is that if there are eight or more individuals in each of the groups being compared, then standard parametric and non-parametric tests should work reasonably well. With a smaller N than that, the best solution is to analyse the data with **randomization tests** instead

---

Ecological validity relates to how well results from a sample can be generalized to the population as a whole. Also called permutation tests, **randomization tests** calculate a test statistic and compare it to the results from repeated randomization of the same data across all groups. This identifies the probability of obtaining a particular pattern of scores when all of the possible combinations of scores are jumbled up.

(Besag and Clifford, 1989; Manly, 1998; Todman and Dugard, 2001).[9] What must be avoided is simply gathering more data on the same animals in the hope that these extra data points increase the sample size: they do not, because you should be using only one mean or median for each animal, or else using separate data points in a repeated measure design. This fault is known as **pseudoreplication** and is discussed more fully by Kuhar (2006).

Randomization tests are designed to identify the probability of finding a particular pattern of scores across two or more groups. Suppose, for example, that you are investigating whether activity increases in an animal that is given a feeding enrichment and, to answer this question, you collect baseline (that is, pre-enrichment) activity data for 20 days and treatment (that is, with enrichment) activity data for a further 20 days. The randomization test then asks into how many different combinations can the resulting forty numbers (that is, the twenty baseline data points and the twenty treatment data points) be put and, out of all of these, what is the probability of getting low values in the baseline group and high values in the treatment group. This test makes no assumptions about the underlying distribution of the data, but it does require that the experimental design includes some randomization. This can be achieved by randomly allocating the treatment among the available observation blocks.

More information on randomization tests in the zoo context is given in Plowman (2006).

14.6 Multi-zoo studies

Many of the research questions that we are interested in asking about zoo animals cannot be answered from single-zoo studies, but require data drawn from across a number of different zoos. Multi-zoo studies of this sort allow much more than a possibility of increasing sample size: they also allow us to start teasing out the effects of a number of different variables, which would remain confounding variables if we were to restrict ourselves to only one zoo. Furthermore, multi-zoo studies are the only real way of finding out about the prevalence of phenomena in which we are interested (such as how widespread is a particular behaviour, a particular parasite infestation, or a particular dietary anomaly), or to confirm whether or not something appearing in our zoo is unusual.

There are, of course, some limitations to multi-zoo research as well. Because zoos are so different, you may end up actually introducing even more confounding variables into the study. There is also the risk that pooling data from animals in different zoos can lead again to lack of independence (Kuhar, 2006), because animals in the same cage have more in common with each other than they do with animals in different cages or in different zoos. There may be practical (and financial) difficulties inherent in visiting a number of zoos, possibly in different parts of the world. Finally, gaining the cooperation of a lot of different zoos is much more of a challenge than working only in your local zoo, and can be both time-consuming and frustrating. This process can, of course, be eased considerably if the project is supported by the relevant zoo association: in the UK and Ireland, for example, BIAZA will scrutinize and support, through its Research Group, multi-zoo projects that have achievable and worthwhile goals (see **Box 14.5**).

There are several ways in which multi-zoo research can be undertaken. Often, they involve a survey of some sort, usually using a questionnaire, and may request current information or historical records. Surveys are widely used to investigate

---

9 Todman and Dugard's (2001) book on randomization tests comes with a useful CD with worked examples of how to carry out these tests.
Pseudoreplication is the term usually used to describe data among which the sample points are not independent. For example, fifty measurements of stride length by a cheetah do not give you a sample size of fifty; the sample size (n) is one (the cheetah).

husbandry problems, but are often not used to best advantage because of poor design (Mellen, 1994). When carried out well, however, they can supply much useful data very quickly. A good example of this approach is seen in a survey by Pickering *et al.* (1992) to determine what factors affect the breeding success of captive flamingos (see **Box 4.3**). They obtained information from forty-four zoos in the UK and Ireland on flock size, egg-laying, and chick-rearing in Chilean *Phoenicopterus chilensis* and Caribbean flamingos *P. ruber ruber*. They found that larger flocks were more likely to breed successfully, and recommended minimum flock sizes of forty birds (Chilean) and twenty birds (Caribbean) to ensure breeding, provided that space allowed these numbers without crowding.

Another way of undertaking a multi-zoo study is to visit each zoo and collect data in person. This is usually the only way of conducting behavioural studies, because quantitative behavioural data are not routinely gathered by zoos and hence none would be available for a survey- or questionnaire-based approach.

An example is the study by Perkins (1992) of the factors that affect activity in orang-utans *Pongo pygmaeus* (**Fig. 14.11**). She observed twenty-nine animals in fourteen enclosures across nine different zoos in the USA and then used **multiple regression** to identify the effects of the different environmental variables that she measured. This allowed her to conclude that large enclosures with a lot of movable objects, together with opportunities for social interaction, were the most favourable environments for promoting activity in these animals. To many people, a result such as this may seem self-evident, but in fact it is not; it may tell us something that we think we already know, but we actually do not know it at all unless it is supported by hard empirical data and, in this case, using data from a number of zoos allowed those data to be collected in a way that would not usually be possible in a single zoo.

Figure **14.11** Behavioural observations of orang-utans in many different zoos—referred to as 'multi-zoo studies'—have shown that they prefer large enclosures with a lot of movable objects and opportunities for social interaction. (Photograph: Ray Wiltshire)

To show just how much a multi-zoo study can achieve, let us look finally at a study by Carlstead *et al.* (1999a; 1999b) on black rhinoceros *Diceros bicornis* (**Fig. 14.12**) in US zoos. This was part of the Methods of Behavioral Assessment (MBA) project, originally set up by researchers from twelve leading US zoos, and designed to assess the behaviour and breeding success of animals across a large number of zoos. In this way, standardized behavioural profiles could be built up and compared across zoos to identify the influence of both individual differences (that is,

Multiple regression is a statistical technique for identifying which of a number of independent variables best predicts the value of a dependent variable: in this example, activity.

Figure **14.12** The black rhinoceros *Diceros bicornis* was the subject of a US multi-zoo study that aimed to identify the reasons for the species' poor reproduction in captivity. (Photograph: Ray Wiltshire)

'personality'; see **Box 4.2**) and captive management regime on breeding success, mortality, behaviour, and welfare. The black rhinoceros was one of the species chosen for this study because it reproduces poorly in captivity. Such a study would not be possible in only one zoo and a survey would not give all of the information that was required. A multi-zoo study using breeding records, keeper assessments of animal traits, enclosure measurements, and behaviours allowed this study to conclude that the most compatible rhinoceros pairs were those with assertive females and submissive males, and that they bred

best in large enclosures with few concrete walls. Such conclusions could only be drawn because the multi-zoo study allowed the comparison of almost thirty pairs of rhinoceros.

14.7 Disseminating the results of zoo research

No matter how good the research may be, it is of greatest value if it is made available to other people. In the case of zoo research, this means not only making it available to other zoo professionals, but also to researchers in other fields and to anyone else who may find it useful or interesting. So how is this done?

The most important way is through publication in **peer-reviewed journals** and there are a number of these that publish zoo research. The advantages of this are that, by being published in these journals, the research has been shown to be of good quality; furthermore, this sort of publication ensures that the results of the research are available to anyone who wishes to see it. By their very nature, however, peer-reviewed journals tend to be mostly concerned with studies that have high ecological validity, which rather restricts the opportunities for much of the zoo work to be acceptable to them. Consequently, a lot of zoo research is published in the so-called 'grey literature'—that is, journals that are not peer-reviewed and which are often zoo community periodicals or internal publications of individual zoos (see **Box 14.4**). This makes the research available to others within the zoo community, but getting access to these sources can be difficult for those who are outside this community. This difficulty is recognized by the WZACS (WAZA, 2005), which urges zoos to publish in the peer-reviewed literature wherever possible.

Ultimately, the research should be published in the most appropriate place, bearing in mind the

Peer-reviewed journals are those that publish papers only once they have been through a rigorous evaluation by independent experts to ensure their quality and scientific value.

Box 14.4 Zoo journals

One of the earliest zoo-related journals was *Zoologica*, published by the New York Zoological Society from 1907 to 1973 (Kisling, 2001), but now no longer published. In the UK, the Zoological Society of London has been publishing the *International Zoo Yearbook* annually since 1960. The journal *Zoo Biology* was first published in 1982 and remains the main peer-reviewed journal for zoo research. In Europe, the journal *Der Zoologische Garten* publishes papers on zoo animal management, mostly in German. Several journals routinely publish the results of research on zoo animals—notably, *Animal Welfare*, *Applied Animal Behaviour Science*, and the *International Journal of Primatology*. Veterinary and health-related research on zoo animals is published in a number of journals, such as *Journal of Zoo and Wildlife Medicine*, *Journal of the American Veterinary Medical Association*, *Journal of Parasitology*, and *Journal of Veterinary Medical Science*.

The magazine *International Zoo News* is not a peer-reviewed journal, but has the advantage of being readily accessible; this publication makes many of its articles and back issues freely available online. Other zoo news, particularly relating to conservation, is published in *Dodo*, the journal of the Jersey Wildlife Preservation Trust, but is harder to come by, because this publication does not have its own website.

Zoo associations have, of course, published their own newsletters from the outset. Some of the more notable of these include the AZA's *Communiqué* and the American Association of Zoo Veterinavians (AAZV) *Journal of Zoo Animal Medicine*. Within Europe, BIAZA and EAZA both produce their own newsletters. The BIAZA Research Group also publishes monthly summaries of zoo research news, available online from the BIAZA website.

Zookeepers, too, have their own journals and magazines: in the UK, for example, the Association of British Zoo and Wild Animal Keepers (ABWAK) produces *Ratel*. These magazines and newsletters, often referred to as the 'grey literature' because the articles in them are not peer-reviewed, are nevertheless a useful source of additional information on zoo research and other activities.

---

intended readership. BIAZA has produced guidelines on getting zoo research published; these include a list of suggested journals (Pankhurst *et al.*, 2008).

Much zoo research is presented and discussed at conferences. Regular research conferences are run by BIAZA, EAZA, and AZA, as well as by organizations such as the Leibniz Institute for Zoo and Wildlife Research (IZW) (see **Box 14.5**), and details of these are usually available at their respective websites. Conferences are a good way of disseminating research findings, because they make the material available for discussion and draw people's attention to research of which they may otherwise be unaware.

14.8 What zoo research still needs to be done?

Let us finish off this chapter by briefly considering some areas of zoo biology in which research still needs to be done. You will have noticed in several chapters that we have said things such as 'very little is known about . . .', so, in this section, we will try to

Box **14.5** Managing and coordinating zoo research

Research in zoos can be achieved on an individual basis, but is much more effective if part of a coordinated effort. Coordination helps avoid duplication of effort, allows better integration of knowledge from different sources, and can help to ease some of the logistical difficulties involved in implementing research.

In the UK and Ireland, BIAZA takes on this coordinating role through its research group. This group publishes a regular newsletter, giving information on current projects, and organizes an annual zoo research symposium, the proceedings of which are published by BIAZA and made available on the BIAZA website. It also lends support to approved projects (which can help to increase response rates in multi-zoo surveys) and publishes a series of guidelines to help researchers in various aspects of zoo research.

Research is also supported by EAZA, which endorses two research meetings that are held in alternate years at Poznan Zoo in Poland and by the Leibniz-Institut für Zoo- und Wildtier-

forschung (Leibniz Institute for Zoo and Wildlife Research (IZW)) in Berlin, respectively. The IZW studies behaviour, evolutionary biology, wildlife diseases, and reproduction in both wild and zoo animals. EAZA and the IZW publish research and hold regular conferences on zoo and wildlife research. EAZA is currently preparing a research strategy, which will place a strong emphasis on **taxon advisory groups (TAGs)** to develop research themes and keep records of research on their species. This means that, in future, one of the first things that a researcher should do is contact the relevant TAG for the species in which they are interested to find out who else is undertaking research on it.

In North America, the AZA includes research in its annual and regional conferences. Proceedings of these are published, the most recent ones (since 2004) being available on the AZA website. AZA also maintains a searchable database of research projects, although, unfortunately, this is only available to AZA members.

recall some of these as a kind of summary of where we think future research should be directed.

We can start with two general, but very important, priorities. Firstly, more zoo research needs to have a more explicit conservation focus. This is, in any case, a legal requirement now for zoos in many European countries including the UK (see sections **3.1** and **10.3**), although, as we have seen, there have been some criticisms of how effective this legislation is. How can this be done? One obvious way is to undertake research on the *ex situ* population, which can then feed back to, and inform, the management

of the *in situ* population. One area in which zoos have great expertise already is in the management of small fragmented populations (Wharton, 2007).

Secondly, much more research should be done on non-mammalian species. In putting this book together, we have often struggled to find suitable examples among the invertebrates, fish, reptiles, amphibians, and even birds, in a literature that is very much dominated by mammalian examples.

The following list represents some areas of zoo biology in which we think that there is scope for worthwhile research projects.

- How does tagging, microchipping, or branding of animals affect their behaviour, physiology, and welfare? This has relevance for the study of wild animals, as well as animals in the zoo.
- Do animals in zoos learn about husbandry events and how does this manifest itself?
- How does seasonality in northern zoos influence the behaviour and physiology of tropical animals?
- What are different animals' preferences and **responses** to different physical **stimuli** in their environment, such as sound (including **ultrasound**), light, humidity, and temperature?
- What do different species require in their housing, in terms of privacy, elevated areas, etc.?
- Is space quality more important than quantity? (The current evidence is almost all from primates.)
- How does the timing of feeding and the way in which the food is presented affect the animals?
- How do differences in group composition and group size affect the behaviour and welfare of the animals?
- How do different species interact with one another in **mixed-species exhibits**?
- How do animals respond to changes in their housing?
- What behaviours do fish, reptiles, and amphibians show that are equivalent to those abnormal behaviours (such as stereotypies) that are sometimes seen in birds and mammals?
- Can **preference testing** be achieved with zoo animals to give us information about their welfare?

- How do non-mammalian species respond to the presence of zoo visitors?
- What are the optimal housing conditions for captive breeding across a range of endangered species?
- What can an analysis of zoo records tell us about **life history traits**, such as tooth eruption patterns, length of gestation, and litter size, in various species?

Finally, it is worth pointing out again that zoos are excellent places in which to research the basic biology of species that would be difficult to study in the wild and about which little is known—and there are a lot of those.

As an example, let us consider five species that have either had their conservation status reclassified or have appeared for the first time in the 2007 IUCN Red List, and see how many times they have appeared in the title of a scientific paper in the 10 years since 1998:

- Western lowland gorilla *Gorilla gorilla gorilla* (537 times);
- Malayan sun bear *Helarctos malayanus* (twenty-three times);
- Rüppell's griffon vulture *Gyps rueppelli* (not at all);
- gharial *Gavialis gangeticus* (five times);
- Banggai cardinal fish *Pterapogon kauderni* (twenty-six times).

And if this makes you want to go and study the Rüppell vulture or the gharial, then take a quick look at ISIS (see section **5.7.1**) to find out which zoos to visit to find them.

Summary

- Research in zoos can be designed to answer applied questions about the maintenance of animals in captivity, or can answer basic questions about the biology of the animals.

- Zoo research is growing in quantity, but is still biased both towards particular topics (behaviour, reproduction) and towards certain taxa (carnivorous mammals, primates, and others).

- Increasingly, zoo research is undertaken by research departments or individual researchers within the zoos.

- Experimental designs in zoo research usually have to contend with difficulties caused by small sample sizes and non-independence of data, but this not problematic for all questions or hypotheses posed.

- Multi-zoo studies can avoid some of these problems and can also answer research questions that could not be answered in single-zoo studies.

- The results of zoo research are published in journals and discussed at conferences, but more needs to be done to make data more accessible.

Questions and topics for further discussion

1. What difficulties of sampling and analysis might you encounter in a research project on zoo animal behaviour?

2. What do you think are neglected areas of zoo research and why?

3. Provide an argument that worthwhile behavioural studies can be undertaken in zoos.

4. What are the advantages and disadvantages of multi-zoo research?

5. 'Small numbers of animals need not be a problem in zoo research.' Discuss this assertion.

Further reading

General principles of biological research are covered in Barnard et al.'s *Asking Questions in Biology: Design, Analysis and Presentation in Practical Work* (1993). Anyone interested in undertaking behavioural research in a zoo should consult a good manual, such as Martin and Bateson's *Measuring Behaviour:* *An Introductory Guide* (2007) or Lehner's *Handbook of Ethological Methods* (1998) before they start.

For a discussion of some of the statistical challenges in zoo research, we recommend Kuhar (2006).

Other sources of further reading are mentioned throughout this chapter.

Websites and other resources

Methods of implementing zoo research are described in the various research guidelines published by BIAZA (Wehnelt et al., 2003; Smith, 2004; Mitchell and Hosey, 2005; Plowman, 2006; Plowman et al., 2006; Pankhurst and Knight, 2008), all of which are available online at www.biaza.org.uk. Also in this series are guidelines on how to publish the results of zoo research (Pankhurst et al., 2008). Copies of BIAZA Research Conference Proceedings and the regular research newsletters can be obtained from this site as well.

Anyone interested in undertaking behavioural research in a zoo should investigate zoos' websites for information about what sort of research it might be possible to undertake there and what special application procedures may apply, because they will need to request permission to conduct their research there first.

Chapter 15 We hope you enjoyed your visit

In much the same way as many **zoos** have a sign near the exit saying 'We hope you enjoyed your visit', we hope that you have enjoyed reading this book and have found it useful. Zoos—and, in particular, zoo educators—set great store by what they hope is the positive impact of a visit to the zoo and on 'getting their message across'. We would like the 'take home' message of this book to be a positive one. We hope that, having read this book, or perhaps only part of it, you will see zoos in a more enlightened way in future: that you will have a greater awareness of what zoos do, of the different points of view within and about the zoo community, and of the many and complex issues with which modern zoos must deal if they are to continue to operate. We would be among the first to acknowledge that there are some poor zoos, in which standards of husbandry and welfare are not as high as they should be, and in which the contribution of the zoo to conservation, or education, or research, is minimal. But we believe that the community of **accredited zoos** in Europe, and in countries such as the USA and Australia (and beyond), takes very seriously its responsibility for raising standards[1] and for providing support to fellow members to continue to become better at what they do.

In this final chapter, we want to revisit some of the themes that we introduced at the start of this book and to consider how well zoos match up to what they say about themselves in their mission statements (**Fig. 15.1**). We saw, in **Chapter 1**, that zoos tend to describe their role in terms of four key words: conservation; education; research; and recreation. But how good are zoos at measuring their success in achieving their stated objectives?

In this chapter, we will look at the extent to which zoos critically evaluate their impact in areas such as conservation, education, and research. How will this sort of evaluation affect the kinds of animals we see in zoos and the way in which these animals are managed? We also want to return to the subject of **sustainability**, on which we have touched only briefly in earlier chapters. Finally, for those readers who are interested in finding a job in a zoo (or who want to develop further an existing zoo career), we have provided a section on training and career opportunities within the zoo world.

The structure of this short concluding chapter is, therefore, as follows.

15.1 Evaluation
15.2 The shape of the collection
15.3 Sustainability
15.4 Careers in zoos

---

1 See Hatchwell *et al.* (2007) for a discussion of possible future changes to the zoo accreditation system, to encourage 'step-wise' schemes that allow zoos to join at an earlier stage and then to be supported through a progressive raising of standards.

Figure **15.1** How well do zoos match up to their mission statements? Zoo and aquarium mission statements often mention conservation and education as key roles, and many zoos provide opportunities for hands-on contact with animals as part of their educational programmes. Here, children at an aquarium are learning about starfish and other marine invertebrates. (Photograph: © Tammy Bryngelson, www.iStockphoto.com)

(a)

(b)

Figure **15.2** There is nothing wrong with zoos emphasizing a good day out in their marketing material, if this allows them to get across other messages about threatened species and conservation after people arrive. (Photographs: (a) Warsaw Zoo; (b) Sheila Pankhurst)

15.1 Evaluation

Zoos are perhaps best at evaluating how good they are at recreation, at providing a good day out (**Fig. 15.2**). 'Gate income' is a very effective indicator of whether or not a zoo is providing what visitors want. There is nothing inherently wrong with zoos wanting their visitors to have a good day out, particularly if this also helps the zoo to get across other messages about conservation, or animal welfare.

Zoos, of course, rely heavily on their visitors to provide income, not only to meet their day-to-day running costs, but also for supporting their conservation work, both *ex situ* and *in situ*. There is a danger that too great a pressure on zoos to focus on conservation, education, and research, with recreation seen as the poor cousin and rather frowned upon as a worthwhile objective in its own right, will result in zoos that are worthy, but dull, and which no one wants to visit. Many museums have found, to their cost, that cataloguing and preserving antiquities, or fossils, or dead insects, may not be enough, on its own, to keep all of their curators in employment. There is even a danger that the World Zoo and Aquarium Conservation Strategy (WZACS) (WAZA, 2005) may direct attention away from other worthwhile areas of activity and research by zoos, such as welfare and **enrichment**.

In recent years, studies such as those of Conway (2003), Miller *et al.* (2004), Balmford *et al.* (2007), and Mace *et al.* (2007) have assessed the achievements of zoos and aquariums in achieving conservation and education goals. Prompted at least in part by these studies, zoos are slowly becoming better at self-evaluation and at recognizing the need to measure achievement against clear objectives. At the Fourth International Conference on Environmental Education (ICEE)[2] (2007), for example, two of the four main recommendations to emerge from the meeting were about how zoos should set and utilize appropriate benchmarks, and then evaluate their performance against these benchmarks. More recently, the European Association of Zoos and Aquaria (EAZA) Education Standards (EAZA, 2008) explicitly require member zoos to evaluate the effectiveness of their educational programmes.

Accredited zoos are also often required by their national or regional zoo association to undertake a process of self-evaluation. To take just one association, the Canadian Association of Zoos and Aquariums (CAZA) requires its member collections '*To establish, maintain and raise standards of operation in the Canadian zoo and aquarium community through a process of self-evaluation, on site inspections and peer review*'. Conservation strategy documents produced by zoos, such as the report from the international amphibian *ex situ* workshop held in Panama in 2006 (Zippel *et al.*, 2006), now include both detailed action plans and a requirement to evaluate achievement of the stated objectives in these plans.

As we touched on in **Chapter 10**, there is a risk that zoos may evaluate the wrong things. It is tempting—and perhaps easier—to measure and report on inputs (funds raised; money spent; number of people employed) rather than to assess impact, or outputs (number of threatened species in **captive breeding programmes**; successful **reintroductions** back into the wild, etc.) (see Mace *et al.*, 2007). The zoo community is not unaware of this risk. During the run-up to the 2004 *Catalysts for Conservation* symposium, for example, zoo professionals collaborated with leading conservation biologists in an independent research group, the Zoo Measures Group, to facilitate the process of measuring the success of zoo conservation initiatives (Leader-Williams *et al.*, 2007). The findings of this group were used in the preparation of some of the papers reported at the symposium and subsequently incorporated into various chapters of the book *Zoos in the 21st Century: Catalysts for Conservation?* (Zimmerman *et al.*, 2007). We also saw in **Chapter 10** that various tools are now available to zoos to help them to achieve a more effective audit of their conservation work. A greater emphasis on the sort of project evaluation proposed by Mace *et al.* (2007) is likely to lead to changes in the shape of zoo collections in the twenty-first century and this is the subject of the next section.

But before we look at how zoo collections are changing, what about evaluation of other aspects of how zoos operate? In **Chapter 7**, for example, there is a description of the systems of **welfare audit** that two UK zoos have initiated, with detailed mechanisms and processes for measuring potential welfare issues and then acting on them, in a planned rather than an ad hoc manner. And looking at the wider role of zoos in modern society, a recent initiative by nine UK zoos, in partnership with the British and Irish Association of Zoos and Aquariums (BIAZA), resulted in the production of *The Manifesto for Zoos* (Regan, 2005). This was an ambitious attempt to

---

2 Confusingly, ICEE is the acronym for a number of organizations. Within the zoo world, ICEE is usually a reference either to the International Conference on Environmental Education, or to the International Conference on Environmental Enrichment. The former is a major intergovernmental conference sponsored by UNESCO and UNEP; these meetings are held once each decade (the first was in Tblisi, Georgia, in 1977). The latter is a conference focused on enrichment and animal welfare, held every 2 years (with the first in 1993).

evaluate not only zoo performance in the field of conservation, but also to assess how zoos can and do contribute to a wider 'public good'. The *Manifesto for Zoos* has sections covering, for example, the economic output of zoos, and zoos as vehicles for local and regional regeneration. The results of this initiative have been watched with interest by the wider zoo community and a number of Australian zoos are now joining forces to produce a similar document (Hatchwell *et al.*, 2007).

15.2 The shape of the collection

As with a visit to a zoo, you may not have found everything you expected to see, or wanted to find out about, in this book. Of course, we hope that the opposite is also true, and that you have found some things in the last fourteen chapters that, perhaps, you did not expect to find, but which you found interesting.

We have seen in this book that there is a bias in zoo collections towards mammals and birds (although, in terms of the total number of different species, invertebrates do quite well, particularly if corals and other marine animals are included). Following this pattern, a majority of the examples and references in this book are about mammals, followed by birds, followed by reptiles and amphibians, and we recognize that, in this book, fish[3] get a very poor look-in (**Fig. 15.3**). This is not an entirely deliberate choice on our part: we struggled to find more than a handful of good peer-reviewed papers about enrichment for reptiles or amphibians, for example, or about the health, **nutrition**, and husbandry of fish in public aquariums.

The same is true if you take a subject-based approach to zoos, rather than a **taxonomic** approach. There is a bias in the literature towards information about animal behaviour in zoos and so, to a certain extent, in this book.

(a)

(b)

Figure **15.3** There is a perception that zoos are largely about mammals and birds, although, in terms of the total number of species held in zoos and aquariums, fish and other marine animals such as corals are often well represented. (Photographs: (a) © Kristian Sekulic, www.iStockphoto.com; (b) Warsaw Zoo)

As we have seen in the previous chapters of this book, however, the zoo world is changing and, in some areas, it is changing quite rapidly. Zoos are keeping more threatened species. The number of non-mammalian species managed as part of coordinated captive breeding programmes more than trebled between 1992–1993 and 2003, although **taxa** such as amphibians, fish, and invertebrates are still under-represented (Leader-Williams *et al.*, 2007). But the success of new aquariums such as The Deep (**Fig. 15.4**)

3 If there is a second edition of this book, we will aim to devote at least one full chapter to fish and aquariums.

or leisure complexes often stay for a weekend, or longer—why not at zoos? The zoo of the future may well have, alongside its animal collection, a separate natural history discovery centre, or perhaps an IMAX cinema showing the best of wildlife films.

15.3 Sustainability

The United Nations has declared the decade 2005–2014 the Decade of Education for Sustainable Development (DESD). Increasingly, zoos see themselves as leading contributors to public education about sustainability (the *Zoos Forum Handbook* has a whole chapter devoted to 'Sustainability initiatives in UK zoos', including education; see Defra, 2007b). But is the zoo industry itself sustainable and, if not, what can it do to become so?

The concept of sustainability applies to zoos at a number of different levels. It can mean economic sustainability: will zoos still attract large numbers of paying visitors in fifty or a hundred years' time? But the word 'sustainability' nowadays is usually taken to mean environmental sustainability, and by far the greatest environmental issue on the agenda is global warming. Can zoos, for example, meet the challenge of becoming 'carbon-neutral' organizations in the way that many big businesses[5] are now pledged to do?

Zoos are certainly making strides in this direction. For example, Indianapolis Zoo in North America launched a new website in 2008 to promote environmentally responsible and sustainable behaviour at the level of individual households (www.mycarbonpledge.com); Chester Zoo in the North of England has as its mission and vision statement 'A diverse, thriving and sustainable natural world', and Auckland Zoo in New Zealand offers

Figure **15.4** Aquariums have not featured as strongly in this book as we would have liked, although the success of newer aquariums such as The Deep in the UK and Océanopolis in France shows that the public is happy to pay to see fish. (Photograph: © Frank Boellmann, www.iStockphoto.com)

in Hull, UK,[4] shows that visitors will come in large numbers to see fish and other marine animals, and demonstrates that the zoo bias towards mammals and birds is not necessarily reflected in what the public is prepared to pay to see. Balmford *et al.* (1996), in a series of censuses of visitors at London Zoo, found that the aquarium and the reptile house were among the most popular exhibits.

What are the other changes that we are likely to see in zoos over the coming years? In terms of the visitor experience, at least fifteen EAZA zoos now have hotels on site and this looks likely to become a future trend. Families visiting theme parks

---

4 When The Deep opened in 2002, the aquarium predicted visitor numbers of 200,000 in its first year. In fact, more than 500,000 visitors came to The Deep in the first 5 months (Garner, 2002).

5 In the UK, the retailer Marks and Spencer has set itself the goal of becoming carbon neutral by 2012.

'sustainability tours' and advice to local businesses on how to reduce their carbon footprint.

We also noted, in **Chapter 10**, that a number of zoos in the UK have now achieved ISO 14001, a challenging, but internationally recognized environmental management standard.

15.3.1 Sustainability in sourcing animals

Good zoos are striving to achieve sustainability in their animal collections, with zoos becoming a growing source of animals *for* the wild, rather than *from* the wild. Accredited zoos have already made great strides to reduce the number of animals taken from the wild, and the goal of zoos within BIAZA, EAZA, AZA, and the Australasian Regional Association of Zoological Parks and Aquaria (ARAZPA) remains the management of self-sustaining populations in captivity (Conway, 2007).

15.3.2 Sustainability in the day-to-day operation of the zoo

So what would a sustainable zoo look like? At the level of the individual zoo, we can imagine that the sustainable zoo a decade or so hence will have some, if not all, of the following characteristics:

- a policy of not sourcing animals from the wild unless from sustainable sources;
- the sourcing of food locally (within 20, or at most 50, miles), both for zoo visitors and for zoo animals;
- energy sourced renewably and locally;
- buildings that are energy-efficient, with high levels of insulation and ground storage heating or cooling systems;
- the use of rainwater and 'greywater' for toilets and for irrigation;
- discounted parking charges for hybrid cars;
- campaigns for better public transport links and provision of bicycle stands;
- greater use of local materials and wood from sustainable sources for buildings (**Fig. 15.5**);

- a zoo gift shop that sells a greater proportion of locally produced (and fair trade) goods;
- one or more recycling centres within the zoo;
- zoo signs that provide information not only about animals and their habitats, but also about sustainability.

In fact, nearly all of these goals are already reality in at least one European zoo. At Apenheul Primate Park in the Netherlands, for example, car parking is free of charge for hybrid cars. The Living Rainforest, UK, has its own biofuel boiler, which runs on locally produced (and carbon-neutral) wood chips. Artis Zoo in Amsterdam operates a sophisticated combined heat and power (CHP) system. At Edinburgh Zoo, a reed-bed system filters waste water. Cologne Zoo in Germany includes references to sustainability on its enclosure signs. Further afield, Perth Zoo in Australia has a renewable energy display, which is part of a larger environmental programme with a grid-connected photovoltaic installation and a small demonstration wind turbine.

And in 2006, BIAZA introduced an annual award for sustainability initiatives in its member zoos.

15.4 Careers in zoos

Jobs in zoos are more varied than many people first imagine. Of course, there are zookeepers and curators, but there are also zoo research officers, zoo education staff, zoo conservation officers, and other support staff.

The following is a brief overview of some of the main jobs in zoos that we have talked about in this book.

15.4.1 Zookeeper

Zookeepers nowadays are very likely to be graduates, or even postgraduates. If you want to become a zookeeper, you also need to demonstrate a real interest in—and, preferably, experience of—handling and caring for animals. This might include, for example, voluntary work at an animal shelter (**Fig. 15.6**).

Figure **15.5** This photograph shows the large and well-used Conservation Education Centre at Marwell Zoo (part of the Marwell Preservation Trust), in Hampshire, UK. The centre was built using environmentally friendly materials, such as wood from sustainable sources and low-emission paints, and makes good use of a variety of energy- and water-saving devices. (Photograph: Sheila Pankhurst)

(a)

(b)

Figure **15.6** If you want to work with animals as a zookeeper, you will need to be able to demonstrate relevant experience, such as voluntary work at an animal shelter. These photographs show (a) keeper Ben Warren with a Pallas kitten *Otocolobus manul*, at Howletts Zoo, UK; and (b) a keeper at Warsaw Zoo with a hand-reared red squirrel *Sciurus vulgaris*. (Photographs: (a) Dave Rolfe, Howletts Zoo; (b) Barbara Zaleweska, Warsaw Zoo)

(a)

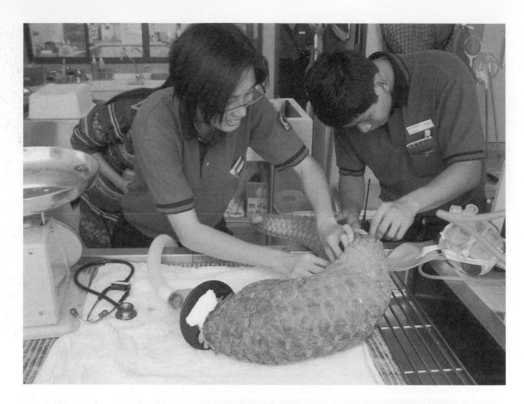

(b)

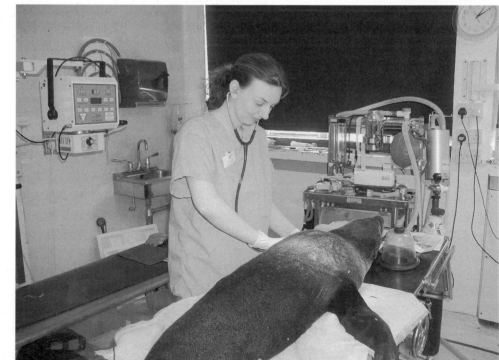

Figure **15.7** Working as a zoo vet is a challenging job: a typical medium-sized zoo may contain around a thousand animals from upwards of 300 species. Here, vets at Singapore Zoo and Bristol Zoo are examining a pangolin (a) and a fur seal pup (b), respectively. (Photographs: (a) Biswajit Guha, Singapore Zoo; (b) Mel Gage, Bristol Zoo)

The websites of keeper associations such as the Association of British Wild Animal Keepers (ABWAK) and the American Association of Zoo Keepers (AAZK) are a good place to start if you are interested in becoming a zookeeper. (The AAZK website has a useful page on zookeeping as a career at `www.aazk.org/zoo_career.php`).

15.4.2 Zoo veterinarian

Becoming a zoo vet (**Fig. 15.7**) can take a long time. Typically, students enrolling on a veterinary medicine course in the UK will study full time for either 5 or 6 years, although some colleges now offer an accelerated veterinary medicine course (usually over 4 years) for students who have already completed a BSc in the biological sciences.

An alternative degree is the BSc in Veterinary Surgery (BVS). Membership of the Royal College of Veterinary Surgeons (RCVS, which entitles the member to add the designation 'MRCVS' after their name) is a legal requirement in the UK before a holder of a BvetMEd or VetMB degree can practise as a veterinarian (see `www.rcvs.org.uk` for further details).

Of course, qualifying as a vet is only the first step towards working with wild animals in a zoo. For qualified veterinarians, the RCVS offers both a postgraduate certificate and diploma courses in animal welfare science, **ethics**, and law.

15.4.3 Zoo veterinary nurse

At the time of writing this book, there are no specific qualifications available in the UK for veterinary nurses working with wild animals in a zoo, although the Zoological Society of London (ZSL) sometimes offers summer work experience placements for veterinary nurses and trainee veterinary nurses. For further details, see `www.zsl.org`.

Summary

- Zoos today understand that it is not enough to state a commitment to conservation and education; their achievements in these areas need to be evaluated against clear objectives.

- As well as striving to become effective centres for conservation, many good zoos are committed to sustainability and want to be seen as organiza- tions that can take a lead in demonstrating sound environmental management.

- The shape of collections will continue to change, with a greater emphasis on the captive breeding of endangered species, and, in particular, species from non-mammalian and non-avian taxa.

Questions and topics for discussion

1. What are the practical obstacles that may prevent a zoo from becoming a 'fully sustainable organization'?

2. Why should zoos keep fewer mammals and more animals from other taxa?

3. How can a zoo evaluate the effectiveness of its educational programmes?

4. To what extent should good zoos be responsible for solving the problems of bad zoos?

5. Discuss how zoos can become better at evaluating the impact of their conservation initiatives, rather than only measuring inputs.

Further reading

In the previous fourteen chapters of this book, we have provided subject-specific suggestions for further reading, along with details of where to find further information on aspects of zoos ranging from legislation to enrichment. In this chapter, we want to do something rather different, and to recommend a few of the books and papers about zoos that have particularly interested, inspired, informed, and entertained us.

First and foremost, we strongly recommend Colin Tudge's thoughtful and well-written *Last Animals at the Zoo* (1992). This book has the subtitle *How Mass Extinction can be Stopped* and Tudge presents a well-argued case for the conservation role of modern zoos.

The second book that we have enjoyed reading and would like to recommend is Gerald Durrell's *The Stationary Ark* (1976): an account of the creation

Figure **15.8** This book began with a giraffe (on the front cover), so it is perhaps appropriate to end with a photograph of the newest addition (in the centre foreground) to the herd of Rothschild giraffes *Giraffa camelopardalis rothschildi* at Woburn Safari Park in Bedfordshire. In many ways, this photograph epitomizes what good zoos are about, with a group of animals of different ages, breeding successfully in captivity, and housed with access to a large outdoor enclosure. (Photograph: Jake Veasey)

of Jersey Zoo. Re-reading this book recently (it was first published more than 30 years ago), we were struck by Durrell's foresight in some of his recommendations about how a modern zoo should operate. In *The Stationary Ark*, Durrell said, for example, that zoo animals should be provided with '*an adequate diet, which is considered interesting by the animal and nutritional by you*' and '*as much freedom from boredom as possible, i.e. plenty of "furniture" in the cages*'. All of this pre-dates the **five freedoms** (Farm Animal Welfare Council, 1992) by some 15 years and the first version of the Secretary of State's Standards of Modern Zoo Practice (SSSMZP) by more than 25 years.

Douglas Adams and Mark Carwardine collaborated on the book *Last Chance to See* (1991), and a BBC radio series of the same name. *Last Chance to See* is an account of Adams' and Cawardine's attempts to see, in the wild, critically endangered species. Better known for books such as *Hitchhiker's Guide to the Galaxy*, Adams (1952–2001) was a committed environmentalist activist. He and his co-author Mark Carwardine contributed a passage from their book *Last Chance to See* to another book, *The Great Ape Project*, written by Paola Cavalieri and the ethical philosopher Peter Singer (1994). The Great Ape Project became, in 1993, a wider campaign for equal moral rights for the great apes.

Susan McCarthy's *Becoming a Tiger: How Baby Animals Learn to Live in the Wild* (2005) is a very readable account of how animals learn, liberally sprinkled with anecdotes and examples. This is a book into which you can dip for information, amusement, and entertainment.

Yann Martel's *Life of Pi* (2001) won the Mann Booker Prize in 2002. Pi, the hero of the book, is the son of a zookeeper in India. He later survives a shipwreck, only to spend 227 days at sea in the company of a Bengal tiger named Richard Parker. Unlikely, yes, but Pi's (Martel's) reflections on the animals in his father's zoo are well worth reading (many of the places described in this work of fiction are book are real, although there is no zoo in the Botanic Gardens at Pondicherry).

We would also like to point readers in the direction of a chapter entitled 'Conclusion: future of zoos' by Matthew Hatchwell and his colleagues in the book *Zoos in the 21ˢᵗ Century: Catalysts for Conservation?* (Hatchwell *et al.*, 2007). Nicole Mazur concludes her book *After the Ark?* (2001) with a thoughtful chapter on 'Sailing into unknown waters'.

And last, but by no means least, we would like to end this book by singling out both a paper and a book chapter by Bill Conway (2003; 2007), for a cogent and clear-sighted view of the role of zoos in the twenty-first century.

Glossary

a **abnormal behaviour** · unusual or rare behaviour that might be pathological

accredited zoo · a zoo that is licensed and a member of a national or regional zoo association (such as BIAZA, EAZA, AZA, ARAZPA, etc.)

activity budget · a quantitative representation of the proportion of its time that an animal spends in each of the behaviours or activities that it performs. This can be in the form of a table or a histogram (see **Fig. 4.33**) (cf. **time budget**).

acute condition · a severe or aggressive illness that often develops quite rapidly (cf. **chronic condition**)

ad libitum (or **ad lib.**) · free access; usually refers to food or water, indicating that the animal can choose to have as much as it wants

adornment · thing such as a tag, bead, ring, collar, or necklace, which is temporarily added to alter an animal's appearance so that they can be recognized

agonistic behaviour · a behaviour that occurs in situations involving conflict or contests between animals. This includes both aggressive and submissive behaviours.

agronomy · the scientific study of crop production (including the study of soils)

allele · an alternative form of a gene. In individuals from sexually reproducing organisms, one allele, or copy, of each gene is inherited from the father and one from the mother. These alternative forms of a gene are responsible for determining traits (such as hair colour), and genes for some traits can come in many different forms or alleles. A stricter definition is 'an alternative form of a gene, found at a single locus (location) on a **chromosome**'.

analgesia · the relief of pain

analgesic drug · a drug used to relieve pain

Animal Health · a UK government agency, created in 2007, which absorbed the State Veterinary Service and other bodies

animal rights · animal rights campaigners generally hold that the rights of animals are equivalent to those of humans and that it is not acceptable to sacrifice the interests of one animal to benefit another (in other words, that humans should not be allowed to exploit non-human animals for their own purposes)

Annex A species · species listed in the first of four annexes to the EU Convention on International Trade in Endangered Species of Wild Fauna and Flora (CITES) legislation. Annex A includes all CITES Appendix I species, as well as some Appendix II and III species for which the EU has adopted stricter measures than CITES.

anthelmintic (or **antihelmintic**) **drug** · a drug used to treat animals infected with parasitic gut worms, such as roundworms

anthropoid ape · this term is rather out of date now (primate **taxonomy** has moved on). It refers to the 'man-like' apes—that is, the tailless primates, such as the chimpanzee, gibbon, gorilla, and orang-utan. Current taxonomy puts the chimpanzee, gorilla, and orang-utan, but not the gibbon, into the family Hominidae, along with humans.

anthropomorphism · the attribution of human qualities to animals (when there is no biologically valid evidence to do so)

anthropozoonotic disease · disease that can be transmitted from humans to non-human animals

Appendix I species · this refers to a species listed in Appendix I of the Convention on International Trade in Endangered Species of Wild Fauna and Flora (CITES). There are four appendices to CITES; Appendix I species are considered to be the most threatened.

approved vet · in the context of zoo licensing and legislation, an approved vet (AV) is a vet listed by the relevant local or national authority on a register of vets able to undertake zoo inspections, etc.

aquaculture · the farming or cultivation of animals or plants that live in water (e.g. fish)

Article 10 certificate · a permit needed by zoos within the EU to authorize the sale or movement of animals listed under Annex A of the EU legislation implementing the Convention on International

Trade in Endangered Species of Wild Fauna and Flora (CITES)

Article 60 certificate · a permit (formerly called an 'Article 30 certificate') that allows zoos to display their Annex A species and also to move them to other zoos within the EU, provided that the other zoos also have an Article 60 certificate

artificial insemination (AI) · a reproductive technology in which semen is collected from one or more male animals prior to being inserted via a catheter into the vagina or another part of the reproductive tract of the female

aversion test · test that studies if an animal is willing to choose to undergo something unfavourable, such as an electric shock, in order to gain access to a resource

b **Balai Directive** · a European Council Directive (92/65/EEC) governing the transport of non-domestic animals by EU member States (*balai* is the French word for broom—this Directive 'sweeps up' animals not covered by other EU legislation)

barren enclosure · enclosure with few, if any, objects or furnishings

basal metabolic rate · the minimum amount of energy usage required to support life processes (cf. **metabolic rate**)

behavioural competence · the ability of an animal to express appropriate behaviour in a given situation

behavioural diversity · a measure of the number and variety of behaviours that a particular individual shows

behavioural ecology · an approach to studying animal behaviour that concentrates on how and why particular behaviours have evolved, and how they influence the individual's fitness

behavioural economic theory · in the context of this book, this is where principles of economic theory, such as elasticity of demand, are applied to animal behaviour

behavioural engineering (or **behavioural enrichment**) · a method of increasing animal activity by requiring them to perform a task for a reward

behavioural husbandry · changes in the captive animal's environment that will impact on its behaviour, e.g. **enrichment** and training

behavioural restriction · the inability of animals to express their full repertoire of behaviours

behaviourism · a scientific discipline that viewed mental events as unknowable and, in its most extreme form, even denied that they existed. This led behaviourists to study only observable behaviour, so that their view of animals was effectively of automata whose behaviour was shaped by **stimuli** through a process of operant conditioning.

Berne Convention · the Convention on the Conservation of European Wildlife and Natural Habitats 1979, which came into force in 1982

binomial name (or **binomial nomenclature**) · the two-part scientific name for a species, comprising the names of the genus and the species. The genus name always starts with a capital letter, and the species name always starts with a lower-case letter. Both parts of the name are italicized (or underlined if handwritten)—e.g. the binomial name of a lion is *Panthera leo*. The binomial system of naming organisms was developed in the eighteenth century by Carolus Linnaeus.

bioavailability · the proportion of a nutrient that is available for physiological functions within the body. The bioavailability of some minerals can be reduced by antagonistic effects with other minerals, where the presence of one blocks the uptake or utilization of another.

biopark (or **BioPark**) · a zoo exhibit in which the animal is shown within a portrayal of its ecosystem (e.g. in a rainforest setting, or a desert setting)

biosecurity · steps to reduce the incidence and spread of disease

birth rate · *See* **fertility**

body condition scoring · a subjective method of measuring or scoring the general body condition of an animal, as an indication of its nutritional status and energetic state. This can be done visually, or by handling the animal (e.g. by feeling the amount of fat over the vertebral processes of the spine).

bovine spongiform encephalopathy (BSE) · a form of **transmissible spongiform encephalopathy (TSE)** that causes a degeneration in the brain and spinal cord of cattle

branding · the marking of an animal's skin through the application of extreme heat or cold

breed and cull · a management strategy within which animals are allowed to breed, but the group size is kept a constant size by the selective culling of individuals with the group

C caecum · the large diverticulum (a sac with just one opening) in the gastrointestinal tract, at the point where the small intestine becomes the large intestine. All that is left of it in humans is the sometimes troublesome appendix.

cafeteria-style feeding (CSF) · a variety of food-stuffs are offered (e.g. a mixture of fresh fruits and vegetables) and animals are allowed to choose what they consume (cf. **complete feed-style feeding**)

callitrichid · a species of small New World monkey. The family *Callitrichidae*, or more correctly now, the sub-family *Callitrichinae*, includes the marmosets and tamarins.

captive breeding programme · the planned and managed breeding in captivity of a threatened species, involving cooperation between zoos within a country or region. Examples of captive breeding programmes include **European Endangered species Programmes (EEPs)** in Europe and **Species Survival Plans (SSPs)** in North America.

capture myopathy · the build-up of lactic acid in muscles during and after capture, which can cause stiffness, paralysis, and even death. Capture myopathy has been reported in a wide range of **taxa** and can, in some cases, cause death almost immediately, or up to a few weeks later.

carnivore · an animal that eats other animals (e.g. the tiger, the shark, the eagle) (cf. **herbivore**; **omnivore**)

Carnivore · a member of an order of mammals, the order Carnivora. This includes meat-eating animals such as lions and tigers, but also plant-eating

giant pandas and omnivorous badgers. (Note the capitalization.)

cartilaginous fish · a fish with a skeleton made of car-tilage rather than bone, of which the **Elasmobranchii** is a sub-class including the skate, the ray, and the shark (cf. **teleost**)

causal (or **proximate**) · explanations of behaviour that attempt to identify its immediate causes, e.g. **stimuli**, motivational changes, hormone levels (cf. **functional**)

cellulose · a carbohydrate macromolecule used extensively by plants in their structure (e.g. as cell walls)

chaining (or **picketing**) · a method of restraining elephants, where one front and the diagonal back leg are chained, so that the elephant is able to move approximately one step forward and back

charismatic megafauna · the animals—nearly always mammals—that have strong popular appeal, such as tigers, pandas, wolves, and elephants. These animals are often (but not always correctly) assumed to be the species that zoo visitors most want to see.

chelonid · a member of the reptilian order Chelonia, the tortoises and turtles

chromosome · a length of DNA that contains genes and other elements. In animals that have cells with nuclei, the chromosomes are paired structures located in the cell nucleus (in humans, for example, there are normally forty-six chromosomes in each body cell, arranged as twenty-three pairs).

chronic condition · a condition that persists over a long period of time and which may require long-term management (e.g. diabetes) (cf. **acute condition**)

chronic wasting disease (CWD) · a form of **trans-missible spongiform encephalopathy (TSE)** that causes small lesions in the brains of infected deer species, such as moose and elk

classical conditioning (or **Pavlovian conditioning**) · a form of **learning** in which the animal learns to associate an existing **response** with a new **stimulus** (e.g. showing food-related behaviours not to food itself, but to the sound of it being prepared) (cf. **instrumental conditioning**)

clinical · of the examination and treatment of sick animals

clinical sign · the visible sign of a possible disease or injury, such as a rash or a swelling

coccidian parasite · a parasite that is protozoan (a small unicellular animal)

cognition · refers to the processes by which animals perceive, process, and store information. Cognition, broadly speaking, includes all of the ways in which animals take in information through the senses, process and retain this information, and act on it.

communicable disease · a disease that can usually (but not always) be passed on, or 'communicated', between animals or humans

complete feed-style feeding · the provision of a processed diet that meets the nutrient requirements of the animal (but does not offer any choice) (cf. **cafeteria-style feeding**)

confounding variable · in studies of animal behaviour, a variable or factor that affects the results of a study, but which cannot be controlled (e.g. the weather, or an unexpected outbreak of disease)

consciousness · the awareness of one's own mental and/or physical actions. Dawkins (2006) offers the definition that '*consciousness refers to a wide range of states in which there is an immediate awareness of thought, memory or sensation*'.

conspecific · another animal of the same species

contrafreeloading · animals actively working for a resource in their enclosure, even when it is freely available

coprophagy · the eating of faeces

corticosteroids · a group of steroid hormones produced by the adrenal cortex (e.g. cortisol)

cortisol · a **corticosteroid** hormone produced by the adrenal cortex and involved in the body's **response** to stress (measurements of circulating cortisol levels are sometimes used as a measure of stress)

d **demography** · the characteristics of a population, such as its age or sex structure, birth and death rates, and so on

dietary drift · where diets are changed incrementally over time by keepers (not always for the better)

digestion · a process within which macromolecules that animals ingest as food are broken down into smaller molecules inside the gastrointestinal (GI) tract

Disinfectant Era · the period from the 1920s onwards when zoo enclosures were designed primarily for ease of cleaning, rather than with regard to the needs of the animals housed within them. This trend for sterile, often minimalist, cages with concrete floors and tiled walls persisted well into the 1960s and 1970s.

displacement behaviour · behaviour that seems to be irrelevant to the **stimulus** situation in which it is performed

dolphinarium · a facility that, strictly speaking, houses only dolphin species, which include the orca, or killer whale, *Orcinus orca* (cf. **oceanarium**)

double-clutching · a process whereby some, or all, of the eggs laid in one clutch are removed and incubated artificially, in the hope that the animal will lay another clutch and thus double its reproductive output

dry matter · the weight of a quantity of a foodstuff without the water content

e **ecosystem exhibit** (or **ecosystem zoo**) · an exhibit or zoo that is designed to represent an entire ecosystem or ecosystems, rather than a single species or a group of animals displayed on **taxonomic** or regional lines (used interchangeably with the term **biopark** by some authors)

effective population · the number of animals that actually breed and contribute offspring to the population

elasmobranch · *See* **cartilaginous fish**

embryo transfer · in which embryos from wild mammals are transferred successfully into a different species for gestation

empirical study · a study based on observation rather than on what has been reported in the literature (data may be obtained from observing animal behaviour, for example, or from observing the results of an experiment)

endogenous stimuli · those **stimuli** that arise from within the animal (cf. **exogenous stimuli**)

endoparasite · a parasite that lives within the body of its host (e.g. a tapeworm) (cf. **exoparasite**)

enrichment · any change to an animal's environment that is implemented to improve the animal's physical fitness and mental well-being (cf. **environmental enrichment**)

environmental challenge · the presence or absence of properties within the environment that may act to overstimulate or understimulate the animal

environmental enrichment · the provision of species-specific opportunities within an animal's environment to enable it to express a diversity of desirable behaviours

epidemiology · the study of the incidence, prevalence, and treatment of disease in the population

ethics · a narrower concept than morality, although the words are often used interchangeably. Ethics, strictly speaking, refers to a branch of philosophy that tries to clarify and analyse the arguments that people use when discussing moral questions.

ethogram · a list, with descriptions, of the behaviours observed in a particular species or group of animals

euphagia · See **nutritional wisdom**

European Endangered species Programme (EEP) · a planned and managed programme for breeding threatened species in settings such as zoos. In Europe, this is achieved via European Endangered species Programmes (EEPs); in North America, it is achieved via Species Survival Plans (SSPs).

euthanasia · the humane killing of an animal if it is sick, for example, or badly injured

evolutionarily significant unit (ESU) · a population that, for conservation purposes, is considered to be distinct—that is, the minimum unit for conservation management. The usefulness of the ESU concept is that it bypasses difficulties over defining species.

exogenous stimuli · those **stimuli** that arise in the environment outside the animal (cf. **endogenous stimuli**)

exoparasite · a parasite that lives on the body of its host (e.g. a flea or a mite) (cf. **endoparasite**)

exotic animal · a very variable term, but generally used to mean a wild, rather than a domesticated, animal (so cats, dogs, and dairy cattle are not exotics). Whether or not an animal is exotic or not depends on your viewpoint and where you are in the world: camels may be regarded as exotics in the UK, but not in Egypt.

ex situ conservation · the conservation of animals outside their natural habitat, e.g. in zoos (cf. **in situ conservation**)

f **F2** · a second-generation zoo animal—that is, a zoo animal that was born in captivity

fecundity · a measure of reproductive potential (often measured as the number of gametes produced)

fertility · the actual reproductive performance of an organism, usually measured as the number of viable offspring. (The term **birth rate** means much the same thing as fertility and the two terms are sometimes used interchangeably.)

fitness · an indication of how well an animal is adapted to its environment. Fitness is usually measured in terms of the number of offspring of an individual animal that survive to reproductive age.

five freedoms · the standards of animal welfare developed by the UK's Farm Animal Welfare Council (1992). Freedom 1, for example, is freedom from hunger and thirst; Freedom 3 is freedom from pain, injury, or disease.

five principles · five basic principles of animal care and management, which form the basis of the standards set out in the UK Secretary of State's Standards of Modern Zoo Practice (SSSMZP). The 'five principles' are derived from the **five freedoms** of animal welfare.

flagship species · a species that can excite public attention, and can therefore also raise public support and perhaps generate funds

flehmen · a behaviour shown by some mammals in which olfactory **stimuli** are detected by taking air into the mouth and, from there, into a specialized receptor—the vomeronasal organ—in the roof of the mouth. The behaviour is usually accompanied by the characteristic facial expression of mouth slightly open and upper lip retracted.

flight distance · the distance at which an animal will not allow an aversive **stimulus** (e.g. a person) to come any closer, usually the point at which the animal will flee

flight restraint (or **wing management**) · physical modifications to a bird's wing that prevent it from flying

fomite · an inanimate object that can spread disease from one individual to another. In zoos, a fomite might be a contaminated bucket or food bowl, or a keeper's boots.

forage · a general term for any kind of fodder made out of herbaceous plants, such as hay, straw, silage, etc.

formal learning · **learning** from structured experiences, such as keeper talks, demonstrations, or short courses (cf. **informal learning**)

founders · individuals from which populations derive

free-ranging · a term used in a variety of ways, but usually to imply that the animals are not restricted in where they go, although, in reality, there will be restrictions at some point

functional (or **ultimate**) · explanations of behaviour that attempt to identify why and how the behaviour has evolved (cf. **causal**)

g **gait scoring** · the subjective scoring of measures such as the rate, direction, and range of movement of an animal, as well as the strength or force of its gait

gallinaceous · (meaning 'chicken-like') birds such as pheasants, partridges, grouse, quails, etc., from the order Galliformes. They are usually ground-feeding birds and many species are hunted as game birds.

gene pool · all of the genes in a given population of a species

genetic diversity · broadly, the number of different genetic characteristics in a population. These genetic characteristics can be the genes, or the chromosomes, or the entire genome (or even the nucleotides that make up genes).

genotype · the genetic constitution of an individual (cf. **phenotype**)

giraffid · a member of the family Giraffidae, the extant members of which comprise the giraffes and the okapi

gut microflora · a general name for the microorganisms that inhabit the digestive tract

h **habituation** · a form of **learning** in which the animal reduces its **response** to a constant or repetitive **stimulus**

ha-ha · a ditch separating two areas, such as an animal enclosure from a visitor area

haemochromatosis · iron-storage disease

haemosiderosis · increased iron storage in the body tissues (benign, rather than disease-causing)

helminth · a species belonging to a range of genera of multicellular parasitic worms that inhabit the gastrointestinal (GI) tract and sometimes other parts of the body. Helminths include: trematodes, or flukes; cestodes, or tapeworms; and nematodes, or roundworms.

herbivore · an animal that eats plants (cf. **carnivore**; **omnivore**)

herpetology · the study of reptiles and amphibians (known generally as **herps**)

herps · a general term used to describe reptiles and amphibians

heterozygous · having two different **alleles** of a gene (cf. **homozygous**)

hispid · meaning covered with stiff, bristly hairs

holism · an ethical viewpoint within which the whole ecosystem—including inanimate objects such as rocks, water, and soil—must be taken into account when debating moral considerations

homeostasis · the processes that maintain the animal's body in a stable state (or state of balance) so that it can carry out necessary survival functions

homozygous · having two identical copies or alleles of a gene (cf. **heterozygous**)

hormone assay · a method of determining the presence and/or amount of a hormone in a given sample

husbandry guidelines (or **husbandry manuals**) · guidance on the day-to-day care of an animal or

species covering, for example, diet, space requirements, etc.

Hygiene Era · *See* **Disinfectant Era**

**hypothalamo-pituitary-adrenal/gonadal
(HPA/HPG) axis** · the hormonal regulatory system
that is common to the reproductive physiology of all
vertebrates

i **immersion exhibit** (or **landscape immersion**) ·
an exhibit within which landscaping includes
the visitor area, so visitors get the sensation that they
are in the animals' habitat (e.g. 'walking through a
rainforest')

imprinting · a form of **learning** in which young
animals learn about their species identity, sex identity, or relatedness to other individuals

inbreeding · referring to when closely related animals
breed with each other

inbreeding depression · a situation in which
inbreeding results in offspring with reduced genetic diversity, because both parents have provided
similar genetic material

induced ovulator · an animal in which the release
of an egg or ovum by the female is stimulated by the
act of copulation with a male

informal learning · the acquisition of knowledge
without formal instruction, e.g. by reading zoo signs
(cf. **formal learning**)

in situ conservation · the conservation of animals
within their natural habitat—that is, 'in the wild'
(cf. **ex situ conservation**)

instrumental conditioning (or **operant conditioning**)
· a form of **learning** in which the animal acquires a
new **response** to an existing **stimulus** (e.g. pressing a
lever in order to get food) (cf. **classical conditioning**)

integrated conservation · the bringing together of
different activities and different agencies in order to
achieve conservation priorities

invasive procedure · a veterinary (medical) procedure that 'invades' the body, such as blood sampling

in vitro fertilization (IVF) · the production of an
embryo in culture in the laboratory or clinic (cf. **in
vivo fertilization**)

in vivo fertilization · the production of an embryo
inside the mother's body (in the case of mammals
and birds) (cf. **in vitro fertilization**)

isozoonosis · a disease that shows very similar pathology in both humans and animals

l **lamellibranch** · a bivalve mollusc belonging
to the class Lamellibranchia, e.g. scallops and
clams

landscape immersion · *See* **immersion exhibits**

latency · the time gap between a **stimulus** or event
and a **response** occurring—e.g. if a novel object
is placed in an animal's cage and the animal first
approaches the object after 3 minutes, the latency to
'approach novel object' is 3 minutes

learning · the process that takes place when an
experience of some sort brings about a relatively
permanent change in the way in which an animal
responds to a situation

life history traits · the characteristics of a species,
such as age at weaning or fledging, age at first reproduction, number and size of offspring, reproductive
lifespan, adult survival, and ageing. In a bird species,
for example, clutch size and re-nesting rate after clutch
failure would both be examples of life history traits.

longevity · a measurement of the length of time
for which an animal lives (also a useful indicator of
how well an animal has been able to survive and
thus succeed)

m **maintenance rations** · referring to when the
amount of energy entering the animal is
equal to the amount leaving the animal

malnutrition · not necessarily a shortage of food per
se, but a deficiency or deficiencies in **nutrients** in
the diet

menagerie · this comes from the French word
ménage, meaning a household or unit of people
living together. The word *ménagerie* was used from
the sixteenth century onwards in France to refer to
the management of a farm or collection of livestock
and gradually came into use to refer to a collection
of exotic animals.

metabolic rate · the amount of **metabolizable energy** expended by an animal over a given period of time—i.e. energy metabolism per unit time

metabolizable energy (ME) · the energy content of the food taken in by an animal; a measure of the energy available to fulfil the animal's metabolic requirements

metapopulation · a collection or network of sub-populations that are connected in some way. The links between the sub-populations are usually immigration and emigration from one sub-population to another. Thus, the groups of animals housed in different zoos can be regarded as sub-populations of a larger metapopulation, which also includes the wild population, if there is transfer of animals between sub-populations.

mixed-species exhibit · an exhibit housing more than one species

monkey chow · food concentrate designed to provide a complete and balanced diet for monkeys

monogamy · a pairing between one male and one female, particularly in those species among which male desertion is not profitable, because females are unable to raise the young alone (e.g. many birds). A distinction needs to be made between social monogamy, and genetic monogamy (the male in a social pair may not always be the father of some, or all, of the female's offspring) (cf. **polygamy**; **promiscuity**)

monogastric · a non-**ruminant** animal with a single-chambered stomach (e.g. bear, human)

morbidity · the incidence of disease, or the number of animals that will become sick

mortality · the death rate from a disease

n **naturalistic enclosure** · an enclosure that attempts to replicate the animals' natural habitat through the appropriate use of land formations and plant life

necropsy · a **post-mortem** examination to try to determine the cause of death

negative reinforcer · a **reinforcer** that involves removal of a **stimulus**, such as the turning off of an electric shock (cf. **positive reinforcer**)

neophila · the attraction to novel objects or new situations (cf. **neophobia**)

neophobia · the avoidance of novel objects or situations (cf. **neophila**)

notching · the making of carefully placed holes or marks in the animal's horn, shell, scales, or ears to act as an identifier

notifiable disease · a disease that must be reported to the relevant authorities. In the UK, a notifiable disease must be reported to the police; in practice, Defra should also be notified. Examples in the UK include rabies, anthrax, and foot-and-mouth disease (FMD).

nutrient · a substance needed for the survival and growth of an organism

nutrition · a process that occurs within organisms of taking in and absorbing **nutrients**, to provide energy and to meet other metabolic needs

nutritional requirements · the particular elements or compounds that an animal must have in its diet in order to satisfy all of its requirements for growth, and for building and maintaining its metabolic machinery

nutritional wisdom (or **euphagia**) · the idea that animals can select a balanced diet by detecting the nutritional quality of the different foods that they encounter

o **obstruction test** · a test within which an obstruction, such as a door or steep ramp, is placed between the animal and a resource in order to determine how hard the animal is willing to work to get around the obstacle to gain the resource

oceanarium · a marine animal park housing species such as seals, sea lions, dolphins, etc., and sometimes also fish, corals, and other marine life (cf. **dolphinarium**)

omnivore · an animal that eats plants and animals (cf. **carnivore**; **herbivore**)

operant conditioning · *See* **instrumental conditioning**

organic · refers to carbon-based compounds, which usually also contain oxygen and hydrogen. These compounds are not necessarily the result of living processes. (The term can also refer to an agricultural system.)

oviparous · meaning egg-laying (cf. **viviparous**)

P **pachyderm** · a non-**ruminant** hoofed animal with a very thick skin, such as a rhinoceros, hippopotamus, or an elephant. As a **taxonomic** term, the word is no longer used.

passive integrated transponder (PIT) · a microchip containing a unique magnetic code, which is usually injected just beneath an animal's skin and is used to code its identity

pathogen · disease-causing agents (from '*patho*' meaning disease), such as bacteria, viruses, fungi, or other organisms such as protozoan parasites. The **prions** that cause **bovine spongiform encephalopathy (BSE)** and other **transmissible spongiform encephalopathies** are pathogenic proteins.

pathology · the study of disease

Pavlovian conditioning · *See* **classical conditioning**

perseveration · an animal's carrying on of an activity in the absence of the appropriate **stimulus**

pest · a species (plant or animal) that occurs where it is unwanted

petting zoo · an area that usually contains domesticated or habituated animals, with which visitors are allowed to come into close contact

phenotype · the product of interaction between an individual's **genotype** and the environment in which it develops; this effectively refers to what the individual looks like and how it behaves

pica · in humans, often used to refer to the consumption of non-food items—from coal to soap—but can also mean cravings for particular food constituents

picketing · *See* **chaining**

pinioning · a form of **flight restraint** involving the amputation of a digit at the wing tip

plane of nutrition · level of nutrition. An animal with a high plane of nutrition has ready access to a high-quality diet, in which food is not in short supply. In the zoo situation, this can sometimes lead to problems such as obesity.

plant secondary metabolite (PSM) · a substance that acts to make a plant distasteful or even toxic, or which may make it difficult for the animal to extract the important **nutrients**. PSMs are so-called because they are synthesized from the products of ordinary or primary metabolism.

polyandry · a mating system—unusual in mammals—in which a female has several male mates or partners in a breeding season. Examples of species among which polyandry is found include the jacanas (a group of wading birds) and several frog species (cf. **polygyny**).

polygamy · a general term for mating systems in which one individual of one sex mates with several individuals of the other sex. It thus includes both **polygyny** and **polyandry** (cf. **monogamy**; **promiscuity**)

polygyny · the mating of one male with many females. It occurs particularly in species among which females are able to raise young with little male input (e.g. many mammals, in which gestation and lactation facilitate male desertion)

population genetics · the study of the change in frequency of **alleles** in a population over time, as a result of processes such as natural selection, genetic drift, mutation and gene flow

positive reinforcement training (PRT) · the use of operant conditioning techniques in which animals are rewarded for exhibiting the behaviours that the trainer wants them to

positive reinforcer · a reward that follows a required **response** (e.g. the provision of food) (cf. **negative reinforcer**)

post-mortem · the examination of a dead animal, or tissues from a dead animal (literally meaning 'after death'). Often used informally to mean **necropsy**.

post-occupancy evaluation (POE) · a technique used to assess how animals use their environment (e.g. their enclosure space)

pre-feeding anticipation (PFA) · the behavioural expression observed prior to an animal being given food, which can occur due to a variety of signals, including circadian cues, visual **stimuli**, or odour

preference test · a test that attempts to measure objectively an animal's preferences (e.g. for different foodstuffs)

preventive (or **preventative**) **medicine** · a branch of veterinary (or human) medicine that tries to prevent disease from occurring. In zoos, preventive medicine includes measures such as regular health checks, quarantine procedures, good enclosure hygiene, etc.

prion · a disease-causing agent or **pathogen** that is not a micro-organism, but a protein. Prions do not contain any genetic material.

promiscuity · a mating system in which both males and females have multiple sexual partners (cf. **monogamy**; **polygamy**)

protozoan · unicellular organism, which includes free-living, as well as parasitic, forms

proximate · *See* **causal**

pseudoreplication · the term usually used to describe data among which the sample points are not independent

psittacine · refers to members of the avian order Psittaciformes, such as parrots, macaws and lories

r andomization test · also called a permutation test, this calculates a test statistic and compares it with the results from repeated randomization of the same data across all groups

regurgitation/reingestion · behaviour shown in some captive gorillas where food is voluntarily brought back up from the stomach to the mouth, or on to a substrate, and is then reswallowed

reinforcer · any event that increases the probability of a particular **response** (*see* **positive reinforcer**)

reintroduction · the intentional movement of an organism into a part of its native range from which it disappeared or became extirpated in historic times as a result of human activity or natural catastrophe

response · the behaviour that an animal shows when stimulated by an event (**stimulus**) in its environment

reversed lighting schedule · a schedule that maintains illuminated conditions during natural darkness and darkness during natural daytime, used to exhibit nocturnal species (those that are active during darkness)

rumen · the first (and usually largest) of the stomach chambers of **ruminant** animals and a primary site for the microbial fermentation of ingested foodstuffs

ruminant · a hoofed animal that 'chews the cud' —that is, regurgitates partially digested food. The stomach of ruminants is chambered.

s catter feed · feeding technique in which the animals' daily food ration, or part of it, is spread out around the enclosure

scientific name · *See* **binomial name**

self-awareness · the knowledge of oneself as an individual, distinct from other individuals. (Along with **theory of mind**, self-awareness is regarded as a higher order cognitive ability and the extent to which animals possess these abilities is controversial.)

self-directed behaviour · behaviour that the animal directs towards its own body, such as scratching and body-shaking

self-injurious behaviour · behaviour that damages one's own body, e.g. biting oneself; some authors include excessive hair or feather plucking within this term

self-medication · in the context of wild animals, whether in the zoo or in their natural habitat, the deliberate ingestion of plants or other material (e.g. soil or minerals), apparently to alleviate the symptoms of disease

sentience · the ability to perceive and respond to the external environment; an alert cognitive state (sentience is often used to imply **consciousness**, but the ability to detect and respond to **stimuli** does not necessarily require consciousness). One of the best and simplest definitions of sentience is having '*feelings that matter*' (Webster, 2006) and the word is often used in connection with animals' ability to perceive pain and to suffer.

signalled predictability · referring to where a signal precedes an event and thus makes it predictable

socialization · a process whereby animals routinely interact with people and become familiar with them. A good simple definition is 'learning to get along with' people.

speciesism · a term coined by Australian philosopher Peter Singer to mean the preferential consideration for human beings over non-human animals

Species Survival Plan (SSP) · a captive breeding programme in North American zoos

species-typical behaviours · the repertoire of behaviours that characterize how a particular species behaves in the wild

spread of participation index (SPI) · a technique that compares the observed frequency of use of different enclosure sectors with the frequency that we would expect, given their relative sizes, thus giving an index of the animal's use of the enclosure

stereotypy · a repetitive, invariant behaviour, which may be the result of frustration, attempts to cope with a suboptimal environment, or a dysfunction of the central nervous system

sternal recumbancy · a mammalian posture involving laying down with the sternum or breast bone on the ground, usually with the legs folded under the body, but with the body still in an upright position

stimulus (pl. **stimuli**) · a change in an animal's environment that may result in it changing its behaviour in an appropriate way. Stimuli can be features in the physical environment (such as light, temperature, sound, etc.), or in the internal environment (e.g. changes in hormone levels), or signals from other animals (e.g. postures, displays, pheromones).

stimulus enhancement · a form of **learning** in which an animal learns an association as a result of having its attention drawn to the relevant **stimuli** by another animal

stocking density · the number of animals maintained in a particular area or volume of space

stockmanship · the skill of managing and caring for animals

strength of motivation test · a type of preference test in which it is assumed that an animal's willingness to 'work' for a resource reflects its level of motivation to gain the resource or to carry out the behaviour

studbook · a computerized database of all of the individual animals in a breeding programme for a given species (usually a threatened species). The aim of studbook management is to reduce inbreeding.

sustainability · the long-term viability of ecosystems, or systems within agriculture, or forestry, or fishing, etc. The term is used to refer to systems that can continue at a set level indefinitely without collapsing. In 1987, the World Commission on Environment and Development developed the following definition (published in the Brundtland Report): '*Sustainable development meets the needs of the present without compromising the ability of future generations to meet their own needs.*'

symbiosis · the close structural and physiological relationship between two or more species, which could not survive independently of the relationship

t **taxon** (pl. **taxa**) · any unit of **taxonomy** or classification, so the term can variously refer to a species, a genus, an order, or any other classification category

Taxon Advisory Group (TAG) · a group comprising keepers, curators, **studbook** managers, etc. with specialist knowledge of a particular **taxon** (e.g. the penguin TAG, or the amphibian TAG). TAGs consider conservation issues, as well as husbandry.

taxonomy · the system of classifying plants and animals

teleost · a bony fish (cf. **cartilaginous fish**; **elasmobranch**)

temporal predictability · referring to whether an event is provided at a fixed time (and thus is predictable), or at different times (so is unpredictable)

testiconid · a mammal in which the testes are located within the abdominal cavity rather than in a scrotal pouch (scrotum)

theory of mind · a theory that refers to an individual's knowledge of another individual's mental processes

tonic immobility · a natural state of paralysis that occurs in some species (e.g. fish, frogs, lizards, birds, rats, and rabbits; see Maser and Gallup, 1974) when they are threatened by an aversive stressor

touch station · a place, particularly in aquariums, where visitors can come into close contact with animals

trace element · element that is found in the body, often as part of an enzyme, and which appears to be necessary for metabolic function, but only in tiny amounts

transmissible spongiform encephalopathies (TSEs) · diseases caused by **prions**, such as **bovine spongiform encephalopathy (BSE)** and **chronic wasting disease (CWD)**

U

ultimate · *See* **functional**

ultrasonography · in relation to medicine, ultrasonography is diagnostic imaging technology that makes use of high-frequency sound waves or **ultrasound** to create an image of an internal part of the body

ultrasound · the sound of a higher frequency than that which is detectable by the human ear and therefore above about 20 kHz (or 20,000 cycles per second)

ungulate · a term that refers to the members of the former taxonomic group *Order Ungulata*. The ungulates are the hoofed mammals, which are now grouped into a number of orders including the *Order Perissodactyla* (the odd-toed ungulates such as horses, tapirs, and rhinoceroses) and the *Order Artiodactyla* (the even-toed ungulates, such as cattle, deer, and giraffes).

utilitarian · a person who believes that the value of something depends on its utility (in other words, how useful it is), meaning that, for the utilitarian, some suffering by animals is justified if the benefit to humans outweighs the cost to the animals (e.g. the use of animal testing during the development of a cure for cancer)

utilitarianism · the idea that we should act so as to produce the greatest good (or utility) for the largest number of individuals

V

vacuum activity · a behaviour that is carried out even if the appropriate **stimuli** are not present

venipuncture · blood sampling, with a needle and syringe, from a vein

viviparous · meaning giving birth to live young (cf. **oviparous**)

W

walk-through exhibit · an exhibit within which visitors and animals may not be separated by barriers

wasting disease (or **wasting syndrome**) · illness as a result of which an animal shows progressive weight loss over a period of time

welfare audit · a process of systematic review of animal records and husbandry parameters, with the aim of identifying whether change is necessary to promote better welfare

'willingness to pay' principle · a principle allowing consumers to promote animal welfare by making informed purchases and by paying more for animals farmed under 'improved' systems

wing management · *See* **flight restraint**

Z

zoo · one definition of a zoo would be a collection of (mostly) wild animals, of different species, that is open to the public for a significant part of the year. Definitions of zoos vary widely, both in relation to national law and in common usage.

zoonosis (pl. **zoonoses**) · a disease that can be transmitted from non-human animals to humans

Zootrition™ · a nutritional software package and database

Other useful glossaries

The glossary provided here is not exhaustive. For other useful glossaries, we recommend the following.

- **Glossary in Alcock (2005)** *Animal Behavior*
  The glossary at the end of this book is primarily to do with animal behaviour, but it covers some of the basic terms relating to mating systems and reproduction.

- The National Human Genome Research Institute (NHGRI) 'Talking Glossary of Genetic Terms', available online at www.genome.gov/10002096 A glossary designed to help those without scientific backgrounds to understand genetic research terms and concepts.

- The Smithsonian National Zoological Park 'Great Apes & Other Primates' glossary, available online at http://nationalzoo.si.edu/Animals/Primates/glossary.cfm

References

a **AATA (Animal Transport Association)** (2007) *AATA Manual for the Transportation of Live Animals* (2nd edn), Houston, TX: AATA.

AAZK (American Association of Zoo Keepers) (2004) *Enrichment Notebook* (CD-ROM format) (3rd edn), Topeka, KS: AAZK.

AAZK (American Association of Zoo Keepers) (2005) *Zoonotic Diseases* (CD-ROM format) (3rd edn), Topeka, KS: AAZK.

Abello, M. T., Colell, M., and Martin, M. (2007) 'Integration of one hand-reared cherry-crowned mangabey *Cercocebus torquatus torquatus* and two hand-reared drills *Mandrillus leucophaeus* into their respective family groups at Barcelona Zoo', *International Zoo Yearbook*, 41: 156–65.

Abeyesinghe, S. M., Nicol, C. J., Hartnell, S. J., and Wathes, C. M. (2005) 'Can domestic fowl, *Gallus gallus domesticus*, show self-control?', *Animal Behaviour*, 70: 1–11.

Adams, D. and Carwardine, M. (1991) *Last Chance to See*, London: Pan Books.

Addessi, E., Stammati, M., Sabbatini, G., and Visalberghi, E. (2005) 'How tufted capuchin monkeys (*Cebus apella*) rank monkey chow in relation to other foods', *Animal Welfare*, 14: 215–22.

Adelman, L. M., Falk, J. H., and James, S. (2000) 'Impact of National Aquarium in Baltimore on visitors' conservation attitudes, behavior and knowledge', *Curator*, 43: 33–61.

Adey, W. and Loveland, K. (2007) *Dynamic Aquaria: Building Living Ecosystems*, San Diego, CA: Academic Press.

Alcock, J. (2005) *Animal Behavior: An Evolutionary Approach* (8th edn), Sunderland, MA: Sinauer Associates.

Alford, P. L., Bloomsmith, M. A., Keeling, M. E., and Beck, T. F. (1995) 'Wounding aggression during the formation and maintenance of captive, multimale chimpanzee groups', *Zoo Biology*, 14: 347–59.

Allchurch, A. F. (2003) 'Yersiniosis in all taxa' in M. E. Fowler and R. E. Miller (eds), *Zoo and Wild Animal Medicine* (5th edn), Philadelphia, PA: Saunders (Elsevier), pp. 724–27.

Allen, M. E. and Ullrey, D. E. (2004) 'Relationships among nutrition and reproduction and relevance for wild animals', *Zoo Biology*, 23: 475–87.

Altman, J. D. (1998) 'Animal activity and visitor learning at the zoo', *Anthrozoös*, 11: 12–21.

Altmann, J. (1974) 'Observational study of behaviour: sampling methods', *Behaviour*, 49: 227–67.

Amato, G., Wharton, D., Zainuddin, Z. Z., and Powell, J. R. (1995) 'Assessment of conservation units for the Sumatran rhinoceros (*Dicerorhinus sumatrensis*)', *Zoo Biology*, 14: 395–402.

Ammerman, C. B. and Goodrich, R. D. (1983) 'Advances in mineral nutrition in ruminants', *Journal of Animal Science (Supplement)*, 57: 519–33.

Amosin, A., Payungporn, S., Theamboonlers, A., Thanawongnuwech, R., Suradhat, S., Pariyothorn, N., Tantilertcharoen, R., Damrongwantanapokin, S., Buranathai, C., Chaisingh, A., Songserm., T., and Poovorawan, Y. (2006) 'Genetic characterization of H5N1 influenza A viruses isolated from zoo tigers in Thailand', *Virology*, 344: 480–91.

Andereck, K. L. and Caldwell, L. L. (1994) 'Variable selection in tourism market segmentation models', *Journal of Travel Research*, 33: 40–6.

Anderson, J. and Chamove, A. (1984) 'Allowing captive primates to forage' in Universities Federation for Animal Welfare (UFAW), *Standards in Laboratory Animal Management*, Wheathampsted: UFAW, pp. 253–356.

Anderson, U. S., Benne, M., Bloomsmith, M., and Maple, T. (2002) 'Retreat space and human visitor density moderate undesirable behaviour in petting zoo animals', *Journal of Applied Animal Welfare Science*, 5: 125–37.

Anderson, U. S., Kelling, A. S., Pressley-Keough, R., Bloomsmith, M. A., and Maple, T. L. (2003) 'Enhancing the zoo visitor's experience by public animal training and oral interpretation at an otter exhibit', *Environment and Behaviour*, 35: 826–41.

Anziani, O., Zimmermann, G., Guglielmone, A., Forchieri, M., and Volpogni, M. (2000) 'Evaluation of insecticide ear tags containing ethion for control of pyrethroid-resistant *Haematobia irritans* (L.) on dairy cattle', *Veterinary Parasitology*, 91: 147–51.

Appleby, M. C. (1999) *What Should We Do About Animal Welfare?*, Oxford: Blackwell Science.

Appleby, M. C. and Hughes, B. O. (1991) 'Welfare of laying hens in cages and alternative systems: environmental, physical and behavioural apsects', *Journal of Wild Poultry Science*, 47: 109–28.

Appleby, M. C. and Hughes, B. O. (1997) *Animal Welfare*, Wallingford, Oxon: CABI Publishing.

Arendt, J. and Skene, D. J. (2005) 'Melatonin as a chronobiotic', *Sleep Medicine Reviews*, 9: 25–39.

ARAZPA (Australasian Regional Association of Zoological Parks and Aquaria) (2008) 'About ARAZPA', available online at `http://www.arazpa.org.au/About-Us/default.aspx` (accessed April 2008).

Armstrong, S. (2004) 'A taste for grass: do zebra (*Equus burchelli*) have a preference for individual grass species?' in C. McDonald (ed.), *Proceedings of the Sixth Annual BIAZA Research Meeting*, 8–9 July, Edinburgh Zoo, Edinburgh, pp. 43–5.

Armstrong, S. and Botzler, R. (2003) *The Animal Ethics Reader*, London: Routledge.

Armstrong, S. and Marples, N. (2005) 'Do captive plains zebra (*E. burchelli*) have a preference for individual grass species?' in T. P. Meehan and M. E. Allen (eds), *Proceedings of the Fourth European Zoo Nutrition Conference*, 20–23 January, Leipzig Zoo, EAZA.

Asa, C. S. (1996) 'Reproductive physiology' in D. G. Kleiman, M. Allen, K. Thompson, and S. Lumpkin (eds), *Wild Mammals in Captivity: Principles and Techniques*, Chicago, IL: University of Chicago Press, pp. 390–417.

Asa, C. S. and Porton, I. J. (2005) *Wildlife Contraception: Issues, Methods and Applications*, Baltimore, MD: John Hopkins University Press.

Ashley, P. J. (2007) 'Fish welfare: current issues in aquaculture', *Applied Animal Behaviour Science*, 104: 199–235.

Asvestas, C. and Reininger, M. (1999) 'Forming a bachelor group of long-tailed macaques (*Macaca fascicularis*)', *Laboratory Primate Newsletter*, 38: 14.

Atkinson, R. L., Atkinson, R. C., Smith, E. E., Bem, D. J., and Nolen-Hoeksema, S. (1996) *Hilgard's Introduction to Psychology* (12th edn), London: Harcourt Brace.

Aujard, F., Seguy, M., Terrien, J., Botalla, R., Blanc, S., and Perret, M. (2006) 'Behavioral thermoregulation in a non-human primate: effects of age and photoperiod on temperature selection', *Experimental Gerontology*, 41: 784–92.

Austin, M., Leader, L., and Reilly, N. (2005) 'Prenatal stress, the hypothalamic-pituitary-adrenal axis, and fetel and infant neurobehaviour', *Early Human Development*, 81: 917–26.

AZA (American Association of Zoos and Aquariums) (1999) *The Collective Impact of America's Zoos and Aquariums*, Silver Spring, MD: AZA.

b Bach, C. (1998) *Birth Date Determination in Australasian Marsupials*, Taronga Zoo, Sydney: ARAZPA.

Baillie, J. M., Hilton-Taylor, C., and Stuart, S. N. (2004) *IUCN Red List of Threatened Species: A Global Species Assessment*, Gland: IUCN.

Baker, K. C. (1997) 'Straw and forage material ameliorate abnormal behaviour in adult chimpanzees', *Zoo Biology*, 16: 225–36.

Baker, K. C. (2000) 'Advanced age influences chimpanzee behavior in small social groups', *Zoo Biology*, 19: 111–19.

Baker, K. C. (2004) 'Benefits of positive human interaction for socially housed chimpanzees', *Animal Welfare*, 13: 239–45.

Balke, J. M. E., Barker, I. K., Hackenberger, M. K., McManamon, R., and Boever, W. J. (1988) 'Reproductive anatomy of three nulliparous female Asian elephants: the development of artificial breeding techniques', *Zoo Biology*, 7: 99–113.

Balmford, A. (2000) 'Separating fact from artefact in analyses of zoo visitor preferences', *Conservation Biology*, 14: 1193–5.

Balmford, A., Leader-Williams, N., and Green, M. J. B. (1995) 'Parks or arks: where to preserve threatened mammals?', *Biodiversity and Conservation*, 4: 595–607.

Balmford, A., Mace, G. M., and Leader-Williams, N. (1996) 'Designing the ark: setting priorities for captive breeding', *Conservation Biology*, 10: 719–27.

Balmford, A., Leader-Williams, N., Mace, G. M., Manica, A., Walter, O., West, C., and Zimmermann, A. (2007) 'Message received? Quantifying the impact of informal conservation education on adults visiting UK zoos' in A. Zimmermann, M. Hatchwell, L. Dickie, and C. West (eds), *Zoos in the 21st Century: Catalysts for Conservation?*, Cambridge: Cambridge University Press, pp. 120–36.

Baratay, E. and Hardouin-Fugier, E. (2004) *Zoo: A History of Zoological Gardens in the West*, London: Reaktion Books.

Barker, D., Fitzpatrick, M. P., and Dierenfeld, E. S. (1998) 'Nutrient composition of selected whole invertebrates', *Zoo Biology*, 17: 123–34.

Barnard, C. and Hurst, J. (1996) 'Welfare by design: the natural selection of welfare criteria', *Animal Welfare*, 56: 405–33.

Barnard, C., Gilbert, F., and McGregor, P. (1993) *Asking Questions in Biology: Design, Analysis and Presentation in Practical Work*, Upper Saddle River, NJ: Prentice Hall.

Barr, S., Laming, P. R., Dick, J. T. A., and Elwood, R. W. (2008) 'Nociception or pain in a decapod crustacean?', *Animal Behaviour*, 75: 745–51.

Barrington-Johnson, J. (2005) *The Zoo: The Story of London Zoo*, London: Robert Hale.

Bartlett, A. D. (1890) *Life Among Wild Beasts in the Zoo*, London: Chapman & Hall.

Bartlett, A. D. (1898) *Wild Animals in Captivity*, London: Chapman & Hall.

Bashaw, M. J. and Maple, T. L. (2001) 'Signs fail to increase zoo visitors' ability to see tigers', *Curator*, 44: 297–304.

Bashaw, M. J., Tarou, L. R., Maki, T. S., and Maple, T. L. (2001) 'A survey assessment of variables related to stereotypy in captive giraffe and okapi', *Applied Animal Behaviour Science*, 73(3): 235–47.

Bashaw, M. J., Bloomsmith, M. A., Marr, M. J., and Maple, T. L. (2003) 'To hunt or not to hunt? A feeding enrichment experiment with captive large felids', *Zoo Biology*, 22(2): 189–98.

Bassett, L. and Buchanan-Smith, H. M. (2007) 'Effects of predictability on the welfare of captive animals', *Applied Animal Behaviour Science, Conservation, Enrichment and Animal Behaviour*, 102: 223–45.

Bassett, L., Buchanan-Smith, H. M., McKinley, J., and Smith, T. E. (2003) 'Effects of training on stress-related behaviour of the common marmoset (*Callithrix jacchus*) in relation to coping with routine husbandry procedures', *Journal of Applied Animal Welfare Science*, 6: 221–33.

Bateson, P. and Bradshaw, E. (1997) 'Physiological effects of hunting red deer (*Cervus elaphus*)', *Proceedings of the Royal Society (Series B: Biological Sciences)*, 264: 1707–14.

Bauert, M. R., Furrer, S. C., Zingg, R., and Steinmetz, H. W. (2007) 'Three years of experience running the Masoala Rainforest ecosystem at Zurich Zoo, Switzerland', *International Zoo Yearbook*, 41: 203–16.

Baxter, E. and Plowman, A. B. (2001) 'The effect of increasing dietary fibre on feeding, rumination and oral stereotypies in captive giraffes (*Giraffa camelopardalis*)', *Animal Welfare*, 10: 281–90.

Beardsworth, A. and Bryman, A. E. (2001) 'The wild animal in late modernity: the case of the Disneyization of zoos', *Tourist Studies*, 1: 83–104.

Beck, B. B., Rapaport, L. G., Stanley-Price, M. R., and Wilson, A. C. (1994) 'Reintroduction of captive born animals' in P. J. S. Olney, G. M. Mace, and A. T. C. Feistner (eds), *Creative Conservation: Interactive Management of Wild and Captive Animals*, London: Chapman & Hall, pp. 265–86.

Beckoff, M. and Byers, J. (1998) *Animal Play: Evolutionary, Comparative and Ecological Perspectives*, Cambridge: Cambridge University Press.

Bell, C. E. (ed.) (2001) *Encyclopedia of the World's Zoos*, Chicago, IL/London: Fitzroy Dearborn.

Benirschke, K., Kumamoto, A. T., and Bogart, M. H. (1981) 'Congenital anomalies in *Lemur variegatus*', *Journal of Medical Primatology*, 10: 38–45.

Bennett, P. (2001) 'Establishing animal germplasm resource banks for wildlife conservation: genetic, population and evolutionary aspects' in P. Watson and W. Holt (eds), *Cryobanking the Genetic Resource: Wildlife Conservation for the Future?*, London: Taylor & Francis, pp. 47–67.

Berg, J. K. (1983) 'Vocalizations and associated behaviours of the African elephant (*Loxodonta africana*) in captivity', *Zeitschrift für Tierpsychologie*, 63: 63–79.

Berge, G. M. (1990) 'Freeze branding of Atlantic halibut', *Aquaculture*, 89: 383–6.

Bergeron, R., Badnell-Waters, A., Lambton, S., and Mason, G. (2006) 'Stereotypic oral behaviour in captive ungulates: foraging, diet and gastrointestinal function' in G. J. Mason (ed.), *Stereotypic Animal Behaviour: Fundamentals and Applications to Welfare*, Wallingford, Oxon: CABI Publishing, pp. 19–57.

Berkson, G., Mason, W., and Saxon, S. (1963) 'Situation and stimulus effect on stereotyped behaviours of chimpanzees', *Journal of Comparative Physiological Psychology*, 56: 786–92.

Bernard, J. B. (1997) *Vitamin D and Ultraviolet Radiation: Meeting Lighting Needs for Captive Animals*, EAZA Nutrition Advisory Group (NAG) Fact Sheet 002 (July 1997), available online at http://www.nagonline.net/Technical%20Papers/NAGFS00297VitD-JONIFEB24,2002MODIFIED.pdf.

Bernard, J. B., Watkins, B., and Ullrey, D. (1989) 'Manifestations of vitamin D deficiency in chicks reared under different artificial lighting regimes', *Zoo Biology*, 8: 349–55.

Bernstein, I. S. (1967) 'Defining the natural habitat' in D. Starck, R. Schneider, and H.-J. Kuhn (eds), *Progress in Primatology*, Stuttgart: Fischer, pp. 177–9.

Bertolino, S., Viano, C., and Currado, I. (2001) 'Population dynamics, breeding patterns and spatial use of the garden dormouse (*Eliomys quercinus*) in an Alpine habitat', *Journal of Zoology*, 253: 513–21.

Besag, J. and Clifford, P. (1989) 'Generalized Monte Carlo significance tests', *Biometrika*, 76: 633–42.

Bestelmeyer, S. V. (1999) 'Behavioural changes associated with introductions of male maned wolves (*Chrysocyon brachyurus*) to females with pups', *Zoo Biology*, 18: 189–97.

BIAZA (British and Irish Association of Zoos and Aquariums) (2007) *Working Together for Wildlife*, London: BIAZA.

Birke, L. (2002) 'Effects of browse, human visitors and noise on the behaviour of captive orang-utans', *Animal Welfare*, 11: 189–202.

Bitgood, S. (2002) 'Environmental psychology in museums, zoos and other exhibition centres' in R. B. Bechtel and A. Churchman (eds), *Handbook of Environmental Psychology*, New York, NY: John Wiley & Sons, pp. 461–80.

Bitgood, S., Patterson, D., and Benefield, A. (1988) 'Exhibit design and visitor behavior: empirical relationships', *Environment and Behavior*, 20: 474–91.

Blaney, E. C. and Wells, D. L. (2004) 'The influence of a camouflage net barrier on the behaviour, welfare and public perceptions of zoo-housed gorillas', *Animal Welfare*, 13: 111–18.

Blasetti, A., Boltani, L., Riviello, M. C., and Visalberghi, E. (1988) 'Activity budgets and use of enclosed space by wild boars (*Sus scrofa*) in captivity', *Zoo Biology*, 7: 69–79.

Blomqvist, L. (1995) 'Three decades of snow leopards *Panthera uncia* in captivity', *International Zoo Yearbook*, 34: 178–85.

Bloomsmith, M. A. and Lambeth, S. P. (1995) 'Effects of predictable versus unpredictable feeding schedules on chimpanzee behavior', *Applied Animal Behaviour Science*, 44: 65–74.

Bloomsmith, M. A. and Lambeth, S. P. (2000) 'Videotapes as enrichment for captive chimpanzees (*Pan troglodytes*)', *Zoo Biology*, 19: 541–51.

Bloomsmith, M. A., Laule, G. E., Alford, P. L., and Thurston, R. H. (1994) 'Using training to moderate chimpanzee aggression during feeding', *Zoo Biology*, 13: 557–66.

Bloomsmith, M. A., Stone, A. M., and Laule, G. E. (1998) 'Positive reinforcement training to enhance the voluntary movement of group-housed chimpanzees within their enclosures', *Zoo Biology*, 17: 333–41.

Bloomsmith, M. A., Jones, M. L., Snyder, R. J., Singer, R. A., Gardner, W. A., Liu, S. C., and Maple, T. L. (2003) 'Positive reinforcement training to elicit voluntary movement of two giant pandas throughout their enclosure', *Zoo Biology*, 22: 323–34.

Bloomsmith, M. A., Marr, M. J., and Maple, T. L. (2007) 'Addressing non-human primate behavioral problems through the application of operant conditioning: is the human treatment approach a useful model?', *Applied Animal Behaviour Science: Conservation, Enrichment and Animal Behaviour*, 102: 205–22.

Bloxam, Q. M. C. and Tonge, S. J. (1995) 'Amphibians: suitable candidates for breeding-release programmes', *Biodiversity and Conservation*, 4: 636–44.

Blunt, W. (1976) *The Ark in the Park: The Zoo in the Nineteenth Century*, London: Hamish Hamilton.

Boakes, E. H., Wang, J., and Amos, W. (2006) 'An investigation of inbreeding depression and purging in captive pedigreed populations', *Heredity*, 98: 172–82.

Boehm, T. and Zufall, F. (2006) 'MHC peptides and the sensory evaluation of genotype', *Trends in Neurosciences*, 29: 100–7.

Boinski, S. (1987) 'Mating patterns in squirrel monkeys (*Saimiri oerstedi*)', *Behavioral Ecology and Sociobiology*, 21: 13–21.

Boinski, S., Gross, T. S., and Davis, J. K. (1999) 'Terrestrial predator alarm vocalizations are a valid monitor of stress in captive brown capuchins (*Cebus apella*)', *Zoo Biology*, 18: 295–312.

Boivin, X., Garel, J. P., Mante, A., and Le Neindre, P. (1998) 'Beef calves react differently to different handlers according to the test situation and their previous interactions with their caretaker', *Applied Animal Behaviour Science*, 55: 245–57.

Boivin, X., Lensink, J., Tallet, C., and Veissier, I. (2003) 'Stockmanship and farm animal welfare', *Animal Welfare*, 12: 479–92.

Bolin, C. A. (2003) 'Leptospirosis' in M. E. Fowler and R. E. Miller (eds), *Zoo and Wild Animal Medicine* (5th edn), Philadelphia, PA: Saunders (Elsevier), pp. 699–702.

Bond, J. C. and Lindburg, D. G. (1990) 'Carcass feeding of captive cheetahs (*Acinonyx jubatus*): the effects of a naturalistic breeding program on oral health and psychological well-being', *Applied Animal Behaviour Science*, 26: 373–82.

Boness, D. (1996) 'Water quality management in aquatic mammal exhibits' in D. Kleiman, M. Allen, K. Thompson, and S. Lumpkin (eds), *Wild Mammals in Captivity*, Chicago, IL: University of Chicago Press, pp. 231–42.

Boogaard, B. K., Oosting, S. J., and Bock, B. B. (2006) 'Elements of societal perception of farm animal welfare: a quantitative study in the Netherlands', *Livestock Science*, 104: 13–22.

Boorer, M. (1972) 'Some aspects of stereotyped patterns of movement exhibited by zoo animals', *International Zoo Yearbook*, 12: 164–8.

Bostock, S. St. C. (1993) *Zoos and Animal Rights*, London: Routledge.

Bowden, C. and Masters, J. (2003) *Textbook of Veterinary Medical Nursing*, New York, NY/Edinburgh: Butterworth-Heinemann.

Bowers, B. B. and Burghardt, G. M. (1992) 'The scientist and the snake: relationships with reptiles' in H. Davis and D. Balfour (eds), *The Inevitable Bond: Examining Scientist–Animal Interactions*, Cambridge: Cambridge University Press, pp. 250–63.

Box, H. (1991) 'Training for life after release: simian primates as examples', *Symposium of the Zoological Society of London*, 62: 111–23.

Braswell, L. D. (1991) 'Exotic animal dentistry', *Compendium on Continuing Education for the Practicing Veterinarian*, 13: 1229–33.

Bremner-Harrison, S., Prodohl, P. A., and Elwood, R. (2004) 'Behavioural trait assessment as a release criterion: boldness predicts early death in a reintroduction programme of captive-bred swift fox (*Vulpes velox*)', *Animal Conservation*, 7: 313–20.

Brent, L. and Stone, A. M. (1996) 'Long-term use of televisions, balls, and mirrors as enrichment for paired and singly caged chimpanzees', *American Journal of Primatology*, 39: 139–45.

Brent, L., Kessel, A. L., and Barrera, H. (1997) 'Evaluation of introduction procedures in captive chimpanzees', *Zoo Biology*, 16: 335–42.

Britt, A. (1998) 'Encouraging natural feeding behaviour in captive bred black and white ruffed lemurs (*Varecia variegata v.*)', *Zoo Biology*, 17: 379–92.

Broad, G. (1996) 'Visitor profile and evaluation of informal education at Jersey Zoo', *Dodo: Journal of the Jersey Wildlife Preservation Trust*, 32: 166–92.

Brodey, P. (1981) 'The LINKS-ZOO: a recreational/educational facility for the future', *International Zoo Yearbook*, 21: 63–8.

Broom, D. M. (1998) 'Stereotypies in animals' in M. Bekoff and C. A. Meaney (eds), *Encyclopedia of Animal Rights and Animal Welfare*, London: Fitzroy Dearborn, p. 256.

Broom, D. M. (2005) 'The effects of land transport on animal welfare', *Revue Scientifique et Technique de L'Office International des Epizooties*, 24: 683–91.

Broom, D. M. and Johnson, K. G. (1993) *Stress and Animal Welfare*, London: Chapman & Hall.

Brooman, S. and Legge, D. (1997) *Law Relating to Animals*, London: Cavendish Publishing.

Brown, C. and Loskutoff, N. (1998) 'A training program for non-invasive semen collection in captive western lowland gorillas (*Gorilla gorilla gorilla*)', *Zoo Biology*, 17: 143–51.

Brown, J. L. (2000) 'Reproductive endocrine monitoring of elephants: an essential tool for assisting captive management', *Zoo Biology*, 19: 347–69.

Brown, J. L. and Wemmer, C. M. (1995) 'Urinary cortisol analysis for monitoring adrenal activity in elephants', *Zoo Biology*, 14: 533–42.

Brown, J. L., Olson, D., Keele, M., and Freeman, E. W. (2004) 'Survey of the reproductive cyclicity status of Asian and African elephants in North America', *Zoo Biology*, 23: 309–21.

Bubier, N. (1996) 'The behavioural priorities of laying hens: the effects of two methods of environmental enrichment on time budgets', *Behavioural Processes*, 374: 239–49.

Buchanan-Smith, H. M., Anderson, D. A., and Ryan, C. W. (1993) 'Responses of cotton-top tamarins (*Saguinus oedipus*) to faecal scents of predators and non-predators', *Animal Welfare*, 2: 17–32.

Buckanoff, H., Frederick, C., and Weston Murphy, H. (2006) 'Hand-rearing a potto *Perodicticus potto* at Franklin Park Zoo, Boston', *International Zoo Yearbook*, 40: 302–12.

Burghardt, G. M. (1995) 'Brain imaging, ethology and the non-human mind', *Behavioural and Brain Sciences*, 18: 339–40.

Burghardt, G. M., Ward, B., and Rosscoe, R. (1996) 'Problem of reptile play: environmental enrichment and play behavior in a captive Nile soft-shelled turtle, *Trionyx triunguis*', *Zoo Biology*, 15: 223–38.

Burke, T. and Bruford, M. W. (1987) 'DNA fingerprinting in birds', *Nature*, 327: 149–52.

Burkhardt, R. W. (2001) 'A man and his menagerie: management of nineteenth-century zoological park by Frédéric Cuvier', *Natural History*, 110: 62–9.

Burley, N. (1985) 'Leg-band color and mortality patterns in captive breeding populations of zebra finches', *The Auk*, 102: 647–51.

Bush, M. (1993) 'Anaesthesia of high-risk animals: giraffe' in M. E. Fowler (ed.), *Zoo and Wild Animal Medicine* (3rd edn), Philadelphia, PA: W. B. Saunders.

Bush, M. (2003) '*Giraffidae*' in M. E. Fowler and R. E. Miller (eds), *Zoo and Wild Animal Medicine* (5th edn), Philadelphia, PA: Saunders (Elsevier).

Bush, M., Montali, R. J., Brownstein, D., James, A. E., and Appel, M. J. G. (1976) 'Vaccine-induced canine distemper in a lesser panda', *Journal of the American Veterinary Medical Association*, 169: 959–60.

Byford, R. L., Craig, M. E., and Crosby, B. L. (1992) 'A review of ectoparasites and their effect on cattle production', *Journal of Animal Science*, 70: 597–602.

C Caine, J. and Melfi, V. (2005) 'A long term study of *Trichuris trichiura* in zoo-housed colobus' in A. Nicklin (ed.), *Proceedings of the Seventh Annual Symposium on Zoo Research*, 7–8 July, Twycross Zoo, London: BIAZA, pp. 56–66.

Caine, N. G. and O'Boyle Jr, V. J. (1992) 'Cage design and forms of play in red-bellied tamarins, *Saguinus labiatus*', *Zoo Biology*, 11: 215–20.

Calle, P. P. (2003) 'Rabies' in M. E. Fowler and R. E. Miller (eds), *Zoo and Wild Animal Medicine* (5th edn), Philadelphia, PA: Saunders (Elsevier), pp. 732–6.

Canfield, P. J. and Cunningham, A. A. (1993) 'Disease and mortality in Australian marsupials held at London Zoo, 1872–1972', *Journal of Zoo and Wildlife Medicine*, 24: 158–67.

Capitanio, J. P. (1999) 'Personality dimensions in adult male rhesus macaques: prediction of behaviors across time and situation', *American Journal of Primatology*, 47: 299–320.

Cardillo, M., Mace, G. M., Jones, K. E., Bielby, J., Bininda-Emonds, O. R. P., Sechrest, W., Orme, C. D. L., and Purvis, A. (2005) 'Multiple causes of high extinction risk in large mammal species', *Science*, 309: 1239–41.

Carlson, N. (2007) *Physiology of Behaviour*, Boston: Pearson Education Inc.

Carlstead, K. (1996) 'Effects of captivity on the behavior of wild mammals' in D. G. Kleiman, M. E. Allen, K. V. Thompson, and S. Lumpkin (eds), *Wild Mammals in Captivity*, Chicago, IL: University of Chicago Press, pp. 317–33.

Carlstead, K. (1998) 'Determining the causes of stereotypic behaviors in zoo carnivores: towards appropriate enrichment strategies' in D. Shepherdson, J. Mellen, and M. Hutchins (eds), *Second Nature: Environmental Enrichment for Captive Animals*, Washington DC: Smithsonian Institute Press, pp. 172–83.

Carlstead, K. and Brown, J. L. (2005) 'Relationships between patterns of fecal corticoid excretion and behavior, reproduction, and environmental factors in captive black (*Diceros bicornis*) and white (*Ceratotherium simum*) rhinoceros', *Zoo Biology*, 24: 215–32.

Carlstead, K. and Seidensticker, J. (1991) 'Seasonal variation in stereotypic pacing in an American black bear *Ursus americanus*', *Behavioural Processes*, 25: 155–61.

Carlstead, K. and Shepherdson, D. (1994) 'Effects of environmental enrichment on reproduction', *Zoo Biology*, 13: 447–58.

Carlstead, K., Brown, J. L., and Seidensticker, J. (1993) 'Behavioral and adrenocortical responses to environmental changes in leopard cats (*Felis bengalensis*)', *Zoo Biology*, 12: 321–31.

Carlstead, K., Mellen, J., and Kleiman, D. G. (1999a) 'Black rhinoceros (*Diceros bicornis*) in US zoos: I. individual behaviour profiles and their relationship to breeding success', *Zoo Biology*, 18: 17–34.

Carlstead, K., Fraser, J., Bennett, C., and Kleiman, D. G. (1999b) 'Black rhinoceros

(*Diceros bicornis*) in US zoos: II. behavior, breeding success, and mortality in relation to housing facilities', *Zoo Biology*, 18: 35–52.

Carlstead, K., Seidensticker, J., and Baldwin, R. (1991) 'Environmental enrichment for zoo bears', *Zoo Biology*, 10: 3–16.

Caro, T. M. (1993) 'Behavioral solutions to breeding cheetahs in captivity: insights from the wild', *Zoo Biology*, 12: 19–30.

Castellote, M. and Fossa, F. (2006) 'Measuring acoustic activity as a method to evaluate welfare in captive beluga whales (*Delphinapterus leucas*)', *Aquatic Mammals*, 32: 325–33.

Cavalieri, P. and Singer, P. (1994) (eds) *The Great Ape Project: Equality Beyond Humanity*, New York: St Martin's Griffin.

Cavigelli, S. A., Yee, J. R., and McClintock, M. K. (2006) 'Infant temperament predicts life span in female rats that develop spontaneous tumors', *Hormones and Behavior*, 50: 454–62.

Caws, C. and Aureli, F. (2003) 'Chimpanzees cope with temporary reduction of escape opportunities', *International Journal of Primatology*, 24: 1077–91.

Ceballos, G., Erhlich, P. R., Soberón, J., Salazar, I., and Fay, J. P. (2005) 'Global mammal conservation: what must we manage?', *Science*, 309: 603–7.

Cerit, H. and Avanus, K. (2007) 'Sex identification in avian species using DNA typing methods', *World's Poultry Science Journal*, 63: 91–99.

Chalmers, K. (2006) *Zoo Keeper Information: Auckland Zoo and its Role in Conservation and Captive Breeding Programmes*, Auckland, New Zealand: Auckland Zoo.

Chamove, A. (1988) 'Assessing the welfare of captive primates: a critique' in Universities Federation for Animal Welfare (UFAW), *Symposium of Laboratory Animal Welfare Research: Primates*, Potters Bar: UFAW, pp. 39–49.

Chamove, A. (1989) 'Environmental enrichment: a review', *Animal Technology*, 40: 155–78.

Chamove, A. and Moodie, E. (1990) 'Are alarming events good for captive monkeys?', *Applied Animal Behaviour Science*, 276: 169–76.

Chamove, A. and Rohrhuber, B. (1989) 'Moving callitrichid monkeys from cages to outside areas', *Zoo Biology*, 8: 151–63.

Chamove, A., Anderson, J., Morgan-Jones, S., and Jones, S. (1982) 'Deep woodchip litter: hygiene, feeding, and behavioural enhancement in eight primate species', *International Journal for the Study of Animal Problems*, 3: 308–18.

Chamove, A. S., Hosey, G. R., and Schaetzel, P. (1988) 'Visitors excite primates in zoos', *Zoo Biology*, 7: 359–69.

Chandroo, K. P., Duncan, I. J. H., and Moccia, R. D. (2004) 'Can fish suffer? Perspectives on sentience, pain, fear and stress', *Applied Animal Behaviour Science*, 86: 225–50.

Chang, T. R., Forthman, D. L., and Maple, T. L. (1999) 'Comparison of confined mandrill (*Mandrillus sphinx*) behaviour in traditional and "ecologically representative" exhibits', *Zoo Biology*, 18: 163–76.

Chastain, B. (2005) 'The defining moment' in *Innovation or replication*, *Proceedings of the Sixth International Symposium on Zoo Design*, Whitley Wildlife Conservation Trust, Paignton, UK.

Cheek, N. J. (1976) 'Sociological perspectives on the Zoological Park Market' in N. H. Check, D. R. Field, and R. J. Burge (eds), *Leisure and Recreation Places*, Ann Arbor, MI: Ann Arbor Science.

Cherfas, J. (1984) *Zoo 2000: A Look Beyond the Bars*, London: BBC.

Cirulli, F., Berry, A., and Alleva, E. (2003) 'Early disruption of the mother–infant relationship: effects on brain plasticity and implications for psychopathology', *Neuroscience and Biobehavioral Reviews*, 27: 73–82.

Clarke, F. and King, A. (2008) 'A critical review of zoo-based olfactory enrichment' in J. Hurst, R. Beynon, S. Roberts, and T. Wyatt (eds), *Chemical Signals in Vertebrates 11*, New York, NY: Springer, pp. 391–8.

Clauss, M. and Dierenfeld, E. S. (2007) 'The nutrition of "browsers" ', in M. E. Fowler and R. E. Miller, *Zoo and Wild Animal Medicine:*

Current Therapy (6th edn), St Louis, MO: Saunders (Elsevier), pp. 444–54.

Clauss, M. and Hatt, J.-M. (2006) 'The feeding of rhinoceros in captivity', *International Zoo Yearbook*, 40: 197–209.

Clauss, M., Kienzle, E., and Wiesner, H. (2003) 'Feeding browse to large zoo herbivores: how much is "a lot", how much is "sufficient"?', in A. Fidgett, M. Clauss, U. Ganslosser, J. M. Hatt, and J. Niijboer (eds), *Zoo Animal Nutrition: Vol. II*, Fürth: Filander Verlag, pp. 17–25.

Clauss, M., Polster, C., Kienzle, E., Weisner, H., Baumgartner, K., von Houwald, F., Streich, W. J., and Dierenfeld, E. (2005) 'Energy and mineral nutrition and water intake in the captive Indian rhinoceros (*Rhinoceros unicornis*)', *Zoo Biology*, 24: 1–14.

Clubb, R. and Mason, G. (2002) *A Review of the Welfare of Zoo Elephants in Europe: A Report Commissioned by the RSPCA*, Oxford: Animal Behaviour Research Group, University of Oxford.

Clubb, R. and Mason, G. (2003) 'Captivity effects on wide-ranging carnivores', *Nature*, 425: 473–4.

Clubb, R. and Mason, G. (2004) 'Pacing polar bears and stoical sheep: testing ecological and evolutionary hypotheses about animal welfare', *Animal Welfare*, 13: 533–40.

Clubb, R. and Mason, G. (2007) 'Natural behavioural biology as a risk factor in carnivore welfare: how analysing species differences could help zoos improve enclosures', *Applied Animal Behaviour Science*, 102: 303–28.

Clubb, R. and Vickery, S. (2006) 'Laboratory stereotypies in carnivores: does pacing stem from hunting, ranging or frustrated escape?' in G. J. Mason (ed.), *Stereotypic Animal Behaviour: Fundamentals and Applications to Welfare*, Wallingford, Oxon: CABI Publishing, pp. 58–84.

Cocks, L. (2007) 'Factors influencing the well-being and longevity of captive female orang-utans', *International Journal of Primatology*, 28: 429–40.

Coe, J. (1985) 'Design and perception: making the zoo experience real', *Zoo Biology*, 4: 197–208.

Coe, J. (1987) 'What's the message? Exhibit design for education' in American Association of Zoological Parks and Aquariums, *Regional Conference Proceedings*, Wheeling, WV: AAZPA, pp. 19–23.

Coe, J. (1989) 'Naturalizing habitats for captive primates', *Zoo Biology*, 8: 117–25.

Coe, J. (1994) 'Landscape immersion: origins and concepts' in J. C. Coe (mod.), *Landscape Immersion Exhibits: How are They Proving as Education Settings? American Zoo and Aquarium Association Convention Proceedings*, Bethesda, MD: AZA, pp. 1–7, available online at `http://www.joncoedesign.com/pub/technical.htm`.

Coe, J. (1996) 'What's the message? Education through exhibit design' in D. Kleiman, M. Allen, K. Thompson, and S. Lumpkin (eds), *Wild Mammals in Captivity: Principles and Techniques*, Chicago, IL: University of Chicago, pp. 167–74.

Coe, J. (1997) 'Entertaining zoo visitors and zoo animals: an integrated approach' in *Proceedings of the American Zoo and Aquarium Association Annual Conference*, Bethesda, MD: AZA, pp. 156–62.

Coe, J. (1999) 'An integrated approach to design: how zoo staff can get the best results from new facilities', *First Annual Rhino Keeper Workshop*, 7–8 May, Disney's Wild Animal Kingdom, Orlando, FL.

Coe, J. (2006) 'Naturalistic enrichment' in *Australasian Regional Association of Zoological Parks and Aquaria Conference Proceedings*, Perth Zoo, ARAZPA, pp. 1–9, available online at `http://www.joncoedesign.com/pub/technical.htm`.

Colahan, H. and Breder, C. (2003) 'Primate training at Disney's Animal Kingdom', *Journal of Applied Animal Welfare Science*, 6: 235–46.

Cole, C. and Townsend, C. (1977) 'Parthenogenetic reptiles: new subjects for laboratory research', *Experientia*, 33: 285–9.

Colman, R. J., McKiernan, S. H., Aiken, J. M., and Weindruch, R. (2005) 'Muscle mass loss in Rhesus monkeys: age of onset', *Experimental Gerontology*, 40: 573–81.

Conservation Breeding Specialist Group (CBSG) of the World Conservation Union (IUCN) (2004) 'Transponders', *CBSG News*, 15(1): 20.

Conte, E. S. (2004) 'Stress and the welfare of cultured fish', *Applied Animal Behaviour Science*, 86: 205–23.

Conway, W. (1986a) 'The consumption of wildlife by man', *Animal Kingdom*, 7: 18–23.

Conway, W. (1986b) 'The practical difficulties and financial implications of endangered species breeding programmes', *International Zoo Yearbook*, 24/25: 210–19.

Conway, W. (2003) 'The role of zoos in the 21st century', *International Zoo Yearbook*, 38: 7–13.

Conway, W. (2007) 'Entering the 21st century' in A. Zimmerman, M. Hatchwell, L. Dickie, and C. West (eds), *Zoos in the 21st Century: Catalysts for Conservation?*, Conservation Biology Series No. 15, Cambridge: Cambridge University Press, pp. 12–21.

Cook, S. and Hosey, G. R. (1995) 'Interaction sequences between chimpanzees and human visitors at the zoo', *Zoo Biology*, 14: 431–40.

Cooper, J. E. and Cooper, M. E. (2007) 'Importance and application of animal law' in J. E. Cooper and M. E. Cooper (eds), *Introduction to Veterinary and Comparative Forensic Medicine*, Oxford: Blackwell Publishing, pp. 42–60.

Cooper, M. E. (2003) 'Zoo legislation', *International Zoo Yearbook*, 38: 81–93.

Cooper, K. A., Harder, J. D., Clawson, D. H., Fredrick, D. L., Lodge, G. A., Peachey, H. C., Spellmire, T. J., and Winstel, D. P. (1990) 'Serum testosterone and musth in captive male African and Asian elephants', *Zoo Biology* 9: 297–306.

Cooper, M. E. and Rosser, A. M. (2002) 'International regulation of wildlife trade: relevant legislation and organisations', *Revue Scientifique et Technique de L'Office International des Epizooties*, 21: 103–23.

Cooper, M. R. and Johnson, A. W. (1998) *Poisonous Plants and Fungi in Britain—Animal and Human Poisoning*, London: HMSO.

Cooper, V. J. and Hosey, G. R. (2003) 'Sexual dichromatism and female preference in *Eulemur fulvus* subspecies', *International Journal of Primatology*, 24: 1177–88.

Coote, T. and Loeve, E. (2003) 'From 61 species to five: endemic tree snails of the Society Islands fall prey to an ill-judged biological control programme', *Oryx*, 37: 91–6.

Coote, T., Clarke, D., Hickman, C. S., Murray, J., and Pearce-Kelly, P. (2004) 'Experimental release of endemic *Partula* species, extinct in the wild, into a protected area of natural habitat on Moorea', *Pacific Science*, 58: 429–34.

Cork, S. C. (2000) 'Iron storage disease in birds', *Avian Pathology*, 29: 7–12.

Cornetto, T. and Estevez, I. (2001) 'Behavior of the domestic fowl in the presence of vertical panels', *Poultry Science*, 80: 1455–62.

Corson-White, E. P. (1932) *Diet in Relation to Degenerative Bone Lesions and Fertility*, Report of the Laboratory and Museum of Comparative Pathology, Zoological Society of Philadelphia. pp. 26–8.

Coulton, L. E., Waran, N. K., and Young, R. J. (1997) 'Effects of foraging enrichment on the behaviour of parrots', *Animal Welfare*, 6: 357–63.

Cousins, D. (2006) 'Review of the use of herb gardens and medicinal plants in primate exhibits in zoos', *International Zoo Yearbook*, 40: 341–50.

Coviello-McLaughlin, G. M. and Starr, S. J. (1997) 'Rodent enrichment devices: evaluation of preference and efficacy', *Contemporary Topics in Laboratory Animal Science*, 36: 66–8.

Cowie, A. (1948) *Pregnancy Diagnosis: A Review*, Reading: Commonwealth Agricultural Bureaux.

Crandall, K. A., Bininda-Emonds, O. R. P., Mace, G. M., and Wayne, R. K. (2000) 'Considering evolutionary processes in conservation biology', *Trends in Ecology and Evolution*, 15: 290–5.

Cranfield, M. R., Graczyk, T. K., and McCuthchan, T. F. (2000) 'ELISA antibody test, PCR and a DNA vaccine for use with avian malaria in African penguins' in *Proceedings of the Annual Meeting of the American Association of Zoo Veterinarians*, 17–21 September, New Orleans, LA, AAZV, p. 39.

Crawshaw, G. (2003) 'Anurans (*Anura, Salienta*): Frogs, Toads' in M. E. Fowler and R. E. Miller (eds), *Zoo and Wild Animal Medicine* (5th edn), Philadelphia, PA: Saunders (Elsevier).

Creel, S., Creel, N. M., Mills, M. G. L., and Monfort, S. L. (1997) 'Rank and reproduction in cooperatively breeding African wild dogs: behavioral and endocrine correlates', *Behavioural Ecology*, 8: 298–306.

Cresswell, W., Lind, J., Quinn, L., Minderman, J., and Whitfield, D. P. (2007) 'Ringing or colour-banding does not increase predation mortality in redshanks *Tringa totanus*', *Journal of Avian Biology*, 38: 309–16.

Crissey, S. (2005) 'The complexity of formulating diets for zoo animals: a matrix', *International Zoo Yearbook*, 39: 36–43.

Critser, J. K., Riley, L. K., and Prather, R. S. (2003) 'Application of nuclear transfer technology to wildlife species' in W. Holt, A. Pickard, J. Rodger, and D. Wildt (2003) *Reproductive Science and Integrated Conservation*, Cambridge: Cambridge University Press, pp. 195–208.

Crockett, C. and Bowden, D. (1994) 'Challenging conventional wisdom for housing monkeys', *Laboratory Animal*, 24: 29–33.

Crockett, C., Bowers, C., Sackett, G., and Bowden, D. (1993a) 'Urinary cortisol responses of longtailed macaques to five cage sizes, tethering, sedation and room change', *American Journal of Primatology*, 30: 55–73.

Crockett, C., Bowers, C., Shimoji, M., Leu, M., Bellanca, R., and Bowden, D. (1993b) 'Appetite and urinary cortisol responses to different cage sizes in female pigtailed macaques', *American Journal of Primatology*, 31: 305 (abstract).

Cronin, M. A. (1993) 'Mitochondrial DNA in wildlife taxonomy and conservation biology: cautionary notes', *Wildlife Society Bulletin*, 21: 339–48.

Csuti, B., Sargent, E. L., and Bechert, U. S. (eds) (2001) *The Elephant's Foot: Prevention and Care of Foot Conditions in Captive Asian and African Elephants*, Ames, IA: Iowa State University Press.

Cueto, G. R., Allekotte, R., and Kravetz, F. O. (2000) 'Scurvy in capybaras bred in captivity in Argentina', *Journal of Wildlife Diseases*, 36: 97–101.

Culik, B., Wilson, R., and Bannasch, R. (1993) 'Flipper-bands on penguins: what is the cost of a life-long commitment?', *Marine Ecology Progress Series*, 98: 209–14.

Cunningham, A. A., Frank, M. J., Croft, P., Clarke, D., and Pearce-Kelly, P. (1997) 'Mortality of captive British wartbiter crickets: implications for reintroduction programs', *Journal of Wildlife Diseases*, 33: 673–6.

d **Dalgetty, G.** (2007) 'Zoo bill receives all-party support', *Toronto Observer*, 15 March, available online at http://www.tobserver.com/CYCLEFEB07/ 15-03-07-DalgettyZoo.html .

Dalley, S. (1993) 'Ancient Mesopotamian gardens and the identification of the Hanging Gardens of Babylon resolved', *Garden History*, 21: 1–13.

Dalton, R. and Buchanan-Smith, H. M. (2005) 'A mixed-species exhibit for Goeldi's monkeys and pygmy marmosets *Callimico goeldii* and *Callithrix pygmaea* at Edinburgh Zoo', *International Zoo Yearbook*, 39: 176–84.

Dantzer, R. (1994) 'Animal welfare methodology and criteria', *Revue Scientifique et Technique de L'Office International des Epizooties*, 13: 277–302.

D'Août, K., Aerts, P., Clercq, D. D., Schoonaert, K., Vereecke, E., and Elsacker, L. V. (2001) 'Studying bonobo (*Pan paniscus*) locomotion using an integrated setup in a zoo environment: preliminary results', *Primatologie*, 4: 191–206.

Da Silva, M. A. M. and da Silva, J. M. C. (2007) 'A note on the relationships between visitor interest and characteristics of the mammal exhibits in Recife Zoo, Brazil', *Applied Animal Behaviour Science*, 105: 223–6.

Davey, G. (2006a) 'Visitor behavior in zoos: a review', *Anthrozoös*, 19: 143–57.

Davey, G. (2006b) 'Relationship between exhibit naturalism, animal visibility and visitor interest in a Chinese zoo', *Applied Animal Behaviour Science*, 96: 93–102.

Davey, G. (2007) 'An analysis of country, socio-economic and time factors on worldwide zoo attendance during a 40-year period', *International Zoo Yearbook*, 41: 217–25.

Davey, G., Henzi, P., and Higgins, L. (2005) 'The influence of environmental enrichment on Chinese visitor behaviour', *Journal of Applied Animal Welfare Science*, 8: 131–40.

Davies, N. B. (1992) *Dunnock Behaviour and Social Evolution*, Oxford: Oxford University Press.

Davis, H. (2002) 'Prediction and preparation: Pavlovian implications of research animals discriminating among humans', *ILAR Journal*, 43: 19–26.

Davis, N., Schaffner, C. M., and Smith, T. E. (2005) 'Evidence that zoo visitors influence HPA activity in spider monkeys (*Ateles geoffroyii rufiventris*)', *Applied Animal Behaviour Science*, 90: 131–41.

Davis, T. and Ovaska, K. (2001) 'Individual recognition of amphibians: effects of toe clipping and fluorescent tagging on the salamander *Plethodon vehiculum*', *Journal of Herpetology*, 35: 217–25.

Dawkins, M. S. (1983) 'Battery hens name their price: consumer demand theory and the measurement of ethological "needs"', *Animal Behaviour*, 31: 1195–205.

Dawkins, M. S. (1988) 'Behavioural deprivation: a central problem in animal welfare', *Applied Animal Behaviour Science*, 20: 209–25.

Dawkins, M. S. (1997) 'D. G. M. Wood-Gush Memorial lecture: Why has there not been more progress in animal welfare research?', *Applied Animal Behaviour Science*, 53: 59–73.

Dawkins, M. S. (2003) 'Behaviour as a tool in the assessment of animal welfare', *Zoology*, 106: 383–7.

Dawkins, M. S. (2006) 'Through animal eyes: what behaviour tells us', *Applied Animal Behaviour Science*, 100: 4–10.

Dayan, A. (1971) 'Comparative neuropathology of aging: studies of the brain of 47 species of vertebrates', *Brain*, 94: 31–42.

Dayrell, E. and Pullen, K. (2003) 'Post-occupancy evaluation of a red river hog (*Potamochoerus porcus*) enclosure' in T. Gilbert (ed.), *Proceedings of the Fifth Annual Symposium on Zoo Research*, 7–8 July, Maxwell Park, Winchester, pp. 226–30.

Deagle, B. and Tollit, D. (2007) 'Quantitative analysis of prey DNA in pinniped faeces: potential to estimate diet composition?', *Conservation Genetics*, 8: 743–7.

Deardorff, T. L. and Throm, R. (1988) 'Commercial blast-freezing of third-stage *Anisakis simplex* larvae encapsulated in salmon and rockfish', *Journal of Parasitology*, 74: 600–3.

De Azevedo, C. S. and Young, R. J. (2006a) 'Behavioural responses of captive-born greater rheas *Rhea americana Linnaeus* (Rheiformes: Rheidae) submitted to antipredator training', *Revista Brasileira de Zoologia*, 23: 186–93.

De Azevedo, C. S. and Young, R. J. (2006b) 'Do captive-born greater rheas *Rhea americana Linnaeus* (Rheiformes: Rheidae) remember antipredator training?', *Revista Brasileira de Zoologia*, 23: 194–201.

Defra (Department for Environment, Food and Rural Affairs) (2002) *Zoo Inspectors' Training Seminar*, 5–7 April, Bath University, available online at http://www.defra.gov.uk/wildlife-countryside/gwd/zoo-inspectors/bath-seminar2002.pdf.

Defra (Department for Environment, Food and Rural Affairs) (2003) *Zoo Licensing Act 1981*, Circular 02/2003, available online at http://www.defra.gov.uk/wildlife-countryside/gwd/govt-circular022003.pdf.

Defra (Department for Environment, Food and Rural Affairs) (2004) *Secretary of State's Standards of Modern Zoo Practice* (rev'd edn), available online at http://www.defra.gov.uk/wildlife-countryside/gwd/zooprac/index.htm.

Defra (Department for Environment, Food and Rural Affairs) (2007a) *Animal Welfare: Protecting Domestic or Captive Animals from Cruelty*, available online at http://www.defra.gov.uk/animalh/welfare/domestic/index.htm.

Defra (Department for Environment, Food and Rural Affairs) (2007b) *Zoos Forum Handbook*, available online at http://www.defra.gov.uk/wildlife-countryside/gwd/zoosforum/handbook/.

Defra (Department for Environment, Food and Rural Affairs) (2008) *Animal Welfare Act 2006*, available online at http://www.defra.gov.uk/animalh/welfare/act/index.htm.

DfES (Department for Education and Skills) (2006) *Learning Outside the Classroom Manifesto*, Nottingham: DfES Publications.

Dembiec, D., Snider, R., and Zanella, A. (2004) 'The effects of transport stress on tiger physiology and behaviour', *Zoo Biology*, 23: 335–46.

Dempsey, J. L. (2004) 'Fruit bats: nutrition and dietary husbandry', *AZA Nutrition Advisory Group Handbook*, Fact Sheet 14, available online at http://www.nagonline.net.

De Passillé, A. M., Rushen, J., Ladewig, J., and Petherick, C. (1996) 'Dairy calves' discrimination of people based on previous handling', *Journal of Animal Science*, 74: 969–74.

De Rouck, M., Kitchener, A. C., Law, G., and Nelissen, M. (2005) 'A comparative study of the influence of social housing conditions on the behaviour of captive tigers (*Panthera tigris*)', *Animal Welfare*, 14: 229–38.

Desmond, T. and Laule, G. (1994) 'Use of positive reinforcement training in the management of species for reproduction', *Zoo Biology*, 13: 471–7.

De Vos, V. (2003) 'Anthrax' in M. E. Fowler and R. E. Miller (eds), *Zoo and Wild Animal Medicine* (5th edn), Philadelphia, PA: Saunders (Elsevier), pp. 696–9.

De Waal, F. B. M. (1982) *Chimpanzee Politics*, London: Jonathan Cape.

De Waal, F. B. M. (1989) 'The myth of a simple relation between space and aggression in captive primates', *Zoo Biology (Supplement)*, 1: 141–8.

Dewsbury, D. A. (1992) 'Studies on rodent–human interactions in animal psychology' in H. Davis and D. Balfour (eds), *The Inevitable Bond: Examining Scientist–Animal Interactions*, Cambridge: Cambridge University Press, pp. 27–43.

Diamond, M. C. (2001) 'Response of the brain to enrichment', *Anais da Academia Brasileira de Ciencias*, 73: 211–20.

Dickie, L. A., Bonner, J. P., and West, C. (2007) '*In situ* and *ex situ* conservation: blurring the boundaries between zoos and the wild' in A. Zimmerman, M. Hatchwell, L. A. Dickie, and C. West (eds), *Zoos in the 21st Century*, Cambridge: Cambridge University Press, pp. 220–35.

Dickinson, H. C. and Fa, J. E. (1997) 'Ultraviolet light and heat source selection in captive spiny-tailed iguanas (*Oplurus cuvieri*)', *Zoo Biology*, 16: 391–401.

Dierenfeld, E. S. (1997a) 'Captive wild animal nutrition: a historical perspective', *Proceedings of the Nutrition Society*, 56: 989–99.

Dierenfeld, E. S. (1997b) 'Orang-utan nutrition' in C. Sodaro (ed.), *Orang-utan SSP Husbandry Manual*, Brookfield, IL: Orang-utan SSP and Brookfield Zoo.

Dierenfeld, E. S. (2005) 'Advancing zoo animal nutrition through global synergy', *International Zoo Yearbook*, 39: 29–35.

Dierenfeld, E. S., Atkinson, S., Craig, A. M., Walker, K. C., Streich, W. J., and Clauss, M. (2005) 'Mineral concentrations in serum/plasma and liver tissue of captive and free-ranging Rhinoceros species', *Zoo Biology*, 24: 51–72.

Dingemanse, N. J., Both, C., Drent, P. J., van Oers, K., and van Noordwijk, A. J. (2002) 'Repeatability and heritability of exploratory behaviour in great tits from the wild', *Animal Behaviour*, 64: 929–38.

Disney, W., Green, J., Forsythe, K., Wiemers, J., and Weber, S. (2001) 'Benefit–cost analysis of animal identification for disease prevention and control', *Revue Scientifique et Technique de L'Office International des Epizooties*, 20: 385–405.

Dobbs, T. and Fry, A. (2008) 'The development and progression of the primate enrichment timetables at Paignton Zoo' in V. Hare (ed.), *Eighth Conference on Environmental Enrichment*, 5–10 August 2007, Vienna: The Shape of Enrichment, Inc. (in press).

Doerfler, R. L. and Peters, K. J. (2006) 'The relativity of ethical issues in animal agriculture related to different cultures and production conditions', *Livestock Science*, 103: 257–62.

Dol, M., Fentener van Vlissingen, M., Kasanmoentalib, S., Visser, T., and Zwart, H. (1999) (eds) *Recognizing the Intrinsic Value of Animals Beyond Animal Welfare*, Assen, the Netherlands: Van Gorcum.

Dolins, F. L. (1999) *Attitudes to Animals: Views in Animal Welfare*, Cambridge: Cambridge University Press.

Dollinger, P. (2007) ' "Balai" Directive of the European Union: difficult veterinary legislation' in M. E. Fowler and R. E. Miller (eds), *Zoo and Wild Animal Medicine* (6th edn), St Louis, MO: Saunders (Elsevier), pp. 68–74.

Donahue, J. and Trump, E. (2006) *The Politics of Zoos: Exotic Animals and their Protectors*, DeKalb, IL: Northern Illinois University Press.

Doncaster, C. P., Dickman, C. R., and MacDonald, D. W. (1990) 'Feeding ecology of red foxes (*Vulpes vulpes*) in the City of Oxford, England', *Journal of Mammalogy*, 71: 188–94.

Donoghue, A. M., Blanco, J. M., Gee, G. F., Kirby, Y. K., and Wildt, D. E. (2003) 'Reproductive technologies and challenges in avian conservation and management' in: W. Holt, A. Pickard, J. Rodger, and D. Wildt (2003) *Reproductive Science and Integrated Conservation*, Cambridge: Cambridge University Press, pp. 321–37.

Donohue, K. C. and Dufty, A. M. (2006) 'Sex determination of red-tailed hawks (*Buteo jamaicensis calurus*) using DNA analysis and morphometrics', *Journal of Field Ornithology*, 77: 74–9.

Dooley, M. and Pineda, M. (2003) 'Patterns of reproduction' in M. H. Pineda and M. P. Dooley (eds), *McDonald's Veterinary Endocrinology and Reproduction*, Oxford: Blackwell Publishing, pp. 377–94.

Douglas-Hamilton, I., Bhalla, S., Wittemyer, G., and Vollrath, F. (2006) 'Behavioural reactions of elephants towards a dying and deceased matriarch', *Applied Animal Behaviour Science*, 100: 87–102.

Duckler, G. (1998) 'An unusal osteological formation in the posterior skulls of captive tigers (*Panthera tigris*)', *Zoo Biology*, 17: 135–42.

Dudink, S., Simonse, H., Marks, I., de Jonge, F. H., and Spruijt, B. M. (2006) 'Announcing the arrival of enrichment increases play behaviour and reduces weaning-stress-induced behaviours of piglets directly after weaning', *Applied Animal Behaviour Science*, 101: 86–101.

Duncan, I. J. H. (1978) 'The interpretation of preference tests in animal behaviour', *Applied Animal Ethology*, 4: 197–200.

Duncan, I. J. H. (1993) 'Welfare is to do with what animals feel', *Journal of Agricultural and Environmental Ethics*, 6: 8–14.

Duncan, I. J. H. (2005) 'Science-based assessment of animal welfare: farm animals', *Revue Scientifique et Technique de L'Office International des Epizooties*, 24: 483–92.

Duncan, I. J. H. (2006) 'The changing concept of animal sentience', *Applied Animal Behaviour Science Sentience in Animals*, 100: 11–19.

Duncan, I. J. H. and Fraser, D. (1997) 'Understanding animal welfare' in M. Appleby and B. Hughes (eds), *Animal Welfare*, Wallingford, Oxon: CABI Publishing, pp. 19–31.

Duncan, M. (2003) 'Fungal diseases in all taxa' in M. E. Fowler and R. E. Miller (eds), *Zoo and Wild Animal Medicine* (5th edn), Philadelphia, PA: Saunders (Elsevier), pp. 727–32.

Durnin, M., Palsbell, P. J., Ryder, O., and McCullough, D. (2007) 'A reliable genetic technique for sex determination of giant panda (*Ailuropoda melanoleuca*) from non-invasively collected hair samples', *Conservation Genetics*, 8: 715–20.

Durrell, G. (1976) *The Stationary Ark*, London: Collins.

e **Eaton, G. G., Kelley, S. T., Axthelm, M. K., Iliffsizemore, S. A., and Shiigi, S. M.** (1994) 'Psychological well-being in paired adult female rhesus (*Macaca mulatta*)', *American Journal of Primatology*, 33: 89–99.

EAZA (European Association of Zoos and Aquaria) (2003) 'From the EAZA Office: fifteen years E(C)AZA', *EAZA Newsletter*, 44: 5–8.

EAZA (European Association of Zoos and Aquaria) (2008) *EAZA Education Standards*, Amsterdam: EAZA.

EC (European Commission) (1999) Council Directive 1999/22/EC of 29 March (1999) relating to the keeping of wild animals in zoos, *Official Journal of the European Communities*, L94/24 (09/04/1999).

EC (European Commission) (2006) *The Convention on Biological Diversity: Implementation in the European Union*, Luxembourg: Office for Official Publications of the European Communities.

Edberg, S. (2004) 'The algae: marine mammal enrichment at Kolmarden Zoo', *The Shape of Enrichment*, 13: 1–3.

EFSA (European Food Safety Authority) (2008) 'Avian influenza', available online at `http://www.efsa.europa.eu/EFSA/KeyTopics/efsa_locale-1178620753812_AvianInfluenza.htm`.

Egliston, K., McMahon, C., and Austin, M. (2007) 'Stress in pregnancy and infant HPA axis function: conceptual and methodological issues relating to the use of salivary cortisol as an outcome measure', *Psychoneuroendocrinology*, 32: 1–32.

Eisenberg, J. and Kleiman, D. (1977) 'The usefulness of behaviour studies in developing captive breeding programmes for mammals', *International Zoo Yearbook*, 17: 81–8.

Ellis, D. and Dein, F. (1996) 'Special techniques: part E flight restraint' in D. Ellis, G. Gee, and C. Mirande, *Cranes: Their Biology, Husbandry and Conservation*, Washington DC/Baraboo, WI: US Department of the Interior, National Biological Service/International Crane Foundation, pp. 241–44.

Elson, H. (2007) 'An investigation into the short-term effects of environmental enrichment on the behaviour of psittacines in captivity', unpublished PhD thesis, Trinity College, University of Dublin.

Elson, H. and Marples, N. (2001) 'Effects of environmental enrichment on parrots in captivity' in S. Wehnelt and C. Hudson (eds), *Proceedings of the Third Annual Symposium on Zoo Research*, 9–10 July, Chester, BIAZA, pp. 1–8.

Embury, A. S. (1992) 'Gorilla rainforest at Melbourne Zoo', *International Zoo Yearbook*, 31: 203–13.

Erickson, G. M., Lappin, A. K., and Vliet, K. A. (2003) 'Comparison of the bite-force performance between long-term captive and wild American alligators (*Alligator mississippiensis*)', *Journal of Zoology*, 262: 21–8.

Erman, A. (1971) *Life in Ancient Egypt*, Mineola, NY: Dover Publications.

Erwin, J. (1979) 'Aggression in captive macaques: interactions of social and spatial factors' in J. Erwin, T. Maple, and G. Mitchell (eds), *Captivity and Behaviour of Primates in Breeding Colonies, Laboratories and Zoos*, New York, NY: Van Norstrand Reinhold, pp. 139–71.

Erwin, J. and Deni, R. (1979) 'Strangers in a strange land: abnormal behaviours or abnormal environments?' in J. Erwin, T. L. Maple, and G. Mitchell (eds), *Captivity and Behaviour of Primates in Breeding Colonies, Laboratories and Zoos*, New York, NY: Van Norstrand Reinhold, pp. 1–28.

Essler, W. and Folkjun, G. (1961) 'Determination of physiological rhythms of unrestrained animals by radio telemetry', *Nature*, 190: 90–1.

Estep, D. Q. and Baker, S. C. (1991) 'The effects of temporary cover on the behavior of socially housed stumptailed macaques (*Macaca arctoides*)', *Zoo Biology*, 10: 465–72.

Estep, D. Q. and Hetts, S. (1992) 'Interactions, relationships, and bonds: the conceptual basis for scientist–animal relations' in H. Davis and D. Balfour (eds), *The Inevitable Bond: Examining Scientist–Animal Interactions*, Cambridge: Cambridge University Press, pp. 6–26.

Ettah, U. (1997) The Impact of Food Preparation on Feeding in *Celebes macaques* at Jersey Zoo, unpublished diploma thesis, Trinity, Jersey: Durrell Wildlife Preservation Trust.

Eurogroup for Animal Welfare (2006) *Report on the Implementation of the EU Zoo Directive*, available online at `http://www.eurogroupanimalwelfare.org/policy/pdf/zooreportmar2006.pdf`.

Evans, J. E., Cuthill, I. C., and Bennett, A. T. D. (2006) 'The effect of flicker from fluorescent lights on mate choice in captive birds', *Animal Behaviour*, 72: 393–400.

Ewbank, R. (1985) 'Behavioral responses to stress in farm animals' in G. Moberg (ed.), *Animal Stress*, Baltimore, MD: Waverly Press Inc., pp. 71–9.

Ewer, R. F. (1968) *Ethology of Mammals*, London: Elek Books.

Exner, C. and Unshelm, J. (1997) 'Climatic condition and airborne contaminants in buildings of wild cats kept in zoos', *Zentralblatt für Hygiene und Umweltmedizin*, 199: 497–512.

f

Fa, J. E. (1989) 'Influence of people on the behaviour of display primates' in E. F. Segal (ed.), *Housing, Care and Psychological Well-Being of Captive and Laboratory Primates*, Park Ridge, IL: Noyes Publications, pp. 270–90.

Fàbregas, M. and Guillén-Salazar, F. (2007) 'Social compatibility in a newly formed all-male group of white crowned mangabeys (*Cercocebus atys lunulatus*)', *Zoo Biology*, 26: 63–9.

Fairbanks, L. A., Newman, T. K., Bailey, J. N., Jorgensen, M. J., Breidenthal, S. E., Ophoff, R. A., Comuzzie, A. G., Martin, L. J., and Rogers, J. (2004) 'Genetic contributions to social impulsivity and aggressiveness in vervet monkeys', *Biological Psychiatry*, 55: 642–7.

Falk, J. H., Reinhard, E. M., Vernon, C. L., Bronnenkant, K., Heimlich, J. E., and Deans, N. L. (2007) *Why Zoos and Aquariums Matter: Assessing the Impact of a Visit to a Zoo or Aquarium*, Silver Springs, MD: AZA.

Farlin, M. and Baumans, V. (2003) 'Environmental enrichment for mice: a hammock in the cage', *Scandinavian Journal of Laboratory Animal Science*, 30: 45–6.

Farm Animal Welfare Council (1992) 'FAWC updates the five freedoms', *Veterinary Record*, 131: 357.

Farmer, H. and Melfi, V. (2008) 'Is it music to their ears?' in V. Hare (ed.), *Eighth Conference on Environmental Enrichment*, 5–10 August 2007, Vienna: The Shape of Enrichment, Inc. (in press).

Farrell, M. A., Barry, E., and Marples, N. (2000) 'Breeding behavior in a flock of Chilean flamingos (*Phoenicopterus chilensis*) at Dublin Zoo', *Zoo Biology*, 19: 227–37.

Faust L. J., Earnhardt, J. E., and Thompson, S. D. (2006) 'Is reversing the decline of Asian elephants in captivity possible? An individual-based modeling approach', *Zoo Biology*, 25: 201–18.

Fekete, J. M., Norcross, J. L., and Newman, J. D. (2000) 'Artificial turf foraging boards as environmental enrichment for pair-housed female squirrel monkeys', *Contemporary Topics in Laboratory Animal Science*, 39: 22–6.

Fenolio, D. B., Graening, G. O., Collier, B. A., and Stout, J. F. (2006) 'Coprophagy in a cave-adapted salamander: the importance of bat guano examined through nutritional and stable isotope analyses', *Proceedings of the Royal Society Series B: Biological Sciences*, 273: 439–43.

Ferner, J. (1979) 'A review of marking techniques for amphibians and reptiles' in Society for the Study of Amphibians and Reptiles, *Herpetological Circular No. 9*, Shoreview, MN: SSAR.

Festa-Bianchet, M. and Apollonio, M. (2003) *Animal Behaviour and Wildlife Conservation*, Washington DC: Island Press.

Fidgett, A. (2005) 'Standardizing nutrition information within husbandry guidelines: the essential ingredients', *International Zoo Yearbook*, 39: 132–8.

Fidgett, A. and Dierenfeld, E. S. (2007) 'Minerals and stork nutrition' in M. E. Fowler and R. E. Miller (eds), *Zoo and Wild Animal Medicine* (6th edn), St Louis, MO: Saunders (Elsevier), pp. 206–13.

Fidgett, A., Clauss, M., Ganslosser, U., Hatt, J. M., and Niijboer, J. (eds) (2003) *Zoo Animal Nutrition: Vol. II*, Fürth: Filander Verlag.

Fidgett, A., Clauss, M., Eulenberger, K., Hatt, J.-M., Hume, I., Janssens, G., and Nijboer, J. (eds) (2006) *Zoo Animal Nutrition: Vol. III*, Fürth: Filander Verlag.

Fiedeldey, A. (1994) 'Wild animals in a wilderness setting: an ecosystemic experience?', *Anthrozoös*, 7: 113–23.

Filadelfi, A. and Castrucci, A. (1996) 'Comparative aspects of the pineal/melatonin system in poikilothermic vertebrates', *Journal of Pineal Research*, 20: 175–86.

Finlay, T. W. and Maple, T. L. (1986) 'A survey of research in American zoos and aquariums', *Zoo Biology*, 5: 261–8.

Finlay, T. W., James, L. R., and Maple, T. L. (1988) 'People's perceptions of animals: the influence of zoo environment', *Environment and Behavior*, 20: 506–28.

Fish, K. D., Sauther, M. L., Loudon, J. E., and Couzzo, F. P. (2007) 'Coprophagy by wild ring-tailed lemurs (*Lemur catta*) in human-disturbed locations adjacent to the Beza Mahafaly Special Reserve, Madagascar', *American Journal of Primatology*, 69: 713–18.

Fisher, J. and Hinde, R. A. (1949) 'The opening of milk bottles by birds', *British Birds*, 42: 347–57.

Fitch, H. and Fagan, D. A. (1982) 'Focal palatine erosion associated with dental malocclusion in captive cheetahs', *Zoo Biology*, 1: 295–310.

Flach, E. (2003) '*Cervidae* and *Tragulidae*' in M. E. Fowler and R. E. Miller (eds), *Zoo and Wild Animal Medicine* (5th edn), Philadelphia, PA: Saunders (Elsevier), pp. 634–49.

Flach, E., Stevenson, M. F., Henderson, G. M. (1990) 'Aspergillosis in gentoo penguins (*Pygoscelis papua*) at Edinburgh Zoo, 1964–1988', *Veterinary Record*, 126: 81–5.

Flammer, K. (2003) 'Chlamydiosis' in M. E. Fowler and R. E. Miller (eds), *Zoo and Wild Animal Medicine* (5th edn), Philadelphia, PA: Saunders (Elsevier), pp. 718–23.

Flecknell, P. and Molony, V. (2003) 'Pain and injury' in M. C. Appleby and B. O. Hughes (eds), *Animal Welfare*, Wallingford, Oxon: CABI Publishing, pp. 63–73.

Flecknell, P. and Waterman-Pearson, A. (2000) *Pain Management in Animals*, London: W. B. Saunders.

Fleetwood, A. J. and Furley, C. W. (1990) 'Spongiform encephalopathy in an eland', *Veterinary Record*, 126: 408–9.

Flesness, N. R. (2003) 'International Species Information System (ISIS): over 25 years of compiling global animal data to facilitate collection and population management', *International Zoo Yearbook*, 38: 53–61.

Fletcher, W. J., Fielder, D. R., and Brown, I. W. (1989) 'Comparison of freeze- and heat-branding techniques to mark the coconut crab *Birgus latro* (Crustacea, Anomura)', *Journal of Experimental Marine Biology and Ecology*, 127: 245–51.

Flew, A. (ed.) (1979) *A Dictionary of Philosophy*, London: Pan.

Fooks, A. R., Brookes, S. M., Johnson, N., McElhinney, L. M., and Hutson, A. M. (2003) 'European bat lyssaviruses: an emerging zoonosis', *Epidemiology and Infection*, 131: 1029–39.

Foose, T. J. (1980) 'Demographic management of endangered species in captivity', *International Zoo Yearbook*, 20: 154–66.

Forthman, D. L., Elder, S. D., Bakeman, R., Kurkowski, T. W., Noble, C. C., and Winslow, S. W. (1992) 'Effects of feeding enrichment on behavior of three species of captive bears', *Zoo Biology*, 11: 187–95.

Forthman-Quick, D. L. (1984) 'An integrative approach to environmental engineering in zoos', *Zoo Biology*, 312: 65–77.

Fowler, M. E. and Miller, R. E. (eds) (1993) *Zoo and Wild Animal Medicine* (3rd edn), Philadelphia, PA: W. B. Saunders.

Fowler, M. E. and Miller, R. E. (eds) (1999) *Zoo and Wild Animal Medicine* (4th edn), Philadelphia, PA: W.B. Saunders.

Fowler, M. E. and Miller, R. E. (eds) (2003) *Zoo and Wild Animal Medicine* (5th edn), St Louis, MO: Saunders (Elsevier).

Fowler, M. E. and Miller, R. E. (eds) (2007) *Zoo and Wild Animal Medicine* (6th edn), St Louis, MO: Saunders (Elsevier).

Frankham, R., Hemmer, H., Ryder, O., Cothran, E., Soulé, M., Murray, N., and Synder, M. (1986) 'Selection of captive populations', *Zoo Biology*, 5: 127–38.

Frankham, R., Ballou, J. D., and Briscoe, D. A. (2002) *Introduction to Conservation Genetics*, Cambridge: Cambridge University Press.

Frankham, R., Ballou, J. D., and Briscoe, D. A. (2004) *A Primer of Conservation Genetics*, Cambridge: Cambridge University Press.

Fraser, A. F. and Broom, D. M. (1990) *Farm Animal Behaviour and Animal Welfare* (3rd edn), London: Baillière Tindall.

Fraser, D. (2008) *Understanding Animal Welfare*, Oxford: Blackwell.

Fraser, D. and Matthews, L. R. (1997) 'Preference and motivation testing' in M. Appleby and B. Hughes (eds), *Animal Welfare*, Wallingford, Oxon: CABI Publishing, pp. 159–72.

Fraser, D. J. and Bernatchez, L. (2001) 'Adaptive evolutionary conservation: towards a unified concept for defining conservation units', *Molecular Ecology*, 10: 2741–52.

Fraser, D. J., Ritchie, J., and Fraser, A. (1975) 'The term "stress" in a veterinary context', *British Veterinary Journal*, 131: 653–62.

Frediani, K. (2008) 'The ethical use of plants in zoos: informing selection choices, uses and management strategies', *International Zoo Yearbook* (in press).

Freeland, W. J. (1991) 'Plant secondary metabolites: biochemical coevolution with herbivores' in R. T. Palo and C. T. Robbins (eds), *Plant Chemical Defenses Against Mammalian Herbivory*, Boca Raton, FL: CRC Press, pp. 61–82.

Freeman, E., Wiess, E., and Brown, J. (2004) 'Examination of the interrelationships of behavior, dominance status, and ovarian activity in captive Asian and African elephants', *Zoo Biology*, 23: 431–48.

Frézard, A. and Le Pape, G. (2003) 'Contribution to the welfare of captive wolves (*Canis lupus lupus*): a behavioral comparison of six wolf packs', *Zoo Biology*, 22: 33–44.

Friend, T. H. and Parker, M. L. (1999) 'The effect of penning versus picketing on stereotypic behavior of circus elephants', *Applied Animal Behaviour Science*, 64: 213–25.

Fry, A. and Dobbs, T. (2005) 'From junk to enrichment: uses for a camera film canister', *The Shape of Enrichment*, 14: 1–3.

Frye, F. L. (1992) *Biomedical and Surgical Aspects of Captive Reptile Husbandry*, Melbourne, FL: Krieger Publishing.

g Galis, F., Wagner, G., and Jackson, E. (2003) 'Why is limb regeneration possible in amphibians but not in reptiles, birds and mammals?', *Evolution and Development*, 5: 208–20.

Garcia, L. S. (1999) *Practical Guide to Diagnostic Parasitology*, Washington, DC: ASM Press.

Garner, R. (2002) 'First "submarium" matches success of Eden Project', *The Independent*, 28 December, 2002.

Garner, J. P. and Mason, G. J. (2002) 'Evidence for a relationship between cage stereotypies and behavioural disinhibition in laboratory rodents', *Behavioural Brain Research*, 136: 83–92.

Garner, J. P., Meehan, C. L., and Mench, J. A. (2003) 'Stereotypies in caged parrots, schizophrenia and autism: evidence for a common mechanism', *Behavioural Brain Research*, 145: 125–34.

Garner, J. P., Meehan, C. L., Famula, T. R., and Mench, J. A. (2006) 'Genetic, environmental, and neighbor effects on the severity of stereotypies and feather picking in orange-winged Amazon parrots (*Amazona amazonica*): an epidemiological study', *Applied Animal Behaviour Science*, 96: 153–68.

Gartrell, B. D., Raidal, S. R., and Jones, S. M. (2003) 'Renal disease in captive swift parrots (*Lathamus discolor*): clinical findings and disease management', *Journal of Avian Medicine and Surgery*, 17: 213–23.

Gascon, C., Collins, J. P., Moore, R. D., Church, D. R., McKay, J. E., and Mendelson III, J. R. (eds) (2007) *Amphibian Conservation Action Plan*, Gland/Cambridge: IUCN/SSC Amphibian Specialist Group.

Gauthier-Clerc, M. and Le Maho, Y. (2001) 'Beyond bird marking with rings', *Ardea (Special issue)*, 89: 221–30.

Geissmann, T. (2007) 'Status reassessment of the gibbons: results of the Asian Primate Red List Workshop 2006', *Gibbon Journal*, 3: 5–15.

Genty, E. and Roeder, J.-J. (2006) 'Self-control: why should sea lions, *Zalophus californianus*, perform better than primates?', *Animal Behaviour*, 72: 1241–7.

Geraci, J. R. (1986) 'Nutrition and nutritional disorders' in M. E. Fowler (ed.), *Zoo and Wild Animal Medicine* (2nd edn), Philadelphia, PA: W. B. Saunders, pp. 760–4.

Gerald, M., Weiss, A., and Ayala, J. (2006) 'Artifical colour treatment mediates aggression among unfamiliar vervet monkeys (*Cercopithecus aethiops*): a model for introducing primates with colourful sexual skin', *Animal Welfare*, 15: 363–9.

Gerhmann, W., Ferguson, G., Odom, T., Roberts, D., and Barcelone, W. (1991) 'Early growth and bone mineralization on the iguanid lizard *Sceloporus occidentalis* in captivity: is vitamin D supplementation or ultraviolet B irradiation necessary?', *Zoo Biology*, 10: 409–16.

Gesualdi, J. (2001) 'North America: licensing and accreditation' in C. E. Bell (ed.), *Encyclopedia of the World's Zoos*, Chicago, IL/London: Fitzroy Dearborn, pp. 883–5.

Gillespie, D., Frye, F. L., Stockham, S. L., and Fredeking, T. (2001) 'Blood values in wild and captive Komodo dragons (*Varanus komodoensis*)', *Zoo Biology*, 19: 495–509.

Gittleman, J. L. (1994) 'Are the pandas successful specialists or evolutionary failures?', *BioScience*, 44: 456–64.

Gittleman, J. L. and Thompson, S. D. (1988) 'Energy allocation in mammalian reproduction', *American Zoologist*, 28: 863–75.

Glander, K. E. (1994) 'Non-human primate self-medication with wild plant foods' in N. Etkin (ed.), *Eating on the Wild Side: The Pharmacologic, Ecologic and Social Implications of Using Noncultigens*, Tucson, AZ: University of Arizona Press, pp. 227–39.

Glatston, A. R. (1998) 'The control of zoo populations with special reference to primates', *Animal Welfare*, 7: 269–81.

Glatston, A. R., Geilvoet-Soeteman, E., Hora-Pecek, E., and Van Hooff, J. A. R. A. M. (1984) 'The influence of the zoo environment on social behaviour of groups of cotton-topped tamarins, *Saguinus oedipus oedipus*', *Zoo Biology*, 3: 241–53.

Glatt, S. E., Francl, K. E., and Scheels, J. L. (2008) 'A survey of current dental problems and treatments of zoo animals', *International Zoo Yearbook*, 42: 206–13.

Goerke, B., Fleming, L., and Creel, M. (1987) 'Behavioral changes of a juvenile gorilla after a transfer to a more naturalistic environment', *Zoo Biology*, 6: 283–95.

Golani, I., Kafkafi, N., and Drai, D. (1999) 'Phenotyping stereotypic behaviour: collective variables, range of variation and predictability', *Applied Animal Behaviour Science*, 65: 191–220.

Golub, M. S., Keen, C. L., and Hendrickx, A. G. (1990) 'Food preference of young Rhesus monkeys fed marginally zinc deficient diets', *Primates*, 32: 49–59.

Gomendio, M., Cassinello, J., and Roldan, E. (2000) 'A comparative study of ejaculate traits in three endangered ungulates with different levels of inbreeding: fluctuating asymmetry as an indicator of reproductive and genetic stress', *Proceedings of the Royal Society of London Series B*, 267: 875–82.

Gomez, J-.C. (2005) 'Species comparative studies and cognitive development', *Trends in Cognitive Sciences Developmental Cognitive Neuroscience*, 9: 118–25.

Goossens, E., Dorny, P., Boomker, J., Vercammen, F., and Vercruysse, J. (2005) 'A 12-month survey of the gastro-intestinal helminths of antelopes, gazelles and giraffids kept at two zoos in Belgium', *Veterinary Parasitology*, 127: 303–12.

Gore, M., Hutchins, M., and Ray, J. (2006) 'A review of injuries caused by elephants in captivity: an examination of predominant factors', *International Zoo Yearbook*, 40: 51–62.

Gosling, S. (2001) 'From mice to men: what can we learn about personality from animal research?', *Psychological Bulletin*, 127: 45–86.

Graczyk, T. K., Cranfield, M. R., Brossy, J. J., Cockrem, J. F., Jouventin, P., and Seddon, P. J. (1995) 'Detection of avian malaria infections in wild and captive penguins', *Journal of the Helminth Society (Washington)*, 62: 135–41.

Graffam, W. S., Irlbeck, N. A., Grandin, T., Mallinckrodt, C., Cambre, R. C., and Phillips, M. (1995) 'Determination of vitamin E status and supplementation for Nyala (*Tragelaphus angasi*)', available online at http://www.nagonline.net/Proceedings/NAG1995/Determination%20of%20Vitamin%20E%20Status...Nyala.pdf.

Graham, S. (1996) 'Issues of surplus animals' in D. G. Kleiman, M. E. Allen, K. V. Thompson, and S. Lumpkin (eds), *Wild Mammals in Captivity: Principles and Techniques*, Chicago, IL: University of Chicago, pp. 290–6.

Grandin, T. (2000) 'Habituating antelope and bison to cooperate with veterinary procedures', *Journal of Applied Animal Welfare Science*, 3: 253–61.

Grech, K. S. (2004) 'Brief summary of the laws pertaining to zoos', Homepage of the Animal Legal and Historical Center, Michigan State University College of Law, available online at http://www.animallaw.info/articles/qvuszoos.htm.

Greenwood, A. (2003) 'Pox disease in all taxa' in M. E. Fowler and R. E. Miller (eds), *Zoo and Wild Animal Medicine* (5th edn), Philadelphia, PA: Saunders (Elsevier), pp. 737–41.

Gregory, N. (2004) *Physiology and Behaviour of Animal Suffering*, Oxford: Blackwell Science.

Griffin, A. S., Blumstein, D. T., and Evans, C. S. (2000) 'Training captive-bred or translocated animals to avoid predators', *Conservation Biology*, 14: 1317–26.

Griffin, A. S., Savani, R., Hausmanis, K., and Lefebvre, L. (2005) 'Mixed species aggregations in birds: zenaida doves, *Zenaida aurita*, respond to the alarm calls of carib grackles, *Quiscalus lugubris*', *Animal Behaviour*, 70: 507–15.

Griffin, D. R. (1992) *Animal Minds*, Chicago, IL: University of Chicago Press.

Grigor, P. N., Hughes, B. O., and Appleby, M. C. (1995) 'Effects of regular handling and exposure to an outside area on subsequent fearfulness and dispersal in domestic hens', *Applied Animal Behaviour Science*, 44: 47–55.

Grindrod, J. A. E. and Cleaver, J. A. (2001) 'Environmental enrichment reduces the performance of stereotypic circling behaviour in captive common seals (*Phoca vitulina*)', *Animal Welfare*, 10: 53–63.

Guilarte, T. R., Toscano, C. D., McGlothan, J. L., and Weaver, S. A. (2003) 'Environmental enrichment reverses cognitive and molecular deficits induced by developmental lead exposure', *Annals of Neurology*, 53: 50–6.

Gwynne, J. A. (2007) 'Inspiration for conservation: moving audiences to care' in A. Zimmerman, M. Hatchwell, L. Dickie, and C. West (eds), *Zoos in the 21st Century: Catalysts for Conservation?*, Conservation Biology Series No. 15, Cambridge: Cambridge University Press, pp. 51–62.

h **Habib, B. and Kumar, S.** (2007) 'Den shifting by wolves in semi-wild landscapes in the Deccan Plateau, Maharashtra, India', *Journal of Zoology*, 272: 259–65.

Hadley, C., Hadley, B., Ephraim, S., Yang, M., and Lewis, M. H. (2006) 'Spontaneous stereotypy and environmental enrichment in deer mice (*Peromyscus maniculatus*): reversibility of experience', *Applied Animal Behaviour Science*, 97: 312–22.

Hagenbeck, C. (1909) *Beasts and Men: Being Carl Hagenbeck's Experiences for Half a Century among Wild Animals* (abridged), H. S. R. Elliot and A. G. Thacker (trans.), London: Longmans.

Hahn, E. (1968) *Zoos*, London: Secker & Warburg.

Haig, S. M. (1998) 'Molecular contributions to conservation', *Ecology*, 79: 413–25.

Halachmi, I., Edan, Y., Maltz, E., Peiper, U. M., Moallem, U., and Brukental, I. (1998) 'A real-time control system for individual dairy cow food intake', *Computers and Electronics in Agriculture*, 20: 131–44.

Halbrooks, R. D., Swango, L. J., Schnurennberger, P. R., Mitchell, F. E., and Hill, E. P. (1981) 'Response of gray foxes to modified live-virus canine distemper vaccines', *Journal of the American Veterinary Medical Association*, 179: 1170–4.

Halliday, T. (1978) *Vanishing Birds: Their Natural History and Conservation*, London: Sidgwick & Jackson.

Halver, J. E. and Hardy, W. (2002) *Fish Nutrition*, London: London Academic Press.

Hamburger, L. (1988) 'Introduction of two young orang-utans *Pongo pygmaeus* into an established family group', *International Zoo Yearbook*, 27: 273–8.

Hamilton, W. D. (1964) 'The genetical evolution of social behaviour', *Journal of Theoretical Biology*, 7: 1–52.

Hammerstrom, F. (1970) *An Eagle in the Sky*, Ames, IA: Iowa State University Press.

Hancocks, D. (1995) 'Lions and tigers and bears, oh no!' in B. G. Norton, M. Hutchins, E. Stevens, and T. L. Maple (eds), *Ethics on the Ark: Zoos, Animal Welfare and Wildlife Conservation*, Washington DC/London: Smithsonian Institute Press, pp. 31–7.

Hancocks, D. (2001) *A Different Nature: The Paradoxical World of Zoos and Their Uncertain Future*, Berkeley, CA: University of California Press.

Hanna, J. (1996) 'Ambassadors of the wild' in M. Nichols (ed.) *Keepers of the Kingdom: The New American Zoo*, New York: Thomasson-Grant and Lickle, pp. 75–82.

Hansen, L. T. and Berthelsen, H. (2000) 'The effect of environmental enrichment on the behaviour of caged rabbits (*Oryctolagus cuniculus*)', *Applied Animal Behaviour Science*, 68: 163–78.

Hansen, S. J. and Møller, S. H. (2001) 'The application of a temperament test to on-farm selection of mink', *Acta Agriculturae Scandinavica, Section A—Animal Sciences*, 51 (Supplement Feb 2001): 93–8.

Hanson, E. (2002) *Animal Attractions: Nature on Display in American Zoos*, Princeton, NJ: Princeton University Press.

Harbone, J. B. (1991) 'The chemical basis of plant defense' in R. T. Palo and C. T. Robbins (eds) *Plant Defense Against Mammalian Herbivores*, Boca Raton, FL: CRC Press, pp. 45–60.

Hare, V. J. (2008) 'Enrichment gone wrong!' in Hare, V. J. (ed.) *Proceedings of the Eighth International Conference on Environmental Enrichment*, 5–10 August 2007, Scheonbrunn Zoo, Vienna, Austria (in press).

Hare, V. J. and Sevenich, M. (1999) 'Is it training or is it enrichment?' in V. J. Hare, K. E. Worley, and K. Myers (eds), *Proceedings of the Fourth Conference on Environmental Enrichment*, 29 August–3 September, Edinburgh: The Shape of Enrichment, Inc., pp. 40–7.

Hare, V. J., Ripsky, D., Battershill, R., Bacon, K., Hawk, K., and Swaisgood, R. R. (2003) 'Giant panda enrichment: meeting everyone's needs', *Zoo Biology*, 22: 401–16.

Harri, M., Mononen, J., Ahola, L., Plyusnina, L., and Rekliä, T. (2003) 'Behavioural and physiological differences between silver foxes selected and not selected for domestic behaviour', *Animal Welfare*, 12: 305–14.

Harris, L. D., Custer, L. B., Soranaka, E. T., Burge, J. R., and Ruble, G. R. (2001) 'Evaluation of objects and food for environmental enrichment of NZW rabbits', *Contemporary Topics in Laboratory Animal Science*, 40: 27–30.

Harrison, R., Ford, S., Young, J., Conley, A., and Freeman, A. (1990) 'Increased milk production versus reproductive and energy status of high-producing dairy cows', *Journal of Dairy Science*, 73: 2749–58.

Harvey, N. C., Farabaugh, S. M., and Druker, B. B. (2002) 'Effects of early rearing experience on adult behavior and nesting in captive Hawaiian crows (*Corvus hawaiiensis*)', *Zoo Biology*, 21: 59–75.

Hastings, B. E., Lowenstine, L. J., and Foster, J. W. (1991) 'Mountain gorillas and measles: ontogeny of a wildlife vaccination program' in R. E. Junge (ed.), *Proceedings of the Annual Meeting of the American Association of Zoo Veterinarians*, 28 September–3 October, Calgary, Alberta, Canada, pp. 198–205.

Hatchwell, M., Rübel, A., Dickie, L. A., West, C., and Zimmermann, A. (2007) 'Conclusion: the future of zoos' in A. Zimmerman, M. Hatchwell, L. Dickie, and C. West (eds), *Zoos in the 21st Century: Catalysts for Conservation?*, Conservation Biology Series No. 15, Cambridge: Cambridge University Press, pp. 343–60.

Hatt, J-.M. (2007) 'Raising giant tortoises' in M. E. Fowler and R. E. Miller (eds), *Zoo and Wild Animal Medicine* (6th edn), St Louis, MO: Saunders (Elsevier), pp. 144–53.

Hatt, J. M. and Clauss, M. (2001) 'Browse silage in zoo animal nutrition: feeding enrichment of browsers during winter', in Abstract Book Second European Zoo Nutrition Conference, 6–9 April, Winchester.

Hayward, J. and Rothenberg, M. (2004) 'Measuring success in the "Congo Gorilla Forest" conservation exhibition', *Curator*, 47: 261–82.

Hayward, M. W. and Kerley, G. I. H. (2005) 'Prey preferences of lions (*Panthera leo*)', *Journal of Zoology*, 267: 309–22.

Hebb, D. O. (1947) 'The effects of early experience on problem solving at maturity', *American Psychology*, 2: 306.

Hediger, H. (1950) *Wild Animals in Captivity* (trans.), London: Butterworth Scientific Publications.

Hediger, H. (1955) *Psychology of Animals in Zoos and Circuses*, London: Butterworth Scientific Publications.

Hediger, H. (1965) 'Man as a social partner of animals and vice versa', *Symposia of the Zoological Society of London*, 14: 291–300.

Hediger, H. (1970) *Man and Animal in the Zoo*, London: Routledge and Kegan Paul.

Heldman, D. R. (ed.) (2003) *Encyclopedia of Agricultural, Food, and Biological Engineering*, New York: Marcel Dekker, Inc.

Heleski, C. R. and Zanella, A. J. (2006) 'Animal science student attitudes to farm animal welfare', *Anthrozoös*, 19: 3–16.

Helme, A., Clayton, N., and Emery, N. (2008) 'Physical and cognitive enrichment for rooks (*Corvus frugilegus*)' in V. Hare (ed.) *Eighth Conference on Environmental Enrichment*, 5–10 August 2007, Vienna: The Shape of Enrichment, Inc. (in press).

Hemdal, J. (2006) *Advanced Marine Aquarium Techniques*, Neptune City, NJ: TFH Publications.

Hemsworth, P. H. (2003) 'Human–animal interactions in livestock production', *Applied Animal Behaviour Science*, 81: 185–98.

Henderson, J. V. and Waran, N. K. (2001) 'Reducing equine stereotypies using an Equiball™', *Animal Welfare*, 10: 73–80.

Herbert, P. and Bard, K. (2000) 'Orang-utan use of vertical space in an innovative habitat', *Zoo Biology*, 19: 239–51.

Herrnstein, R. J. (1979) 'Acquisition, generalization, and discrimination reversal of a natural concept', *Journal of Experimental Psychology: Animal Behaviour Processes*, 5: 116–29.

Hesterman, H., Gregory, N., and Boardman, W. (2001) 'Deflighting procedures and their welfare implications in captive birds', *Animal Welfare*, 10: 405–19.

Hiby, E. F., Rooney, N. J., and Bradshaw, J. W. S. (2006) 'Behavioural and physiological responses of dogs entering re-homing kennels', *Physiology and Behavior*, 89: 385–91.

Hildebrandt, T., Göritz, F., and Hermes, R. (2006) 'Ultrasonography: an important tool in captive breeding management in elephants and rhinoceroses', *European Journal of Wildlife Research*, 52: 23–7.

Hinshaw, K. C., Amand, W. B., and Tinkelman, C. L. (1996) 'Preventive medicine' in D. G. Kleiman, M. E. Allen, K. V. Thompson, and S. Lumpkin (eds), *Wild Mammals In Captivity: Principles and Techniques*, Chicago, IL/London: University of Chicago Press, pp. 16–24.

HMSO (2002) *The Zoo Licensing Act 1981 (Amendment) (England and Wales) Regulations 2002*, SI 2002/3080 (© Crown Copyright 2002), London: Her Majesty's Stationery Office.

Hoage, R. J., and Deiss, W. A. (1996) *New Worlds, New Animals: From Menagerie to Zoological Park in the Nineteenth Century*, Baltimore, MD/London: John Hopkins University Press.

Hoff, M. P., Powell, D. M., Lukas, K. E., and Maple, T. L. (1997) 'Individual and social behaviour of lowland gorillas in outdoor exhibits compared with indoor holding areas', *Applied Animal Behaviour Science*, 54: 359–70.

Hogan, E. S., Houpt, K. A., and Sweeney, K. (1988) 'The effect of enclosure size on social interactions and daily activity patterns of the captive Asiatic wild horse (*Equus przewalskii*)', *Applied Animal Behaviour Science*, 21: 147–68.

Hogan, L. A. and Tribe, A. (2007) 'Prevalence and cause of stereotypic behaviour in common wombats (*Vombatus ursinus*) residing in Australian zoos', *Applied Animal Behaviour Science*, 105: 180–91.

Höhn, M., Kronschnabel, M., and Gansloßer, U. (2000) 'Similarities and differences in activities and agonistic behaviour of male Eastern grey kangaroos (*Macropus giganteus*) in captivity and the wild', *Zoo Biology*, 19: 529–39.

Holst, B., and Dickie, L. A. (2007) 'How do national and international regulations and policies influence the role of zoos and aquariums in conservation?' in A. Zimmerman, M. Hatchwell, L. Dickie, and C. West (eds), *Zoos in the 21st Century: Catalysts for Conservation?*, Conservation Biology Series No. 15, Cambridge: Cambridge University Press, pp. 22–33.

Holt, W., Pickard, A., Rodger, J., and Wildt, D. (eds) (2003) *Reproductive Science and Integrated Conservation*, Cambridge: Cambridge University Press.

Holzer, D. and Scott, D. (1997) 'The long-lasting effects of early zoo visits', *Curator*, 40: 255–7.

Honess, P. E. and Marin, C. M. (2006) 'Enrichment and aggression in primates', *Neuroscience and Biobehavioral Reviews*, 30: 413–36.

Honess, P. E., Johnson, P. J., and Wolfensohn, S. E. (2004) 'A study of behavioural responses of non-human primates to air transport and re-housing', *Laboratory Animals*, 38: 119–32.

Hooper, K. J. and Newsome, J. T. (2004) 'Proactive compliance: the team program approach to revitalizing primate enrichment', *Contemporary Topics*, 43: 37–8.

Hope, K. and Deem, S. L. (2006) 'Retrospective study of morbidity and mortality of captive jaguars

(*Panthera onca*) in North America: 1982–2002', *Zoo Biology*, 25: 501–12.

Hörnicke, H. (1981) 'Utilization of caecal digesta by caecotrophy (soft faeces ingestion) in the rabbit', *Livestock Production Science*, 8: 361–6.

Horwich, R. H. (1989) 'Use of surrogate parental models and age periods in a successful release of hand-reared sandhill cranes', *Zoo Biology*, 8: 379–90.

Hosey, G. R. (1989) 'Behavior of the Mayotte lemur, *Lemur fulvus mayottensis*, in captivity', *Zoo Biology*, 8: 27–36.

Hosey, G. R. (1997) 'Behavioural research in zoos: academic perspectives', *Applied Animal Behaviour Science*, 51: 199–207.

Hosey, G. R. (2000) 'Zoo animals and their human audiences: what is the visitor effect?', *Animal Welfare*, 9: 343–57.

Hosey, G. R. (2005) 'How does the zoo environment affect the behaviour of captive primates?', *Applied Animal Behaviour Science*, 90: 107–29.

Hosey, G. R. (2008) 'A preliminary model of human–animal relationships in the zoo', *Applied Animal Behaviour Science*, 109: 105–27.

Hosey, G. R. and Druck, P. L. (1987) 'The influence of zoo visitors on the behaviour of captive primates', *Applied Animal Behaviour Science*, 18: 19–29.

Hosey, G. R. and Skyner, L. J. (2007) 'Self-injurious behaviour in zoo primates', *International Journal of Primatology*, 28: 1431–7.

Hosey, G. R., Jacques, M., and Pitts, A. (1997) 'Drinking from tails: social learning of a novel behaviour in a group of ring-tailed lemurs (*Lemur catta*)', *Primates*, 38: 415–22.

Houts, L. (1999) 'Supplemental carcass feeding for zoo carnivores', *The Shape of Enrichment*, 8: 1–3.

Howard, B. R. (1992) 'Health risks of housing small psittacines in galvanized wire mesh cages', *Journal of American Veterinary Medical Association*, 235: 469–83.

Howard, J. and Allen, M. E. (2007) 'Nutritional factors affecting semen quality in felids' in M. E. Fowler and R. E. Miller (eds), *Zoo and Wild Animal Medicine* (6th edn), St Louis, MO: Saunders (Elsevier), pp. 272–83.

Howell, S., Matevia, M., Fritz, J., Nash, L., and Maki, S. (1993) 'Pre-feeding agonism and seasonality in captive groups of chimpanzees (*Pan troglodytes*)', *Animal Welfare*, 2: 153–63.

HSE (Health and Safety Executive) (2006) *Managing Health and Safety in Zoos*, available online at http://www.hse.gov.uk/pubns/web15.pdf.

Hu, H. and Jiang, Z. (2002) 'Trial release of Père David's deer *Elaphurus davidianus* in the Dafeng Reserve, China', *Oryx*, 36: 196–9.

Huffman, M. A. and Hirata, S. (2004) 'An experimental study of leaf swallowing in captive chimpanzees: insights into the origin of a self-medicative behavior and the role of social learning', *Primates*, 45: 113–18.

Hunt, R. D., Garcia, F. G., Hegsted, D. M., and Kaplinsky, N. (1967) 'Vitamins D_2 and D_3 in New World primates: influence on calcium absorption', *Science*, 157: 943–5.

Hunt, S., Cuthill, I., Swaddle, J., and Bennett, A. (1997) 'Ultraviolet vision and band-colour preferences in female zebra finches, *Taeniopygia guttata*', *Animal Behaviour*, 54: 1383–92.

Hunter Jr., M. L. (1995) *Fundamentals of Conservation Biology*, Oxford: Blackwell Science.

Huntingford, F. and Turner, A. (1987) *Animal Conflict*, London: Chapman & Hall.

Hurme, K., Gonzalez, K., Halvorsen, M., Foster, B., Moore, D., and Chepko-Sade, B. D. (2003) 'Environmental enrichment for dendrobatid frogs', *Journal of Applied Animal Welfare Science*, 6: 285–99.

Hutchings, K. and Mitchell, H. (2003) 'A comparison of the behaviour of captive lemurs subjected to different causes of disturbance at Marwell Zoological Park' in T. C. Gilbert (ed.), *Proceedings of the Fifth Annual Symposium on Zoo Research*, 7–8 July, Winchester: BIAZA, pp. 139–43.

Hutchings, M., Hancocks, D., and Calip, T. (1978) 'Behavioural engineering in the zoo: a critique', *International Zoo News*, 25: 18–23.

Hutchins, D. M., Willis, K., and Wiese, R. J. (1995) 'Strategic collection planning: theory and practice', *Zoo Biology*, 14: 5–25.

Hutchins, M. (2001) 'Research: overview' in C. E. Bell (ed.), *Encyclopedia of the World's Zoos*, Chicago, IL/London: Fitzroy Dearborn, pp. 1076–80.

Hutchins, M. (2006) 'Variation in nature: its implications for zoo elephant management', *Zoo Biology*, 25: 161–71.

Hutchins, M. and Wiese, R. (1991) 'Beyond genetic and demographic management: the future of the SSP and related AAZPA conservation efforts', *Zoo Biology*, 10: 285–92.

Hutchins, M., Smith, G. M., Mead, D. C., Elbin, S., and Steenberg, J. (1991) 'Social behaviour of Matschie's tree kangaroos (*Dendrolagus matschiei*) and its implications for captive management', *Zoo Biology*, 10: 147–64.

I

IATA (International Air Transport Association) (2007) *Live Animal Regulations*, Montreal: IATA.

ICEE (International Conference on Environmental Education) (2007) *Environmental Education Towards a Sustainable Future: Partners for the Decade of Education for Sustainable Development*, Fourth International Conference on Environmental Education, 24–28 November, Ahmedabad, India.

Ickes, B. R., Pham, T. M., Sanders, L. A., Albeck, D. S., Mohammed, A. H., and Granholm, A. C. (2000) 'Long-term environmental enrichment leads to regional increases in neurotrophin levels in rat brain', *Experimental Neurology*, 164: 45–52.

Inglis, I. R. and Ferguson, N. (1986) 'Starlings search for food rather than eat freely-available, identical food', *Animal Behaviour*, 34: 614–17.

Inglis, I. R., Forkman, B., and Lazarus, J. (1997) 'Free food or earned food? A review and fuzzy model of contrafreeloading', *Animal Behaviour*, 53: 1171–91.

Ings, R., Waran, N. K., and Young, R. J. (1997a) 'Attitude of zoo visitors to the idea of feeding live prey to zoo animals', *Zoo Biology*, 16: 343–7.

Ings, R., Waran, N. K., and Young, R. J. (1997b) 'Effect of wood-pile feeders on the behaviour of captive bush dogs (*Speothos venaticus*)', *Animal Welfare*, 6: 145–52.

Isaza, R. (2003) 'Tuberculosis in all taxa' in M. E. Fowler and M. E. Miller (eds), *Zoo and Wild Animal Medicine* (5th edn), Philadelphia, PA: Elsevier (Saunders), pp. 689–96.

ISIS (International Species Information System) (2004) *User Manual for SPARKS: Single Population Analysis and Records Keeping System Version 1.5*, Eagen, MN: ISIS.

IUCN (International Union for the Conservation of Nature and Natural Resources) (2006) *Red Data List of Threatened Species*, available online at http://www.iucnredlist.org (accessed June 2007).

IUCN (International Union for the Conservation of Nature and Natural Resources) (2007) *Red Data List of Threatened Species*, available online at http://www.iucnredlist.org (accessed 20 April 2008).

IUCN/SSC (International Union for the Conservation of Nature and Natural Resources/Species Survival Commission) (1998) *IUCN Guidelines for Re-Introductions*, prepared by the IUCN/SSC Re-introduction Specialist Group, Gland/Cambridge: IUCN.

IUDZG/CBSG (International Union of Directors of Zoological Gardens/Conservation Breeding Specialist Group (The World Zoo Organisation and the Captive Breeding Specialist Group of the IUCN/SSC) (1993) *The World Zoo Conservation Strategy: The Role of the Zoos and Aquaria of the World in Global Conservation*, Chicago, IL: Chicago Zoological Society.

J

Jackson, D. (1996) 'Horticultural philosophies in zoo exhibit design' in D. Kleiman, M. Allen, K. Thompson, and S. Lumpkin (eds), *Wild Mammals in Captivity: Principles and Techniques*, Chicago, IL: University of Chicago, pp. 175–9.

Jackson, S. and Wilson, R. P. (2002) 'The potential costs of flipper-bands to penguins', *Functional Ecology*, 16: 141–8.

Jadavji, N. M., Kolb, B., and Metz, G. A. (2006) 'Enriched environment improves motor function in intact and unilateral dopamine-depleted rats', *Neuroscience*, 140: 1127–38.

Jago, N., Dorey, N., and Melfi, V. (2006) 'Training Goeldi's monkeys (*Callomico goeldii*) for weighting and crating' in N. Dorey (ed.) *Proceedings of the Animal Behavior Management Alliance Annual Conference*, 5–10 March, San Diego Zoo, San Diego Zoo's Wild Animal Park and Sea World Adventure Park, San Diego, pp. 113–15.

Jamieson, D. (1995) 'Zoos revisited' in B. G. Norton, M. Hutchins, E. Stevens, and T. L. Maple (eds), *Ethics on the Ark: Zoos, Animal Welfare and Wildlife Conservation*, Washington DC/London: Smithsonian Institute Press, pp. 52–66.

Jenkins, M. (2003) 'Prospects for biodiversity', *Science*, 302: 1175–7.

Jennison, G. (2005) *Animals for Show and Pleasure in Ancient Rome*, Philadelphia, PA: University of Pennsylvania Press.

Jeppesen, L. L., Heller, K. E., and Bildsoe, M. (2004) 'Stereotypies in female farm mink (*Mustela vison*) may be genetically transmitted and associated with higher fertility due to effects on body weight', *Applied Animal Behaviour Science*, 86: 137–43.

Johnson, A. J., Pessier, A. P., Wellehan, J. F. X., Brown, R., and Jacobson, E. R. (2005) 'Identification of a novel herpesvirus from a California desert tortoise (*Gopherus agassizii*)', *Veterinary Microbiology*, 111: 107–16.

Johnson, L. (2000) 'Sexing mammalian sperm for production of offspring: the state-of-the-art', *Animal Reproduction Science*, 60/61: 93–107.

Jones, G., Coe, J., and Paulson, D. (1976) *Long-Range Plan for Woodland Park Zoological Gardens*, Seattle, WA: Jones & Jones for the Seattle Department of Parks and Recreation.

Jones, M. and Pillay, N. (2004) 'Foraging in captive hamadryas baboons: implications for enrichment', *Applied Animal Behaviour Science*, 88: 101–10.

Jones, T. A., Hawrylak, N., Klintsova, A. Y., and Greenough, W. T. (1998) 'Brain damage, behavior, rehabilitation, recovery, and brain plasticity', *Mental Retardation and Developmental Disabilities Research Reviews*, 4: 231–7.

Judge, P. G. and de Waal, F. B. M. (1997) 'Rhesus monkey behaviour under diverse population densities: coping with long-term crowding', *Animal Behaviour*, 54: 643–62.

Kawai, M. (1965) 'Newly-acquired pre-cultural behavior of the natural troop of Japanese monkeys on Koshima Islet', *Primates*, 6: 1–30.

Kawakami, K., Takeuchi, T., Yamaguchi, S., Ago, A., Nomura, M., Gonda, T., and Komemushi, S. (2003) 'Preference of guinea pigs for bedding materials: wood shavings versus paper cutting sheet', *Experimental Animals*, 52: 11–15.

Keane, C. and Marples, N. (2003) 'The effects of zoo visitors on gorilla behaviour' in T. C. Gilbert (ed.), *Proceedings of the Fifth Annual Symposium on Zoo Research*, 7–8 July, Winchester: BIAZA, pp. 144–54.

Keeling, C. H. (1984) *Where the Lion Trod: A Study of Forgotten Zoological Gardens*, London: Clam Productions.

Kellert, S. (1979) 'Zoological parks in American society' in *AAZPA Annual Proceedings*, Wheeling, WV: AAZPA, pp. 88–126.

Kelley, J. L. and Magurran, A. E. (2006) 'Captive breeding promotes aggression in an endangered Mexican fish', *Biological Conservation*, 133: 169–77.

Kells, A., Dawkins, M. S., and Borja, M. C. (2001) 'The effect of a "freedom food" enrichment on the behaviour of broilers on commercial farms', *Animal Welfare*, 10: 347–56.

Kerridge, F. J. (1996) 'Behavioural enrichment of ruffed lemurs (*Varecia variegata*) based upon a wild–captive comparison of their behaviour', PhD thesis, Manchester: University of Manchester (Bolton Institute).

Kerridge, F. J. (2005) 'Environmental enrichment to address behavioral differences between wild and captive black-and-white ruffed lemurs (*Varecia variegata*)', *American Journal of Primatology*, 66: 71–84.

Kertesz, P. (1993) *A Colour Atlas of Veterinary Dentistry and Oral Surgery*, London: Wolfe.

Kesler, D., Lopes, I., and Haig, S. (2006) 'Sex determination of Pohnpei Micronesian Kingfishers using morphological and molecular genetic techniques', *Journal of Field Ornithology*, 77: 229–32.

Ketz-Riley, C. J. (2003) 'Salmonellosis and shigellosis' in M. E. Fowler and M. E. Miller (eds), *Zoo and Wild Animal Medicine* (5th edn), Philadelphia, PA: Elsevier (Saunders), pp. 686–9.

Kiley-Worthington, M. (1989) 'Ecological, ethological, and ethically sound environments for animals: towards symbiosis', *Journal of Agricultural Ethics*, 2: 323–47.

Kilner, R. (1998) 'Primary and secondary sex ratio manipulation by zebra finches', *Animal Behaviour*, 56: 155–64.

Kinkel, L. (1989) 'Lasting effects of wing tags on ring-billed gulls', *The Auk*, 106: 619–24.

Kirkwood, J. K. (2001a) 'United Kingdom: legislation' in C. E. Bell (ed.), *Encyclopedia of the World's Zoos*, Chicago, IL/London: Fitzroy Dearborn, pp. 1281–3.

Kirkwood, J. K. (2001b) 'United Kingdom: licensing' in C. E. Bell (ed.), *Encyclopedia of the World's Zoos*, Chicago, IL/London: Fitzroy Dearborn, pp. 1284–5.

Kirkwood, J. K. and Cunningham, A. A. (1994) 'Epidemiologic observations on spongiform encephalopathies in captive wild animals in the British Isles', *Veterinary Record*, 135: 296–303.

Kirkwood, J. K., Wells., G. A. H., Wilesmith, J. W., Cunningham, A. A., and Jackson, S. I. (1990) 'Spongiform encephalopathy in an Arabian oryx (*Oryx leucoryx*) and a greater kudu (*Tragelaphus strepsiceros*)', *Veterinary Record*, 127: 418–20.

Kirkwood, J. K., Cunningham, A. A., Flach, E. J., Thornton, S. M., and Wells, G. A. H. (1995) 'Spongiform encephalopathy in another captive cheetah (*Acinonyx jubatus*): evidence for variation in susceptibility or incubation periods between species?', *Journal of Zoo and Wildlife Medicine*, 26: 577–82.

Kisling, V. N. (2001) *Zoo and Aquarium History: Ancient Animal Collections to Zoological Gardens*, New York, NY/London: CRC Press.

Kitchener, A. C. and Macdonald, A. (2004) 'The longevity legacy: the problem of old mammals in zoos' in B. Hiddinga (ed.), *Proceedings of the EAZA Conference*, 21–25 September, Kolmarden, Amsterdam: EAZA Executive Office, pp. 132–7.

Klasing, K. C. (1998) *Comparative Avian Nutrition*, Wallingford, Oxon: CABI Publishing.

Kleiman, D. G. (1992) 'Behaviour research in zoos: past, present and future', *Zoo Biology*, 1110: 301–12.

Kleiman, D. G. (1994) 'Animal behaviour studies and zoo propagation programs', *Zoo Biology*, 1310: 411–12.

Kleiman, D. G., Beck, B. B., Dietz, J. M., Dietz, L. A., Ballou, J. D., and Coimbra-Filho, A. C. (1986) 'Conservation program for the golden lion tamarins: captive research and management, ecological studies, educational strategies and reintroduction' in K. Benirshke (ed.), *Primates: The Road to Self-Sustaining Populations*, New York, NY: Springer, pp. 959–79.

Kleiman, D. G., Beck, B. B., Dietz, J. M., and Dietz, L. A. (1991) 'Costs of reintroduction and criteria for success: accounting and accountability in the golden lion tamarin conservation program' in J. H. W. Gipps (ed.) 'Beyond captive breeding: re-introducing endangered animals to the wild', *Symposium of the Zoological Society of London*, 62: 125–42.

Kleiman, D. G., Stanley Price, M. R., and Beck, B. B. (1994) 'Criteria for reintroductions' in P. J. S. Olney, G. M. Mace, and A. T. C. Feistner (eds), *Creative Conservation: Interactive Management of Wild and Captive Animals*, London: Chapman & Hall, pp. 287–303.

Kleiman, D. G., Allen, M. E., Thompson, K. V., and Lumpkin, S. (eds) (1996) *Wild Mammals in Captivity: Principles and Techniques*, Chicago, IL/London: University of Chicago Press.

Knight, K., Pearson, R., and Melfi, V. (2005) 'Does the provision of carcasses compromise the health of zoo-housed carnivores?' in A. Nicklin (ed.) *Proceedings of the Seventh Annual BIAZA Research Meeting*, 7–8 July, Twycross Zoo, Warwickshire BIAZA, pp. 194–8.

Knowles, J. M. (1985) 'Wild and captive populations: triage, contraception and culling', *International Zoo Yearbook*, 24–25: 206–10.

Koene, P. and Duncan, I. J. H. (2001) 'From environmental requirement to environmental enrichment: from animal suffering to animal pleasure' in M. Hawkins (ed.), *Proceedings of the Fifth Annual Conference on Environmental Enrichment*, 4–9 November, Sydney, The Shape of Enrichment, Inc., p. 36.

Kohn, B. (1994) 'Zoo animal welfare', *Revue Scientifique et Technique de L'Office International Des Epizooties*, 13: 233–45.

Kolter, N. and Kolter, P. (1998) *Museum Strategy and Marketing: Designing Missions, Building Audiences, Generating Revenue and Resources*, San Francisco, CA: Jossey-Bass.

Kratochvil, H. and Schwammer, H. (1997) 'Reducing acoustic disturbances by aquarium visitors', *Zoo Biology*, 16: 349–53.

Krebs, J. R. and Davies, N. B. (1993) *An Introduction to Behavioural Ecology* (3rd edn), Oxford: Blackwell Scientific.

Kreger, M. D. and Mench, J. A. (1995) 'Visitor–animal interactions at the zoo', *Anthrozoös*, 8: 143–58.

Kuhar, C. W. (2006) 'In the deep end: pooling data and other statistical challenges in zoo and aquarium research', *Zoo Biology*, 25: 339–52.

Kuhar, C. W. (2008) 'Group differences in captive gorillas' reaction to large crowds', *Applied Animal Behaviour Science*, 110: 377–85.

Kuhar, C. W., Bettinger, T., and Laudenslager, M. (2005) 'Salivary cortisol and behaviour in an all-male group of western lowland gorillas (*Gorilla. g. gorilla*)', *Animal Welfare*, 14: 187–93.

Kuhar, C. W., Stoinski, T. S., Lukas, K. E., and Maple, T. L. (2006) 'Gorilla Behavior Index

revisited: age, housing and behavior', *Applied Animal Behaviour Science*, 96: 315–26.

Kusuda, S., Ikoma, M., Morikaku, K., Koizumi, J., Kawaguchi, Y., Kobayashi, K., Matsui, K., Nakamura, A., Hashikawa, H., Kobayashi, K., Ueda, M., Kaneko, M., Akikawa, T., Shibagaki, S., and Doi, O. (2007) 'Estrous cycle based on blood progesterone profiles and changes in vulvar appearance in Malayan tapirs (*Tapirus indicus*)', *Journal of Reproduction and Development*, 53: 1283–9.

Kyle, D. G. (2001) *Spectacles of Death in Ancient Rome*, London: Routledge.

Ladewig, J. (1987) 'Endocrine aspects of stress: evaluation of stress reactions in farm animals' in P. Wiepkema and P. van Adrichem (eds), *Biology of Stress in Farm Animals: An Integrative Approach*, Dordrecht: Martinus Nijhoff, pp. 13–25.

Laikre, L. (1999) 'Conservation genetics of Nordic carnivores: lessons from zoos', *Hereditas*, 130: 203–16.

Laikre, L. and Ryman, N. (1991) 'Inbreeding depression in a captive wolf (*Canis lupus*) population', *Conservation Biology*, 5: 33–40.

Lair, S., Barker, I. K., Mehren, K. G., and Williams, E. S. (2002) 'Epidemiology of neoplasia in captive black-footed ferrets (*Mustela nigripes*) 1986–1996', *Journal of Zoo and Wildlife Medicine*, 33: 204–13.

Lamb, M. E. and Hwang, C. P. (1982) 'Maternal attachment and mother–neonate bonding: a critical review' in M. E. Lamb and A. L. Brown (eds), *Advances in Developmental Psychology*, Hillsdale, NJ: Laurence Erlbaum, pp. 1–39.

Lamberski, N. (2003) 'Psittaciformes (parrots, macaws, lories)' in M. E. Fowler and M. E. Miller (eds), *Zoo and Wild Animal Medicine* (5th edn), Philadelphia, PA: Elsevier (Saunders), pp. 187–210.

Lamikanra, O., Imam, S. H., and Ukuku, D. (2005) *Produce Degradation*, Boca Raton, FL: CRC Press.

Landgkilde, T. and Shine, R. (2006) 'How much stress do researchers inflict on their study animals? A case study using a scincid lizard, *Eulamprus heatwolei'*, *Journal of Experimental Biology*, 209: 1035–43.

Landolfi, J. A., Baktiar, O. K., Poynton, S. L., and Mankowski, J. L. (2003) 'Hepatic *Calodium hepaticum* (Nematoda) infection in a zoo colony of black-tailed prairie dogs (*Cynomys ludovicianus*)', *Journal of Zoo and Wildlife Medicine*, 34: 371–4.

Lane, J. (2006) 'Can non-invasive glucocorticoid measures be used as reliable indicators of stress in animals?', *Animal Welfare*, 15: 331–42.

Lanza, R. P., Cibelli, J. B., Diaz, F., Morales, C. T., Farin, P. W., Farin, C. E., Hammer, C. J., West, M. D., and Damiani, P. (2000) 'Cloning of an endangered species (*Bos gaurus*) using interspecies nuclear transfer', *Cloning*, 2: 9–90.

Latham, N. R. and Mason, G. J. (2008) 'Maternal deprivation and the development of stereotypic behaviour', *Applied Animal Behaviour Science*, 110: 84–108.

Lauer, J. (1976) *Saqqara: The Royal Cemetery of Memphis, Excavations and Discoveries since 1850*, London: Thames & Hudson.

Laule, G. E. and Desmond, T. (1998) 'Positive reinforcement training as an enrichment strategy' in D. J. Shepherdson, J. D. Mellen, and M. Hutchins (eds), *Second Nature: Environmental Enrichment for Captive Animals*, Washington DC: Smithsonian Institution Press, pp. 302–13.

Laule, G. E., Bloomsmith, M. A., and Schapiro, S. J. (2003) 'The use of positive reinforcement training techniques to enhance the care, management, and welfare of primates in the laboratory', *Journal of Applied Animal Welfare Science*, 6: 163–73.

Laurenson, M. (1993) 'Early maternal behavior of wild cheetahs: implications for captive husbandry', *Zoo Biology*, 12: 31–43.

Lay Jr, D. C., Friend, T. H., Bowers, C. L., Grissom, K. K., and Jenkins, O. C. (1992) 'A comparative physiological and behavioral study of freeze and hot-iron branding using dairy cows', *Journal of Animal Science*, 70: 1121–5.

Leader-Williams, N., Balmford, A., Linkie, M., Mace, G. M., Smith, R. J., Stevenson, M., Walter, O.,

West, C., and Zimmermann, A. (2007) 'Beyond the ark: conservation biologists' views of the achievements of zoos in conservation' in A. Zimmerman, M. Hatchwell, L. Dickie, and C. West (eds), *Zoos in the 21st Century: Catalysts for Conservation?*, Conservation Biology Series No. 15, Cambridge: Cambridge University Press, pp. 236–54.

Le Boeuf, B. J. (1974) 'Male–male competition and reproductive success in elephant seals', *American Zoology*, 14: 163–76.

Leck, C. (1980) 'Establishment of new population centers with changes in migration patterns', *Journal of Field Ornithology*, 51: 168–73.

Lees, C. (1993) 'Managing harems in captivity: the Sulawesi crested macaque (*Macaca nigra*)', MSc dissertation, University of Kent, UK.

Lehmann, J. and Boesch, C. (2004) 'To fission or to fusion: effects of community size on wild chimpanzee (*Pan troglodytes verus*) social organisation', *Behavioral Ecology and Sociobiology*, 56: 207–16.

Lehner, P. N. (1998) *Handbook of Ethological Methods* (2nd edn), Cambridge: Cambridge University Press.

Leipold, H. W. (1980) 'Congenital defects of zoo and wild mammals: a review' in R. J. Montali and G. Migaki (eds), *The Comparative Pathology of Zoo Animals*, Washington DC: Smithsonian Institute, pp. 457–70.

Lemasson, A., Gautier, J.-P., and Hausberger, M. (2005) 'A brief note on the effects of the removal of individuals on social behaviour in a captive group of Campbell's monkeys (*Cercopithecus campbelli campbelli*): a case study', *Applied Animal Behaviour Science*, 91: 289–96.

Lemm, J., Steward, S., and Schmidt , T. (2005) 'Reproduction of the critically endangered Anegada island iguana *Cyclura pinguis* at San Diego Zoo', *International Zoo Yearbook*, 39: 141–52.

LeNeindre, P., Boivin, X., and Boissy, A. (1996) 'Handling of extensively kept animals', *Applied Animal Behaviour Science*, 49: 73–81.

Leus, K. (2006) *EAZA Adaptation of AZA Studbook Analysis and Population Management Handbook*, Amsterdam/Antwerp: EAZA/Royal Zoological Society of Antwerp.

LeVan, N. F., Estevez, I., and Stricklin, W. R. (2000) 'Use of horizontal and angled perches by broiler chickens', *Applied Animal Behaviour Science*, 65: 349–65.

Lewis, J. C. M., Fitzgerald, A. J., Gulland, F. M. D., Hawkey, C. M., Kertesz, P., Kirkwood, J. K., and Kock, R. A. (1989) 'Observations on the treatment of necrobacillosis in wallabies', *British Veterinary Journal*, 145: 394–6.

Lewke, R. and Stroud, R. (1974) 'Freeze-branding as a method of marking snakes', *Copeia*, 4: 997–1000.

Lindberg, J., Björnerfeldt, S., Saetre, P., Svartberg, K., Seehuus, B., Bakken, M., Vilà, C., and Jazin, E. (2005) 'Selection for tameness has changed brain gene expression in silver foxes', *Current Biology*, 15: 915–16.

Lindburg, D. G. (1988) 'Improving the feeding of captive felines through application of field data', *Zoo Biology*, 7: 211–18.

Line, S., Morgan, K., Markowitz, H., and Strong, S. (1989a) 'Heart rate and activity of rhesus monkeys in response to routine events', *Laboratory Primate Newsletter*, 28: 9–12.

Line, S., Morgan, K., Markowitz, H., and Strong, S. (1989b) 'Influence of cage size on heart rate and behaviour in rhesus monkeys', *American Journal of Veterinary Research*, 40: 1523–6.

Line, S., Morgan, K., Markowitz, H., and Strong, S. (1990) 'Increased cage size does not alter heart rate or behaviour in female rhesus monkeys', *American Journal of Primatology*, 20: 107–13.

Line, S., Markowitz, H., Morgan, K., and Strong, S. (1991) 'Effect of cage size and environmental enrichment on behavioural and physiological responses of rhesus macaques to the stress of daily events' in M. Novak and A. Petto (eds), *Through a Looking Glass: Issues in Psychological Well-Being in Captive Nonhuman Primates*, Washington DC: American Psychology Association, pp. 160–79.

Lintzenich, B. A. and Ward, A. M. (1997) 'Hay and pellet ratios: considerations in feeding ungulates', Fact Sheet 006, *AZA Nutrition Advisory Group Handbook*, available online at `http://www.nagonline.net`.

Little, K. A. and Sommer, V. (2002) 'Change of enclosure in langur monkeys: implications for the evaluation of environmental enrichment', *Zoo Biology*, 21: 549–59.

Liu, J. Chen, Y., Guo, L., Gu, B., Liu, H., Hou, A., Liu, X., Sun, L., and Liu, D. (2006) 'Stereotypic behavior and fecal cortisol level in captive giant pandas in relation to environmental enrichment', *Zoo Biology*, 25: 445–59.

Loeske, E. B., Kruuk, L. E. B., Clutton-Brock, T. H., Albon, S. D., Pemberton, J. M., and Guinness, F. E. (1999) 'Population density affects sex ratio variation in red deer', *Nature*, 399: 460–61.

Lombardi, J. (1998) *Comparative Vertebrate Reproduction*, London: Kluwer.

Loskutoff, N. M. (2003) 'Role of embryo technologies in genetic management and conservation of wildlife' in W. Holt, A. Pickard, J. Rodger, and D. Wildt (2003) *Reproductive Science and Integrated Conservation*, Cambridge: Cambridge University Press, pp. 183–94.

Lovell, T. (1998) *Nutrition and Feeding of Fish*, London: Kluwer.

Lozano, G. A. (1998) 'Parasitic stress and self-medication in wild animals' in A. P. Møller, M. Milinski, and P. J. B. Slater (eds), *Advances in the Study of Behavior, Vol. 27: Stress and Behavior*, London: Academic Press, pp. 291–317.

Lucentini, L., Caporali, S., Palomba, A., Lancioni, H., and Panara, F. (2006) 'A comparison of conservative DNA extraction methods from fins and scales of freshwater fish: a useful tool for conservation genetics', *Conservation Genetics*, 7: 1009–12.

Ludes, E. and Anderson, J. (1996) 'Comparison of the behaviour of captive white-faced capuchin monkeys (*Cebus capucinus*) in the presence of four kinds of deep litter', *Applied Animal Behaviour Science*, 49: 293–303.

Ludes-Fraulob, E. and Anderson, J. R. (1999) 'Behaviour and preferences among deep litters in captive capuchin monkeys (*Cebus capucinus*)', *Animal Welfare*, 8: 127–34.

Lukas, K. E. (1999) 'A review of nutritional and motivational factors contributing to the performance of regurgitation and reingestion in captive lowland gorillas (*Gorilla gorilla gorilla*)', *Applied Animal Behaviour Science*, 63: 237–49.

Lukas, K. E., Marr, M. J., and Maple, T. L. (1998) 'Teaching operant conditioning at the zoo', *Teaching of Psychology*, 25: 112–16.

Lukas, K. E., Hoff, M. P., and Maple, T. L. (2003) 'Gorilla behavior in response to systematic alternation between zoo enclosures', *Applied Animal Behaviour Science*, 81: 367–86.

Luttrell, L., Acker, L., Urben, M., and Reinhardt, V. (1994) 'Training a large troop of rhesus macaques to co-operate during catching: analysis of the time investment', *Animal Welfare*, 3: 135–40.

Lutz, C. K. and Novak, M. A. (1995) 'Use of foraging racks and shavings as enrichment tools for groups of rhesus monkeys (*Macaca mulatta*)', *Zoo Biology*, 14: 463–74.

Lyles, A. M. and May, R. M. (1987) 'Problems in leaving the ark', *Nature*, 326: 245–6.

Lyons, J., Young, R. J., and Deag, J. M. (1997) 'The effects of physical characteristics of the environment and feeding regime on the behavior of captive felids', *Zoo Biology*, 16: 71–83.

m **MacDonald, D.** (1996) 'Social behaviour of captive bush dogs (*Speothos venaticus*)', *Journal of Zoology*, 239: 525–43.

Macdonald, D. (2001) *The New Encyclopedia of Mammals*, Oxford: Oxford University Press.

Mace, G. M. (1986) 'Genetic management of small populations', *International Zoo Yearbook*, 24/25: 167–74.

Mace, G. M. (2004) 'The role of taxonomy in species conservation', *Philosophical Transactions of the Royal Society B*, 359: 711–19.

Mace, G. M., Balmford, A., Leader-Williams, N., Manica, A., Walter, O., West, C., and Zimmerman, A. (2007) 'Measuring conservation success: assessing zoos' contribution' in A. Zimmerman, M. Hatchwell, L. Dickie, and C. West (eds), *Zoos in the 21st Century: Catalysts for Conservation?*, Conservation Biology Series No. 15, Cambridge: Cambridge University Press, pp. 322–42.

Maestripieri, D. (2000) 'Measuring temperament in rhesus macaques: consistency and change in emotionality over time', *Behavioural Processes*, 49: 167–71.

Maestripieri, D. (2001) 'Is there mother–infant bonding in primates?', *Developmental Review*, 21: 93–120.

MAFF (Ministry of Agriculture, Fisheries and Food) (1986) *Manual of Veterinary Parasitological Laboratory Techniques*, Technical Bulletin 18, London: Her Majesty's Stationery Office.

Magin, C., Johnson, T., Groombridge, B., Jenkins, M., and Smith, H. (1994) 'Species extinction, endangerment and captive breeding' in P. J. S. Olney, G. M. Mace, and A. T. C. Feistner (eds), *Creative Conservation: Interactive Management of Wild and Captive Animals*, London: Chapman & Hall, pp. 3–31.

Mairéad, A., Farrell, E. B., and Marples, N. (2000) 'Breeding behavior in a flock of Chilean flamingos (*Phoenicopterus chilensis*) at Dublin Zoo', *Zoo Biology*, 19: 227–37.

Maki, S. and Bloomsmith, M. (1989) 'Uprooted trees facilitate the psychological well-being of captive chimpanzees', *Zoo Biology*, 8: 79–87.

Mallapur, A. and Chelan, R. (2002) 'Environmental influences on stereotypy and the activity budget of Indian leopards (*Panthera pardus*) in four zoos in southern India', *Zoo Biology*, 21: 585–95.

Mallapur, A. and Choudhury, B. C. (2003) 'Behavioural abnormalities in captive non-human primates', *Journal of Applied Animal Welfare Science*, 6: 275–84.

Mallapur, A., Qureshi, Q., and Chellam, R. (2002) 'Enclosure design and space utilization by Indian leopards (*Panthera pardus*) in four zoos in southern India', *Journal of Applied Animal Welfare Science*, 5: 111–24.

Mallapur, A., Sinha, A., and Waran, N. (2005) 'Influence of visitor presence on the behaviour of captive lion-tailed macaques (*Macaca silenus*) housed in Indian zoos', *Applied Animal Behaviour Science*, 94: 341–52.

Mallapur, A., Sinha, A., and Waran, N. (2007) 'A world survey of husbandry practices for lion-tailed macaques *Macaca silenus* in captivity', *International Zoo Yearbook*, 41: 166–75.

Mallet, J. (2001) 'Mimicry: an interface between psychology and evolution', *Proceedings of the National Academy of Sciences USA*, 98: 8928–30.

Mallinson, J. (1995) 'Zoo breeding programmes: balancing conservation and animal welfare', *Dodo: Journal of the Jersey Wildlife Preservation Trust*, 3110: 66–73.

Malmkvist, J. and Hansen, S. W. (2001) 'The welfare of farmed mink (*Mustela vison*) in relation to behavioural selection: a review', *Animal Welfare*, 10: 41–52.

Manly, B. F. J. (1998) *Randomization, Bootstrap and Monte Carlo Methods in Biology* (2nd edn), London: Chapman & Hall.

Mann, J. R. (1988) 'Full-term development of mouse eggs fertilized by a spermatozoan microinjected under the zona pellucida', *Biology of Reproduction*, 38: 1077–83.

Manning, A. and Dawkins, M. S. (1998) *An Introduction to Animal Behaviour* (5th edn), Cambridge: Cambridge University Press.

Maple, T. and Finlay, T. (1989) 'Applied primatology in the modern zoo', *Zoo Biology (Supplement)*, 112: 101–16.

Marcellini, D. L and Jenssen, T. A. (1988) 'Visitor behavior in the National Zoo's reptile house', *Zoo Biology*, 7: 329–38.

Margulis, S. W. and Westhus, E. J. (2008) 'Evaluation of different observational sampling regimes for use in zoological parks', *Applied Animal Behaviour Science*, 110: 363–76.

Margulis, S. W., Hoyos, C., and Anderson, M. (2003) 'Effect of felid activity on zoo visitor interest', *Zoo Biology*, 22: 587–99.

Maria, G. A. (2006) 'Public perception of farm animal welfare in Spain', *Livestock Science*, 103: 250–6.

Marker-Kraus, L. (1997) 'History of the cheetah: *Acinonyx jubatus* in zoos 1829–1994', *International Zoo Yearbook*, 35: 27–43.

Marker-Kraus, L. and Grisham, J. (1993) 'Captive breeding of cheetahs in North American Zoos: 1987–1991', *Zoo Biology*, 12: 5–18.

Markowitz, H. (1982) *Behavioural Enrichment in the Zoo*, New York, NY: Van Nostrand Reinhold.

Markowitz, H. and Woodworth, G. (1978) 'Experimental analysis and control of group behaviour' in H. Markowitz and V. J. Stevens (eds), *Behavior of Captive Wild Animals*, Chicago, IL: Nelson-Hall, pp. 107–31.

Markowitz, H., Schmidt, M., and Moody, A. (1978) 'Behavioural engineering and animal health in the zoo', *International Zoo Yearbook*, 18: 190–5.

Markowitz, H., Aday, C., and Gavazzi, A. (1995) 'Effectiveness of acoustic prey: environmental enrichment for a captive African leopard (*Panthera pardus*)', *Zoo Biology*, 14: 371–9.

Maroldo, G. K. (1978) 'Zoos worldwide as settings for psychological research', *American Psychologist*, 33: 1000–4.

Marqués H., Navidad, G., Baucells, M., and Albanell, E. (2001) 'Animals' nutritional wisdom: pros and cons of cafeteria-style feeding', *Zoo Nutrition News (EAZA)*, 2: 23–5.

Marriner, L. M. and Drickamer, L. C. (1994) 'Factors influencing stereotyped behaviour of primates in a zoo', *Zoo Biology*, 13: 267–75.

Marris, E. (2007) 'Linnaeus at 300: the species and the specious', *Nature*, 446: 250–3.

Martel, Y. (2002) *Life of Pi*, Edinburgh: Canongate.

Martin, J. E. (2002) 'Early life experiences: activity levels and abnormal behaviours in resocialised chimpanzees', *Animal Welfare*, 11: 419–36.

Martin, J. E. (2005) 'The effects of rearing conditions on grooming and play behaviour in captive chimpanzees', *Animal Welfare*, 14: 125–33.

Martin, P. and Bateson, P. (2007) *Measuring Behaviour: An Introductory Guide* (3rd edn), Cambridge: Cambridge University Press.

Masefield, W. (1999) 'Forage preferences and enrichment in a group of captive Livingstone's fruit bats *Pteropus livingstonii*', *Dodo. Journal of the Jersey Wildlife Preservation Trust*, 35: 48–56.

Maser, J. and Gallup, G. J. (1974) 'Tonic immobility in the chicken: catalepsy potentiation by uncontrollable shock and alleviation by imipramine', *Psychosomatic Medicine*, 36: 199–205.

Mason, G. J. (1991) 'Stereotypies: a critical review', *Animal Behaviour*, 41: 1015–37.

Mason, G. J. (2006) 'Stereotypic behaviour in captive animals: fundamentals, and implications for welfare and beyond' in G. J. Mason (ed.), *Stereotypic Animal Behaviour: Fundamentals and Applications to Welfare*, Wallingford, Oxon: CABI Publishing, pp. 325–56.

Mason, G. J. and Latham, N. R. (2004) 'Can't stop, won't stop: is stereotypy a reliable animal welfare indicator?', *Animal Welfare*, 13: 557–69.

Mason, G. J. and Littin, K. E. (2003) 'The humaneness of rodent pest control', *Animal Welfare*, 12: 1–37.

Mason, G. J. and Rushen, J. (eds) (2006) *Stereotypic Animal Behaviour: Fundamentals and Applications to Welfare*, Wallingford, Oxon: CABI Publishing.

Mason, G. J., Cooper, J., and Clarebrough, C. (2001) 'Frustrations of fur-farmed mink', *Nature*, 410: 35–6.

Mason, P. (2000) 'Zoo tourism: the need for more research', *Journal of Sustainable Tourism*, 8: 333–9.

Mather, J. (1992) 'Underestimating the octopus' in H. Davis and D. Balfour (eds), *The Inevitable Bond: Examining Scientist–Animal Interactions*, Cambridge: Cambridge University Press, pp. 240–9.

Mathews, F., Orrors, M., McLaren, G., Gelling, M., and Foster, R. (2005) 'Keeping fit on the ark: assessing the suitability of captive-bred animals for release', *Biological Conservation*, 121: 569–577.

Matthews, K. (1998) *Behavioural Studies of Sulawesi Crested Black Macaques (*Macaca nigra*): Managing Males in Captivity*, Trinity, Jersey: Durrell Wildlife Conservation Trust.

May, H. Y. and Mercier, A. J. (2006) 'Responses of crayfish to a reflective environment depend on dominant status', *Canadian Journal of Zoology*, 84: 1104–11.

May, R. M. and Lyles, A. M. (1987) 'Living Latin binomials', *Nature*, 326: 642–3.

Mayor, J. (1984) 'Hand-feeding an orphaned scimitar-horned oryx *Oryx dammah* calf after its integration with the herd', *International Zoo Yearbook*, 23: 243–8.

Mayr, E. (1942) *Systematics and the Origin of Species*, New York, NY: Columbia University Press.

Mazur, N. A. (2001) *After the Ark? Environmental Policy Making and the Zoo*, Melbourne: Melbourne University Press.

McCann, C., Buchanan-Smith, H. M., Jones-Engel, L., Farmer, F., Prescott, M., Fitch-Snyder, H., Taylor, S., Buchanan-Smith, H. M., Jones-Engel, L., Farmer, F., Prescott, M., Fitch-Snyder, H., and Taylor, S. (2007) *IPS International Guidelines for the Acquisition, Care and Breeding of Non-Human Primates* (2nd edn), available online at `http://www.internationalprimatologicalsociety.org`.

McCann, C. M. and Rothman, J. M. (1999) 'Changes in nearest-neighbor associations in a captive group of western lowland gorillas after the introduction of five hand-reared infants', *Zoo Biology*, 18: 261–78.

McCarthy, S. (2005) *Becoming a Tiger: How Baby Animals Learn to Live in the Wild*, New York: Harper Perennial.

McCormick, A. E. (1983) 'Canine distemper in African cape hunting dogs (*Lycaon pictus*): possibly vaccine-induced', *Journal of Zoo Animal Medicine*, 14: 66–71.

McCormick, W. (2003) 'How enriching is training?' in T. C. Gilbert (ed.) *Proceedings of the Fifth Annual Symposium on Zoo Research*, 7–8 July, Winchester: BIAZA, pp. 9–19.

McCormick, W., Melfi, V., and Muller, C. (2006) 'Lemurs pumping iron' in T. P. Meehan and M. E. Allen (eds), *Proceedings of the Fourth European Zoo Nutrition Conference*, 20–23 January, Leipzig Zoo: EAZA, p. 42.

McDonald, P., Edwards, R. A., Greenhalgh, J. F. D., and Morgan, C. A. (2002) *Animal Nutrition*, Harlow/London: Prentice Hall (Pearson Education).

McDougall, P. T., Réale, D., Sol, D., and Reader, S. M. (2006) 'Wildlife conservation and animal temperament: causes and consequences of evolutionary change for captive, reintroduced and wild populations', *Animal Conservation*, 9: 39–48.

McDowell, L. R. (1989) *Vitamins in Animal Nutrition: Comparative Aspects to Human Nutrition*, San Diego, CA: Academic Press.

McDowell, L. R. (1992) *Minerals in Animal and Human Nutrition*, San Diego, CA: Academic Press.

McDowell, L. R. (2003) *Minerals in Animal and Human Nutrition* (2nd edn), Amsterdam: Elsevier Science.

McFarland, D. (1999) *Animal Behaviour: Psychobiology, Ethology and Evolution* (3rd edn), Harlow: Longman.

McGrew, W. C., Brennan, J. A., and Russel, J. (1986) 'An artificial gum-tree for marmosets (*Callithrix jacchus*)', *Zoo Biology*, 5: 45–50.

McKay, S. (2003) 'Personality profiles of the cheetah in the UK and Ireland, in relation to environmental factors and performance variables' T. C. Gilbert (ed.) *Proceedings of the Fifth Annual Symposium on Zoo Research*, 7–8 July, Winchester: BIAZA, pp. 177–89.

McKenna, V., Travers, W., and Wray, J. (eds) (1987) *Beyond the Bars: The Zoo Dilemma*, Wellingborough: Thorsons Publishing Group.

McKenzie, S., Chamove, S., and Feistner, A. (1986) 'Floor-coverings and hanging screens alter arboreal monkey behaviour', *Zoo Biology*, 5: 339–48.

McKinley, J., Buchanan-Smith, H. M., Bassett, L., and Morris, K. (2003) 'Training common marmosets (*Callithrix jacchus*) to cooperate during routine laboratory procedures: ease of training and time investment', *Journal of Applied Animal Welfare Science*, 6: 209–20.

McPhee, M. (2002) 'Intact carcasses as enrichment for large felids: effects on on- and off-exhibit behaviours', *Zoo Biology*, 21: 37–47.

McPhee, M., Foster, J., Sevenich, M., and Saunders, C. (1998) 'Public perceptions of behavioral enrichment: assumptions gone awry', *Zoo Biology*, 17: 525–34.

McWilliams, D. A. (2005a) 'Nutrition research on calcium homeostasis I: lizards (with recommendations)', *International Zoo Yearbook*, 39: 77–84.

McWilliams, D. A. (2005b) 'Nutrition research on calcium homeostasis II: freshwater turtles (with recommendations)', *International Zoo Yearbook*, 39: 85–98.

Measey, G., Gower, D., Oommen, O., and Wilkinson, M. (2001) 'Permanent marking of a fossorial caecilian, *Gegeneophis ramaswamii* (Amphibia: Gymnophiona: Caeciliidae)', *Journal of South Asian Natural History*, 5: 141–7.

Meehan, C. L. and Mench, J. A. (2002) 'Environmental enrichment affects the fear and exploratory responses to novelty of young Amazon parrots', *Applied Animal Behaviour Science*, 79: 75–88.

Meehan, C. L. and Mench, J. A. (2007) 'The challenge of challenge: can problem solving opportunities enhance animal welfare?', *Applied Animal Behaviour Science*, 102: 246–61.

Meehan, C. L., Garner, J. P., and Mench, J. A. (2003) 'Isosexual pair housing improves the welfare of young Amazon parrots', *Applied Animal Behaviour Science*, 81: 73–88.

Meehan, C. L., Garner, J. P., and Mench, J. A. (2004) 'Environmental enrichment and development of cage stereotypy in orange-winged Amazon parrots (*Amazona amazonica*)', *Developmental Psychobiology*, 44: 209–16.

Melfi, V. A. (2001) 'Identification and evaluation of the captive environmental factors that affect the behaviour of Sulawesi crested black macaques (*Macaca nigra*)', unpublished PhD thesis, Department of Zoology, University of Dublin, Trinity College Dublin, Ireland.

Melfi, V. A. (2005) 'The appliance of science to zoo-housed primates', *Applied Animal Behaviour Science*, 90: 97–106.

Melfi, V. A. and Feistner, A. T. C. (2002) 'A comparison of the activity budgets of wild and captive Sulawesi crested black macaques (*Macaca nigra*)', *Animal Welfare*, 11: 213–22.

Melfi, V. and Knight, K. (2008) 'Public perceptions of carnivore feeding methods: preliminary results from an international study' in A. Hartley (ed.) *Proceedings of the Annual BIAZA Research Meeting*, Whipsnade Zoo (in press).

Melfi, V. and Poyser, F. (2008) '*Trichuris* burdens in zoo-housed *Colobus guereza*', *International Journal of Primatology*, 28: 1449–56.

Melfi, V. A. and Thomas, S. (2005) 'Can training zoo-housed primates compromise their conservation? A case study using Abyssinian colobus monkeys (*Colobus guereza*)', *Anthrozoös*, 18: 304–17.

Melfi, V. A., Garcia, L., Dicks, J., Bowers, C., Hendy, M., Chapman, J., and Bemment, N. (2004, unpublished) 'Wire hay racks: simple but effective enrichment', data presented at the BIAZA Elephant TAG meeting, Chester Zoo.

Melfi, V. A., McCormick, W., and Gibbs, A. (2004a) 'A preliminary assessment of how zoo visitors evaluate animal welfare according to enclosure style and the expression of behaviour', *Anthrozoös*, 17: 98–108.

Melfi, V. A., Uwakaneme, C., and Rees, M. (2004b) 'Crocodile environmental enrichment: as necessary as monkey puzzles!', *BIAZA Research News*, 5: 2–3.

Melfi, V. A., Bowkett, A., Plowman, A. B., and Pullen, K. (2007) 'Do zoo designers know enough about animals?' in A. B. Plowman and S. Tonge (eds), *Innovation or Replication: Proceedings of the Sixth International Symposium on Zoo Design*, 9–14 May, Paignton, pp. 119–27.

Mellen, J. D. (1991) 'Factors influencing reproductive success in small captive exotic felids (*Felis* spp.): a multiple regression analysis', *Zoo Biology*, 10: 95–110.

Mellen, J. D. (1992) 'Effects of early rearing experience on subsequent adult sexual behavior using domestic cats (*Felis catus*) as a model for exotic small felids', *Zoo Biology*, 11: 17–32.

Mellen, J. D. (1994) 'Survey and interzoo studies used to address husbandry problems in some zoo vertebrates', *Zoo Biology*, 13: 459–70.

Mellen, J. D. and Ellis, S. (1996) 'Animal learning and husbandry training techniques' in D. G. Kleiman, M. E. Allen, K. V. Thompson, and S. Lumpkin (eds), *Wild Mammals in Captivity*, Chicago, IL/London: University of Chicago Press, pp. 88–99.

Mellen, J. D. and MacPhee, M. S. (2001) 'Philosophy of environmental enrichment: past, present, and future', *Zoo Biology*, 20: 211–26.

Meller, C. L., Croney, C. C., and Shepherdson, D. (2007) 'Effects of rubberized flooring on Asian elephant behavior in captivity', *Zoo Biology*, 26: 51–61.

Mello, I., Nordensten, L., and Amundin, M. (2005) 'Reactions of three bottlenose dolphin dams with calves to other members of the group in connection with nursing', *Zoo Biology*, 24: 543–55.

Melnick, D. and Pearl, M. (1987) 'Cercopithecines in multimale groups: genetic diversity and population structure' in B. Smuts, D. Cheney, R. Seyfarth, R. Wrangham, and T. Struhsaker (eds), *Primate Societies*, Chicago, IL: Chicago University Press, pp. 121–34.

Mench, J. A. (1998) 'Thirty years after Brambell: whither animal welfare science?', *Journal of Applied Animal Welfare Science*, 1: 91–102.

Mensink, M. and Shafternaar, W. (1998) 'When bad things happen to bats: the occurrence of a *Lyssavirus* in a closed population of Egyptian frugivorous bats (*Rousettus aegyptiacus*) at Rotterdam Zoo', *Proceedings of the Annual Meeting of the European Association of Zoo and Wildlife Veterinarians*, Chester, pp. 147–51.

Menzel, C. (1991) 'Cognitive aspects of foraging in Japanese monkeys', *Animal Behaviour*, 41: 397–402.

Meretsky, V. J., Snyder, N. F. R., Beissinger, F. R., Clendenen, D. A., and Wiley, J. W. (2000) 'Demography of the Californian condor: implications for reestablishment', *Conservation Biology*, 14: 957–67.

Mete, A., Hendriks, H. G., Klaren, P. H. M., Dorrestein, G. M., van Dijk, J. E., and Marx, J. J. M. (2003) 'Iron metabolism in mynah birds (*Gracula religiosa*) resembles human hereditary haemochromatosis', *Avian Pathology*, 32: 625–32.

Meyer-Holzapfel, M. (1968) 'Abnormal behaviour in zoo animals' in M. W. Fox (ed.), *Abnormal Behavior in Animals*, Philadelphia, PA: W. B. Saunders, pp. 476–503.

Midgley, M. (1983) *Animals and Why They Matter*, Athens, GA: The University of Georgia Press.

Mikota, S.K. (2007) 'Tuberculosis in elephants', in M. E. Fowler and R. E. Miller (eds), *Zoo and Wild Animal Medicine* (6th edn), St Louis, MO: Saunders (Elsevier), pp. 355–64.

Mikota, S. K., Larsen, R. S., and Montali, R. J. (2000) 'Tuberculosis in elephants in North America', *Zoo Biology*, 19: 393–403.

Millam, J. R., Kenton, B., Jochim, L., Brownback, T., and Brice, A. T. (1995) 'Breeding orange-winged Amazon parrots in captivity', *Zoo Biology*, 14: 275–84.

Millar, J. S. and Hickling, G. J. (1990) 'Fasting endurance and the evolution of mammalian body size', *Functional Ecology*, 4: 5–12.

Miller, B., Conway, W., Reading, R. P., Wemmer, C., Wildt, D., Kleiman, D., Monfort, S., Rabinowitz, A., Armstrong, B., and Hutchins, M. (2004) 'Evaluating the conservation mission of zoos, aquariums, botanical gardens and natural history museums', *Conservation Biology*, 18: 86–93.

Miller, H. (2006) 'Cloacal and buccal swabs are a reliable source of DNA for microsatellite genotyping of reptiles', *Conservation Genetics*, 7: 1001–3.

Millman, S. T. and Duncan, I. J. H. (2000) 'Strain differences in aggressiveness of male domestic fowl in response to a male model', *Applied Animal Behaviour Science*, 66: 217–33.

Millman, S. T., Duncan, I. J. H., Stauffacher, M., and Stookey, J. M. (2004) 'The impact of applied ethologists and the International Society for Applied Ethology in improving animal welfare', *Applied Animal Behaviour Science, International Society for Applied Ethology Special Issue: A Selection of Papers from the 36th ISAE International Congress*, 86: 299–311.

Millspaugh, J. J. and Washburn, B. E. (2003) 'Within-sample variation of fecal glucocorticoid measurements', *General and Comparative Endocrinology*, 132: 21–6.

Millspaugh, J. J. and Washburn, B. E. (2004) 'Use of fecal glucocorticoid metabolite measures in conservation biology research: considerations for application and interpretation', *General and Comparative Endocrinology*, 138: 189–99.

Misslin, R. and Cigrang, M. (1986) 'Does neophobia necessarily imply fear or anxiety?', *Behavioral Proceedings*, 12: 45–50.

Mistlberger, R. E. (1994) 'Circadian food-anticipatory activity: formal models and physiological mechanisms', *Neuroscience and Biobehavioral Reviews*, 18: 171–95.

Mitchell, G. and Gomber, J. (1976) 'Moving laboratory rhesus monkeys (*Macaca mulatta*) to unfamiliar home cages', *Primates*, 17: 543–7.

Mitchell, G., Herring, F., Obradovich, S., Tromborg, C., Dowd, B., Neville, L., and Field, L. (1991a) 'Effects of visitors and cage changes on the behaviours of mangabeys', *Zoo Biology*, 10: 417–23.

Mitchell, G., Obradovich, S. D., Herring, F. H., Dowd, B., and Tromborg, C. (1991b) 'Threats to observers, keepers, visitors, and others by zoo mangabeys (*Cercocebus galeritus chrysogaster*)', *Primates*, 32: 515–22.

Mitchell, G., Herring, F., and Obradovich, S. (1992a) 'Like threaten like in mangabeys and people', *Anthrozoös*, 5: 106–12.

Mitchell, G., Tromborg, C. T., Kaufman, J., Bargabus, S., Simoni, R., and Geissler, V. (1992b) 'More on the "influence" of zoo visitors on the behaviour of captive primates', *Applied Animal Behaviour Science*, 35: 189–98.

Mitchell, H. and Hosey, G. (2005) *Zoo Research Guidelines: Studies of the Effects of Human Visitors on Zoo Animal Behaviour*, London: BIAZA.

Moberg, G. P. and Mench, J. A. (eds) (2000) *The Biology of Animal Stress: Basic Principles and Implications for Animal Welfare*, Wallingford, Oxon: CABI Publishing.

Moe, M. (1993) *The Marine Aquarium Reference: Systems and Invertebrates*, Plantation, FL: Green Turtle Publications.

Molloy, L. and Hart, J. A. (2002) 'Duiker food selection: palatability trials using natural foods in the Ituri Forest, Democratic Republic of Congo', *Zoo Biology*, 21: 149–59.

Molony, V. and Kent, J. E. (1997) 'Assessment of acute pain in farm animals using behavioral and physiological measurements', *Journal of Animal Science*, 75: 266–72.

Montaudouin, S. and Le Pape, G. (2004) 'Comparison of the behaviour of European brown bears (*Ursus arctos arctos*) in six different parks, with particular attention to stereotypies', *Behavioural Processes*, 67: 235–44.

Montaudouin, S. and Le Pape, G. (2005) Comparison between 28 zoological parks: stereotypic and social behaviours of captive brown bears (*Ursus arctos*)', *Applied Animal Behaviour Science*, 92: 129–41.

Moodie, E. M. and Chamove, A. S. (1990) 'Brief threatening events beneficial for captive tamarins?', *Zoo Biology*, 9: 275–86.

Moore, B. D., Marsh, K. J., Wallis, I. R., and Foley, W. J. (2005) 'Taught by animals: how understanding diet selection leads to better zoo diets', *International Zoo Yearbook*, 39: 43–61.

Moore, T. L., Killiany, R. J., Herndon, J. G., Rosene, D. L., and Moss, M. B. (2006) 'Executive system dysfunction occurs as early as middle-age in the rhesus monkey', *Neurobiology of Aging*, 27: 1484–93.

Moran, G. and Sorensen, L. (1984) 'The behavioural researcher and the zoological park', *Applied Animal Behaviour Science*, 13: 143–55.

Morgan, J. M. and Hodgkinson, M. (1999) 'The motivation and social orientation of visitors attending a contemporary zoological park', *Environment and Behavior*, 31: 227–39.

Morgan, K. N. and Tromborg, C. T. (2007) 'Sources of stress in captivity', *Applied Animal Behaviour Science*, 102: 262–302.

Moritz, C. (1994) 'Defining "evolutionarily significant units" for conservation', *Trends in Ecology and Evolution*, 9: 373–5.

Mormede, P., Andanson, S., Auperin, B., Beerda, B., Guemene, D., Malmkvist, J., Manteca, X., Manteuffel, G., Prunet, P., van Reenen, C. G., Richard, S., and Veissier, I. (2007) 'Exploration of the hypothalamic-pituitary-adrenal function as a tool to evaluate animal welfare', *Physiology and Behavior*, 92: 317–39.

Morris, D. (1964) 'The response of animals to a restricted environment', *Symposium of the Zoological Society of London*, 13: 99–118.

Morris, D. (1969) *The Human Zoo*, London: Jonathon Cape.

Morton, A. C. (1990) 'Captive breeding of butterflies and moths II: conserving genetic variation and managing biodiversity', *International Zoo Yearbook*, 30: 89–97.

Moyle, M. (1989) 'Vitamin D and UV radiation: guidelines for the herpetoculturist' in M. Uricheck (ed.), *Proceedings of the 13th International Herpetological Symposium on Captive Propagation and Husbandry*, Thurmont, MD: Zoological Consortium Inc., pp. 61–70.

Murphy, J. B. (2007) *Herpetological History of the Zoo and Aquarium World*, Melbourne, FL: Krieger.

Murray, J., Murray, E., Johnson, M. S., and Clarke, B. (1988) 'The extinction of *Partula* on Moorea', *Pacific Science*, 42: 150–3.

Myers, O. E., Saunders, C. D., and Birjulin, A. A. (2004) 'Emotional dimensions of watching zoo animals: an experience sampling study building on insights from psychology', *Curator*, 47: 299–321.

N Nash, L. T. (1986) 'Dietary, behavioral, and morphological aspects of gummivory in primates', *American Journal of Physical Anthropology*, 29: 113–37.

Nash, L. and Chilton, S. (1986) 'Space or novelty? Effects of altered cage size on galago behaviour', *American Journal of Primatology*, 10: 37–50.

Nelson, R. (2000) *An Introduction to Behavioral Endocrinology*, Sunderland, MA: Sinauer Associates.

Neptune, D. and Walz, D. (2005) 'Thinking outside the cardboard box: taking enrichment to the next level' in N. Clum, S. Silver, and P. Thomas (eds), *Seventh Conference on Environmental Enrichment*, 31 July–5 August, New York, The Shape of Enrichment, Inc., pp. 90–5.

Newberry, R. (1995) 'Environmental enrichment: increasing the biological relevance of captive environments.', *Applied Animal Behavioural Science*, 44: 229–43.

Nicholls, N. (2003) 'Development of a method to determine behavioural need in Mediterranean tortoise *Testudo hermanni*', *Federation Research Newsletter*, 4: 3.

Nieuwenhuijsen, K. and de Waal, F. (1982) 'Effects of spatial crowding on social behavior in a chimpanzee colony', *Zoo Biology*, 1: 5–28.

Nijboer, J. and Hatt, J. (2000) *Zoo Animal Nutrition, Vol. I*, Fürth: Filander Verlag.

Nimon, A. J. and Dalziel, F. R. (1992) 'Cross-species interaction and communication: a study method applied to captive siamang (*Hylobates syndactylus*) and long-billed corella (*Cacatua tenuirostris*) contacts with humans', *Applied Animal Behaviour Science*, 33: 261–72.

Noë, R. and Bshary, R. (1997) 'The formation of red colobus-diana monkey associations under predation pressure from chimpanzees', *Proceedings of the Royal Biological Society*, 264: 253–9.

Nogge, G. (1997) 'Introduction: zoo research—the role of the EAZA Research Committee', *Applied Animal Behaviour Science*, 51: 195–7.

Norcup, S. (2001) *European Endangered Species Programme Studbook for Sulawesi Crested Black Macaques (*Macaca nigra*)* (4th edn), Trinity, Jersey: Durrell Wildlife Conservation Trust.

Norris, R. (2005) 'Transport of animals by sea', *OIE Scientific and Technical Review*, 24: 673–81.

Norton, B. G., Hutchins, M., Stevens, E., and Maple, T. L. (eds) (1995) *Ethics on the Ark: Zoos, Animal Welfare and Wildlife Conservation*, Washington DC/London: Smithsonian Institute Press.

Novak, M. A. (2003) 'Self-injurious behaviour in rhesus monkeys: new insights into its etiology, physiology, and treatment', *American Journal of Primatology*, 59: 3–19.

Novak, M. A., Musante, A., Munroe, H., O'Neill, P. L., Price, C., and Suomi, S. J. (1993) 'Old, socially housed rhesus monkeys manipulate objects', *Zoo Biology*, 12: 285–98.

NRC (National Research Council) (1982) *Nutrient Requirements of Mink and Foxes*, Washington DC: National Academy Press.

NRC (National Research Council) (1998) *Nutrient Requirements of Swine*, Washington DC: National Academy Press.

NRC (National Research Council) (2001) *Nutrient Requirements of Dairy Cattle*, Washington DC: National Academy Press.

NRC (National Research Council) (2006) *Nutrient Requirements of Cats and Dogs*, Washington DC: National Academy Press.

O Oaks, J. L., Gilbert, M., Virani, M. Z., Watson, R. T., Meteyer, C. U., Rideout, B. A., Shivaprasad, H. L., Ahmed, S., Chaudhry, M. J. I., Arshad, M., Mahmood, S., Ali, A., and Khan, A. A. (2004) 'Diclofenac residues as the cause of vulture population decline in Pakistan', *Nature*, 427: 630–3.

Oates, J. (1989) 'Food distribution and foraging behaviour' in B. Smuts, D. Cheney, R. Seyfarth,

R. Wrangham, and T. Struhsaker (eds), *Primate Societies*, Chicago, IL: Chicago University Press, pp. 197–209.

O'Brien, J. (2006) 'Effects of conspecific playback recordings on a pair of Toco toucans', *The Shape of Enrichment*, 15: 3–5.

O'Brien, S., Roelke, M., Marker, L., Newman, A., Winkler, C., Meltzer, D., Colly, L., Evermann, J., Bush, M., and Wildt, D. (1985) 'Genetic basis for species vulnerability in the cheetah', *Science*, 227: 1428–34.

O'Brien, T. and Kinnaird, M. (1997) 'Behaviour, diet and movement of the Sulawesi crested black macaque', *International Journal of Primatology*, 18: 321–51.

O'Connor, K. I. (2000) 'Mealworm dispensers as environmental enrichment for captive Rodrigues fruit bats (*Pteropus rodricensis*)', *Animal Welfare*, 9: 123–37.

O'Donovan, D., Hindle, J. E., McKeown, S., and O'Donovan, S. (1993) 'Effect of visitors on the behaviour of female cheetahs *Acinonyx jubatus*', *International Zoo Yearbook*, 32: 238–44.

Oftedal, O. T. and Allen, M. E. (1996) 'Nutrition and dietary evaluation in zoos' in D. Kleiman, M. E. Allen, K. V. Thompson, and S. Lumpkin (eds), *Wild Mammals in Captivity: Principles and Techniques*, Chicago, IL/London: University of Chicago Press, pp. 109–16.

Oftedal, O. T., Baer, D. J., and Allen, M. E. (1996) 'The feeding and nutrition of herbivores' in D. Kleiman, M. E. Allen, K. V. Thompson, and S. Lumpkin (eds), *Wild Mammals in Captivity: Principles and Techniques*, Chicago, IL/London: University of Chicago Press, pp. 129–38.

Ogden, J., Lindburg, D., and Maple, T. (1993) 'Preference for structural environmental features in captive lowland gorillas', *Zoo Biology*, 1215: 381–95.

Olney, P. (1975) 'Walk-through avaries' in A. Michelmore (ed.), *Proceedings of the First International Symposium on Zoo Design and Construction*, 13–15 May, Paignton, pp. 130–5.

Olney, P. (1980) 'The London Zoo and Whipsnade Zoo' in L. S. Zuckerman (ed.), *Great Zoos of the World: Their Origins and Significance*, London: Weidenfeld & Nicolson, pp. 37–59.

Olney, P. J. S., Mace, G. M., and Feistner, A. T. C. (eds) (1994) *Creative Conservation: Interactive Management of Wild and Captive Animals*, London: Chapman & Hall.

O'Regan, H. J. (2001) 'Morphological effects of captivity in big cat skulls' in S. Wehnelt and C. Hudson (eds), *Proceedings of the Third Annual Symposium on Zoo Research*, 9–10 July, Chester: BIAZA, pp. 18–22.

O'Regan, H. J. and Kitchener, A. C. (2005) 'The effects of captivity on the morphology of captive, domesticated and feral mammals', *Mammal Review*, 35: 215–30.

Ortega, J., Franco, R., Adams, B. A., Ralls, K., and Maldonado, J. E. (2004) 'A reliable, non-invasive method for sex determination in the endangered San Joaquin kit fox (*Vulpes macrotis mutica*) and other canids', *Conservation Genetics*, 5: 715–18.

Osawa, R., Blanchard, W. H., and O'Callaghan, P. G. (1993) 'Microbiological studies of the intestinal microflora of the koala, *Phascolarctos cinereus*', *Australian Journal of Zoology*, 41: 611–20.

Ostrowski, S., Bedin, E., Lenain, D. M., and Abuzinada, A. H. (1998) 'Ten years of Arabian oryx conservation breeding in Saudi Arabia: achievements and regional perspectives', *Oryx*, 32: 209–22.

Ott-Joslin, J. E. (1993) 'Zoonotic diseases of nonhuman primates' in M. E. Fowler (ed.), *Zoo and Wild Animal Medicine: Current Therapy* (3rd edn), Philadelphia, PA: W. B. Saunders, pp. 358–73.

Overmier, J. B., Patterson, J., and Wielkiewics, R. M. (1980) 'Environmental contingencies as sources of stress in animals' in S. Levine and H. Ursin (eds) *Coping and Health*, New York: Plenum Press, pp. 1–38.

Owen, M., A., Swaisgood, R., R., Czekala, N., M. and Lindburg, D. G. (2005) 'Enclosure choice and well-being in giant pandas: is it all about control?', *Zoo Biology*, 24: 475–81.

Owen, M. A., Swaisgood, R. R., Czekala, N. M., Steinman, K., and Lindburg, D. G. (2004) 'Monitoring stress in captive giant pandas (*Ailuropoda melanoleuca*): behavioural and hormonal responses to ambient noise', *Zoo Biology*, 23: 147–64.

P Packard, J., Babbitt, K., Hannon, P., and Grant, W. (1990) 'Infanticide in captive collared peccaries (*Tayassu tajacu*)', *Zoo Biology*, 9(1): 49–53.

Pankhurst, S. J. (1998) *The Social Organisation of the Mara (*Dolichotis patagonum*) at Whipsnade Wild Animal Park*, unpublished PhD thesis, Cambridge: University of Cambridge.

Pankhurst, S. J. and Knight, K. (2008) *Zoo Research Guidelines: Behavioural Profiling of Zoo Animals*, London: BIAZA (in press).

Pankhurst, S. J., Plumb, A., and Walter, O. (2008) *Zoo Research Guidelines: Getting Zoo Research Published*, London: BIAZA.

Parker, M., Goodwin, D., Redhead, E., and Mitchell, H. (2006) 'The effectiveness of environmental enrichment on reducing stereotypic behaviour in two captive vicugna (*Vicugna vicugna*)', *Animal Welfare*, 15: 59–62.

Pearce, J. M. (1997) *Animal Learning and Cognition* (2nd edn), Hove: Psychology Press.

Pearce-Kelly, P., Mace, G. M., and Clarke, D. (1995) 'The release of captive-bred snails (*Partula taeniata*) into a semi-natural environment', *Biodiversity and Conservation*, 4: 645–63.

Pearce-Kelly, P., Jones, R., Clarke, D., Walker, C., Atkin, P., and Cunningham, A. A. (1998) 'The captive rearing of threatened *Orthoptera*: a comparison of the conservation potential and practical considerations of two species' breeding programmes at the Zoological Society of London', *Journal of Insect Conservation*, 2: 201–10.

Pearson, G. L. (1977) 'Vaccine-induced canine distemper virus in black-footed ferrets', *Journal of the American Veterinary Medicine Association*, 170: 103–9.

Peeters, E. and Geers, R. (2006) 'Influence of provision of toys during transport and lairage on stress responses and meat quality of pigs', *Animal Science*, 82: 591–5.

Pennisi, E. (2001) 'Zoo's new primate exhibit to double as research lab', *Science*, 293: 1247.

Perkins, L. A. (1992) 'Variables that influence the activity of captive orang-utans', *Zoo Biology*, 11(3): 177–86.

Peterson, R. O., Jacobs, A. K., Drummer, T. D., Mech, L. D., and Smith, D. W. (2002) 'Leadership behavior in relation to dominance and reproductive status in gray wolves, *Canis lupus*', *Canadian Journal of Zoology*, 80(8): 1405–12.

Pickard, A. (2003) 'Reproductive and welfare monitoring for the management of *ex situ* populations' in W. Holt, A. Pickard, J. Rodger, and D. Wildt (eds), *Reproductive Science and Integrated Conservation*, Cambridge: Cambridge University Press, pp. 132–46.

Pickering, S., Creighton, E., and Stevens-Wood, B. (1992) 'Flock size and breeding success in flamingos', *Zoo Biology*, 11: 229–34.

Picq, J.-L. (2007) 'Aging affects executive functions and memory in mouse lemur primates', *Experimental Gerontology*, 42(3): 223–32.

Pierce, J. (2005) 'A system for mass culture of upside-down jellyfish *Cassiopea* spp as a potential food item for medusivores in captivity', *International Zoo Yearbook*, 39: 62–9.

Pinder, N. J. and Barkham, J. P. (1978) 'An assessment of the contribution of captive breeding to the conservation of rare mammals', *Biological Conservation*, 13: 187–245.

Pizzi, R. (2004) 'Slap me with a dead penguin: what we can learn from 1001 penguin post-mortems at Edinburgh Zoo' in C. Macdonald (ed.), *Proceedings of the Sixth Annual Symposium on Zoo Research*, 8–9 July, Edinburgh: BIAZA, pp. 253–55.

Plowman, A. B. (2003) 'A note on a modification of the spread of participation index allowing for unequal zones', *Applied Animal Behaviour Science*, 83: 331–6.

Plowman, A. B. (ed.) (2006) *Zoo Research Guidelines: Statistics for Typical Zoo Datasets*, London: BIAZA.

Plowman, A. B. and Knowles, L. (2003) 'Overcoming habituation in an enrichment programme for tigers' in: V. J. Hare, K. E. Worley, and B. Hammond (eds) *Proceedings of the Fifth International Conference on Environmental Enrichment*, 4–9 November 2001, Taronga Zoo, Sydney, Australia. San Diego, CA: The Shape of Enrichment Inc., pp. 263–8.

Plowman, A. B. and Turner, I. (2001a) 'A survey and database of browse use for mammals in UK and Irish zoos' in S. Wehnelt and C. Hudson (eds), *Proceedings of the Third Annual Symposium on Zoo Research*, 9–10 July, Chester: BIAZA, pp. 50–5.

Plowman, A. B. and Turner, I. (2001b) *Database of Browse Used in Federation Zoos*, CD-ROM, London: BIAZA.

Plowman, A. B. and Turner, I. (2006) 'A survey and database of browse use for mammals in UK and Irish zoos' in A. Fidgett, M. Claus, K. Eulenberger, J.-M. Hatt, I. Hume, G. Janssens, and J. Nijboer (eds), *Zoo Animal Nutrition, Vol. III*, Fürth: Filander Verlag, pp. 193–7.

Plowman, A. B., Jordan, N. R., Anderson, N., Condon, E., and Fraser, O. (2005) 'Welfare implications of captive primate population management: behavioural and psycho-social effects of female-based contraception, oestrus and male removal in hamadryas baboons (*Papio hamadryas*)', *Applied Animal Behaviour Science*, 90: 155–65.

Plowman, A. B., Hosey, G., and Stevenson, M. (2006) *Zoo Research Guidelines: Surveys and Questionnaires*, London: BIAZA.

Plowman, A., Green, K., and Taylor, L. (2008) 'Should zoo food be chopped?' in A. Fidgett (ed.) *Proceedings of the Fifth European Zoo Nutrition Conference*, 24–27 January 2008, Chester Zoo, Chester (in press).

Pochon, V. (1998) 'Mixed species exhibit for Eastern black-and-white colobus and patas monkeys', *International Zoo Yearbook*, 36: 69–73.

Polakowski, W. (1987) *Zoo Design: The Reality of Wild Illusions*, Ann Arbour, MI: University of Michigan, School of Natural Sciences.

Pond, W. G., Church, D. C., Pond, K., and Schoknecht, P. A. (2005) *Basic Animal Nutrition and Feeding* (5th edn), New York: John Wiley & Sons.

Poole, T. (1999) *UFAW Handbook on the Care and Management of Laboratory Animals* (7th edn), Oxford: Blackwell Science.

Pope, C. E. (2000) 'Embryo technology in conservation efforts for endangered felids', *Theriogenology*, 53: 163–74.

Popp, J. W. (1984) 'Interspecific aggression in mixed ungulate species exhibits', *Zoo Biology*, 3: 211–19.

Pounds, J. A., Bustamante, A. R., Coloma, L. A., Consuegra, J. A., Fogden, M. P., Foster, P. N., La Marca, E., Masters, K. L., Merino-Viteri, A., Puschendorf, R., Ron, S. R., Sànchez-Azofeifa, G. A., Still, C. J., and Young, B. E. (2006) 'Widespread amphibian extinctions from epidemic disease driven by global warming', *Nature*, 439: 161–7.

Povey, K. D. and Rios, J. (2002) 'Using interpretive animals to deliver affective messages in zoos', *Journal of Interpretation Research*, 7: 19–28.

Powell, D. (1995) 'Preliminary evaluation of environmental enrichment techniques for African lions (*Panthera leo*)', *Animal Welfare*, 4: 361–71.

Powell, D. M., Carlstead, K., Tarou, L. R., Brown, J. L., and Monfort, S. L. (2006) 'Effects of construction noise on behavior and cortisol levels in a pair of captive giant pandas (*Ailuropoda melanoleuca*)', *Zoo Biology*, 25: 391–408.

Preston, D. J. (1983) 'Jumbo, king of elephants', *Natural History*, 92: 80–3.

Price, D. (2000) 'Psychological and neural mechanisms of the affective dimension of pain', *Science*, 288: 1769–72.

Price, E. and Caldwell, C. A. (2007) 'Artificially generated cultural variation between two groups of captive monkeys, *Colobus guereza kikuyuensis*', *Behavioural Processes*, 74: 13–20.

Price, E. C., McGivern, A.-M., and Ashmore, L. (1991) 'Vigilance in a group of free-ranging cotton-top tamarins *Saguinus oedipus*', *Dodo: Journal of the Jersey Wildlife Preservation Trust*, 27: 41–9.

Price, E. C., Ashmore, L. A., and McGivern, A.-M. (1994) 'Reactions of zoo visitors to free-ranging monkeys', *Zoo Biology*, 13: 355–73.

Price, E. E. and Stoinski, T. S. (2007) 'Group size: determinants in the wild and implications for the captive housing of wild mammals in zoos', *Applied Animal Behaviour Science*, 103: 255–64.

Price, E. O. (1984) 'Behavioural aspects of animal domestication', *Quarterly Review of Biology*, 59: 1–32.

Provenza, F. D., Scott, C. B., Phy, T. S., and Lynch, J. J. (1996) 'Preference of sheep for foods varying in flavors and nutrients', *Journal of Animal Science*, 74: 2355–61.

Prusky, G. T., Reidel, C., and Douglas, R. M. (2000) 'Environmental enrichment from birth enhances visual acuity but not place learning in mice', *Behavioural Brain Research*, 114: 11–15.

Pryor, K. (2002) *Don't Shoot the Dog! The New Art of Teaching and Training*, Dorking: Ringpress Books.

Puan, C. L. and Zakaria, M. (2007) 'Perception of visitors towards the role of zoos: a Malaysian perspective', *International Zoo Yearbook*, 41: 226–32.

Pullen, K. (2005) 'Preliminary comparisons of male/male interactions within bachelor and breeding groups of western lowland gorillas (*Gorilla gorilla gorilla*)', *Applied Animal Behaviour Science*, 90: 143–53.

Pusey, A. and Packer, C. (1987) 'Dispersal and philopatry' in B. Smuts, D. Cheney, R. Seyfarth, R. Wrangham, and T. Stuhsaker (eds), *Primate Societies*, Chicago, IL: University of Chicago Press, pp. 250–66.

q **Quinn, H. and Quinn, H.** (1993) 'Estimated number of snake species that can be managed by Species Survival Plans in North America', *Zoo Biology*, 12: 243–56.

r **Rabb, G. B.** (1994) 'The changing roles of zoological parks in conserving biological diversity', *American Zoologist*, 34: 159–64.

Rabb, G. B. and Saunders, C. D. (2005) 'The future of zoos and aquariums: conservation and caring', *International Zoo Yearbook*, 39: 1–26.

Rabb, G. B. and Sullivan, T. A. (1995) 'Coordinating conservation: global networking for species survival', *Biodiversity and Conservation*, 4: 536–43.

Rabin, L. A. (2003) 'Maintaining behavioural diversity in captivity for conservation: natural behaviour management', *Animal Welfare*, 12: 85–94.

Rachels, J. (1976) 'Do animals have a right to liberty?' in P. Singer and T. Regan (eds), *Animal Rights and Human Obligations*, Englewood Cliffs, New Jersey: Prentice Hall, pp. 205–23.

Radcliffe, R. W., Czekala, N. M., and Osofsky, S. A. (1997) 'Combined serial ultrasonography and fecal progestin analysis for reproductive evaluation of the female white rhinoceros (*Ceratotherium simum simum*): preliminary results', *Zoo Biology*, 16: 445–6.

Radford, M. (2001) *Animal Welfare Law in Britain: Regulation and Responsibility*, Oxford: Oxford University Press.

Ralls, K., Lundrigan, B., and Kranz, K. (1987) 'Mother–young relationships in captive ungulates: spatial and temporal patterns', *Zoo Biology*, 6: 11–20.

Ramirez, K. (ed.) (1999) *Animal Training: Successful Animal Management Through Positive Reinforcement*, Chicago, IL: Shedd Aquarium Society.

Raphael, B. L. (2003) 'Chelonians (Turtles, Tortoises)' in M. E. Fowler and M. E. Miller (eds), *Zoo and Wild Animal Medicine* (5th edn), Philadelphia, PA: Elsevier (Saunders), pp. 48–58.

Raven, P. H. (2002) 'Science, sustainability, and the human prospect', *Science*, 297: 954–8.

Rawlins, C. G. C. (1985) 'Zoos and conservation: the last 20 years', *Symposia of the Zoological Society of London*, 54: 59–69.

Reade, L. S. and Waran, N. K. (1996) 'The modern zoo: how do people percieve zoo animals?', *Applied Animal Behavioural Sciences*, 4710: 109–18.

Reading, R. P. and Miller, B. J. (2007) 'Attitudes and attitude change among zoo visitors' in A. Zimmermann, M. Hatchwell, L. Dickie, and C. West (eds), *Zoos in the 21st Century: Catalysts for Conservation?*, Cambridge: Cambridge University Press, pp. 63–91.

Reaka-Kudka, M. L., Wilson, D. E., and Wilson, E. O. (eds) (1996) *Biodiversity II: Understanding and Protecting our Biological Resources*, Washington DC: Joseph Henry Press.

Réale, D. and Festa-Bianchet, M. (2003) 'Predator-induced natural selection on temperament in bighorn ewes', *Animal Behaviour*, 65: 463–70.

Redig, P. T. and Ackermann, J. (2000) 'Raptors' in T. N. Tully, M. P. C. Lawton, and G. M. Dorrestein (eds), *Avian Medicine*, Oxford: Butterworth-Heinemann, pp. 180–214.

Redshaw, M. E. and Mallinson, J. J. C. (1991) 'Learning from the wild: improving the psychological and physical well-being of captive primates', *Dodo: Journal of the Jersey Wildlife Preservation Trust*, 2712: 18–26.

Reed, H. J., Wilkins, L. J., Austin, S. D., and Gregory, N. G. (1993) 'The effect of environmental enrichment during rearing on fear reactions and depopulation trauma in adult caged hens', *Applied Animal Behaviour Science*, 36: 39–46.

Rees, P. A. (2004) 'Low environmental temperature causes an increase in stereotypic behaviour in captive Asian elephants (*Elephas maximus*)', *Journal of Thermal Biology*, 29: 37–43.

Rees, P. A. (2005a) 'The EC Zoos Directive: a lost opportunity to implement the Convention on Biological Diversity', *Journal of International Wildlife Law and Policy*, 8: 51–62.

Rees, P. A. (2005b) 'Will the EC Zoos Directive increase the conservation value of zoo research?', *Oryx*, 39: 128–31.

Regal, P. (1980) 'Temperature and light requirements of captive reptiles' in J. Murphy and J. Collins (eds), *Reproductive Biology and Diseases in Captive Reptiles*, St Louis, MO: Society for the Study of Amphibians and Reptiles, pp. 79–91.

Regan, J. (2005) *The Manifesto for Zoos*, Manchester: John Regan Associates Ltd.

Regan, T. (1983) *The Case for Animal Rights*, Berkeley, CA: University of California Press.

Regan, T. (1995) 'Are zoos morally defensible?' in B. G. Norton, M. Hutchins, E. Stevens, and T. L. Maple (eds), *Ethics on the Ark: Zoos, Animal Welfare and Wildlife Conservation*, Washington DC/London: Smithsonian Institute Press, pp. 38–51.

Rehling, M. (2001) 'Octopus enrichment techniques' in M. Hawkins, K. E. Worley, and B. Hammond (eds), *Fifth International Conference on Environmental Enrichment*, 4–9 November, Sydney, The Shape of Enrichment Inc., pp. 94–8.

Reinhardt, V. (1994a) 'Caged Rhesus macaques voluntarily work for ordinary food', *Primates*, 35: 95–8.

Reinhardt, V. (1994b) 'Safe pair formation technique for previously single-caged Rhesus macaques', available online at http://www.awionline.org/Lab_animals/biblio/tou-safe.htm.

Reinhardt, V. (1995) 'Arguments for single-caging of Rhesus macaques: are they justified?', *Animal Welfare Information Center Newsletter*, 6: 1–2.

Reinhardt, V. (2003) 'Working with rather than against macaques during blood collection', *Journal of Applied Animal Welfare Science*, 6: 189–97.

Reinhardt, V. and Reinhardt, A. (2000) 'Social enhancement for adult non-human primates in research laboratories: a review', *Lab Animal*, 29: 34–41.

Reinhardt, V. and Roberts, A. (1997) 'Effective feeding enrichment for non-human primates: a brief review', *Animal Welfare*, 6: 265–72.

Reinhardt, V., Liss, C., and Stevens, C. (1996) 'Space requirement stipulations for caged non-human primates in the United States: a critical review', *Animal Welfare*, 5: 361–72.

Reinhardt, V., Bryant, D., Kurth, B., Lynch, R., Asvestas, C., Byrum, R., Claire, M. S., and Seelig, D. (1998) 'Discussion: a plea for pair-housing of adult macaques', *Laboratory Primate Newsletter*, 37: 4.

Reiss, D. (2005) 'Enriching animals while enriching science: providing choice and control to dolphins' in N. Clum, S. Silver, and P. Thomas (eds), *Seventh Conference on Environmental Enrichment*, 31 July– 5 August, New York, The Shape of Enrichment, Inc., pp. 26–31.

Rendall, D. and Taylor, L. T. (1991) 'Female sexual behavior in the absence of male–male competition in captive Japanese macaques (*Macaca fuscata*)', *Zoo Biology*, 10: 319–28.

Renner, M. J. and Lussier, J. P. (2002) 'Environmental enrichment for the captive spectacled bear (*Tremarctos ornatus*)', *Pharmacology Biochemistry and Behavior*, 73: 279–83.

Rhoads, D. L. and Goldsworthy, R. J. (1979) 'The effects of zoo environments on public attitudes toward endangered wildlife', *International Journal of Environmental Studies*, 13: 283–7.

Richman, L. K. (2007) 'Elephant herpesviruses' in M. E. Fowler and R. E. Miller (eds), *Zoo and Wild Animal Medicine* (6th edn), St Louis, MO: Saunders (Elsevier), pp. 349–54.

Robbins, C. T. (1993) *Wildlife Feeding and Nutrition* (2nd edn), New York, NY: Academic Press.

Roberts, R. L., Roytburd, L. A., and Newman, J. D. (1999) 'Puzzle feeders and gum feeders as environmental enrichment for common marmosets', *Contemporary Topics in Laboratory Animal Science*, 38: 27–31.

Robertson, D. R. (1982) 'Fish faeces as fish food on a Pacific coral-reef', *Marine Ecology Progress Series*, 7: 253–65.

Robinson, M. H. (1989) 'The zoo that is not: education for conservation', *Conservation Biology*, 3: 213–15.

Robinson, M. H. (1996a) 'Foreword' in R. J. Hoage and W. A. Deiss (eds), *New Worlds, New Animals:*

From Menagerie to Zoological Park in the Nineteenth Century, Baltimore, MD/London: John Hopkins University Press, pp. vii–xi.

Robinson, M. H. (1996b) 'The BioPark concept and the exhibition of mammals' in D. G. Kleiman, M. E. Allen, K. V. Thompson, and S. Lumpkin (eds), *Wild Mammals in Captivity: Principles and Techniques*, Chicago, IL/London: University of Chicago Press, pp. 161–6.

Roca, A. L., Georgiadis, N., Pecon-Slattery, J., and O'Brien, S. J. (2001) 'Genetic evidence for two species of elephant in Africa', *Science*, 293: 1473–7.

Rodda, G., Bock, B., Burghardt, G., and Rand, A. (1988) 'Techniques for identifying individual lizards at a distance reveal influences of handling', *Copeia*, 1988(4): 905–13.

Roeder, J.-J. (1980) 'Marking behaviour and olfactory recognition in genets (*Genetta genetta* L.; in Carnivora-Viverridae)', *Behaviour*, 72: 200–10.

Roeder, J.-J. (1983) 'Études des interactions sociales entre mâle et femelle chez la genette (*Genetta genetta* L.): relations entre marquage olfactif et agression', *Zeitschrift für Tierpsychologie*, 61: 293–310.

Roeder, J.-J. (1984) 'Ontogenèse des systèmes de communication chez la genette (*Genetta genetta* L.)', *Behaviour*, 90: 259–301.

Rollin, B. E. (1992) *Animal Rights and Human Morality*, New York: Prometheus Books.

Ross, S. R. (2006) 'Issues of choice and control in the behaviour of a pair of captive polar bears (*Ursus maritimus*)', *Behavioural Processes*, 73: 117–20.

Ross, S. R. and Lukas, K. E. (2006) 'Use of space in a non-naturalistic environment by chimpanzees (*Pan troglodytes*) and lowland gorillas (*Gorilla gorilla gorilla*)', *Applied Animal Behaviour Science*, 96: 143–52.

Rothfels, N. (2002) *Savages and Beasts: The Birth of the Modern Zoo*, Baltimore, MD/London: John Hopkins University Press.

Rothschild, B. M., Rothschild, C., and Woods, R. J. (2001) 'Inflammatory arthritis in canids: spondyloarthropy', *Journal of Zoo Wildlife Medicine*, 32: 58–64.

Roush, R. S., Burkhardt, R., Converse, L., Dreyfus, T. A., Garrison, C., Porter, T. A., Snowdon, C. T., and Ziegler, T. E. (1992) 'Comment on "Are alarming events good for captive monkeys?"', *Applied Animal Behaviour Science*, 33: 291–3.

Rowland, D. L., Helgeson, V. S., and Cox, C. C. (1984) 'Temporal patterns of parturition in mammals in captivity', *Chronobiologica*, 11: 31–9.

Rushen, J. (1993) 'The "coping" hypothesis of stereotypic behaviour', *Animal Behaviour*, 45: 613–15.

Rushen, J. (2003) 'Changing concepts of farm animal welfare: bridging the gap between applied and basic research', *Applied Animal Behaviour Science, International Society for Applied Ethology Special Issue: A Selection of Papers from the ISAE International Congresses, 1999–2001*, 81: 199–214.

Rushen, J. and Depassillé, A. M. B. (1992) 'The scientific assessment of the impact of housing on animal welfare: a critical review', *Canadian Journal of Animal Science*, 72: 721–43.

Russel, A. (1984) 'Body condition scoring of sheep', *In Practice*, 6: 91–3.

Russow, L.-M. (2002) 'Ethical implications of the human–animal bond in the laboratory', *ILAR Journal*, 43: 33–7.

Rutherford, K. (2002) 'Assessing pain in animals', *Animal Welfare*, 11: 31–53.

Ryan, S., Thompson, S., Roth, A., and Gold, K. (2002) 'Effects of hand-rearing on the reproductive success of western lowland gorillas in North America', *Zoo Biology*, 21: 389–401.

Ryder, O. A. (1986) 'Species conservation and systematics: the dilemma of subspecies', *Trends in Ecology and Evolution*, 1: 9–10.

Ryder, O. A. and Feistner, A. T. C. (1995) 'Research in zoos: a growth area in conservation', *Biodiversity and Conservation*, 4: 671–7.

S Sainsbury, D. (1998) *Animal Health* (2nd edn), London: Blackwell Science.

Sales, G. D., Milligan, S. R., and Khirnykh, K. (1999) 'Sources of sound in the laboratory animal environment: a survey of the sounds produced by procedures and equipment', *Animal Welfare*, 8: 97–115.

Sambrook, T. D. and Buchanan-Smith, H. M. (1997) 'Control and complexity in novel object enrichment', *Animal Welfare*, 6: 207–16.

Sanderson, S. (2007) 'Appendix A: Contraception' in V. Melfi (ed.) *Proceedings of the Eighth European Endangered Species Programme (EEP) Studbook for Sulawesi Crested Black Macaque (Macaca nigra)*, Paignton Zoo Environmental Park, pp. 51–4.

Sanderson, S., Fidgett, A. L., and Fletcher, E. (2004) 'The effect of food presentation on the mortality rates and reproductive success of a colony of Rodrigues fruit bats (*Pteropus rodricensis*)' in *Proceedings of the European Association of Zoo and Wildlife Veterinarians Fifth Scientific Meeting*, 19–23 May, Ebeltoft: EAZWV, pp. 13–19.

Sannen, A., Van Elsacker, L., and Eens, M. (2004) 'Effect of spatial crowding on aggressive behaviour in a bonobo colony', *Zoo Biology*, 23: 383–95.

Saskia, J. and Schmid, H. (2002) 'Effect of feeding boxes on the behaviour of stereotyping Amur tigers (*Panthera tigris altaica*) in the Zurich Zoo, Zurich, Switzerland', *Zoo Biology*, 21: 573–84.

Savage, A., Rice, J. M., Brangan, J. M., Martini, D. P., Pugh, J. A., and Miller, C. D. (1994) 'Performance of African elephants (*Loxodonta africana*) and California sea lions (*Zalophus californianus*) on a two-choice object discrimination task', *Zoo Biology*, 13: 69–75.

Savastano, G., Hanson, A., and McCann, C. (2003) 'The development of an operant conditioning training program for New World primates at the Bronx Zoo', *Journal of Applied Animal Welfare Science*, 6: 247–61.

Schaaf, C. (1984) 'Animal behaviour and the captive management of wild mammals: a personal view', *Zoo Biology*, 310: 373–7.

Schafer, E. H. (1968) 'Hunting parks and animal enclosures in Ancient China', *Journal of the Economic and Social History of the Orient*, 11: 318–43.

Schaffer, N., Beehler, B., Jeyendran, R. S., and Balke, B. (1990) 'Methods of semen collection in an ambulatory greater one-horned rhinoceros (*Rhinoceros unicornis*)', *Zoo Biology*, 9: 211–21.

Schaffner, C. M. and Smith, T. E. (2005) 'Familiarity may buffer the adverse effects of relocation on marmosets (*Callithrix kuhlii*): preliminary evidence', *Zoo Biology*, 24: 93–100.

Schapiro, S. J., Bloomsmith, M. A., Suarez, S. A., and Porter, L. M. (1997) 'A comparison of the effects of simple versus complex environmental enrichment on the behaviour of group-housed, subadult rhesus macaques', *Animal Welfare*, 6: 17–28.

Schapiro, S. J., Perlman, J. E., and Boudreau, B. A. (2001) 'Manipulating the affiliative interactions of group-housed rhesus macaques using positive reinforcement training techniques', *American Journal of Primatology*, 55: 137–49.

Schapiro, S. J., Bloomsmith, M. A., and Laule, G. E. (2003) 'Positive reinforcement training as a technique to alter non-human primate behaviour: quantitative assessments of effectiveness', *Journal of Applied Animal Welfare Science*, 6: 175–87.

Scharmann, C. and van Hooff, J. (1986) 'Reproductive strategies of the orang-utan: new data and a reconsideration of existing sociosexual models', *International Journal of Primatology*, 7: 265–87.

Schiml, P. A., Mendoza, S. P., Saltzman, W., Lyons, D. M., and Mason, W. A. (1996) 'Seasonality in squirrel monkeys (*Saimiri sciureus*): social facilitation by females', *Physiology and Behavior*, 60: 1105–13.

Schmid, J., Heistermann, M., Gansloßer, U., and Hodges, J. K. (2001) 'Introduction of foreign female Asian elephants (*Elephas maximus*) into an existing group: behavioural reactions and changes in cortisol levels', *Animal Welfare*, 10: 357–72.

Schmidt-Nielsen, K. (1997) *Animal Physiology: Adaptation and Environment*, Cambridge: Cambridge University Press.

Schmitt, D. L. (2003) 'Proboscidea (Elephants)' in M. E. Fowler and M. E. Miller (eds), *Zoo and Wild Animal Medicine* (5th edn), Philadelphia, PA: Elsevier (Saunders), pp. 541–50.

Schulte-Hostedde, A. I., Zinner, B., Millar, J. S., and Hickling, G. J. (2005) 'Restitution of mass-size residuals: validating body condition indices', *Ecology*, 86: 155–63.

Schwabe, C. W. (1984) *Veterinary Medicine and Human Health* (3rd edn), Baltimore, MD: Williams & Wilkins.

Schwaibold, U. and Pillay, N. (2001) 'Stereotypic behaviour is genetically transmitted in the African striped mouse *Rhabdomys pumilio*', *Applied Animal Behaviour Science*, 74: 273–80.

Schwartzkopf-Genswein, K. S., Huisma, C., and McAllister, T. A. (1999) 'Validation of a radio frequency identification system for monitoring the feeding patterns of feedlot cattle', *Livestock Production Science*, 60: 27–31.

Schwartzkopf-Genswein, K. S., Stookey, J. M., and Welford, R. (1997) 'Behavior of cattle during hot-iron and freeze branding and the effects on subsequent handling ease', *Journal of Animal Science*, 75: 2064–72.

Schwitzer, C. and Kaumanns, W. (2001) 'Body weights of ruffed lemurs (*Varecia variegata*) in European zoos, with reference to the problem of obesity', *Zoo Biology*, 20: 261–9.

Scott, L., Pearce, P., Fairhall, S., Muggleton, N., and Smith, J. (2003) 'Training non-human primates to cooperate with scientific procedures in applied biomedical research', *Journal of Applied Animal Welfare Science*, 6: 199–207.

Scruton, D. M. and Herbert, J. (1972) 'The reaction of groups of captive talapoin monkeys to the introduction of male and female strangers of the same species', *Animal Behaviour*, 20: 463–73.

Seal, U. S. (1991) 'Life after extinction' in J. H. W. Gipps (ed.) 'Beyond captive breeding: re-introducing endangered animals to the wild', *Symposium of the Zoological Society of London*, 62: 39–55.

Seebeck, J. and Booth, R. (1996) 'Eastern barred bandicoot recovery: the role of the veterinarian in the management of endangered species', *Australian Veterinary Journal*, 73: 81–3.

Segovia, G., Yague, A. G., Garcia-Verdugo, J. M., and Mora, F. (2006) 'Environmental enrichment promotes neurogenesis and changes the extracellular concentrations of glutamate and GABA in the hippocampus of aged rats', *Brain Research Bulletin*, 70: 8–14.

Seidensticker, J. and Doherty, J. (1996) 'Integrating animal behaviour and exhibit design' in D. Kleiman, M. Allen, K. Thompson, and S. Lumpkin (eds), *Wild Mammals in Captivity*, Chicago, IL: Chicago University Press, pp. 180–90.

Sellinger, R. L. and Ha, J. C. (2005) 'The effects of visitor density and intensity on the behaviour of two captive jaguars (*Panthera onca*)', *Journal of Applied Animal Welfare Science*, 8: 233–44.

Seltzer, L. J. and Ziegler, T. E. (2007) 'Non-invasive measurement of small peptides in the common marmoset (*Callithrix jacchus*): a radiolabeled clearance study and endogenous excretion under varying conditions', *Hormones and Behaviour*, 51: 436–42.

Selye, H. (1973) 'The evolution of the stress concept', *American Scientist*, 61: 692–9.

Semple, S. (2002) 'Analysis of research projects conducted in Federation collections to 2000', *Federation Research Newsletter*, 3: 3.

Seres, M., Aureli, F., and de Waal, F. B. M. (2001) 'Successful formation of a large chimpanzee group out of two preexisting subgroups', *Zoo Biology*, 20: 501–15.

Shackleton-Bailey, D. R. (ed. and trans.) (2004) *Cicero, Epistulae ad Familiares Vol. 2, 47–43 BC*, Cambridge Classical Texts and Commentaries No. 17, Cambridge: Cambridge University Press.

Shackley, M. (1996) *Widlife Tourism*, London: Routledge.

Shannon, G. (2005) 'The effects of sexual dimorphism on the movements and foraging ecology of the African elephant', unpublished PhD thesis, School of Biological and Conservation Sciences, Durban, University of KwaZulu-Natal.

Shepherdson, D. (1991) 'A wild time at the zoo: practical enrichment for zoo animals', *Annual Conference of the American Association of Zoological Parks and Aquariums*, San Diego, CA: AAZPA, pp. 413–20.

Shepherdson, D. (1994) 'The role of environmental enrichment in the captive breeding and reintroduction of endangered species' in P. J. S. Olney, G. M. Mace, and A. T. C. Feistner (eds), *Creative Conservation: Interactive Management of Wild and Captive Animals*, London: Chapman & Hall, pp. 167–77.

Shepherdson, D. (1998) *Second Nature: Environmental Enrichment for Captive Animals*, Washington DC: Smithsonian Books.

Shepherdson, D. J., Carlstead, K. C., Mellen, J., and Seidensticker, J. (1993) 'The influence of food presentation on the behavior of small cats in confined environments', *Zoo Biology*, 12: 203–16.

Shepherdson, D., Carlstead, K. C., and Wielebnowski, N. (2004) 'Cross-institutional assessment of stress responses in zoo animals using longitudinal monitoring of faecal corticoids and behaviour', *Animal Welfare*, 13: 105–13.

Sheppard, C. (1995) 'Propagation of endangered birds in US institutions: how much space is there?', *Zoo Biology*, 14: 197–210.

Sheppard, C. and Dierenfeld, E. (2002) 'Iron storage disease in birds: speculation on etiology and implications for captive husbandry', *Journal of Avian Medicine and Surgery*, 16: 192–7.

Sherrill, J., Spelman, L. H., Reidel, C. L., and Montali, R. J. (2000) 'Common cuttlefish (*Sepia officinalis*) mortality at the National Zoological Park: implications for clinical management', *Journal of Zoo and Wildlife Medicine*, 31: 523–31.

Sherwin, C. (2001) 'Can invertebrates suffer? Or, how robust is argument-by-analogy?', *Animal Welfare*, 10: 103–18.

Shi, Y. and Yokoyama, S. (2003) 'Molecular analysis of the evolutionary significance of ultraviolet vision in vertebrates', *Proceedings of the National Academy of Sciences USA*, 100: 8308–13.

Shine, C., Shine, N., Shine, R., and Slip, D. (1988) 'Use of subcaudal scale anomalies as an aid in recognizing individual snakes', *Herpetological Review*, 19: 79.

Sibley, C. G. and Monroe Jr, B. L. (1990) *Distribution and Taxonomy of the Birds of the World*, New Haven, CT/London: Yale University Press.

Sibley, C. G. and Monroe Jr, B. L. (1993) *Supplement to Distribution and Taxonomy of the Birds of the World*, New Haven, CT/London: Yale University Press.

Signal, T. D. and Taylor, N. (2006) 'Attitudes to animals in the animal protection community compared to a normative community sample', *Society and Animals*, 14: 265–74.

Sigursdon, C. J. and Miller, M. W. (2003) 'Other animal prion diseases', *British Medical Bulletin*, 66: 199–212.

Silk, J. (1989) 'Social behaviour in evolutionary perspective' in B. Smuts, D. Cheney, R. Seyfarth, R. Wrangham, and T. Strusaker (eds), *Primate Societies*, Chicago, IL: Chicago University Press, pp. 318–29.

Silver, S. C., Ostro, L. E. T., Yeager, C. P., and Dierenfeld, E. S. (2000) 'Phytochemical and mineral components of foods consumed by black howler monkeys (*Alouatta pigra*) at two sites in Belize', *Zoo Biology*, 19: 95–109.

Simeone, A., Wilson, R. P., Knauf, G., Knauf, W., and Schützendübe, J. (2002) 'Effects of attached data-loggers on the activity budgets of captive Humboldt penguins', *Zoo Biology*, 21: 365–73.

Simmonds, M. P. (2006) 'Into the brains of whales', *Applied Animal Behaviour Science*, 100: 103–16.

Singer, P. (1990) *Animal Liberation*, Cambridge: Cambridge University Press.

Singer, P. and Regan, T. (1999) *Animal Rights and Human Obligations*, Upper Saddle River, NJ: Prentice Hall.

Skinner, B. (1938) *The Behavior of Organisms: An Experimental Analysis*, New York, NJ: Appleton-Century.

Sklan, D., Melamed, D., and Friedman, A. (1994) 'The effects of varying levels of dietary vitamin A on immune response in the chick', *Poultry Science*, 73: 843–7.

Slocombe, K. E. and Zuberbühler, K. (2005) 'Functionally referential communication in a chimpanzee', *Current Biology*, 15: 1779–84.

Smith, A., Lindburg, D. G., and Vehrencamp, S. (1989) 'Effect of food preparation on feeding behavior of lion-tailed macaques', *Zoo Biology*, 8: 57–65.

Smith, E. J., Partridge, J. C., Parsons, K. N., White, E. M., Cuthill, I. C., Bennett, A. T. D., and Church, S. C. (2002) 'Ultraviolet vision and mate choice in the guppy (*Poecilia reticulata*)', *Behavioural Ecology*, 13: 11–19.

Smith, S. (2006) 'Environmental enrichment plan for elasmobranchs at Shark Bay, Sea World, Gold Coast, Australia', *Proceedings of the First Australasian Regional Environmental Enrichment Conference*, The Royal Melbourne Zoological Gardens, Australia.

Smith, T. (2004) *Zoo Research Guidelines: Monitoring Stress in Zoo Animals*, London: BIAZA.

Sneddon, I. A., Beattie, V. E., Dunne, L., and Neil, W. (2000) 'The effect of environmental enrichment on learning in pigs', *Animal Welfare*, 9: 373–83.

Sneddon, L. (2003) 'The evidence for pain in fish: the use of morphine as an analgesic', *Applied Animal Behaviour Science*, 83: 153–62.

Sodhi, N. S., Bickford, D., Diesmos, A. C., Lee, T. M., Koh, L. P., Brook, B. W., Sekercloglu, C. H., and Bradshaw, C. J. A. (2008) 'Measuring the meltdown: drivers of global amphibian extinction and decline', *Public Library of Science (PLoS) ONE*, 3: e1636.

Soloman, N. G. and French, J. A. (2007) *Cooperative Breeding in Mammals*, Cambridge: Cambridge University Press.

Sommerfeld, R., Bauert, M., Hillmann, E., and Stauffacher, M. (2006) 'Feeding enrichment by self-operated food boxes for white-fronted lemurs (*Eulemur fulvus albifrons*) in the Masoala exhibit of the Zurich Zoo', *Zoo Biology*, 25: 145–54.

Soulé, M., Gilpin, M., Conway, W., and Foose, T. (1986) 'The millenium ark: how long a voyage, how many staterooms, how many passengers?', *Zoo Biology*, 5: 101–13.

Southwick, C. H. (1967) 'An experimental study of intragroup agonistic behaviour in rhesus monkeys (*Macaca mulatta*)', *Behaviour*, 28: 182–209.

Sowell, B. F., Bowman, J. G.P., Branine, M. E., and Hubbert, M. E. (1998) 'Radio frequency technology to measure feeding behavior and health of feedlot steers', *Applied Animal Behaviour Science*, 59: 277–84.

Spalton, J. A., Brend, S. A., and Lawrence, M. W. (1999) 'Arabian oryx reintroduction in Oman: successes and setbacks', *Oryx*, 33: 168–75.

Speakman, J. (2005) 'Review: body size, energy metabolism and lifespan', *Journal of Experimental Biology*, 218: 1717–30.

Spelman, L. H. (1999) 'Vermin control' in M. E. Fowler and M. E. Miller (eds), *Zoo and Wild Animal Medicine: Current Therapy* (4th edn), Philadelphia, PA: Elsevier (Saunders), pp. 114–20.

Spelman L. H., Osborn K. G., and Anderson, M. P. (1989) 'Pathogenesis of hemosiderosis in lemurs: role of dietary iron, tannin, and ascorbic acid', *Zoo Biology*, 8: 239–51.

Spencer, W. and Spencer, J. (2006) *Management Guideline Manual for Invertebrate Live Food Species*, Amsterdam, The Netherlands: EAZA Terrestrial Invertebrate TAG.

Spinelli, J. and Markowitz, H. (1985) 'Prevention of cage-associated distress', *Lab Animal*, 14: 19–28.

Spinka, M. (2006) 'How important is natural behaviour in animal farming systems?', *Applied Animal Behaviour Science*, 100: 117–28.

Spotte, S. (1992) *Captive Seawater Fishes: Science and Technology*, New York, NY: John Wiley & Sons.

Spraker, T. R. (2003) 'Spongiform encephalopathy' in M. E. Fowler and M. E. Miller (eds), *Zoo and Wild Animal Medicine: Current Therapy* (5th edn), Philadelphia, PA: Elsevier (Saunders) , pp. 741–5.

Spring, S. E., Clifford, J. O., and Tomko, D. L. (1997) 'Effect of environmental enrichment devices on behaviors of single- and group-housed squirrel monkeys (*Saimiri sciureus*)', *Contemporary Topics in Laboratory Animal Science*, 36: 72–5.

Springer, M. S. and de Jong, W. W. (2001) 'Phylogenetics: which mammalian supertree to bark up?', *Science*, 291: 1709–11.

Stacey, P. B. and Koenig, W. D. (1990) *Cooperative Breeding in Birds: Long-term Studies of Ecology and Behavior*, Cambridge: Cambridge University Press.

Stanley, M. E. and Aspey, W. P. (1984) 'An ethometric analysis in a zoological garden: modification of ungulate behavior by the visual presence of a predator', *Zoo Biology*, 3: 89–109.

Stanley Price, M. R. (1991) 'A review of mammal reintroductions, and the role of the reintroduction specialist group of IUCN/SSC' in J. H. W. Gipps (ed.), 'Beyond captive breeding: re-introducing endangered animals to the wild', *Symposium of the Zoological Society of London*, 62: 9–25.

Stanley Price, M. R. and Fa, J. E. (2007) 'Reintroductions from zoos: a conservation guiding light or a shooting star?' in A. Zimmerman, M. Hatchwell, L. Dickie, and C. West (eds), *Zoos in the 21st Century: Catalysts for Conservation?*, Conservation Biology Series No. 15, Cambridge: Cambridge University Press, pp. 155–77.

Steele, K. M., Linn, M. J., Schoepp, R. J., Komar, N., Geisbert, T. W., Manduca, R. M., Calle, P. P., Raphael, B. L., Clippinger, T. L., Larsen, T., Smith, J., Lanciotti, R. S., Panella, N. A., and McNamara, T. S. (2000) 'Pathology of fatal West Nile virus infections in native and exotic birds during the 1999 outbreak in New York City, New York', *Veterinary Pathology*, 37: 208–24.

Steiner, A. (2007) 'Foreword' in A. Zimmerman, M. Hatchwell, L. A. Dickie, and C. West (eds), *Zoos in the 21ˢᵗ Century*, Cambridge: Cambridge University Press, pp. xi–xii.

Sterling, E., Lee. J., and Wood, T. (2007) 'Conservation education in zoos: an emphasis on behavioral change' in A. Zimmerman, M. Hatchwell, L. Dickie, and C. West (eds), *Zoos in the 21st Century: Catalysts for Conservation?*, Conservation Biology Series No. 15, Cambridge: Cambridge University Press, pp. 37–50.

Sternicki, T., Szablewski, P., and Szwaczkowski, T. (2003) 'Inbreeding effects on lifetime in David's deer (*Elaphurus davidianus*, Milne Edwards 1866) population', *Journal of Applied Genetics*, 44: 175–83.

Stevens, C. E. and Hume, I. D. (1995) *Comparative Physiology of the Vertebrate Digestive System* (2nd edn), Cambridge: Cambridge University Press.

Stevens, E. F. (1991) 'Flamingo breeding: the role of group displays', *Zoo Biology*, 10: 53–63.

Stevens, E. F. and Pickett, C. (1994) 'Managing the social environments of flamingos for reproductive success', *Zoo Biology*, 13: 501–7.

Stevenson, M. (1983) 'The captive environment: its effect on exploratory and related behavioural responses in wild animals' in J. Archer and L. Birke (eds) *Exploration in Animals and Man*, New York: Van Nostrand Rheinhold, pp. 176–97.

Stevenson, M. (2005) *Management Guidelines for the Welfare of Zoo Animals: Elephant*, Report for the Federation of Zoological Gardens of Great Britain, London: FZG.

Stevenson, M. (2008) *BIAZA Questionnaire: What It Is For and What It Achieves*, London: BIAZA.

Stevenson, M. and Walter, O. (2002) *Management Guidelines for the Welfare of Zoo Animals: Elephants* (Loxodonta africana *and* Elephas maxismus), London: BIAZA.

St Louis, V., Barlow, J., and Sweerts, J. (1989) 'Toenail-clipping: a simple technique for marking individual nidicolous chicks', *Journal of Field Ornithology*, 60: 211–15.

Stoinski, T. S., Beck, B., Bowman, M., and Lehnhardt, J. (1997) 'The Gateway zoo program: a recent initiative in golden lion tamarin reintroduction' in J. Wallis (ed.), 'Primate conservation: the role of zoological parks', *American Society of Primatologists, Special Topics in Primatology*, 1: 29–41.

Stoinski, T. S., Lukas, K. E., and Maple, T. L. (1998) 'A survey of research in North American zoos and aquariums', *Zoo Biology*, 17: 167–80.

Stoinski, T. S., Daniel, E., and Maple, T. L. (2000) 'A preliminary study of the behavioral effects of feeding enrichment on African elephants', *Zoo Biology*, 19: 485–93.

Stoinski, T. S., Hoff, M. P., Lukas, K. E., and Maple, T. L. (2001) 'A preliminary behavioral comparison of two captive all-male gorilla groups', *Zoo Biology*, 20: 27–40.

Stoinski, T. S., Allen, M. T., Bloomsmith, M. A., Forthman, D. L., and Maple, T. L. (2002) 'Educating zoo visitors about complex environmental issues: should we do it and how?', *Curator*, 45: 129–43.

Stolba, A. and Wood-Gush, D. (1984) 'The identification of behavioural key features and their incorporation into a housing design for pigs', *Annales de Recherches Veterinaires*, 15: 287–98.

Stone, R. (2003) 'Foreword' in W. Holt, A. Pickard, J. Rodger, and D. Wildt (eds), *Reproductive Science and Integrated Conservation*, Cambridge: Cambridge University Press, pp. xiii–xv.

Storms, T. N., Clyde, V. L., Munson, L., and Ramsay, E. C. (2003) 'Blastomycosis in non-domestic felids', *Journal of Zoo and Wildlife Medicine*, 34: 231–8.

Straughan, R. (2003) *Ethics, Morality and Crop Biotechnology*, Report sponsored by the Biotechnology and Biological Sciences Research Council (BBSRC), available online at `http://www.bbsrc.ac.uk/organisation/policies/position/public_interest/animal_biotechnology.pdf`.

Strouhal, E. (1992) *Life in Ancient Egypt*, Cambridge: Cambridge University Press.

Stuart, S. N. (1991) 'Re-introductions: to what extent are they needed?' in J. H. W. Gipps, (ed.), 'Beyond captive breeding: re-introducing endangered animals to the wild', *Symposium of the Zoological Society of London*, 62: 27–37.

Stuart, S. N., Chanson, J. S., Cox, N. A., Young, B. E., Rodrigues, A. S. L., Fischman, D. L., and Waller, R. W. (2004) 'Status and trends of amphibian declines and extinctions worldwide', *Science*, 306: 1783–6.

Sunquist, F. (1995) 'End of the ark?', *International Wildlife*, 25: 23–9.

Sutherland, W., Newton, I., and Green, R. (2004) *Bird Ecology and Conservation: A Handbook of Techniques.* Oxford: Oxford University Press.

Swain, D. L., Wilson, L. A., and Dickinson, J. (2003) 'Evaluation of an active transponder system to monitor spatial and temporal location of cattle within patches of a grazed sward', *Applied Animal Behaviour Science*, 84: 185–95.

Swaisgood, R. R. (2007) 'Current status and future directions of applied behavioral research for animal welfare and conservation', *Applied Animal Behaviour Science*, 102: 139–62.

Swaisgood, R. R. and Shepherdson, D. J. (2005) 'Scientific approaches to enrichment and stereotypies in zoo animals: what's been done and where should we go next?', *Zoo Biology*, 24: 499–518.

Swaisgood, R. R. and Shepherdson, D. J. (2006) 'Environmental enrichment as a strategy for mitigating stereotypies in zoo animals: a literature review and meta-analysis' in G. J. Mason and J. Rushen (eds), *Stereotypic Animal Behaviour: Fundamentals and Applications to Welfare*, Wallingford: CABI Publishing, pp. 256–85.

Swaisgood, R. R., White, A. M., Zhou, X. P., Zhang, H. M., Zhang, G. Q., Wei, R. P., Hare, V. J., Tepper, E. M., and Lindburg, D. G. (2001) 'A quantitative assessment of the efficacy of an environmental enrichment programme for giant pandas', *Animal Behaviour*, 61: 447–57.

Swan, G., Naidoo, V., Cuthbert, R., Green, R. E., Pain, D. J., Swarup, D., Prakash, V., Taggart, M., Bekker, L., Das, D., Diekmann, D., Diekmann, M.,
Killian, E., Mehar, G., Chandra Patra, R., Saini, M., and Wolter, K. (2006) 'Removing the threat of diclofenac to critically endangered Asian vultures', *Public Library of Science (PLoS) Biology*, 4: e66.

Swanson, W. F., Johnson, W. E., Cambre, R. C., Citino, S. B., Quigley, K. B., Brousset, D. M., Morais, R. N., Moreira, N., O'Brien, S. J., and Wildt, D. E. (2003) 'Reproductive status of endemic felid species in Latin American zoos and implications for *ex situ* conservation', *Zoo Biology*, 22: 421–41.

Swenson, J., Wallin, K., Ericsson, G., Cederlund, G., and Sandegren, F. (1999) 'Effects of ear-tagging with radiotransmitters on survival of moose calves', *Journal of Wildlife Management*, 63: 354–8.

Sykes, J. B. (ed.) (1977) *Concise Oxford Dictionary of Current English* (6th edn), Oxford: Oxford University Press.

t **Taber, A. and Macdonald, D. W.** (1992) 'Spatial organisation and monogamy in the mara, *Dolichotis patagonum*', *Journal of Zoology*, 227: 417–38.

Taberlet, T. and Bouvet, J. (1991) 'A single plucked feather as a source of DNA for bird genetic studies', *Auk*, 108: 959–60.

Tarou, L. R., Bashaw, M. J., and Maple, T. L. (2000) 'Social attachment in giraffe: response to social separation', *Zoo Biology*, 19: 41–51.

Tarou, L. R., Bashaw, M. J., and Maple, T. L. (2003) 'Failure of a chemical spray to significantly reduce stereotypic licking in a captive giraffe', *Zoo Biology*, 22: 601–7.

Tarou, L. R., Kuhar, C. W., Adcock, D., Bloomsmith, M. A., and Maple, T. L. (2004) 'Computer-assisted enrichment for zoo-housed orang-utans (*Pongo pygmaeus*)', *Animal Welfare*, 13: 445–53.

Taylor, A. C. (2003) 'Assessing the consequences of inbreeding for population fitness: past challenges and future prospects' in W. Holt, A. Pickard, J. Rodger, and D. Wildt (eds) *Reproductive Science and Integrated Conservation*, Cambridge: Cambridge University Press, pp. 67–81.

Taylor J. J. (1984) 'Iron accumulation in avian species in captivity', *Dodo: Journal of the Jersey Wildlife Preservation Trust*, 21: 126–31.

Terranova, C. J. and Coffman, B. S. (1997) 'Body weights of wild and captive lemurs', *Zoo Biology*, 16: 17–30.

Testa, J. W. and Rothery, P. (1992) 'Effectiveness of various cattle ear tags as markers for Weddell seals', *Marine Mammal Science*, 8: 344–53.

Thiermann, A. and Babcock, S. (2005) 'Animal welfare and international trade', *Science and Technology Review*, 24: 747–55.

Thodberg, K., Jensen, K. H., Herskin, M. S., and Jorgensen, E. (1999) 'Influence of environmental stimuli on nest building and farrowing behaviour in domestic sows', *Applied Animal Behaviour Science*, 63: 131–44.

Thomas, C. D., Cameron, A., Green, R. E., Bakkenes, M., Beaumont, L. J., Collingham, Y. C., Erasmus, B. F. N., Ferreira de Siqueira, M., Grainger, A., Hannah, L., Hughes, L., Huntley, B., van Jaarsveld, A. S., Midgley, G. F., Miles, L., Ortega-Huerta, M. A., Peterson, A. T., Phillips, O. L., and Williams, S. E. (2004) 'Extinction risk from climate change', *Nature*, 427: 145–8.

Thomas, L. W., Kline, C., Duffelmeyer, J., Maclaughlin, K., and Doherty, J. G. (1986) 'The hand-rearing and social reintegration of a Californian sealion', *International Zoo Yearbook*, 24: 279–85.

Thomas, P. R. and Powell, D. M. (2006) 'Birth and simultaneous rearing of two litters in a pack of captive African wild dogs (*Lycaon pictus*)', *Zoo Biology*, 25: 461–77.

Thomas, R. (2005) 'Internal drive vs external directive: the delivery of conservation through zoo-based research—a response to Rees', *Oryx*, 39: 134.

Thomas, R., Bartlett, L., Marples, N., Kelly, D., and Cuthill, I. (2004) 'Prey selection by wild birds can allow novel and conspicuous colobus morphs to spread in prey populations', *Oikos*, 106: 285–94.

Thomas, W. and Maruska, E. (1996) 'Mixed-species exhibits with mammals' in D. Kleiman, M. Allen, K. Thompson, and S. Lumpkin (eds), *Wild Mammals in Captivity: Principles and Techniques*, Chicago, IL: University of Chicago, pp. 204–11.

Thompson, C. B., Holter, J. B., Hayes, H. H., Silver, H., and Urban Jr, W. E. (1973) 'Nutrition of white-tailed deer I: energy requirements of fawns', *The Journal of Wildlife Management*, 37: 301–11.

Thompson, K., Roberts, M., and Rall, W. (1995) 'Factors affecting pair compatibility in captive kangaroo rats, *Dipodomys heermanni*', *Zoo Biology*, 14: 317–30.

Tinbergen, N. (1963) 'On aims and methods of ethology', *Zeitschrift für Tierpsychologie*, 20: 410–33.

Todman, J. B. and Dugard, P. (2001) *Single-Case and Small-n Experimental Designs*, Mahwah, NJ: Lawrence Erlbaum Associates.

Tofield, S., Coll, R. K., Vyle, B., and Bolstad, R. (2003) 'Zoos as a source of free choice learning', *Research in Science and Technological Education*, 21: 67–99.

Toone, W. D. and Wallace, M. P. (1994) 'The extinction in the wild and reintroduction of the California condor (*Gymnogyps californianus*)' in P. J. S. Olney, G. M. Mace, and A. T. C. Feistner (eds) *Creative Conservation: Interactive Management of Wild and Captive Animals*, London: Chapman & Hall, pp. 411–19.

Travis, D. (2007) 'West Nile virus in birds and mammals' in M. E. Fowler and R. E. Miller (eds), *Zoo and Wild Animal Medicine* (6th edn), St Louis, MO: Saunders (Elsevier), pp. 2–9.

Traylor-Holzer, K. and Fritz, P. (1985) 'Utilization of space by adult and juvenile groups of captive chimpanzees (*Pan troglodytes*)', *Zoo Biology*, 4: 115–27.

Tribe, A. and Booth, R. (2003) 'Assessing the role of zoos in wildlife conservation', *Human Dimensions of Wildlife*, 8: 65–74.

Trivers, R. L. (1971) 'The evolution of reciprocal altruism', *Quarterly Review of Biology*, 46: 35–57.

Trivers, R. L. (1972) 'Parental investment and sexual selection' in B. Campbell (ed.), *Sexual Selection and the Descent of Man*, Chicago, IL: Aldine, pp. 35–57.

Troyer, K. (1984a) 'Diet selection and digestion in *Iguana iguana*: the importance of age and nutrient requirements', *Oecologia*, 61: 201–7.

Troyer, K. (1984b) 'Behavioral acquisition of the hindgut fermentation system by hatchling *Iguana iguana*', *Behavioral Ecology and Sociobiology*, 14: 189–93.

Tudge, C. (1992) *Last Animals at the Zoo: How Mass Extinction Can Be Stopped*, Washington DC/London: Island Press/Hutchinson Radius.

Tunnicliffe, S. D., Lucas, A. M., and Osborne, J. (1997) 'School visits to zoos and museums: a missed educational opportunity?', *International Journal of Science Education*, 19: 1039–56.

Turley, S. K. (1999) 'Exploring the future of the traditional UK zoo', *Journal of Vacation Marketing*, 5: 340–55.

Tyler, S. (1979) 'Time sampling: a matter of convention', *Animal Behaviour*, 27: 801–10.

U **Ullrey, D. E.** (2003) 'Metabolic bone diseases' in M. E. Fowler and R. E. Miller (eds), *Zoo and Wild Animal Medicine* (5th edn), Philadelphia, PA: Elsevier (Saunders), pp. 749–56.

UN (United Nations) (1973) *Convention on International Trade in Endangered Species of Wild Fauna and Flora (CITES)*, Washington DC: UNEP.

UN (United Nations) (1992) 'Convention on Biodiversity (CBD)', *International Environmental Laws: Multilateral Treaties*, 992: 1–43.

V **Valutis, L. L. and Marzuluff, J. M.** (1999) 'The appropriateness of puppet-rearing birds for reintroduction', *Conservation Biology*, 13: 584–91.

Van der Berg, L., van der Borg, J., and van der Meer, J. (1995) *Urban Tourism: Performance and Strategies in Eight European Cities*, Aldershot: Ashgate Publishing.

Van Gelder, R. (1991) 'A big pain', *Natural History*, 100: 22–7.

Van Hoek, C. S. and King, C. E. (1997) 'Causation and influence of environmental enrichment on feather picking of the crimson-bellied conure (*Pyrrhura perlata perlata*)', *Zoo Biology*, 16: 161–72.

Van Keulen-Kromhout, G. (1978) 'Zoo enclosures for bears: their influence on captive behaviour and reproduction', *International Zoo Yearbook*, 18: 177–86.

Van Linge, J. (1992) 'How to out-zoo the zoo', *Tourism Management*, 13: 114–17.

Van Oers, K., Drent, P. J., de Goede, P., and van Noordwijk, A. J. (2004) 'Realized heritability and repeatability of risk-taking behaviour in relation to avian personalities', *Proceedings of the Royal Society Series B: Biological Sciences*, 271: 65–73.

Van Praag, H., Kempermann, G., and Gage, F. H. (2000) 'Neural consequences of environmental enrichment', *Nature Reviews Neuroscience*, 1: 191–8.

Vargas-Ashby, H. and Pankhurst, S. (2007) 'Effects of feeding enrichment on the behaviour and welfare of captive Waldrapps (Northern bald ibis *Geronticus eremita*)', *Animal Welfare*, 16: 369–74.

Vasey, N. and Tattersall, I. (2002) 'Do ruffed lemurs form a hybrid zone? Distribution and discovery of *Varecia*, with systematic and conservation implications', *American Museum Novitates*, 3376: 1–26.

Veasey, J. (2006) 'Concepts in the care and welfare of captive elephants', *International Zoo Yearbook*, 40: 63–79.

Veasey, J. S., Waran, N. K., and Young, R. J. (1996a) 'On comparing the behaviour of zoo-housed animals with wild conspecifics as a welfare indicator', *Animal Welfare*, 5: 13–24.

Veasey, J. S., Waran, N. K., and Young, R. J. (1996b) 'On comparing the behaviour of zoo housed animals with wild conspecifics as a welfare indicator, using the giraffe (*Giraffa camelopardalis*) as a model', *Animal Welfare*, 5: 139–53.

Vehrs, K. L. (1996) 'Summary of United States wildlife regulations applicable to zoos' in D. G. Kleiman, M. E. Allen, K. V. Thompson, and S. Lumpkin (eds), *Wild Mammals in Captivity: Principles and Techniques*, Chicago/London: University of Chicago Press, pp. 593–9.

Veltman, K. and van der Zanden, R. (2000) Biological control: fighting pests with pests. *De Harpij*, 19: 5–7 (in Dutch with English summary).

Verderber, S., Gardner, L., Islam, D., and Nakanishi, L. (1988) 'Elderly persons' appraisal of the zoological environment', *Environment and Behavior*, 20: 492–507.

Vevers, G. (1976) *London's Zoo*, London: Bodley Head.

Vickery, S. and Mason, G. (2004) 'Stereotypic behavior in Asiatic black and Malayan sun bears', *Zoo Biology*, 23: 409–30.

Vickery, S. S. and Mason, G. J. (2005) 'Stereotypy and perseverative responding in caged bears: further data and analyses', *Applied Animal Behaviour Science*, 91: 247–60.

Vié, J. C. (1996) 'Reproductive biology of captive Arabian oryx *Oryx leucoryx* in Saudi Arabia', *Zoo Biology*, 15: 371–81.

Vignes, S., Newman, J. D., and Roberts, R. L. (2001) 'Mealworm feeders as environmental enrichment for common marmosets', *Contemporary Topics in Laboratory Animal Science*, 40(3): 26–9.

Vining, J. (2003) 'The connection to other animals and caring for nature', *Research in Human Ecology*, 10(2): 87–99.

Visalberghi, E. and Anderson, J. (1993) 'Reasons and risks associated with manipulating captive primates' social environments', *Animal Welfare*, 212: 3–15.

Visalberghi, E. and Vitale, A. F. (1990) 'Coated nuts as an enrichment device to elicit tool use in tufted capuchins (*Cebus apella*)', *Zoo Biology*, 9: 65–71.

Vogelnest, L. and Ralph, H. K. (1997) 'Chemical immobilisation of giraffe to facilitate short procedures', *Australian Veterinary Journal*, 75: 180–2.

Voipio, H. M., Nevalainen, T., Halonen, P., Hakumaki, M., and Bjork, E. (2006) 'Role of cage material, working style and hearing sensitivity in perception of animal care noise', *Laboratory Animals*, 40: 400–9.

Vonk, J. and MacDonald, S. E. (2002) 'Natural concepts in a juvenile gorilla (*Gorilla gorilla gorilla*) at three levels of abstraction', *Journal of the Experimental Analysis of Behaviour*, 78: 315–32.

W Waiblinger, S., Boivin, X., Pedersen, V., Tosi, M.-V., Janczak, A. M., Visser, E. K., and Jones, R. B. (2006) 'Assessing the human–animal relationship in farmed species: a critical review', *Applied Animal Behaviour Science*, 101: 185–242.

Waits, L. P., Talbot, S. L., Ward, R. H., and Shields, G. F. (1998) 'Mitochondrial DNA phylogeography of the North American brown bear and implications for conservation', *Conservation Biology*, 12: 408–17.

Waitt, C. and Buchanan-Smith, H. M. (2001) 'What time is feeding? How delays and anticipation of feeding schedules affect stump-tailed macaque behaviour', *Applied Animal Behaviour Science*, 75: 75–85.

Waitt, C., Buchanan-Smith, H. M., and Morris, K. (2002) 'The effects of caretaker–primate relationships on primates in the laboratory', *Journal of Applied Animal Welfare Science*, 5: 309–19.

Walker, S. (2001) 'Africa: national legislation and licensing' in C. E. Bell (ed.), *Encyclopedia of the World's Zoos*, Chicago, IL/London: Fitzroy Dearborn, pp. 14–15.

Ward, P. I., Mosberger, N., Kistler, C., and Fischer, O. (1998) 'The relationship between popularity and body size in zoo animals', *Conservation Biology*, 12: 1408–11.

Ward, S. and Melfi, V. (2004) 'The influence of stockmanship on the behaviour of black rhinoceros (*Diceros bicornis*)' in C. Macdonald (ed.), *Proceedings of the Sixth Annual Symposium on Zoo Research*, 8–9 July, Edinburgh: BIAZA, pp. 160–9.

Wardzynski, C., Arne, P., and Millemann, Y. (2005) 'Methods of restraint for zoo mammals', *Point Veterinaire*, 36: 46.

Warwick, C. (1990) 'Reptilian ethology in captivity: observations of some problems and an evaluation of their aetiology', *Applied Animal Behaviour Science*, 26: 1–13.

Washio, K., Misawa, S., and Ueda, S. (1989) 'Individual identification of non-human primates using DNA fingerprinting', *Primates*, 30: 217–21.

Wasser, S. K., Hunt, K. E., Brown, J. L., Cooper, K., Crockett, C. M., Bechert, U., Millspaugh, J. J., Larson, S., and Monfort, S. L. (2000) 'A generalized fecal glucocorticoid assay for use in a diverse array of non-domestic mammalian and avian species', *General and Comparative Endocrinology*, 120: 260–75.

Wasserman, F. E. and Cruikshank, W. W. (1983) 'The relationship between time of feeding and aggression in a group of captive hamadryas baboons', *Primates*, 24: 432–5.

Waterhouse, M. and Waterhouse, H. (1971) 'Population density and stress in zoo monkeys', *The Ecologist*, 1: 19–21.

Wathes, C. and Charles, D. (eds) (1994) *Livestock Housing*, Wallingford, Oxon: CABI Publishing.

Watson, J. B. (1928) *The Psychological Care of Infant and Child*, London: Allen.

Watson, P. F. and Holt, W. V. (2001) 'Organizational issues concerning the establishment of a genetic resource bank', in P. F. Watson and W. Holt (eds), *Cryobanking the Genetic Resource: Wildlife Conservation for the Future?*, London: Taylor & Francis, pp. 113–22.

Watters, J. V. and Meehan, C. L. (2007) 'Different strokes: can managing behavioral types increase post-release success?', *Applied Animal Behaviour Science*, 102: 364–79.

Watts, J. M. and Stookey, J. M. (1999) 'Effects of restraint and branding on rates and acoustic parameters of vocalization in beef cattle', *Applied Animal Behaviour Science*, 62: 125–35.

Watts, P. C., Buley, K. R., Sanderson, S., Boardman, W., Ciofi, C., and Gibson, R. (2006) 'Parthenogenesis in Komodo dragons', *Nature*, 444: 1021–2.

Wayne, R. K., Bruford, M. W., Girman, D., Rebholz, W. E. R., Sunnucks, P., and Taylor, A. C. (1994) 'Molecular genetics of endangered species' in P. J. S. Olney, G. M. Mace, and A. T. C. Feistner (eds), *Creative Conservation: Interactive Management of Wild and Captive Animals*, London: Chapman & Hall, pp. 92–117.

WAZA (World Association of Zoos and Aquariums) (1999) *Code of Ethics*, Liebefeld-Berne: WAZA.

WAZA (World Association of Zoos and Aquariums) (2005) *Building a Future for Wildlife: The World Zoo and Aquarium Conservation Strategy*, Berne: WAZA.

WAZA (World Association of Zoos and Aquariums) (2006) *Understanding Animals and Protecting Them: About the World Zoo and Aquarium Strategy*, Liebefeld-Berne: WAZA.

WAZA (World Association of Zoos and Aquariums) (2008) 'About WAZA', available online at http://www.waza.org/home/index.php?main=home.

Weary, D. M., Niel, L., Flower, F. C., and Fraser, D. (2006) 'Identifying and preventing pain in animals', *Applied Animal Behaviour Science*, 100: 64–76.

Webster, J. (1994) *Animal Welfare: A Cool Eye Towards Eden*, Oxford: Blackwell Science Ltd.

Webster, J. (2006) 'Animal sentience and animal welfare: what is it to them and what is it to us?', *Applied Animal Behaviour Science*, 100: 1–3.

Wechsler, B. (1991) 'Stereotypies in polar bears', *Zoo Biology*, 10: 177–88.

Wehnelt, S. and Wilkinson, R. (2005) 'Research, conservation and zoos: the EC Zoos Directive—a response to Rees', *Oryx*, 39: 132–3.

Wehnelt, S., Hosie, C., Plowman, A., and Feistner, A. (2003) *Zoo Research Guidelines: Project Planning and Behavioural Observations*, London: BIAZA.

Wehnelt, S., Bird, S., and Lenihan, A. (2006) 'Chimpanzee forest exhibit at Chester Zoo', *International Zoo Yearbook*, 40: 313–22.

Wei, F., Feng, Z., Wang, Z., Zhou, A. and Hu, J. (1999) 'Use of the nutrients in bamboo by the red panda (*Ailurus fulgens*)', *Journal of Zoology*, 248: 535–41.

Weipkema, P. and Koolhaas, J. (1993) 'Stress and animal welfare', *Animal Welfare*, 26: 195–218.

Weiss, E. and Wilson, S. (2003) 'The use of classical and operant conditioning in training Aldabra tortoises (*Geochelone gigantea*) for venipuncture and other husbandry issues', *Journal of Applied Animal Welfare Science*, 6: 33–8.

Weller, S. H. and Bennett, C. L. (2001) 'Twenty-four hour activity budgets and patterns of behavior in captive ocelots (*Leopardus pardalis*)', *Applied Animal Behaviour Science*, 71: 67–79.

Wells, D. L. (2005) 'A note on the influence of visitors on the behaviour and welfare of zoo-housed gorillas', *Applied Animal Behaviour Science*, 93: 13–17.

Wells, D. L. and Egli, J. M. (2004) 'The influence of olfactory enrichment on the behaviour of captive black-footed cats, *Felis nigripes*', *Applied Animal Behaviour Science*, 85: 107–19.

Wemelsfelder, F. (1999) 'The problem of animal subjectivity and its consequences for the scientific measurement of animal suffering' in F. Dolins (ed.), *Attitudes to Animals: Views in Animal Welfare*, Cambridge: Cambridge University Press, pp. 37–53.

Wemelsfelder, F. and Birke, L. (1997) 'Environmental challenge' in M. C. Appleby and B. O. Hughes (eds) *Animal Welfare*, Wallingford, Oxon: CABI Publishing, pp. 35–47.

Wendeln, M. C., Runkle, J. R., and Kalko, E. K. V. (2000) 'Nutritional values of 14 fig species and bat feeding preferences in Panama', *Biotropica*, 32: 489–501.

Westneat, D. F. and Stewart, I. R. K. (2003) 'Extra-pair paternity in birds: causes, correlates, and conflict', *Annual Review of Ecology, Evolution, and Systematics*, 34: 365–96.

Wharton, D. (2007) 'Research by zoos' in A. Zimmerman, M. Hatchwell, L. A. Dickie,

and C. West (eds), *Zoos in the 21st Century*, Cambridge: Cambridge University Press, pp. 178–91.

Wheater, R. (1995) 'World Zoo Conservation Strategy: a blueprint for zoo development', *Biodiversity and Conservation*, 4: 544–52.

Wheler, C. L. and Fa, J. E. (1995) 'Enclosure utilization and activity of round island geckos (*Phelsuma guentheri*)', *Zoo Biology*, 14: 361–9.

Whitaker, B. R. (1999) 'Preventive medicine programs for fish' in M. E. Fowler and R. E. Miller (eds), *Zoo and Wild Animal Medicine: Current Therapy* (4th edn), Philadelphia, PA: W. B. Saunders, pp. 163–81.

White, B. C., Houser, L. A., Fuller, J. A., Taylor, S., and Elliott, J. L. L. (2003) 'Activity-based exhibition of five mammalian species: evaluation of behavioral changes', *Zoo Biology*, 22: 269–85.

White, P. A. (2005) 'Maternal rank is not correlated with cub survival in the spotted hyena, *Crocuta crocuta*', *Behavioural Ecology*, 16: 606–13.

Whitehead, G. K. (1972) *Deer of the World*, London: Constable.

Whitehead, M. (1995) 'Saying it with genes, species and habitats: biodiversity education and the role of zoos', *Biodiversity and Conservation*, 4: 664–70.

Whiten, A., Goodall, J., McGrew, W. C., Nishida, T., Reynolds, V., Sugiyama, Y. , Tutin, C. E. G., Wrangham, R. W., and Boesch, C. B. (1999) 'Chimpanzee cultures', *Nature*, 399: 682–5.

Whitford, H. L. and Young, R. J. (2004) 'Trends in the captive breeding of threatened and endangered birds in British zoos, 1988–1997', *Zoo Biology*, 23: 85–9.

Whiting, M. J., Stuart-Fox, D. M., O'Connor, D., Firth, D., Bennett, N. C., and Blomberg, S. P. (2006) 'Ultraviolet signals ultra-aggression in a lizard', *Animal Behaviour*, 72: 353–63.

Whitney, R. and Wickings, E. (1987) 'Macaques and other old world simians' in T. Poole (ed.), *The UFAW Handbook on the Care and Management of Laboratory Animals*, New York, NY: Churchill Livingston, pp. 599–627.

Wielebnowski, N. (1996) 'Reassessing the relationship between juvenile mortality and genetic monomorphism in captive cheetahs', *Zoo Biology*, 15: 353–69.

Wielebnowski, N. (1998) 'Contributions of behavioral studies to captive management and breeding of rare and endangered mammals' in T. Caro (ed.), *Behavioral Ecology and Conservation Biology*, New York, NY/London: Oxford University Press, pp. 130–62.

Wielebnowski, N. (1999) 'Behavioral differences as predictors of breeding status in captive cheetahs', *Zoo Biology*, 18: 335–49.

Wielebnowski, N., Fletchall, N., Carlstead, K., Busso, J., and Brown, J. (2002) 'Non-invasive assessment of adrenal activity associated with husbandry and behavioral factors in the North American clouded leopard population', *Zoo Biology*, 21: 77–98.

Wiese, R., Willis, K., Lacy, R., and Ballou, J. (2003) *AZA Studbook Analysis and Population Management Handbook*, Bethesda, MD: AZA.

Wiesner, C. S. and Iben, C. (2003) 'Influence of environmental humidity and dietary protein on pyramidal growth of carapaces in African spurred tortoises (*Geochelone sulcata*)', *Journal of Animal Physiology and Animal Nutrition*, 87: 66–74.

Wiggs, R. B. and Lobprise, H. B. (1997) 'Exotic animal oral disease and dentistry' in R. B. Wiggs and H. B. Lobprise (eds), *Veterinary Dentistry: Principles and Practice*, Oxford: Blackwells, pp. 538–58.

Wilcken, J. and Lees, C. (1998) *Managing Zoo Populations: Compiling and Analysing Studbook Data*, Sydney: ARAZPA.

Wildt, D., Ellis, S., Janssens, D., and Buff, J. (2003) 'Toward more efficient reproductive science for conservation' in W. Holt, A. Pickard, J. Rodger, and D. Wildt (eds), *Reproductive Science and Integrated Conservation*, Cambridge: Cambridge University Press, pp. 2–20.

Wilkinson, R. (2000) 'An overview of captive-management programmes and regional collection planning for parrots', *International Zoo Yearbook*, 37: 36–58.

Williams, E. S. (2003) 'Plague' in M. E. Fowler (ed.), *Zoo and Wild Animal Medicine* (3rd edn), Philadelphia, PA: Elsevier (Saunders), pp. 705–9.

Williams, E. S., Yuill, T., Artois, M., Fischer, J., and Haigh, J. A. (2002) 'Emerging infectious diseases in wildlife', *OEI Scientific and Technical Review*, 21: 139–57.

Williams, L. E. and Abee, C. R. (1988) 'Aggression with mixed age-sex groups of Bolivian squirrel monkeys following single animal introductions and new group formations', *Zoo Biology*, 7: 139–45.

Willmer, P., Stone, G., and Johnston, I. (2000) *Environmental Physiology of Animals*, Oxford: Blackwell Science.

Wilson, A. C. and Stanley Price, M. R. (1994) 'Reintroduction as a reason for captive breeding' in P. J. S. Olney, G. M. Mace, and A. T. C. Feistner (eds), *Creative Conservation: Interactive Management of Wild and Captive Animals*, London: Chapman & Hall, pp. 243–64.

Wilson, D. E. and Reader, D. M. (eds) (2005) *Mammal Species of the World*, Baltimore, MD/London: John Hopkins University Press.

Wilson, E. O. (1975) *Sociobiology: The New Synthesis*, Cambridge, MA: Belknap Press.

Wilson, E. O. (1984) *Biophilia*, Cambridge, MA: Harvard University Press.

Wilson, E. O. (ed.) (1988) *BioDiversity*, Washington DC: National Academy Press.

Wilson, M., Kelling, A., Poline, L., Bloomsmith, M., and Maple, T. (2003) 'Post-occupancy evaluation of Zoo Atlanta's Giant Panda Conservation Center: staff and visitor reactions', *Zoo Biology*, 22: 365–82.

Wilson, M. L., Bloomsmith, M. A., and Maple, T. L. (2004) 'Stereotypic swaying and serum cortisol concentrations in three captive African elephants (*Loxodonta africana*)', *Animal Welfare*, 13: 39–43.

Wilson, S. C., Mitlohner, F. M., Morrow-Tesch, J., Dailey, J. W., and McGlone, J. J. (2002) 'An assessment of several potential enrichment devices for feedlot cattle', *Applied Animal Behaviour Science*, 76: 259–65.

Wilson, S. F. (1982) 'Environmental influences on the activity of captive apes', *Zoo Biology*, 115: 201–9.

WIN (Wildlife Information Network) (2008) 'Gateway to WILDPRO', available online at http://www.wildlifeinformation.org.

Winne, C., Willson, J., Andrews, K., and Reed, R. (2006) 'Efficacy of marking snakes with disposable medical cautery units', *Herpetological Review*, 31: 52–4.

Winter, Y., Lopez, J., and von Helversen, O. (2003) 'Ultraviolet vision in a bat', *Nature*, 425: 612–14.

Wishart, G. J. (2001) 'The cryopreservation of germplasm in domestic and non-domestic birds' in P. F. Watson and W. V. Holt (eds) *Cryobanking the Genetic Resource: Wildlife Conservation for the Future?*, London: Taylor & Francis, pp. 179–200.

Wojciechowski, S. (2001) 'Is enrichment still good the next day? Overcoming the challenges of providing daily enrichment to multiple animal groups in a colony-type situation' in M. Hawkins, K. E. Worley and B. Hammond (eds), *Fifth International Conference on Environmental Enrichment*, 4–9 November, Sydney, The Shape of Enrichment Inc., pp. 211–20.

Wolf, R. and Tymitz, B. (1981) 'Studying visitor perceptions of zoo environments: a naturalistic view', *International Zoo Yearbook*, 21: 49–53.

Wolfe, B. A. (2003) 'Toxoplasmosis' in M. E. Fowler and M. E. Miller (eds), *Zoo and Wild Animal Medicine* (5th edn), Philadelphia, PA: Elsevier (Saunders), pp. 745–9.

Wolters, S. and Zuberbuhler, K. (2003) 'Mixed-species associations of Diana and Campbell's monkeys: the costs and benefits of a forest phenomenon', *Behaviour*, 140: 371–85.

Wood, W. (1998) 'Interactions among environmental enrichment, viewing crowds and zoo chimpanzees', *Zoo Biology*, 17: 211–30.

Woodfine, T., Gilbert, T., and Engel, H. (2005) 'A summary of past and present initiatives for the conservation and reintroduction of addax and scimitar-horned oryx in North Africa' in B. Hiddinga (ed.), *Proceedings of the EAZA Conference*, 21–25 September, Kolmarden, Amsterdam: EAZA Executive Office, pp. 208–11.

Woolcock, D. (2000) 'Husbandry and management of kea *Nestor notabilis* at Paradise Park, Hayle', *International Zoo Yearbook*, 37: 146–52.

Woollard, S. (1998) 'The development of zoo education', *International Zoo News*, 45: 422–6.

Woollard, S. (1999) 'A review of zoo education in the United Kingdom and Ireland', *International Zoo News*, 46: 20–4.

Woollard, S. (2001) 'Teachers' evaluation of zoo education', *International Zoo News*, 48: 240–5.

Woolverton, W., Ator, N., Beardsley, P., and Carroll, M. (1989) 'Effects of environmental conditions on the psychological well-being of primates: a review of literature', *Life Sciences*, 4414: 901–17.

Wormell, D., Brayshaw, M., Price, E., and Herron, S. (1996) 'Pied tamarins *Saguinus bicolor bicolor* at the Jersey Wildlife Preservation Trust: management, behaviour and reproduction', *Dodo: Journal of the Wildlife Preservation Trust*, 32: 76–97.

WRI/IUCN/UNEP/FAO/UNESCO (World Resources Institute/The World Conservation Union/United Nations Environment Programme in consultation with the Food and Agriculture Organization and the United Nations Education, Scientific and Cultural Organization) (1992) *Global Biodiversity Strategy: Guidelines for Action to Save, Study and Use Earth's Biotic Wealth Sustainably and Equitably*, Washington DC: WRI.

Wyatt, T. D. (2003) *Pheromones and Animal Behaviour*, Cambridge: Cambridge University Press.

X Xiao, J., Wang, K., and Wang, D. (2005) 'Diurnal changes of behavior and respiration of Yangtze finless porpoises (*Neophocaena phocaenoides asiaeorientalis*) in captivity', *Zoo Biology*, 24: 531–41.

Xu, Y., Fang, S., and Li, Z. (2007) 'Sustainability of the South China tiger: implications of inbreeding depression and introgression', *Conservation Genetics*, 8: 1199–207.

Y

Yalowitz, S. S. (2004) 'Evaluating visitor conservation research at the Monterey Bay Aquarium', *Curator*, 47: 283–98.

Yates, K. and Plowman, A. (2004) 'Hoof overgrowth in Hartmann's mountain zebra is a consequence of diet, substrate, and behaviour' in C. Macdonald (ed.), *Proceedings of the Sixth Annual Symposium on Zoo Research*, 8–9 July, Edinburgh: BIAZA, pp. 305–12.

Young, R. J. (1998) 'Behavioural studies of guenons *Cercopithecus* spp at Edinburgh Zoo', *International Zoo Yearbook*, 36: 49–56.

Young, R. J. (2003) *Environmental Enrichment for Captive Animals*, Oxford: Blackwell Science.

Young, R. J. and Cipreste, C. F. (2004) 'Applying animal learning theory: training captive animals to comply with veterinary and husbandry procedures', *Animal Welfare*, 13: 225–32.

Z

Zhang, G. Q., Swaisgood, R. R., Wei, R. P., Zhang, H. M., Han, H. Y., Li, D. S., Wu, L. F., White, A. M., and Lindburg, D. G. (2000) 'A method for encouraging maternal care in the giant panda', *Zoo Biology*, 19: 53–63.

Ziegler, G. (1995) 'An alternative to processed meat diets: carcass feeding at Wildlife Safari', *The Shape of Enrichment*, 4: 1–5.

Ziegler, T. (2002) 'Selected mixed-species exhibits of primates in German zoological gardens', *Primate Reports*, 64: 7–71.

Zimmermann, A. and Wilkinson, R. (2007) 'The conservation mission in the wild: zoos as conservation NGOs?' in A. Zimmerman, M. Hatchwell, L. Dickie, and C. West (eds), *Zoos in the 21st Century: Catalysts for Conservation?*, Conservation Biology Series No. 15, Cambridge: Cambridge University Press, pp. 303–22.

Zimmerman, A., Hatchwell, M., Dickie, L., and West, C. (2007) *Zoos in the 21st Century: Catalysts for Conservation?*, Conservation Biology series No. 15, Cambridge: Cambridge University Press.

Zimmermann, M. (1986) 'Behavioural investigations of pain in animals' in I. J. H. Duncan and V. Molony (eds), *Assessing Pain in Farm Animals*, Luxembourg: Commission of the European Communities, pp. 16–27.

Zippel, K. (2005) 'Zoos play a vital role in amphibian conservation', *AmphibiaWeb*, available online at `http://amphibiaweb.org/declines/zoo/index.html`.

Zippel, K., Lacy, R., and Byers, O. (eds) (2006) *CBSG/WAZA Amphibian Ex Situ Conservation Planning Workshop Final Report*, Apple Valley, MN: IUCN/SSC Conservation Breeding Specialist Group.

ZSL (Zoological Society of London) Living Conservation (2005) *Annual Report of the Zoological Society of London 2004–2005*, London: ZSL.

Index

Entries have been indexed to page number, with figure references in *italic*, and table and box references in **bold**.